A-Level Year 1/AS

Pure Mathematics

Exam Board: Edexcel

When it comes to AS-Level Pure Maths, all you need is love. No, hang on — all you need is this incredible CGP Student Book. We always mix those up.

But there's still plenty in here to fall in love with. Crystal-clear study notes, tips, examples, a huge helping of practice questions and a whole practice exam... with fully worked answers for everything included at the back.

All in all, it's the best Student Book you can buy — and it includes a free Online Edition to read on your PC, Mac or tablet! Dreamy.

How to get your free Online Edition

Go to **cgpbooks.co.uk/extras** and enter this code...

1109 8845 0226 5523

This code will only work once. If someone has used this book before you, they may have already claimed the Online Edition.

Contents

About this Book

In this book you'll find...

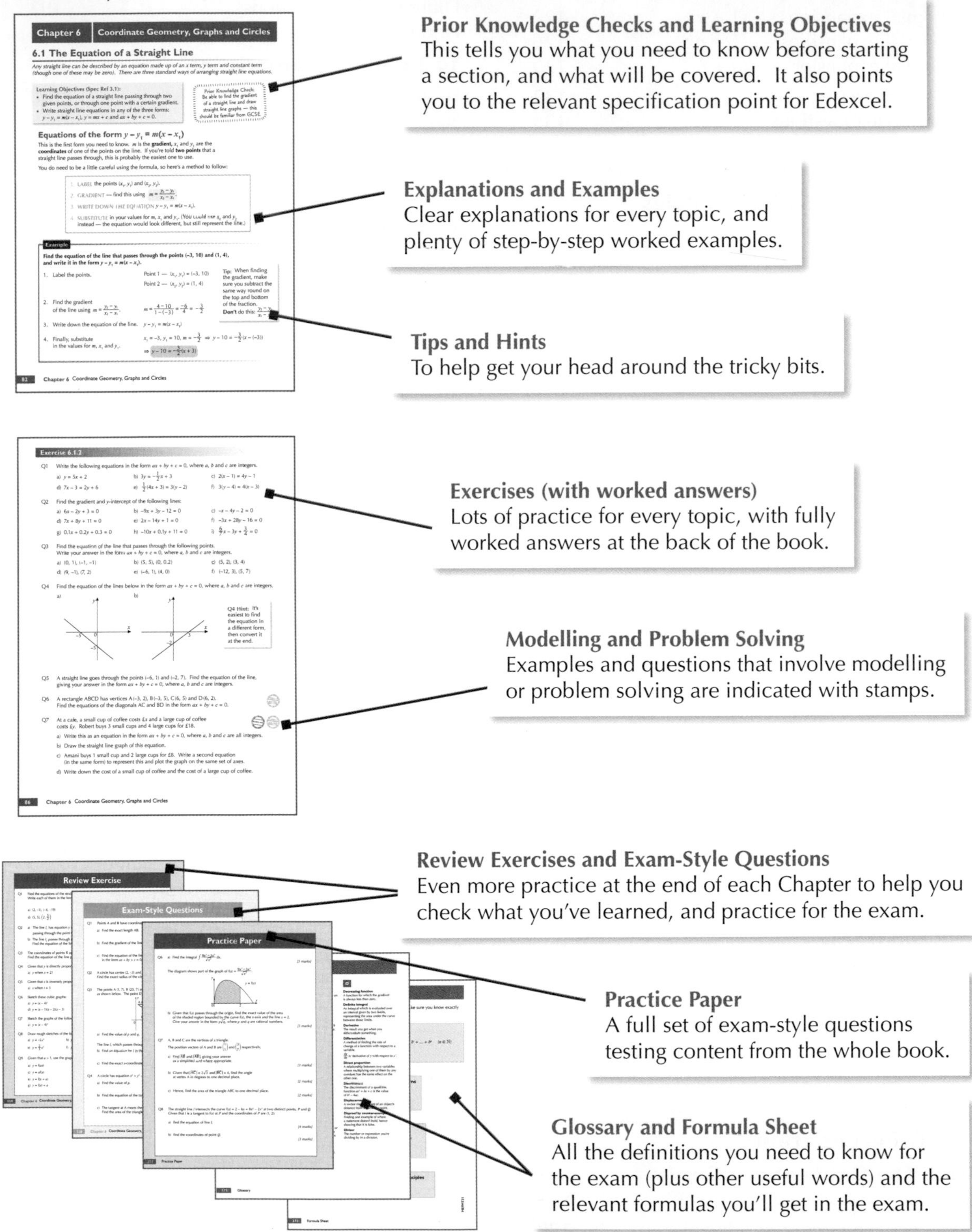

About this Course

This book covers the Pure content in AS-level Maths...

The Edexcel **AS-level** Mathematics course has **two** exam papers:

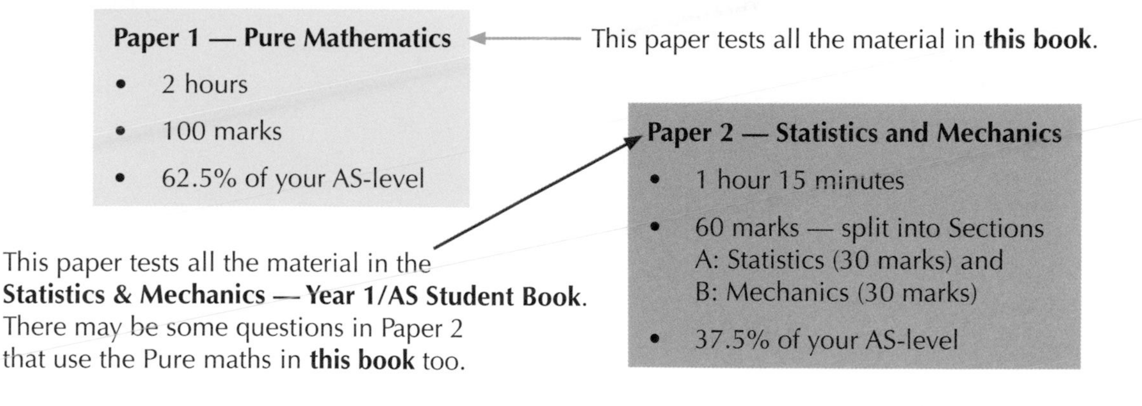

Paper 1 — Pure Mathematics

- 2 hours
- 100 marks
- 62.5% of your AS-level

This paper tests all the material in **this book**.

Paper 2 — Statistics and Mechanics

- 1 hour 15 minutes
- 60 marks — split into Sections A: Statistics (30 marks) and B: Mechanics (30 marks)
- 37.5% of your AS-level

This paper tests all the material in the **Statistics & Mechanics — Year 1/AS Student Book**. There may be some questions in Paper 2 that use the Pure maths in **this book** too.

... And half of the Pure content in A-level Maths.

If you're studying for the Edexcel **A-level** in Mathematics, you'll sit **three** exam papers:

Paper 1 — Pure Mathematics 1
Paper 2 — Pure Mathematics 2

- 2 hours each
- 100 marks each
- 33.33% of your A-level each

These two papers test the same material. Start off with the material covered in **this book**, and the rest of the content is in the **Pure Mathematics — Year 2 Student Book**.

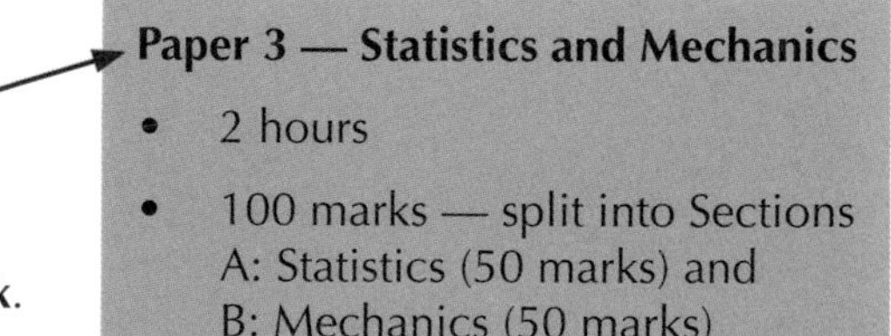

Paper 3 — Statistics and Mechanics

- 2 hours
- 100 marks — split into Sections A: Statistics (50 marks) and B: Mechanics (50 marks)
- 33.33% of your A-level

This paper tests the material in the **Statistics & Mechanics — Year 1/AS Student Book** and the **Statistics & Mechanics — Year 2 Student Book**. There may be some questions in Paper 3 that use the Pure maths in **this book** and the **Pure Mathematics – Year 2 Student Book** too.

Formulas, Tables and Data Sets

With each exam paper you'll have a **formula booklet** which contains **formulas** and **statistical tables** that you might need to use. The relevant ones for the material in this book are on p.378.

You'll also be working with a **large data set** throughout your course. This will only be used in Paper 2 of AS-level and Paper 3 of A-level.

More information and practice on using the data set can be found in the **Statistics & Mechanics — Year 1/AS Student Book** and the **Statistics & Mechanics — Year 2 Student Book**.

Tip: Although you don't have to learn these formulas off by heart, it's important that you practise using them, and also know which formulas are **not** given to you.

Published by CGP

Editors:
Sarah George, Samuel Mann, Alison Palin, David Ryan, Ben Train, Ruth Wilbourne, Dawn Wright.

Contributors:
Katharine Brown, Vijesh Chauhan, Mick Coe, Eva Cowlishaw, John Fletcher, Daniel Goldberg, Aleksander Goodier, Allan Graham, Phil Harvey, Alan Mason, Andy Pierson, Lauren Reynolds, Rosemary Rogers, Andy Smith, Simon Thornhill, Janet West.

ISBN: 978 1 78908 361 3

Cover design by emc design ltd.

With thanks to Janet Dickinson, Liam Dyer and Glenn Rogers for the proofreading.
With thanks to Emily Smith for the copyright research.

Printed by Elanders Ltd, Newcastle upon Tyne.
Clipart from Corel®

Chapter 1 Modelling and Problem Solving

1.1 Modelling

A mathematical model is a mathematical description of a real-life situation. Modelling involves simplifying the situation so that you can understand its behaviour and predict what is going to happen.

Learning Objectives (Spec Ref OT3):
- Understand the mathematical modelling process.
- Describe some examples of mathematical modelling.
- Understand why mathematical modelling involves making assumptions.
- Understand the process of refining a mathematical model.

Using mathematical models

Modelling in maths generally boils down to using an **equation** or a set of equations to **predict** what will happen in real life. You'll meet them in all areas of this course. For example:

- In algebra, formulas involving exponentials are used to model things like population growth (see pages 177-180).
- In mechanics, equations are used to model how the speed and acceleration of a moving object changes over time.
- In statistics, probability distributions are models which are used to predict the probability of a particular outcome in a trial.

Example 1

A company predicts that its latest product will sell 500 units in its first month on sale, and that sales will gradually increase by an average of 250 units per month.

Write an equation to model the predicted sales, s, of the product during its mth month on sale.

s is the mth term of an arithmetic sequence with first term $a = 500$ and common difference $d = 250$.

So $s = a + (m - 1)d = 500 + 250(m - 1) = 250 + 250m = 250(m + 1)$

This is a model, because it's based on a prediction of how the sales will go. The actual sales might end up being much higher or lower if the company has got its predictions wrong.

Modelling assumptions

Models are always **simplifications** of the real-life situation. When you construct a model, you have to make **assumptions** — you **ignore** or **simplify** some factors that affect the real-life outcome to keep the maths simpler. There are many reasons for making assumptions — a factor might only have a small effect or be hard to predict, or you might not have enough data to model it accurately. For example:

- A population growth model might ignore the fact that the population will eventually run out of **food**, because that won't happen in the **time period** you're modelling.
- A model for the speed of a moving object might ignore some of the forces acting upon it (e.g. **air resistance**), because including them would make the maths **more complicated**, or because you might want a **general result** for objects of all shapes and sizes.

Example 2

Leon owns a gooseberry farm. This week, he had 5 workers picking fruit, and they picked a total of 1000 punnets of gooseberries. Leon wants to hire more workers for next week. He predicts that next week, if the number of workers on his farm is w, the farm will produce p punnets of gooseberries, where $p = 200w$. Suggest three assumptions Leon has made in his model.

Leon's model predicts that the mean number of punnets produced per worker next week will be the same as this week. That means he's assumed all the conditions next week will be the same as the conditions this week. For example:

- Leon has assumed all the new workers he employs next week will work at the same speed on average as the ones he employed this week.
- He has assumed that the weather next week will be good enough to allow each worker to work as many hours as this week.
- He has assumed that there will be enough gooseberries to fill 200 punnets per worker, however many workers he employs.

Tip: Don't forget to link your answers back to the original context of the question.

Refining models

An important part of the modelling process is **refining** a model. This usually happens after the model has been **tested** by comparing it with real-world outcomes, or if you find out some **extra information** that affects the model.

Refining a model usually means changing some of the **assumptions**. For example:

- You might adjust a population growth model if you found that **larger populations** were more susceptible to **disease**, so grew more slowly.
- You might decide to refine a model for the speed of an object to take into account the **friction** from the surface the object is travelling over.

Tip: You could be asked to criticise or evaluate a model — e.g. you might need to assess if any assumptions are unrealistic. There's more on this on page 179.

Example 2 (cont.)

Leon discovers that the weather forecast for next week is bad, and his workers are only likely to be able to pick gooseberries for half the number of hours they did this week. How should he refine his model?

Leon's original model was $p = 200w$, based on all the workers next week picking at the same average weekly rate as this week. If his workers can only pick for half the time, they can only pick half as many gooseberries. So the refined model would be $p = 200w \div 2 \Rightarrow p = 100w$

The model is refined because there's new information. He might also refine his model at the end of next week, e.g. if he found that his new workers were a lot slower or faster than his current ones.

Modelling is one of the **overarching themes** of the AS and A-Level Maths courses. It could come up within any other topic — you might be given a model to use, or be asked to create one yourself.

Throughout this book, examples and questions involving modelling are marked with stamps like this:

1.2 Problem Solving

It's not always obvious what you need to do to answer a question — 'problem solving' questions are those tricky ones where you have to work out for yourself exactly what maths you need to do.

Learning Objectives (Spec Ref OT2.1-2.3 & OT2.6):
- Understand the problem solving cycle.
- Apply problem solving skills to maths questions.

The problem solving cycle

Like modelling, problem solving is an overarching theme that will come up throughout the course. Whenever maths is used to solve a real-life problem, the process used can be described using a **problem solving cycle**. The basic cycle looks like this:

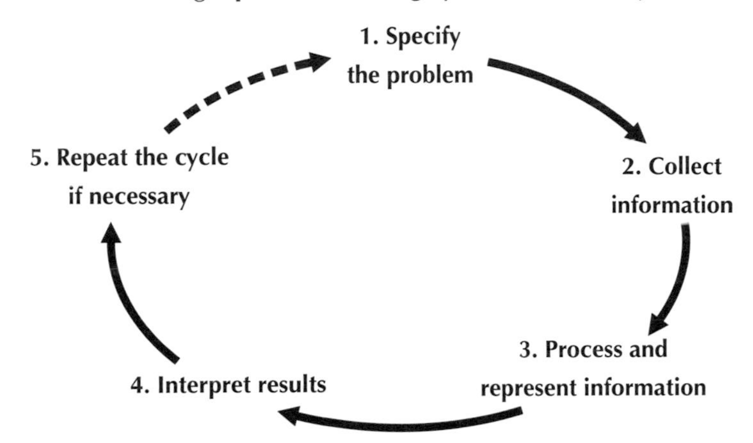

Tip: Problem solving questions include: questions that don't have 'scaffolding' (i.e. they're not broken down into parts a), b), c) etc.), 'wordy' questions with a real-life context, questions that use more than one area of maths, and questions that test your understanding of the maths as well as being able to use it.

These steps apply to pretty much anything you can use maths for.

1. Specify the problem
   The **problem** is the actual question you want to answer. It could be anything from "What's the probability I roll a 6 on this dice? to "How long will it take this ball to fall 10 m?" to "What's the angle of elevation from here to the top of that lighthouse?"

2. Collect information
   You'll need some **numbers** to solve the problem. This step might involve taking measurements, carrying out surveys or looking up data.

3. Process and represent information
   Once you've got the information you need, you can do the **calculations**. Representing the information might involve drawing a **graph** or **diagram**.

4. Interpret results
   Once you've done the calculation, you might need to **interpret** it to work out what the numerical answer means in terms of the original question.

5. Repeat the cycle if necessary
   You might decide to **repeat** the process. E.g. if you collected your information via a survey, you might want to repeat it with a larger or more representative sample. If your calculations involved modelling, you might want to **refine** your model (see previous page).

Problem solving in maths questions

You can apply a version of the problem solving cycle to any maths question where it's not immediately obvious what you're supposed to do.

1. **Specify the problem**
   The first thing to do is work out what the question is actually asking. The question might be phrased in an unusual way to try to throw you, or it might be written in a 'wordy' context, where you need to turn the words into maths.
2. **Collect information**
   Write down what you know. All the information you need to answer the question will either be given in the question somewhere, or it'll require facts that you should already know.
3. **Process and represent information**
   When you know what you're trying to find out, and what you already know, you can do the calculation to answer the question.
4. **Interpret results**
   Don't forget to give your answer in terms of the original context. The result of your calculation won't necessarily be the final answer.

Tip: When you're doing a maths question, it's unlikely you'll need to repeat the problem solving cycle once you've found the answer — just be aware that it's part of the real-world problem solving process.

Example 3

Armand cuts out a semicircle of cardboard from a rectangular sheet of cardboard measuring 20 cm by 40 cm and throws the rest away. The cardboard he throws away has an area of 398.08 cm². How long is the straight side of the semicircle? (Use $\pi = 3.14$.)

1. **Specify the problem.** What are you trying to find?
   The length of the straight side of a semicircle is the diameter of the full circle, which is twice the radius. So that's what you're looking for here.
2. **Collect information.** What do you know?
   The total area of the sheet of cardboard is 20 cm × 40 cm. 398.08 cm² was thrown away and the rest is the area of the semicircle. Area of a semicircle $= \frac{1}{2} \times$ area of a circle $= \frac{1}{2}\pi r^2$
3. **Process and represent information.** Do the maths.
   Area of semicircle $= 20 \times 40 - 398.08 = 800 - 398.08 = 401.92$ cm²
   So $401.92 = \frac{1}{2}\pi r^2 = \frac{1}{2} \times 3.14 \times r^2 = 1.57 \times r^2$
   $\Rightarrow r^2 = 401.92 \div 1.57 = 256 \Rightarrow r = \sqrt{256} = 16$ cm $\Rightarrow d = 2r = 2 \times 16 = 32$ cm
4. **Interpret results.** Give the answer in the context of the question.
   The length of the straight side of the semicircle is 32 cm.

Tip: You wouldn't normally do so much writing for a problem solving question — it's written out like this here as an example of how the problem solving thought process might work.

PROBLEM SOLVING

Throughout the rest of this book, if you see a stamp like this, it means there's an example that shows some of these problem solving skills in action, or a question where you'll probably need them.

Chapter 2 Proof

2.1 Proof

Mathematical proof is a logical argument that shows that a statement is true (or false). There are a few different types of proof that you need to be familiar with, but first there's a bit of notation to learn...

Learning Objectives (Spec Ref OT1.1-1.3):

- Be comfortable with mathematical notation for sets, functions, logical arguments and equivalence.
- Show that sums and products of integers are odd or even.
- Use proof by exhaustion and proof by deduction to show that statements are true.
- Use disproof by counter-example to show that statements are false.

Prior Knowledge Check: Be able to apply problem-solving skills — see p.3-4.

Notation

Set notation

In maths, a '**set**' is just a **collection** of objects or numbers (called **elements**) — a set is often represented by a capital letter. To show something is a set you put **curly brackets** around it — e.g. A = {0, 1, 2}.

- You can write out the **complete list** of elements: {1, 3, 5, 7, 9, 11, 13, 15, 17, 19}
- Or you can write out the **rule** that connects the elements: {odd numbers between 0 and 20}
- The rule can also be written out using numbers and symbols: $\{x: x < 0\}$. In words, this means "the set of values of x such that x is less than 0".
- A set containing **no elements** is called the **empty set**, written as: $\varnothing = \{\}$.

If you have more than one set you can talk about their **union** and **intersection**.

The **union** of two sets A and B is the combination of both sets — so it contains **all the elements** of **both sets**. It is written $A \cup B$.

The **intersection** of two sets A and B is a smaller set that contains only the elements present in **both** sets. It is written $A \cap B$.

Tip: Have a look at p.63-65 to see how set notation is used to show solutions to inequalities.

So if A = {1, 3, 5, 7, 9} and B = {1, 4, 9, 16}, then $A \cup B$ = {1, 3, 4, 5, 7, 9, 16} and $A \cap B$ = {1, 9}.

Function notation

- A **function** of x, written **f(x)**, takes a value of x, does something to it, and then outputs another value — it's just like an **equation** of x, but y is replaced with **f(x)**. You'll see function notation used a lot in the chapters on quadratics (Chapter 4) and differentiation (Chapter 10).
- For example, the function $f(x) = x^2 + 1$ takes a value of x, squares it and then adds 1. So if $x = 4$, $f(4) = (4)^2 + 1 = 17$.

Logical notation

The symbols $\Rightarrow$ and $\Leftrightarrow$ crop up all over the place. These are logic symbols — they show when one thing **implies** another.

'$p \Rightarrow q$' means 'p implies q'. You can read it as '**if** p, **then** q' — so **if** p is true, **then** q must also be true — e.g. $x = 2 \Rightarrow x^2 = 4$.

'$p \Leftrightarrow q$' means 'p implies q and q implies p'. This means that p is true **if and only if** (or **iff**) q is true — e.g. $x^2 = 4 \Leftrightarrow x = \pm 2$.

Tip: You might also see an arrow like this $\Leftarrow$. It works in the same way as the other implication arrow — it just goes the opposite way. E.g. $p \Leftarrow q$ means 'q implies p'.

Equivalence notation

There are three variations on the equals sign that you also need to know.

$\neq$ means **not equal to** — e.g. $\sin 90° \neq \cos 90°$

$\approx$ means **approximately equal to** — e.g. $1 \div 3 \approx 0.33$.

$\equiv$ is the **identity symbol**. It means that two things are **identically equal** to each other. So $(a + b)(a - b) \equiv a^2 - b^2$ is true for **all values** of a and b (unlike an equation like $a^2 = 9$, which is only true for certain values of a).

Tip: Identities crop up again in trigonometry — see Chapter 8.

Proof

Simple proofs — odd and even numbers

Before you get onto the trickier proofs, there are some nice simple proofs about **odd** and **even** numbers that are really useful. But first you need to know these 'proper' definitions for them:

Any **even** number can be written as $\mathbf{2a}$, where a is an integer.

Any **odd** number can be written as $\mathbf{2b + 1}$, where b is an integer.

Tip: Integers are just whole numbers — they can be positive, negative or 0.

In the proofs below, $\mathbf{2j + 1}$ and $\mathbf{2k + 1}$ represent any two **odd numbers**, and $\mathbf{2l}$ and $\mathbf{2m}$ represent any two **even numbers** (where j, k, l and m are integers).

Example 1

a) Prove that the sum of two even numbers is even.

Proof: $2l + 2m = 2(l + m) =$ even so even + even = even

Tip: It can be written as $2 \times$ something, so it's divisible by two, i.e. it's an even number.

b) Prove that the sum of an odd number and an even number is odd.

Proof: $(2j + 1) + (2l) = 2j + 2l + 1 = 2(j + l) + 1 =$ odd, so odd + even = odd

c) Prove that the product of two odd numbers is odd.

Proof: $(2j + 1)(2k + 1) = 4jk + 2j + 2k + 1 = 2(2jk + j + k) + 1 =$ odd, so odd $\times$ odd = odd

Tip: You can prove that e.g. the product of two even numbers is even in a similar way.

Proof by deduction

A **proof by deduction** is when you use **known facts** to build up your argument and show a statement **must** be true.

Example 2

A definition of a rational number is 'a number that can be written as a quotient of two integers, where the denominator is non-zero'.

Use this definition to prove that the following statement is true: ***"The product of two rational numbers is always a rational number."***

Tip: A quotient is what you get when you divide one number by another.

1. Take **any two** rational numbers and call them $\boldsymbol{a}$ and $\boldsymbol{b}$.
2. By the **definition** of rational numbers you can write them in the form $a = \frac{p}{q}$ and $b = \frac{r}{s}$, where p, q, r and s are all integers, and q and s are non-zero.
3. So the **product** of a and b is $ab = \frac{p}{q} \times \frac{r}{s} = \frac{pr}{qs}$.
4. pr and qs are the products of integers, so they must also be integers, and because q and s are non-zero, qs must also be non-zero.
5. We've shown that ab is a quotient of two integers and has a non-zero denominator, so by definition, $\boldsymbol{ab}$ **is rational**. Hence the original statement is **true**.

Proof by exhaustion

In **proof by exhaustion** you break things down into two or more **cases**. You have to make sure that your cases cover **all possible situations**, then prove separately that the statement is true for **each case**.

Example 3

Prove the following statement: ***"For any integer x, the value of f(x) = $x^3 + x + 1$ is an odd integer."***

1. To prove the statement, split the situation into **two cases** that cover all possible situations: (i) x is an **even number**, and (ii) x is an **odd number**.

   (i) If x is an even integer, then it can be written as $x = 2n$ for some integer n.

   $f(2n) = (2n)^3 + 2n + 1 = 8n^3 + 2n + 1 = 2(4n^3 + n) + 1$

   n is an integer $\Rightarrow$ $(4n^3 + n)$ is an integer
   $\Rightarrow$ $2(4n^3 + n)$ is an even integer
   $\Rightarrow$ $2(4n^3 + n) + 1$ is an odd integer
   So f(x) is **odd** when x is **even**.

   (ii) If x is an odd integer, then it can be written as $x = 2m + 1$, for some integer m.

   $$\begin{aligned} f(2m + 1) &= (2m + 1)^3 + 2m + 1 + 1 \\ &= (8m^3 + 12m^2 + 6m + 1) + 2m + 1 + 1 \\ &= 8m^3 + 12m^2 + 8m + 3 \\ &= 2(4m^3 + 6m^2 + 4m + 1) + 1 \end{aligned}$$

   m is an integer $\Rightarrow$ $(4m^3 + 6m^2 + 4m + 1)$ is an integer
   $\Rightarrow$ $2(4m^3 + 6m^2 + 4m + 1)$ is an even integer
   $\Rightarrow$ $2(4m^3 + 6m^2 + 4m + 1) + 1$ is an **odd integer**
   So f(x) is **odd** when x is **odd**.

2. You have shown that f(x) is odd when x is even and when x is odd. As any integer x must be either odd or even, you have shown that f(x) is odd for any integer x, so the statement is true.

Disproof by counter-example

Disproof by **counter-example** is the easiest way to show a mathematical statement is **false**.
All you have to do is find **one case** where the statement doesn't hold.

Example 4

Disprove the following statement: *"For any pair of integers x and y, if $x > y$, then $x^2 + x > y^2 + y$."*

To **disprove** the statement, it's enough to find just **one example** of x and y where $x > y$, but $x^2 + x \leq y^2 + y$.

You might have to try a few different numbers before you come up with an example that doesn't work.

E.g. Let $x = 2$ and $y = -4$.
Then $2 > -4 \Rightarrow x > y$
But $x^2 + x = 2^2 + 2 = 6$ and $y^2 + y = (-4)^2 + (-4) = 12$,
so $x^2 + x < y^2 + y$

So when $x = 2$ and $y = -4$, the first part of the statement holds, but the second part doesn't.
So the statement is **not true**.

Exercise 2.1.1

PROBLEM SOLVING

Q1 a) Prove that the sum of two odd numbers is even.

b) Prove that the product of two even numbers is even.

c) Prove that the product of an odd number and an even number is even.

Q1 Hint: These proofs are similar to the ones on p.6.

Q2 Prove that when an odd number is subtracted from an even number, the result is always odd.

Q3 By finding a counter-example, disprove the following statement:
"If p is a non-zero integer, then $\frac{1}{p^2} < \frac{1}{p}$."

Q4 Prove that, for any integer x, $(x + 5)^2 + 3(x - 1)^2$ is always divisible by four.

Q4 Hint: To show a number is divisible by 4, you need to be able to write it as $4 \times$ an integer.

Q5 Prove by exhaustion that the product of any three consecutive integers is even.

Q6 Disprove the following statement: "$n^2 - n - 1$ is a prime number for any integer $n > 2$."

Q7 Disprove the following: $\sqrt{x^2 + y^2} < x + y$.

Q8 Prove that the sum of two rational numbers is also a rational number.

Q9 Disprove the following statement:
"When one rational number is divided by another rational number, the result is always rational."

Q8-9 Hint: Use the definition of rational numbers from p.7 — and have a look at p.14-15 for more on adding fractions.

Q10 a) Prove the following statement: "For any integer n, $n^2 - n - 1$ is always odd."

b) Hence prove that $(n^2 - n - 2)^3$ is always even.

Exam-Style Questions

Q1 Prove the following statement:
"For any integer value of x, $f(x) = 2x^2 + 2x + 3$ is odd."

[2 marks]

Q2 Disprove the following statement: "$x^3 > x$, when $x \neq 0$."

[2 marks]

Q3 Prove that $a^m \times a^n = a^{m+n}$ for all positive integers m and n.

[3 marks]

Q4 Prove that the sum of the squares of two consecutive odd integers is an even number.

[4 marks]

Q5 Prove that raising an even number to an odd power gives an even number.

[2 marks]

Q6 Prove that the difference between any two rational numbers is also a rational number.

[3 marks]

Q7 Prove algebraically that $(4n + 1)^2 - (2n - 1)$ is an even number.

[3 marks]

Q8 An isosceles triangle has two identical angles, each measuring $y°$, and another angle measuring $x°$.
Given that y is an odd number, prove algebraically that x is an even number.

[3 marks]

Q9 Disprove the following statement:
"$ax = bx$ implies that $a = b$ for all rational numbers a, b and x."

[2 marks]

Chapter 3 Algebra

3.1 Algebraic Expressions

This chapter will cover some of the basic algebra skills that you'll need again and again throughout the course — the good news is you should have seen a lot of it before.

Learning Objectives (Spec Ref 2.6):

- Use and expand brackets.
- Identify common factors and take them outside the brackets.
- Simplify complicated expressions including algebraic fractions.

Expanding brackets

Single brackets

When you've got just **one set of brackets** multiplied by a single number or letter, multiply each term in the brackets by the term outside the brackets.

$$a(b + c + d) = ab + ac + ad$$

Double brackets

For **two sets** of brackets multiplied together (where there are **two terms** in each), multiply **each term** in one set of brackets by **each term** in the other. You should **always** get **four terms** from multiplying out double brackets (though sometimes two of the terms will **combine**).

$$(a + b)(c + d) = ac + ad + bc + bd$$

Tip: Remember **FOIL**: **F**irst **O**utside **I**nside **L**ast as a rule for multiplying out double brackets. It's just an easy way to remember it.

Squared brackets

Squared brackets are just a **special case** of double brackets where both brackets are the **same**. Write them out as two sets of brackets until you're comfortable with it.

$$(a + b)^2 = (a + b)(a + b) = a^2 + ab + ba + b^2 = a^2 + 2ab + b^2$$

A common **mistake** is to write $(a + b)^2 = a^2 + b^2$ — remember that $(a + b)^2$ is actually $(a + b)(a + b)$.

Long brackets

Long brackets are brackets with **many terms**. Just like with double brackets, you need to multiply each term in the first set of brackets by each term in the second — you just do it with more terms.

Write out the expression again with each term from the first set of brackets separately multiplied by the second set of brackets — always use this middle step so you don't get confused by all the terms.

$$(x + y + z)(a + b + c + d) = x(a + b + c + d) + y(a + b + c + d) + z(a + b + c + d)$$

Then multiply out each of these single brackets, **one at a time**.

Many brackets

When you've got **many sets** of brackets multiplied together, multiply them out **two at a time**, treating each set of two as double brackets or long brackets.

Multiply out the first **two** sets of brackets...

$$(a + b)(c + d)(e + f) = (ac + ad + bc + bd)(e + f)$$

...then multiply out the remaining **two sets**.

$$= ac(e + f) + ad(e + f) + bc(e + f) + bd(e + f)$$

Now multiply out each of these single brackets, **one at a time**.

Tip: Once you've multiplied out the first pair, the resulting terms may cancel or simplify — making the second step easier.

Examples

Expand $3xy(x^2 + 2x - 8)$.

Multiply each term inside the brackets by the bit outside — separately.

$$(3xy \times x^2) + (3xy \times 2x) + (3xy \times (-8))$$
$$= (3x^3y) + (6x^2y) + (-24xy)$$
$$= 3x^3y + 6x^2y - 24xy$$

Expand $(2y^2 + 3x)^2$.

Either write it as two sets of brackets, multiply it out and collect like terms...

$$(2y^2 + 3x)(2y^2 + 3x)$$
$$= 2y^2 \cdot 2y^2 + 2y^2 \cdot 3x + 3x \cdot 2y^2 + 3x \cdot 3x$$
$$= 4y^4 + 6xy^2 + 6xy^2 + 9x^2$$
$$= 4y^4 + 12xy^2 + 9x^2$$

...or do it in one go, using $(a + b)^2 = a^2 + 2ab + b^2$.

$$(2y^2 + 3x)^2 = (2y^2)^2 + 2(2y^2)(3x) + (3x)^2$$
$$= 4y^4 + 12xy^2 + 9x^2$$

Expand $(2x^2 + 3x - 6)(4x^3 + 6x^2 + 3)$.

1. Multiply each term in the first set of brackets by the whole second set of brackets.

$$(2x^2 + 3x - 6)(4x^3 + 6x^2 + 3)$$
$$= 2x^2(4x^3 + 6x^2 + 3) + 3x(4x^3 + 6x^2 + 3) + (-6)(4x^3 + 6x^2 + 3)$$

2. Now multiply out each of these sets of brackets and simplify it all.

$$= (8x^5 + 12x^4 + 6x^2) + (12x^4 + 18x^3 + 9x) + (-24x^3 - 36x^2 - 18)$$
$$= 8x^5 + 24x^4 - 6x^3 - 30x^2 + 9x - 18$$

Expand $(2x + 5)(x + 2)(x - 3)$.

1. Multiply the first two sets of brackets.

$$(2x + 5)(x + 2)(x - 3) = (2x^2 + 4x + 5x + 10)(x - 3)$$
$$= (2x^2 + 9x + 10)(x - 3)$$

2. Now multiply the long bracket by the final set of brackets.

$$= 2x^2(x - 3) + 9x(x - 3) + 10(x - 3)$$

3. Expand the single brackets and simplify.

$$= (2x^3 - 6x^2) + (9x^2 - 27x) + (10x - 30)$$
$$= 2x^3 + 3x^2 - 17x - 30$$

Exercise 3.1.1

Q1 Expand the brackets in these expressions:

a) $5(x + 4)$ b) $a(4 - 2b)$ c) $-2(x^2 + y)$

d) $6mn(m + 1)$ e) $-4ht(t^2 - 2ht - 3h^3)$ f) $7z^2(2 + z)$

g) $4(x + 2) + 3(x - 5)$ h) $p(3p^2 - 2q) + (q + 4p^3)$ i) $7xy(x^2 + z^2)$

Q2 Expand and simplify:

a) $(x + 5)(x - 3)$ b) $(2z + 3)(3z - 2)$ c) $(u + 8)^2$

d) $(ab + cd)(ac + bd)$ e) $(10 + f)(2f^2 - 3g)$ f) $(7 + q)(7 - q)$

g) $(2 - 3w)^2$ h) $(4rs^2 + 3)^2$ i) $(5k^2l - 2kn)^2$

Q3 Expand and simplify the following expressions:

a) $(l + 5)(l^2 + 2l + 3)$ b) $(2 + q)(3 - q + 4q^2)$ c) $(m + 1)(m + 2)(m - 4)$

d) $(r + s)^3$ e) $(3x + 2)(x - 4)(2x + 1)$ f) $(4 + x + y)(1 - x - y)$

g) $(j + 2k - 3)(j^2 + 2j + 1)$ h) $(2c^2 - cd + d)(2d - c - 5c^2)$ i) $(2f^3 - 4f - 1)(f^2 + 3f + 2)$

Q4 The volume of a cylinder is given by πr^2h, where r is the radius and h is the height. A certain cylinder has a radius of 3 times its height (h cm), minus 2 cm. Find the volume of the cylinder in terms of h. There should be no brackets in your answer. PROBLEM SOLVING

Q5 Carole's garden is a square with sides of length x metres. Mark's garden is a rectangle. One side of the rectangle is 3 metres longer than the side of the square and the other is twice as long as the side of the square, plus an extra metre. Find the difference in area between Mark's garden and Carole's. Give your answer as a simplified expression in x. PROBLEM SOLVING

Factorising

Common Factors

The **factors** of a term are all the bits that **multiply together** to make it up — if something is a factor of a term, the term will be **divisible** by it.

For example, consider the term $\mathbf{12xy^2}$ — it has many factors including:

- All the **factors of 12** — 1, 2, 3, 4, 6 and 12.
- The variables x and y (and also y^2).
- Any combinations of these multiplied together e.g. $3xy$, $12y^2$, $6x$ etc.

Tip: The definition of a **term** is a collection of numbers, letters and brackets all multiplied or divided together.

Example 1

Find all the factors of $6x$.

1. A good way to do this is to break it up as much as you can: $6x = 1 \times 2 \times 3 \times x$
2. Now list all possible combinations of 1, 2, 3 and x: $1,\ 2,\ 3,\ 6,\ x,\ 2x,\ 3x,\ 6x$

1 is always a factor. The term itself is also a factor.

A factor which is in every term of an expression is called a **common factor**. They can be '**taken out**' and put outside brackets — when you've taken out **all** possible factors, an expression is **factorised**.

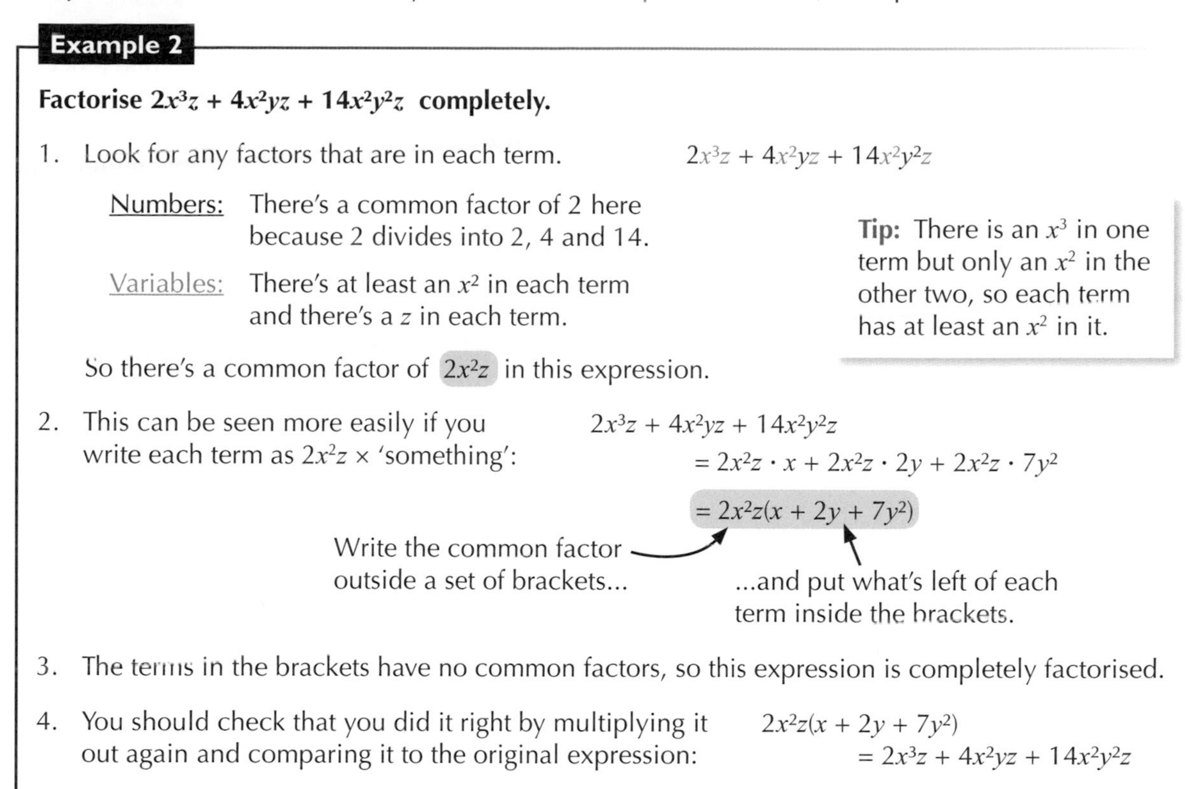

Example 2

Factorise $2x^3z + 4x^2yz + 14x^2y^2z$ completely.

1. Look for any factors that are in each term. $2x^3z + 4x^2yz + 14x^2y^2z$

   Numbers: There's a common factor of 2 here because 2 divides into 2, 4 and 14.

   Variables: There's at least an x^2 in each term and there's a z in each term.

   So there's a common factor of $2x^2z$ in this expression.

   Tip: There is an x^3 in one term but only an x^2 in the other two, so each term has at least an x^2 in it.

2. This can be seen more easily if you write each term as $2x^2z$ × 'something':

   $2x^3z + 4x^2yz + 14x^2y^2z$
   $= 2x^2z \cdot x + 2x^2z \cdot 2y + 2x^2z \cdot 7y^2$
   $= 2x^2z(x + 2y + 7y^2)$

   Write the common factor outside a set of brackets...

   ...and put what's left of each term inside the brackets.

3. The terms in the brackets have no common factors, so this expression is completely factorised.

4. You should check that you did it right by multiplying it out again and comparing it to the original expression: $2x^2z(x + 2y + 7y^2) = 2x^3z + 4x^2yz + 14x^2y^2z$

It's not just numbers and variables that you need to look for — you can sometimes take out **whole sets of brackets** as factors of an expression. You might be asked to write an expression as a **product of factors** — this means write all its factors multiplied together, which is exactly the **same** as **factorising**.

Example 3

Express $(y + a)^2(x - a)^3 + (x - a)^2$ as a product of factors.

1. This can be written $(y + a)^2(x - a)(x - a)^2 + (x - a)^2$.
2. $(x - a)^2$ is a common factor — so write the common factor outside a set of brackets and put what's left of each term inside the brackets: $(x - a)^2[(y + a)^2(x - a) + 1]$
3. The two terms in the brackets share no common factors so the expression is factorised.

Difference of two squares

If you expand brackets of the form $(a - b)(a + b)$, the 'ab' terms cancel and you're left with one square minus another:

$$(a - b)(a + b) = a^2 + ab - ba - b^2 = a^2 + ab - ab - b^2 = a^2 - b^2$$

This result is called the **difference of two squares**: $a^2 - b^2 = (a - b)(a + b)$

Watch out for it when factorising — if you spot that an expression is just 'something squared' minus 'something else squared', you can use this result to rewrite the expression as a pair of brackets.

Example 4

a) **Factorise $x^2 - 36y^2$.**

36 is a square number so $36y^2$ can be written as a square:

$x^2 - 36y^2 = x^2 - 6^2y^2$
$= x^2 - (6y)^2$
$= (x - 6y)(x + 6y)$

This is a difference of two squares.

b) **Factorise $x^2 - 5$.**

5 isn't a square number but you can still write it as a square:

$x^2 - 5 = x^2 - (\sqrt{5})^2$
$= (x - \sqrt{5})(x + \sqrt{5})$

Tip: Any number can be written as the square of its square root (see p.20).

Exercise 3.1.2

Q1 Factorise the following expressions completely:

a) $9k + 15l$ b) $u^2 - uv$ c) $2x^2y - 12xy^2$ d) $f^2g^2 - fg$

e) $p^3 + 3pq^3 + 2p$ f) $mnp^2 + 7m^2np^3$ g) $2ab^4 + 3a^3b^2 - 4ab$ h) $36xyz - 8x^2z^2 + 20y^2z^2$

Q2 Write the following expressions as products of factors:

a) $x^2 - y^2$ b) $9a^2 - 4b^2$ c) $25x^2 - 49z^2$ d) $a^2c - 16b^2c$

e) $y^2 - 2$ f) $m^2 - 11$ g) $4x^2 - 3$ h) $7p^2 - 13$

Q3 Express the following as the product of factors.

a) $(4 - z)^2(2 - z) + p(2 - z)$ b) $(r - d)^3 + 5(r - d)^2$

c) $(b + c)^5(a + b) - (b + c)^5$ d) $l^2m(a - 2x) + rp^2(2x - a)$

Q3d) Hint: Remember that $(b - a) = -(a - b)$

Q4 Simplify each expression, leaving your answer in its factorised form.

a) $(p + q)^2 + 2q(p + q)$ b) $2(2x - y)^2 - 6x(2x - y)$ c) $(l + w + h)^2 - l(l + w + h)$

Q5 Simplify these expressions by expanding brackets, factorising or both.

a) $(m + 5)(m^2 - 5m + 25)$ b) $(p - 2q)(p^2 + 2pq + 4q^2)$

c) $(u - v)(u + v) - (u + v)^2$ d) $(c + d)^3 - c(c + d)^2 - d(c + d)^2$

Algebraic fractions

The rules for **algebraic fractions** should be familiar from working with numerical fractions.

Adding and subtracting

If you're **adding fractions** together that all have the same **denominator**, you just add the **numerators**.

$$\frac{a}{x} + \frac{b}{x} + \frac{c}{x} \equiv \frac{a+b+c}{x}$$

x is the **common denominator**

Tip: This equals sign with 3 lines $\equiv$ means it's true for all values of a, b, c or x — this is called an **identity** (see page 6).

If the fractions you want to add don't have a common denominator you can 'find' one — **rewrite** the fractions so that the denominators are the same by multiplying **top** and **bottom** by the same thing.

Example 1

a) Express $\frac{1}{2x} - \frac{1}{3x} + \frac{1}{5x}$ as a single fraction.

1. Rewrite the fractions so that their **denominators** are all the same. You want something that all of these denominators divide into.

2. **30** is the lowest number that 2, 3 and 5 all go into and each denominator contains an x, so make the common denominator $\mathbf{30x}$. Multiply the top and bottom lines of each fraction by whatever makes the bottom line $30x$.

3. Then, just add the numerators.

$$\frac{1}{2x} - \frac{1}{3x} + \frac{1}{5x} = \frac{1}{2x} \cdot \frac{15}{15} - \frac{1}{3x} \cdot \frac{10}{10} + \frac{1}{5x} \cdot \frac{6}{6}$$
$$= \frac{15}{30x} - \frac{10}{30x} + \frac{6}{30x}$$
$$= \frac{15 - 10 + 6}{30x} = \frac{11}{30x}$$

b) Simplify $\frac{3}{x+2} + \frac{5}{x-3}$.

1. Again, the first step is to rewrite the fractions so that they have a **common denominator**. You need an expression that both $(x + 2)$ and $(x - 3)$ divide into — you can get one by multiplying the denominators together to give a common denominator of $(x + 2)(x - 3)$.

2. Multiply the top and bottom lines of each fraction by whatever makes the bottom line the same as the common denominator.

3. Once the denominators are all the same, you can just add the numerators and simplify.

$$\frac{3(x-3)}{(x+2)(x-3)} + \frac{5(x+2)}{(x+2)(x-3)}$$
$$= \frac{3(x-3) + 5(x+2)}{(x+2)(x-3)}$$
$$= \frac{3x - 9 + 5x + 10}{(x+2)(x-3)}$$
$$= \frac{8x + 1}{(x+2)(x-3)}$$

c) Simplify $\frac{3}{2x^2} + \frac{6}{5x}$.

1. Finding a common denominator here is a bit more tricky. You could still **multiply** the two denominators together to get $10x^3$ — but this wouldn't give the **simplest** one.

- You have $2x^2 = 2 \times x \times x$ and $5x = 5 \times x$ and you need to find a term which **both** of them divide into. You must include each different factor **at least once** in your term — but some **more than once**.
- The different factors are **2**, **5** and x so you need **at least one** of each — there are **two** x's in the first denominator so you'll need an x^2, but you **don't** need another for the x in the second denominator since this is accounted for by multiplying by the x^2. So the common denominator is $2 \times 5 \times x \times x = \mathbf{10x^2}$.

Tip: Finding any old common denominator is easy — just multiply all the denominators together. But if you're careful and don't include any bits twice, you'll have a lot less simplifying to do at the end.

2. As before, multiply the top and bottom lines of each fraction by whatever makes the bottom line the same as the common denominator.

3. Then add the numerators and simplify by factorising.

$$\frac{3}{2x^2} + \frac{6}{5x} = \frac{3 \times 5}{2x^2 \times 5} + \frac{6 \times 2x}{5x \times 2x}$$
$$= \frac{15}{10x^2} + \frac{12x}{10x^2}$$
$$= \frac{15 + 12x}{10x^2} = \frac{3(5 + 4x)}{10x^2}$$

Simplifying

Algebraic fractions can sometimes be simplified by cancelling **factors** that appear in both the numerator and denominator. You can do this in **two ways** — use whichever method you prefer, but make sure you understand the ideas behind both.

Example 2

Simplify $\frac{ax + ay}{az}$.

You can **factorise**, then cancel...

$$\frac{ax+ay}{az} = \frac{a(x+y)}{az} = \frac{\cancel{a}(x+y)}{\cancel{a}z} = \frac{x+y}{z}$$

Factorise the top line.

Cancel the 'a'.

... or **split into two fractions**, then cancel.

$$\frac{ax+ay}{az} = \frac{ax}{az} + \frac{ay}{az}$$

This is the rule from page 14 for adding fractions, but backwards.

$$= \frac{\cancel{a}x}{\cancel{a}z} + \frac{\cancel{a}y}{\cancel{a}z} = \frac{x}{z} + \frac{y}{z} = \frac{x+y}{z}$$

Exercise 3.1.3

Q1 Express each of these as a single fraction.

a) $\frac{x}{3} + \frac{x}{4}$ b) $\frac{2}{t} + \frac{13}{t^2}$ c) $\frac{1}{2p} - \frac{1}{5q}$

d) $\frac{2}{3h} + \frac{1}{2h} - \frac{3}{4h}$ e) $\frac{ab}{c} + \frac{bc}{a} + \frac{ca}{b}$ f) $\frac{2}{mn} - \frac{3m}{n} + \frac{n^2}{m}$

g) $\frac{2}{ab^3} - \frac{9}{a^3b}$ h) $\frac{1}{x} + \frac{2x}{y} + \frac{4}{x^2}$ i) $2 + \frac{a^2}{b} - \frac{2b}{a^2}$

Q2 Express the following as single fractions in their simplest form.

a) $\frac{5}{y-1} + \frac{3}{y-2}$ b) $\frac{7}{r-5} - \frac{4}{r+3}$ c) $\frac{8}{p} - \frac{1}{p-3}$

d) $\frac{w}{2(w-2)} + \frac{3w}{w-7}$ e) $\frac{z+1}{z+2} - \frac{z+3}{z+4}$ f) $\frac{1}{q+1} + \frac{3}{q-2}$

g) $\frac{x}{x+z} + \frac{2z}{x-z}$ h) $\frac{y}{2x+3} - \frac{2y}{3-x}$ i) $\frac{5}{r-4} + \frac{3}{r} - \frac{r}{r+1}$

Q3 Simplify these expressions.

a) $\frac{2x+10}{6}$ b) $\frac{6a-12b-15c}{3}$ c) $\frac{np^2-2n^2p}{np}$

d) $\frac{4st+6s^2t+9s^3t}{2t}$ e) $\frac{10yz^3-40y^3z^3+60y^2z^3}{10z^2}$ f) $\frac{12cd-6c^2d+3c^3d^2}{12c^2de}$

g) $\frac{2x+x^2y-x^2}{x^2+3x}$ h) $\frac{2w^3+14w^2}{w^2-49}$ i) $\frac{4g^2-4h^2}{g^2+gh}$

Q4 Maya has a ribbon of length 10 cm and Hal has a ribbon of length 15 cm. Maya cuts her ribbon into x equal pieces. Hal cuts his into three more pieces than Maya. What is the total length of one of Maya's pieces and one of Hal's pieces laid end to end? Give your answer as a single fraction involving x.

PROBLEM SOLVING

3.2 Laws of Indices

The laws of indices are just a set of simple rules for manipulating expressions involving indices (powers). They'll be used regularly throughout this course for simplifying expressions, equations and formulas.

Learning Objective (Spec Ref 2.1):

- Use the laws of indices to simplify expressions.

Laws of indices

You should already know that the expression x^n just means ***n* lots** of x multiplied together. The n is called the **index** or power of x. So when you square a number (e.g. x^2), the index or power is 2.

Tip: Index and power mean the same thing — we'll use power for the rest of the section.

OK, now you're ready to see the laws. Here are the first **three** you need to know:

If you multiply two numbers, you **add** their powers:

$$a^m \times a^n = a^{m+n}$$

If you divide two numbers, you **subtract** their powers:

$$\frac{a^m}{a^n} = a^{m-n}$$

If you have a power to the power of something else, you **multiply** the powers together:

$$(a^m)^n = a^{mn}$$

There are also laws for manipulating **fractional** and **negative** powers...

$$a^{\frac{1}{m}} = \sqrt[m]{a} \qquad a^{-m} = \frac{1}{a^m} \qquad a^{\frac{m}{n}} = \sqrt[n]{a^m} = (\sqrt[n]{a})^m$$

Tip: $\sqrt[m]{a}$ is the mth root of a.

...and one simple law for **zero** powers, which works for any non-zero number or letter.

$$a^0 = 1$$

Now, let's see the laws in action in some worked examples.

Example 1

Simplify the following:

a) **(i)** a^2a **(ii)** $x^{-2} \cdot x^5$ **(iii)** $(a+b)^2(a+b)^5$ **(iv)** $ab^3 \cdot a^2b$

$a^m \times a^n = a^{m+n}$

(i) $a^2a = a^{2+1} = a^3$

(ii) $x^{-2} \cdot x^5 = x^{-2+5} = x^3$

(iii) $(a+b)^2(a+b)^5 = (a+b)^{2+5} = (a+b)^7$

(iv) $ab^3 \cdot a^2b = a^{1+2}b^{3+1} = a^3b^4$

Add the powers of a and b separately.

Tip: Note that $x = x^1$.

b) **(i)** $(x^2)^3$ **(ii)** $\{(a+b)^3\}^4$ **(iii)** $(ab^2)^4$

$(a^m)^n = a^{mn}$

(i) $(x^2)^3 = x^6$

(ii) $\{(a+b)^3\}^4 = (a+b)^{12}$

(iii) $(ab^2)^4 = a^4(b^2)^4 = a^4b^8$

This power of 4 applies to both bits inside the brackets.

Tip: For part b) (iii), remember that:
$(ab^2)^4 = (ab^2) \cdot (ab^2) \cdot (ab^2) \cdot (ab^2)$
$= a \cdot a \cdot a \cdot a \cdot b^2 \cdot b^2 \cdot b^2 \cdot b^2$
$= a^4 \cdot (b^2)^4 = a^4b^8$

c) **(i)** $\frac{x^{\frac{3}{4}}}{x}$ **(ii)** $\frac{x^3y^2}{xy^3}$

> **Tip:** For part c) (i), you could simplify further to $\frac{1}{\sqrt[4]{x}}$.

(i) $\frac{x^{\frac{3}{4}}}{x} = x^{\frac{3}{4}-1} = x^{-\frac{1}{4}} = \frac{1}{x^{\frac{1}{4}}}$ $\quad \frac{a^m}{a^n} = a^{m-n}$

(ii) $\frac{x^3y^2}{xy^3} = x^{3-1}y^{2-3} = x^2y^{-1} = \frac{x^2}{y}$

Subtract the powers of x and y separately.

d) **(i)** $4^{\frac{1}{2}}$ **(ii)** $125^{\frac{1}{3}}$

(i) $4^{\frac{1}{2}} = \sqrt{4} = 2$ $\quad a^{\frac{1}{m}} = \sqrt[m]{a}$

(ii) $125^{\frac{1}{3}} = \sqrt[3]{125} = 5$

e) **(i)** $9^{\frac{3}{2}}$ **(ii)** $16^{\frac{3}{4}}$

(i) $9^{\frac{3}{2}} = (9^{\frac{1}{2}})^3 = (\sqrt{9})^3 = 3^3 = 27$ $\quad a^{\frac{m}{n}} = \sqrt[n]{a^m} = (\sqrt[n]{a})^m$

(ii) $16^{\frac{3}{4}} = (16^{\frac{1}{4}})^3 = (\sqrt[4]{16})^3 = 2^3 = 8$

It's often easier to work out the root first, then raise it to the power.

f) **(i)** 2^{-3} **(ii)** $(x + 1)^{-1}$

(i) $2^{-3} = \frac{1}{2^3} = \frac{1}{8}$ $\quad a^{-m} = \frac{1}{a^m}$

(ii) $(x+1)^{-1} = \frac{1}{x+1}$

g) **(i)** 2^0 **(ii)** $(a + b)^0$

(i) $2^0 = 1$ $\quad a^0 = 1$

(ii) $(a+b)^0 = 1$

Example 2

Express $\frac{(7^{\frac{1}{3}})^6 \times (7^{-1})^4}{(7^{-4})^{-2}}$ as 7^k, where k is an integer.

1. Use the $(a^m)^n = a^{mn}$ rule to get rid of the brackets.
2. Combine the powers of 7 on the top of the fraction using the $a^m \times a^n = a^{m+n}$ rule.
3. Get rid of the fraction using the $\frac{a^m}{a^n} = a^{m-n}$ rule.

$$\frac{(7^{\frac{1}{3}})^6 \times (7^{-1})^4}{(7^{-4})^{-2}} = \frac{7^{\frac{6}{3}} \times 7^{-1\times4}}{7^{-4\times-2}}$$
$$= \frac{7^2 \times 7^{-4}}{7^8}$$
$$= \frac{7^{2-4}}{7^8}$$
$$= \frac{7^{-2}}{7^8}$$
$$= 7^{-2-8}$$
$$= 7^{-10}$$

> **Tip:** You could also write this as $\frac{1}{7^{10}}$.

Exercise 3.2.1

Q1 Simplify the following, leaving your answer as a power:

a) 10×10^4 b) $y^{-1} \times y^{-2} \times y^7$ c) $5^{\frac{1}{2}} \times 5^3 \times 5^{\frac{-3}{2}}$ d) $6^5 \div 6^2$

e) $3^4 \div 3^{-1}$ f) $\frac{6^{11}}{6}$ g) $\frac{r^2}{r^6}$ h) $(3^2)^3$

i) $(k^{-2})^5$ j) $(z^4)^{-\frac{1}{8}}$ k) $(8^{-6})^{-\frac{1}{2}}$ l) $\frac{p^5q^4}{p^4q}$

m) $\frac{c^{-1}d^{-2}}{c^2d^4}$ n) $(ab^2)^2$ o) $\frac{12yz^{-\frac{1}{2}}}{4yz^{\frac{1}{2}}}$ p) $(mn^{\frac{1}{2}})^4$

Q2 Evaluate:

a) $4^{\frac{1}{2}} \times 4^{\frac{3}{2}}$ b) $\frac{2^3 \times 2}{2^5}$ c) $\frac{7^5 \times 7^3}{7^6}$ d) $\frac{6^4}{6^{\frac{5}{4}} \times 6^{\frac{3}{4}}}$

e) $(3^2)^5 \div (3^3)^3$ f) $(4^{-\frac{1}{2}})^2 \times (4^{-3})^{-\frac{1}{3}}$ g) $\frac{(2^{\frac{1}{2}})^6 \times (2^{-2})^{-2}}{(2^{-1})^{-1}}$ h) 1^0

i) $\left(\frac{4}{5}\right)^0$ j) $(-5.726324)^0$ k) 8.374936^1 l) $\frac{(3^3)^2}{(9^{\frac{1}{2}})^4 \times (9^{\frac{1}{4}})^8}$

Q3 Express the following as negative or fractional powers or both:

a) $\frac{1}{p}$ b) $\frac{5}{y^4}$ c) $\sqrt{q}$ d) $\sqrt{r^3}$

e) $\sqrt[4]{s^5}$ f) $\frac{1}{\sqrt[3]{t}}$ g) $\left(\frac{1}{\sqrt[3]{x}}\right)^4$ h) $\frac{\sqrt{z}}{z^3}$

Q4 Evaluate:

a) $9^{\frac{1}{2}}$ b) $8^{\frac{1}{3}}$ c) $4^{\frac{3}{2}}$ d) $27^{-\frac{1}{3}}$

e) $16^{-\frac{3}{4}}$ f) $125^{\frac{2}{3}}$ g) $81^{\frac{1}{4}}$ h) $64^{\frac{1}{2}} \times 64^{-\frac{1}{3}}$

Q5 Clare thinks of a number, x.
She takes the cube root of the number, and then raises the result to the power 6.
Finally, she divides by the number she thought of.
What number does she have now?

PROBLEM SOLVING

Q6 If $p = \frac{1}{16}q^2$, write the following expressions in terms of q:

PROBLEM SOLVING

a) $p^{\frac{1}{2}}$ b) $2p^{-1}$ c) $p^{\frac{1}{2}} \div 2p^{-1}$

d) p^2q e) $\frac{4p}{q^3}$ f) $\frac{q^2}{4p^2}$

Q7 $a = 8^{-3}$, $b = 64^3$ and $c = 8^{\frac{1}{2}}$.
Write $\frac{a^{-5}}{b^{\frac{1}{2}} \times c^4}$ as a single power of 8.

PROBLEM SOLVING

Q8 Find the value of x for each of the following:

PROBLEM SOLVING

a) $4^x = \sqrt[3]{16}$ b) $9^x = \frac{1}{3}$ c) $\sqrt{5} \times 5^x = \frac{1}{25}$

d) $(16^x)^2 = \frac{1}{4}$ e) $x^{-3} = -8$ f) $\sqrt{100^x} = 0.001$

3.3 Surds

This section will cover how to simplify expressions containing square roots. There are laws for simplifying these expressions just like the ones for powers.

Learning Objectives (Spec Ref 2.2):
- Simplify expressions containing surds.
- Rationalise denominators.

The laws of surds

Put $\sqrt{2}$ into a calculator and you'll get 1.414213562... But if you square 1.414213562, you get 1.999999999. No matter how many decimal places you use, you'll never get exactly 2. This is because $\sqrt{2}$ is an **irrational number** — its decimal expansion **continues forever**.

The only way to express a number like this **exactly** is to leave it as a root. Numbers like $\sqrt{2}$ that can only be written exactly using roots are called **surds**. The number $\sqrt{3}$ is a surd because it can't be written exactly without a root — $\sqrt{9}$ is **not** a surd because it can be simplified to 3.

Tip: A rational number is a number that can be expressed as $\frac{p}{q}$ where p and q are integers and $q \neq 0$. An irrational number is just one which is not rational.

There are three rules you need to know to be able to use surds properly:

$$\sqrt{ab} = \sqrt{a}\sqrt{b}$$

$$\sqrt{\frac{a}{b}} = \frac{\sqrt{a}}{\sqrt{b}}$$

$$a = (\sqrt{a})^2 = \sqrt{a}\sqrt{a}$$

Simplifying surds usually just means making the number in the $\sqrt{\ }$ sign smaller, or getting rid of a fraction inside the $\sqrt{\ }$ sign.

Tip: Remember that $\sqrt{x} = x^{\frac{1}{2}}$ (see p.17).

Examples

a) Simplify $\sqrt{12}$

$$\sqrt{12} = \sqrt{4 \times 3} = \sqrt{4} \times \sqrt{3} = 2\sqrt{3}$$

($\sqrt{ab} = \sqrt{a}\sqrt{b}$)

b) Simplify $\sqrt{\frac{3}{16}}$

$$\sqrt{\frac{3}{16}} = \frac{\sqrt{3}}{\sqrt{16}} = \frac{\sqrt{3}}{4}$$

($\sqrt{\frac{a}{b}} = \frac{\sqrt{a}}{\sqrt{b}}$)

c) Find $(2\sqrt{5} + 3\sqrt{6})^2$

Multiply out the squared brackets: $(2\sqrt{5} + 3\sqrt{6})^2 = (2\sqrt{5} + 3\sqrt{6})(2\sqrt{5} + 3\sqrt{6})$

$$= (2\sqrt{5})^2 + (2 \times (2\sqrt{5}) \times (3\sqrt{6})) + (3\sqrt{6})^2$$

$$= (2^2 \times \sqrt{5}^2) + (2 \times 2 \times 3 \times \sqrt{5} \times \sqrt{6}) + (3^2 \times \sqrt{6}^2)$$

$$= 20 + 12\sqrt{30} + 54$$

$$= 74 + 12\sqrt{30}$$

$= 4 \times 5 = 20$

$= 12\sqrt{5}\sqrt{6} = 12\sqrt{30}$

$= 9 \times 6 = 54$

Tip: Remember $(a + b)^2 = a^2 + 2ab + b^2$.

d) Express $\sqrt{63} - \sqrt{28}$ in the form $k\sqrt{x}$ where k and x are integers.

1. Try to write both numbers as 'a square number' × x. Here, x is 7.

$\sqrt{63} - \sqrt{28} = \sqrt{9 \times 7} - \sqrt{4 \times 7}$
$= \sqrt{9}\sqrt{7} - \sqrt{4}\sqrt{7}$
$= 3\sqrt{7} - 2\sqrt{7}$
$= \sqrt{7}$ ← In this case, k is just 1.

2. Square root the square numbers to leave two integer multiples of $\sqrt{7}$, which you can subtract to simplify.

Tip: An integer is just a positive or negative whole number, including 0.

Exercise 3.3.1

Q1 Simplify the following surds:

a) $\sqrt{8}$ b) $\sqrt{24}$ c) $\sqrt{50}$ d) $\sqrt{63}$

e) $\sqrt{72}$ f) $\sqrt{\frac{5}{4}}$ g) $\sqrt{\frac{7}{100}}$ h) $\sqrt{\frac{11}{9}}$

Q2 Evaluate the following, giving your answers as either a whole number or a surd.

a) $2\sqrt{3} \times 4\sqrt{3}$ b) $\sqrt{5} \times 3\sqrt{5}$ c) $(\sqrt{7})^2$ d) $2\sqrt{2} \times 3\sqrt{5}$

e) $(2\sqrt{11})^2$ f) $5\sqrt{8} \times 2\sqrt{2}$ g) $4\sqrt{3} \times 2\sqrt{27}$ h) $2\sqrt{6} \times 5\sqrt{24}$

i) $\frac{6}{\sqrt{11}} \times \sqrt{44}$ j) $2\sqrt{18} \times \frac{5}{\sqrt{8}}$ k) $\frac{\sqrt{10}}{6} \times \frac{12}{\sqrt{5}}$ l) $\frac{\sqrt{12}}{3} \times \frac{2}{\sqrt{27}}$

Q3 Express the following in the form $k\sqrt{x}$, where k and x are integers and x is as small as possible.

a) $\sqrt{20} + \sqrt{5}$ b) $\sqrt{32} - \sqrt{8}$

c) $\sqrt{27} + 4\sqrt{3}$ d) $2\sqrt{8} - 3\sqrt{2}$

e) $3\sqrt{10} + \sqrt{250}$ f) $4\sqrt{27} + 2\sqrt{48} + 5\sqrt{108}$

Q3 Hint: To add surds, the $\sqrt{x}$ bit must be the same in each term.

Q4 Expand the following expressions, giving your answers in their simplest form.

a) $(1 + \sqrt{2})(2 + \sqrt{2})$ b) $(3 + 4\sqrt{3})(2 - \sqrt{3})$ c) $(\sqrt{11} + 2)(\sqrt{11} - 2)$

d) $(9 - 2\sqrt{5})(9 + 2\sqrt{5})$ e) $(\sqrt{3} + 2)^2$ f) $(3\sqrt{5} - 4)^2$

Q5 A parallelogram has base length $2\sqrt{6}$ cm and vertical height $\sqrt{3}$ cm.
Find the area of the parallelogram.
Give your answer in the form $p\sqrt{2}$ cm², where p is an integer.

PROBLEM SOLVING

Q6 A rectangular swimming pool is $3\sqrt{28}$ m long and $\sqrt{63}$ m wide.
What is the perimeter of the swimming pool?
Give your answer in the form $s\sqrt{7}$ m, where s is an integer.

PROBLEM SOLVING

Q7 Triangle ABC is right-angled with angle ABC = 90°.
Side AC has length $5\sqrt{2}$ cm and side AB has length $\sqrt{2}$ cm.
Find the length of side BC in the form $k\sqrt{3}$ cm, where k is an integer.

PROBLEM SOLVING

Rationalising the denominator

Surds are pretty complicated — they're probably the last thing you want on the bottom of a fraction. You can remove surds from the denominators of fractions by **rationalising the denominator**.

To rationalise the denominator you multiply **top and bottom** of the fraction by an **expression** that will get rid of surds in the denominator. (Multiplying a fraction by the same thing on the top and bottom doesn't change its value.)

Examples

a) **Show that $\frac{9}{\sqrt{3}} = 3\sqrt{3}$.**

To get rid of the surd, multiply the top and bottom by $\sqrt{3}$.

$$\frac{9}{\sqrt{3}} = \frac{9 \times \sqrt{3}}{\sqrt{3} \times \sqrt{3}} = \frac{9\sqrt{3}}{3} = 3\sqrt{3}$$

You can cancel 3 from top and bottom

b) **Rationalise the denominator of $\frac{1}{1 + \sqrt{2}}$.**

1. If the denominator is of the form $a + \sqrt{b}$, multiply top and bottom by $a - \sqrt{b}$ — just change the sign in front of the surd.
2. Using the difference of two squares rule (see page 13), the surds in the denominator cancel each other out.
3. Multiply top and bottom by –1 to give the answer in its simplest form.

$$\frac{1}{1+\sqrt{2}} \times \frac{1-\sqrt{2}}{1-\sqrt{2}} = \frac{1-\sqrt{2}}{(1+\sqrt{2})(1-\sqrt{2})}$$
$$= \frac{1-\sqrt{2}}{1^2 - \sqrt{2} + \sqrt{2} - \sqrt{2}^2}$$
$$= \frac{1-\sqrt{2}}{1-2}$$
$$= \frac{1-\sqrt{2}}{-1}$$
$$= -1 + \sqrt{2}$$

c) **Rationalise the denominator of $\frac{7 + \sqrt{5}}{3 + \sqrt{5}}$.**

1. Multiply top and bottom by $3 - \sqrt{5}$.
2. Expand the brackets on the top and bottom and simplify. The surds on the bottom cancel each other out.
3. Cancel 4 from the top and bottom of the fraction to give the answer in its simplest form.

$$\frac{7+\sqrt{5}}{3+\sqrt{5}} \times \frac{3-\sqrt{5}}{3-\sqrt{5}} = \frac{(7+\sqrt{5})(3-\sqrt{5})}{(3+\sqrt{5})(3-\sqrt{5})}$$
$$= \frac{(7 \times 3) - 7\sqrt{5} + 3\sqrt{5} - (\sqrt{5})^2}{3^2 - 3\sqrt{5} + 3\sqrt{5} - (\sqrt{5})^2}$$
$$= \frac{21 - 4\sqrt{5} - 5}{9 - 5}$$
$$= \frac{16 - 4\sqrt{5}}{4}$$
$$= 4 - \sqrt{5}$$

Exercise 3.3.2

Q1 Simplify the following, giving your answers in the form $p\sqrt{q}$, where q is an integer and p is an integer or fraction:

a) $\frac{6}{\sqrt{3}}$ b) $\frac{21}{\sqrt{7}}$ c) $\frac{30}{\sqrt{5}}$

d) $\sqrt{45} + \frac{15}{\sqrt{5}}$ e) $\frac{\sqrt{54}}{3} - \frac{12}{\sqrt{6}}$ f) $\frac{\sqrt{300}}{5} + \frac{30}{\sqrt{12}}$

g) $\frac{1}{\sqrt{18}} - \frac{1}{\sqrt{2}}$ h) $\frac{1}{\sqrt{28}} + \frac{3}{\sqrt{7}}$ i) $\frac{2}{\sqrt{72}} - \frac{5}{\sqrt{8}}$

Q2 Express the following in the form $a + b\sqrt{k}$, where a, b and k are integers:

a) $\frac{4}{1+\sqrt{3}}$ b) $\frac{8}{-1+\sqrt{5}}$ c) $\frac{18}{\sqrt{10}-4}$

d) $\frac{\sqrt{6}}{2-\sqrt{6}}$ e) $\frac{3}{5+2\sqrt{7}}$ f) $\frac{6}{3\sqrt{2}-4}$

Q3 Express the following in the form $p + q\sqrt{r}$, where r is an integer, and p and q are integers or fractions:

a) $\frac{\sqrt{2}+1}{\sqrt{2}-1}$ b) $\frac{\sqrt{5}+3}{\sqrt{5}-2}$ c) $\frac{3-\sqrt{3}}{4+\sqrt{3}}$

d) $\frac{3\sqrt{5}-1}{2\sqrt{5}-3}$ e) $\frac{\sqrt{2}+\sqrt{3}}{3\sqrt{2}-\sqrt{3}}$ f) $\frac{2\sqrt{7}-\sqrt{5}}{\sqrt{7}+2\sqrt{5}}$

g) $\frac{2\sqrt{2}+\sqrt{3}}{\sqrt{3}-\sqrt{12}}$ h) $\frac{6-4\sqrt{2}}{5\sqrt{2}-\sqrt{8}}$ i) $\frac{\sqrt{3}+3}{\sqrt{5}+\sqrt{15}}$

Q4 Express the following in the form $k(\sqrt{x} \pm \sqrt{y})$, where x and y are integers and k is an integer or fraction.

a) $\frac{4}{\sqrt{7}-\sqrt{3}}$ b) $\frac{24}{\sqrt{11}-\sqrt{17}}$ c) $\frac{2}{\sqrt{13}+\sqrt{5}}$

d) $\frac{\sqrt{5}}{\sqrt{6}+\sqrt{3}}$ e) $\frac{\sqrt{3}}{\sqrt{21}-3\sqrt{5}}$ f) $\frac{3\sqrt{2}}{2\sqrt{3}+\sqrt{20}}$

Q5 Find the reciprocal of $4\sqrt{3} - 2\sqrt{5}$. Give your answer in the form $k(\sqrt{x} \pm \sqrt{y})$, where x and y are integers and k is an integer or fraction. PROBLEM SOLVING

Q6 Solve the equation $12 = \sqrt{3}z$ giving your answer in the form $k\sqrt{3}$ where k is an integer. PROBLEM SOLVING

Q7 Solve the equation $8 = (\sqrt{5} - 1)x$ giving your answer in the form $a + b\sqrt{5}$ where a and b are integers. PROBLEM SOLVING

Q8 Solve the equation $5 + \sqrt{7} = (3 - \sqrt{7})y$ giving your answer in the form $p + q\sqrt{7}$ where p and q are integers. PROBLEM SOLVING

Q9 A rectangle has an area of $(2 + \sqrt{2})$ cm² and a width of $(3\sqrt{2} - 4)$ cm. Find the length of the rectangle. Give your answer in the form $a + b\sqrt{2}$ where a and b are integers. PROBLEM SOLVING

Review Exercise

Q1 Remove the brackets and simplify the following expressions:

a) $(a + b)(a - b)$ b) $(p + q)(p + q)$ c) $35xy + 25y(5y + 7x) - 100y^2$

d) $(x + 3y + 2)(3x + y + 7)$ e) $(c + 2d)(c - d)(2d - 3c)$ f) $[(s - 2t)(s + 2t)]^2$

Q2 The length of the longest side of a cuboid-shaped box is x cm.
The shortest side of the box is 10 cm shorter than the longest side,
and the other side is 8 cm shorter than the longest side.
Find an expression for the total volume of 5 of these boxes, in terms of x.
Your final answer should contain no brackets.

PROBLEM SOLVING

Q3 Fully factorise each of the following expressions:

a) $2x^2y + axy + 2xy^2$ b) $a^2x + a^2b^2x^2$ c) $16y + 8yx + 56x$

d) $24s + 60st + 15s^2t^2$ e) $27c - 9c^3d - 45cd^2$ f) $x^2y^2z^2 + x^3y^3 - x^2yz$

Q4 Write each of the following expressions as a product of factors:

a) $(x + 1) - y(x + 1)$ b) $z^4 + z^2(3 - z)$ c) $(x + y)^2 + x(x + y)$

d) $x(x - 2) + 3(2 - x)$ e) $25 - x^4$ f) $9b^2c^4 - 4c^2d^6$

Q5 Put the following expressions over a common denominator:

a) $\frac{2x}{3} + \frac{y}{12} + \frac{x}{5}$ b) $\frac{5}{xy^2} - \frac{2}{x^2y}$ c) $\frac{1}{x} + \frac{x}{x+y} + \frac{y}{x-y}$

d) $\frac{a}{b} + \frac{4}{a} - \frac{7}{a^2}$ e) $3x - \frac{4}{3xy}$ f) $\frac{2s}{t^2} + \frac{5}{2t} - \frac{t}{s^2}$

Q6 Simplify these expressions:

a) $\frac{2a}{b} - \frac{a}{2b}$ b) $\frac{2p}{p+q} + \frac{2q}{p-q}$ c) $\frac{c+d}{(c-d)^2} + \frac{1}{c+d}$

d) $\frac{1}{1+x} - \frac{1-x}{2x^2}$ e) $\frac{2k}{k^2-1} + \frac{k^2}{k-1}$ f) $\frac{4}{z+1} + \frac{2}{y+z} - \frac{6}{y-1}$

Q7 The diagram on the right shows part of a garden.
The combined area of the lawn and flower bed is $3x^2$ m².
The area of the flower bed is x^2 m².
Show that y can be expressed as:

$$y = \frac{x^2(2x-15)}{(x+6)(x-3)}$$

$(x + 6)$ m

y m

Lawn

$(x - 3)$ m

Flower bed

PROBLEM SOLVING

Review Exercise

Q8 Simplify:

a) $x^3 \cdot x^5$ b) $a^7 \cdot a^8$ c) $\frac{x^8}{x^2}$

d) $(a^2)^4$ e) $(xy^2) \cdot (x^3yz)$ f) $\frac{a^2b^4c^6}{a^3b^2c}$

Q9 Simplify:

a) $g^2 \times g^{-5}$ b) $p^4r^2 \div p^5r^{-6}$ c) $(k^{\frac{1}{3}})^6$

d) $(mn^8 \times m^4n^{-11})^{-2}$ e) $s^4t^3 \times \left(\frac{1}{s^2t^5}\right)^{-3}$ f) $\frac{a^2}{b^2c} \times \frac{b^6}{a^4c^{-2}} \div \frac{c^2}{a^3b}$

Q10 Work out the following:

a) $16^{\frac{1}{2}}$ b) $8^{\frac{1}{3}}$ c) $81^{\frac{3}{4}}$

d) x^0 e) $49^{-\frac{1}{2}}$ f) $\frac{1}{27^{-\frac{2}{3}}}$

Q11 Simplify:

a) $\sqrt{28}$ b) $\sqrt{\frac{5}{36}}$ c) $\sqrt{18}$ d) $\sqrt{\frac{9}{16}}$

Q12 Simplify the following expressions by writing them in the form $k\sqrt{x}$, where k and x are integers and x is as small as possible.

a) $\sqrt{3} - \sqrt{12}$ b) $3\sqrt{5} + \sqrt{45}$ c) $\sqrt{7} + \sqrt{448}$

d) $\sqrt{52} + \sqrt{117}$ e) $4\sqrt{150} + \sqrt{54} - \sqrt{5}\sqrt{120}$

Q13 The diagram on the right shows a shape which has been made by cutting a small square from one corner of a larger square.
The area of the larger square was 1920 cm².
The area of the smaller square was 1080 cm².
Find the value of a. Give your answer in the form $k\sqrt{x}$, where k and x are integers and x is as small as possible.

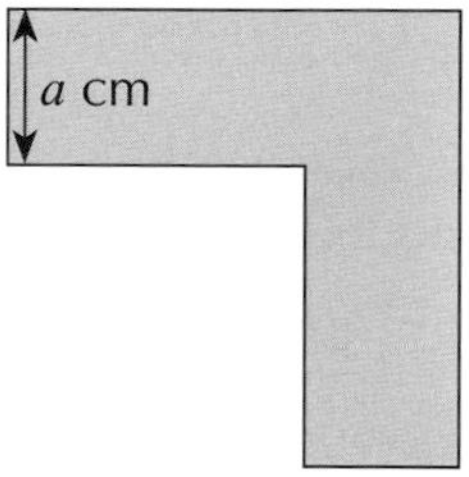

Q14 Find $(6\sqrt{3} + 2\sqrt{7})^2$.

Q15 Show that: a) $\frac{8}{\sqrt{2}} = 4\sqrt{2}$ b) $\frac{\sqrt{2}}{2} = \frac{1}{\sqrt{2}}$

Q16 Rationalise the denominator of $\frac{2}{3 + \sqrt{7}}$.

Q17 Write the following in the form $p + q\sqrt{r}$, where r is an integer, and p and q are integers or fractions:

a) $\frac{11 + \sqrt{13}}{5 - \sqrt{13}}$ b) $\frac{2\sqrt{7} + 9}{3 - \sqrt{7}}$ c) $\frac{3\sqrt{5} + \sqrt{15}}{\sqrt{60} - \sqrt{20}}$

Exam-Style Questions

Q1 Fully simplify $\frac{(3a^2b^2)^2 \times (2a^2b)^2}{(8a^6b^{-3})^{\frac{1}{3}}}$.

[2 marks]

Q2 Find x such that:

a) $9^x = 3$

[1 mark]

b) $9^{3x} \cdot 81^{2x-1} = 27$

[3 marks]

Q3 Show that $6^{\frac{1}{3}} + 6^{\frac{1}{3}} + 6^{\frac{1}{3}} = 2^p 3^q$, where p and q are constants to be found.

[2 marks]

Q4 Express $(3\sqrt{5} - 5\sqrt{3})^2$ in the form $a(b + \sqrt{c})$ where a, b and c are integers.

[3 marks]

Q5 Simplify $\frac{\sqrt{x}}{\sqrt{x} + \sqrt{y}} + \frac{\sqrt{y}}{\sqrt{x} - \sqrt{y}}$.

[3 marks]

Q6 Solve the simultaneous equations:

$$3x + 2y = 23 + 5\sqrt{5}$$
$$\sqrt{5}x + y = 12 + 4\sqrt{5}$$

[7 marks]

Q7 Three numbers are given as $5 - 2\sqrt{m}$, $10 + 9\sqrt{m}$ and $n\sqrt{m}$, where m and n are positive constants.

a) The mean of the three numbers is $5 + 4\sqrt{m}$. What is the value of n?

[2 marks]

b) The range of the three numbers is $5 + 11\sqrt{m}$. What range of values can m take?

[5 marks]

Q8 A circular pond has circumference 10π m.
A circular path of width x m is going to be built around the pond.
There is 20π m^2 of surfacing material available for the path.

Find the maximum possible value of x, giving x in the form $a\sqrt{b} \pm c$, where a, b and c are integers to be found.

[6 marks]

Chapter 4 Quadratics and Cubics

4.1 Quadratic Equations

In this section, you'll learn three methods for solving quadratic equations — factorising, completing the square and the quadratic formula. They should also help you to sketch graphs of quadratic functions.

Learning Objectives (Spec Ref 2.3):

- Solve quadratic equations by factorising.
- Solve quadratic equations by using the quadratic formula.
- Solve quadratic equations by completing the square.
- Solve quadratic equations involving a function of x.

Prior Knowledge Check: Be able to expand brackets (see p.10) and work with surds (see p.20-23).

Factorising a quadratic

Quadratic equations are equations of the general form: $ax^2 + bx + c = 0$
where a, b and c are constants (i.e. numbers) and $a \neq 0$.

Factorising a quadratic means putting it into two brackets called **factors** — the **solutions** to the equation can be easily worked out from these factors. There are **two cases** that you need to know: when $a = 1$, and when $a \neq 1$.

Factorising when $a = 1$

Fortunately, there's a step-by-step method you can follow when factorising this sort of quadratic:

To factorise a quadratic with $a = 1$:

1. Rearrange into the standard $ax^2 + bx + c$ form.
2. Write down the two brackets: $(x \quad)(x \quad)$
3. Find two numbers that multiply to give 'c' and add / subtract to give 'b' (ignoring signs).
4. Put the numbers in the brackets and choose their signs.

Tip: All quadratics can be rearranged into this standard form — but not all will factorise. Methods of solving quadratics that don't factorise are covered later in this chapter.

This will all make more sense once you've seen a worked example...

Example 1

Solve $x^2 - 8 = 2x$ by factorising.

1. Rearrange into standard $ax^2 + bx + c = 0$ form.

   Subtract $2x$ from both sides to give... $x^2 - 2x - 8 = 0$
   So $a = 1$, $b = -2$, $c = -8$.

2. Write down the two brackets with x's in: $x^2 - 2x - 8 = (x \quad)(x \quad)$

   Since $a = 1$, you know that there will be an x in each bracket, which will multiply together to give x^2.

Tip: Be careful with the values of b and c — don't let the minus signs catch you out.

3. Find two numbers that **multiply** together to make c but which also **add or subtract** to give b. You can ignore any minus signs for now.

1 and 8 multiply to give 8 — and add / subtract to give 9 and 7.
2 and 4 multiply to give **8** — and add / subtract to give 6 and **2**.

These are the values for c and b you're after — so this is the right combination: **2** and **4**.

4. Find the signs. So far you've got: $x^2 - 2x - 8 = (x \quad 2)(x \quad 4)$

Now all you have to do is put in the plus or minus signs.
It must be +2 and –4 because
$2 \times (-4) = -8$ and
$2 + (-4) = 2 - 4 = -2$

$x^2 - 2x - 8 = (x + 2)(x - 4)$

Tip: If two things multiplied together give a negative answer, they must have opposite signs. So if c is negative, then the signs must be opposite.

5. Now that you've factorised using the step by step method — you can use the factors to solve the equation: $(x + 2)(x - 4) = 0$

The factors (brackets) multiply to give 0, so one of them **must** be 0. $\Rightarrow x + 2 = 0$ or $x - 4 = 0$

Don't forget this last step. The factors aren't the answer. → $\Rightarrow x = -2$ or $x = 4$

Example 2

Solve $x^2 + 4x - 21 = 0$ by factorising.

1. It's already in the standard form, so start by writing down the brackets: $x^2 + 4x - 21 = (x \quad)(x \quad)$

2. Find the numbers.

1 and 21 multiply to give 21 — and add / subtract to give 22 and 20.
3 and 7 multiply to give **21** — and add / subtract to give 10 and **4**.

These are the values you need, so 3 and 7 are the right numbers, so: $x^2 + 4x - 21 = (x \quad 3)(x \quad 7)$

3. c is negative so we must need opposite signs.
The signs must be –3 and +7
because $7 - 3 = 4$ and $7 \times (-3) = -21$

$x^2 + 4x - 21 = (x - 3)(x + 7)$

4. Solve the equation to find x.
If two things multiply together to give 0, one of them must be equal to 0.

$(x - 3)(x + 7) = 0$

$\Rightarrow x = 3$ or $x = -7$

Factorising when $a \neq 1$

The basic method's the same as before — but it can be a bit more awkward.

<u>To factorise a quadratic with $a \neq 1$</u>:

1. Rearrange into the standard $ax^2 + bx + c$ form.
2. Write down the two brackets, but instead of just having x in each, you need two things that will multiply to give ax^2:
   $$(nx \quad)(mx \quad)$$
   where n and m are two numbers that multiply to give a.
3. Find two numbers that multiply to give 'c' but which will give you bx when you multiply them by nx and mx, and then add / subtract them.
4. Put the numbers in the brackets and choose their **signs**.

Tip: In practice, this third step is a case of working through all possible cases until you get it right.

Again, a worked example will help.

Example 3

Factorise $3x^2 + 4x - 15$.

1. This quadratic's already in the standard form so you don't need to rearrange it.
2. As before, write down two brackets — but instead of just having x in each, you need two things that will multiply to give $3x^2$.

   It's got to be $3x$ and x here. $3x^2 + 4x - 15 = (3x \quad)(x \quad)$

3. Work out the numbers. You need to find two numbers that multiply together to make 15 — but which will give you $4x$ when you multiply them by x and $3x$, and then add / subtract them.

   It's a good idea to write out the brackets for each possible number combination — it makes it easier to see if you've got the right numbers.

   $(3x \quad 1)(x \quad 15) \Rightarrow x$ and $45x$
   which then add or subtract to give $46x$ and $44x$.

   $(3x \quad 15)(x \quad 1) \Rightarrow 15x$ and $3x$
   which then add or subtract to give $18x$ and $12x$.

   $(3x \quad 3)(x \quad 5) \Rightarrow 3x$ and $15x$
   which then add or subtract to give $18x$ and $12x$.

   This is the value you're after — so this is the right combination.

   $(3x \quad 5)(x \quad 3) \Rightarrow 5x$ and $9x$
   which then add or subtract to give $14x$ and $\mathbf{4x}$.

4. Add the signs.
   You know the brackets must be like these: $(3x \quad 5)(x \quad 3) = 3x^2 + 4x - 15$
   'c' is negative — that means the signs in the brackets are opposite.
   You've only got two choices for the signs of the numbers, -5 and 3 or 5 and -3.
   If you're unsure which it is, just multiply each case out to see which is right.
   The numbers must be -5 and $+3$
   since $9x - 5x = 4x$ and $-5 \times 3 = -15$. $(3x - 5)(x + 3) = 3x^2 + 4x - 15$

Exercise 4.1.1

Q1 Factorise the following expressions.

a) $x^2 - 6x + 5$
b) $x^2 - 3x - 18$
c) $x^2 + 22x + 121$
d) $x^2 - 12x$
e) $y^2 - 13y + 42$
f) $x^2 + 51x + 144$
g) $x^2 - 121$
h) $x^2 - 35x + 66$

Q1 Hint: If b or c is zero, use the factorising methods from Chapter 3.

Q2 Solve the following equations.

a) $x^2 - 2x - 8 = 0$
b) $2x^2 + 2x - 40 = 0$
c) $p^2 + 21p + 38 = 0$
d) $x^2 - 15x + 54 = 0$
e) $x^2 + 18x = -65$
f) $x^2 - x = 42$
g) $x^2 + 1100x + 100\,000 = 0$
h) $3x^2 - 3x - 6 = 0$

Q2 Hint: Look out for questions where the equation can be simplified before factorising — for example by dividing through by a number.

Q3 Factorise the following expressions.

a) $4x^2 - 4x - 3$
b) $2x^2 + 23x + 11$
c) $7x^2 - 19x - 6$
d) $-x^2 - 5x + 36$
e) $6x^2 - 7x - 3$
f) $2x^2 - 2$
g) $3x^2 - 3$
h) $-x^2 + 9x - 14$

Q4 Solve the following equations.

a) $-5x^2 - 22x + 15 = 0$
b) $32x^2 + 60x + 13 = 0$
c) $5a^2 + 12a = 9$
d) $8x^2 + 22x + 15 = 0$
e) $4q^2 + 6 = 11q$
f) $24y^2 + 23y - 12 = 0$

Q5 Solve $(x - 1)(x - 2) = 37 - x$.

Q6 $f(x) = -x^2 + 7x + 30$.
Find the x-coordinates of the point or points at which the graph of $f(x)$ meets the x-axis.

PROBLEM SOLVING

Q7 $f(x) = (x - 8)(x + 10)$ and $g(x) = (3x + 2)(x - 11)$.
Find the x-coordinates of the points of intersection of the two functions.

Q7 Hint: Solve for x where $f(x) = g(x)$.

PROBLEM SOLVING

Q8 In a scientific experiment, the temperature, T °C, is modelled by the equation $T = -2h^2 + 13h - 20$, where h is the time in hours from the start of the experiment.
Find both times at which the temperature is 0 °C.

MODELLING

Q9 Factorise $x^2 + 6xy + 8y^2$.

Q10 The perimeter of a square is $(2z + 5)$ cm.
Find the possible values of z if the area of the square is $\frac{9z}{2}$ cm^2.

PROBLEM SOLVING

Q11 The area of the triangle shown on the right is $(9x + 3)$ m^2.
By solving a quadratic equation to find x, find the exact length of the hypotenuse.

$(5x - 3)$ m
$(x + 2)$ m

PROBLEM SOLVING

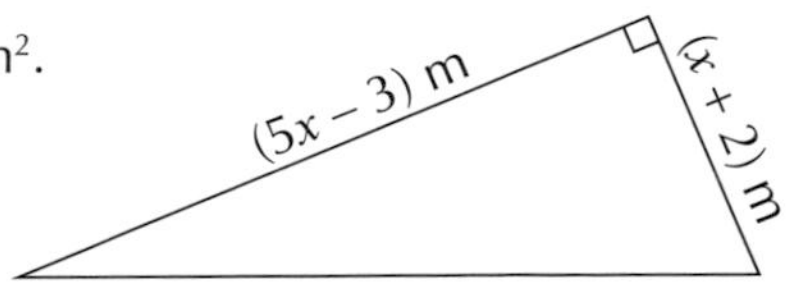

The quadratic formula

You should now be comfortable with solving quadratics by factorising.
But there are two important points to bear in mind:

- The expression **won't** always factorise.
- Sometimes factorising is so messy that it's **easier** to just use other methods.

So if the question doesn't tell you to factorise, **don't assume** it will factorise.

Example 1

Solve $6x^2 + 87x - 144 = 0$.

This will actually factorise, but there are 2 possible bracket forms to try: $(6x \quad)(x \quad)$ or $(3x \quad)(2x \quad)$

And for each of these, there are 8 possible ways of making 144 to try.

If you tried to factorise this example, you'd be going all day. Luckily, there's a formula which will work out the **solutions** of a quadratic equation, even when you can't factorise — it's known as **the quadratic formula**.

If $ax^2 + bx + c = 0$ then:

$$x = \frac{-b \pm \sqrt{b^2 - 4ac}}{2a}$$

Example 2

Solve the quadratic equation $3x^2 - 4x = 8$, leaving your answer in surd form.

The mention of surds in the answer suggests that the quadratic will be too hard to factorise, so we'll use the quadratic formula instead.

Tip: If the question asks you to give your answer in surd form or as a decimal, that's a big hint to use the quadratic formula instead of trying to factorise.

1. Get the equation in the standard $ax^2 + bx + c = 0$ form.

$$3x^2 - 4x = 8$$
$$\Rightarrow 3x^2 - 4x - 8 = 0$$

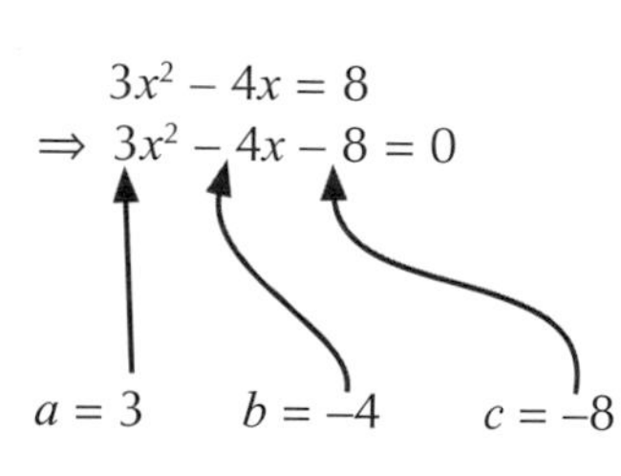

2. Write down the coefficients a, b and c — making sure you don't forget minus signs.

$a = 3 \qquad b = -4 \qquad c = -8$

3. Very carefully, plug these numbers into the formula. It's best to write down each stage as you do it.

If any of the coefficients in your quadratic equation are negative, be especially careful. There are a couple of minus signs in the formula which can catch you out if you're not paying attention.

$$x = \frac{-b \pm \sqrt{b^2 - 4ac}}{2a}$$

$$x = \frac{-(-4) \pm \sqrt{(-4)^2 - 4 \times 3 \times (-8)}}{2 \times 3}$$

These minus signs multiply together to get a plus.

$$x = \frac{4 \pm \sqrt{16 + 96}}{6}$$

4. Simplify your answer as much as possible, using the rules of surds.

$$x = \frac{4 \pm \sqrt{112}}{6}$$

$$x = \frac{4 \pm \sqrt{16}\sqrt{7}}{6}$$

$$x = \frac{4 \pm 4\sqrt{7}}{6}$$

$$x = \frac{2 \pm 2\sqrt{7}}{3}$$

Tip: See p.20 for a reminder of these rules.

The ± sign means that you actually have two different expressions for x, which you get by replacing the ± with + and –.
Doing this gives you the two solutions to the quadratic equation.

$$x = \frac{2 + 2\sqrt{7}}{3} \quad \text{or} \quad x = \frac{2 - 2\sqrt{7}}{3}$$

Example 3

Solve the quadratic equation $2x^2 = 4x + 3$, leaving your answer in the form $p \pm q\sqrt{r}$ where p, q and r are whole numbers or fractions.

Rearranging $2x^2 = 4x + 3$ you get $2x^2 - 4x - 3 = 0$ and so $a = 2$, $b = -4$ and $c = -3$.

So plugging these values into the quadratic formula, you get:

$$x = \frac{-b \pm \sqrt{b^2 - 4ac}}{2a}$$

$$x = \frac{-(-4) \pm \sqrt{(-4)^2 - 4 \times 2 \times (-3)}}{2 \times 2}$$

$$x = \frac{4 \pm \sqrt{16 + 24}}{4} = \frac{4 \pm \sqrt{40}}{4} = \frac{4 \pm 2\sqrt{10}}{4} = \frac{2 \pm \sqrt{10}}{2} = \frac{2}{2} \pm \frac{1}{2}\sqrt{10} = 1 \pm \frac{1}{2}\sqrt{10}$$

Using a calculator

You can also use a **graphical calculator** to solve quadratic equations. There are two different ways to do this:

- Use the calculator to **plot** the graph of the quadratic, then work out where it crosses the x-axis — these values of x are the solutions to $f(x) = 0$ (there's more about this on pages 41-45).
- Some calculators will allow you to **solve** quadratic equations directly — enter the values for a, b and c, and it'll give you the solutions.

Tip: Different calculators work in different ways — make sure you know how to do this on your calculator.

In each case, you'll have to make sure your equation is in the **standard form** ($ax^2 + bx + c$). Be careful though — sometimes your calculator will only give you one solution (even if there are two), and it won't usually give answers in surd form.

Exercise 4.1.2

Q1 Solve the following equations using the quadratic formula, giving your answers in surd form where necessary.

a) $x^2 - 4x = -2$

b) $x^2 - 2x - 44 = 0$

c) $x^2 + 3x = 12$

d) $x^2 - 14x + 42 = 0$

e) $4x^2 + 4x - 1 = 0$

f) $-x^2 + 4x - 3 = 0$

g) $x^2 - \frac{5}{6}x + \frac{1}{6} = 0$

h) $x^2 - 2\sqrt{11}x + 11 = 0$

Q1 Hint: Have a go at solving these equations using a calculator too.

Q2 a) Multiply out $(x - 2 + \sqrt{5})(x - 2 - \sqrt{5})$.

b) Solve the equation $x^2 - 4x - 1 = 0$ using the quadratic formula.

c) How does your answer to b) relate to the expression given in a)?

Q3 The roots of the equation $x^2 + 8x + 13 = 0$ can be written in the form $x = A \pm \sqrt{B}$ where A and B are integers. Find A and B.

Q4 Solve the following equations, giving your answers in surd form where necessary.

a) $x^2 + x + \frac{1}{4} = 0$

b) $x^2 - \frac{7}{4}x + \frac{2}{3} = 0$

c) $25x^2 - 30x + 7 = 0$

d) $60x - 5 = -100x^2 - 3$

e) $2x(x - 4) = 7 - 3x$

f) $(3x - 5)(x + 2) = 3x - 2$

Q5 One of the roots of the equation $kx^2 + 4x - 2 = 0$ is $x = \frac{\sqrt{14}}{5} - \frac{2}{5}$.
Find k, and state the exact value of the other root.

Q6 A shopper leaves a supermarket at 2 pm and walks home, which is 1 km away. The walk can be modelled by the equation $d = 2t^2 - 5t + 1$, where d is their distance from home in km, and t is the time taken in hours. At what time, to the nearest minute, will they arrive home?

Completing the square

You could be asked to **solve** a quadratic equation by **completing the square** so you need to know this method just as well as the others.
And what's more — it gives you loads of **useful information** about the quadratic.

> Completing the square just means writing a quadratic expression $ax^2 + bx + c$ in the form: $a(x + \text{something})^2 + d$
>
> - Basically, the '**square**' is this bit: $a(x + \text{something})^2$
>   The 'something' is chosen so that it will produce the correct x^2 and x terms when the square is multiplied out.
> - But this square won't always give the right constant term — so you need to '**complete**' it by adding a number to the square to make it the **same** as the original quadratic: $a(x + \text{something})^2 + d$

The method can seem complicated at first, but is actually very simple when you get it.
As always, working through examples is the best way to learn it.

When $a = 1$

We'll start with the slightly easier case of $a = 1$...

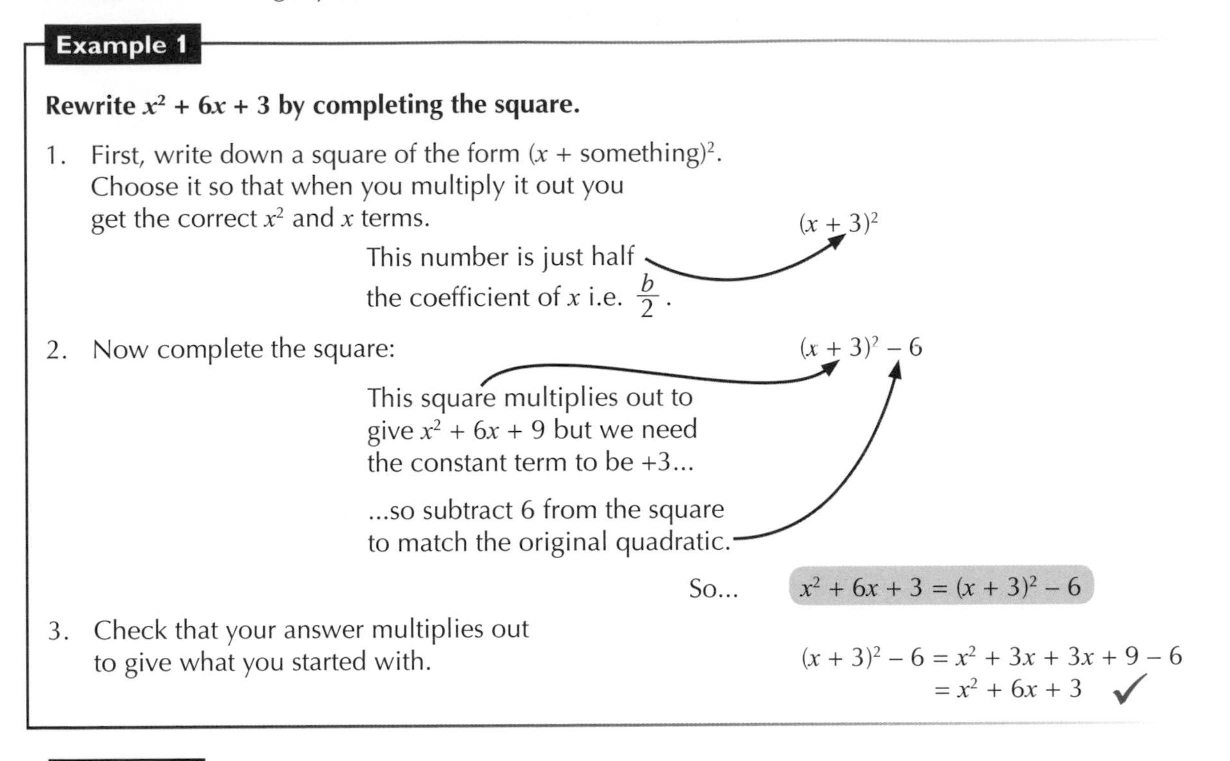

Example 1

Rewrite $x^2 + 6x + 3$ by completing the square.

1. First, write down a square of the form $(x + \text{something})^2$. Choose it so that when you multiply it out you get the correct x^2 and x terms. $(x + 3)^2$

   This number is just half the coefficient of x i.e. $\frac{b}{2}$.

2. Now complete the square: $(x + 3)^2 - 6$

   This square multiplies out to give $x^2 + 6x + 9$ but we need the constant term to be +3...

   ...so subtract 6 from the square to match the original quadratic.

   So... $x^2 + 6x + 3 = (x + 3)^2 - 6$

3. Check that your answer multiplies out to give what you started with.

$$(x + 3)^2 - 6 = x^2 + 3x + 3x + 9 - 6$$
$$= x^2 + 6x + 3 \checkmark$$

Example 2

Rewrite $x^2 - 5x - 1$ by completing the square.

1. Again, start by writing down the square. This example has a negative coefficient of x — so make sure you have a minus sign in the brackets. $\left(x - \frac{5}{2}\right)^2$

   Remember, this is just $\frac{b}{2}$.

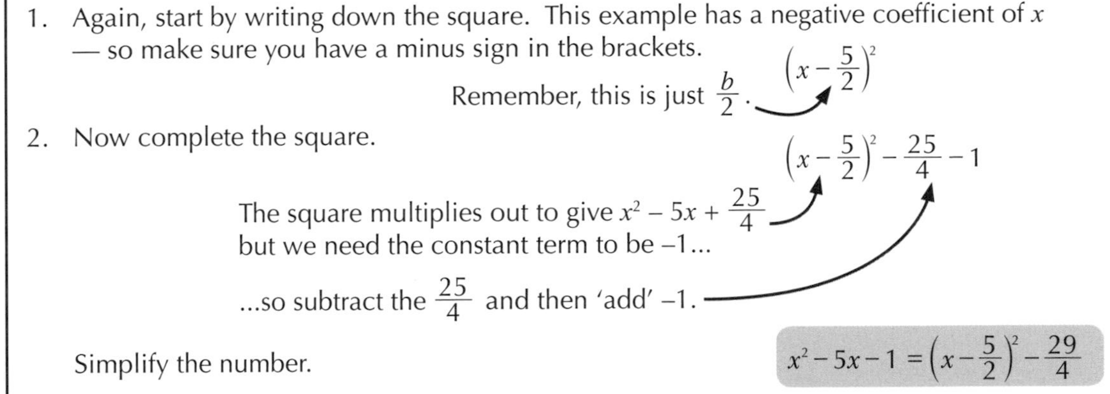

2. Now complete the square. $\left(x - \frac{5}{2}\right)^2 - \frac{25}{4} - 1$

   The square multiplies out to give $x^2 - 5x + \frac{25}{4}$ but we need the constant term to be -1...

   ...so subtract the $\frac{25}{4}$ and then 'add' -1.

   Simplify the number. $x^2 - 5x - 1 = \left(x - \frac{5}{2}\right)^2 - \frac{29}{4}$

   (You can always find the number that completes the square by subtracting off the number term you get from the bracket and adding on the number term from the original quadratic.)

3. Check your answer.

$$\left(x - \frac{5}{2}\right)^2 - \frac{29}{4}$$
$$= x^2 - \frac{5}{2}x - \frac{5}{2}x + \frac{25}{4} - \frac{29}{4}$$
$$= x^2 - 5x - 1 \checkmark$$

When $a \neq 1$

It's a little more complicated in cases where a is not 1. You have to put a outside of the squared bracket, and allow for this when choosing the number to go inside the bracket — basically by dividing by a.

Tip: The formula for completing the square for the general quadratic $ax^2 + bx + c$ is: $a\left(x+\frac{b}{2a}\right)^2 + \left(c - \frac{b^2}{4a}\right)$

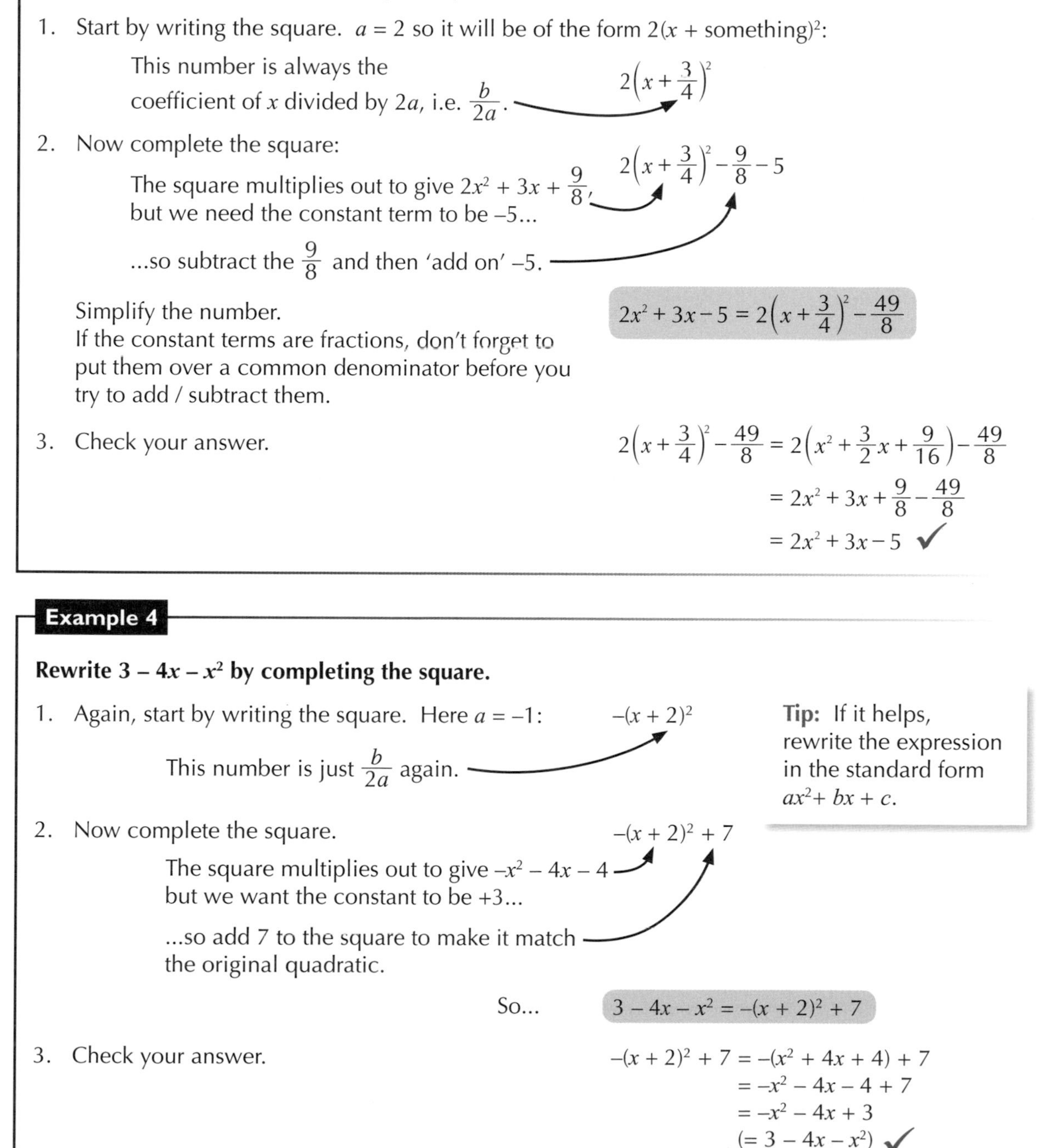

Example 3

Rewrite $2x^2 + 3x - 5$ by completing the square.

1. Start by writing the square. $a = 2$ so it will be of the form $2(x + \text{something})^2$:

   This number is always the coefficient of x divided by $2a$, i.e. $\frac{b}{2a}$. $2\left(x+\frac{3}{4}\right)^2$

2. Now complete the square: $2\left(x+\frac{3}{4}\right)^2 - \frac{9}{8} - 5$

   The square multiplies out to give $2x^2 + 3x + \frac{9}{8}$, but we need the constant term to be –5...

   ...so subtract the $\frac{9}{8}$ and then 'add on' –5.

   Simplify the number.
   If the constant terms are fractions, don't forget to put them over a common denominator before you try to add / subtract them.

   $$2x^2 + 3x - 5 = 2\left(x+\frac{3}{4}\right)^2 - \frac{49}{8}$$

3. Check your answer.

$$2\left(x+\frac{3}{4}\right)^2 - \frac{49}{8} = 2\left(x^2 + \frac{3}{2}x + \frac{9}{16}\right) - \frac{49}{8}$$
$$= 2x^2 + 3x + \frac{9}{8} - \frac{49}{8}$$
$$= 2x^2 + 3x - 5 \checkmark$$

Example 4

Rewrite $3 - 4x - x^2$ by completing the square.

1. Again, start by writing the square. Here $a = -1$: $-(x + 2)^2$

   This number is just $\frac{b}{2a}$ again.

Tip: If it helps, rewrite the expression in the standard form $ax^2 + bx + c$.

2. Now complete the square. $-(x + 2)^2 + 7$

   The square multiplies out to give $-x^2 - 4x - 4$ but we want the constant to be +3...

   ...so add 7 to the square to make it match the original quadratic.

   So... $3 - 4x - x^2 = -(x + 2)^2 + 7$

3. Check your answer.

$$-(x + 2)^2 + 7 = -(x^2 + 4x + 4) + 7$$
$$= -x^2 - 4x - 4 + 7$$
$$= -x^2 - 4x + 3$$
$$(= 3 - 4x - x^2) \checkmark$$

Once you've completed the square, a quadratic equation becomes very easy to **solve**:

- Take the **constant term** to the other side of the equals sign.
- Take the square root of both sides — don't forget the **negative** square root.
- **Rearrange** to find the solutions.

Example 5

Solve $3 - 4x - x^2 = 0$ by completing the square.

From Example 4, you can write $3 - 4x - x^2$ as $-(x + 2)^2 + 7$ by completing the square.
So now all you need to do is set this equal to zero and rearrange.

Tip: When you take the square root of something, you need to put a ± sign in front of the √ sign.

$$-(x + 2)^2 + 7 = 0$$

Take the constant to the other side. → $-(x + 2)^2 = -7$

$$(x + 2)^2 = 7$$

Take a square root — don't forget the ± sign. → $x + 2 = \pm\sqrt{7}$

Subtract 2 from both sides. → $x = -2 \pm\sqrt{7}$

So $x = -2 + \sqrt{7}$ or $x = -2 - \sqrt{7}$

Exercise 4.1.3

Q1 Solve the following equations, leaving your answer in surd form where appropriate:

a) $(x + 4)^2 = 25$ b) $(2x + 5)^2 = 9$ c) $(5x - 3)^2 = 21$

Q1 Hint: In these questions you don't need to complete the square — they'll just give you practice at the 'solving' bit.

Q2 Rewrite the following expressions in the form $p(x + q)^2 + r$:

a) $x^2 + 6x + 8$ b) $x^2 + 8x - 10$ c) $x^2 - 3x - 10$

d) $x^2 - 20x + 15$ e) $x^2 - 2mx + n$ f) $x^2 + 6tx + s$

g) $3x^2 - 12x + 7$ h) $2x^2 - 4x - 3$ i) $6x^2 + 30x - 20$

j) $-x^2 - 9x + 9$ k) $4x^2 - 22x + 5$ l) $-3x^2 + 9x + 1$

Q3 Solve the following equations by completing the square:

a) $x^2 - 6x - 16 = 0$ b) $p^2 - 10p = 200$ c) $x^2 + 2x + k = 0$

d) $x^2 + 4x - 8 = 0$ e) $4x^2 + 24x - 13 = 0$ f) $9x^2 + 18x = 16$

g) $2x^2 - 12x + 9 = 0$ h) $2x^2 - 12x - 54 = 0$ i) $5x^2 + 10x = 1$

j) $-3x^2 - 18x + 2 = 0$ k) $3x^2 + 2x = \frac{7}{6}$ l) $5x^2 - 3x + \frac{2}{5} = 0$

Q4 By completing the square, show that the expression $3x^2 - 12x + 14$ is positive for all x. PROBLEM SOLVING

Q5 By completing the square, show that the solutions to $ax^2 + bx + c = 0$ are $x = \frac{-b \pm \sqrt{b^2 - 4ac}}{2a}$. PROBLEM SOLVING

Quadratics involving functions of x

Sometimes you'll be asked to solve an equation that doesn't look like a quadratic — it might involve different powers of x, or functions like $\sin x$ or e^x. However, as long as it's in the form **a(something)^2 + b(something) + c**, you can solve it like a normal quadratic. Instead of '$x =$ ', you'll be left with 'something = ', which you'll then have to solve.

The way to solve these equations is:

1. Identify the function of x, and replace it with a different letter, say u.
2. Solve the resulting quadratic equation for u.
3. Replace u with the original function, and solve it to find x.

This will make more sense with a worked example:

Tip: There's more on solving trig equations in Chapter 8, and on solving exponential equations in Chapter 9.

Example

PROBLEM SOLVING

Solve the equation $x^{\frac{2}{3}} + 3x^{\frac{1}{3}} - 40 = 0$.

1. To treat this as a quadratic, you need to find a function of x (u) that allows you to rewrite the equation in the form $au^2 + bu + c$.

   If you take $u = x^{\frac{1}{3}}$, then u^2 would be $x^{\frac{2}{3}}$ — this fits with the equation.
   So the equation can be written as:
   $u^2 + 3u - 40 = 0$, where $u = x^{\frac{1}{3}}$.

2. Now solve this quadratic — luckily it factorises.

   $u^2 + 3u - 40 = 0 \Rightarrow (u + 8)(u - 5) = 0$
   So $u = -8$ or $u = 5$.

3. Finally, substitute $x^{\frac{1}{3}}$ back into the equations and solve for x.

   $x^{\frac{1}{3}} = -8$, so $x = (-8)^3 = -512$ or
   $x^{\frac{1}{3}} = 5$, so $x = 5^3 = 125$

Exercise 4.1.4

Q1 Find an expression for u, in terms of x, that allows you to write each equation below in the form $au^2 + bu + c = 0$. You do **not** need to solve the resulting equations.

a) $2x + 4x^{\frac{1}{2}} - 7 = 0$ b) $e^x(e^x - 6) = 8$ c) $5^x + 5^{2x} = 4$ d) $2\cos^2 x + 3 = 5\cos x$

Q2 a) Rewrite the equation $x^2 + 6x + 7$ in the form $(x + m)^2 + n$.

b) Hence solve the equation $(2x + 1)^2 + 6(2x + 1) + 7 = 0$. Leave your answers in surd form.

Q3 Find all four solutions to the equation $x^4 - 17x^2 + 16 = 0$. PROBLEM SOLVING

Q4 Solve the equations: PROBLEM SOLVING

a) $\dfrac{3}{(5x+2)^2} + \dfrac{1}{5x+2} = 10$ b) $3x + \sqrt{x} = 14$

Q5 An ornithologist predicts the increase in population of a bird species in a coastal area using the equation: $p = 5y^3 - y^6$, where p is the population growth in 1000s of birds, and y is the time from now in 100s of years. Using this model, what is the shortest time it would take, to the nearest year, for there to be 1000 more birds in the area? MODELLING

4.2 Quadratic Functions and Roots

The roots of a quadratic function f(x) are just the solutions to the equation f(x) = 0.
But you don't actually need to solve the equation to find out how many roots there are.

Learning Objectives (Spec Ref 2.3):
- Identify the number of real roots of a quadratic function.
- Calculate the discriminant of a quadratic function.
- Use the discriminant to solve problems involving quadratics with unknown coefficients.

Prior Knowledge Check:
Be able to solve equations and inequalities — this should be familiar from GCSE.

The roots of a quadratic function

Quadratic functions are just functions of the form $f(x) = ax^2 + bx + c$. Their graphs all have the same **general shape**, no matter what the values of a, b and c are. This general shape is called a **parabola**. Parabolas are either '**u**'-shaped or '**n**'-shaped:

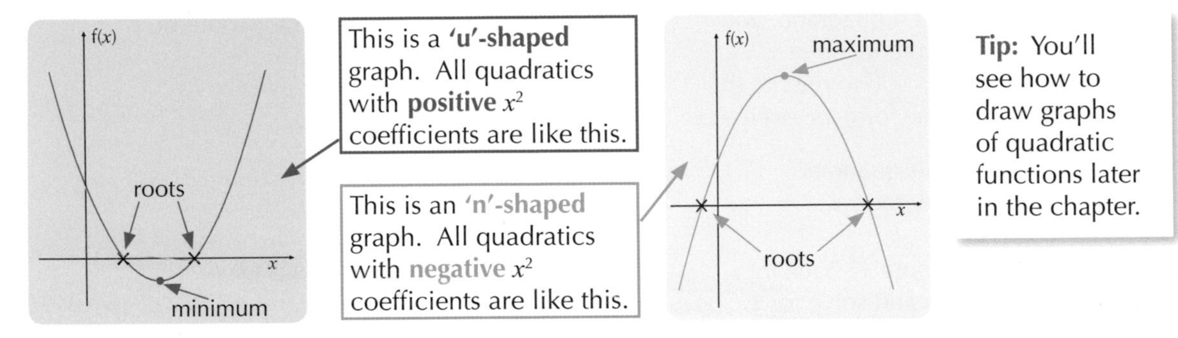

Tip: You'll see how to draw graphs of quadratic functions later in the chapter.

The **roots** of a quadratic function are the values of x where the function f(x) is equal to **zero** — i.e. where the graph **crosses the x-axis**. They are the **solutions** to the quadratic equation f(x) = 0. The functions above each have 2 roots because their graphs cross the x-axis twice.

A quadratic function may have **0**, **1** or **2 roots**. You'll see two methods for finding out which it is — **completing the square** and using the **discriminant**.

Using the completed square

If you've already **completed the square**, you can easily work out the number of roots by examining the completed square.

The function will look like this: $\mathbf{f(x) = p(x + q)^2 + r}$

The key to this method is remembering that anything squared is ≥ 0.

So, let's assume for now that p is positive:

- Since p is **positive**, the graph will be **u-shaped** and have a minimum.
- The smallest value that f(x) can take will occur when the bracket is 0 (since the square is ≥ 0). At that point f(x) is just r, and x must be $-q$.
- So the minimum is $\mathbf{(-q\ ,\ r)}$.
  This also tells you that the graph of f(x) has a **line of symmetry** at $\mathbf{x = -q}$.

Now the **value of r** tells us the number of roots — the number of times the graph crosses the x-axis depends on whether the minimum is above, below or on the axis.

- If $r < 0$, the minimum is below the x-axis, so the graph must cross the axis twice — meaning there are **two roots**. (Picture the u-shaped graph.)
- If $r > 0$, the graph is always above the x-axis — so there are **no roots**.
- If $r = 0$, the minimum point is on the x-axis, so there's **one root**.

$$\mathbf{f}(x) = p(x + q)^2 + r$$

Next, we'll see what happens when p is negative:

- Since p is **negative**, the graph will be **n-shaped** and have a **maximum**.
- And also because p is negative, the highest value of $p(x + q)^2$ will be when the bracket is 0. At that point f(x) is just r, and x is $-q$.
- So the maximum is **$(-q, r)$**. Again, this tells you that the graph has a **line of symmetry** at **$x = -q$**.

Tip: The coordinates of the maximum are actually just the same as those we found for the minimum: $(-q, r)$.

Look at the **value of r** to work out the number of roots.

- If $r < 0$, the graph is always below the x-axis — so there are **no roots**.
- If $r > 0$, the maximum is above the x-axis, so the graph must cross the axis twice, meaning there are **two roots**.
- If $r = 0$, the maximum point is on the x-axis, so there's **one root**.

Let's see what this all means for a few functions.

Tip: Remember, real numbers are just all the rational and irrational numbers. Look at p.41 for an explanation of what is meant by 'real' roots.

Two real roots

$y = x^2 - 6x + 8$

- The completed square is $(x - 3)^2 - 1$.
- The minimum is $(3, -1)$ which is below the x-axis.
- So there are two roots ($x = 2$ and $x = 4$).

One real root

$y = x^2 - 6x + 9$

- The completed square is $(x - 3)^2$.
- The minimum is $(3, 0)$, so the graph just touches the x-axis.
- $x = 3$ is the only root.

Tip: When you're factorising a quadratic equation, if both factors come out the same, in this case $(x - 3)(x - 3)$, the function has one root. We call this one **repeated** root.

No real roots

$y = x^2 - 6x + 10$

- The completed square is $(x - 3)^2 + 1$.
- The minimum is $(3, 1)$ which is above the x-axis.
- So the graph never touches the x-axis, and there are no roots.

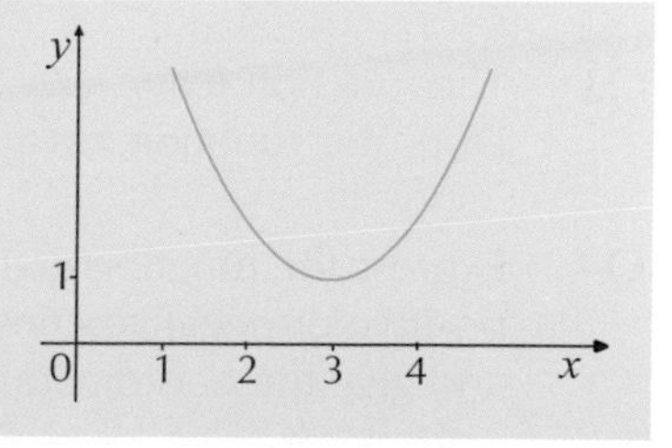

All the different cases we've covered can actually be summarised in these three simple rules:

> For a quadratic function of the form $f(x) = p(x + q)^2 + r$:
> - If p and r have **different signs**, the function has **two** real roots.
> - If $r = 0$ then the function has **one** real root.
> - If p and r have the **same sign**, the function has **no** real roots.

Example

How many real roots does the quadratic function $f(x) = x^2 + 4x + 7$ have?

Completing the square, you can rewrite the function in the form $f(x) = p(x + q)^2 + r$:

$f(x) = (x + 2)^2 + 3$

$p = 1$ and $r = 3$ are of the same sign, so the function has no real roots.

You can see why this works using the following argument:

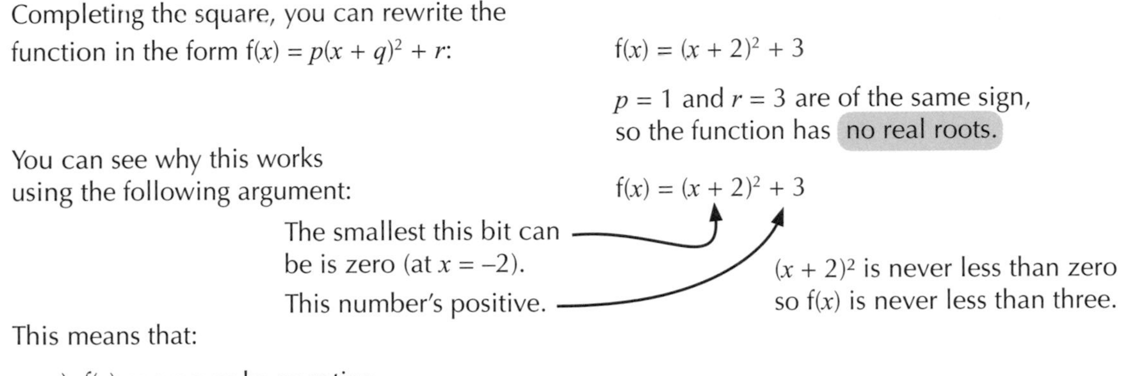

$f(x) = (x + 2)^2 + 3$

The smallest this bit can be is zero (at $x = -2$).

This number's positive.

$(x + 2)^2$ is never less than zero so $f(x)$ is never less than three.

This means that:

a) $f(x)$ can never be negative.

b) The graph of $f(x)$ never crosses the x-axis — so there are no real roots.

Exercise 4.2.1

Q1 How many real roots does each quadratic function have?

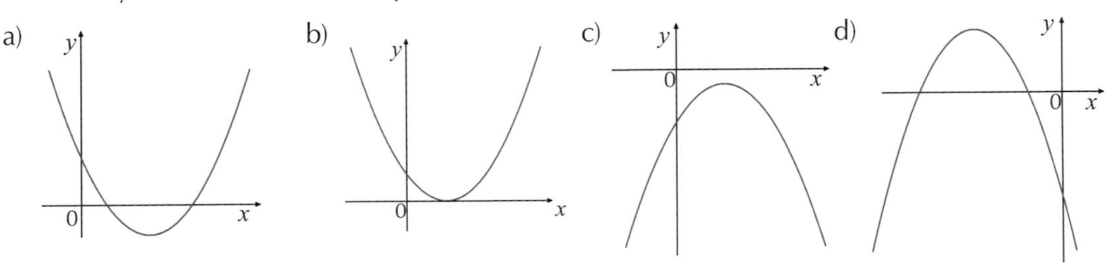

Q2 Express $f(x) = x^2 + 6x + 10$ in the form $f(x) = (x + q)^2 + r$, where q and r are positive or negative constants. Using your answer, state whether $f(x)$ has any real roots and give the equations of any lines of symmetry of the graph of $f(x)$.

Q2 Hint: Complete the square.

Q3 The function $f(x) = -x^2 - 7x - 6$ can be expressed in the form $f(x) = -\left(x + \frac{7}{2}\right)^2 + \frac{25}{4}$. Does this function have any real roots? Explain your answer.

Q4 Express the function $g(x) = 4x^2 - 3x - 5$ in the form $g(x) = p(x + q)^2 + r$, where p, q and r are positive or negative constants. Use your answer to find the number of real roots $g(x)$ has, and give the coordinates of the turning point of the graph of $g(x)$.

Q5 The function $f(x) = -2x^2 + 3x + k$ has one real root.

a) Find the value of k.

b) Give the equation of the line of symmetry of the graph of $f(x)$.

c) State the coordinates of the turning point of the graph of $f(x)$ and whether it is a maximum or a minimum.

PROBLEM SOLVING

Using the discriminant

Remember the **quadratic formula** for solving an equation of the form $ax^2 + bx + c = 0$:

$$x = \frac{-b \pm \sqrt{b^2 - 4ac}}{2a}$$

The $b^2 - 4ac$ bit is called the **discriminant**.

- If the discriminant is **positive**, the formula will give you **two** different values for x — when you **add** and **subtract** the $\sqrt{b^2 - 4ac}$ bit.
- If it's **zero**, you'll only get **one** value for x, since adding and subtracting zero gets the same value.
- If it's **negative**, you don't get any (real) values for x because you can't take the square root of a negative number.

Tip: In some areas of maths, you can actually take the square root of negative numbers and get 'imaginary' or 'complex' numbers. That's why we say no 'real' roots.

To picture what this means, recall the examples from page 39:

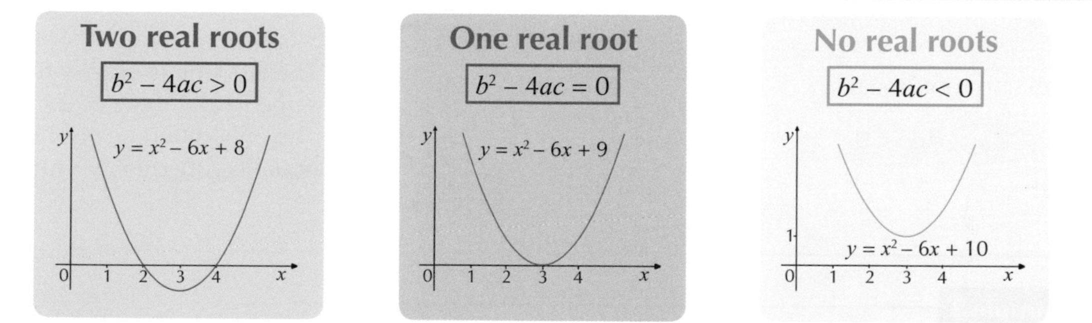

When working out the **discriminant**, the first thing to do is work out what ***a***, ***b*** and ***c*** are. Make sure you get them the right way round — it's easy to get mixed up if the quadratic's in a different order.

Example 1

Find the discriminant of $15 - x - 2x^2$.
How many real roots does the function $f(x) = 15 - x - 2x^2$ have?

1. First, identify a, b and c by rewriting as $-2x^2 - x + 15$. (Don't make the mistake of writing $a = 15$, $b = -1$ and $c = -2$ — this would be the function $15x^2 - x - 2$.)

   $a = -2$, $b = -1$ and $c = 15$

2. Then put these values into the formula for the discriminant:

   $b^2 - 4ac = (-1)^2 - (4 \times -2 \times 15)$
   $= 1 + 120 = 121$

3. The discriminant is > 0, so $15 - x - 2x^2$ has two distinct real roots.

You may need to work with a quadratic where one or more of a, b and c are given in terms of an **unknown**. This means that you'll end up with an equation or inequality for the discriminant in terms of the unknown — you might have to solve it to find the **value** or **range of values** of the unknown.

Example 2

Find the range of values of k for which the function $f(x) = 3x^2 + 2x + k$:
a) has 2 distinct roots, b) has 1 root, c) has no real roots.

1. First, decide what a, b and c are. $a = 3, b = 2, c = k$

2. Then work out what the discriminant is. $b^2 - 4ac = 2^2 - 4 \times 3 \times k$
$= 4 - 12k$

3. Solve to find k for the different values of the discriminant:

a) Two distinct roots means:

$b^2 - 4ac > 0$

$\Rightarrow 4 - 12k > 0$
$\Rightarrow 4 > 12k$
$\Rightarrow k < \frac{1}{3}$

b) One root means:

$b^2 - 4ac = 0$

$\Rightarrow 4 - 12k = 0$
$\Rightarrow 4 = 12k$
$\Rightarrow k = \frac{1}{3}$

c) No roots means:

$b^2 - 4ac < 0$

$\Rightarrow 4 - 12k < 0$
$\Rightarrow 4 < 12k$
$\Rightarrow k > \frac{1}{3}$

The working is exactly the same in all three cases. The only difference is the equality / inequality symbol.

Tip: The discriminant often comes up in exam questions — but sometimes they'll be sneaky and not actually tell you that's what you have to find. Any question that mentions **roots** of a quadratic will probably mean that you need to find the **discriminant**.

Example 3

PROBLEM SOLVING

The equation $kx^2 + 12x + 9k = 0$ has two distinct roots.
Find the range of possible values for k.

1. First, decide what a, b and c are. $a = k, b = 12, c = 9k$

2. Then work out what the discriminant is. $b^2 - 4ac = 12^2 - 4 \times k \times 9k$
$= 144 - 36k^2$

3. Two distinct roots means:

$b^2 - 4ac > 0$

$\Rightarrow 144 - 36k^2 > 0$
$\Rightarrow 36k^2 < 144$
$\Rightarrow k^2 < 4$
$\Rightarrow -2 < k < 2$

Tip: You'll learn more about quadratic inequalities in Chapter 5 — but here you just need to notice that if $k^2 < 4$, then k must be between 2 and –2.

Exercise 4.2.2

Q1 Find the discriminant, and hence the number of real roots, for each of the following:

a) $x^2 + 8x + 15$

b) $x^2 + 2\sqrt{3}x + 3$

c) $(2x + 1)(5x - 3)$

d) $-3x^2 - \frac{11}{5}x - \frac{2}{5}$

e) $9x^2 + 20x$

f) $\frac{19}{16}x^2 - 4$

Q1 Hint: Make sure the equation is written in the form $ax^2 + bx + c$ before trying to calculate the discriminant.

Q2 The discriminant of the equation $15x^2 + bx = 2$ is 169, where b is a positive number. Find all possible values of b.

PROBLEM SOLVING

Q3 The equation $0 = ax^2 + 7x + \frac{1}{4}$ has one real root. Find a.

PROBLEM SOLVING

Q4 Determine the number of real roots of the following equations, without solving them:

a) $13x^2 + 8x + 2 = 0$

b) $\frac{x^2}{3} + \frac{5}{2}x + 3 = 0$

c) $4 - \frac{x}{3} - \frac{x^2}{2} = 0$

Q5 Find the range of values of p for which $x^2 - 12x + 27 + p = 0$ has two distinct real roots.

PROBLEM SOLVING

Q6 Find the range of values of q for which $10x^2 - 10x + \frac{q}{2} = 0$ has two distinct real roots.

PROBLEM SOLVING

Q7 The equation $2x^2 + (10p + 1)x + 5 = 0$ has no real roots.
Show that p satisfies: $p(5p + 1) < \frac{39}{20}$

PROBLEM SOLVING

Q8 Find the range of values of k for which $-2x^2 - 2x + k = 0$ has:

PROBLEM SOLVING

a) two distinct roots.

b) one real root.

c) no roots.

Q9 The equation $x^2 + (k + 5)x + \frac{k^2}{4} = 0$, where k is a constant, has no real roots.

PROBLEM SOLVING

a) Show that k satisfies $10k + 25 < 0$.

b) Find the range of possible values of k.

Q10 a) Find the discriminant of $\left(k - \frac{6}{5}\right)x^2 + \sqrt{k}x + \frac{5}{4}$.

PROBLEM SOLVING

b) For what values of k would the equation $\left(k - \frac{6}{5}\right)x^2 + \sqrt{k}x + \frac{5}{4} = 0$ have:

(i) one real root?

(ii) no real roots?

(iii) two distinct real roots?

Q11 For what values of m would the equation $\left(\frac{m}{\sqrt{3}} + 1\right)x + \frac{m}{2}x^2 + \frac{m}{6} = 0$ have:

PROBLEM SOLVING

a) one real root?

b) no real roots?

c) two distinct real roots?

Q12 A primary school teacher marks out a rectangular space in the playground with a single rope. The rectangle's length is twice its width, w metres.
She then reduces the length by 5 m and increases the width by 5 m.

MODELLING

a) Write an expression for the area of the new rectangle in terms of w.

b) The new rectangle has an area of 25 m^2.

(i) By evaluating the discriminant, show that there are two real solutions for the value of w.

(ii) Explain why there might not be two possible values for w in practice.

4.3 Quadratic Graphs

Using the methods you've learnt for finding roots of quadratic functions, you'll be able to draw the graph of any quadratic function at all.

Learning Objective (Spec Ref 2.3 & 2.7):
- Sketch graphs of quadratic functions accurately.

Prior Knowledge Check: You should be familiar with graph sketching and the shape of quadratics from GCSE.

Sketching a quadratic graph

There are two pieces of information you **always need** to know about a quadratic function before you can sketch it.

- The **shape** — u-shaped or n-shaped.
- The coordinates of the **points of intersection** with the x- and y-axes.

Sometimes, there will be two **different** graphs which have the same points of intersection and shape — in this case you'll need to work out the location of the **vertex point** (maximum or minimum) to decide which graph is right.

Tip: The vertex of a quadratic graph is just the point where the graph changes direction. It is either a maximum point or a minimum point depending on the shape of the graph.

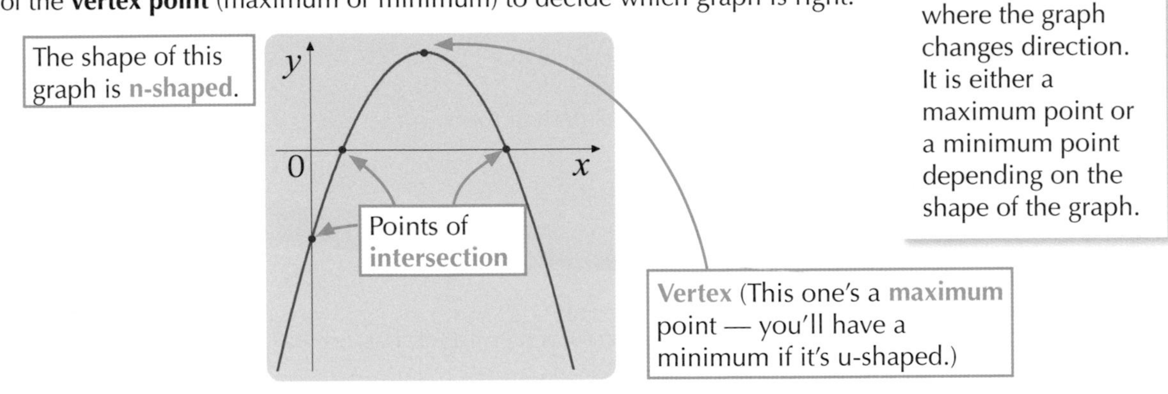

Shape

The first thing you need to decide is the **shape** of the graph — look at the coefficient of the x^2 term.

- If the coefficient of x^2 is **positive** — the graph will be **u-shaped**.
- If the coefficient of x^2 is **negative** — the graph will be **n-shaped**.

Intercepts

The next bit of information you need is where the graph **intersects the axes** — set x or y equal to zero and work out the other coordinate. If you're sketching the function $y = ax^2 + bx + c$:

- To find the **y-intercept** — let $x = 0$ and calculate the value of y.
- To find the **x-intercepts** — let $y = 0$ and solve the equation $0 = ax^2 + bx + c$ to find the value or values of x.

Don't forget the x-intercepts correspond to the roots of the quadratic function — bear in mind that there may be only **one** root, or **no** roots.

Tip: Use one of the methods of solving quadratics from earlier in the chapter to work out the x-intercepts — they're just the solutions of the equation.

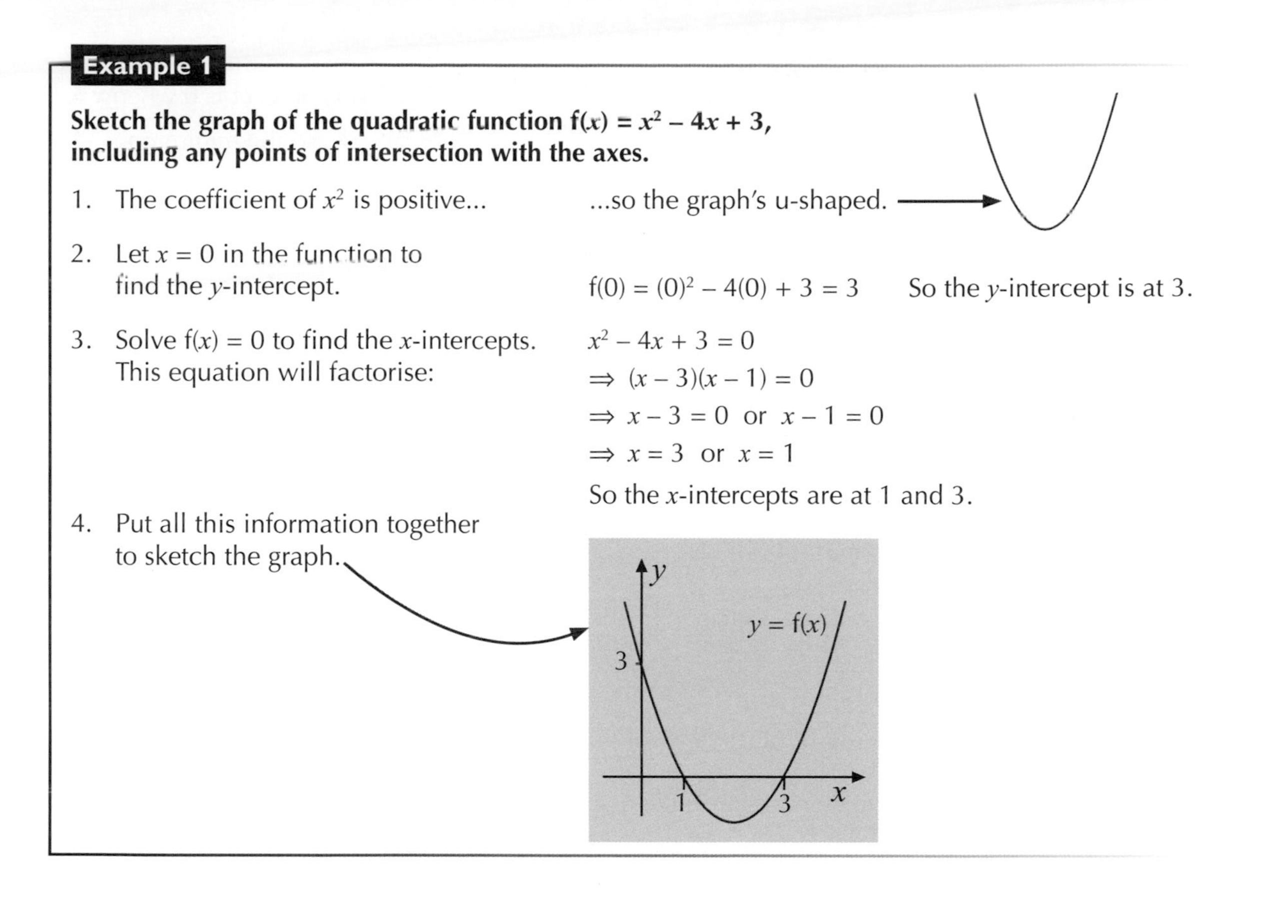

Example 1

Sketch the graph of the quadratic function $f(x) = x^2 - 4x + 3$, including any points of intersection with the axes.

1. The coefficient of x^2 is positive... ...so the graph's u-shaped.
2. Let $x = 0$ in the function to find the y-intercept. $f(0) = (0)^2 - 4(0) + 3 = 3$ So the y-intercept is at 3.
3. Solve $f(x) = 0$ to find the x-intercepts. This equation will factorise:

$x^2 - 4x + 3 = 0$

$\Rightarrow (x - 3)(x - 1) = 0$

$\Rightarrow x - 3 = 0$ or $x - 1 = 0$

$\Rightarrow x = 3$ or $x = 1$

So the x-intercepts are at 1 and 3.

4. Put all this information together to sketch the graph.

Vertex points

You'll sometimes need to find the minimum or maximum of the graph — which one it is depends on whether your graph is u-shaped or n-shaped. One way to find the vertex is to **complete the square** and then interpret this. You actually did this back on pages 35-36 when finding the number of roots of a quadratic. Have a look back at those pages to remind yourself of the method.
But here's the key result you need...

- A graph with an equation of the form $y = p(x + q)^2 + r$ has a vertex at $(-q, r)$.
- If $p > 0$, the graph is u-shaped, so the vertex is a minimum.
- If $p < 0$, the graph is n-shaped, so the vertex is a maximum.

This comes from the fact that a square is always positive and so can never be less than 0.

Example 2

a) Find the vertex of the graph of $y = f(x)$, where $f(x) = 3x^2 - 6x - 7$, stating whether it is a maximum or minimum.

1. As it's a quadratic function and the coefficient of x^2 is positive... ...it's a **u-shaped** graph so it has a minimum.
2. Completing the square gives: $f(x) = 3x^2 - 6x - 7 = 3(x - 1)^2 - 10$.

This is a square — it can never be negative. The smallest it can be is 0.

3. When the squared bit is zero, $f(x)$ reaches its minimum value. So find the value of x that makes the squared bit zero.

$f(x) = 3(x-1)^2 - 10$ ← This bracket is 0 when $x = 1$...

$f(1) = 3(1-1)^2 - 10$

$f(1) = 3(0)^2 - 10 = -10$ ← ...so the minimum is –10.

The vertex is (1, –10).

Tip: f(1) means using $x = 1$ in the function.

b) Find where the graph of $y = f(x)$ crosses the axes and hence sketch the graph.

1. $y = f(x)$ crosses the y-axis when $x = 0$ which gives: $y = 3(0)^2 - 6(0) - 7 = -7$

2. $y = f(x)$ crosses the x-axis when $f(x) = 0$ so... $3x^2 - 6x - 7 = 0$

Complete the square. ⟶ $\Rightarrow 3(x-1)^2 - 10 = 0$

Solve it to find where $y = f(x)$ crosses the x-axis. ⟶ $\Rightarrow (x-1)^2 = \frac{10}{3}$

$$\Rightarrow x - 1 = \pm\sqrt{\frac{10}{3}}$$

$$\Rightarrow x = 1 \pm\sqrt{\frac{10}{3}}$$

So $y = f(x)$ crosses the x-axis when $x = 1 + \sqrt{\frac{10}{3}}$ or $x = 1 - \sqrt{\frac{10}{3}}$

3. Now use this information to sketch the graph...

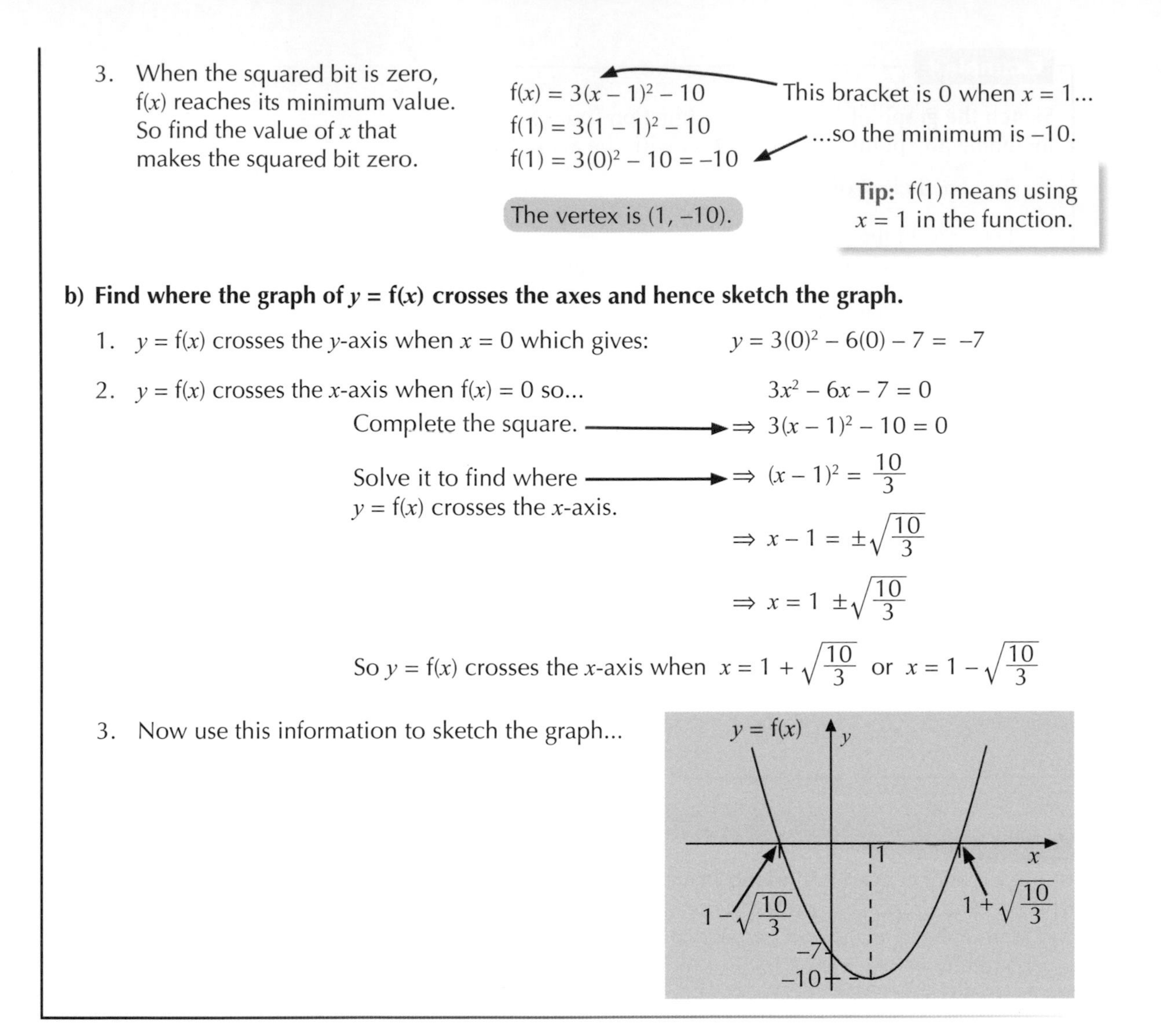

If a function has **no real roots**, the shape and the axis intercepts won't be **enough** to draw the graph. You'll **have** to find the coordinates of the **vertex point**, even if the question doesn't ask you to.

Example 3

Sketch the graph of the equation $y = 2x^2 - 4x + 3$, showing any intersection points with the axes.

1. The coefficient of x^2 here is positive... ...so the graph's u-shaped.

2. Now find the places where the graph crosses the axes (both the y-axis and the x-axis).

   (i) Put $x = 0$ to find where it meets the y-axis. $y = 2x^2 - 4x + 3$

   $y = (2 \times 0)^2 - (4 \times 0) + 3$

   so $y = 3$ is where it crosses the y-axis.

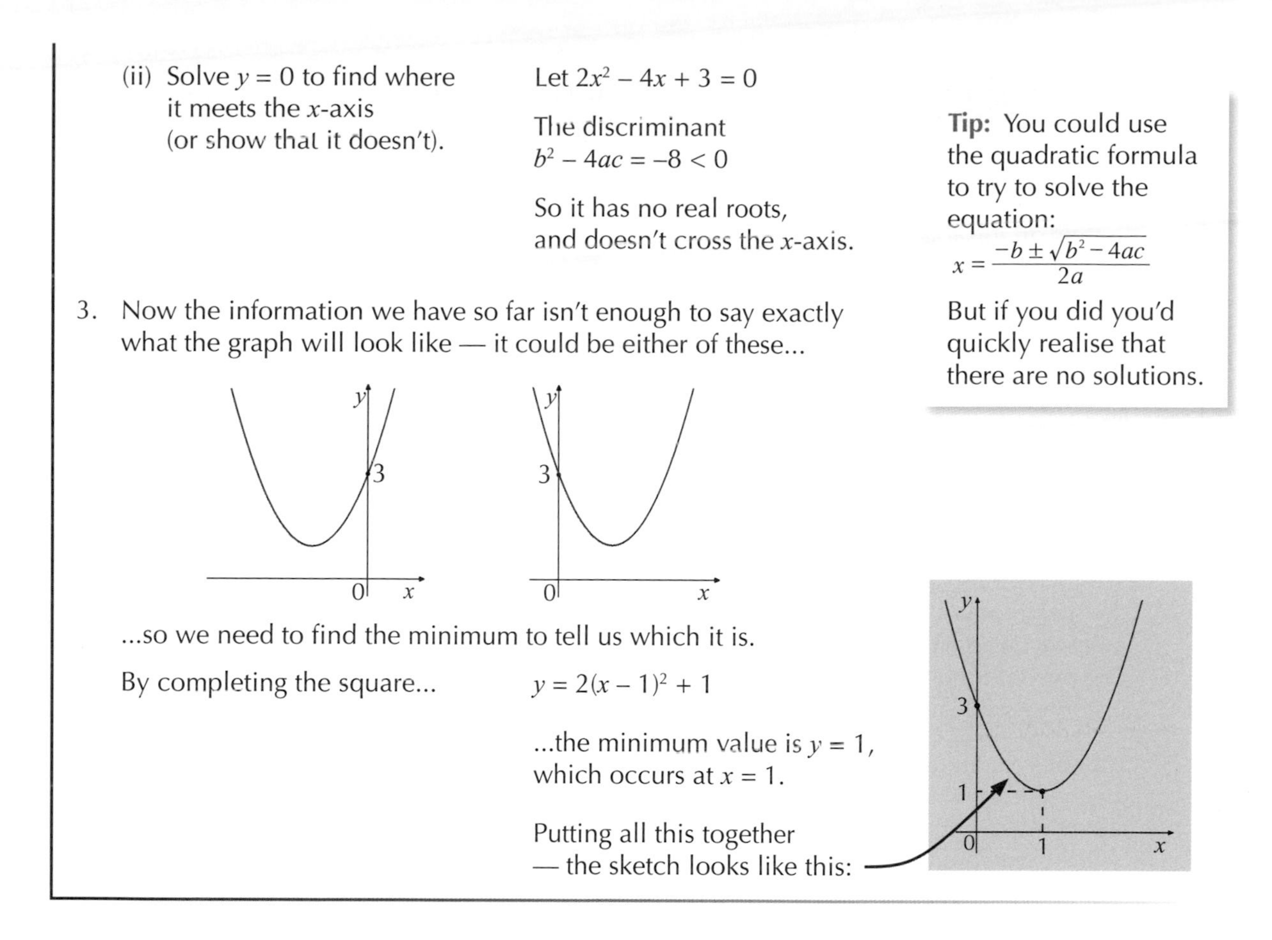

(ii) Solve $y = 0$ to find where it meets the x-axis (or show that it doesn't).

Let $2x^2 - 4x + 3 = 0$

The discriminant $b^2 - 4ac = -8 < 0$

So it has no real roots, and doesn't cross the x-axis.

Tip: You could use the quadratic formula to try to solve the equation:
$x = \frac{-b \pm \sqrt{b^2 - 4ac}}{2a}$
But if you did you'd quickly realise that there are no solutions.

3. Now the information we have so far isn't enough to say exactly what the graph will look like — it could be either of these...

...so we need to find the minimum to tell us which it is.

By completing the square... $y = 2(x - 1)^2 + 1$

...the minimum value is $y = 1$, which occurs at $x = 1$.

Putting all this together — the sketch looks like this:

So we've seen a couple of examples of finding the vertex by completing the square. But if you've already worked out the roots, and found that there are **one** or **two** real roots, you can work out the vertex more easily like this:

If the function has **two distinct roots** — use symmetry of quadratic graphs

- The graph of a quadratic function is **symmetrical**, so the x-coordinate of the vertex is **halfway** between the roots of the function.
- Work out the x-value halfway between the two roots and put it into the function to find the corresponding y-value of the vertex.

If the function has **one root** — the vertex is at the root

If a function has one root, then its graph just **touches** the x-axis at the root — this point will always be the vertex.

You can also find the vertex of a quadratic function using differentiation — see p.199-200.

Example 4

A rocket is launched from the ground. Its height, h m, is modelled by the equation $h = 30t - 5t^2$, where t is the time in seconds. Sketch a graph to show the rocket's flight, and hence find its maximum height above the ground and the total duration of its flight.

MODELLING

Don't let the different letters put you off — y has been replaced by h, and x has been replaced by t.

1. The coefficient of t^2 is negative so the graph is n-shaped.

2. Now find the places where the graph crosses the axes.

Putting $t = 0$ gives $h = 0$ as the h-intercept.

Putting $h = 0$ gives:
$30t - 5t^2 = 0 \Rightarrow 5t(6 - t) = 0$

$\Rightarrow t = 0$ and $t = 6$ as the t-intercepts

3. Putting all this together, the sketch looks like this:

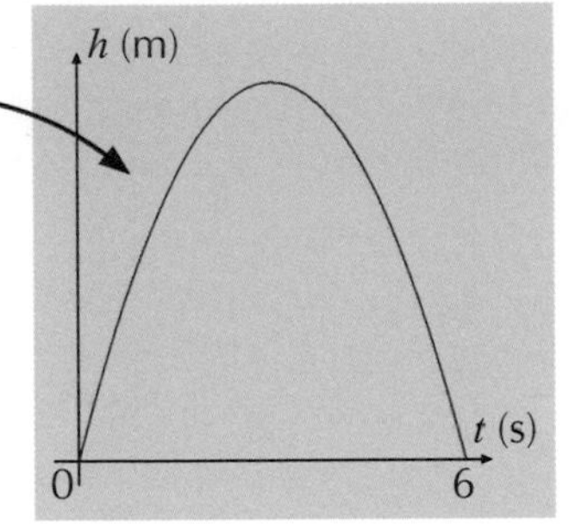

Now use your graph to answer the questions.

Tip: Just ignore the parts of the graph where t and h are less than 0 — time and height can't be negative in this example.

4. First find the maximum height of the rocket — to do this, you need to find the vertex of the graph.
You know the two roots of the equation and you know the graph is symmetrical, so you can work out the maximum by finding the t-value halfway between the two roots.
The maximum height of the rocket is just the h-value at the maximum point.

Halfway between 0 and 6

The maximum point is at $t = 3$.
$t = 3$ gives $h = 30(3) - 5(3)^2$
$= 90 - 45 = 45$ m.

So the coordinates of the maximum point are (3, 45).

5. The total duration of the flight is the time from when the rocket was launched to when it hit the ground — i.e. the two times at which $h = 0$.

These points are $t = 0$ and $t = 6$, so the flight lasted for $6 - 0 =$ 6 s.

Exercise 4.3.1

Q1 Sketch the following graphs on the same set of axes, indicating the x-intercepts of each.

a) $y = x^2 - 1$ b) $y = x^2 - 9$

Q2 a) Factorise the expression $f(x) = x^2 - 10x + 9$.

b) Use your answer to a) to sketch the graph of $f(x)$, showing the points where it crosses both axes.

c) Sketch the graph of $-f(x)$ on the same axes.

Q2c) Hint:
Remember that the graph of $y = -f(x)$ is just $y = f(x)$ reflected in the x-axis (see page 102 for more on this).

Q3 For each of the following quadratic functions:

(i) Describe its shape.

(ii) Find the y-intercept.

(iii) Find the number of real roots.

(iv) Find the values of x at which the graph intersects the x-axis — if it does.

(v) Find the coordinates of the vertex.

(vi) Sketch the graph of the function, marking on all the information you've found.

a) $y = -x^2 + 2x + 1$ b) $y = x^2 - 7x + 15$ c) $y = 2x^2 + 4x - 9$ d) $y = -x^2 + 4x - 7$

Q4 The graph on the right shows the quadratic function $y = f(x)$.

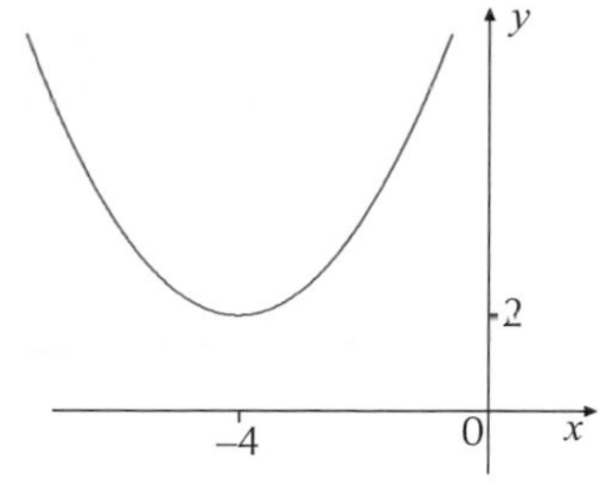

a) $f(x)$ can be written in the form $f(x) = (x + q)^2 + r$, where q and r are integers. Use the graph to find the values of q and r.

b) Copy the sketch, and on the same axes, sketch the function $g(x) = (x + 4)^2$.

c) How many real roots does each function have?

Q5 a) Complete the square of the expression $x^2 - 6x + 5$.

b) Use part a) to solve the equation $x^2 - 6x + 5 = 0$

c) Draw a graph of $y = x^2 - 6x + 5$ showing any intersections with the axes and marking the vertex.

Q6 Sketch the following graphs, showing any intersections with the axes:

a) $y = x^2 - 2x + 1$ b) $y = x^2 + x - 1$ c) $y = x^2 - 8x + 18$

d) $y = -x^2 + 3$ e) $2x^2 + 5x + 2$ f) $2x^2 - 5x - 1$

Q7 a) What are the roots of the quadratic function shown in the graph on the right?

b) The quadratic can be written in the form $y = -x^2 + px + q$ where p and q are integers. Use your answer to part a) to find p and q.

Q8 Sketch the graph of $y = f(x)$, where $f(x) = (x + 3)^2$, showing the coordinates of the vertex and any intersection points with the axes.

Q9 Sketch the graph of $y = g(x)$, where $g(x) = x^2 - 2x - 15$, showing the coordinates of the vertex and any intersection points with the axes.

Q10 A tiddlywink is fired from the ground. Its height, h cm, is modelled by the equation $h = 24t - 16t^2$, where t is the time in seconds.

MODELLING

a) Sketch a graph to show the tiddlywink's flight.

b) Find the maximum height that the tiddlywink reaches.

Q11 A theme park is designing a new roller coaster that starts on a raised platform then goes underground. The first 10 seconds of the roller coaster's vertical path are modelled by the equation $h = 0.25t^2 - 2.5t + 4$, where h is the height in metres above the ground and t is the time in seconds.

MODELLING

a) Sketch a graph showing height of the roller coaster during the first 10 seconds of the ride.

Q11a) Hint: You can multiply through by 4 to give integer coefficients.

b) For the first 10 seconds of the ride, use your graph to find:
(i) the height of the raised platform. (ii) the lowest point of the roller coaster.
(iii) how long the roller coaster is underground for.

Q12 The height of an aeroplane stunt, h m, is modelled by the equation $h = 0.5t^2 - 13t + 100$, where t is the time in seconds. The stunt is completed when the aeroplane returns to the starting height.

MODELLING

a) Sketch a graph of the aeroplane's flight showing the time taken to complete the stunt.

b) Find the minimum height that the aeroplane reaches.

4.4 Factorising Cubics

So far you've seen lots of ways to solve quadratics. Cubic equations (ones with an x^3 term) are a bit trickier. Luckily, there are some useful tools to help, such as the Factor Theorem and algebraic division.

Learning Objectives (Spec Ref 2.6):
- Factorise cubic expressions.
- Use the Factor Theorem to find factors of polynomials.
- Carry out simple algebraic division.

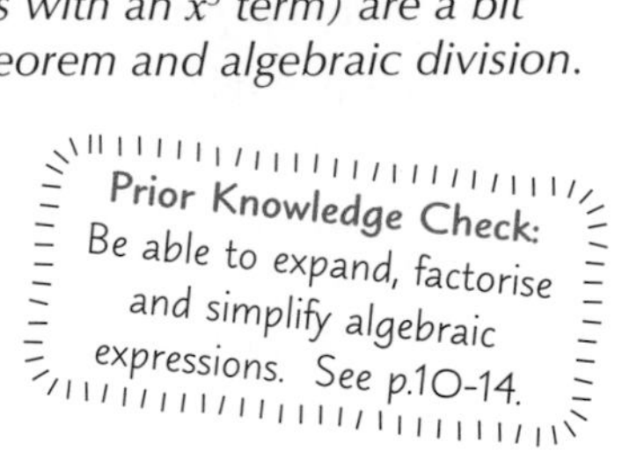

Factorising a cubic (when x is a factor)

Factorising a **cubic** means exactly what it meant with a quadratic — writing it as a product of **factors** in **brackets**. Let's start with factorising cubics that have an x in **every term**.

For a cubic of the form $ax^3 + bx^2 + cx$, **take out x** as your first factor as follows:

$$ax^3 + bx^2 + cx = x(ax^2 + bx + c)$$

Now you can just factorise the **quadratic** inside the brackets to get the other factors using the methods given earlier in the chapter. Once you've factorised, you can solve a cubic equation just as you would solve a quadratic.

Tip: Cubic equations can have one, two or three real solutions. So unlike quadratics, they always have at least one real root.

Example 1

Factorise and solve the following cubic equations.

a) $x^3 - 2x^2 - 24x = 0$

Start by taking out a factor of x... $x^3 - 2x^2 - 24x = x(x^2 - 2x - 24)$ — This quadratic will factorise.

Put in the x's... $= x(x \quad)(x \quad)$

... work out the numbers... $= x(x \quad 6)(x \quad 4)$

... then choose the signs. $= x(x - 6)(x + 4)$

$\Rightarrow x(x - 6)(x + 4) = 0$

So either $x = 0$, $x - 6 = 0$ or $x + 4 = 0$

$\Rightarrow x = 0, x = 6 \text{ or } x = -4$

b) $-x^3 - 2x^2 + 3x = 0$

If the x^3 coefficient is negative, take out a factor of $-x$... $-x^3 - 2x^2 + 3x = -x(x^2 + 2x - 3)$ — ...but don't forget these signs change.

$= -x(x \quad)(x \quad)$

Factorise the quadratic in the brackets using the normal method.

$= -x(x \quad 3)(x \quad 1)$

$= -x(x + 3)(x - 1)$

$\Rightarrow -x(x + 3)(x - 1) = 0$

So either $x = 0$, $x + 3 = 0$ or $x - 1 = 0$

$\Rightarrow x = 0, x = -3 \text{ or } x = 1$

Example 2

Solve the cubic equation $-4x^3 - 4x^2 + x = 0$.

1. The first thing you need to do is factorise. The x^3 coefficient is negative, so take out a factor of $-x$.

$$-4x^3 - 4x^2 + x = 0$$
$$-x(4x^2 + 4x - 1) = 0$$

Now $-x(4x^2 + 4x - 1) = 0$ so either $x = 0$ or $4x^2 + 4x - 1 = 0$.
$x = 0$ is one solution and solving the quadratic equation will give the other two.

2. This quadratic $4x^2 + 4x - 1$ won't factorise so use the quadratic formula.

$$x = \frac{-b \pm \sqrt{b^2 - 4ac}}{2a}$$
$$= \frac{-4 \pm \sqrt{4^2 - 4 \times 4 \times (-1)}}{2 \times 4}$$
$$= \frac{-4 \pm \sqrt{32}}{8} = \frac{-4 \pm 4\sqrt{2}}{8}$$
$$= -\frac{1}{2} \pm \frac{1}{2}\sqrt{2}$$

So the solutions are $x = 0$, $x = -\frac{1}{2} + \frac{1}{2}\sqrt{2}$ and $x = -\frac{1}{2} - \frac{1}{2}\sqrt{2}$.

Exercise 4.4.1

Q1 Factorise the following cubic expressions:

a) $x^3 + 5x^2 + 6x$
b) $x^3 + 6x^2 - 7x$
c) $x^3 - 18x^2 + 81x$
d) $x^3 + 7x^2 + 10x$
e) $-x^3 + 4x^2 - 3x$
f) $x^3 + 4x^2 + 3x$
g) $x^3 + 2x^2 - 35x$
h) $x^3 - 6x^2 - 16x$
i) $-x^3 - 3x^2 + 4x$
j) $2x^3 + 15x^2 + 25x$
k) $2x^3 - 7x^2 + 6x$
l) $4x^3 + 13x^2 - 12x$
m) $x^3 - \frac{4}{25}x$
n) $x^3 - 49x$
o) $x^3 - \frac{9}{4}x$

Q2 Solve the following cubic equations:

a) $-x^3 + 2x^2 + 24x = 0$
b) $x^3 - \frac{7}{9}x^2 + \frac{10}{81}x = 0$
c) $2x^3 + 9x^2 + 4x = 0$
d) $3x^3 - 3x^2 + 4x = 0$
e) $4x - x^3 = 0$
f) $5x^3 + 7x^2 - 3x = 0$
g) $3x^3 + 26x^2 - 9x = 0$
h) $x^3 + \frac{2}{3}x^2 - \frac{8}{9}x = 0$
i) $x^2(4x + 3) = x$
j) $2x^3 + 8x^2 = -3x$

Q2 Hint: Some of the quadratics may not factorise — use the quadratic formula to get the remaining solutions.

Q3 a) Factorise the expression $f(x) = -x^3 + 36x$

b) Use your answer to part a) to write down all of the roots of $f(x)$.

Q4 Show that the equation $2x^3 + 3x = x^2$ has only one real solution.

The Factor Theorem

When x **isn't** a factor of the cubic, factorising it becomes a lot trickier. First, you'll need to **find a linear factor** (i.e. of the form $(ax + b)$), and then you need to **divide** the cubic by that factor (there are a couple of ways of doing this, which are covered later in this section).

Before you can do that, you'll need to learn the **Factor Theorem**. The Factor Theorem gives you a quick way of finding a **factor** of a polynomial, **without** having to do the division.

The **Factor Theorem** states:

> If $f(x)$ is a polynomial, and $f(a) = 0$, then $(x - a)$ is a factor of $f(x)$.

This also works the other way round — if $(x - a)$ is a factor of $f(x)$, then $\mathbf{f(a) = 0}$.

Multiples of a factor are also factors — e.g. if $(x - 3)$ is a factor of $f(x)$, then $2(x - 3) = (2x - 6)$ is a factor of $f(x)$ too as both give the root $x = 3$.

Remember, a root is just a value of x that makes $f(x) = 0$. So if you know the roots of $f(x)$, you also know the factors of $f(x)$ — and vice versa.

Tip: A polynomial is an algebraic expression made up of the sum of constant terms and variables raised to positive integer powers, such as quadratics and cubics.

Example 1

Use the Factor Theorem to show that $(x - 2)$ is a factor of $(x^3 - 5x^2 + x + 10)$.

1. $a = 2$, so work out $f(a)$: $f(a) = f(2) = 8 - (5 \times 4) + 2 + 10 = 0$
2. $f(2) = 0$, so that means $(x - 2)$ divides into $x^3 - 5x^2 + x + 10$ **exactly**. So $(x - 2)$ must be a **factor** of $x^3 - 5x^2 + x + 10$.

If you're given factors of a polynomial, you can also use the Factor Theorem to find any **unknown coefficients** in the original polynomial.

Example 2

PROBLEM SOLVING

$(x + 1)$ and $(x - 3)$ are factors of $f(x) = x^3 + bx + c$.
Use the Factor Theorem to find the values of b and c.

1. $(x + 1)$ and $(x - 3)$ are factors, so using the Factor Theorem, $f(-1) = 0$ and $f(3) = 0$.
2. Use the fact that $f(-1) = 0$ to form an equation in b and c.
   $f(-1) = 0 \Rightarrow (-1)^3 + b(-1) + c = 0$
   $\Rightarrow -1 - b + c = 0$
   Rearranging gives $c = 1 + b$ (equation 1)
3. Use the fact that $f(3) = 0$ to form an equation in b and c.
   $f(3) = 0 \Rightarrow (3)^3 + b(3) + c = 0$
   $\Rightarrow 27 + 3b + c = 0$
   Rearranging gives $c = -27 - 3b$ (equation 2)
4. Sub (2) into (1) to get:
   $1 + b = -27 - 3b$
   $4b = -28 \Rightarrow b = -7$
5. Sub b into (1) to get:
   $c = 1 - 7 = -6$
   So $f(x) = x^3 - 7x - 6$.

If you have a polynomial with a factor in the form of $(ax - b)$ you can use this extension of the Factor Theorem:

If f(x) is a polynomial, and $f\left(\frac{b}{a}\right) = 0$, then $(ax - b)$ is a factor of f(x).

Tip: Note, $\frac{b}{a}$ is just the value of x that would make the bracket 0.

This also means that if $(ax - b)$ is a factor of f(x), then $f\left(\frac{b}{a}\right) = 0$.

Example 3

Show that $(2x + 1)$ is a factor of $f(x) = 2x^3 - 3x^2 + 4x + 3$.

1. $a = 2$ and $b = -1$, so work out $f\left(\frac{b}{a}\right)$. $\quad f\left(\frac{b}{a}\right) = f\left(-\frac{1}{2}\right) = 2\left(-\frac{1}{8}\right) - 3\left(\frac{1}{4}\right) + 4\left(-\frac{1}{2}\right) + 3 = 0$
2. $f\left(-\frac{1}{2}\right) = 0$, so $(2x + 1)$ must be a factor of $2x^3 - 3x^2 + 4x + 3$.

Or you could show $(2x + 1)$ is a factor using the method from the previous page:

1. Notice that $2x + 1 = 0$ when $x = -\frac{1}{2}$. So plug this value of x into f(x).
2. If $f\left(-\frac{1}{2}\right) = 0$, then the Factor Theorem says that $\left(x + \frac{1}{2}\right)$ is a factor.
3. A multiple of a factor is also a factor, so this means that $2\left(x + \frac{1}{2}\right) = (2x + 1)$ is also a factor.
4. $f\left(-\frac{1}{2}\right) = 2\left(-\frac{1}{8}\right) - 3\left(\frac{1}{4}\right) + 4\left(-\frac{1}{2}\right) + 3 = 0$
5. So, by the Factor Theorem, $(2x + 1)$ is a factor of f(x).

Just one more useful thing to mention about polynomials and factors:

If the **coefficients** in a polynomial **add up to 0**, then $(x - 1)$ is a **factor**.

This works for all polynomials — there are no exceptions.
For example, $(x - 1)$ is a factor of $4x^3 + 2x^2 - 5x - 1$, as $4 + 2 - 5 - 1 = 0$.

Tip: If you put $x = 1$ into a polynomial f(x), x^2, x^3 etc. are all just 1, so f(1) is the sum of the coefficients.

Example 4

Factorise the polynomial $f(x) = 6x^2 - 7x + 1$.

1. The coefficients (6, –7 and 1) add up to 0, which means $f(1) = 0$, and so $(x - 1)$ is a factor.
2. Then just factorise it like any quadratic to get this: $\quad f(x) = 6x^2 - 7x + 1 = (6x - 1)(x - 1)$

Exercise 4.4.2

Q1 Use the Factor Theorem to show that:

a) $(x - 1)$ is a factor of $x^3 - x^2 - 3x + 3$

b) $(x + 1)$ is a factor of $x^3 + 2x^2 + 3x + 2$

c) $(x + 2)$ is a factor of $x^3 + 3x^2 - 10x - 24$

d) $(x - 3)$ is a factor of $x^3 + 2x^2 - 9x - 18$

Q2 Use the Factor Theorem to show that:

a) $(2x - 1)$ is a factor of $2x^3 - x^2 - 8x + 4$

b) $(3x - 2)$ is a factor of $3x^3 - 5x^2 - 16x + 12$

Q3 Use the Factor Theorem to show that:

a) $(5x + 1)$ is a factor of $5x^3 - 44x^2 + 61x + 14$ b) $(1 - 2x)$ is a factor of $-2x^3 + 3x^2 + 11x - 6$

Q4 a) Use the Factor Theorem to show that $(x - 3)$ is a factor of $x^3 - 2x^2 - 5x + 6$.

b) Show, by adding the coefficients, that $(x - 1)$ is also a factor of this cubic.

Q5 $f(x) = 3x^3 - 5x^2 - 58x + 40$. Use the Factor Theorem to show that the following are factors of $f(x)$:

a) $(x + 4)$ b) $(3x - 2)$ c) $(x - 5)$

Q6 $(x - 2)$ is a factor of the cubic $2x^3 - 7x^2 + px + 20$. Find the value of p. PROBLEM SOLVING

Q7 $(x - 3)$ is a factor of the cubic $qx^3 - 4x^2 - 7qx + 12$. Find the value of q. PROBLEM SOLVING

Q8 The polynomial $f(x) = x^3 + cx^2 + dx - 2$ has factors $(x - 1)$ and $(x - 2)$.
Using the Factor Theorem, find the values of c and d. PROBLEM SOLVING

Q9 A cuboid is made from a number of centimetre cube blocks.
The volume, V cm³, of the cuboid can be modelled by the function:
$V(x) = 6x^3 + 37x^2 + 37x + 10$, where x is a positive integer number of blocks. MODELLING

a) Use the Factor Theorem to show that the cuboid has a length, width and height of $(3x + 2)$ blocks, $(2x + 1)$ blocks and $(x + 5)$ blocks.

b) Find the smallest possible volume of the cuboid.

Factorising a cubic (when x isn't a factor)

You can use the **Factor Theorem** to factorise cubics that **don't** have an x in every term. First, you need to use the Factor Theorem (and a bit of trial and error) to find one of the factors of the cubic:

- First, add up the coefficients to check if $(x - 1)$ is a factor.
- If that doesn't work, keep trying small numbers (find $f(-1)$, $f(2)$, $f(-2)$, $f(3)$, $f(-3)$ and so on) until you find a number that gives you zero when you put it in the cubic. Call that number k. $(x - k)$ is a factor of the cubic.

Once you've found one of the factors, here's how to **factorise the cubic**:

1. Write down the **factor** you know $\boldsymbol{(x - k)}$, and another set of brackets: $(x - k)(\qquad)$.
2. In the brackets, put the $\boldsymbol{x^2}$ **term** needed to get the right x^3 term.
3. In the brackets, put in the **constant** (that when multiplied by k gives the constant in the cubic).
4. Put in $\boldsymbol{nx}$ as the $\boldsymbol{x}$ **term** and then **multiply** to find the x^2 terms.
5. **Equate the coefficients** of the x^2 terms you've just found with the coefficient of x^2 from the question, then **solve** to find n. Check that it gives you the correct x term as well.
6. **Factorise** the quadratic you've found — if that's possible.
If the quadratic can't be factorised, just leave it as it is.

Example 1

$f(x) = x^3 + 6x^2 + 5x - 12$ Factorise $f(x)$ and find all the solutions of $f(x) = 0$.

1. Check to see if the coefficients add up to 0.

   $1 + 6 + 5 - 12 = 0$
   They do, so $(x - 1)$ is a factor.

2. Then factorise your cubic.

   $(x - 1)(x^2 \quad)$
   $(x - 1)(x^2 \quad + 12)$
   $(x - 1)(x^2 + nx + 12)$

   Work out the term that gives x^3 when multiplied by x — this is x^2.

   The term that gives -12 when multiplied by -1 is 12.

   Multiply out and equate the x^2 terms.

   $nx^2 - x^2 = 6x^2$ ← From the question.
   $n - 1 = 6 \Rightarrow n = 7$

3. Check that this gives you the right x term.

   $(x - 1)(x^2 + 7x + 12)$
   $12x - 7x = 5x$ ✓

4. So you have:

   $x^3 + 6x^2 + 5x - 12 = (x - 1)(x^2 + 7x + 12)$

5. You can factorise this quadratic.

   $x^2 + 7x + 12 = (x + 3)(x + 4)$

   The final factorisation is:

   $x^3 + 6x^2 + 5x - 12 = (x - 1)(x + 3)(x + 4)$

6. So the solutions of $f(x) = 0$ are:

   $x = 1$, $x = -3$ and $x = -4$

Example 2

$f(x) = 2x^3 - 3x^2 - 12x + 20$. Factorise $f(x)$ and find all the solutions of $f(x) = 0$.

1. Check to see if the coefficients add up to 0. They don't, so...

   $2 - 3 - 12 + 20 = 7$

2. ...use trial and error for values of x, until you find a number that gives you a value of zero.
   The coefficients don't add up to zero so you don't need to do f(1), as you already know that $(x - 1)$ isn't a factor.

   $f(-1) = (2 \times -1) - (3 \times 1) - (12 \times -1) + 20 = 27$ ✗
   $f(2) = (2 \times 8) - (3 \times 4) - (12 \times 2) + 20 = 0$ ✓
   $\Rightarrow (x - 2)$ is a factor

3. Factorise the cubic.

   $2x^3 - 3x^2 - 12x + 20) = (x - 2)(2x^2 + x - 10)$
   $= (x - 2)(x - 2)(2x + 5)$
   $= (x - 2)^2(2x + 5)$

4. One of the factors is repeated, so there are only 2 solutions.

   The solutions of $f(x) = 0$ are: $x = 2$ and $x = -\frac{5}{2}$

Example 3

$f(x) = 4x^3 + 9x^2 - 30x - 8$. Given that $x = -\frac{1}{4}$ is one solution of $f(x) = 0$, fully factorise $f(x)$ and find all the other solutions of $f(x) = 0$.

1. If $-\frac{1}{4}$ is a root, that means $\left(x+\frac{1}{4}\right)$ is a factor.
2. Any multiple of $\left(x+\frac{1}{4}\right)$ is a factor, so: $(4x + 1)$ is also a factor.
3. Then factorise your cubic: $f(x) = (4x + 1)(x^2 + 2x - 8)$

   $f(x) = (4x + 1)(x - 2)(x + 4)$
4. So the solutions of $f(x) = 0$ are: $x = -\frac{1}{4}$, $x = 2$ and $x = -4$

Tip: You could also use $\left(x+\frac{1}{4}\right)$ as the factor, but it's a lot harder to find the quadratic when you have fractions in the expression.

Exercise 4.4.3

Q1 Factorise the following:

a) $x^3 - 3x^2 + 2x$

b) $2x^3 + 3x^2 - 11x - 6$

c) $x^3 - 3x^2 + 3x - 1$

d) $x^3 - 3x^2 + 4$

e) $x^3 - x^2 - 7x + 7$

f) $x^3 + 2x^2 - 5x - 6$

Q2 Find all solutions to $f(x) = 0$, where:

a) $f(x) = x^3 - 3x^2 - 33x + 35$

b) $f(x) = x^3 - 28x + 48$

Q3 $f(x) = x^3 + 4x^2 - 8$

a) Write $f(x)$ as the product of a linear factor and a quadratic factor.

b) Find the solutions of $f(x) = 0$.
Give your answers in surd form where appropriate.

Q3b) Hint: The mention of surds suggests that you'll need to use the quadratic formula.

Q4 Find the roots of the cubic equation $x^3 - 2x^2 - x + 2 = 0$.

Q5 Find the roots of the cubic equation $x^3 - x^2 - 3x + 3 = 0$.

Q6 $f(x) = 6x^3 + 37x^2 + 5x - 6$. Use the fact that $(3x - 1)$ is one factor of $f(x)$ to fully factorise $f(x)$.

Q7 $f(x) = x^3 - px^2 + 17x - 10$, where $(x - 5)$ is a factor of $f(x)$.

a) Find the value of p.

b) Factorise $f(x)$.

c) Find the solutions to $f(x) = 0$.

PROBLEM SOLVING

Q8 Factorise the following cubic equations:

a) $3x^3 + 2x^2 - 7x + 2$

b) $5x^3 - 13x^2 + 4x + 4$

Q9 Factorise and solve the cubic equation $4x^3 - 7x = -3$.

Q10 Show that $x = 2$ is the only real root of the cubic equation $2x^3 - x^2 - 2x - 8 = 0$.

Algebraic division

Once you've found one linear factor of a cubic, you can use **algebraic long division** to find the quadratic factor.

The best way to explain how this works is with a worked example:

Tip: This is an alternative method for factorising cubics — use the method on p.54 if you prefer.

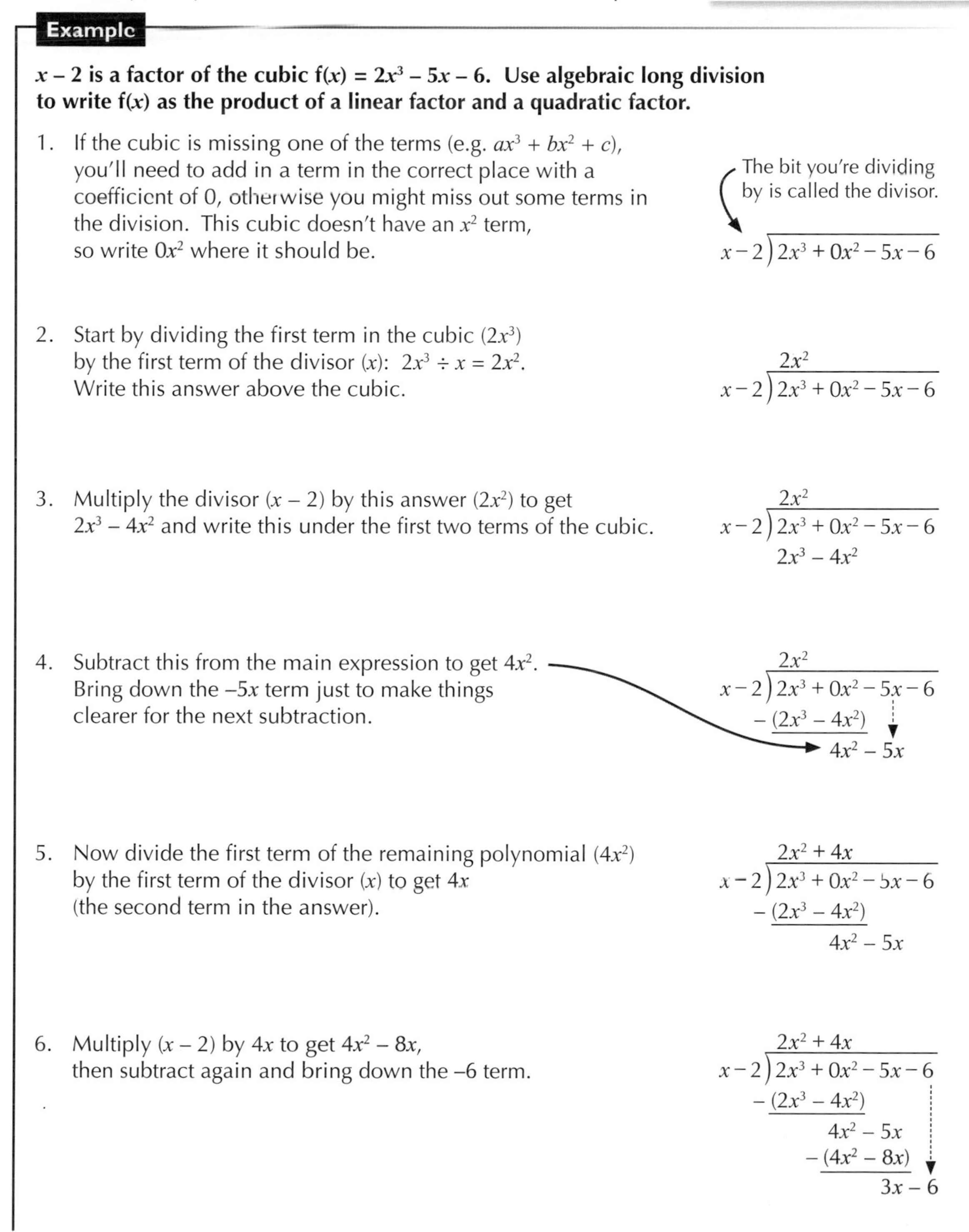

Example

$x - 2$ is a factor of the cubic $f(x) = 2x^3 - 5x - 6$. Use algebraic long division to write f(x) as the product of a linear factor and a quadratic factor.

1. If the cubic is missing one of the terms (e.g. $ax^3 + bx^2 + c$), you'll need to add in a term in the correct place with a coefficient of 0, otherwise you might miss out some terms in the division. This cubic doesn't have an x^2 term, so write $0x^2$ where it should be.

   The bit you're dividing by is called the divisor.

   $x - 2\overline{)2x^3 + 0x^2 - 5x - 6}$

2. Start by dividing the first term in the cubic ($2x^3$) by the first term of the divisor (x): $2x^3 \div x = 2x^2$. Write this answer above the cubic.

   $2x^2$
   $x - 2\overline{)2x^3 + 0x^2 - 5x - 6}$

3. Multiply the divisor ($x - 2$) by this answer ($2x^2$) to get $2x^3 - 4x^2$ and write this under the first two terms of the cubic.

   $2x^2$
   $x - 2\overline{)2x^3 + 0x^2 - 5x - 6}$
   $2x^3 - 4x^2$

4. Subtract this from the main expression to get $4x^2$. Bring down the $-5x$ term just to make things clearer for the next subtraction.

   $2x^2$
   $x - 2\overline{)2x^3 + 0x^2 - 5x - 6}$
   $-\underline{(2x^3 - 4x^2)}$
   $4x^2 - 5x$

5. Now divide the first term of the remaining polynomial ($4x^2$) by the first term of the divisor (x) to get $4x$ (the second term in the answer).

   $2x^2 + 4x$
   $x - 2\overline{)2x^3 + 0x^2 - 5x - 6}$
   $-\underline{(2x^3 - 4x^2)}$
   $4x^2 - 5x$

6. Multiply ($x - 2$) by $4x$ to get $4x^2 - 8x$, then subtract again and bring down the -6 term.

   $2x^2 + 4x$
   $x - 2\overline{)2x^3 + 0x^2 - 5x - 6}$
   $-\underline{(2x^3 - 4x^2)}$
   $4x^2 - 5x$
   $-\underline{(4x^2 - 8x)}$
   $3x - 6$

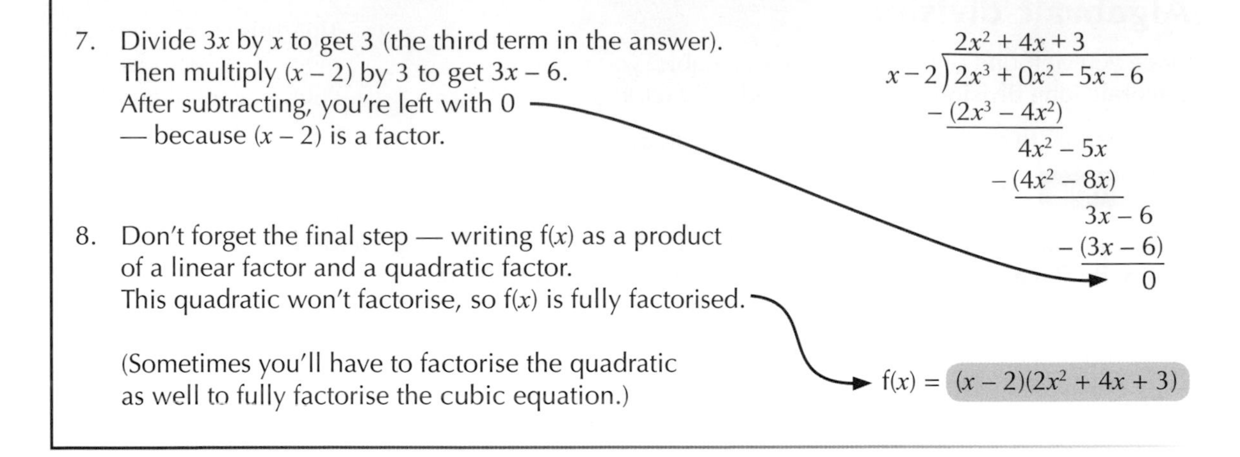

Exercise 4.4.4

Q1 Use algebraic long division and the given factors to fully factorise the following cubic equations:

a) $x^3 - 2x^2 - 15x + 36$, factor: $(x - 3)$

b) $x^3 - x^2 - 11x - 10$, factor: $(x + 2)$

c) $2x^3 + 11x^2 - 23x - 14$, factor: $(x - 2)$

d) $x^3 + 10x^2 + 31x + 30$, factor: $(x + 5)$

Q2 Write $x^3 - 5x + 4$ as the product of a linear factor and a quadratic factor using long division.

Q2, 4 Hint: Remember to add in any missing terms, giving them a coefficient of 0.

Q3 $f(x) = x^3 + 2x^2 - 7x - 2$. Use algebraic long division to express $f(x)$ in the form $(x - 2)g(x)$, where $g(x)$ is a quadratic.

Q4 $f(x) = x^3 - 7x - 6$ and $f(-2) = 0$.
Use algebraic long division to find all the solutions of $f(x) = 0$.

Q5 Write $x^3 + x^2 - 12$ as the product of a linear factor and a quadratic factor using long division.

Q6 a) Using algebraic long division, find the remainder when you divide $f(x) = x^3 - 2x^2 + x - 1$ by $(x + 3)$.

b) Write $f(x)$ in the form $(x + 3)g(x)$ + remainder, where $g(x)$ is a quadratic.

Q6 Hint: When dividing by something that isn't a factor of the cubic, you'll end up with a non-zero constant as the answer to the final subtraction step. This is the remainder.

Q7 a) Use algebraic division to find the remainder when $f(x) = x^3 - 8x^2 + 20x - 3$ is divided by $(x - 2)$.

b) Hence find the solutions to $f(x) - 13 = 0$.

PROBLEM SOLVING

Q8 Use algebraic division to show that $x = 5$ is the only real solution to $x^3 - 15x^2 + 75x = 125$.

PROBLEM SOLVING

Q9 The speed of a car is recorded as it accelerates and decelerates to a stop, over the time period $0 \leq t \leq T$, and modelled as $S = -t^3 + 2t^2 + 13t + 10$, where S is the speed in miles per hour, t is the time in minutes, and T is the time at which the car stops. Given that $(t + 1)$ is a factor of the function, use algebraic division to find T.

PROBLEM SOLVING MODELLING

Review Exercise

Q1 Factorise the following expressions:

a) $x^2 + 2x + 1$ b) $x^2 - 13x + 30$ c) $x^2 - 4$

d) $3 + 2x - x^2$ e) $2x^2 - 7x - 4$ f) $5x^2 + 7x - 6$

Q2 Solve the following equations:

a) $x^2 - 3x + 2 = 0$ b) $x^2 + x - 12 = 0$ c) $2 + x - x^2 = 0$

d) $x^2 + x - 16 = x$ e) $3x^2 - 15x - 14 = 4x$ f) $4x^2 - 1 = 0$

g) $6x^2 - 11x + 9 = 2x^2 - x + 3$ h) $3x^2 + 10x - 8 = 2 - x - 3x^2$ i) $4 - 9x^2 = 0$

Q3 Solve these quadratic equations, leaving your answers in surd form where necessary.

a) $3x^2 - 7x + 3 = 0$ b) $2x^2 - 6x - 2 = 0$ c) $x^2 + 4x + 6 = 12$

Q4 a) Rewrite the expression $x^2 + 6x + 7$ in the form $(x + m)^2 + n$.

b) Hence solve the equation $(2x + 1)^2 + 6(2x + 1) + 7 = 0$. Leave your answers in surd form.

PROBLEM SOLVING

Q5 Find all four solutions to the equation $x^4 - 17x^2 + 16 = 0$.

PROBLEM SOLVING

Q6 If the quadratic equation $x^2 + kx + 4 = 0$ has two distinct real roots, what are the possible values of k?

Q7 Rewrite these quadratics by completing the square. Then state their maximum or minimum value and the value of x where this occurs. Also, say if and where their graphs cross the x-axis.

a) $x^2 - 4x - 3$ b) $x^2 + 5x + 8$ c) $3 - 3x - x^2$

d) $2x^2 - 4x + 11$ e) $4x^2 - 28x + 48$ f) $14 + 12x - 3x^2$

Q8 How many roots do these quadratic equations have? Sketch the graph of each quadratic function.

a) $x^2 - 2x - 3 = 0$ b) $x^2 - 6x + 9 = 0$ c) $2x^2 + 4x + 3 = 0$

Q9 a) Find the coordinates of the points of intersection of the graphs of $y = f(x)$ and $y = g(x)$, where $f(x) = 5x^3 - 13x^2 + 6x$ and $g(x) = -5x^3 + 7x^2 + 6x$.

b) Express $f(x) = 5x^3 - 13x^2 + 6x$ as the product of three factors.

Q10 a) Show that the x-coordinates of the points where the curves $y = x(x - 6)^2$ and $y = -x(2x - 31)$ intersect are given by the solutions to the equation $x^3 - 10x^2 + 5x = 0$.

PROBLEM SOLVING

b) Find the x-coordinates of the points where the two curves meet. Where appropriate, express your answers in surd form.

Review Exercise

Q11 A car travels through a multistorey car park. It starts above ground level, then goes underground. The first 10 seconds of the car's vertical path are modelled by the equation $h = 0.25t^2 - 2.75t + 6$, where h is the height in metres above the ground and t is the time in seconds.

MODELLING

a) Sketch a graph showing the height of the car during the first 10 seconds.

b) For the first 10 seconds of the car's journey, use your graph to find:

(i) the initial height of the car.

(ii) the lowest point of the car.

(iii) how long the car is underground for.

Q12 Solve the following cubic equations:

a) $x^3 - 4x^2 = 0$ b) $x^3 + 5x^2 - 6x = 0$ c) $x^3 - 6x^2 + 9x = 0$

d) $2x^3 + 5x^2 + 15x = x^3 - 3x^2$ e) $2x^3 + 20x^2 + 12x = 9x^3 - 20x^2$ f) $6x^3 - 5x^2 - 4x = 0$

Q13 Which of the following are factors of $f(x) = x^5 - 4x^4 + 3x^3 + 2x^2 - 2$?

a) $x - 1$ b) $x + 1$ c) $x - 2$ d) $2x - 2$

PROBLEM SOLVING

Q14 $f(x) = (x + 5)(x - 2)(x - 1) + k$. If $(x + 2)$ is a factor of $f(x)$, find the value of k.

PROBLEM SOLVING

Q15 Use long division to divide the cubics below. In each case state the quotient and remainder.

a) $x^3 - x^2 - 3x + 3$ by $(x + 3)$

b) $x^3 - 3x^2 - 5x + 6$ by $(x - 2)$

c) $x^3 + 2x^2 + 3x + 2$ by $(x + 2)$

Q15 Hint: The quotient is the quadratic part of the answer.

Q16 Write the following functions $f(x)$ in the form $f(x) = (x + 2)g(x) + \text{remainder}$, where $g(x)$ is a quadratic:

a) $f(x) = 3x^3 - 4x^2 - 5x - 6$ b) $f(x) = x^3 + 2x^2 - 3x + 4$ c) $f(x) = 2x^3 + 6x - 3$

Q17 Find the values of c and d so that $2x^4 + 3x^3 + 5x^2 + cx + d$ is exactly divisible by $(x - 2)(x + 3)$.

PROBLEM SOLVING

Q18 Given that $(x - 3)$ is a factor of the cubic $f(x) = x^3 - 9x^2 + 7x + 33$, find the exact solutions of $f(x) = 0$.

PROBLEM SOLVING

Q19 Find the roots of $f(x) = 0$ where $f(x) = x^3 + 6x^2 + 11x + 6$.

PROBLEM SOLVING

Q20 a) Write $f(x) = x^3 - 5x^2 - 2$ in the form $f(x) = (x + 2)g(x) + \text{remainder}$, where $g(x)$ is a quadratic.

PROBLEM SOLVING

b) Hence show that $x = -2$ is the only solution to $f(x) + 30 = 0$.

Exam-Style Questions

Q1 Consider the function $f(x) = x^2 - 2x - 14$.

a) (i) Rewrite the function in the form $f(x) = (x + a)^2 + b$, where a and b are integers to be found.

[2 marks]

(ii) Hence, or otherwise, find the exact solutions to the equation $f(x) = 0$.

[2 marks]

b) Sketch the graph of $y = f(x)$, labelling all intersections with the axes.

[3 marks]

Q2 The equation $2kx^2 + 5x + k = 0$ has repeated roots. Find the possible values for k. Give your answers in simplified surd form.

[5 marks]

Q3 a) Factorise and solve the equation $2x^2 - 5x - 3 = 0$.

[2 marks]

b) Hence, or otherwise, solve the equation $2x - 5\sqrt{x} = 3$.

[3 marks]

Q4 For the function $f(x) = x^3 + x^2$, show that the curve with equation $y = f(x)$ intersects the x-axis more than once. State the coordinates of the points where $y = f(x)$ intersects the axes.

[3 marks]

Q5 a) Use the Factor Theorem to show that $(x^2 - 1)$ is a factor of $f(x) = 2x^3 + 3x^2 - 2x - 3$.

[3 marks]

b) Sketch the graph of $y = f(x)$, including all intersections with the axes.

[4 marks]

Exam-Style Questions

Q6

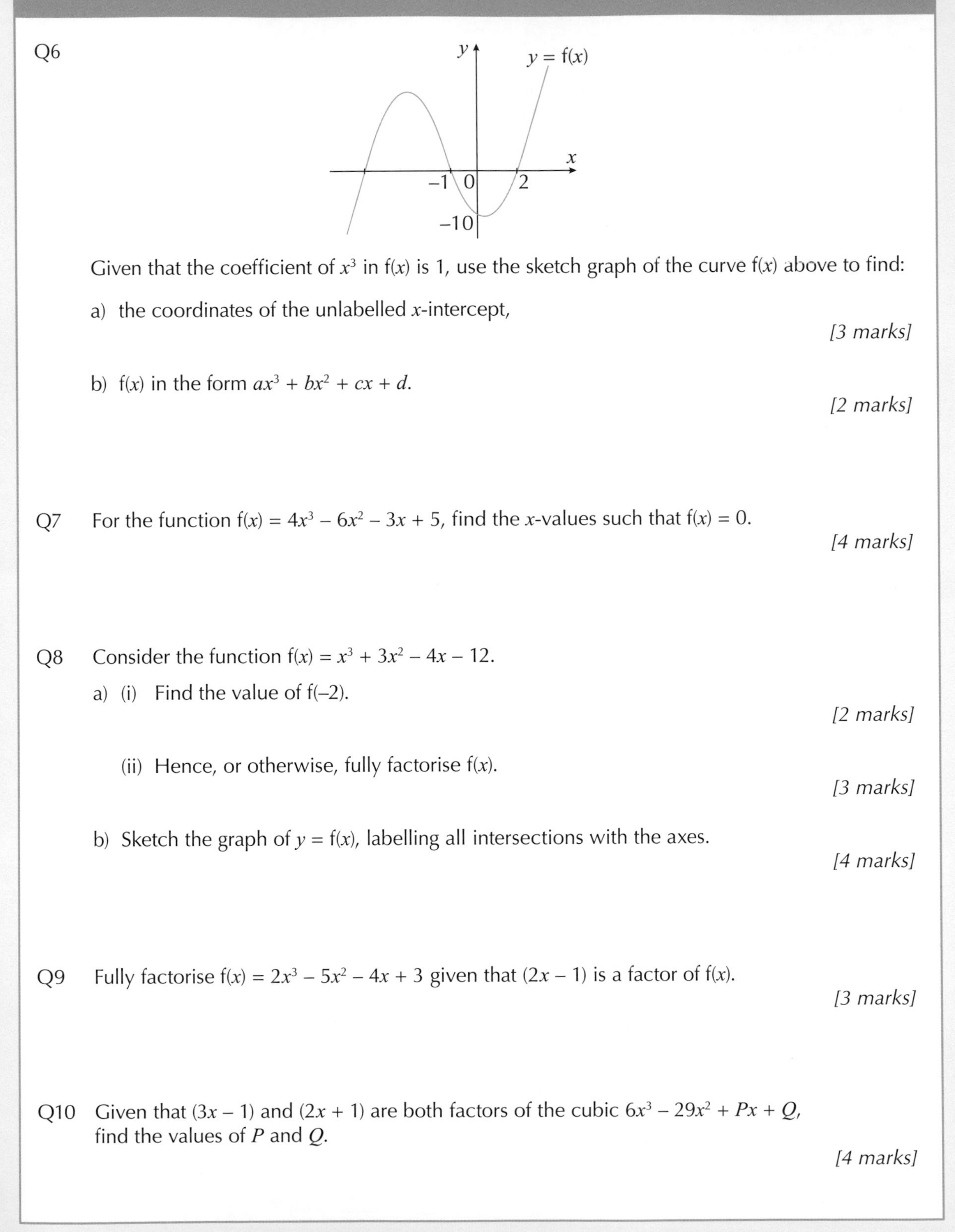

Given that the coefficient of x^3 in f(x) is 1, use the sketch graph of the curve f(x) above to find:

a) the coordinates of the unlabelled x-intercept,

[3 marks]

b) f(x) in the form $ax^3 + bx^2 + cx + d$.

[2 marks]

Q7 For the function $f(x) = 4x^3 - 6x^2 - 3x + 5$, find the x-values such that $f(x) = 0$.

[4 marks]

Q8 Consider the function $f(x) = x^3 + 3x^2 - 4x - 12$.

a) (i) Find the value of $f(-2)$.

[2 marks]

(ii) Hence, or otherwise, fully factorise f(x).

[3 marks]

b) Sketch the graph of $y = f(x)$, labelling all intersections with the axes.

[4 marks]

Q9 Fully factorise $f(x) = 2x^3 - 5x^2 - 4x + 3$ given that $(2x - 1)$ is a factor of f(x).

[3 marks]

Q10 Given that $(3x - 1)$ and $(2x + 1)$ are both factors of the cubic $6x^3 - 29x^2 + Px + Q$, find the values of P and Q.

[4 marks]

Chapter 5 Inequalities and Simultaneous Equations

5.1 Inequalities

Solving an inequality is very similar to solving an equation. But when multiplying or dividing both sides of an inequality, you've got to make sure that you keep the inequality sign pointing the right way.

Learning Objectives (Spec Ref 2.5):
- Solve linear inequalities, e.g. $ax + b > cx + d$.
- Solve quadratic inequalities, e.g. $ax^2 + bx + c \geq 0$.
- Give solutions of inequalities in set notation.
- Represent inequalities and their solutions graphically.

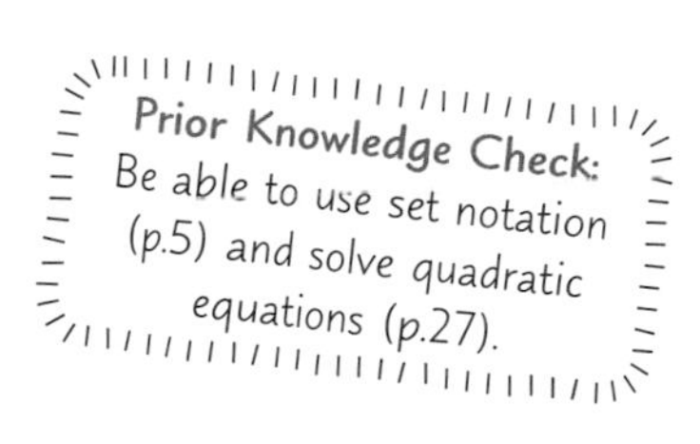

Linear inequalities

Solving where the inequality sign doesn't change direction

Solve inequalities like you solve equations — anything you do to one side, you do to the other.

- If you **add** or **subtract** something from both sides of an inequality, the inequality sign **doesn't** change direction.
- Multiplying or dividing both sides of an inequality by a **positive** number **doesn't** affect the direction of the inequality sign.

You might have to give your answers in **set notation** (see page 5 for a reminder of what this means) or in **interval notation**. For example, $1 < x \leq 2$ can be written as $(1, 2]$ — the round bracket means 1 is not included and the square bracket means 2 is included. Intervals such as $x \geq 3$ are written as $[3, \infty)$.

Example 1

Find the set of values for x which satisfy:

a) $x - 3 < -1 + 2x$

1. Add 1 to both sides of the inequality — this leaves the inequality sign pointing in the same direction.
2. Then subtract x from both sides — this also doesn't affect the direction of the inequality sign.

$$x - 3 + 1 < -1 + 2x + 1$$
$$x - 2 < 2x$$
$$x - 2 - x < 2x - x$$
$$-2 < x$$

b) $2(4x + 1) \geq 2x + 17$

1. First, expand the brackets.
2. Subtract 2 from both sides, then subtract $2x$ from both sides.
3. Finally, divide both sides by 6 — the inequality sign is pointing in the same direction.

$$8x + 2 \geq 2x + 17$$
$$8x + 2 - 2 \geq 2x + 17 - 2$$
$$8x \geq 2x + 15$$
$$8x - 2x \geq 2x + 15 - 2x$$
$$6x \geq 15$$
$$\frac{6x}{6} \geq \frac{15}{6}$$
$$x \geq \frac{5}{2}$$

Solving where the inequality sign does change direction

Multiplying or dividing an inequality by a **negative** number **changes** the direction of the inequality sign.

Example 2

Find the set of values of x for which $4 - 3x \leq 16$.

1. Subtract 4 from both sides. $4 - 3x - 4 \leq 16 - 4$
   $-3x \leq 12$
2. Then divide both sides by −3 — but change the direction of the inequality sign. $\frac{-3x}{-3} \geq \frac{12}{-3}$
   $x \geq -4$

Tip: Dividing by −1 is the same as swapping everything from one side to the other. So when you divide by any negative number the sign changes.

Example 3

Find the set of values of x for which $\frac{2-4x}{3} > \frac{5-3x}{4}$. Give your answer in set notation.

1. Multiply both sides by 12 to remove the fractions, and expand the brackets. $4(2 - 4x) > 3(5 - 3x)$
   $8 - 16x > 15 - 9x$
2. Subtract 8 from both sides, then add $9x$ to both sides. $-16x > 7 - 9x$
   $-7x > 7$
3. Divide both sides by −7 and change the direction of the inequality sign. $x < -1$
4. Write the final answer in set notation. $\{x : x < -1\}$

Finding the solution to two inequalities

You may be given two inequalities and be asked to find a solution which satisfies **both** of them.

Example 4

Find the set of values for x which satisfy both the inequalities $x - 5 < -3 + 2x$ and $2x > 4x - 6$.

1. Solve both inequalities separately:

   $x - 5 < -3 + 2x$ | $2x > 4x - 6$
   $x - 2 < 2x$ | $2x + 6 > 4x$
   $-2 < x$ | $6 > 2x$
   | $3 > x$

2. Show both solutions on a number line. Each line has an open circle at the end to show that this number isn't equal to x. Solutions with a ≤ or ≥ sign end with a filled-in circle (●→) to show that number is a possible value of x.

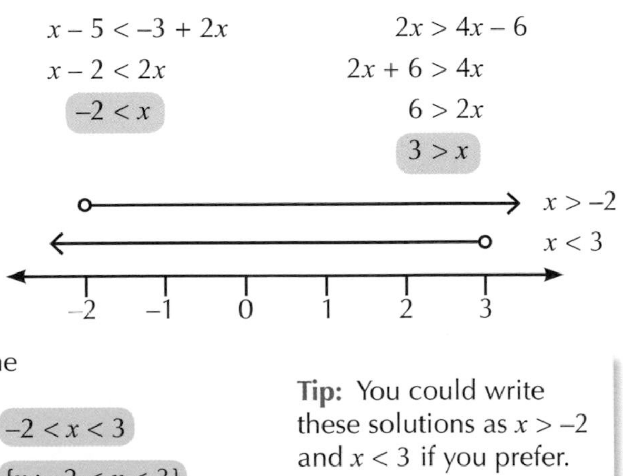

3. Look where the two lines overlap to find the set of values that satisfy **both** inequalities. The lines overlap between −2 and 3, so: $-2 < x < 3$

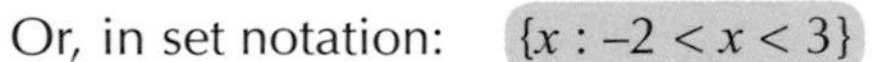
Or, in set notation: $\{x : -2 < x < 3\}$

Tip: You could write these solutions as $x > -2$ and $x < 3$ if you prefer.

Exercise 5.1.1

Q1 Find the set of values for x which satisfy:

a) $2x - 1 < x + 4$ b) $4 - 3x \geq 10 - 5x$ c) $5x + 7 > 3x + 1$

d) $3 - 2x \leq 5x - 4$ e) $9 - x \geq 7x + 5$ f) $12x - 9 \leq 4x + 11$

g) $3x - 6 > 6 - 3x$ h) $-4x < 16 - 7x$

Q2 Hint: Multiply out the brackets first.

Q2 Find the set of values for x which satisfy the inequalities below. Give your answers in set notation.

a) $2(x + 3) > 3(x + 2)$ b) $5(1 + 3x) \leq 7$ c) $12 \geq 2(5 - 2x)$

Q3 Find the set of values for x which satisfy:

a) $\frac{6-5x}{2} < \frac{4-8x}{3}$ b) $\frac{3x-1}{4} \geq 2x$ c) $\frac{x-2}{2} - \frac{2x+3}{3} < 7$

Q4 Find the set of values for x which satisfy the inequalities below. Give your answers in set notation.

a) $-5 < 2x - 3 < 15$ b) $-5 \leq 4 - 3x < 19$ c) $5 \leq 7 + 6x \leq 11$

Q5 Solve the following inequalities, and represent the solutions on a number line:

a) $2x \geq 3 - x$ b) $5x - 1 < 3x + 5$

c) $2x + 1 \geq 3x + 2$ d) $3(x - 3) \leq 5(x - 1)$

e) $9 - x \leq 3 - 4x$ f) $\frac{2(x-3)}{3} + 1 < \frac{2x-1}{2}$

Q5 Hint: Look at the inequality sign to decide the direction of your arrow and whether the line should end with an open or filled-in circle.

Q6 a) Find the set of values of x for which $7 \leq 3x - 2 < 16$.

b) Show your solution to part a) on a number line.

Q7 Find the set of values for x which satisfy both $4 - 2x < 10$ and $3x - 1 < x + 7$. Give your answer in set notation.

Q8 Find the values of x which satisfy both inequalities:

a) $2x \geq 3x - 5$ and $3x - 2 \geq x - 6$ b) $5x + 1 \leq 11$ and $2x - 3 < 5x - 6$

c) $2x - 1 \leq 3x - 5$ and $5x - 6 > x + 22$ d) $3x + 5 < x + 1$ and $6x - 1 \geq 3x + 5$

Quadratic inequalities

When solving inequalities, it's important that you **don't divide** or **multiply** by **variables** (anything you don't know the value of, e.g. x or y).

- The variable might be **negative**, so the inequality sign may end up pointing in the wrong direction.
- The variable could be equal to **zero** — you can't divide something by zero.

Tip: If you have an x on the bottom of a fraction (e.g. $\frac{2}{x} < 1$), you have to multiply both sides by x^2 (as this is always positive). So $\frac{2}{x} < 1$ would become $2x < x^2$.

Example 1

Simplify the quadratic inequality $36x < 6x^2$.

1. Start by dividing by 6 — dividing by 6 is okay because 6 is definitely positive.

$36x < 6x^2$

$\Rightarrow 6x < x^2$

2. It's tempting to divide both sides by x now, but x could be negative (or zero). Instead, subtract $6x$ from both sides.

$0 < x^2 - 6x$ which is... $x^2 - 6x > 0$

In general, the best way to **solve** a **quadratic inequality** is to do the following:

1. Rewrite the inequality with zero on one side.
2. Sketch the graph of the quadratic function.
3. Use the graph to find the solution.

Example 2

Find the values of x which satisfy $-x^2 + 2x + 4 \geq 1$.

1. First rewrite the inequality with zero on one side.

$-x^2 + 2x + 3 \geq 0$

2. Find where the graph crosses the x-axis (i.e. where $y = 0$), by factorising to find the roots — there's more on factorising quadratics on pages 27-29.

$-x^2 + 2x + 3 = 0$

$\Rightarrow x^2 - 2x - 3 = 0$

$\Rightarrow (x + 1)(x - 3) = 0$

$\Rightarrow x = -1$ and $x = 3$

3. Then you need to draw the graph of $y = -x^2 + 2x + 3$. (See pages 44-48 for more on drawing quadratic functions.) The coefficient of x^2 is negative, so the graph is n-shaped.

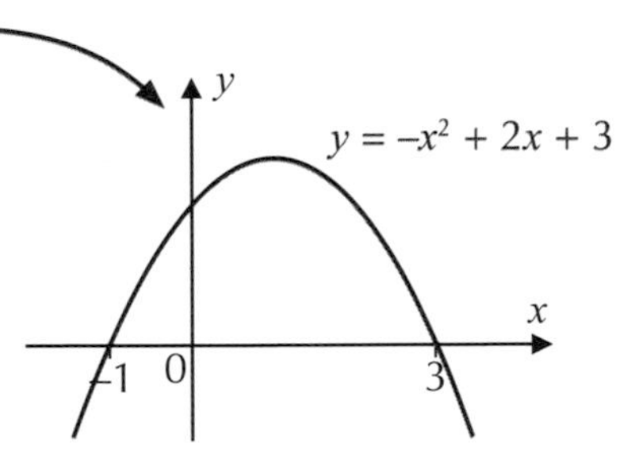

4. Now you're trying to solve the inequality $-x^2 + 2x + 3 \geq 0$, so you're interested in when the graph is **positive or zero**, i.e. when it's **above the x-axis**.

5. From the graph, this is when x is between -1 and 3 (including those points).

6. So the solution is: $-1 \leq x \leq 3$

Example 3

Find the values of x which satisfy $2x^2 + 2x - 5 > 3x - 2$. Give your answer in set notation.

1. First rewrite the inequality with zero on one side.

$2x^2 - x - 3 > 0$

2. Factorise the quadratic equation to find where it crosses the x-axis.

$2x^2 - x - 3 = 0$

$\Rightarrow (2x - 3)(x + 1) = 0$

$\Rightarrow x = \frac{3}{2}$ and $x = -1$

3. Then draw the graph of $y = 2x^2 - x - 3$.
   The coefficient of x^2 is positive, so the graph is u-shaped.

4. Now you're trying to solve $2x^2 - x - 3 > 0$,
   so you need to say when the graph is **positive**.

5. Looking at the graph, there are two parts of the x-axis where this is true — when x is less than -1 and when x is greater than $\frac{3}{2}$.

6. So the solution is $x < -1$ or $x > \frac{3}{2}$.

7. In set notation, this is: $\{x : x < -1\} \cup \left\{x : x > \frac{3}{2}\right\}$

Example 1 revisited

1. On the last page you had to simplify $\mathbf{36x < 6x^2}$.

$$36x < 6x^2$$
$$\Rightarrow \quad 6x < x^2$$
$$\Rightarrow \quad 0 < x^2 - 6x$$

2. To **solve** this, draw the graph of $y = x^2 - 6x$.
   This can be factorised as $y = x(x - 6)$,
   so the roots are $x = 0$ and $x = 6$.

3. You're looking for when it is positive — i.e. where the graph is above the x-axis. This occurs when x is less than 0 and when x is greater than 6.

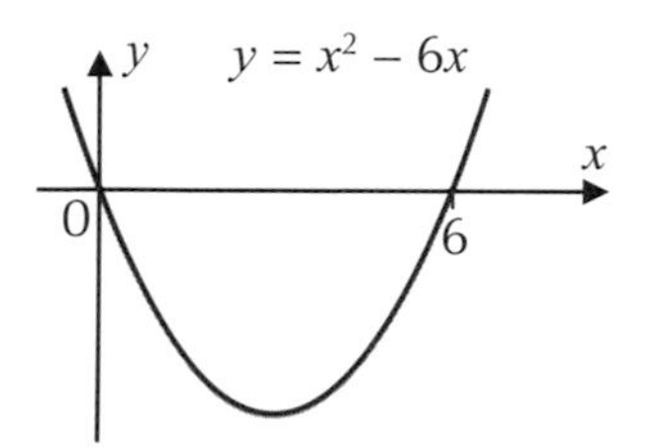

4. So the solution is: $x < 0$ or $x > 6$

   (If you had divided the inequality $6x < x^2$ by x, you would have missed the $x < 0$ part of the solution.)

You may be asked to find the set of values for x which satisfy **both** a quadratic inequality and a linear inequality. To do this, you just work out the solution of each inequality separately and then use a **graph** to help you find the solution that satisfies both.

Example 4

Find the set of values of x which satisfy:

a) $5x - 10 > 4x - 7$

Solve in the usual way.

$$5x - 10 > 4x - 7$$
$$\Rightarrow \quad 5x > 4x + 3$$
$$\Rightarrow \quad x > 3$$

b) $2x^2 - 11x + 5 < 0$

1. You've already got zero on one side, so just factorise the quadratic to find where the graph crosses the x-axis:

$$2x^2 - 11x + 5 = 0$$
$$\Rightarrow \quad (2x - 1)(x - 5) = 0$$
$$\Rightarrow \quad x = \frac{1}{2} \text{ and } x = 5$$

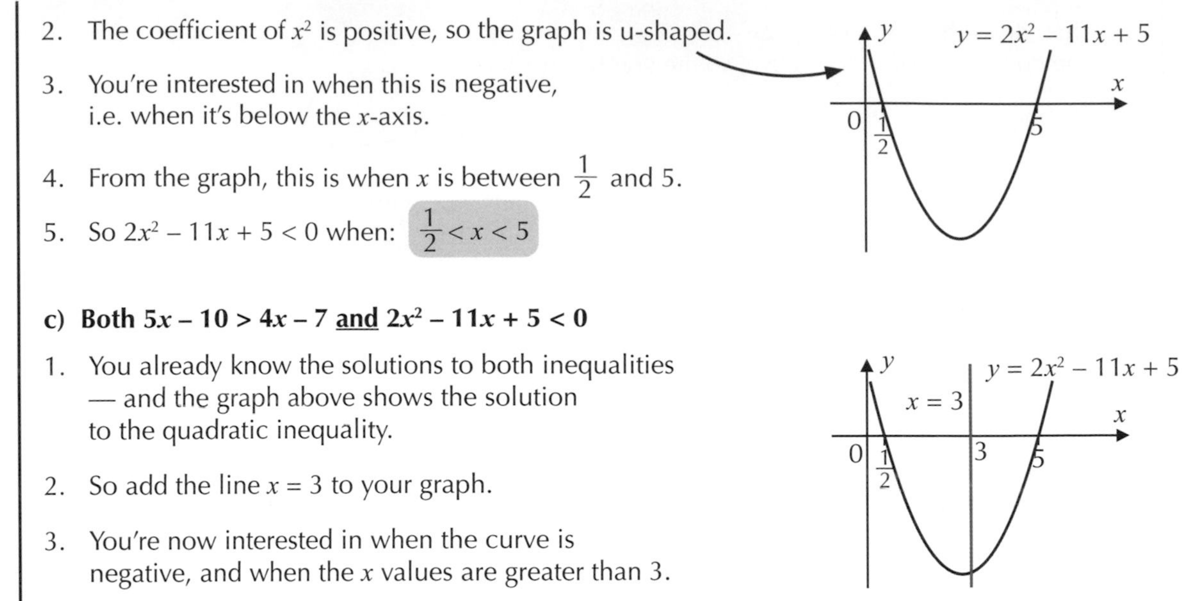

2. The coefficient of x^2 is positive, so the graph is u-shaped.
3. You're interested in when this is negative, i.e. when it's below the x-axis.
4. From the graph, this is when x is between $\frac{1}{2}$ and 5.
5. So $2x^2 - 11x + 5 < 0$ when: $\frac{1}{2} < x < 5$

c) Both $5x - 10 > 4x - 7$ and $2x^2 - 11x + 5 < 0$

1. You already know the solutions to both inequalities — and the graph above shows the solution to the quadratic inequality.
2. So add the line $x = 3$ to your graph.
3. You're now interested in when the curve is negative, and when the x values are greater than 3.
4. So both inequalities are satisfied when: $3 < x < 5$

Solving a quadratic inequality to find k

On page 42, you came across a quadratic containing an unknown constant (k) and used the formula for the **discriminant** to form a **linear inequality** in terms of k. The example below is similar, but it results in a **quadratic inequality**.

Example 5

PROBLEM SOLVING

The equation $kx^2 + (k + 3)x + 4 = 0$ has two distinct real solutions. Show that $k^2 - 10k + 9 > 0$, and find the set of values of k which satisfy this inequality.

1. The original equation has two distinct real solutions, so the discriminant must be greater than 0. (The discriminant is the $b^2 - 4ac$ bit of the quadratic formula — see page 41 for more.)
2. Then put these values into the formula for the discriminant.

$a = k$, $b = (k + 3)$ and $c = 4$

$$b^2 - 4ac > 0 \Rightarrow (k+3)^2 - (4 \times k \times 4) > 0$$
$$\Rightarrow k^2 + 6k + 9 - 16k > 0$$
$$\Rightarrow k^2 - 10k + 9 > 0$$

3. Now, to find the set of values for k, you have to factorise the quadratic.

$k^2 - 10k + 9 = (k - 1)(k - 9)$

4. So, the graph of the quadratic will cross the horizontal axis at $k = 1$ and $k = 9$, and it's u-shaped.
5. Sketching the graph, you can see that the quadratic is > 0 when: $k < 1$ or $k > 9$
6. Or in set notation: $\{k : k < 1\} \cup \{k : k > 9\}$

y
$y = k^2 - 10k + 9$
k
0
1
9

Exercise 5.1.2

Q1 Use the graphs given to solve the following quadratic inequalities:

a) $x^2 + 2x - 3 < 0$

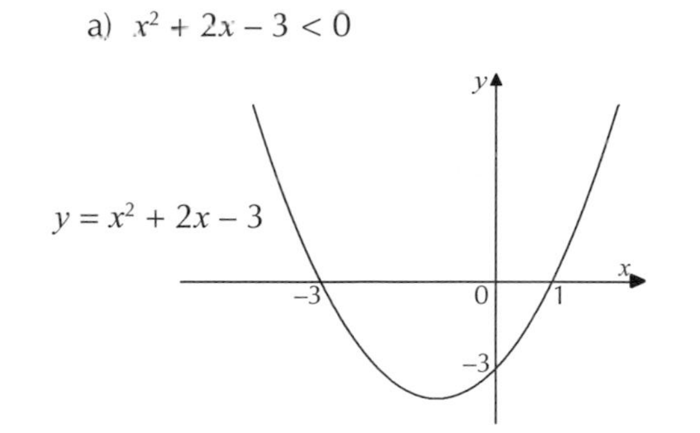

b) $4x - x^2 < 0$

$y = 4x - x^2$

c) $2x^2 \geq 5 - 9x$

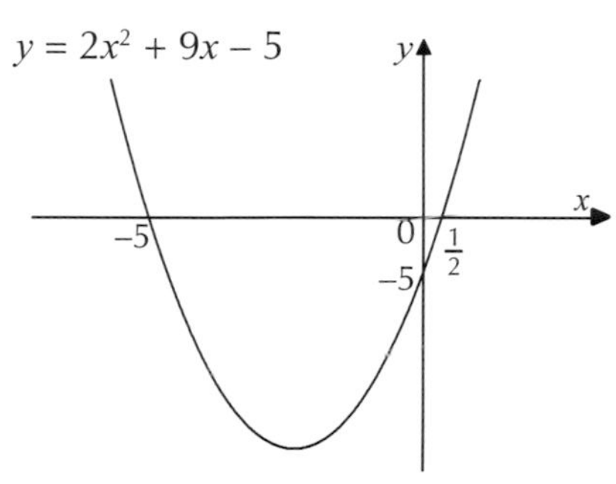

d) $x^2 - 2x - 5 > 0$

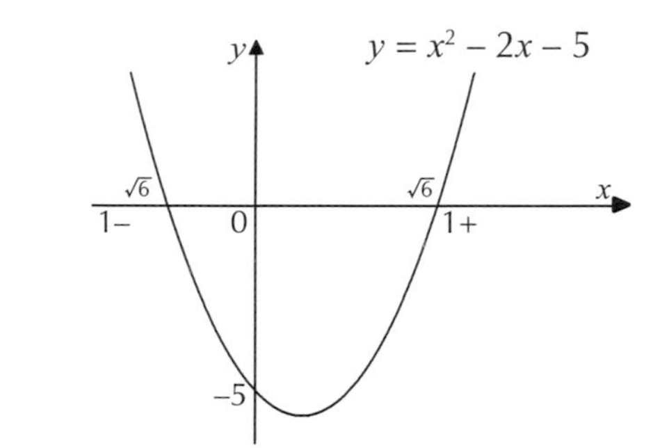

Q2 Use the graphs given to help you solve the quadratic inequalities below. Give your answers in set notation.

a) $x^2 \leq 4$

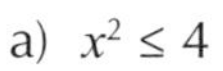

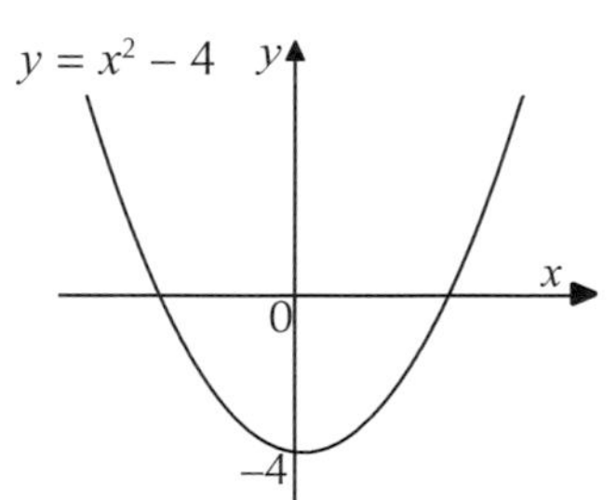

b) $13x < 3x^2 + 4$

$y = -3x^2 + 13x - 4$

c) $x^2 + 4 < 6x$

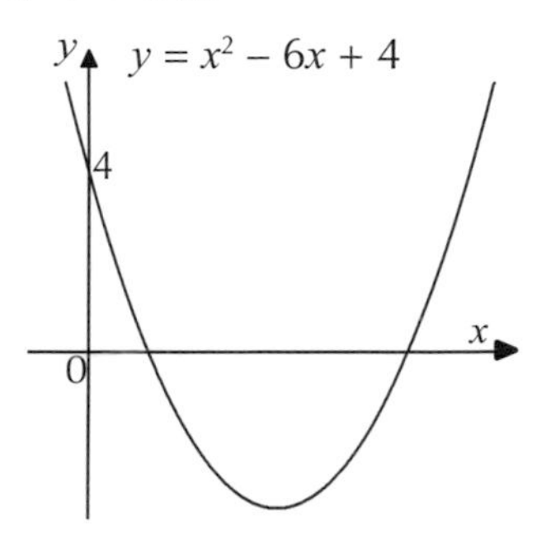

d) $7x > 4 - 2x^2$

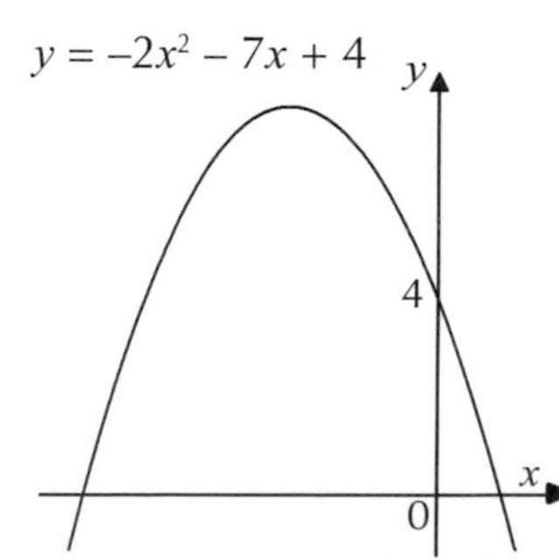

Q3 Find the ranges of values of x which satisfy the following quadratic inequalities. Include a sketch of the graph for each answer.

a) $x^2 + 5x - 6 \geq 0$ b) $x^2 - 3x + 2 < 0$ c) $6 - 5x > 6x^2$

d) $x^2 - 5x + 24 \leq 5x + 3$ e) $36 - 4x^2 \leq 0$ f) $x^2 - 6x + 3 > 0$

g) $x^2 - x + 3 > 0$ h) $6 \geq 5x^2 + 13x$ i) $2x^2 > 3(x + 3)$

j) $(x + 4)^2 \leq 5x$ k) $x^2 + 5x < \frac{1}{2}$ l) $\frac{3}{4}x^2 \geq 1 + \frac{1}{4}x$

Q4 Find the values of x which satisfy the following inequalities, giving your answers in set notation:

a) $\frac{1}{x} > 5$ b) $7 > \frac{3}{x}$

c) $-5 > \frac{2}{x}$ d) $-\frac{6}{x} > 1$

Q4 Hint: See the tip on p.65 for a hint on how to solve these.

Q5 Use the following graphs to find the values of x which satisfy the corresponding inequalities. Give your answers in interval notation, where appropriate.

a) $x^2 - 6x - 7 < 0$ and $x \leq 4$ b) $-2x^2 + 19x - 30 > 0$ and $x > 5$

c) $x^2 - x \leq 56$ and $\frac{1}{x} + \frac{1}{x^2} > 0$ d) $4x \leq \frac{x^2}{3}$ and $5x - 2 < 4x + 8$

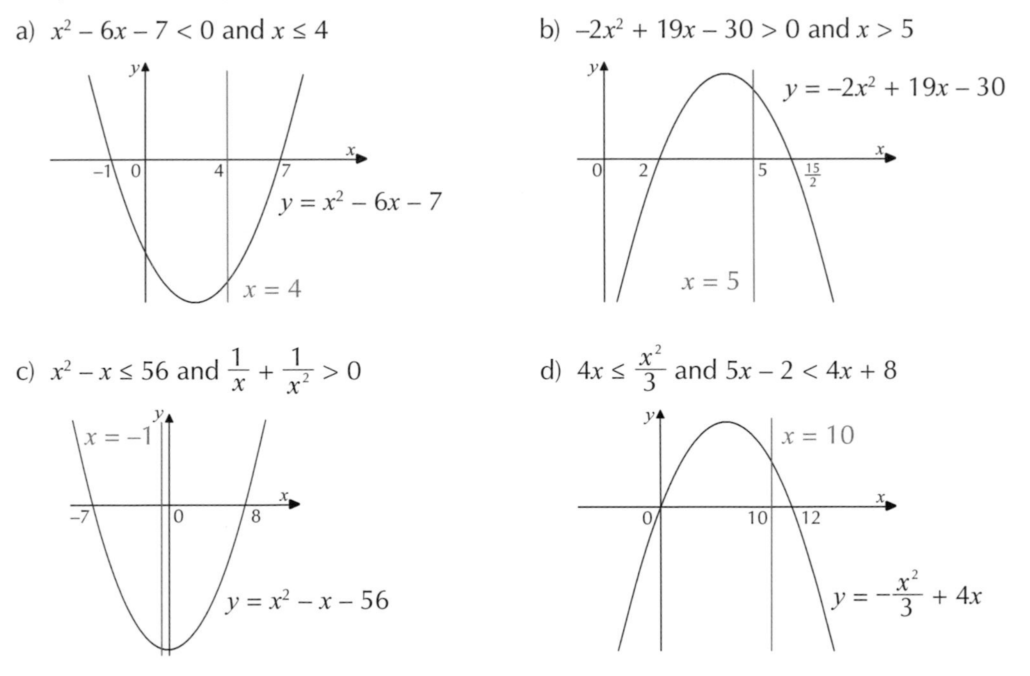

Q6 A rectangular office is to be built, measuring $(x - 9)$ metres wide and $(x - 6)$ metres long. Given that at least 28 m^2 of floor space is required, find the set of possible values of x.

Q7 a) Find the set of values of k for which $kx^2 - 6x + k = 0$ has two distinct real solutions.

b) Find the set of values for k which gives the equation $x^2 - kx + k = 0$ no real roots. Give your answer in set notation.

Q8 Find the values of x which satisfy both $4(3 - x) \geq 13 - 5x$ and $7x + 6 \geq 3x^2$.

Graphing inequalities

You've seen on the last few pages how **graphs** can be used to solve quadratic inequalities. You can also show **regions** on a graph that satisfy **more than one inequality** in **two variables** (x and y) — whether they're linear or quadratic. The method has four steps:

1. **Write each inequality as an equation.**
   Just put = wherever you have an inequality sign, and **rearrange** into the form "$y = ...$". You might have to **split up** any inequalities that are of the form $a < x < b$ into two separate bits — i.e. $a = x$ and $x = b$.

2. **Draw the graph for each equation.**
   If the original inequality was < or >, draw a **dotted line**, and draw a **solid line** for ≤ or ≥.

3. **Work out which side of each line you want.**
   Look back at each inequality and **substitute** in the coordinates of a point to see whether or not it **satisfies** the inequality (usually the **origin** is an easy point to use). If it does, you want the side of the line that the point is on, and if not, you want the other side.

Tip: If the origin lies on one of the lines, you might have to use a different point — but the method is the same.

4. **Label the correct region.**
   Once you know which side of each line you need, shade the **other side** (the side that does **not** satisfy the inequality). Once you've done this for each inequality, **label** the **unshaded** area. This will be the region that satisfies **all** of the inequalities. Make sure you read the question **carefully** — you could be asked to shade the region that **satisfies** the inequalities rather than the regions that don't.

Example 1

Draw the following inequalities on a graph and label the region that satisfies all three:

$$2x + y > 4 \qquad x - y < 1 \qquad y \leq 3$$

1. Change the inequalities into equations: $y = 4 - 2x$ $\quad y = x - 1$ $\quad y = 3$

2. Plot the lines on your graph, using dotted lines for the first two, and a solid line for $y = 3$.

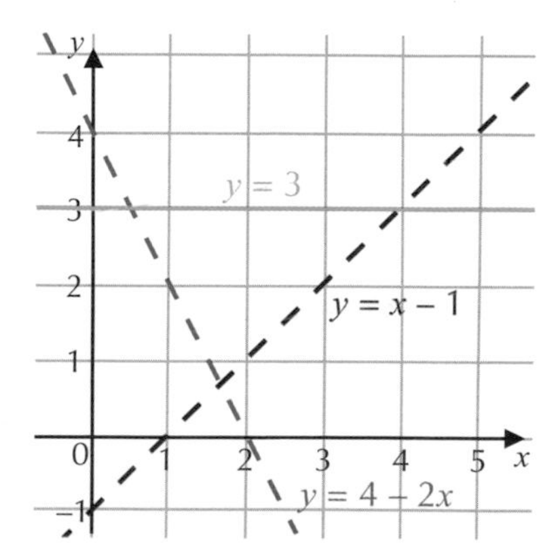

3. Decide which side of each line you want. Try the origin (0, 0) in each inequality.
   - $2x + y > 4 \Rightarrow 0 + 0 > 4$ which is **false**. The origin is on the **wrong** side of the line, so shade this side.
   - $x - y < 1 \Rightarrow 0 - 0 < 1$ which is **true**. The origin is on the **correct** side of the line, so shade the other side.
   - $y \leq 3 \Rightarrow 0 \leq 3$ which is **true**. The origin is on the **correct** side of the line, so shade the other side.

4. Finally, the shaded graph looks like this.

   Check your answer by testing a point — for example, the point (2, 2) lies in the unshaded region and satisfies all three inequalities here. Don't forget to label the correct region — this one is labelled R.

This unshaded region is the area you want — it satisfies all three inequalities.

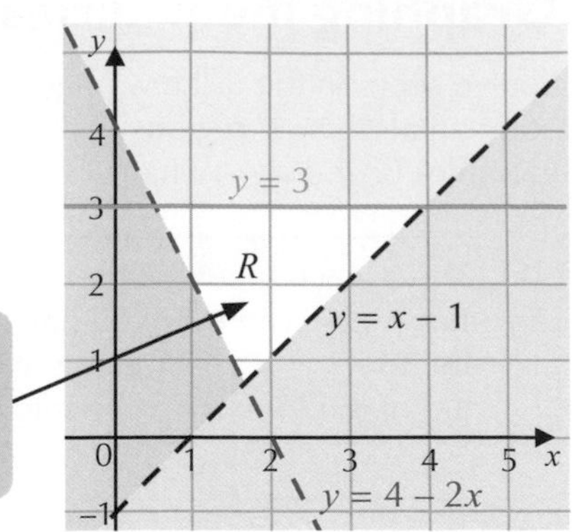

Example 2

Draw and label the region that satisfies the following inequalities:

$$y > x^2 - x - 2 \qquad y \geq 4 + 7x - 2x^2$$

1. Write out the inequalities as equations.

   $y = x^2 - x - 2 = (x + 1)(x - 2)$ $\qquad$ $y = 4 + 7x - 2x^2 = (4 - x)(1 + 2x)$

2. Draw the graphs of these equations, using a **dotted line** for $y = x^2 - x - 2$ and a **solid line** for $y = 4 + 7x - 2x^2$.

   Try and sketch the quadratics as accurately as you can (by finding the x- and y-intercepts and vertices — see pages 44-48).

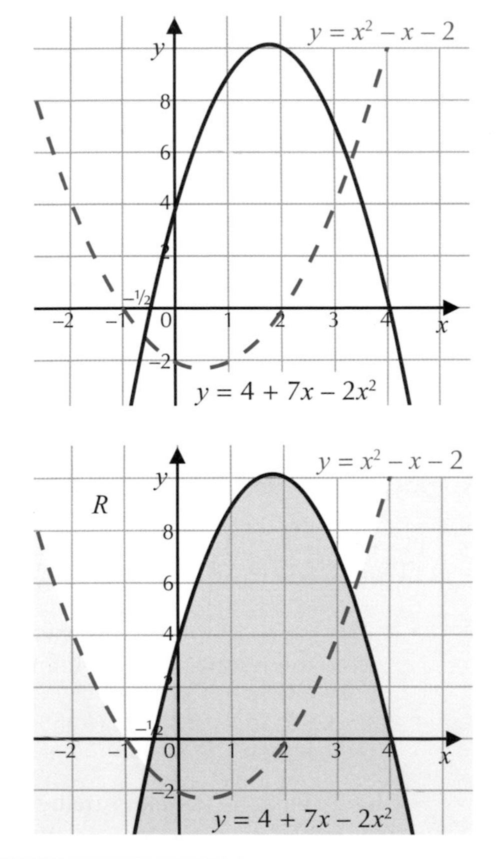

3. Try the origin in each inequality:
   - $y > x^2 - x - 2 \Rightarrow 0 > 0 - 0 - 2$ which is **true**. The origin is on the **correct** side of the line, so shade the other side.
   - $y \geq 4 + 7x - 2x^2 \Rightarrow 0 \geq 4 + 0 - 0$ which is **false**. The origin is on the **wrong** side of the line, so shade this side.

4. So you get an unshaded region that looks like this:

   Notice that the unshaded region is not enclosed — it continues upwards to infinity.

Exercise 5.1.3

Q1 Work out whether the following statements are true or false:

a) The point (2, 4) is in the region that satisfies $3y > 8x - 3$.

b) The point (–3, –5) lies outside the region that satisfies $4y + x^2 \leq 3$.

c) The point (8, –4) is in the region that satisfies $y^2 + (x + 6)^2 \geq 68$.

d) The point (1, 3) is in the region that satisfies $x + 2y > 4$ and $3x^2 > 20 - 4y$.

e) The point $\left(\frac{1}{2}, \frac{3}{2}\right)$ lies outside the region that satisfies $y^2 < 10 - 8x^2$ and $3x + 4y \geq 6$.

Q2 Give the inequalities that define the shaded regions on the following graphs.

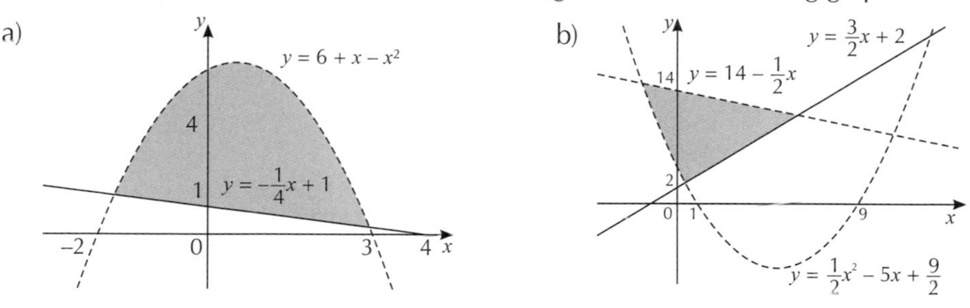

Q3 Draw and shade the regions that satisfy the following sets of inequalities:

a) $x + y < 5, \quad 2x + y \geq 4, \quad x + 2y > 6$

b) $x \leq 4, \quad y \leq 7, \quad x + y > 4$

c) $y > x^2, \quad x - y \geq -3$

d) $y - 2 \leq x^2, \quad 2x^2 - y < 2$

e) $4x^2 > y - 5$, $3x + 5y \leq 40$

f) $2y + 4x^2 < 6x + 10$, $5y > 2x + 5$

Q3 Hint: The question asks you to shade the region that satisfies the inequalities, not the region that doesn't.

Q4 Regions A and B are described by the sets of inequalities below:

A: $x + 2y \leq 12, \quad 2y - 3x \leq 4, \quad y \geq 2$ $\qquad$ B: $x \geq 3, \quad 2x \leq y + 9, \quad x + 3y \leq 15$

Which region has the greater area?

PROBLEM SOLVING

Q5 A bakery is running out of ingredients and wants to see how many sponge cakes and baguettes they can make. A cake requires 1 lb of flour and 3 eggs, and a baguette requires 2 lb of flour and 1 egg. The bakery can make x cakes and y baguettes. They have 24 lb of flour remaining, which can be represented by the inequality $x + 2y \leq 24$.

MODELLING

a) They only have 42 eggs left. Use this information to form another inequality in x and y.

b) On a graph, draw and label the region that satisfies both of these inequalities.

c) A customer requests 8 cakes and 10 baguettes for a fête. Can the bakery meet their order? If not, what ingredient(s) do they not have enough of?

Q5b) Hint: Note that x and y can't be negative (since they can't make a negative number of cakes/baguettes). So $x \geq 0$ and $y \geq 0$ have to be satisfied as well.

5.2 Simultaneous Equations

Solving simultaneous equations just means finding the answers to two equations at the same time — i.e. finding values for x and y for which both equations are true.

Learning Objectives (Spec Ref 2.4):

- Solve two linear simultaneous equations using elimination or substitution.
- Solve simultaneous equations where one is linear and one is quadratic using substitution.

Prior Knowledge Check: Be able to expand brackets and identify common factors — see page 10.

Simultaneous equations — both linear

Solving by elimination

Simultaneous equations are just a pair of equations containing two unknown quantities, often x and y.

This is how simultaneous equations are often shown:

$$3x + 5y = -4$$
$$-2x + 3y = 9$$

But they'll look different sometimes, maybe like this:

$$4 + 5y = -3x$$
$$-2x = 9 - 3y$$

You can solve two linear simultaneous equations in two different ways — by **substitution** (see page 76) and by **elimination.**

Before you can use the elimination method, you need to **rearrange** the equations as '$ax + by = c$'.

$$\begin{aligned} 4 + 5y &= -3x \\ -2x &= 9 - 3y \end{aligned} \quad \longrightarrow \quad \begin{aligned} 3x + 5y &= -4 \\ -2x + 3y &= 9 \end{aligned}$$

The elimination method involves **four** steps:

1. Match the coefficients
   Multiply the equations by numbers that will make either the x's or the y's **match** in the two equations (ignoring minus signs).

Tip: Multiply each equation by the number that gives the lowest common multiple (LCM) of the coefficients.

2. Eliminate to find one variable
   If the coefficients are the **same** sign, you'll need to **subtract** one equation from the other. If the coefficients are **different** signs, you need to **add** the equations.

3. Find the other variable (that you eliminated)
   When you've found one variable, put its value into one of the **original equations** so you can find the **other** variable.

4. Check your answer
   Put the values you found into the **other original equation**.

Tip: You should always check your answer to make sure you've worked out x and y correctly. Then pat yourself on the back if you have. You deserve it.

Example

Solve the simultaneous equations $3x + 5y = -4$ and $-2x + 3y = 9$.

1. Number your equations 1 and 2 so you know which one you're working with.

   ① $3x + 5y = -4$

   ② $-2x + 3y = 9$

2. Match the coefficients:
   To get the x's to match, multiply the first equation by 2 and the second by 3.

   ①×2 $6x + 10y = -8$ → ③

   ②×3 $-6x + 9y = 27$ → ④

   Number the new equations.

3. Eliminate to find one variable:
   Add the new equations together to eliminate the x's.

   ③+④ $19y = 19$

   $y = 1$

4. Find the variable you eliminated:
   $y = 1$, so substitute that value for y into one of the original equations to find x.

   $y = 1$ in ① $\Rightarrow 3x + 5 = -4$

   $3x = -9$

   $x = -3$

5. So the solution is $x = -3, y = 1$

6. Check your answer:
   Put these values into the other equation.

   $-2x + 3y = 9$

   $x = -3$

   $y = 1$

   $\Rightarrow -2 \times (-3) + 3 \times 1 = 6 + 3 = 9$

   If these two numbers are the same, then the values you've got for the variables are right.

If you drew the **graph** of each equation you'd get two straight lines.

- The point where these two lines **intersect** gives the **solution** to the two simultaneous equations.
- For the last example, the graph of the two lines $3x + 5y = -4$ and $-2x + 3y = 9$ would look like this:

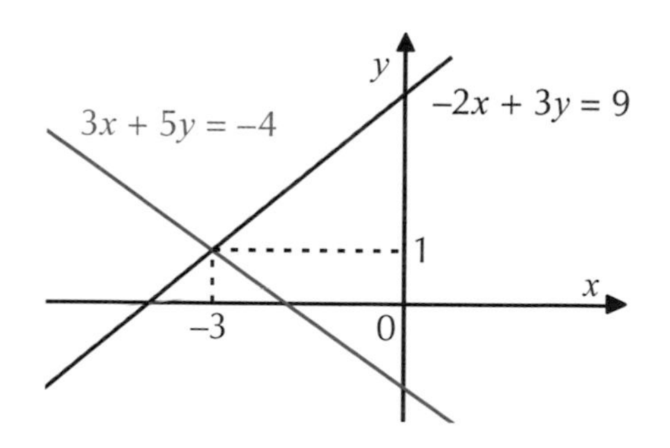

Tip: See Chapter 6 for more on straight-line graphs and their equations.

- The two lines intersect at the point (–3, 1), which is the same as the answer worked out above.
- However, **not all** simultaneous equations have solutions that work in both equations — for example, no values of x and y satisfy both $2x + 3y = 5$ and $4x + 6y = 7$. This would be obvious if you **sketched the graphs** — the lines are **parallel** so they never intersect.

If you're asked to find the coordinates of the point of intersection, you need to find them using an **algebraic method**, rather than reading off the graph.

Exercise 5.2.1

Q1 Solve the following simultaneous equations:

a) $2x - 3y = 3$
$x + 3y = 6$

b) $3x + 2y = 7$
$7x - y = -12$

c) $4x + 3y = -4$
$6x - 4y = 11$

d) $7x - 6y = 4$
$11x + 9y = -6$

e) $6x + 2y - 8 = 0$
$4x + 3 = -3y$

f) $2x + 18y - 21 = 0$
$-14y = 3x + 14$

g) $2x + 16y = 10$
$64y - 5 + 3x = 0$

h) $4x - 3y = 15$
$5y - 12 = 9x$

Q1 Hint: Rearrange the equations so they're in the form $ax + by = c$.

Q2 Find the point of intersection of each pair of straight lines.

a) $y = 2x - 3$
$y = \frac{1}{2}x + 3$

b) $y = -\frac{2}{3}x + 7$
$y = \frac{1}{2}x + \frac{21}{2}$

c) $x + 2y + 5 = 0$
$3x - 5y - 7 = 0$

d) $2x - 3y = 7$
$5x - \frac{15}{2}y = 9$

e) $8x = -3y + 10$
$9y = 3 - 6x$

f) $7x - 5y = 15$
$2x - 9 = 3y$

g) $6x + 3y = 10$
$-9 = 8y - 4x$

h) $10y = 3 - x$
$5y = 6x + 5$

i) $\frac{7}{3}x = 2 + \frac{5}{3}y$
$y = \frac{3}{4}x + \frac{1}{3}$

j) $\frac{3}{4}y = \frac{9}{5}x - 10$
$-\frac{3}{5}y + \frac{3}{2}x + 10 = 0$

Q2 Hint: Remember, not all simultaneous equations have solutions (see previous page).

Q3 Three roads on a map are modelled (in (x, y) coordinates) by the following equations: MODELLING

A: $5x + 2y = -11$ B: $2x = y + 1$ C: $5y = 13 + x$

Signposts are placed at each intersection of the roads. Find the coordinates of each signpost.

Simultaneous equations — if one is not linear

Solving by substitution

Elimination is great for simple equations, but it won't always work. Sometimes one of the equations has not just x's and y's in it — but bits with x^2 and y^2 as well. When one of the equations has quadratic terms, you can **only** use the **substitution** method. The substitution method involves **four** steps:

1. **Isolate variable in linear equation**
— rearrange the linear equation to get x or y on its own.
2. **Substitute into the quadratic equation**
— to get a quadratic equation in just one variable.
3. **Solve to get values for one variable**
— either by factorising or using the quadratic formula.
4. **Stick the values in the linear equation** —
to find corresponding values for the other variable.

Tip: Always check your answer at the end too, by putting the values back into the original equations.

Example 1

Solve the simultaneous equations $-x + 2y = 5$ and $x^2 + y^2 = 25$.

1. Start by labelling the two equations. Here the linear equation is labelled ①, and the equation with quadratic terms is labelled ②.

   ① $-x + 2y = 5$

   ② $x^2 + y^2 = 25$

   Tip: The linear equation is the one with only x's and y's in. The quadratic is the one with x^2 or y^2 terms.

2. Rearrange the linear equation so that either x or y is on its own on one side of the equals sign.

   ① $-x + 2y = 5$

   $\Rightarrow x = 2y - 5$

3. Substitute this expression into the quadratic...

   Sub into ②: $x^2 + y^2 = 25$

   $\Rightarrow (2y - 5)^2 + y^2 = 25$

4. ...and then rearrange this into the form $ax^2 + bx + c = 0$, so you can solve it — either by factorising or using the quadratic formula.

   $\Rightarrow (4y^2 - 20y + 25) + y^2 = 25$

   $\Rightarrow 5y^2 - 20y = 0$

   $\Rightarrow 5y(y - 4) = 0$

   $\Rightarrow y = 0$ or $y = 4$

5. Put both these values back into the linear equation to find corresponding values for x:

   When $y = 0$:

   ① $-x + 2y = 5$

   $\Rightarrow x = -5$

   When $y = 4$:

   ① $-x + 2y = 5$

   $-x + 8 = 5$

   $\Rightarrow x = 3$

6. So solving these simultaneous equations has produced a **pair** of solutions. You'll often (but not always) get a pair of solutions if one of the equations is quadratic.

   $x = -5, y = 0$ and $x = 3, y = 4$

7. Now, **check your answers** by putting each set of values back into the original equations.

   $x = -5, y = 0$:

   $-(-5) + 2 \times 0 = 5$ ✓

   $(-5)^2 + 0^2 = 25$ ✓

   $x = 3, y = 4$:

   $-(3) + 2 \times 4 = 5$ ✓

   $3^2 + 4^2 = 25$ ✓

8. The equation $x^2 + y^2 = 25$ is actually a circle about the origin with radius 5 and the linear equation is just a standard straight line.

   So by solving the simultaneous equations you're finding the two points where the line passes through the circle — the points $(-5, 0)$ and $(3, 4)$.

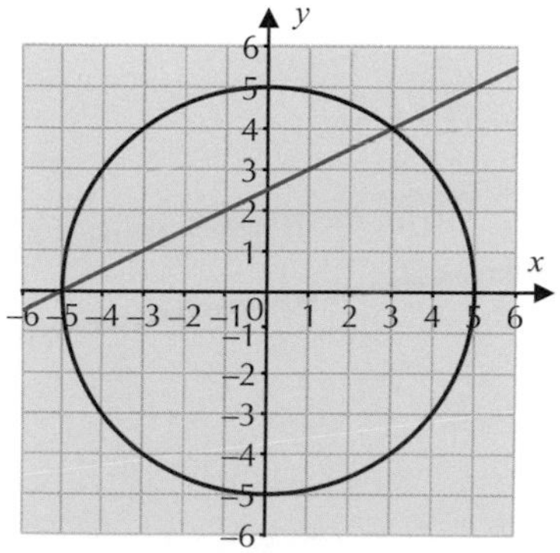

Example 2

Find any points of intersection of the following graphs:

a) $y = x^2 - 4x + 5$ and $y = 2x - 4$

1. Label the two equations: ① $y = x^2 - 4x + 5$ ② $y = 2x - 4$

2. Substitute ② in ①: $2x - 4 = x^2 - 4x + 5$

3. Rearrange and solve:
$$x^2 - 6x + 9 = 0$$
$$(x - 3)^2 = 0$$
$$x = 3$$
This is a double root — i.e. you only get 1 solution from the quadratic equation.

4. In Equation ② this gives: $y = 2 \times 3 - 4 \Rightarrow y = 2$

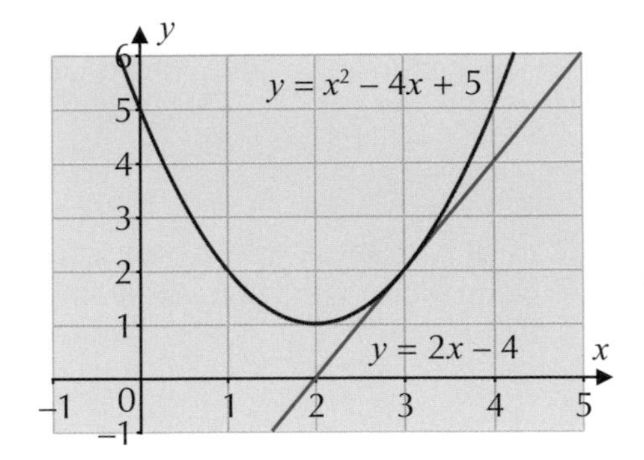

5. So there's one solution: $x = 3, y = 2$

6. Since the equations have only one solution, the two graphs only meet at one point: (3, 2). So the straight line is actually a **tangent** to the curve.

b) $y = x^2 - 4x + 5$ and $y = 2x - 5$

1. Label the two equations: ① $y = x^2 - 4x + 5$ ② $y = 2x - 5$

2. Substitute ② in ①: $2x - 5 = x^2 - 4x + 5$

3. Rearrange and try to solve with the quadratic formula (since it won't factorise). Start by finding the discriminant (see p.41):
$$x^2 - 6x + 10 = 0$$
$$b^2 - 4ac = (-6)^2 - 4 \times 1 \times 10$$
$$= 36 - 40 = -4$$

4. Since $b^2 - 4ac < 0$, the quadratic has no real roots.

So the simultaneous equations have no solutions.

5. This means the graphs never meet:

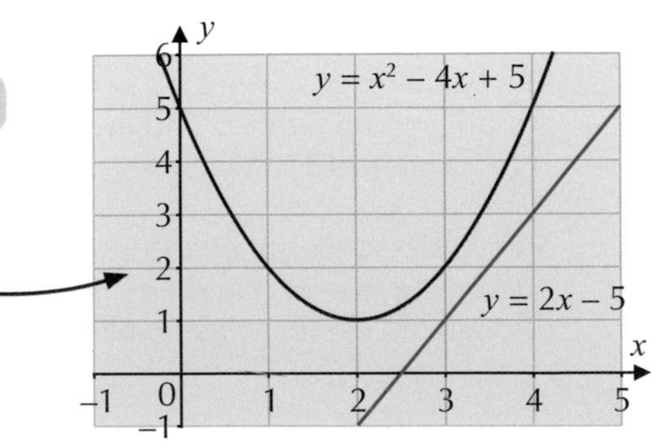

Exercise 5.2.2

Q1 Solve the following simultaneous equations using substitution:

a) $y = 4x + 3$
$2y - 3x = 1$

b) $5x + 2y = 16$
$2y - x - 4 = 0$

Q2 Solve the following simultaneous equations:

a) $y = 2x + 5$
$y = x^2 - x + 1$

b) $y = 2x^2 - 3$
$y = 3x + 2$

c) $2x^2 - xy = 6$
$y - 3x + 7 = 0$

d) $xy = 6$
$2y - x + 4 = 0$

e) $y = x^2 - 2x - 3$
$y + x + 8 = 0$

f) $y = 2x^2 - 3x + 5$
$5x - y = 3$

g) $2x^2 + 3y^2 + 18x = 347$
$4x + y = 7$

h) $2y = 2x^2 + x + 1$
$y + 2x = 2$

i) $x^2 + 4x = 4y + 40$
$12y + 5x + 30 = 0$

j) $y - x = x + 2$
$\frac{1}{4}y^2 + 25 = 3x^2 + 11x$

Q2 Hint: For these sets of equations you'll need to do some rearranging.

Q3 Find the points of intersection between the lines and curves on the graphs below.

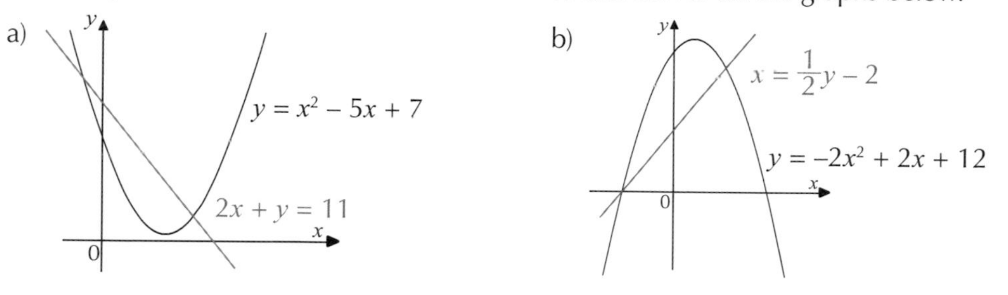

Q4 Find the points of intersection of the following curves and straight lines:

a) $y = \frac{1}{2}x^2 + 4x - 8$
$y = 4 + \frac{3}{2}x$

b) $y = 2x^2 + x - 6$
$5x - y + 10 = 0$

c) $x^2 + y^2 = 50$
$x + 2y = 5$

d) $2x^2 - y + 3x + 1 = 0$
$y - x - 5 = 0$

e) $3x^2 + 9x + 1 = 6y$
$2x + 3y = \frac{11}{2}$

f) $4x - y + 10 = 0$
$2y - 19 = 4x^2 + 8x$

Q5 a) Solve the simultaneous equations $x^2 + y^2 = 10$ and $x - 3y + 10 = 0$.

b) Say what your answer to part a) means geometrically.

Q5b) Hint: You don't need to draw a graph, just describe what your answer to part a) means.

Q6 Without drawing the graphs, determine whether the following curves and lines intersect at one or two points, or do not intersect at all:

PROBLEM SOLVING

a) $y = x^2 + 6x - 7$ and $y = 2x - 3$

b) $3x^2 + 4y^2 + 6x = 9$ and $x + 2y = 3$

c) $xy + 2x - y = 8$ and $x + y = 1$

Review Exercise

Q1 Solve:

a) $7x - 4 > 2x - 42$
b) $12y - 3 \leq 4y + 4$
c) $9y - 4 \geq 17y + 2$
d) $x + 6 < 5x - 4$
e) $4x - 2 > x - 14$
f) $7 - x \leq 4 - 2x$
g) $11x - 4 < 4 - 11x$
h) $1 + 10y \geq 7y - 12$
i) $8y - 6 \leq 6 - 8y$

Q2 Find the set of values for x that satisfy the following inequalities:

a) $3x^2 - 5x - 2 \leq 0$
b) $x^2 + 2x + 7 > 4x + 9$
c) $3x^2 + 7x + 4 \geq 2(x^2 + x - 1)$
d) $x^2 + 3x - 1 \geq x + 2$
e) $2x^2 > x + 1$
f) $3x^2 - 12 < x^2 - 2x$
g) $3x^2 + 6x \leq 2x^2 + 3$
h) $(x + 2)(x - 3) \geq 8 - 3x^2$

Q3 Draw and shade the region which satisfies each of the following sets of inequalities.

a) $8 \leq y - x, \quad y < 12 - x, \quad 9x + 2y < -4$
b) $x + 3y > 15, \quad 3x + y < 12, \quad 4y \leq x + 36$
c) $10y + 10x > x^2, \quad y < -x^2 + 8x - 12$

Q4 Solve these sets of simultaneous equations:

a) $3x - 4y = 7$ and $-2x + 7y = -22$
b) $2x - 3y = \frac{11}{12}$ and $x + y = -\frac{7}{12}$
c) $2x + 3y = 8$ and $6y = 5 - 4x$
d) $11y = 9x + 4$ and $3x - 2y = 7$
e) $\frac{1}{2}x + \frac{1}{3}y = 50$ and $x + 4y = 25$
f) $x + 4y = \frac{1}{4}$ and $y + 2x = \frac{1}{5}$

Q5 Find where the following lines meet:

a) $y = 3x - 4$ and $y = 7x - 5$
b) $y = 13 - 2x$ and $7x - y - 23 = 0$
c) $2x - 3y + 4 = 0$ and $x - 2y + 1 = 0$
d) $5x - 7y = 22$ and $3y - 4x - 13 = 0$
e) $9 - 8y = \frac{2}{3}x$ and $\frac{1}{3}x + \frac{2}{3}y = 10$
f) $24x + 15y = 2$ and $18x + 36y = 5$

Q6 Find, where possible, the solutions to these sets of simultaneous equations. Interpret your answers geometrically.

a) $y = x^2 - 7x + 4$ and $2x - y - 10 = 0$
b) $y = 30 - 6x + 2x^2$ and $y = 2(x + 11)$
c) $2x^2 + 2y^2 - 3 = 0$ and $y = x + 4$
d) $4y + 3x = 8$ and $2y - 2x^2 - 4x = 7$
e) $\frac{1}{4}x^2 + 3x + 15 = 4y$ and $2y = 3x + 3$
f) $(x - 3)^2 + (y + 4)^2 = 25$ and $x - 7y = 6$

Q7 Without drawing the graphs, decide whether the curve $y = x^2 - 2x - 3$ and the line $y = 3x + 11$ intersect at one or two points, or do not intersect at all.

Exam-Style Questions

Q1 a) Find the coordinates of the points where the curve $y = 3x^2 + 7x + 15$ and the line $y = 6x + 25$ intersect.

[5 marks]

b) Hence find the range of values for x where $3x^2 + 7x + 15 > 6x + 25$.

[2 marks]

Q2 Show that the curve C, $3x^2 + 5y = 14$, and the line l, $5x - 10y = -38$, do not intersect.

[4 marks]

Q3 The straight line $y = 2x + 7$ intersects the circle $(x - 9)^2 + (y - 5)^2 = 160$ at two points. Find the coordinates of the points of intersection.

[5 marks]

Q4 Solve the inequality $2x^2 - 5x - 3 > 0$, giving your solution in set notation.

[4 marks]

Q5 Consider the equations: $2x + ay = 4$
$3x - 2y = 1$, where a is constant

a) Assuming a solution exists, find x and y as fractions in terms of a.

[4 marks]

b) Find the value of a for which no solution for the equations exists. Explain your reasoning.

[2 marks]

c) What is the geometrical significance of the value of a in part b)?

[2 marks]

Q6 The line with equation $y = 2x + k$ is a tangent to the curve with equation $y = x^2 - 4x + 6$. Find the value of k and the point on the curve at which the line is a tangent.

[6 marks]

Q7 a) On a set of axes, sketch the graph of $y = x^2$, where $-2 \leq x \leq 4$.

[1 mark]

b) Sketch the graph of $y = 2x + 3$ on the same set of axes, labelling all axis intercepts.

[2 marks]

c) Shade in the region on your sketch that is represented by the set $\{(x, y): y \geq x^2\} \cap \{(x, y): y \leq 2x + 3\}$.

[1 mark]

Chapter 6 Coordinate Geometry, Graphs and Circles

6.1 The Equation of a Straight Line

Any straight line can be described by an equation made up of an x term, y term and constant term (though one of these may be zero). There are three standard ways of arranging straight line equations.

Learning Objectives (Spec Ref 3.1):

- Find the equation of a straight line passing through two given points, or through one point with a certain gradient.
- Write straight line equations in any of the three forms: $y - y_1 = m(x - x_1)$, $y = mx + c$ and $ax + by + c = 0$.

Prior Knowledge Check: Be able to find the gradient of a straight line and draw straight line graphs — this should be familiar from GCSE.

Equations of the form $y - y_1 = m(x - x_1)$

This is the first form you need to know. m is the **gradient,** x_1 and y_1 are the **coordinates** of one of the points on the line. If you're told **two points** that a straight line passes through, this is probably the easiest one to use.

You do need to be a little careful using the formula, so here's a method to follow:

1. LABEL the points (x_1, y_1) and (x_2, y_2).
2. GRADIENT — find this using $m = \frac{y_2 - y_1}{x_2 - x_1}$.
3. WRITE DOWN THE EQUATION $y - y_1 = m(x - x_1)$.
4. SUBSTITUTE in your values for m, x_1 and y_1. (You could use x_2 and y_2 instead — the equation would look different, but still represent the line.)

Example

Find the equation of the line that passes through the points (–3, 10) and (1, 4), and write it in the form $y - y_1 = m(x - x_1)$.

1. Label the points.

   Point 1 — $(x_1, y_1) = (-3, 10)$

   Point 2 — $(x_2, y_2) = (1, 4)$

2. Find the gradient of the line using $m = \frac{y_2 - y_1}{x_2 - x_1}$.

   $m = \frac{4 - 10}{1 - (-3)} = \frac{-6}{4} = -\frac{3}{2}$

3. Write down the equation of the line.

   $y - y_1 = m(x - x_1)$

4. Finally, substitute in the values for m, x_1 and y_1.

   $x_1 = -3$, $y_1 = 10$, $m = -\frac{3}{2} \Rightarrow y - 10 = -\frac{3}{2}(x - (-3))$

   $\Rightarrow y - 10 = -\frac{3}{2}(x + 3)$

Tip: When finding the gradient, make sure you subtract the same way round on the top and bottom of the fraction.
Don't do this: $\frac{y_2 - y_1}{x_1 - x_2}$

Equations of the form $y = mx + c$

This form for the straight line equation is probably the most popular — it's certainly the easiest form to make sense of. m is the **gradient** of the line, and c is the **y-intercept** (where it crosses the y-axis).

When in $y = mx + c$ form, you can simply read off the values of m and c — and together these give you a fairly good idea of what the graph will look like.

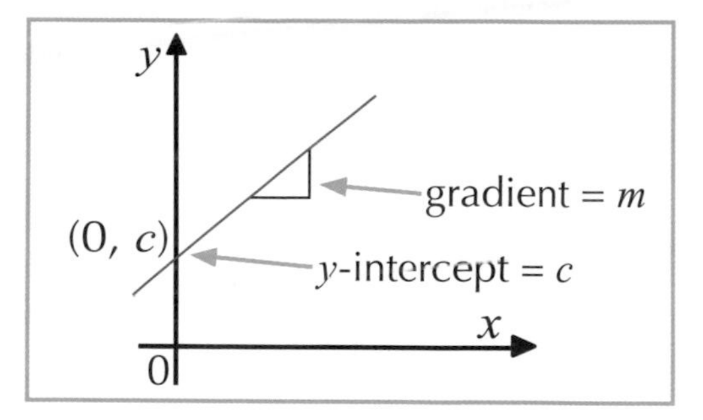

Tip: This diagram shows a straight line with positive values for m and c. A negative value for m would make the graph slope downwards.

As well as being easy to interpret, it's fairly easy to find the equation of a line in $y = mx + c$ form. Here are a couple of examples — you're given different information in each case.

Example 1

A straight line has a gradient of –2 and passes through the point (3, 1). Find the equation of the line.

1. To find c, sub in the values of m, x and y given in the question.

   $y = mx + c$

   $1 = (-2 \times 3) + c \Rightarrow 7 = c$

2. Then put your values of m and c into the equation $y = mx + c$.

   $y = -2x + 7$

Tip: Put the x- and y-values of your given point into your final equation to check it's right.

Example 2

Find the equation of the straight line that passes through the points (–18, 16) and (10, 2).

1. Start by finding m.

   $m = \frac{2-16}{10-(-18)} = \frac{-14}{28} = -\frac{1}{2}$

2. Write down the equation with $m = -\frac{1}{2}$:

   $y = -\frac{1}{2}x + c$

3. Using one of the given points, substitute in values for x and y — this will find c.

   $x = 10$, $y = 2$ gives: $2 = -\frac{1}{2}(10) + c \Rightarrow c = 7$

4. So the equation is:

   $y = -\frac{1}{2}x + 7$

The method in Example 2 is very similar to that on the previous page — find the gradient, then put in the x and y values of one of the points.

Exercise 6.1.1

Q1 Give the gradient and y-intercept of the following straight lines:

a) $y = -4x + 11$ b) $y = 4 - x$ c) $y = 1.7x - 2.3$

Q2 Give equations for the following straight lines in the form $y = mx + c$:

a) gradient -3, y-intercept $(0, 2)$ b) gradient 5, y-intercept $(0, -3)$

c) gradient $\frac{1}{2}$, y-intercept $(0, 6)$ d) gradient 0.8, y-intercept $(0, 1.2)$

e) gradient -0.4, y-intercept $(0, -7)$ f) gradient $-\frac{5}{3}$, y-intercept $(0, \frac{1}{2})$

Q3 Use the information in the diagrams to the find the equation of each straight line in the form $y = mx + c$.

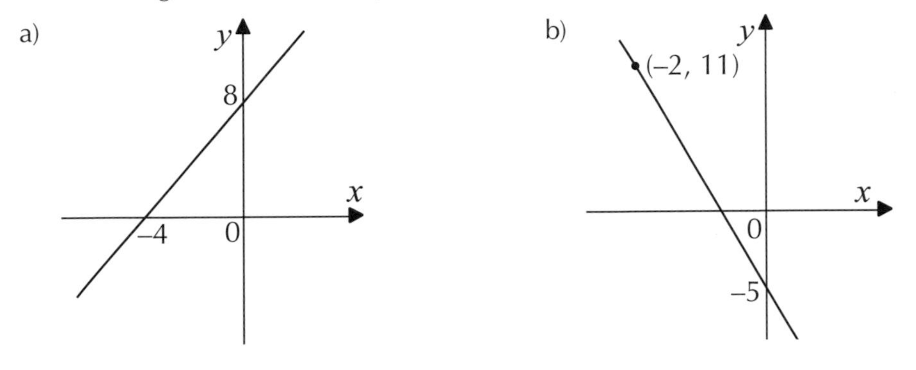

Q4 Find the equations of the lines that pass through the following pairs of points. Give each answer in these forms: (i) $y - y_1 = m(x - x_1)$ (ii) $y = mx + c$.

a) $(2, -2)$, $(6, 10)$ b) $(-1, -6)$, $(3, 4)$ c) $(4, 1)$, $(0, -3)$

d) $(12, -3)$, $(14, 1)$ e) $(5, 7)$, $(-2, 5)$ f) $(-3, 6)$, $(4, -2)$

Q5 A straight line l goes through the points $(-0.3, 0.65)$ and $(0.2, -1.85)$.

a) Find the gradient of the line l.

b) Find the equation of l in the form $y = mx + c$.

Q6 Find the equation of the straight line which passes through the point $(-4, -3)$ and has a gradient of $\frac{1}{4}$. Give your answer in the form $y = mx + c$.

Q7 A line with gradient $-\frac{2}{5}$ passes through the point $(-8, 2)$. Find the equation of the line in the form $y - y_1 = m(x - x_1)$ and in the form $y = mx + c$.

Q8 A straight line has gradient 3 and passes through the point $(2, -7)$. State which of the following coordinates are points on the line.

PROBLEM SOLVING

a) $(1, -10)$ b) $(-2, -7)$ c) $(5, 2)$

d) $(0.5, 2.5)$ e) $(7, 8)$ f) $(0, -12)$

Q9 A straight line passes through the points (6, 6) and (–1, 20).
Which of the following coordinates are also on the line?

a) (2, 14) b) (–4, –24) c) (10, –2)

d) (26, –34) e) (–34, 88) f) (2.5, 13)

Q10 The distance travelled by a car is modelled by a straight line graph. At time $t = 0$ hours, it starts at distance $d = 0$ kilometres, and its speed (the gradient) is a constant 32 km/h throughout the journey.

MODELLING PROBLEM SOLVING

a) Give the equation of the line in the form $d = mt + c$.

b) How long does it take the car to travel a distance of 9.6 km? Give your answer in minutes.

c) Give one criticism of this model.

Q10 Hint: In this question, x and y have been replaced with t and d respectively.

Equations of the form $ax + by + c = 0$

This is the last form you need to know for straight line equations. In this form, a, b and c are **integers**.

This form doesn't involve m, so it's not as easy to work with. If you're asked to give an equation in this form, it's often easiest just to find it in one of the previous two forms, then rearrange it at the end.

It's important to remember that a, b and c are integers, so you must get rid of any fractions.

Example 1

Find the gradient and y-intercept of the line $7x + 3y - 6 = 0$.

1. The easiest way to answer this question is to rearrange the equation into the form $y = mx + c$.
   $7x + 3y - 6 = 0$
   $\Rightarrow 3y = -7x + 6 \Rightarrow y = -\frac{7}{3}x + 2$
2. Now compare the equation with $y = mx + c$:
   $m = -\frac{7}{3}$ and $c = 2$
3. So for the line $7x + 3y - 6 = 0$:
   gradient $= -\frac{7}{3}$, y-intercept is (0, 2)

Example 2

Find the equation of the line that passes through the point (2, –15) and has gradient $-\frac{3}{2}$, giving your answer in the form $ax + by + c = 0$, where a, b and c are integers.

1. Start by finding the equation in one of the easier forms. We'll use $y - y_1 = m(x - x_1)$ this time, but $y = mx + c$ would be just as easy...
   $m = -\frac{3}{2}$ gives: $y - y_1 = -\frac{3}{2}(x - x_1)$
2. Now sub in $x_1 = 2$ and $y_1 = -15$:
   $y + 15 = -\frac{3}{2}(x - 2)$
3. Rearrange into $ax + by + c = 0$ form:
   $y + \frac{3}{2}x + 15 - 3 = 0 \Rightarrow y + \frac{3}{2}x + 12 = 0$
   $\Rightarrow 3x + 2y + 24 = 0$

Exercise 6.1.2

Q1 Write the following equations in the form $ax + by + c = 0$, where a, b and c are integers.

a) $y = 5x + 2$
b) $3y = -\frac{1}{2}x + 3$
c) $2(x - 1) = 4y - 1$
d) $7x - 3 = 2y + 6$
e) $\frac{1}{2}(4x + 3) = 3(y - 2)$
f) $3(y - 4) = 4(x - 3)$

Q2 Find the gradient and y-intercept of the following lines:

a) $6x - 2y + 3 = 0$
b) $-9x + 3y - 12 = 0$
c) $-x - 4y - 2 = 0$
d) $7x + 8y + 11 = 0$
e) $2x - 14y + 1 = 0$
f) $-3x + 28y - 16 = 0$
g) $0.1x + 0.2y + 0.3 = 0$
h) $-10x + 0.1y + 11 = 0$
i) $\frac{6}{7}x - 3y + \frac{3}{4} = 0$

Q3 Find the equation of the line that passes through the following points.
Write your answer in the form $ax + by + c = 0$, where a, b and c are integers.

a) (0, 1), (–1, –1)
b) (5, 5), (0, 0.2)
c) (5, 2), (3, 4)
d) (9, –1), (7, 2)
e) (–6, 1), (4, 0)
f) (–12, 3), (5, 7)

Q4 Find the equation of the lines below in the form $ax + by + c = 0$, where a, b and c are integers.

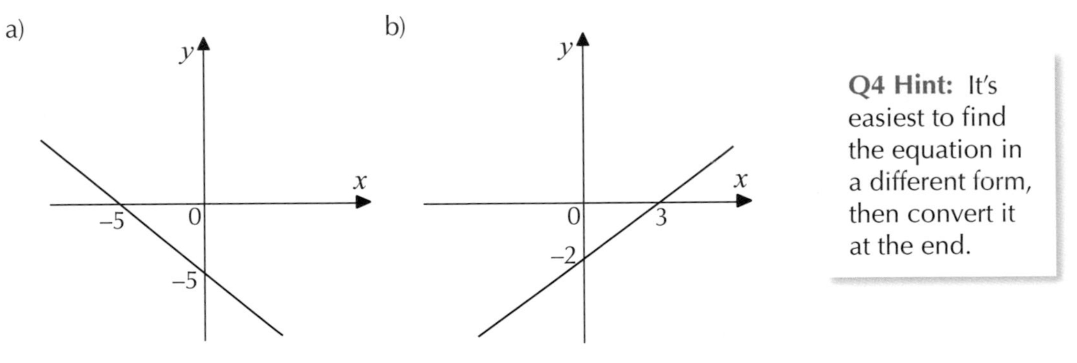

Q4 Hint: It's easiest to find the equation in a different form, then convert it at the end.

Q5 A straight line goes through the points (–6, 1) and (–2, 7). Find the equation of the line, giving your answer in the form $ax + by + c = 0$, where a, b and c are integers.

Q6 A rectangle ABCD has vertices A (–3, 2), B (–3, 5), C (6, 5) and D (6, 2).
Find the equations of the diagonals AC and BD in the form $ax + by + c = 0$.

PROBLEM SOLVING

Q7 At a cafe, a small cup of coffee costs £x and a large cup of coffee costs £y. Robert buys 3 small cups and 4 large cups for £18.

MODELLING PROBLEM SOLVING

a) Write this as an equation in the form $ax + by + c = 0$, where a, b and c are all integers.

b) Draw the straight line graph of this equation.

c) Amani buys 1 small cup and 2 large cups for £8. Write a second equation (in the same form) to represent this and plot the graph on the same set of axes.

d) Write down the cost of a small cup of coffee and the cost of a large cup of coffee.

6.2 Parallel and Perpendicular Lines

You can work out the equation of a line if it's parallel or perpendicular to a line you already know the equation of and you know a point the line passes through — first, you need to find the gradient.

Learning Objectives (Spec Ref 3.1):
- Find the equation of a line parallel or perpendicular to a given line through a given point.
- Know the conditions for two straight lines to be parallel or perpendicular to each other.

Prior Knowledge Check: You should be familiar with rearranging straight line equations into $y = mx + c$ form from GCSE.

Parallel lines

Parallel lines have **equal gradient** — that's what makes them parallel. So when finding the equation of a line parallel to a line with a given equation, you know the gradient will be the same for both.

Example 1

Find the line parallel to $y = \frac{3}{4}x - \frac{7}{4}$ that:

a) has a y-intercept of (0, 4)

1. Parallel lines have the same gradient.
   Compare $y = \frac{3}{4}x - \frac{7}{4}$ with $y = mx + c$: $m = \frac{3}{4}$
   So the gradient of the parallel line is also $\frac{3}{4}$.
2. You then need to find c — this is just the given y-intercept:
   $c = 4$
3. Put m and c into the equation $y = mx + c$:
   $y = \frac{3}{4}x + 4$

b) passes through the point (3, –1).

1. The gradient is $\frac{3}{4}$, so the equation has the form: $y = \frac{3}{4}x + c$
2. You then need to find c. We know that the line passes through point (3, –1), so stick $x = 3$ and $y = -1$ into the equation to find c:
   $-1 = \frac{3}{4}(3) + c \Rightarrow c = -\frac{13}{4}$
3. So the equation of the line is:
   $y = \frac{3}{4}x - \frac{13}{4}$

Example 2

Find the line parallel to $2x - 8y + 11 = 0$ that passes through the point (3, –1).
Give your equation in the form $ax + by + c = 0$, where a, b and c are integers.

1. First, put the given line in a more useful form, i.e. $y = mx + c$.
   $2x - 8y + 11 = 0 \Rightarrow -8y = -2x - 11$
   $\Rightarrow y = \frac{1}{4}x + \frac{11}{8}$
2. Use the gradient of the given line, $\frac{1}{4}$, and the given point (3, –1) to find the constant c.
   When $x = 3$, $y = -1$:
   $y = \frac{1}{4}x + c \Rightarrow -1 = \frac{1}{4}(3) + c \Rightarrow c = -\frac{7}{4}$
3. Write the equation in the correct form:
   $y = \frac{1}{4}x - \frac{7}{4} \Rightarrow x - 4y - 7 = 0$

You may be asked whether two lines are parallel — to do this you need to **compare** the gradients. This is easiest when both equations are in the **same form** — so one or both equations may need **rearranging**.

Example 3

Line l_1 is given by the equation $y = \frac{1}{2}x + 6$ and line l_2 is given by the equation $3x + 6y - 1 = 0$. Find out whether the lines are parallel.

1. To compare the gradients you want both lines in the form $y = mx + c$. So rearrange line l_2 into this form:

   $3x + 6y - 1 = 0 \Rightarrow 6y = -3x + 1$
   $\Rightarrow y = -\frac{3}{6}x + \frac{1}{6} \Rightarrow y = -\frac{1}{2}x + \frac{1}{6}$

2. Then compare the two equations:

   $y = \frac{1}{2}x + 6$ (Line l_1) $\quad y = -\frac{1}{2}x + \frac{1}{6}$ (Line l_2)

3. You're only concerned about the gradient, so look at the bit before the x.

   Line l_1 has a gradient of $\frac{1}{2}$.
   Line l_2 has a gradient of $-\frac{1}{2}$.
   So the lines l_1 and l_2 are NOT parallel.

Exercise 6.2.1

Q1 State which of the following straight lines are parallel to $y = -3x - 1$.

a) $2y = -6x + 2$ b) $y - 3x - 1 = 0$ c) $6y + 18x = 7$
d) $\frac{1}{3}(y + 1) = x$ e) $-9y - 2 = 27x$ f) $4y = 12x$

Q2 Find the equations of the parallel lines shown in blue.
Write them in the form $ax + by + c = 0$, where a, b and c are integers.

Q3 State whether the following pairs of lines are parallel.

a) $y = 2x + 1,\ y + \frac{1}{2}x = 1$ b) $2x - 3y + 1 = 0,\ y = \frac{2}{3}x + 2$
c) $-5x + 4y + 3 = 0,\ 8y = 10x$ d) $3x - 4y + 7 = 0,\ 16x + 12y - 3 = 0$

Q4 Line A passes through the point (4, 3) and is parallel to the line $2x - 4y + 3 = 0$.
Find the equation of line A in the form: a) $y = mx + c$, b) $ax + by + c = 0$.

Q5 Find the equations of the lines which are parallel to each of the following lines and pass through the points given. Give your answers in the form $ax + by + c = 0$, where a, b, and c are integers.

a) $y = 2x - 1$, (2, 1) b) $5x + y - 11 = 0$, (3, –1) c) $3y = \frac{1}{3}x + 2$, (–6, 2)
d) $x - \frac{1}{4}y + 1 = 0$, (–6, –5) e) $x - y = 13$, (0, 0) f) $100 = y + \frac{1}{5}x$, (50, 50)
g) $0.5x + 2.2y - 12 = 0$, (4, 8) h) $3(x + 1) - 2(y - 1) = 4$, (–2, 2) i) $\frac{y - 3x}{2} = \frac{4 + y}{3}$, (2, 3)

Perpendicular lines

Finding the equations of **perpendicular** lines (or '**normals**') is just as easy as finding the equations of parallel lines — you just need to know one key fact:

The gradients of perpendicular lines **multiply to give –1**.

Which means:

Gradient of the perpendicular line = **–1 ÷ the gradient of the other one.**

Tip: Remember, 'perpendicular' just means 'at right angles'.

So if a line has a gradient of m, a line perpendicular to it will have a gradient of $-\frac{1}{m}$.

Example 1

Find the equation of the line perpendicular to $y = \frac{1}{3}x - 1$ that passes through (–2, 4).

1. Use the gradient rule:

   Tip: Remember, to divide by a fraction, turn it upside down and then multiply by it.

   Gradient of perpendicular line
   = –1 ÷ gradient of the other one
   $= -1 \div \frac{1}{3} = -3$
   So: $y = -3x + c$

2. To find c, put the coordinates (–2, 4) into the equation:

   $4 = (-3) \times (-2) + c$
   $\Rightarrow c = 4 - 6 = -2$

3. So the equation of the line is:

   $y = -3x - 2$

Example 2

Find the equation of the line perpendicular to $7x - 3y + 5 = 0$ that passes through the point (–3, –11).

1. Start by converting the equation into a more useful form:

   $7x - 3y + 5 = 0 \Rightarrow -3y = -7x - 5$
   $\Rightarrow y = \frac{7}{3}x + \frac{5}{3}$, so the gradient is $\frac{7}{3}$.

2. Now use the gradient rule:

   Gradient of perpendicular line $= -1 \div \frac{7}{3} = -\frac{3}{7}$
   So we have $y = -\frac{3}{7}x + c$

3. Substitute in (–3, –11) to find c:

   $-11 = -\frac{3}{7}(-3) + c \Rightarrow c = -11 - \frac{9}{7} = -\frac{86}{7}$

4. So the perpendicular line has equation:

   $y = -\frac{3}{7}x - \frac{86}{7} \Rightarrow 3x + 7y + 86 = 0$

You can use the fact that the gradients of perpendicular lines **multiply** to give **–1** to work out whether two lines are perpendicular.

Example 3

PROBLEM SOLVING

Show that the line $2x + 5y + 3 = 0$ is perpendicular to $y = \frac{5}{2}x + 5$.

1. To work out if they are perpendicular, first find the gradient of both lines.

Rearrange $2x + 5y + 3 = 0$ into $y = mx + c$ form:

$2x + 5y + 3 = 0 \Rightarrow 5y = -2x - 3$

$\Rightarrow y = -\frac{2}{5}x - \frac{3}{5}$

So the gradient is $-\frac{2}{5}$.

Comparing $y = \frac{5}{2}x + 5$ to $y = mx + c$, its gradient is $\frac{5}{2}$.

2. The two lines are perpendicular if the gradients of the two lines multiply together to make –1:

$-\frac{2}{5} \times \frac{5}{2} = -1$

So the two lines are perpendicular.

Example 4

The points A (2, 5) and B (6, 0) lie on the line l_1.
The line l_2 is perpendicular to l_1 and passes through point A.

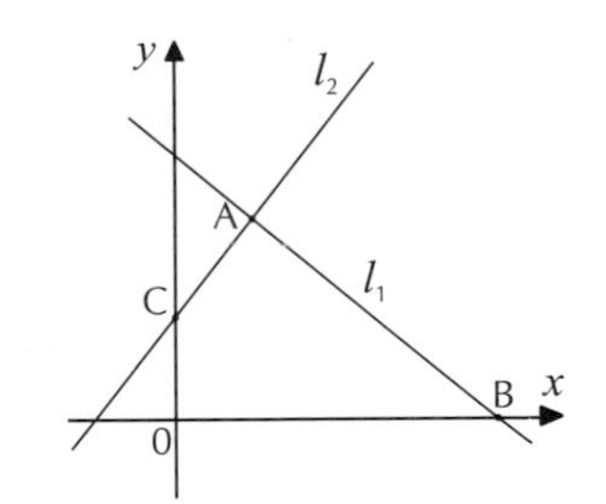

a) Find an equation for l_2 in the form $ax + by + c = 0$, where a, b and c are integers.

1. To find the equation of l_2 we need to find its gradient. We know two points on the line l_1, so work out its gradient first.

$(x_1, y_1) = (2, 5)$, $(x_2, y_2) = (6, 0)$

$m = \frac{0 - 5}{6 - 2} = -\frac{5}{4}$

2. The gradient of a perpendicular line is $-1 \div$ the other one. So the gradient of l_2 is:

$m = -1 \div -\frac{5}{4} \Rightarrow m = \frac{4}{5}$

So the equation of l_2 is: $y = \frac{4}{5}x + c$

3. To find c, put the coordinates of A (2, 5) into the equation.

$5 = \frac{4}{5} \times 2 + c \Rightarrow c = \frac{17}{5}$

4. So the equation of the line l_2 is:

$y = \frac{4}{5}x + \frac{17}{5} \Rightarrow 4x - 5y + 17 = 0$

b) Find the coordinates of point C.

At C, $x = 0$.
So put $x = 0$ into the equation $y = \frac{4}{5}x + \frac{17}{5}$:

$y = \frac{4}{5} \times 0 + \frac{17}{5} = \frac{17}{5}$

So the coordinates of point C are $\left(0, \frac{17}{5}\right)$.

Exercise 6.2.2

Q1 Find the equations of the dotted lines. Give your answers in the form $y = mx + c$

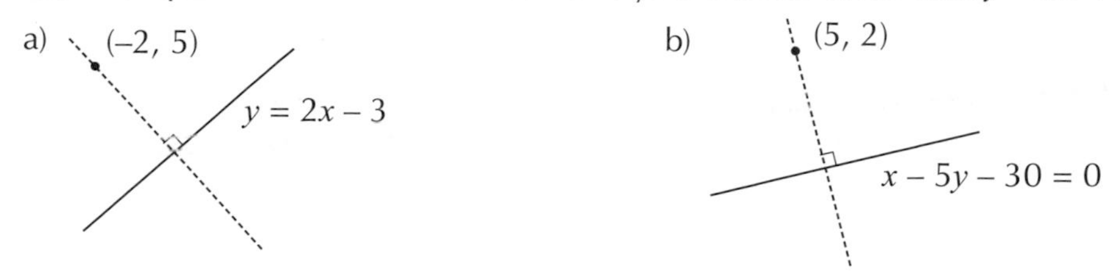

Q2 Find the equations of the lines perpendicular to each of the following that pass through the points given. Give your answers in the form $ax + by + c = 0$, where a, b, and c are integers.

a) $y = \frac{1}{4}x - 1$ $(-1, 2)$

b) $2x + 3y - 1 = 0$ $(-3, -1)$

c) $5x - 10y + 1 = 0$ $(6, -5)$

d) $y = \frac{3}{2}x + 2$ $(2, 1)$

e) $-4x + 21y = 2$ $(0.5, 7)$

f) $5(2 - x + 3y) = 2$ $(-5, -1)$

g) $7y + 1 = \frac{2x-3}{8}$ $(7, 8)$

h) $y = 2(2x + 0.1y) + 1$ $(3, 4.4)$

Q3 Work out which of the following pairs of lines are perpendicular.

a) $y = \frac{4}{3}x - 2$ and $3x + 4y - 1 = 0$

b) $y = \frac{3}{2}x - 1$ and $3x + 2y - 3 = 0$

c) $4x - y + 3 = 0$ and $2x + 8y + 1 = 0$

d) $3x - 5y + 10 = 0$ and $15x + 6y - 4 = 0$

Q4 Given P = (4, –2) and Q = (–1, 7), find the equation of the line that is perpendicular to PQ and passes through the point (2, 5). Give your answer in the form $ax + by + c = 0$.

Q5 Triangle ABC has vertices at A(0, 2), B(4, 3) and C(5, –1). PROBLEM SOLVING

a) Find the equations of the lines AB, BC and AC in the form $y = mx + c$.

b) What type of triangle is ABC? Explain why.

Q6 A quadrilateral PQRS has vertices at P(1, –1), Q(0, 2), R(3, 3) and S(4, 0). PROBLEM SOLVING

a) Find the equations of the lines PR and QS in the form $y = mx + c$.

b) Hence, or otherwise, show that PQRS is a square.

Q7 Line A passes through the point (a, b) and is perpendicular to the line $3x - 2y = 6$. Find an equation of line A in terms of a and b. PROBLEM SOLVING

Q8 The perpendicular bisector of a line segment AB is the line that is perpendicular to AB, passing through its midpoint. Find the equation of the perpendicular bisector of the line AB, where A = (1, 4) and B = (5, 2). PROBLEM SOLVING

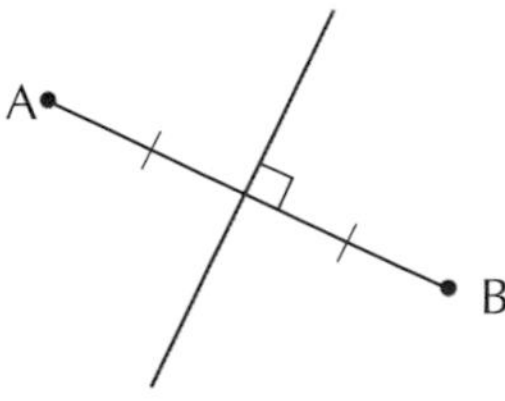

6.3 Proportion

Proportion is just a way of describing how two variables are related. There are a few different types of proportion, but the main one you need to know about is direct proportion.

Learning Objectives (Spec Ref 2.7):

- Interpret direct proportion and convert a statement of proportion into an algebraic equation.
- Use proportion to find values and calculate the constant of proportionality.

Prior Knowledge Check: Be able to write direct and inverse proportion statements as algebraic equations — you will have seen this at GCSE.

Direct proportion

If two variables are in **direct proportion**, it means that changing one variable will change the other by the same scale factor. So doubling one variable will result in the other variable doubling, and the same for tripling, halving, etc. In fact, multiplying or dividing by **any** constant will have the same effect on both.

To say that "y is directly proportional to x", you can write: $y \propto x$

This is equivalent to writing: $y = kx$

where k is a constant. k is sometimes called the **constant of proportionality**.

If you compare this equation to $y = mx + c$, you can see that the graph of two variables in direct proportion is a straight line. The gradient of the line is the constant of proportionality, k, and the y-intercept is 0 — i.e. the line passes through the origin (0, 0). Any time you have two variables in direct proportion, the graph will be a straight line through the origin.

Example 1

The circumference of a circle, C, is directly proportional to its radius, r.

a) What is the constant of proportionality linking C and r?

$C \propto r$, which is the same as saying $C = kr$.
Compare this with the formula for circumference.

Circumference of a circle $= 2\pi r$
So $kr = 2\pi r \Rightarrow k = 2\pi$

b) A circle with a radius of p cm has a circumference of 13 cm. Find the circumference of a circle with a radius of $2.5p$ cm.

1. You could use the formula to find p, but you don't need to.

2. The radius of the second circle is 2.5 times the size of the original, so the circumference will be 2.5 times the original as well.

So $C = 2.5 \times 13 =$ 32.5 cm

c) Sketch the graph of C against r.

The equation of the line is $C = 2\pi r$, which is a straight line through the origin with gradient 2π:

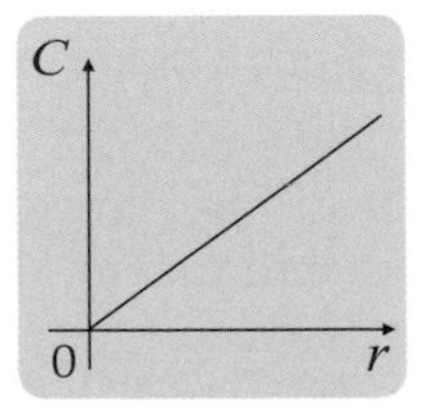

Example 2

The amount of money that Tina earns, y, is directly proportional to the number of hours that she works, x. Given that y = £91 when $x = 7$:

MODELLING

a) Find the constant of proportionality.

$y \propto x$, so first write this as $y = kx$, then substitute $y = 91$ and $x = 7$:

$y = kx \Rightarrow 91 = k \times 7 \Rightarrow k = 91 \div 7 =$ **13**

b) Find how much she would earn if she worked for 4 hours.

Substitute $x = 4$ into your equation $y = kx$:

$y = 13x \Rightarrow y = 13 \times 4 =$ **£52**

c) How many hours does she need to work to earn £104?

Substitute $y = 104$ into the equation:

$104 = 13x \Rightarrow x = 104 \div 13 =$ **8 hours**

Proportion can also be used for relationships that are **not linear**. These are not 'direct' proportion, but they do work in a similar way. For example:

$$y \propto \frac{1}{x} \Rightarrow y = \frac{k}{x}$$
$$V \propto r^3 \Rightarrow V = kr^3$$
$$\omega \propto \frac{1}{\sqrt{m}} \Rightarrow \omega = \frac{k}{\sqrt{m}}$$

Tip: When $y \propto \frac{1}{x}$, this is known as **inverse proportion** — when one doubles, the other halves (and so on). Similarly, "y is inversely proportional to x^2" means $y \propto \frac{1}{x^2}$.

You might see these written as e.g. "y is proportional to the cube root of x" instead of $y \propto \sqrt[3]{x}$. In any case, you can always replace the $\propto$ with "$= k \times$" to convert a proportion statement into an equation.

Exercise 6.3.1

Q1 Given that $y \propto x$ in each case, find the value of a if:

a) $y = 24$ when $x = 8$ and $y = a$ when $x = 5$ b) $y = 28$ when $x = 7$ and $y = 96$ when $x = a$

Q2 Find the value of a, given that x and y are inversely proportional:

a) $y = 3$ when $x = 6$ and $y = a$ when $x = 9$ b) $y = 12$ when $x = 12$ and $y = 36$ when $x = a$

Q3 For each equation below, explain whether $y \propto x$:

PROBLEM SOLVING

a) $y = 7x + 2$

b) $y = ax - bx$ (a and b constants)

c) $y = 2x + 2x^2 - 2 - x - 2x^2$

d) $y = (x + 3)^2 - (x - 3)^2$

Q4 Given that y is proportional to x^2 ($x > 0$), and that $y = 40$ when $x = 4$:

a) Find the value of y when x is 2.

b) Find the exact value of x when y is 45.

Q5 Prove that if $y \propto x$ and $y \propto z$, then $x \propto z$.

PROBLEM SOLVING

Q6 Siobhan is modelling the motion of a sliding box. She finds that the frictional force on the box is directly proportional to its mass. If the frictional force, F, is 15 N when its mass, m, is 12 kg, estimate the frictional force when the mass of the box is increased to 18 kg.

MODELLING

6.4 Curve Sketching

Being able to sketch the graph of a curve is an important skill — it can help you get your head round tricky questions. Usually, you only need a rough sketch — so knowing the basic shapes of these graphs will do.

Learning Objectives (Spec Ref 2.7):
- Sketch the graphs of simple cubic and quartic functions.
- Sketch graphs of reciprocal functions and those with negative powers.
- Know what the term asymptote means.

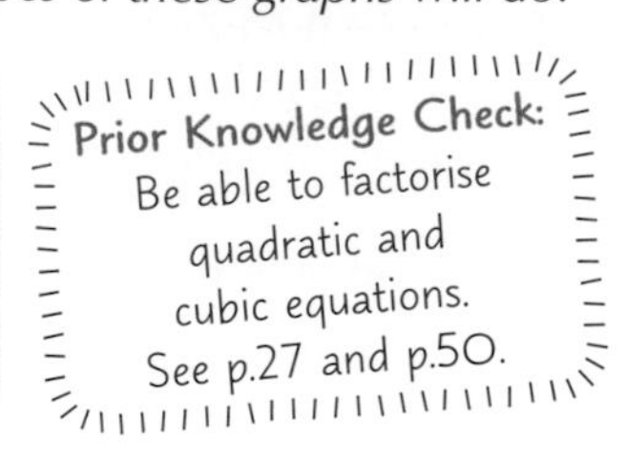

Cubic and quartic functions

Cubic functions are **polynomials** that have an x^3 term in them as the highest power of x. Similarly, quartics go up to x^4. Here are the graphs of some cubic and quartic functions:

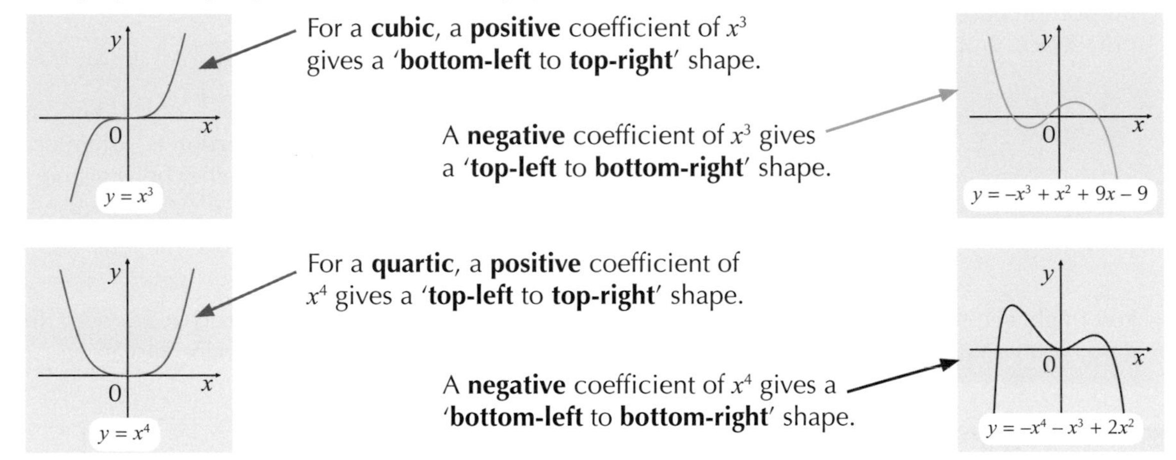

For a **cubic**, a **positive** coefficient of x^3 gives a '**bottom-left** to **top-right**' shape.

A **negative** coefficient of x^3 gives a '**top-left** to **bottom-right**' shape.

For a **quartic**, a **positive** coefficient of x^4 gives a '**top-left** to **top-right**' shape.

A **negative** coefficient of x^4 gives a '**bottom-left** to **bottom-right**' shape.

In general, for any graph in the form $y = kx^n$ (when n is positive):

EVEN

When n is EVEN, you get a u-shape or an n-shape.

If k **is POSITIVE**, you get a u-shape above the x-axis.

And if k **is NEGATIVE**, you get an n-shape below the x-axis.

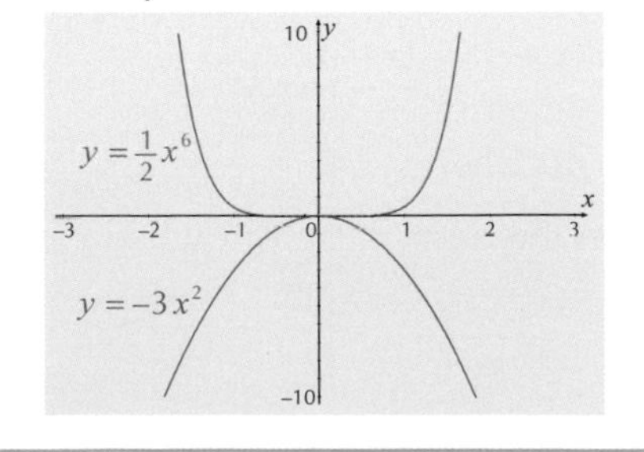

ODD

When n is ODD, you get a 'corner-to-corner' shape.

If k **is POSITIVE**, you get a 'bottom-left to top-right' shape.

And if k **is NEGATIVE**, you get a 'top-left to bottom-right' shape.

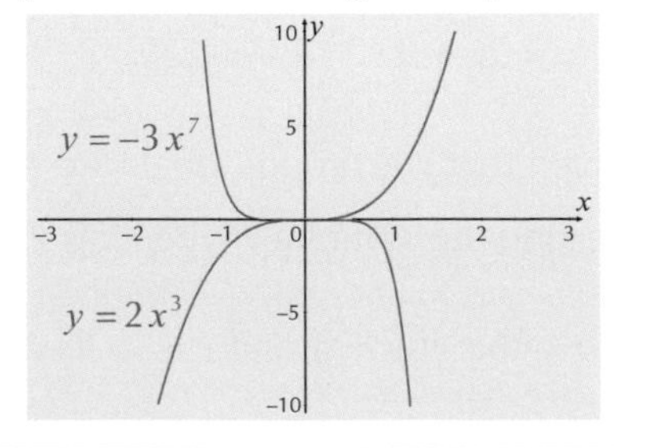

This is the basic shape for $y = kx^n$ — if the polynomial has other terms, the graph might have 'dips' in it. A constant term gives the y-intercept, just like in the equation of a straight line.

You may be asked to sketch some simple cubic and quartic graphs.
The key to this is finding where the graph **crosses the axes**.

- To find where a graph crosses the y-axis, just set $\boldsymbol{x = 0}$ and find the value of y.
- The easiest way to find where it crosses the x-axis is to **factorise** the polynomial — it crosses the x-axis when **each bracket** is set equal to **0**.

When you're sketching cubics and quartics, watch out for any **repeated roots**.
Repeated roots occur when a **factor** is repeated (i.e. when a bracket is squared, cubed, etc.).

- A **squared** bracket means it's a **double root**, and the graph will only **touch** the x-axis, not cross it, at this point (see part b) of the example below).
- A **cubed** bracket means a **triple root**, which still crosses the x-axis, but **flattens out** as it does so (see part c) on the next page).

Examples

Sketch the graphs of the following cubic functions.

a) $\mathbf{f(x) = x(x - 1)(2x + 1)}$

1. Expand the brackets to find the highest power of x and its coefficient.

   $f(x) = x(x - 1)(2x + 1) = 2x^3 - x^2 - x$

   The coefficient of x^3 is 2, i.e. positive, so the graph will have a bottom-left to top-right shape.

2. Find where the graph crosses the y-axis, i.e. where $x = 0$:

   $f(0) = 0(-1)(1) = 0 \Rightarrow$ y-intercept $= 0$

3. Find where the graph crosses the x-axis, i.e. where $f(x) = 0$:

   $f(x) = x(x - 1)(2x + 1) = 0 \Rightarrow x = 0, 1 \text{ and } -\frac{1}{2}$

4. So the curve crosses the x-axis three times and will look like this:

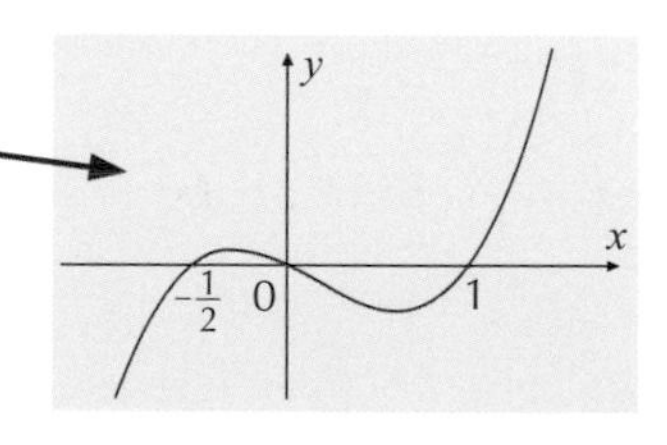

b) $\mathbf{g(x) = (x - 3)^2(x + 1)(x - 1)}$

1. Multiply out the brackets.

   $g(x) = (x - 3)^2(x + 1)(x - 1) = (x^2 - 6x + 9)(x^2 - 1)$
   $= x^4 - 6x^3 + 8x^2 + 6x - 9$

   The coefficient of x^4 is positive, so the graph will have 'top-left to top-right' shape.

2. Find where it meets both axes:

   $g(0) = (0 - 3)^2 \times (0 + 1) \times (0 - 1) = -9$
   So the y-intercept is $(0, -9)$

   $(x - 3)^2(x + 1)(x - 1) = 0 \Rightarrow x = 3, x = 1 \text{ and } x = -1$

3. So the graph meets the x-axis three times and looks like this:

4. The $(x - 3)$ bracket is squared, so the quartic has a double root at $x = 3$ — the graph touches the x-axis here, but doesn't cross it.

c) $h(x) = (2 - x)^3$

1. Multiply out the brackets.

$h(x) = 8 - 12x + 6x^2 - x^3$.

The coefficient of x^3 is negative, so the graph will have a top-left to bottom-right shape.

2. Find where it crosses both axes:

$h(0) = (2 - 0)^3 = 8 \Rightarrow$ y-intercept is $(0, 8)$

$(2 - x)^3 = 0 \Rightarrow x = 2$ — this is a triple root because the bracket is cubed.

3. So the graph looks like this:

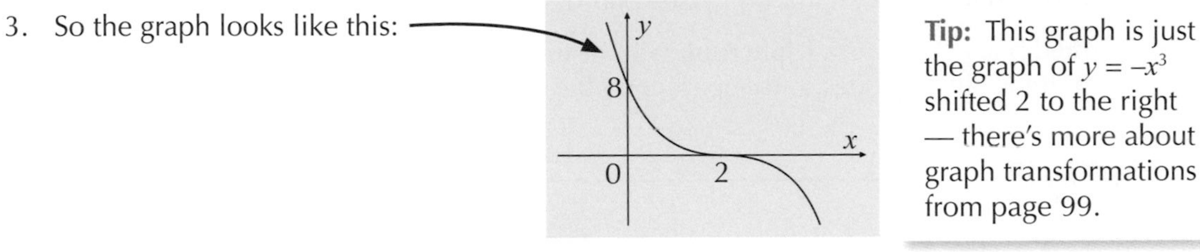

Tip: This graph is just the graph of $y = -x^3$ shifted 2 to the right — there's more about graph transformations from page 99.

Exercise 6.4.1

Q1 The diagram shows four graphs A, B, C and D.
State which graph would represent each of the following functions.

a) $y = -1.5x^4$ b) $y = 0.5x^3$ c) $y = 2x^6$ d) $y = -3x^3$

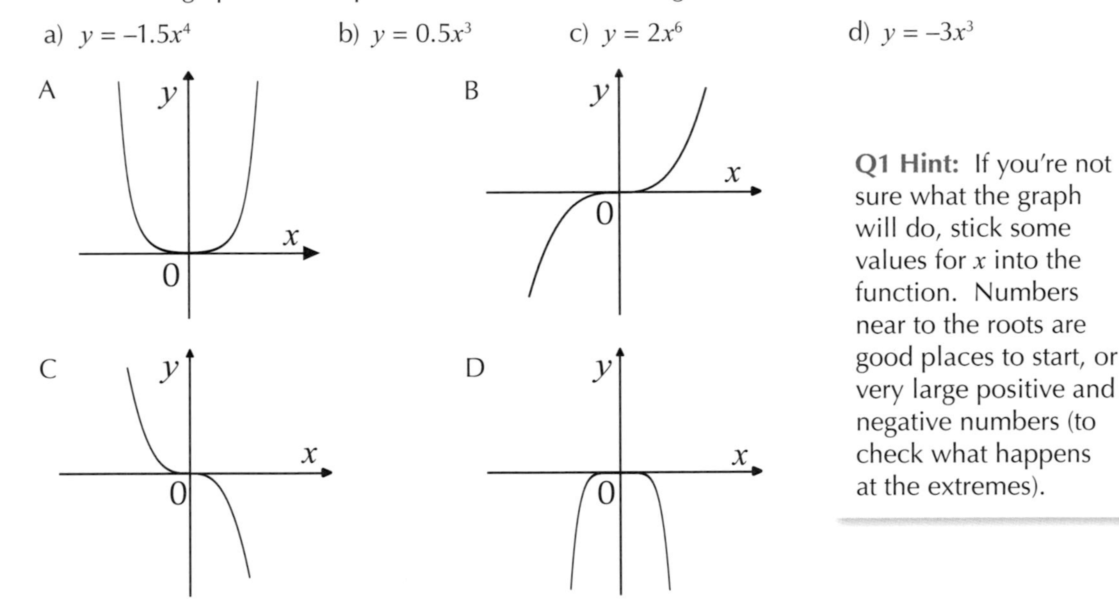

Q1 Hint: If you're not sure what the graph will do, stick some values for x into the function. Numbers near to the roots are good places to start, or very large positive and negative numbers (to check what happens at the extremes).

For Questions 2-4, sketch the graphs of the functions, showing clearly where they meet the x-axis.

Q2 a) $y = x(x + 2)(x - 3)$ b) $y = (x + 1)(2x - 1)(x - 3)$ c) $y = x(x + 1)(2 - x)$

Q3 a) $y = x^2(2x - 5)$ b) $y = x(5 - x)^2$ c) $y = (1 - x)(2 - x)^2$

Q4 a) $y = -5x^2(3x - 2)$ b) $y = (7 - x)(9 - 2x)(3 - x)$ c) $y = (4 + x)^3$

Q5 a) Factorise completely $x^3 - 7x^2 + 12x$.

b) Use your answer to part a) to sketch the graph of $y = x^3 - 7x^2 + 12x$, showing clearly where the graph meets the coordinate axes.

Q6 Sketch the graphs of these functions, showing clearly where the graph meets the coordinate axes.

a) $y = x^3 - 16x$ b) $y = 2x^3 - 12x^2 + 18x$ c) $y = -3x^2 - x^3$

Q7 Sketch the graph of $y = \mathrm{f}(x)$, where:

a) $\mathrm{f}(x) = 3x(x - 4)^2(2x - 1)$ b) $\mathrm{f}(x) = -3x^2(2x - 7)^2$ c) $\mathrm{f}(x) = (4 - x)(x + 2)^3$

Q8 Sketch the graphs of the following quartics:

a) $y = x^2(x^2 - 9x + 14)$ b) $y = (x + 1)(2 - 3x)(4x^2 - 9)$ c) $y = (x - 5)(2x^3 + 5x^2 - 3x)$

Q9 Sketch the graph of $y = 4x^3 + 4x^2 - 5x - 3$, showing where the graph meets the axes. PROBLEM SOLVING

Q10 a) Given that $x = 1$ is a root of the equation $y = (x + 1)(2x^3 - 5x^2 + x + 2)$, factorise y fully. PROBLEM SOLVING

b) Hence, sketch the graph of y, showing the points where the graph meets the x- and y- axes.

Q11 Sketch the graph of $y = (x - 3)(x^3 - 7x^2 + 14x - 8)$, given that its roots are all positive integers. PROBLEM SOLVING

Reciprocal functions and negative powers

Reciprocal functions are those of the form $y = \frac{k}{x}$, where k is a constant.
The graph of a reciprocal function always has **asymptotes**.

> An **asymptote** of a curve is a **line** which the curve gets infinitely close to, but **never** touches.

You need to be able to sketch the graphs of reciprocal functions. Here are some examples:

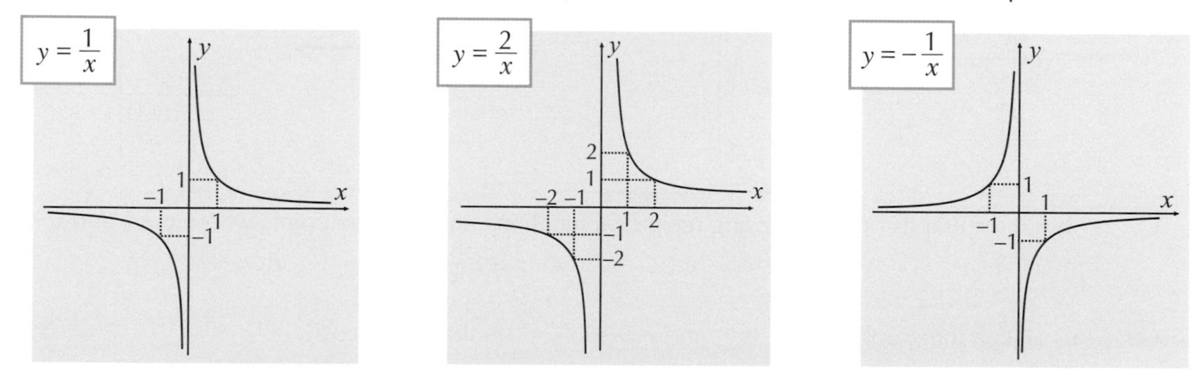

The graph never crosses the x- or y-axes, as you can't divide by zero. So the y-axis is a vertical asymptote ($x = 0$) and the x-axis is a horizontal asymptote ($y = 0$). For reciprocal graphs of the form $y = \frac{1}{x+a} + b$, the asymptotes will be at $y = b$ and $x = -a$ (see Example 2 on page 100).

The function $y = \frac{k}{x}$ can also be written as $y = kx^{-1}$. In general, **negative powers** of x (i.e. functions of the form $y = kx^{-n}$, such as $y = 2x^{-3}$) can also be written in the form $y = \frac{k}{x^n}$.
You need to be familiar with functions written in either form.

For any graph in the form $y = \frac{k}{x^n}$ or kx^{-n}:

EVEN

When n is EVEN, you get a graph with two bits next to each other.

If k **is POSITIVE**, both parts of the graph are above the x-axis

And if k **is NEGATIVE**, the graph is below the x-axis.

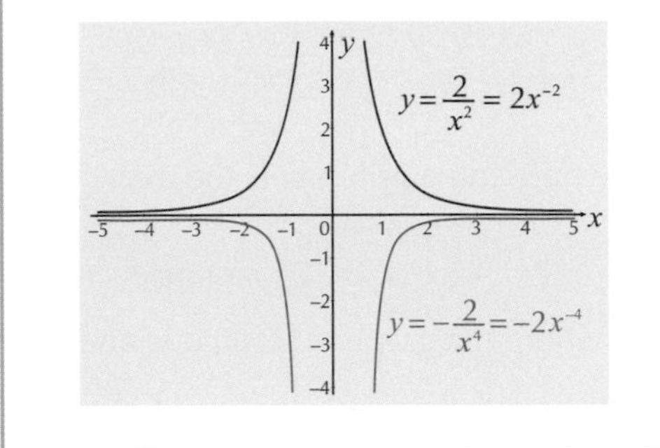

ODD

When n is ODD, you get a graph with two bits opposite each other.

If k **is POSITIVE**, the graph is in the top-right and bottom-left quadrants.

And if k **is NEGATIVE**, it's in the top-left and the bottom-right.

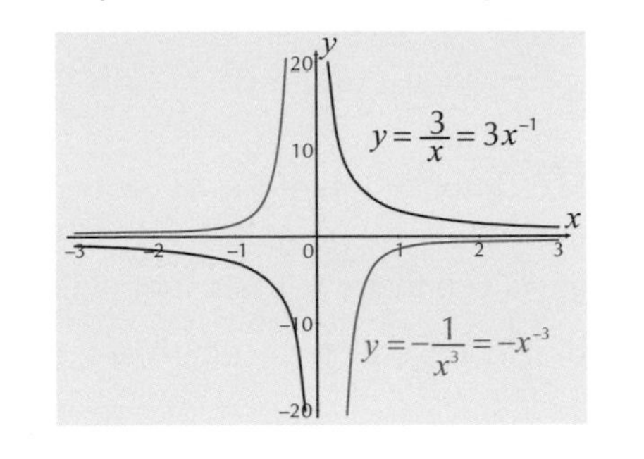

Exercise 6.4.2

Q1 The diagram shows four graphs A, B, C and D.
State which graph would represent each of the following functions:

a) $y = x^{-2}$ b) $y = -3x^{-3}$ c) $y = -\frac{3}{x^4}$ d) $y = 2x^{-5}$

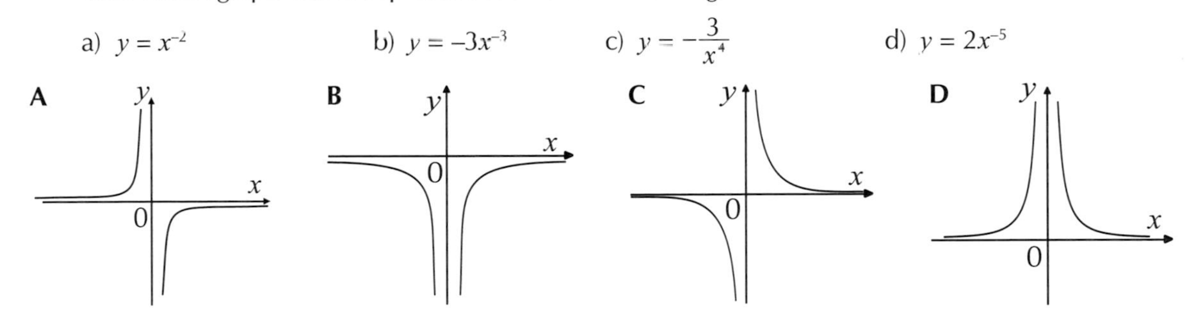

Q2 Sketch the graphs of the following reciprocal functions, showing the points where $x = 1$ and $x = -1$:

a) $y = 1.5x^{-5}$ b) $y = 7x^{-2}$ c) $y = -\frac{6}{x}$ d) $y = -1.2x^{-4}$

Q3 a) Sketch the graphs of $y = 3x^{-2}$ and $y = -x^3 - 2x^2$ on the same axes.

b) Find the number of real roots of $3x^{-2} = -x^3 - 2x^2$.

PROBLEM SOLVING

Q4 a) Use graph paper to draw the graphs of $y = -\frac{2}{x}$ and $y = 4 - x^2$ on the same axes for $-3 \leq x \leq 3$. Use a scale of 2 cm for 1 unit.

b) Use your answer to part a) to estimate the solutions to $-\frac{2}{x} = 4 - x^2$.

PROBLEM SOLVING

6.5 Graph Transformations

If you have a function f(x), you can transform its graph in three different ways — by translating it, stretching it or reflecting it.

Learning Objectives (Spec Ref 2.8):

- Know the effect of the transformations $y = f(x) + a$, $y = f(x + a)$, $y = af(x)$, $y = f(ax)$.
- Apply any of these transformations to quadratic, cubic and reciprocal functions.
- Given the graph of a function f(x), sketch the graph resulting from any of the above transformations.

Prior Knowledge Check: You should be familiar with vector notation from GCSE, and be able to use function notation. See p.5.

Translations

Translating the graph of a function means moving it either **horizontally** or **vertically**. The shape of the graph itself doesn't change, it just moves. There are two types of translation:

$y = f(x) + a$

Adding a number to the **whole function** translates the graph in the **y-direction**.

- If $a > 0$, the graph goes **upwards**.
- If $a < 0$, the graph goes **downwards**.

This can be described by a **column vector**: $\begin{pmatrix} 0 \\ a \end{pmatrix}$ or $\begin{bmatrix} 0 \\ a \end{bmatrix}$.

$y = f(x + a)$

Writing '$x + a$' instead of 'x' means the graph moves **sideways** ("translated in the **x-direction**").

- If $a > 0$, the graph goes to the **left**.
- If $a < 0$, the graph goes to the **right**.

As a **column vector**, this would be $\begin{pmatrix} -a \\ 0 \end{pmatrix}$ or $\begin{bmatrix} -a \\ 0 \end{bmatrix}$.

Example 1

Shown to the right is the graph of $y = f(x)$, where $f(x) = x(x + 2)(x - 2)$.

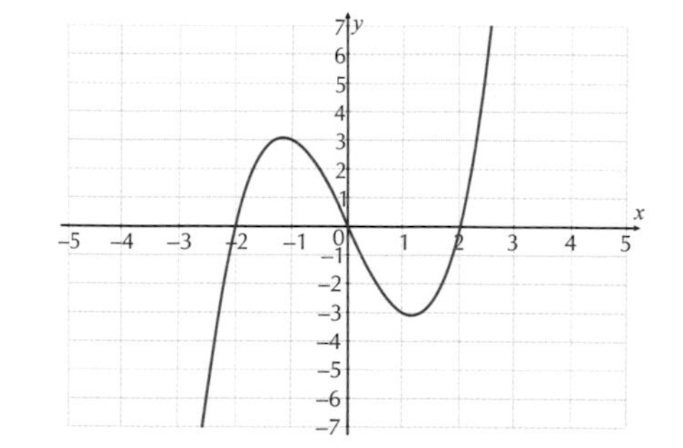

a) Sketch the graph $y = f(x) + 2$.

2 has been added to the whole function, i.e. $a = 2$. So the graph will be translated 2 units in the y-direction, i.e. shifted upwards by 2.

This is a translation by the vector $\begin{pmatrix} 0 \\ 2 \end{pmatrix}$.

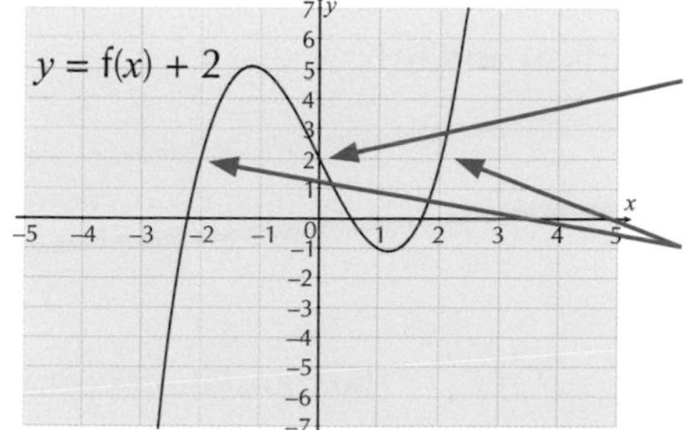

The point (0, 0) on f(x) has become the point (0, 2).

The other roots of f(x), (–2, 0) and (2, 0) have become (–2, 2) and (2, 2).

Tip: When sketching a transformed graph, you need to show what happens to its key points, e.g. where it crosses the axes, max/min points, etc.

b) Sketch the graph $y = f(x - 1)$.

Here, it's of the form $y = f(x + a)$ so it's a translation in the x-direction.

$a = -1$ which is negative, so it's a translation to the right by 1 unit.

The column vector for this translation is $\begin{pmatrix} 1 \\ 0 \end{pmatrix}$.

1 is added to the x-coordinate of every point.

E.g. $(-2, 0)$ becomes $(-1, 0)$.

Example 2

Given that $f(x) = \frac{1}{x}$:

a) Sketch the graph of $y = f(x) + 2$ and state the equations of the asymptotes.

1. First, sketch the graph of $f(x) = \frac{1}{x}$: It has asymptotes at $x = 0$ and $y = 0$.

2. The graph $y = f(x) + 2$ is a translation of the graph upwards by 2:

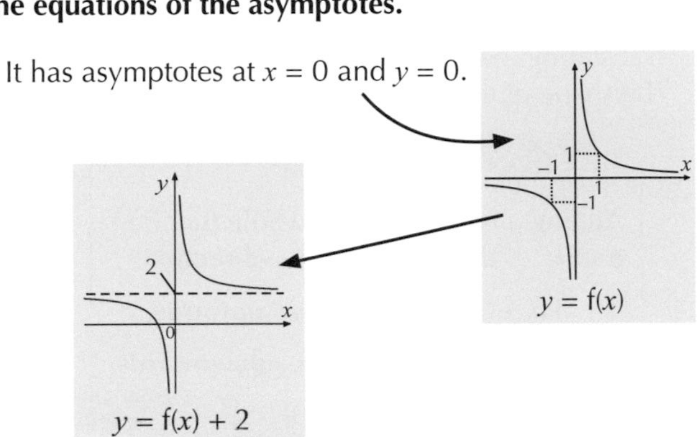

3. The asymptotes of $y = f(x)$ are also translated upwards by 2. The horizontal asymptote at $y = 0$ becomes $y = 2$ but the vertical asymptote is still $x = 0$ because the graph has only moved upwards.

b) Sketch the graph of $y = f(x + 2)$ and state the equations of the asymptotes.

The graph of $f(x + 2)$ is a translation to the left by 2.

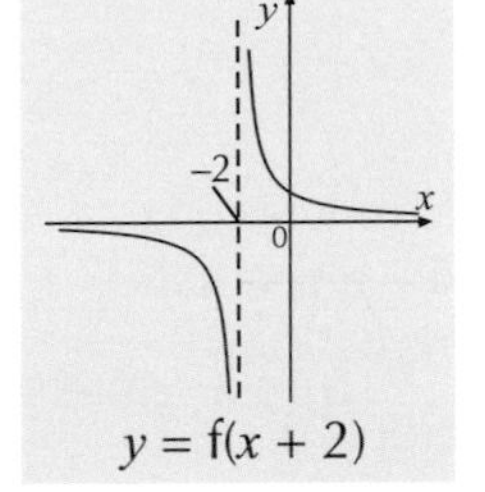

This time, the horizontal asymptote remains at $y = 0$ but the vertical asymptote moves to $x = -2$.

c) What column vectors describe the translations in parts a) and b)?

1. In part a), the graph is translated by 2 units in the positive y-direction: The translation vector is 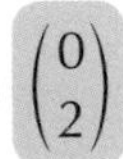 $\begin{pmatrix} 0 \\ 2 \end{pmatrix}$.

2. For part b), the translation is in the negative x-direction (again by 2 units): The translation vector is $\begin{pmatrix} -2 \\ 0 \end{pmatrix}$.

Exercise 6.5.1

Q1 The diagram on the right shows the graph of $y = \mathrm{f}(x)$. The curve has a maximum at (2, 4) and meets the x-axis at (0, 0) and (5, 0).

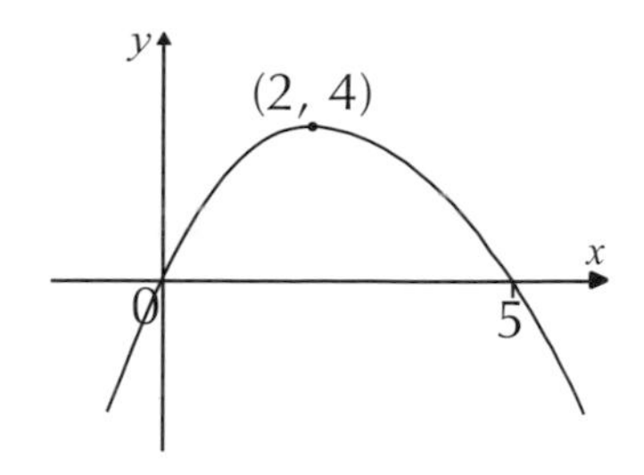

a) Sketch the graph of $y = \mathrm{f}(x) + 2$, labelling the coordinates of the maximum and where the curve meets the y-axis.

b) Sketch the graph of $y = \mathrm{f}(x + 2)$, labelling the points where the curve meets the x-axis and the maximum.

Q2 If $\mathrm{g}(x) = -\frac{2}{x}$, sketch these graphs and write down the equations of the asymptotes for each.

a) $y = \mathrm{g}(x)$ b) $y = \mathrm{g}(x + 3)$ c) $y = \mathrm{g}(x) + 3$

Q3 Given that $y = x^2(x - 4)$, use a column vector to describe the translation that gives the graph of:

a) $x^2(x - 4) + 1$ b) $y = (x - 2)^2(x - 6)$ c) $y = x(x + 4)^2$

Q4 Explain how the graph of $y = x^3 + 3x + 7$ can be translated to give the graph of $y = x^3 + 3x + 2$. Include a column vector in your answer.

Q5 The graph of $y = x^2 - 3x + 7$ is translated by the vector $\begin{pmatrix} -1 \\ 0 \end{pmatrix}$. Write down the equation of the new graph. Give your answer in as simple a form as possible.

Q6 The diagram shows the graph of $y = \mathrm{f}(x)$. The graph has a maximum at (5, 3), crosses the x-axis at (3, 0) and (6, 0) and crosses the y-axis at (0, –1).

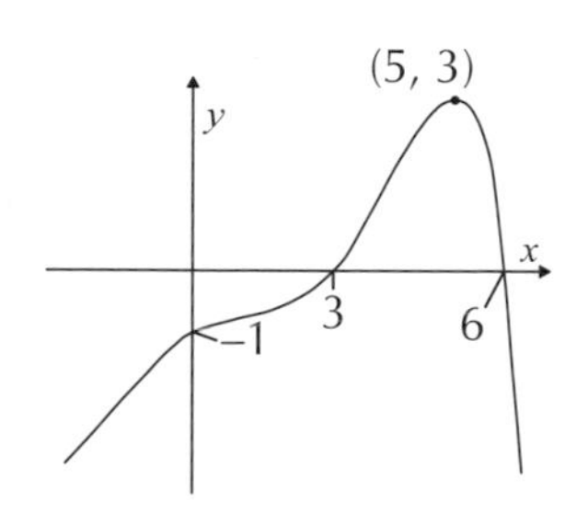

a) Sketch the graph of $y = \mathrm{f}(x) - 2$.

b) Label the coordinates of the maximum and the point where the graph meets the y-axis.

Q7 a) Sketch the graph of $y = (x - 1)(2x - 3)(4 - x)$ and label the points where the graph meets the coordinate axes.

b) The graph in part a) is translated by the vector $\begin{pmatrix} 2 \\ 0 \end{pmatrix}$.

Give the equation of the translated graph in its simplest form.

c) On separate axes, sketch the graph of the equation from part b), labelling all the points where the graph meets the x-axis.

Q8 a) The graph of $\mathrm{f}(x) = x^2 + 5$ is translated to the left by 3 to give the graph of $\mathrm{g}(x)$. Find the equation of $\mathrm{g}(x)$ in its simplest form.

b) The graph of $\mathrm{g}(x)$ is then translated downwards by 4 to give the graph of $\mathrm{h}(x)$. Find the equation of $\mathrm{h}(x)$ in its simplest form.

Q9 Give the equations of the asymptotes of the following graphs:

a) $y = \frac{1}{x} - 4$ b) $y = \frac{1}{x + 3}$ c) $y = \frac{1}{x - 1} + 7$

Stretches and reflections

The graph of a function can be stretched, squashed or reflected by **multiplying** the whole function or the x's in the function by a number. The result you get depends on what you multiply and whether the number is positive or negative.

$y = af(x)$

Multiplying the **whole function** by a, stretches the graph **vertically** (i.e. parallel to the y-axis) by a scale factor of a.

- If $a > 1$ or $a < -1$, the graph is **stretched**.
- If $-1 < a < 1$, the graph is **squashed**.
- If a is **negative**, the graph is **also reflected** about the **x-axis**.

For every point on the graph, the x-coordinate stays the same and the y-coordinate is multiplied by a.

Tip: Don't describe a transformation as a "squash" in the exam — call it a "stretch with a scale factor of...".

Example 1

The diagram shows the graph of a function f(x).

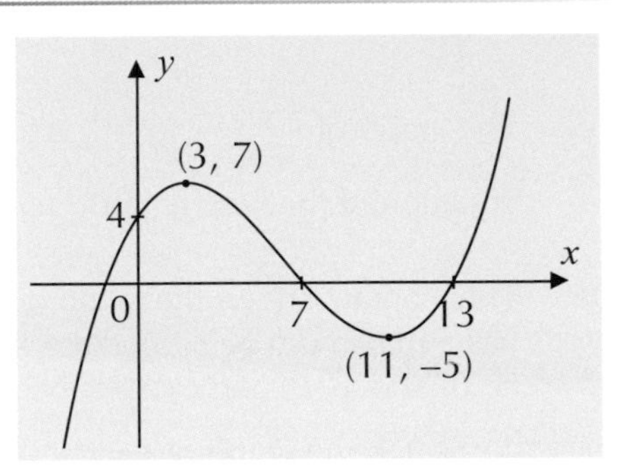

a) Sketch the graph $y = \frac{1}{3}f(x)$.

1. The graph will be stretched vertically by a scale factor of $\frac{1}{3}$.
2. As $\frac{1}{3}$ is less than 1, the 'stretch' will actually be a squash.
3. The diagram gives 5 key points on the graph — (0, 4), (3, 7), (7, 0), (11, –5) and (13, 0). You need to show where each of these points has moved to on the transformed graph.

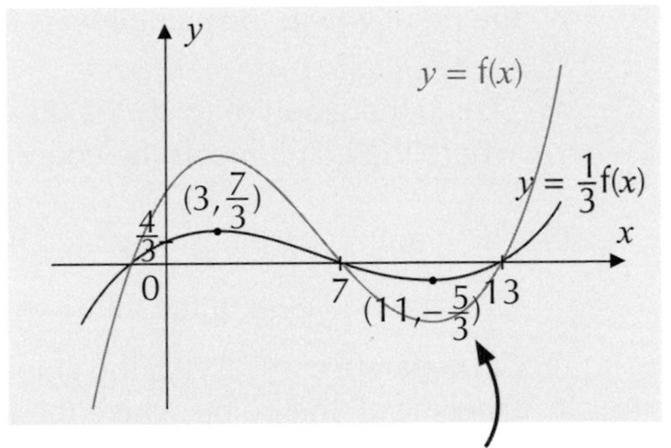

The x-coordinates don't change — just multiply the y-coordinates by a.

Tip: The graph still crosses the x-axis at the same points as the original graph — this is true for all $y = af(x)$ transformations.

b) Sketch the graph $y = -2f(x)$.

1. Here the whole function has been multiplied by -2.
2. So the graph will be stretched vertically by a factor of 2, but also reflected in the x-axis because of the minus sign.
3. Again, you need to show what has happened to each key point.

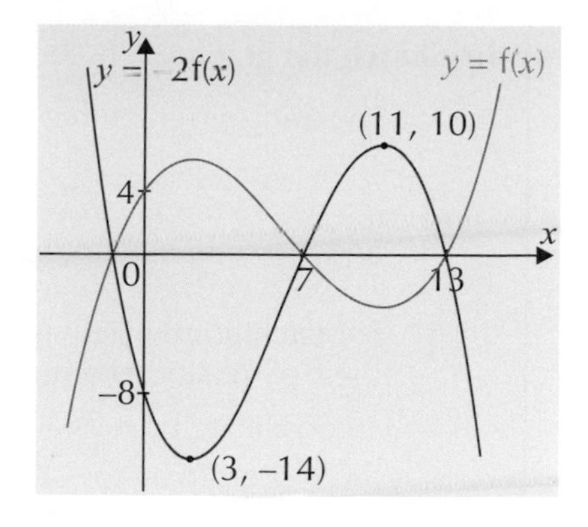

So that's vertical stretches covered. Next up are horizontal stretches:

$y = f(ax)$

Writing 'ax' instead of 'x' stretches the graph **horizontally** by a scale factor of $\frac{1}{a}$.

- If $a > 1$ or $a < -1$, the graph is **squashed**.
- If $-1 < a < 1$, the graph is **stretched**.
- **Negative** values of a **reflect** the basic shape in the **y-axis**.

For these transformations, the y-coordinate of each point stays the same and the x-coordinate is multiplied by $\frac{1}{a}$.

Notice that a being bigger or smaller than 1 has the **opposite effect** for horizontal stretches compared to vertical stretches.

Example 2

The diagram to the right shows the graph of $y = f(x)$ again.

a) Sketch the graph $y = f\left(\frac{1}{2}x\right)$.

1. The transformation has the form $y = f(ax)$, so it's a horizontal stretch.
2. $\frac{1}{2}$ is positive and between -1 and 1, so the graph will be stretched horizontally by a scale factor of 2.
3. For each point given, the x-coordinate is multiplied by 2 but the y-coordinate doesn't change.
4. The graph looks like this:

   This time, the y-intercept doesn't change, but the two x-intercepts do.

b) Sketch the graph $y = f(-3x)$.

1. The transformation has the form $y = f(ax)$, so it's a horizontal stretch.

2. $a = -3$, so the graph will be 'stretched' by a scale factor of $\frac{1}{3}$, i.e. squashed, and also reflected in the y-axis since a is negative.

3. For each point given, find (and label) the new position by multiplying the x-coordinate by $-\frac{1}{3}$ (and leaving the y-coordinate the same.)

4. So the graph looks like this:

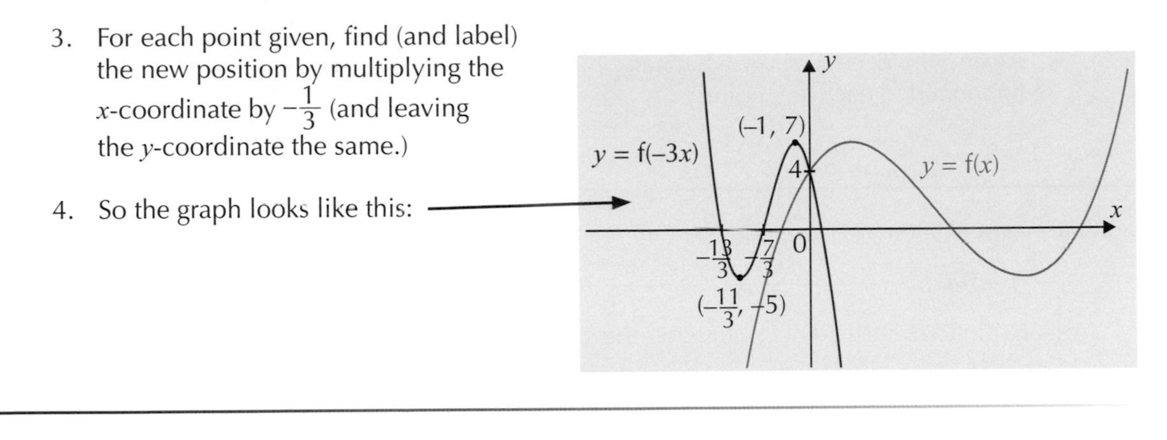

Exercise 6.5.2

Q1 The diagram below shows the graph of $y = g(x)$. The graph has a minimum at (–2, –3), a maximum at (2, 3) and crosses the x-axis at (0, 0), (–4, 0) and (4, 0).

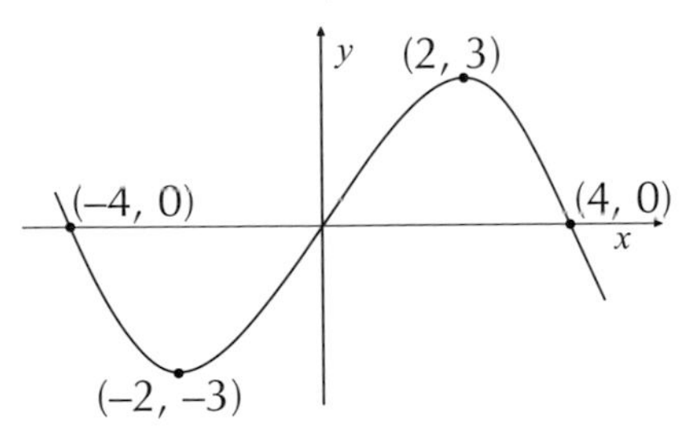

Sketch the graphs of these functions, labelling clearly the coordinates of any maximums, minimums and intersections with the axes:

a) $y = 2g(x)$ b) $y = g(2x)$ c) $y = -2g(x)$ d) $y = g(-2x)$

Q2 The diagram shows the graph of f(x) and Graph A of a function that is a transformation of f(x).

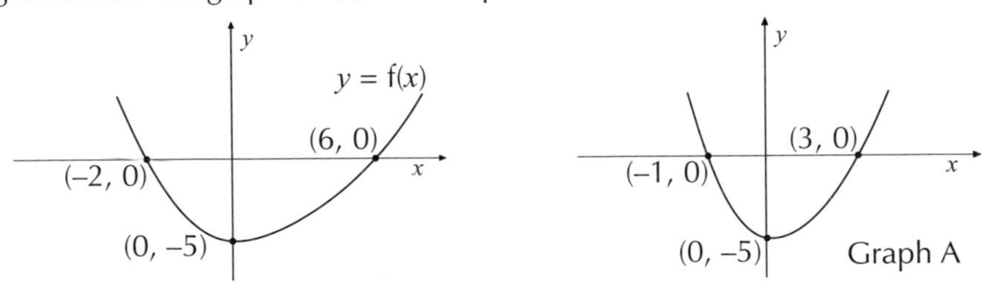

Which of these equations gives the transformed graph?

a) $y = 2f(x)$ b) $y = f(2x)$ c) $y = f(0.5x)$ d) $y = 0.5f(x)$

Q3 The diagram shows the graph of f(x) and Graph A of a function that is a transformation of f(x).

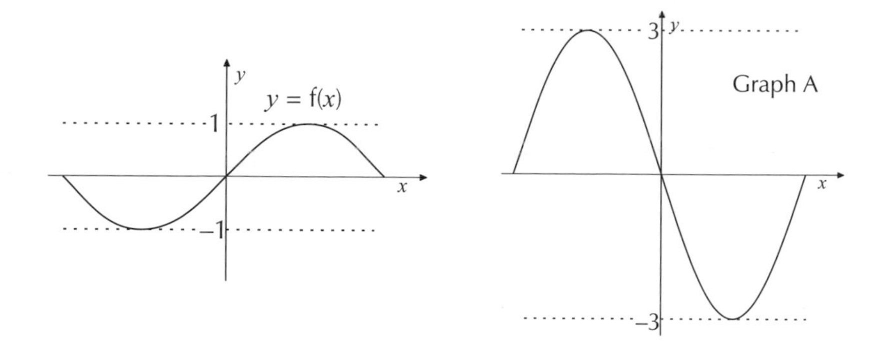

Which of these equations could give the transformed graph?

a) $y = 3f(x)$ b) $y = -3f(x)$ c) $y = f(-3x)$ d) $y = f(3x)$

Q4 Given that $f(x) = x^3 - x$, sketch the graphs of the following functions:

a) $y = f(x) + 2$ b) $y = f(x - 2)$ c) $y = f(-2x)$ d) $y = -2f(x)$

e) $y = f(2x)$ f) $y = -0.2f(x)$ g) $y = f(0.2x)$ h) $y = 0.2f(x)$

Q5 Describe clearly the transformation that is required to take the graph of $y = x^3 + 2x + 4$ to the graph of $y = 3x^3 + 6x + 12$.

PROBLEM SOLVING

Q6 Describe clearly the transformation that is required to take the graph of $y = x^2 + x + 4$ to the graph of $y = 4x^2 - 2x + 4$.

PROBLEM SOLVING

Q7 The diagram shows the graph of $y = e^x$. Use this to sketch the graphs below, labelling clearly where the transformation takes the point (0, 1) to.

PROBLEM SOLVING

a) $y = e^{x-3}$ b) $y = e^x + 1$

c) $y = 0.1e^x$ d) $y = -4e^x$

e) $y = -\frac{1}{2}e^x$ f) $y = e^{7x}$

y
(0, 1)
0
x

Q8 The graph of $f(x) = x^2 - 3x + 3$ is transformed to give the graph of g(x). Give the equation of g(x) in its simplest form if the transformation that takes the graph of f(x) to the graph of g(x) is:

PROBLEM SOLVING

a) vertical stretch, scale factor 2 b) translation by $\begin{pmatrix} -3 \\ 0 \end{pmatrix}$

c) horizontal stretch, scale factor 0.25

Q9 a) Sketch the graph of $f(x) = x^2 - 6x - 7$ showing clearly the coordinates of any maximum or minimum points and where the curve meets the coordinate axes.

PROBLEM SOLVING

b) Write down the equation of the graph obtained by stretching the graph of f(x) vertically with a scale factor of –2.

c) Sketch the graph with equation you found in part b) showing clearly the coordinates of any maximum or minimum points and where the curve meets the coordinate axes.

Q9a) Hint: Complete the square — see pages 33-36 and 44-45 for a recap.

6.6 Circles

If you've survived maths this long, you'll be very familiar with circles. Now we're going to have a look at using equations to describe a circle. An equation of a circle can tell you its radius and where its centre is.

Learning Objectives (Spec Ref 3.2):

- Know and find the equation of a circle with radius r and centre (a, b), in the form: $(x - a)^2 + (y - b)^2 = r^2$.
- Find the radius and coordinates of the centre of a circle, given the equation of the circle.
- Know the three main circle properties, and apply them to find lengths, angles and coordinates.

Prior Knowledge Check:
Be able to complete the square of a function. See p.33-36.
You should also be familiar with finding lengths and midpoints of line segments from GCSE.

The equation of a circle

Circles with centre (0, 0)

The diagram to the right shows a circle centred on the origin (0, 0) and with radius r.

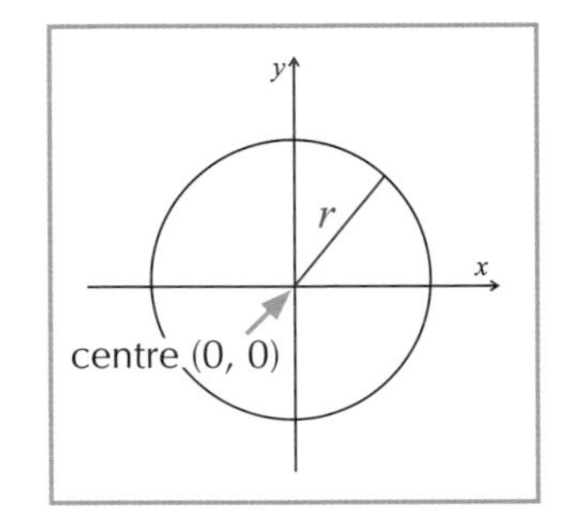

You can describe a circle centred on the origin, with radius r, using the equation: $\mathbf{x^2 + y^2 = r^2}$

Even though you're dealing with circles, you get the equation above using **Pythagoras' theorem**. Here's how:

- The **centre** of the circle is at the origin, labelled **C**.
- **A** is any point on the circle, and has the coordinates (x, y).
- **B** lies on the x-axis and has the same x-coordinate as A.
- So the length of line **CB** $= x$, and **AB** $= y$.

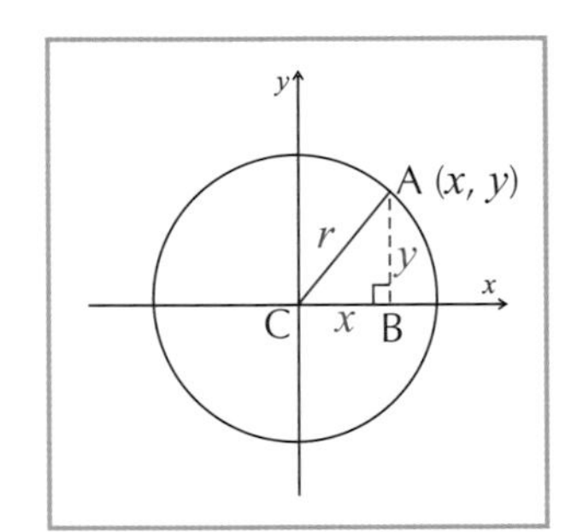

Therefore, using Pythagoras' theorem to find the radius r, we get:

$$CB^2 + AB^2 = r^2$$

Which is... $\mathbf{x^2 + y^2 = r^2}$

If you're given the equation of a circle in the form $x^2 + y^2 = r^2$, you can work out the radius of the circle by taking the square root of the r^2 part.

Example

A circle has the equation $x^2 + y^2 = 4$. Find the radius of the circle.

1. Compare the equation $x^2 + y^2 = 4$ to $x^2 + y^2 = r^2$. Equating the two equations gives: $r^2 = 4$
2. Take the positive square root of r^2 to find r. Ignore the negative square root as the radius will be positive. $\Rightarrow r = 2$

Circles with centre (a, b)

Unfortunately, circles aren't always centred at the origin. This means we need a general equation for circles that have a centre somewhere else — the point (a, b).

The general equation for circles with **radius r** and **centre (a, b)** is: $(x - a)^2 + (y - b)^2 = r^2$

Notice that if the circle had a centre at (0, 0), then you'd get $a = 0$ and $b = 0$, so you'd just get $x^2 + y^2 = r^2$ (the equation for a circle centred at the origin).

The example below shows how you get the equation of a circle when the coordinates of the centre and the value of the radius are given.

Example 1

Find the equation of the circle with centre (6, 4) and radius 3.

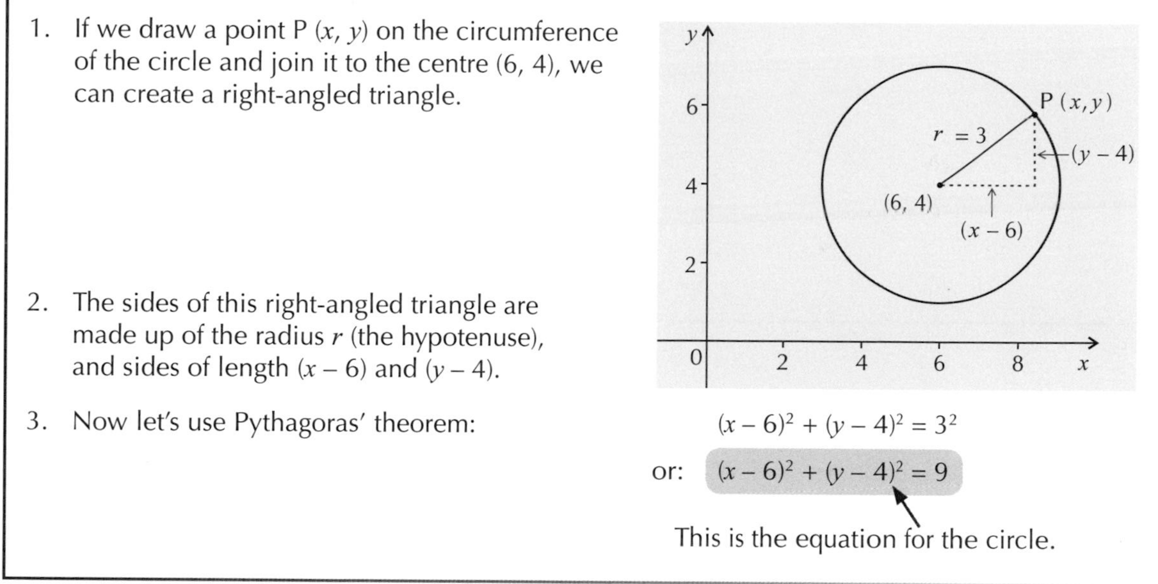

1. If we draw a point P (x, y) on the circumference of the circle and join it to the centre (6, 4), we can create a right-angled triangle.
2. The sides of this right-angled triangle are made up of the radius r (the hypotenuse), and sides of length $(x - 6)$ and $(y - 4)$.
3. Now let's use Pythagoras' theorem:

$$(x - 6)^2 + (y - 4)^2 = 3^2$$

or: $(x - 6)^2 + (y - 4)^2 = 9$

This is the equation for the circle.

Example 2

What is the centre and radius of the circle with equation $(x - 2)^2 + (y + 3)^2 = 16$?

1. Compare $(x - 2)^2 + (y + 3)^2 = 16$ with the general form:

$(x - a)^2 + (y - b)^2 = r^2$

So $a = 2$, $b = -3$ and $r = \sqrt{16} = 4$.

The centre (a, b) is $(2, -3)$ and the radius (r) is 4.

2. On a set of axes, the circle would look like this:

$(x - 2)^2 + (y + 3)^2 = 16$

If you're given the centre and radius of a circle and you're asked to find the equation of the circle, just put the values of a, b and r into the equation $(x - a)^2 + (y - b)^2 = r^2$.

Example 3

Write down the equation of the circle with centre (–4, 2) and radius 6.

1. The question says, 'Write down...', so you know you don't need to do much working.

   The centre of the circle is (–4, 2), so $a = -4$ and $b = 2$.
   The radius is 6, so $r = 6$.

2. Using the general equation for a circle $(x - a)^2 + (y - b)^2 = r^2$ you can write:

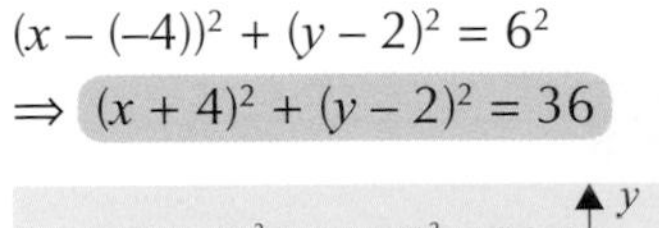

   $(x - (-4))^2 + (y - 2)^2 = 6^2$
   $\Rightarrow (x + 4)^2 + (y - 2)^2 = 36$

3. On a set of axes, the circle would look like this:

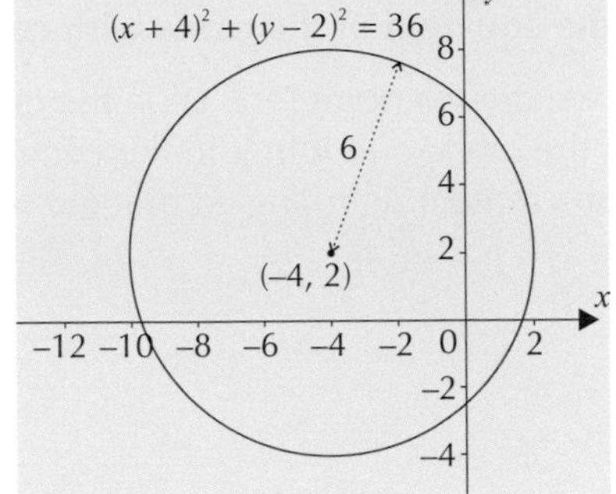

Exercise 6.6.1

Q1 Find an equation for a circle with centre (0, 0) and radius:

a) 5 b) 7 c) $\sqrt{23}$ d) $3\sqrt{2}$

Q2 Find the equation for each of the following circles:

a) centre (2, 5), radius 3 b) centre (–3, 2), radius 5 c) centre (–2, –3), radius 7

d) centre (3, 0), radius 4 e) centre (–1, 3), radius 7 f) centre (5, 4), radius $\sqrt{11}$

g) centre (–7, 5), radius $\sqrt{27}$ h) centre (–10, 7), radius 11 i) centre (8, 0), radius $\sqrt{17}$

Q3 Write down the equation of each of the following circles:

a)

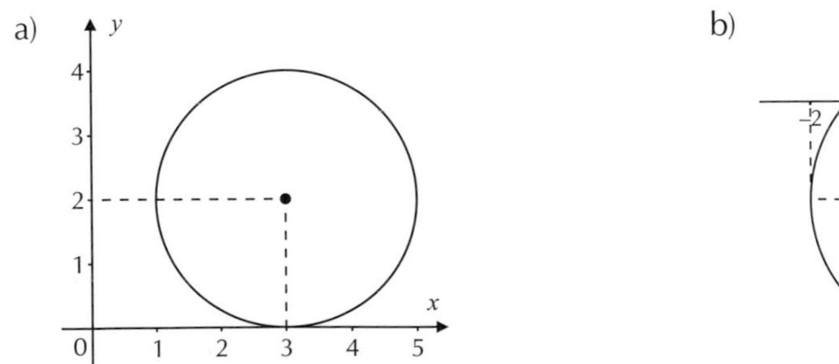

b)

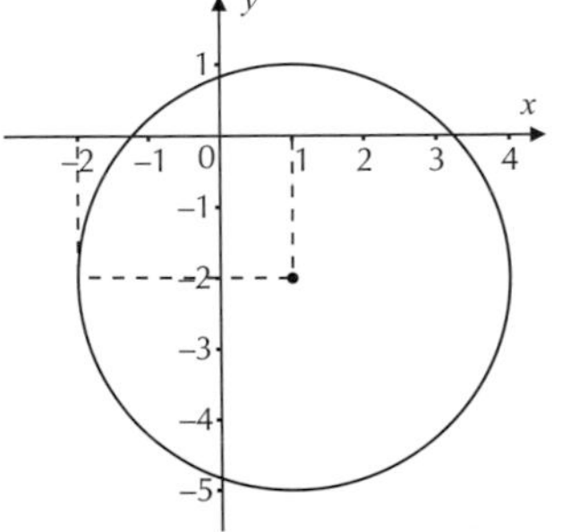

Q4 Find the centre and radius of the circles with the following equations:

a) $(x-1)^2+(y-5)^2=4$
b) $(x-3)^2+(y-5)^2=64$
c) $(x-3)^2+(y+2)^2=25$
d) $(x-6)^2+(y-4)^2=20$
e) $(x+8)^2+(y+1)^2=27$
f) $x^2+(y-12)^2=147$

Q5 A circle has centre (5, 3) and radius 8. Find an equation for the circle.

Q6 A circle has centre (3, 1) and a radius of $\sqrt{31}$. Find an equation for the circle.

Q7 Find the centre and radius of the circle with equation $x^2+(y+3)^2=18$.

Q8 A circle has radius $\sqrt{5}$ and centre (–3, –2). Find an equation for the circle.

Q9 The equation of the circle C is $(x+9)^2+(y-3)^2=45$.

a) Find the coordinates of the centre of the circle.

b) Find the radius of the circle and give your answer in the form $p\sqrt{5}$.

Q10 A circle has centre (6, 3) and a radius of $4\sqrt{5}$.
Identify which coordinates below give points that lie on this circle.

PROBLEM SOLVING

a) (–2, 7)
b) (10, –3)
c) (14, –1)
d) (2, 5)

Q11 Circle C has its centre at (3, 5) and passes through the point (10, 4).
Find the exact radius of the circle and hence give the equation of C.

PROBLEM SOLVING

Q12 A circle has centre (–2, 9) and goes through the point (–11, 7).
Find an equation for the circle.

PROBLEM SOLVING

Q13 A circle has its centre at point A and passes through the origin.
Find the equation of the circle if:

PROBLEM SOLVING

a) A = (1, 1)
b) A = (–7, 13)
c) A = (8, –6)
d) A = (14, 22)

Q14 Points A (–3, –10) and B (7, –2) lie on a circle C. The line AB is a diameter of C.

PROBLEM SOLVING

a) Find: (i) the length of the line AB, (ii) the midpoint of the line AB.

b) Hence, or otherwise, write down the equation of circle C.

Rearranging circle equations

Sometimes you'll be given an equation for a circle that doesn't look much like $(x-a)^2+(y-b)^2=r^2$ — e.g. $x^2+y^2+8x+6y+3=0$.

The general form of this type of equation is: $x^2+y^2+2fx+2gy+c=0$

In this form, you can't immediately tell what the **radius** is or where the **centre** is. All you'll need to do is a bit of **rearranging** to get the equation into the form $(x-a)^2+(y-b)^2=r^2$. To do this, you'll normally have to **complete the square** — this is shown in the examples on the next page.

Example 1

The equation of a circle is $x^2 + y^2 - 6x + 4y + 4 = 0$. Find the centre of the circle and the radius.

1. You need to get the equation $x^2 + y^2 - 6x + 4y + 4 = 0$ into the form $(x - a)^2 + (y - b)^2 = r^2$. To do this, complete the square. So, first rearrange the equation to group the x's and the y's together:

   $x^2 - 6x + y^2 + 4y + 4 = 0$

2. Complete the square for the x- and the y-terms — write down the squared bracket that will produce the x^2 and x terms, then subtract the number term that the bracket will also produce.

   $(x - 3)^2 - 9 + (y + 2)^2 - 4 + 4 = 0$

3. Then rearrange to get it into the form $(x - a)^2 + (y - b)^2 = r^2$.

   $(x - 3)^2 + (y + 2)^2 = 9$

4. This is the recognisable form, so use this equation to find:

   Centre: $(3, -2)$, radius $= \sqrt{9} = 3$

Example 2

PROBLEM SOLVING

a) Show that the circle with equation $x^2 + y^2 + 2fx + 2gy + c = 0$, where f, g and c are constants, has its centre at $(-f, -g)$ and a radius of $\sqrt{f^2 + g^2 - c}$.

1. Rearrange $x^2 + y^2 + 2fx + 2gy + c = 0$ into the form $(x - a)^2 + (y - b)^2 = r^2$, and complete the square for the x-terms and y-terms.

   $x^2 + y^2 + 2fx + 2gy + c = 0$
   $\Rightarrow x^2 + 2fx + y^2 + 2gy + c = 0$
   $\Rightarrow (x + f)^2 - f^2 + (y + g)^2 - g^2 + c = 0$
   $\Rightarrow (x + f)^2 + (y + g)^2 = f^2 + g^2 - c$

2. Now compare this to the general equation of a circle (given above) to find a, b and r.

   So $a = -f$, $b = -g$ and $r^2 = f^2 + g^2 - c$, i.e. the circle has its centre at $(-f, -g)$ and its radius is $\sqrt{f^2 + g^2 - c}$.

b) Use this result to find the centre and radius of the circle given by the equation $x^2 + y^2 - 5x - 5y + 10 = 0$.

1. Write $x^2 + y^2 - 5x - 5y + 10 = 0$ in the form $x^2 + y^2 + 2fx + 2gy + c = 0$:

   $x^2 + y^2 + 2\left(-\frac{5}{2}x\right) + 2\left(-\frac{5}{2}y\right) + 10 = 0$

2. So $f = g = -\frac{5}{2}$ and $c = 10$. Use the results from part a) to work out the centre and radius:

   Centre of the circle $= (-f, -g) = \left(\frac{5}{2}, \frac{5}{2}\right)$

   $$\text{Radius} = \sqrt{f^2 + g^2 - c}$$
   $$= \sqrt{\left(-\frac{5}{2}\right)^2 + \left(-\frac{5}{2}\right)^2 - 10}$$
   $$= \sqrt{\frac{25}{4} + \frac{25}{4} - 10} = \sqrt{\frac{50}{4} - \frac{40}{4}}$$
   $$= \sqrt{\frac{10}{4}} = \sqrt{\frac{5}{2}} \left(\text{or } \frac{\sqrt{10}}{2}\right)$$

Exercise 6.6.2

Q1 A circle has the equation $x^2 + y^2 + 2x - 4y - 3 = 0$.

a) Find the centre of the circle.

b) Find the radius of the circle. Give your answer in the form $k\sqrt{2}$.

Q2 A circle has the equation $x^2 + y^2 - 3x + 1 = 0$.

a) Find the coordinates of the centre of the circle.

b) Find the radius of the circle. Simplify your answer as much as possible.

Q3 For each of the following circles find the radius and the coordinates of the centre.

a) $x^2 + y^2 + 2x - 6y - 6 = 0$

b) $x^2 + y^2 - 2y - 4 = 0$

c) $x^2 + y^2 - 6x - 4y = 12$

d) $x^2 + y^2 - 10x + 6y + 13 = 0$

e) $x^2 + y^2 + 14x - 8y - 1 = 0$

f) $x^2 + y^2 - 4x + y = 3.75$

g) $x^2 + y^2 + 2x + 3y - 1.25 = 0$

h) $(x - 2)^2 + y^2 + 2x + 4y - 12 = 0$

Q4 A circle has centre (0, –3) and radius $\sqrt{10}$. Find the equation of the circle in the form $x^2 + y^2 + 2fx + 2gy + c = 0$ and give the values of f, g and c.

Q5 A circle has centre (–4, 2) and passes through the origin. Find the equation of the circle in the form $x^2 + y^2 + 2fx + 2gy + c = 0$ and give the values of f, g and c.

Q6 Find the equation for each of the circles below in the form $x^2 + y^2 + 2fx + 2gy + c = 0$ and give the values of f, g and c.

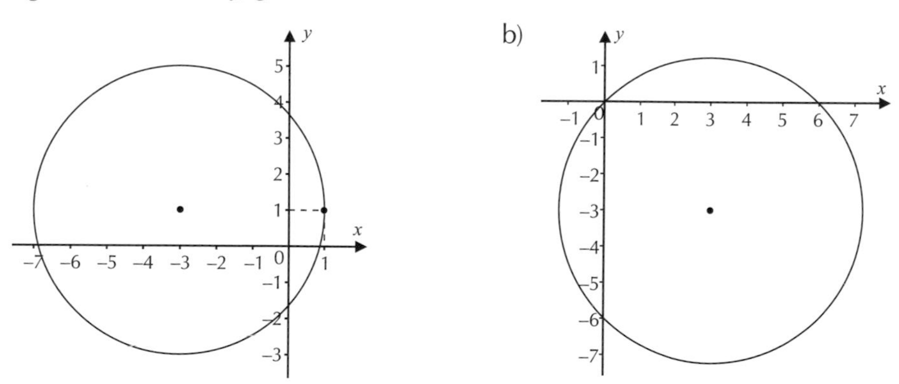

Q7 A circle has the equation $2x^2 + 2y^2 + 16x - 8y = 2$.
Find the coordinates of the centre and the exact radius of the circle.

Q8 A circle has the equation $x^2 + y^2 + 6ax - 7y - \frac{3}{4} = 0$, where a is a constant.

PROBLEM SOLVING

a) Find the centre and radius of the circle in terms of a.

b) Given that the point $\left(3, -\frac{1}{2}\right)$ lies on the circle, find the value of a.

Using circle properties

Here's a reminder of some of the most useful circle properties. Although it might not be obvious when you first look at a question, these rules could help you answer some tricky-sounding circle questions.

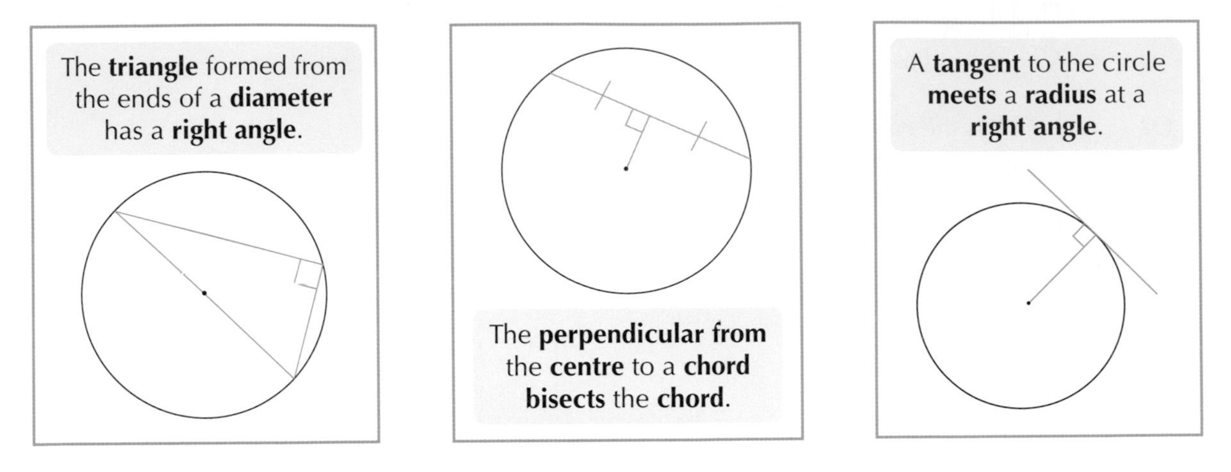

Make sure you're happy with what these terms mean — a chord is a line joining two points which lie on the circumference of a circle, and 'bisecting' just means dividing into two equal parts.

Example 1

The circle shown is centred at C. Points A and B lie on the circle. Point B has coordinates (6, 3). The midpoint, M, of the line AB has coordinates (4, 4). Line *l* passes through both C and M.

Find an equation for the line *l*.

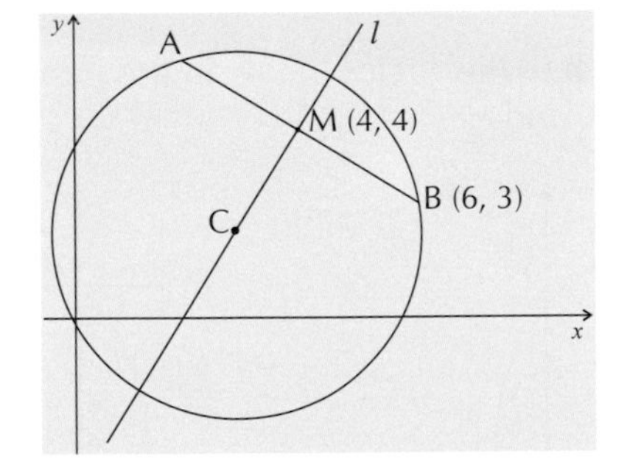

Tip: In the exam, after a question like this, they might then give you a bit more info and ask you to work out the equation for the circle.

1. AB is a chord. l goes through the centre of the circle and bisects the chord. So we can say that the line l is perpendicular to the chord.
2. We know two points on AB, so start by finding its gradient: $\frac{3-4}{6-4} = -\frac{1}{2}$
3. The gradients of perpendicular lines multiply to give –1 (see page 89), so the gradient of l will be –1 divided by the gradient of the chord. $\frac{-1}{-\frac{1}{2}} = 2$
4. Then substitute the gradient, 2, and the point on l that you know, (4, 4), into one of the equations for a straight line to work out the equation.

$$y - y_1 = m(x - x_1)$$
$$\Rightarrow y - 4 = 2(x - 4)$$
$$\Rightarrow y - 4 = 2x - 8$$
$$\Rightarrow y = 2x - 4$$

Example 2

Point A (6, 4) lies on a circle with the equation $x^2 + y^2 - 4x - 2y - 20 = 0$.

a) Find the centre and radius of the circle.

1. Get the equation into the form $(x - a)^2 + (y - b)^2 = r^2$. First, rearrange so that the x's and y's are together: $x^2 - 4x + y^2 - 2y - 20 = 0$
2. Then complete the square for the x-terms and the y-terms, and rearrange to get it into the form $(x - a)^2 + (y - b)^2 = r^2$. $(x - 2)^2 - 4 + (y - 1)^2 - 1 - 20 = 0$ $\Rightarrow (x - 2)^2 + (y - 1)^2 = 25$
3. Read off the centre and radius. Centre: (2, 1), radius = 5

b) Find the equation of the tangent to the circle at A.

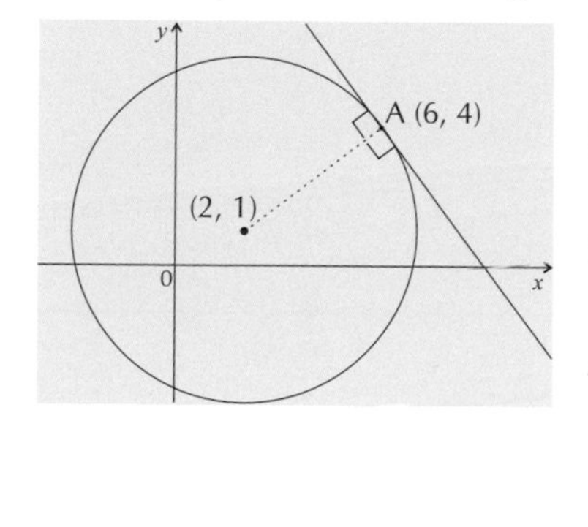

1. The tangent is at right angles to the radius at (6, 4).
2. Gradient of radius at (6, 4): $\frac{4-1}{6-2} = \frac{3}{4}$
3. So gradient of tangent: $\frac{-1}{\frac{3}{4}} = -\frac{4}{3}$
4. Using $y - y_1 = m(x - x_1)$: $y - 4 = -\frac{4}{3}(x - 6)$
   $\Rightarrow 3y - 12 = -4x + 24$
   $\Rightarrow 4x + 3y - 36 = 0$

Example 3

The points A (–2, 4), B (n, –2) and C (5, 5) all lie on the circle shown below. AB is a diameter of the circle. Show that $n = 6$.

PROBLEM SOLVING

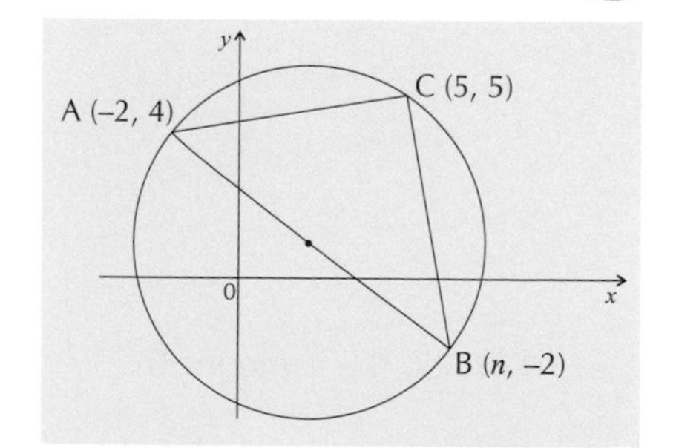

The line AB is a diameter of the circle. So the angle ACB is an angle in a semicircle and must be a right angle. This means the lines AC and BC are perpendicular to each other. The gradients of two perpendicular lines multiply to give –1, so use this to work out n.

1. First, find the gradient of AC: $m_1 = \frac{5-4}{5-(-2)} = \frac{1}{7}$
2. Then find the gradient of BC: $m_2 = \frac{-2-5}{n-5} = \frac{-7}{n-5}$
3. Use the gradient rule that $m_1 \times m_2 = -1$. $\frac{1}{7}\left(\frac{-7}{n-5}\right) = -1$ ← Cancel the 7s
   $\Rightarrow \frac{-1}{n-5} = -1 \Rightarrow 1 = n - 5 \Rightarrow n = 6$ as required

A circle that passes through all three vertices of a triangle is called the **circumcircle** of the triangle. Working out the equation of a circumcircle involves using the properties of circles, perpendicular lines and lots of algebra.

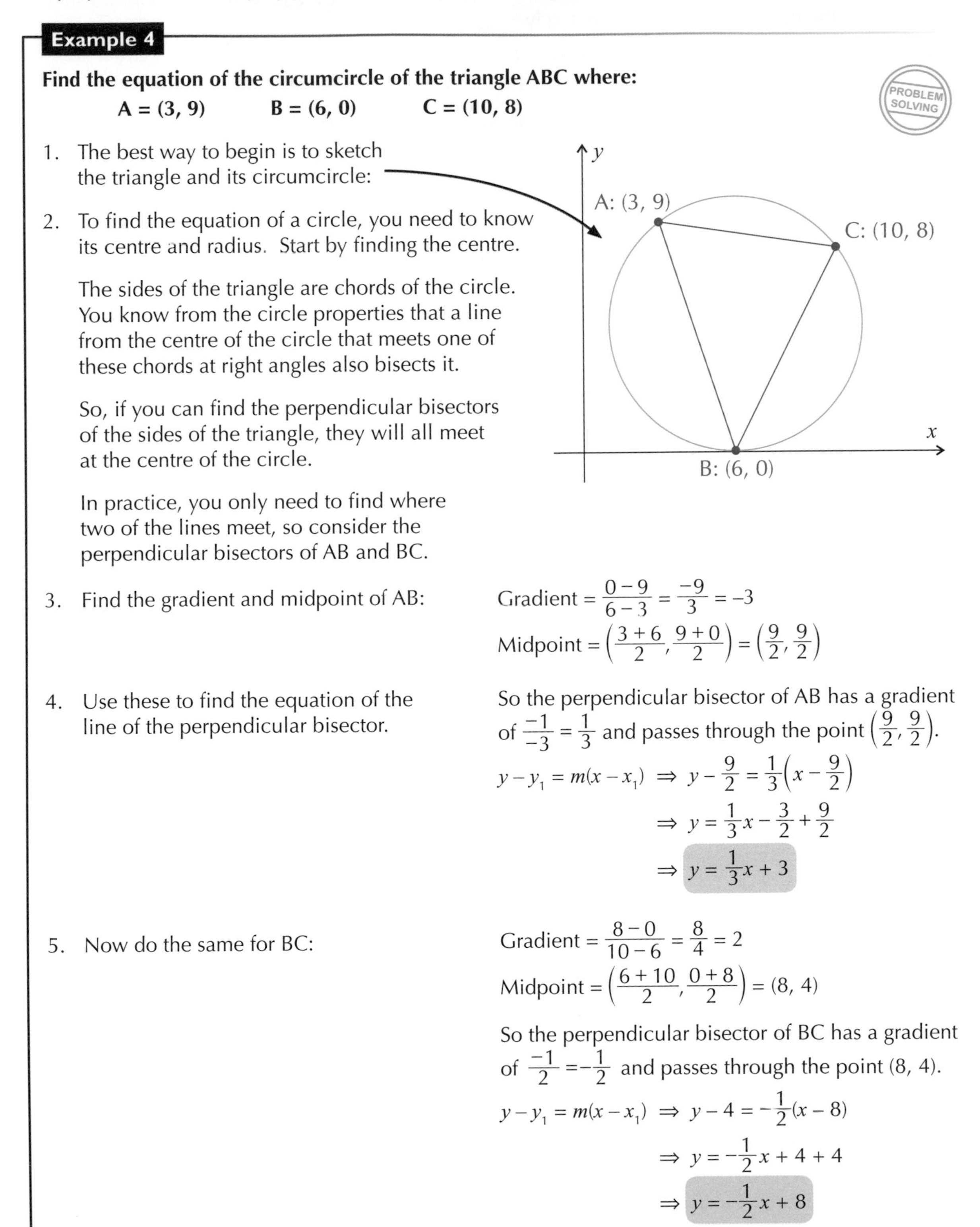

Example 4

Find the equation of the circumcircle of the triangle ABC where:

A = (3, 9) B = (6, 0) C = (10, 8)

1. The best way to begin is to sketch the triangle and its circumcircle:
2. To find the equation of a circle, you need to know its centre and radius. Start by finding the centre.

   The sides of the triangle are chords of the circle. You know from the circle properties that a line from the centre of the circle that meets one of these chords at right angles also bisects it.

   So, if you can find the perpendicular bisectors of the sides of the triangle, they will all meet at the centre of the circle.

   In practice, you only need to find where two of the lines meet, so consider the perpendicular bisectors of AB and BC.

3. Find the gradient and midpoint of AB:

   Gradient $= \frac{0-9}{6-3} = \frac{-9}{3} = -3$

   Midpoint $= \left(\frac{3+6}{2}, \frac{9+0}{2}\right) = \left(\frac{9}{2}, \frac{9}{2}\right)$

4. Use these to find the equation of the line of the perpendicular bisector.

   So the perpendicular bisector of AB has a gradient of $\frac{-1}{-3} = \frac{1}{3}$ and passes through the point $\left(\frac{9}{2}, \frac{9}{2}\right)$.

   $y - y_1 = m(x - x_1) \Rightarrow y - \frac{9}{2} = \frac{1}{3}\left(x - \frac{9}{2}\right)$

   $\Rightarrow y = \frac{1}{3}x - \frac{3}{2} + \frac{9}{2}$

   $\Rightarrow y = \frac{1}{3}x + 3$

5. Now do the same for BC:

   Gradient $= \frac{8-0}{10-6} = \frac{8}{4} = 2$

   Midpoint $= \left(\frac{6+10}{2}, \frac{0+8}{2}\right) = (8, 4)$

   So the perpendicular bisector of BC has a gradient of $\frac{-1}{2} = -\frac{1}{2}$ and passes through the point (8, 4).

   $y - y_1 = m(x - x_1) \Rightarrow y - 4 = -\frac{1}{2}(x - 8)$

   $\Rightarrow y = -\frac{1}{2}x + 4 + 4$

   $\Rightarrow y = -\frac{1}{2}x + 8$

6. With these two lines, you can find the centre of the circle by seeing where they cross.

7. Now find the x-coordinate of the centre by solving the simultaneous equations:

$$\frac{1}{3}x + 3 = -\frac{1}{2}x + 8$$
$$\Rightarrow \frac{1}{3}x + \frac{1}{2}x = 5 \Rightarrow 2x + 3x = 30$$
$$\Rightarrow 5x = 30 \Rightarrow x = 6$$

8. Substitute this value of x into one of the equations to find y: $y = \frac{1}{3}(6) + 3 = 2 + 3 = 5$, so the centre of the circle is $(6, 5)$.

9. The radius is the distance from the centre to a point on the edge. The distance from the centre (6, 5) to point B (6, 0) is easy to read off — you can see that the radius is 5.

Tip: You could use Pythagoras' theorem to find the radius from either of the other points.

10. Put the values for the centre and radius into the standard equation of a circle.

$$(x - 6)^2 + (y - 5)^2 = 25$$

Exercise 6.6.3

Q1 The circle shown below has the equation $(x - 3)^2 + (y - 1)^2 = 10$.
The line shown is a tangent to the circle and touches it at point A (4, 4).

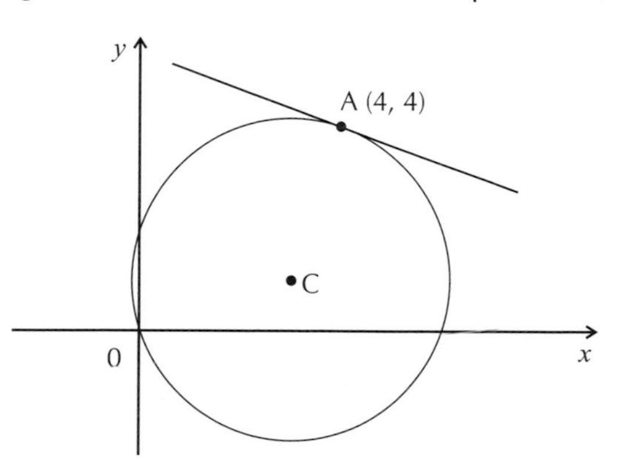

a) Find the centre of the circle, C.

b) Work out the gradient of the radius at (4, 4).

c) Find the equation of the tangent at A in the form $ax + by = c$.

Q2 Find the equation of the tangent to each circle below at the point given.
Give your answers in the form $ax + by + c = 0$, where a, b and c are integers.

a) $(x - 3)^2 + (y + 4)^2 = 37$, $(2, -10)$

b) $x^2 + y^2 + 8x - 9 = 0$, $(-7, 4)$

c) $x^2 + y^2 - 6x + 10y = 7$, $(8, -1)$

d) $(x - 8)^2 + (y + 2)^2 = 13$, $(6, -5)$

e) $(x - 4)^2 + (y + 2)^2 = 26$, $(9, -3)$

f) $x^2 + y^2 + 16x - 6y + 28 = 0$, $(-11, 9)$

g) $(x + 3)^2 + (y + 1)^2 = 13$, $(-5, -4)$

h) $x^2 + y^2 - 14x - 12y = 61$, $(-4, 1)$

Q3 A circle has the equation $(x + 1)^2 + (y - 2)^2 = 13$. The circle passes through the point A $(-3, -1)$.
Find the equation of the tangent at A in the form $ax + by + c = 0$.

Q4 The circle C has the equation $x^2 + y^2 + 2x - 7 = 0$.
Find an equation of the tangent to the circle at the point $(-3, 2)$.

Q5 A circle has the equation $x^2 + y^2 + 2x + 4y = 5$. The point A $(0, -5)$ lies on the circle.
Find the tangent to the circle at A in the form $ax + by = c$.

Q6 The circle C has equation $(x - 2)^2 + (y - 1)^2 = 100$. The point A $(10, 7)$ lies on the circle.
Find an equation of the tangent at A.

Q7 A circle with centre $(-2, 4)$ passes through the point A $(n, 1)$.
Given that the tangent to the circle at A has a gradient of $\frac{5}{3}$, find the value of n.

PROBLEM SOLVING

Q8 A circle has centre $(3, -1)$ and goes through the point $(4, 3)$.

PROBLEM SOLVING

a) Find the equation of the circle.

Line l is the tangent to the circle at the point $(4, 3)$.

b) Show that the area of the triangle formed by the line l and the positive x- and y-axes is 32 units².

Q9 The circle shown has centre C. Points A and B lie on the circle. Point A has coordinates $(-3, 7)$.
The midpoint of the line AB, M, has coordinates $(-1, 1)$. Line l passes through both C and M.

PROBLEM SOLVING

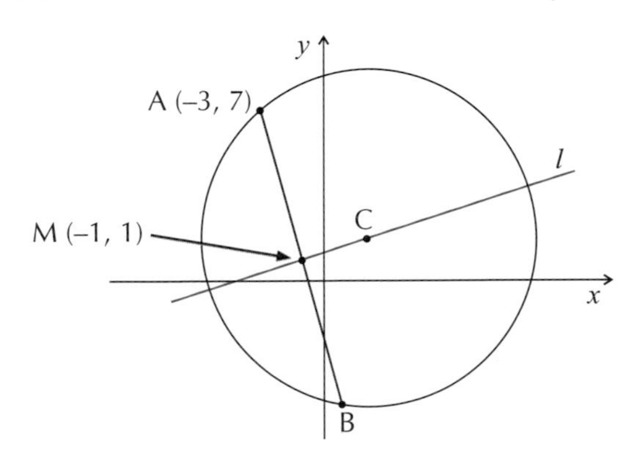

a) Use the information above to find an equation for the line l.

b) The coordinates of C are $(2, 2)$. Find an equation for the circle.

Q10 The points A $(2, a)$, B $(4, -1)$ and C $(0, 1)$ lie on a circle, and AB is a diameter of the circle. PROBLEM SOLVING

a) Find the value of a.

b) Find the equation of the circle.

Q11 The points A $(-2, 12)$, B $(4, 14)$ and C $(8, 2)$ all lie on the circle shown below. PROBLEM SOLVING

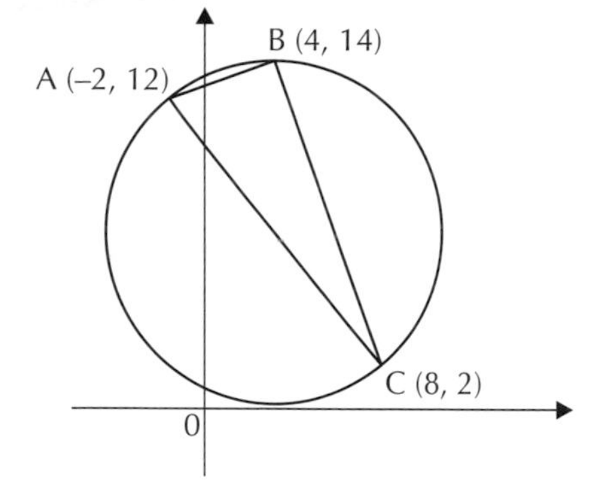

a) Prove that the line AC is a diameter of the circle.

b) Hence find the equation of the circle.

Q12 Points P and Q lie on a circle as shown on the diagram below. P has the coordinates $(-8, 6)$ and Q has the coordinates $(-3, 7)$. M is the midpoint of PQ and C is the centre of the circle. PROBLEM SOLVING

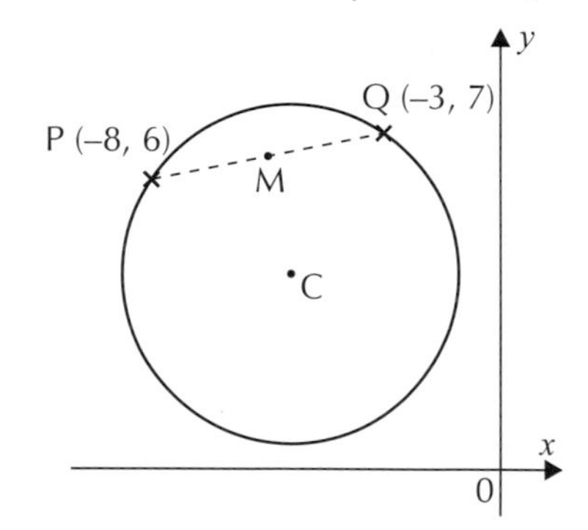

Find the equation, in the form $ax + by + c = 0$, of the line passing through M and C.

Q13 A triangle ABC has vertices A = $(2, 11)$, B = $(6, 5)$ and C = $(-9, 0)$. PROBLEM SOLVING
Find the equation of the circumcircle of ABC.

Q14 Find the equation of the circumcircle of the triangle XYZ where: PROBLEM SOLVING

a) X = $(8, 2)$, Y = $(-4, -4)$, Z = $(2, 8)$

b) X = $(2, 6)$, Y = $(5, 9)$, Z = $(-5, 9)$

c) X = $(14, 14)$, Y = $(22, 6)$, Z = $(22, 10)$

d) X = $(17, -14)$, Y = $(-7, 10)$, Z = $(-11, -2)$

Q15 A triangle ABC has vertices with coordinates A $(2, -2)$, B $(11, 1)$ and C $(10, -6)$. PROBLEM SOLVING

a) Find the equation of the circumcircle of the triangle ABC.

A fourth point D $(3, -5)$ also lies on the circumcircle of ABC.

b) Show that BD is a diameter of the circumcircle.

Review Exercise

Q1 Find the equations of the straight lines that pass through the following pairs of points.
Write each of them in the forms: (i) $y - y_1 = m(x - x_1)$, (ii) $y = mx + c$,
(iii) $ax + by + c = 0$, where a, b and c are integers.

a) $(2, -1), (-4, -19)$
b) $\left(0, -\frac{1}{3}\right), \left(5, \frac{2}{3}\right)$
c) $(8, 7), (-7, -2)$
d) $(5, 5), \left(2, \frac{5}{2}\right)$
e) $(1.3, 2), (1.8, 0)$
f) $(4.6, -2.3), (-5.4, -0.3)$

Q2 a) The line l_1 has equation $y = \frac{3}{2}x - \frac{2}{3}$. Find the equation of the line parallel to l_1, passing through the point with coordinates $(4, 2)$.

b) The line l_2 passes through the point $(6, 1)$ and is perpendicular to $2x - y - 7 = 0$. Find the equation of the line l_2.

Q3 The coordinates of points R and S are $(1, 9)$ and $(10, 3)$ respectively.
Find the equation of the line perpendicular to RS, passing through the point $(1, 9)$.

Q4 Given that y is directly proportional to x, and that $x = 4.5$ when $y = 3$, find:

a) y when $x = 21$
b) y when $x = -3$
c) x when $y = 58$

Q5 Given that s is inversely proportional to t^3, and that $s = 18$ when $t = 6$, find:

a) s when $t = 3$
b) s when $t = 0.5$
c) t when $s = 486$

Q6 Sketch these cubic graphs:

a) $y = (x - 4)^3$
b) $y = (3 - x)(x + 2)^2$
c) $y = (1 - x)(x^2 - 6x + 8)$
d) $y = (x - 1)(x - 2)(x - 3)$
e) $y = 3x^3 - 6x^2$
f) $y = x^3 - x^2 - 12x$

Q7 Sketch the graphs of the following quartic functions:

a) $y = (x - 4)^4$
b) $y = -(x + 1)^2(x - 1)^2$
c) $y = x^2(x^2 - 16)$

Q8 Draw rough sketches of the following curves:

a) $y = -2x^4$
b) $y = \frac{7}{x^2}$
c) $y = -5x^3$
d) $y = -\frac{2}{x^5}$
e) $y = \frac{2}{3}x^5$
f) $y = -4x^{-4}$
g) $y = \frac{x^2}{2}$
h) $y = \frac{4}{5x^6}$

Q9 Given that $a > 1$, use the graph of $f(x)$ to sketch the graph of:

a) $y = f(ax)$
b) $y = f\left(\frac{1}{a}x\right)$
c) $y = af(x)$
d) $y = \frac{1}{a}f(x)$
e) $y = f(x + a)$
f) $y = f(x - a)$
g) $y = f(x) + a$
h) $y = f(x) - a$

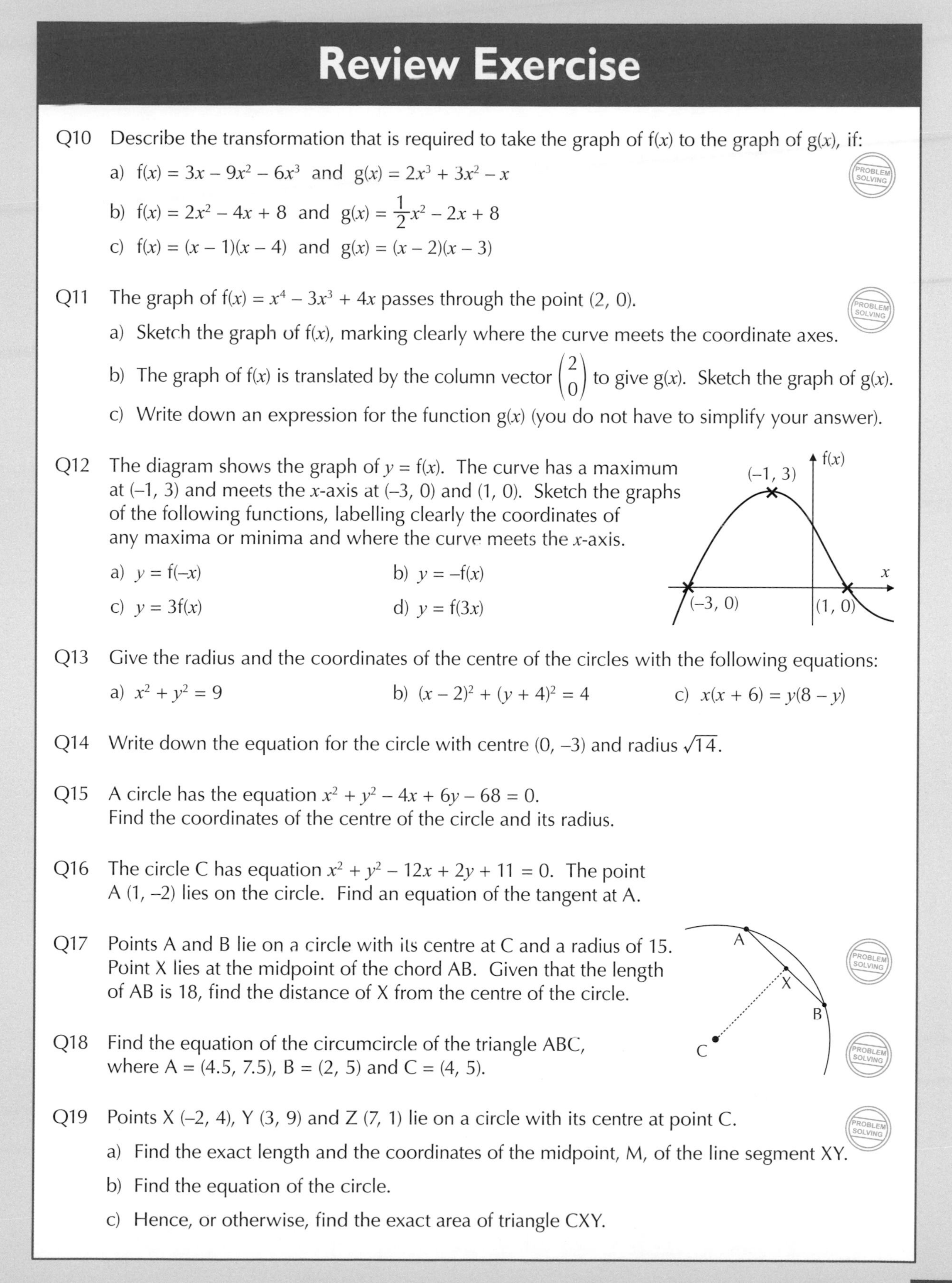

Review Exercise

Q10 Describe the transformation that is required to take the graph of f(x) to the graph of g(x), if:

PROBLEM SOLVING

a) $f(x) = 3x - 9x^2 - 6x^3$ and $g(x) = 2x^3 + 3x^2 - x$

b) $f(x) = 2x^2 - 4x + 8$ and $g(x) = \frac{1}{2}x^2 - 2x + 8$

c) $f(x) = (x - 1)(x - 4)$ and $g(x) = (x - 2)(x - 3)$

Q11 The graph of $f(x) = x^4 - 3x^3 + 4x$ passes through the point (2, 0).

PROBLEM SOLVING

a) Sketch the graph of f(x), marking clearly where the curve meets the coordinate axes.

b) The graph of f(x) is translated by the column vector $\begin{pmatrix} 2 \\ 0 \end{pmatrix}$ to give g(x). Sketch the graph of g(x).

c) Write down an expression for the function g(x) (you do not have to simplify your answer).

Q12 The diagram shows the graph of $y = f(x)$. The curve has a maximum at (–1, 3) and meets the x-axis at (–3, 0) and (1, 0). Sketch the graphs of the following functions, labelling clearly the coordinates of any maxima or minima and where the curve meets the x-axis.

a) $y = f(-x)$

b) $y = -f(x)$

c) $y = 3f(x)$

d) $y = f(3x)$

Q13 Give the radius and the coordinates of the centre of the circles with the following equations:

a) $x^2 + y^2 = 9$

b) $(x - 2)^2 + (y + 4)^2 = 4$

c) $x(x + 6) = y(8 - y)$

Q14 Write down the equation for the circle with centre (0, –3) and radius $\sqrt{14}$.

Q15 A circle has the equation $x^2 + y^2 - 4x + 6y - 68 = 0$.
Find the coordinates of the centre of the circle and its radius.

Q16 The circle C has equation $x^2 + y^2 - 12x + 2y + 11 = 0$. The point A (1, –2) lies on the circle. Find an equation of the tangent at A.

Q17 Points A and B lie on a circle with its centre at C and a radius of 15. Point X lies at the midpoint of the chord AB. Given that the length of AB is 18, find the distance of X from the centre of the circle.

PROBLEM SOLVING

Q18 Find the equation of the circumcircle of the triangle ABC, where A = (4.5, 7.5), B = (2, 5) and C = (4, 5).

PROBLEM SOLVING

Q19 Points X (–2, 4), Y (3, 9) and Z (7, 1) lie on a circle with its centre at point C.

PROBLEM SOLVING

a) Find the exact length and the coordinates of the midpoint, M, of the line segment XY.

b) Find the equation of the circle.

c) Hence, or otherwise, find the exact area of triangle CXY.

Exam-Style Questions

Q1 Points A and B have coordinates (–2, 4) and (4, –10) respectively.

a) Find the exact length AB.

[2 marks]

b) Find the gradient of the line segment AB.

[2 marks]

c) Find the equation of the line that passes through points A and B, giving your answer in the form $ax + by + c = 0$, where a, b and c are integers to be found.

[3 marks]

Q2 A circle has centre (2, –3) and radius r. Its equation is $x^2 + y^2 + ax + by - 4 = 0$. Find the exact radius of the circle and the values of a and b.

[4 marks]

Q3 The points A (1, 7), B (20, 7) and C (p, q) are the vertices of a triangle ABC, as shown below. The point D (8, 2) is the midpoint of AC.

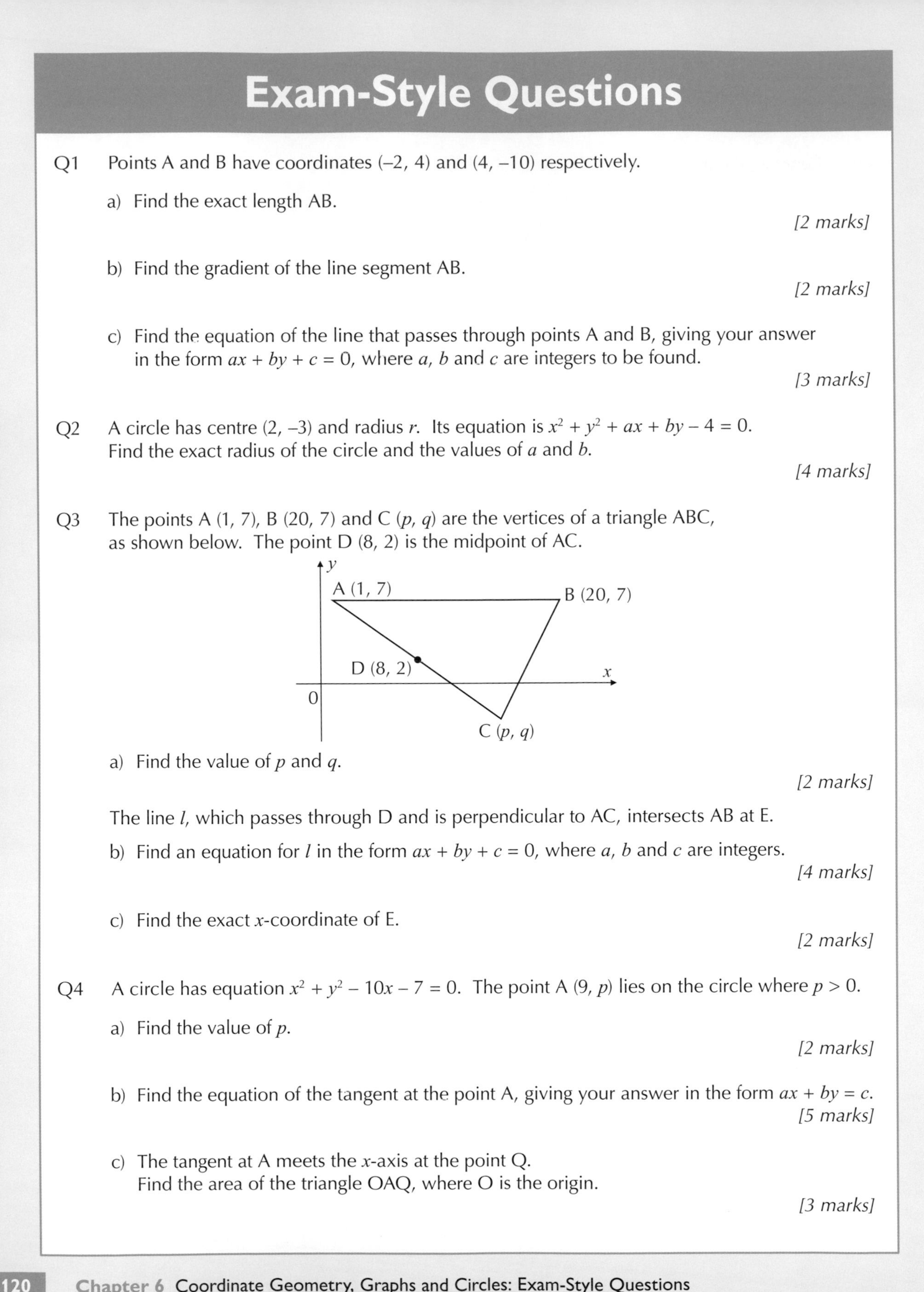

a) Find the value of p and q.

[2 marks]

The line l, which passes through D and is perpendicular to AC, intersects AB at E.

b) Find an equation for l in the form $ax + by + c = 0$, where a, b and c are integers.

[4 marks]

c) Find the exact x-coordinate of E.

[2 marks]

Q4 A circle has equation $x^2 + y^2 - 10x - 7 = 0$. The point A $(9, p)$ lies on the circle where $p > 0$.

a) Find the value of p.

[2 marks]

b) Find the equation of the tangent at the point A, giving your answer in the form $ax + by = c$.

[5 marks]

c) The tangent at A meets the x-axis at the point Q.
Find the area of the triangle OAQ, where O is the origin.

[3 marks]

Exam-Style Questions

Q5 The diagram below shows the trapezium ABCD, in which sides AB and CD are parallel. The point C has coordinates (6, 6) and the point D has coordinates (–4, 2).

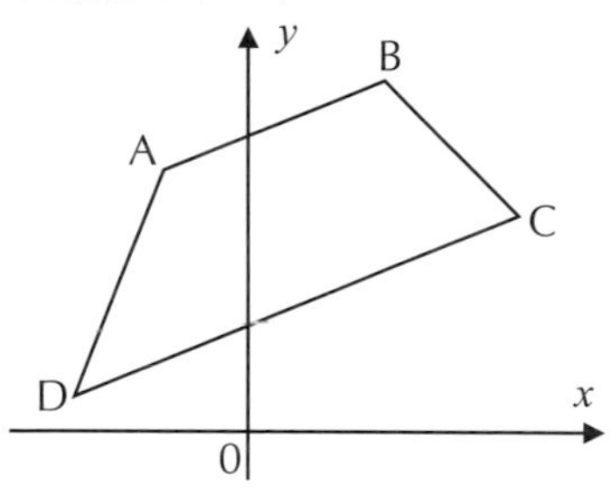

a) Find the equation of the straight line passing through C and D, giving your answer in the form $ax + by + c = 0$, where a, b and c are integers.

[3 marks]

b) Given that the gradient of BC is –1, find an equation of the line BC.

[2 marks]

c) Point A has coordinates (–2, 7). Find the coordinates of the point B.

[4 marks]

d) Show that $\angle DBC = 90°$.

[2 marks]

Q6 A circle with centre C has equation $x^2 + y^2 - 4x + 12y + 15 = 0$.

a) Find the coordinates of the centre point and the radius of circle C.

[4 marks]

b) Explain why the circle C lies entirely below the x-axis.

[2 marks]

c) The point $(-1, k)$ lies on the circle. Find the possible values of k.

[3 marks]

Q7 A triangle XYZ has vertices with coordinates X (7, 6), Y (–1, –10) and Z (13, –12). Find the equation of the circumcircle of the triangle XYZ.

[8 marks]

Q8 Two points, A and B, are the points of intersection between a circle with centre C (2, 4) and radius 5, and the line $x + 2y = 15$. M is the midpoint of the chord AB.

a) Show that the length CM is $\sqrt{5}$.

[7 marks]

b) Hence, find the area of the triangle ABC.

[3 marks]

Chapter 7 The Binomial Expansion

7.1 Binomial Expansions

Binomials are just polynomials that only have two terms (the 'bi' and 'poly' bits come from the Greek words for 'two' and 'many'). So binomial expansion is all about multiplying out brackets with two terms.

Learning Objectives (Spec Ref 4.1):

- Use Pascal's triangle to find the coefficients of a binomial expansion.
- Find binomial coefficients using factorials and using the notation $\binom{n}{r}$ or ${}^nC_r$.
- Use the formula to expand binomials of the form $(1 + x)^n$, $(1 + ax)^n$ and $(a + bx)^n$.

Prior Knowledge Check: Be able to expand brackets. See p.10.

Binomial expansions — $(1 + x)^n$

Pascal's triangle

A **binomial expansion** is what you get when you **multiply out the brackets** of a polynomial with two terms, like $(1 + x)^5$ or $(2 - 3x)^8$. It would take ages to multiply out a bracket like this by hand if the power was really big — fortunately, binomial expansions **follow a pattern**:

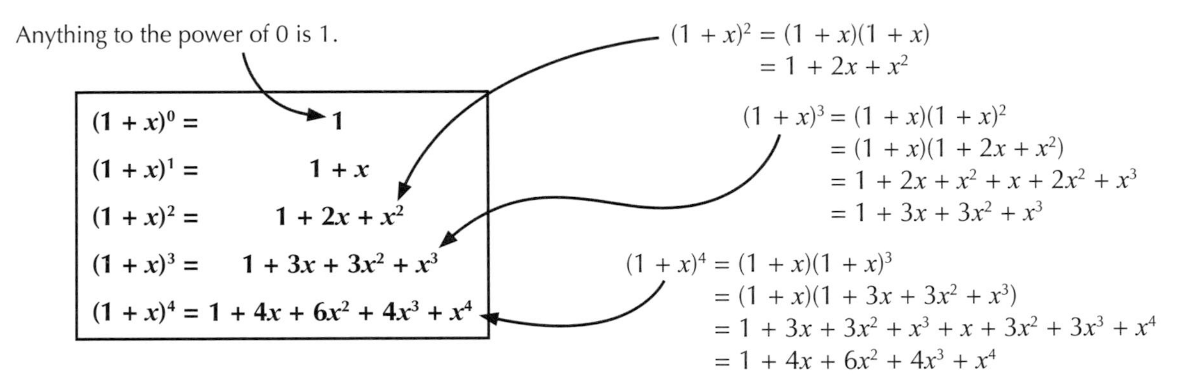

A French man called Blaise Pascal spotted the pattern in the **coefficients** and wrote them down in a **triangle**, so it's imaginatively known as '**Pascal's triangle**'. Each number is the **sum** of the two above it — and the triangle is symmetrical, so you only need to work out the first half of the coefficients:

1

1 1

1 2 1

1 3 3 + 1

1 4 6 4 1

The next line is: **1 5 10 10 5 1**, so $(1 + x)^5 = 1 + 5x + 10x^2 + 10x^3 + 5x^4 + x^5$.

If you're expanding a binomial with a power that's not too huge, writing out a quick **Pascal's triangle** is a good way to **find the coefficients**. Make sure you go down to the **correct row** — you need **one more row** than the **power** you're raising the bracket to. You usually write the expansion in increasing powers of x, from x^0 to x^n.

Example 1

Find the binomial expansion of $(1 + x)^6$.

Draw **Pascal's triangle** — you're raising the bracket to the **power of 6**, so go down to the **7th row**.

Write the answer out, using the **coefficients** from the 7th row, and increasing the power of x:

$(1 + x)^6 = 1 + 6x + 15x^2 + 20x^3 + 15x^4 + 6x^5 + x^6$

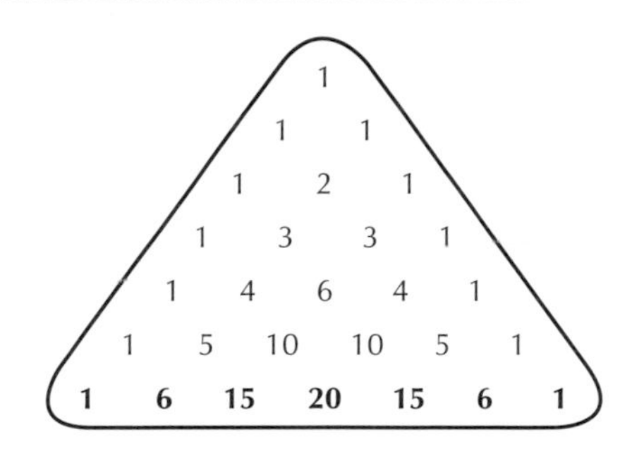

The binomial formula

For expansions with higher powers you don't need to write out Pascal's triangle — use the binomial formula instead:

$$(1 + x)^n = 1 + \frac{n}{1}x + \frac{n(n-1)}{1 \times 2}x^2 + \frac{n(n-1)(n-2)}{1 \times 2 \times 3}x^3 + \ldots + x^n$$

At first glance this looks a bit awful, but each term follows a pattern:

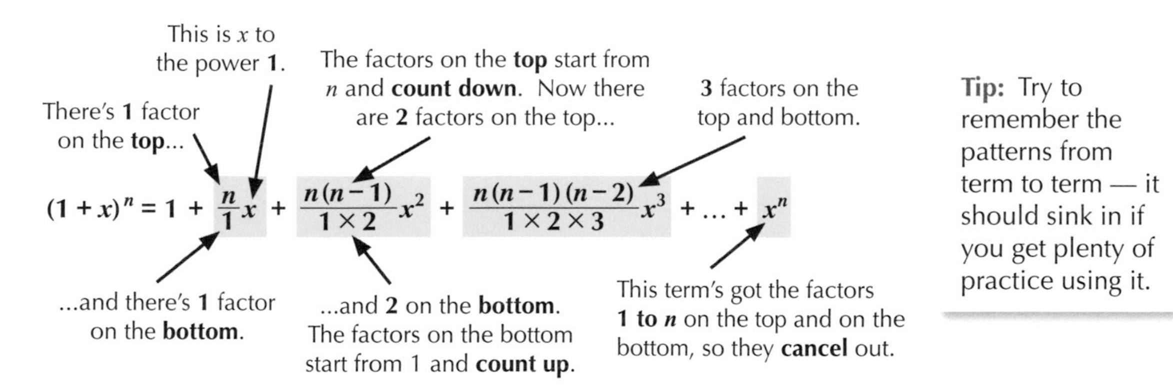

Tip: Try to remember the patterns from term to term — it should sink in if you get plenty of practice using it.

Once you get **halfway** along, the **factors** on the top and bottom start to **cancel**, and the coefficients repeat themselves (they're symmetrical):

$$(1 + x)^5 = 1 + \frac{5}{1}x + \frac{5 \times 4}{1 \times 2}x^2 + \frac{5 \times 4 \times \cancel{3}}{1 \times 2 \times \cancel{3}}x^3 + \frac{5 \times \cancel{4} \times \cancel{3} \times \cancel{2}}{1 \times \cancel{2} \times \cancel{3} \times \cancel{4}}x^4 + \frac{\cancel{5} \times \cancel{4} \times \cancel{3} \times \cancel{2} \times \cancel{1}}{\cancel{1} \times \cancel{2} \times \cancel{3} \times \cancel{4} \times \cancel{5}}x^5$$

$$= 1 + \frac{5}{1}x + \frac{5 \times 4}{1 \times 2}x^2 + \frac{5 \times 4}{1 \times 2}x^3 + \frac{5}{1}x^4 + x^5$$

$$= 1 + 5x + 10x^2 + 10x^3 + 5x^4 + x^5$$

Example 2

a) Expand $(1 + x)^{20}$, giving the first four terms only.

1. The binomial formula is: $(1+x)^n = 1 + \frac{n}{1}x + \frac{n(n-1)}{1 \times 2}x^2 + \frac{n(n-1)(n-2)}{1 \times 2 \times 3}x^3 + \ldots + x^n$

2. You're looking for $(1 + x)^{20}$, so here, $n = 20$.

$$(1+x)^{20} = 1 + \frac{20}{1}x + \frac{20 \times 19}{1 \times 2}x^2 + \frac{20 \times 19 \times 18}{1 \times 2 \times 3}x^3 + \ldots + x^{20}$$

Make sure you write down your working or you won't be able to spot if you've made a mistake.

$$(1+x)^{20} = 1 + \frac{20}{1}x + \frac{\overset{10}{\cancel{20}} \times 19}{1 \times \cancel{2}}x^2 + \frac{\overset{10}{\cancel{20}} \times 19 \times \overset{6}{\cancel{18}}}{1 \times \cancel{2} \times \cancel{3}}x^3 + \ldots$$

$$= 1 + 20x + (10 \times 19)x^2 + (10 \times 19 \times 6)x^3 + \ldots$$

3. The first 4 terms are: $= 1 + 20x + 190x^2 + 1140x^3 + \ldots$

b) What is the term in x^7 in this expansion? Give your answer in its simplest form.

1. The term in x^7 has 7 factors on the top and bottom of the coefficient.

2. In this expansion $n = 20$, so on the top you count down from 20.

3. Term in $x^7 = \frac{\cancel{20} \times 19 \times \cancel{18} \times 17 \times 16 \times 15 \times \cancel{14}}{1 \times \cancel{2} \times \cancel{3} \times \cancel{4} \times \cancel{5} \times \cancel{6} \times \cancel{7}}x^7$

$$= (19 \times 17 \times 16 \times 15)x^7 = 77\,520x^7$$

There's some **notation** you need to know that will make writing out the binomial formula a bit easier.

Factorials

- The product on the **bottom** of each binomial coefficient is $\mathbf{1 \times 2 \times \ldots \times r}$, where r is the **power x is raised to** in that term.

- This product can be written as a **factorial: $r! = 1 \times 2 \times \ldots \times r$**
E.g. in the binomial expansion of $(1 + x)^{20}$, the coefficient of the term in x^3 is: $\frac{20 \times 19 \times 18}{1 \times 2 \times 3} = \frac{20 \times 19 \times 18}{3!}$

Tip: $r! = 1 \times 2 \times \ldots \times r$ is said 'r factorial'. By convention, $0! = 1$.

- In fact, you can write the whole coefficient using factorials. For example, the coefficient of x^3 above is:

$$\frac{20 \times 19 \times 18}{1 \times 2 \times 3} = \frac{20 \times 19 \times 18 \times 17 \times \ldots \times 2 \times 1}{1 \times 2 \times 3 \times 1 \times 2 \times \ldots \times 17}$$

$$= \frac{20 \times 19 \times \ldots \times 1}{(1 \times 2 \times 3)(1 \times 2 \times \ldots \times 17)}$$

$$= \frac{20!}{3!17!}$$

Tip: Here you've multiplied top and bottom by 17! — this is just so you can write the factors on the top as a factorial (you need to multiply all the way down to 1 to do this).

- For a general binomial expansion of $(1 + x)^n$, the coefficient of x^r is:

$$\frac{n \times (n-1) \times \ldots \times (n-(r-1))}{1 \times 2 \times \ldots \times r}$$

$$= \frac{n \times (n-1) \times \ldots \times (n-(r-1)) \times (n-r) \times \ldots \times 2 \times 1}{1 \times 2 \times \ldots \times r \times 1 \times 2 \times \ldots \times (n-r)}$$

$$= \frac{n \times (n-1) \times \ldots \times 1}{(1 \times 2 \times \ldots \times r)(1 \times 2 \times \ldots \times (n-r))}$$

$$= \frac{n!}{r!(n-r)!}$$

Tip: The two numbers on the bottom of the factorial fraction always add up to the number on the top.

- So each term in a **binomial expansion** of $(1 + x)^n$ is of the form:

$$\frac{n(n-1)(n-2)\dots(n-(r-1))}{1\times2\times3\times\dots\times r}x^r = \frac{n!}{r!\,(n-r)!}x^r$$

where n is the power you're raising the **bracket** to,
and r is the power of x in the **term** the coefficient belongs to.

${}^nC_r$ notation

There are a couple of even **shorter ways** of writing the **binomial coefficients**:

$$\frac{n!}{r!\,(n-r)!} = \binom{n}{r} = {}^nC_r$$

Tip: This is in the formula booklet.

The C in ${}^nC_r$ stands for 'choose' — you say these coefficients 'n choose r', e.g. $\binom{3}{2}$ is '3 choose 2'.

Going back to the coefficient of x^3 in the expansion of $(1 + x)^{20}$:

$$\frac{20\times19\times18}{1\times2\times3}x^3 = \frac{20!}{3!17!}x^3 = \binom{20}{3}x^3 = {}^{20}C_3x^3$$

The binomial formula can be written using any of these notations, and you need to be familiar with all of them.

$$(1+x)^n = 1 + \frac{n!}{1!(n-1!)}x + \frac{n!}{2!(n-2)!}x^2 + \frac{n!}{3!(n-3)!}x^3 + \dots + x^n$$

$$(1+x)^n = 1 + \binom{n}{1}x + \binom{n}{2}x^2 + \binom{n}{3}x^3 + \dots + x^n$$

$$(1+x)^n = 1 + {}^nC_1x + {}^nC_2x^2 + {}^nC_3x^3 + \dots + x^n$$

Tip: These coefficients will always come out as a whole number that's at least 1. If they don't, you've gone wrong somewhere.

Tip: ${}^nC_0 = {}^nC_n = 1$.

Most **calculators** will have an '**nCr**' button for finding binomial **coefficients**. To use it, put in n, press '**nCr**', then put in r. This is particularly handy if you're just looking for a **specific term** in a **binomial expansion** and you don't want to write the whole thing out. (If you get confused between which number is n and which is r, remember that n is always greater than or equal to r.)

Example 3

Find the 6th term of the expansion of $(1 + x)^8$.

1. You're raising the **bracket** to the power of **8**, so $n = 8$.
2. The **6th term** is the x^5 term (the first term is $1 = x^0$, the second is the $x = x^1$ term), so $r = 5$.
3. Put '**8 nCr 5**' into your calculator: ${}^8C_5 = 56$.
4. The question asks for the **whole term** (not just the coefficient) so the answer is: $56x^5$

Exercise 7.1.1

Q1 Use Pascal's triangle to expand $(1 + x)^4$.

Q2 Use your calculator to work out the following:

a) ${}^6C_2$ b) $\binom{12}{5}$ c) $\frac{30!}{4!26!}$ d) ${}^8C_8$

Q3 Without using a calculator, work out the following:

a) $\frac{9!}{4!5!}$ b) ${}^{10}C_3$ c) $\frac{15!}{11!4!}$ d) $\binom{8}{6}$

Q4 Find the first 4 terms, in ascending powers of x, of the binomial expansion of $(1 + x)^{10}$. Give each term in its simplest form.

Q5 Write down the full expansion of $(1 + x)^6$.

Q6 Find the first 4 terms in the expansion of $(1 + x)^7$.

Q7 Find the first three terms, in ascending powers of x, of the binomial expansion of:

a) $(1 + x)^{11}$ b) $(1 + x)^{12}$ c) $(1 + x)^{15}$ d) $(1 + x)^{30}$

Q8 The coefficient of x^{12} in the expansion of $(1 + x)^{17}$ is a. Find the value of a.

Q9 For the expansion of $(1 + x)^5 + (2 - 3x)$:

PROBLEM SOLVING

a) Find the term in x.

b) Write down the coefficient of x^4.

Binomial expansions — $(1 + ax)^n$

When the **coefficient of x** in your binomial **isn't 1** (e.g. $(1 + 2x)^6$)
you have to substitute the **whole x term** (e.g. $2x$) into the **binomial formula**:

$$(1 + ax)^n = 1 + \binom{n}{1}(ax) + \binom{n}{2}(ax)^2 + \binom{n}{3}(ax)^3 + ... + (ax)^n$$

When a is **–1** (i.e. $(1 - x)^n$) the formula looks just like the formula for $(1 + x)^n$, but the **signs** of the terms **alternate**:

$$\begin{aligned}(1 - x)^n &= (1 + (-x))^n \\ &= 1 + \frac{n}{1}(-x) + \frac{n(n-1)}{1 \times 2}(-x)^2 + \frac{n(n-1)(n-2)}{1 \times 2 \times 3}(-x)^3 + ... + (-x)^n \\ &= 1 - \frac{n}{1}x + \frac{n(n-1)}{1 \times 2}x^2 - \frac{n(n-1)(n-2)}{1 \times 2 \times 3}x^3 + ... \pm x^n\end{aligned}$$

Tip: The sign of the last term is plus if n is even and minus if n is odd.

So for **$(1 - x)^n$** you just use the usual binomial **coefficients**, but with **alternating signs**:

$$(1 - x)^n = 1 - \binom{n}{1}x + \binom{n}{2}x^2 - \binom{n}{3}x^3 + ... \pm x^n$$

Example 1

a) What is the term in x^5 in the expansion of $(1 - 3x)^{12}$?

1. The general binomial formula for $(1 - 3x)^n$ is:

$$(1-3x)^n = (1+(-3x))^n$$
$$= 1 + \frac{n}{1}(-3x) + \frac{n(n-1)}{1\times 2}(-3x)^2 + \frac{n(n-1)(n-2)}{1\times 2\times 3}(-3x)^3 + \ldots + (-3x)^n$$

2. So when $n = 12$: $(1-3x)^{12} = 1 + \frac{12}{1}(-3x) + \frac{12\times 11}{1\times 2}(-3x)^2 + \frac{12\times 11\times 10}{1\times 2\times 3}(-3x)^3 + \ldots + (-3x)^{12}$

3. The term in x^5 is: $\frac{12!}{5!7!}(-3x)^5 = \binom{12}{5}(-3)^5x^5 = (792 \times -243)x^5 = -192\,456x^5$

b) Find the coefficient of x^2 in the expansion of $(1 + 6x)^4(1 - 2x)^6$.

PROBLEM SOLVING

1. To find the x^2 term in the combined expansion, you need to find all the **terms up to x^2** in both expansions (as these will form the x^2 term when they're multiplied) and then **multiply** together.

$$(1 + 6x)^4 = 1 + \binom{4}{1}(6x) + \binom{4}{2}(6x)^2 + \ldots$$
$$= 1 + 24x + 216x^2 + \ldots$$
$$(1 - 2x)^6 = 1 + \binom{6}{1}(-2x) + \binom{6}{2}(-2x)^2 + \ldots$$
$$= 1 - 12x + 60x^2 - \ldots$$

2. Multiply the two separate expansions together:

$$(1 + 6x)^4(1 - 2x)^6 = (1 + 24x + 216x^2 + \ldots)(1 - 12x + 60x^2 - \ldots)$$
$$= 1 - 12x + 60x^2 - \ldots$$
$$+ 24x - 288x^2 + 1440x^3 - \ldots$$
$$+ 216x^2 - 2592x^3 + 12\,960x^4 - \ldots$$
$$= 1 + 12x - 12x^2 + (\textit{higher power terms}) + \ldots$$

3. So the coefficient of x^2 is: -12

Approximations

You can use a binomial expansion to expand **longer expressions** or to find **approximations** of a number raised to a power.

Approximating usually involves taking a really small value for x so that you can **ignore high powers** of x (because they'll be really, really small).

For example, if you're asked to approximate 1.001^9, then you just stick $x = 0.001$ into the expansion of $(1 + x)^9$. Because 0.001 is small, 0.001^2 is really small and adding on really small terms won't make much difference, so just the **first few terms** will give a good approximation.

Example 2

a) Expand $(1 + 2x)^7$ to find the first 4 terms in ascending powers of x.

You can find the binomial coefficients using any of the methods from pages 122-125 — just pick whichever one you prefer.

$$(1 + 2x)^7 = 1 + {}^7C_1(2x) + {}^7C_2(2x)^2 + {}^7C_3(2x)^3 + \ldots$$
$$= 1 + 7(2x) + 21(4x^2) + 35(8x^3) + \ldots$$
$$= 1 + 14x + 84x^2 + 280x^3 + \ldots$$

b) When x is small, x^3 and higher powers can be ignored. Hence show that for small x: $(2 - x)(1 + 2x)^7 \approx 2 + 27x + 154x^2$

1. Multiply your expansion of $(1 + 2x)^7$ through by $(2 - x)$:
2. You only need to include the terms up to the one in x^2 as the question says to ignore terms in x^3 and above.

$$(2 - x)(1 + 2x)^7 \approx (2 - x)(1 + 14x + 84x^2)$$
$$= 2 + 28x + 168x^2 - x - 14x^2 - 84x^3$$
$$= 2 + 27x + 154x^2 - 84x^3$$
$$\approx 2 + 27x + 154x^2 \text{ as required}$$

Example 3

a) Find the first 3 terms of the expansion of $\left(1 - \frac{x}{4}\right)^9$.

Use the **formula**, but replace x with $\left(-\frac{x}{4}\right)$:

$$\left(1 - \frac{x}{4}\right)^9 = 1 + \binom{9}{1}\left(-\frac{x}{4}\right) + \binom{9}{2}\left(-\frac{x}{4}\right)^2 + \ldots$$
$$= 1 - 9\left(\frac{x}{4}\right) + 36\left(\frac{x^2}{16}\right) - \ldots = 1 - \frac{9}{4}x + \frac{9}{4}x^2 - \ldots$$

b) Use your expansion to estimate $(0.998)^9$.

1. Write $(0.998)^9$ in the same form as the expansion.
2. Substitute $x = 0.008$ into the expansion — the first three terms are enough as 0.008^3 and higher powers will be very small.

$(0.998)^9 = (1 - 0.002)^9 = \left(1 - \frac{x}{4}\right)^9$ when $x = 0.008$.

$$(0.998)^9 = \left(1 - \frac{0.008}{4}\right)^9 \approx 1 - \frac{9}{4}(0.008) + \frac{9}{4}(0.008)^2$$
$$= 1 - 0.018 + 0.000144$$
$$= 0.982144$$

Exercise 7.1.2

Q1 Find the full expansions of:

a) $(1 + 3x)^4$ b) $(1 - x)^4$ c) $(1 - x)^6$ d) $(1 - 2x)^5$

e) $(1 - 4x)^3$ f) $(1 - 5x)^5$ g) $(1 + 2x)^6$ h) $(1 + x)^9 - (1 - x)^9$

Q2 What is the term in x^4 in the expansion of $(1 - 2x)^{16}$?

Q3 Find the first 3 terms in the expansion of $(1 + x)^3(1 - x)^4$.

Q3 Hint: Go up to x^2 in both expansions, then multiply the expansions together.

Q4 Find the coefficient of x^3y^2 in the expansion of $(1 + x)^5(1 + y)^7$. PROBLEM SOLVING

Q5 Find the coefficient of x^3 in the expansion of $(1 + 4x)^4(1 - 6x)^3$. PROBLEM SOLVING

Q6 a) Find the first 4 terms, in ascending powers of x, of the binomial expansion of $(1 + kx)^8$, where k is a non-zero constant.

b) If the fourth term of the above expansion is $448x^3$, find the value of k.

Q7 In the expansion of $(1 - kx)^6$, the coefficient of x^2 is 135. Use this information to find the value of k, given that k is positive. PROBLEM SOLVING

Q8 If x is small, so that x^2 and higher powers can be ignored, show that $(1 + x)(1 - 3x)^6 \approx 1 - 17x$.

PROBLEM SOLVING

Q9 a) Find, in their simplest form, the first 5 terms in the expansion of $\left(1 + \frac{x}{2}\right)^{12}$, in ascending powers of x.

b) Use the expansion to work out the value of 1.005^{12} to 7 d.p.

Q10 Marc finds the expansion of $(1 + x)^{10}$ up to the term in x^3. He wants to substitute $x = 1.6$ into this expression to approximate 2.6^{10}. Explain why this will not be a good approximation.

PROBLEM SOLVING

Q11 In the expansion of $(1 + 7x)^n$, the coefficient of the term in x^2 is 490.

PROBLEM SOLVING

a) Given that n is positive, find the value of n.

b) Find the term in x^3.

Binomial expansions — $(a + b)^n$

When your binomial is of the form $\boldsymbol{(a + b)^n}$ (e.g. $(2 + 3x)^7$, where $a = 2$ and $b = 3x$) you can use a slightly **different formula**:

$$(a+b)^n = a^n + \binom{n}{1}a^{n-1}b + \binom{n}{2}a^{n-2}b^2 + \ldots + \binom{n}{n-1}ab^{n-1} + b^n$$

Tip: The powers of a decrease (from n to 0) as the powers of b increase (from 0 to n). The sum of the powers of a and b in each term is always n.

This formula is in the **formula booklet** and you don't need to know the proof, but seeing where it comes from might make things a bit clearer. You can find it from the binomial formula you've already seen:

- First rearrange so the binomial is in a form you can work with.

$$(a+b)^n = \left(a\left(1+\frac{b}{a}\right)\right)^n = a^n\left(1+\frac{b}{a}\right)^n$$

- You expand this by putting '$\frac{b}{a}$' into the binomial formula for $(1 + x)^n$, just like in the previous section.

$$= a^n\left(1 + \binom{n}{1}\left(\frac{b}{a}\right) + \binom{n}{2}\left(\frac{b}{a}\right)^2 + \ldots + \binom{n}{n-1}\left(\frac{b}{a}\right)^{n-1} + \left(\frac{b}{a}\right)^n\right)$$

$$= a^n\left(1 + \binom{n}{1}\frac{b}{a} + \binom{n}{2}\frac{b^2}{a^2} + \ldots + \binom{n}{n-1}\frac{b^{n-1}}{a^{n-1}} + \frac{b^n}{a^n}\right)$$

- **Multiply** through by $\boldsymbol{a^n}$.

$$= a^n + \binom{n}{1}a^{n-1}b + \binom{n}{2}a^{n-2}b^2 + \ldots + \binom{n}{n-1}ab^{n-1} + b^n$$

This is a general formula that works for any a and b, including 1 and x. So given **any binomial,** you can pop your values for $\boldsymbol{a}$, $\boldsymbol{b}$ and $\boldsymbol{n}$ into this formula and you'll get the **expansion**.

Example 1

Give the first three terms, in ascending powers of x, of the expansion of $(4 - 5x)^7$.

Use the formula above, with $\boldsymbol{a = 4}$, $\boldsymbol{b = -5x}$ and $\boldsymbol{n = 7}$.

$$(4 - 5x)^7 = (4 + (-5x))^7 = 4^7 + \left(\binom{7}{1} \times 4^6 \times (-5x)\right) + \left(\binom{7}{2} \times 4^5 \times (-5x)^2\right) + \ldots$$

$$= 16\,384 + (7 \times 4096 \times -5x) + (21 \times 1024 \times 25x^2) + \ldots$$

$$= 16\,384 - 143\,360x + 537\,600x^2 + \ldots$$

Tip: Be careful with b here — there's a minus sign that might catch you out.

Your other option with expansions of $(a + b)^n$ is to **factorise** the binomial so you get $a^n\left(1+\frac{b}{a}\right)^n$, then plug $\frac{b}{a}$ into the **original binomial formula** (as you did with $(1 + ax)^n$ expansions in the last section).

Example 2

PROBLEM SOLVING

Find the coefficient of x^4 in the expansion of $(2 + 5x)^7$.

1. Factorise the bracket: $(2 + 5x) = 2\left(1 + \frac{5}{2}x\right)$, so $(2 + 5x)^7 = 2^7\left(1 + \frac{5}{2}x\right)^7$

   So the expansion $(2 + 5x)^7$ is the same as the expansion of $\left(1 + \frac{5}{2}x\right)^7$ multiplied by 2^7.

2. Find the coefficient of x^4 in the expansion of $\left(1 + \frac{5}{2}x\right)^7$.

   The term is $\binom{7}{4} \times \left(\frac{5}{2}x\right)^4 = \frac{7 \times 6 \times 5 \times 4}{1 \times 2 \times 3 \times 4} \times \frac{5^4}{2^4}x^4$

   $= 35 \times \frac{5^4}{2^4}x^4 = \frac{21875}{16}x^4$

   So the coefficient is $\frac{21875}{16}$

3. Multiply this by 2^7 to get the coefficient of x^4 in the original binomial:

   $2^7 \times \frac{21875}{16} = 175\,000$

You can find an **unknown** in a binomial expansion if you're given information about the coefficients:

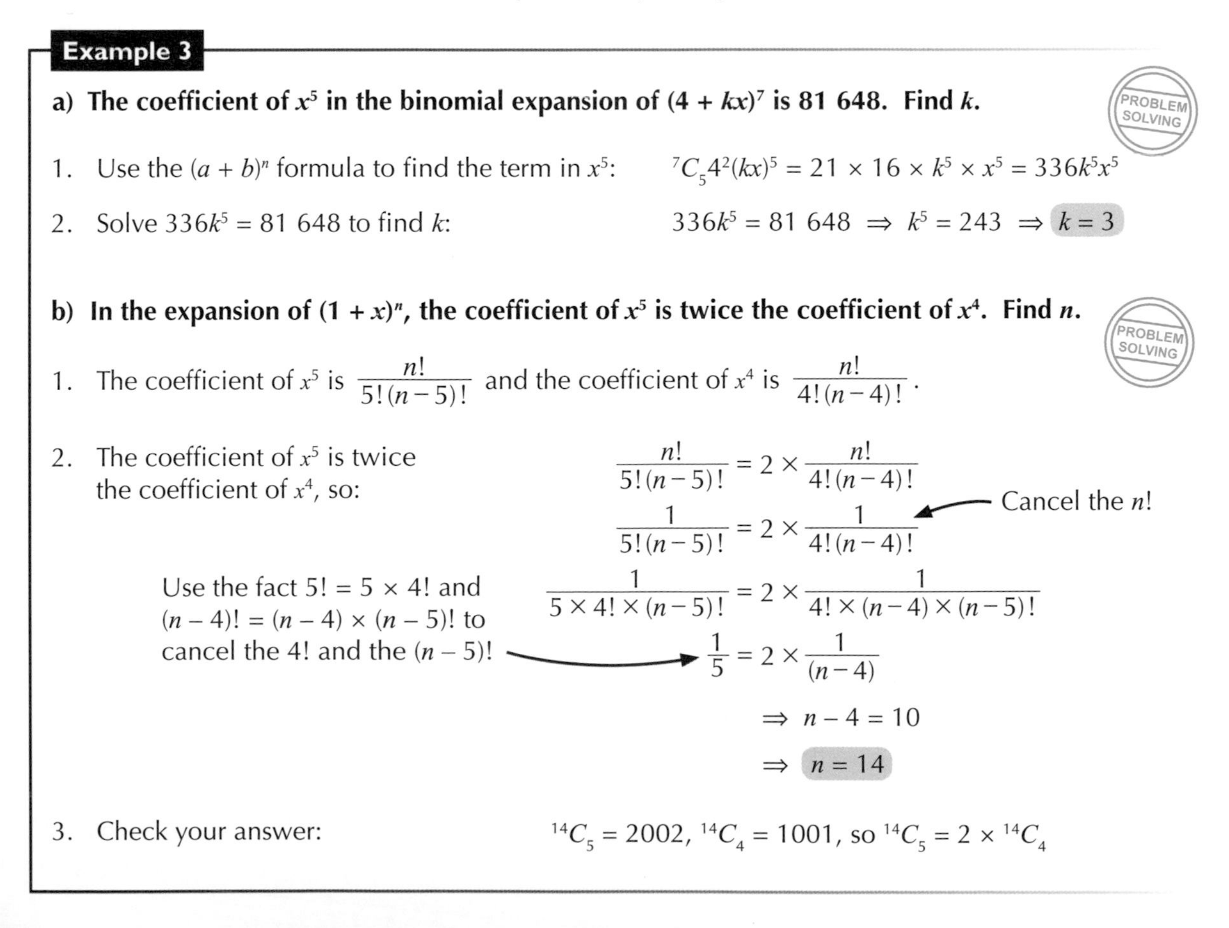

Example 3

PROBLEM SOLVING

a) The coefficient of x^5 in the binomial expansion of $(4 + kx)^7$ is 81 648. Find k.

1. Use the $(a + b)^n$ formula to find the term in x^5: ${}^7C_5 4^2(kx)^5 = 21 \times 16 \times k^5 \times x^5 = 336k^5x^5$

2. Solve $336k^5 = 81\,648$ to find k: $336k^5 = 81\,648 \Rightarrow k^5 = 243 \Rightarrow k = 3$

PROBLEM SOLVING

b) In the expansion of $(1 + x)^n$, the coefficient of x^5 is twice the coefficient of x^4. Find n.

1. The coefficient of x^5 is $\frac{n!}{5!(n-5)!}$ and the coefficient of x^4 is $\frac{n!}{4!(n-4)!}$.

2. The coefficient of x^5 is twice the coefficient of x^4, so:

$$\frac{n!}{5!(n-5)!} = 2 \times \frac{n!}{4!(n-4)!}$$

$$\frac{1}{5!(n-5)!} = 2 \times \frac{1}{4!(n-4)!}$$ ← Cancel the $n!$

Use the fact $5! = 5 \times 4!$ and $(n - 4)! = (n - 4) \times (n - 5)!$ to cancel the $4!$ and the $(n - 5)!$ →

$$\frac{1}{5 \times 4! \times (n-5)!} = 2 \times \frac{1}{4! \times (n-4) \times (n-5)!}$$

$$\frac{1}{5} = 2 \times \frac{1}{(n-4)}$$

$$\Rightarrow n - 4 = 10$$

$$\Rightarrow n = 14$$

3. Check your answer: ${}^{14}C_5 = 2002$, ${}^{14}C_4 = 1001$, so ${}^{14}C_5 = 2 \times {}^{14}C_4$

Exercise 7.1.3

Q1 Find the first 4 terms of the binomial expansion of $(3 + x)^6$.

Q2 Find the full expansion of:

a) $(2 + x)^4$ b) $(2 + 2x)^4$ c) $(2 - 2x)^4$ d) $\left(\frac{1}{2} + \frac{1}{2}x\right)^4$

Q3 In the expansion of $(1 + \lambda x)^8$, the coefficient of x^5 is 57 344.

PROBLEM SOLVING

a) Work out the value of λ. b) Find the first 3 terms of the expansion.

Q4 a) Find the first 5 terms in the expansion of $(2 + x)^8$.

b) Use this expansion to find an approximation for 2.01^8 to 5 d.p.

Q4b) Hint: This is the same method as the approximation example for $(1 + ax)^n$ expansions on page 126.

Q5 Find the first 4 terms in the expansion of $(3 + 5x)^7$.

Q6 Use the terms up to x^3 of the expansion of $(3 - 2x)^7$ to give an approximate value to 2 d.p. of:

a) 2.998^7 b) 2.8^7 c) 2.94^7 d) 3.002^7

Q7 Find the first 5 terms, in ascending order, of $(1 + x)(3 + 2x)^6$.

Q8 Find the coefficient of x^2 in the following binomial expansions:

PROBLEM SOLVING

a) $(4 + 2x)^5$ b) $(2 - 5x)^8$ c) $2x\left(1 + \frac{1}{2}x\right)^7$ d) $(5 + x)(3x - 1)^9$

Q9 Find the full expansion of:

a) $(1 + 3x)(2 - x)^5$ b) $(1 + 3x)^5(2 - x)$ c) $(1 + 3x)^2(2 - x)^4$ d) $(1 + 3x)^3(2 - x)^3$

Q10 The term in x^2 for the expansion of $(2 - 5x)^n$ is $150x^2$ when n is positive.

PROBLEM SOLVING

a) What is the value of n? b) What is the term in x^3?

Q11 In the expansion of $(a + 3x)^8$, the coefficient of x^2 is $\frac{32}{27}$ times bigger than the coefficient of x^5. What is the value of a?

PROBLEM SOLVING

Q12 In the expansion of $(1 + 2x)^5(3 - x)^4$, what is the coefficient of x^3?

PROBLEM SOLVING

Q13 In the expansion of $(1 + x)^n$, the coefficient of x^3 is 3 times larger than the coefficient of x^2.

PROBLEM SOLVING

a) Calculate the value of n.

b) If the coefficient of x^2 is $a \times$ (the coefficient of x), what is a?

Q14 In the expansion of $(2 + \mu x)^8$, where μ is a constant, the coefficient of x^2 is 87 808. What are the possible values of μ?

PROBLEM SOLVING

Q15 Find the coefficient of x^2 in the expansion of $[(x + 2)^3(x + 3)^2]^2$.

PROBLEM SOLVING

Review Exercise

Q1 Find the first 3 terms, in ascending powers of x, of the binomial expansion of:

a) $(1 + x)^{40}$ b) $(1 - x)^{20}$ c) $(1 + 3x)^{20}$ d) $(2 + 3x)^{10}$

Q2 a) Find the first 4 terms of the expansion of $\left(1 + \frac{x}{3}\right)^9$ in ascending powers of x, giving each term in its simplest form.

b) Use your expansion to estimate the value of $(1.003)^9$ to 6 decimal places.

Q3 a) Find the first 5 terms, in ascending powers of x, of the expansion of $(1 + ax)^8$, where a is a non-zero constant.

b) Given that the coefficient of x^2 in this expansion is double the coefficient of x^3, find the value of a and the coefficient of x.

PROBLEM SOLVING

Q4 Find the first 3 terms, in ascending powers of x, of the binomial expansion of $(4 - 5x)^7$. Give each term in its simplest form.

Q5 Find the coefficient of x^2 in the expansion of $(2 + 3x)^5$.

Q6 a) Find the first 5 terms, in ascending powers of x, of the expansion of $\left(3 + \frac{x}{4}\right)^{11}$. Give each term in its simplest form.

b) Use your expansion to find an estimate for the value of $(3.002)^{11}$ to 3 decimal places.

Q7 a) Find the first 3 terms, in ascending powers of x, of the binomial expansion of $(2 + kx)^{13}$, where k is a non-zero constant.

b) Given that the coefficient of x in this expansion is $\frac{1}{6}$ of the coefficient of x^2, find the value of k.

PROBLEM SOLVING

Q8 Use the terms up to x^3 of the binomial expansion of $(1 - 3x)^8$ to approximate the value of 0.97^8 to 3 decimal places.

Q9 a) Find the first four terms, in ascending powers of x, of the binomial expansion of $\left(\frac{x}{3} - 2\right)^6$.

b) Given that terms in x^3 or higher powers can be ignored, find an approximation of:

PROBLEM SOLVING

(i) $4x\left(\frac{x}{3} - 2\right)^6$ (ii) $(3 - x)\left(\frac{x}{3} - 2\right)^6$ (iii) $\left(\frac{x}{3} - 2\right)^6(x + 1)^{16}$

Q10 a) Find the first 5 terms, in ascending powers of x, of the expansion of $(1 + x)^{15}$.

b) Explain why this would not give an accurate estimate of the value of 2.01^{15}.

Q11 a) Approximate the value of 2.5^6 using the terms up to x^3 of the following expansions:

(i) $(2 + x)^6$ (ii) $(3 - x)^6$

b) Given that the actual value of 2.5^6 is 244.14 (2 d.p.), which expansion provides the best approximation?

Exam-Style Questions

Q1 a) Given that $\binom{37}{33} = \frac{37!}{33!k!}$, write down the value of k.

[1 mark]

b) In the expansion of $(1 - x)^{37}$, the coefficient of x^{33} is p. Find the value of p.

[2 marks]

Q2 Find the value of n, given that $\frac{(n-4)!}{34 \times 35} = 33!$

[2 marks]

Q3 a) Expand $(2 + 2x)^5$ in ascending powers of x, up to and including the term in x^3.

[4 marks]

b) Use your answer to part a) to estimate the value of 2.04^5, giving your answer correct to 3 decimal places.

[2 marks]

Q4 Expand $(x - x^{-1})^3$, fully simplifying each term in your answer.

[4 marks]

Q5 The coefficient of x^2 in the binomial expansion of $(1 - 3x)^n$ is 495. Find the value of n.

[4 marks]

Q6 a) Expand $(5 + 2x)^4$ in ascending powers of x.

[4 marks]

b) Hence or otherwise expand $(5 - 2x)^4$ in ascending powers of x.

[2 marks]

c) Hence show that $(5 + 2\sqrt{3})^4 + (5 - 2\sqrt{3})^4$ is an integer and state its value.

[3 marks]

Q7 a) Find the first three terms, in ascending powers of x, in the binomial expansion of $\left(1 - \frac{x}{2}\right)^8$. Give each term in its simplest form.

[3 marks]

The function f is defined by

$$f(x) = \left(1 - \frac{x}{2}\right)^8 (p + qx),$$

where p and q are constants.

b) In the binomial expansion of $f(x)$, the x term has coefficient –26 and the x^2 term has coefficient 50. Find the values of p and q.

[6 marks]

Chapter 8 Trigonometry

8.1 The Sine and Cosine Rules

In this section you'll see how SOHCAHTOA and the sine and cosine rules can be used to find the length of each side and the size of each angle in a triangle, as well as its area. You might have seen some of this at GCSE, so parts of this chapter will just be a recap.

Learning Objectives (Spec Ref 5.1):

- Find the values of sin, cos and tan of 30°, 60° and 45° without a calculator.
- Use the unit circle to find values of sin and cos.
- Know the sine and cosine rules and use them on any triangle.
- Find the area of a triangle using the formula $\frac{1}{2}ab \sin C$.

Prior Knowledge Check: You should be able to find lengths and angles in triangles from GCSE.

Trig values from triangles

You need to know the values of **sin**, **cos** and **tan** at 30°, 60° and 45°, and there are two **triangles** that can help you remember them. It may seem like a long-winded way of doing it, but once you know how to do it, you'll always be able to work them out — even without a calculator.

The idea is you draw the triangles below, putting in their angles and side lengths. Then you can use them to work out special trig values like **sin 45°** or **cos 60°** with exact values instead of the decimals given by calculators.

First, make sure you can remember SOHCAHTOA: $\sin = \frac{\text{opp}}{\text{hyp}}$ $\cos = \frac{\text{adj}}{\text{hyp}}$ $\tan = \frac{\text{opp}}{\text{adj}}$

These are the two triangles that you'll use:

Half an equilateral triangle with sides of length 2:

You can work out the height using Pythagoras' theorem: height = $\sqrt{2^2 - 1^2} = \sqrt{3}$.

Then you can use the triangle to work out sin, cos and tan of 30° and 60°.

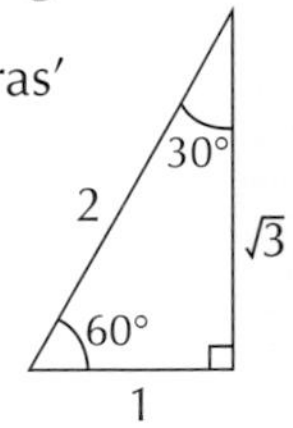

$\sin 30° = \frac{1}{2}$ $\sin 60° = \frac{\sqrt{3}}{2}$

$\cos 30° = \frac{\sqrt{3}}{2}$ $\cos 60° = \frac{1}{2}$

$\tan 30° = \frac{1}{\sqrt{3}}$ $\tan 60° = \sqrt{3}$

Right-angled triangle with two sides of length 1:

You can work out the hypotenuse using Pythagoras' theorem: hypotenuse = $\sqrt{1^2 + 1^2} = \sqrt{2}$.

Then you can use the triangle to work out sin, cos and tan of 45°.

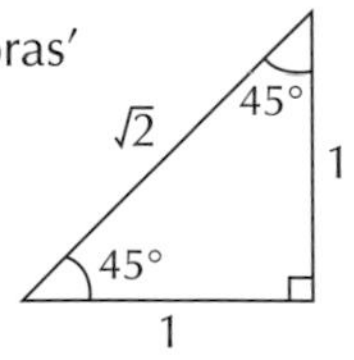

$\sin 45° = \frac{1}{\sqrt{2}}$

$\cos 45° = \frac{1}{\sqrt{2}}$

$\tan 45° = 1$

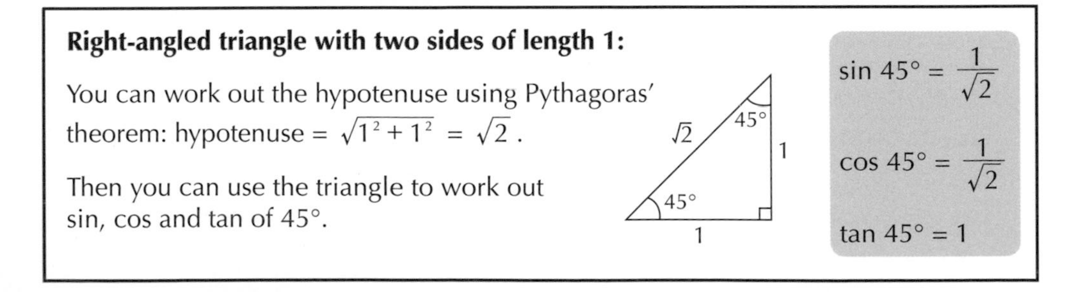

Trig values from the unit circle

You can also find trig values from the **unit circle** — a circle with **radius 1**, centred on the **origin**. Take a point on the unit circle and make a right-angled triangle:

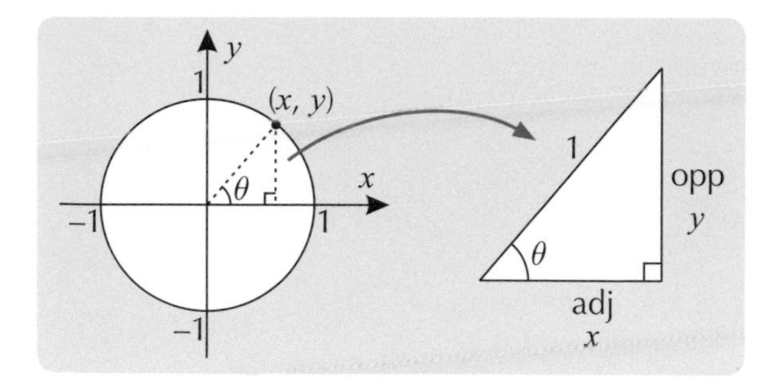

The hypotenuse is always the radius — so it's always 1. From this you know that:

$$\sin\theta = \frac{\text{opp}}{1} = \text{opp} = y$$

$$\cos\theta = \frac{\text{adj}}{1} = \text{adj} = x$$

So on the unit circle, the y-coordinate is **sin θ** and the x-coordinate is **cos θ**.

For any point on the unit circle, the coordinates are $(\cos\theta, \sin\theta)$, where θ is the angle measured from the **positive** x-axis in an **anticlockwise** direction.

This is true for **all** values of θ, including values greater than 90°.

You can use the unit circle to easily find sin and cos of **0°** and **90°** — they're just **points** on the **axes**. E.g. at point (0, 1), cos 90° = 0 and sin 90° = 1.

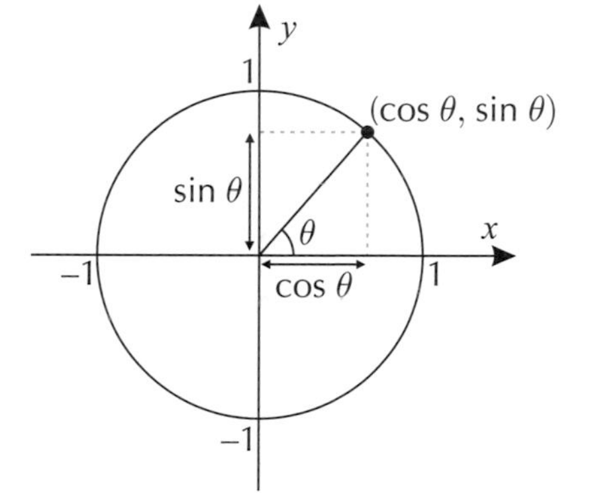

Example 1

The coordinates of a point on the unit circle, given to 3 s.f., are shown on the diagram to the right. Find θ to the nearest degree.

1. The point is on the unit circle, so you know that the coordinates are $(\cos\theta, \sin\theta)$.
   So $\cos\theta = 0.788$ and $\sin\theta = 0.616$
2. You only need one of these to find the value of θ.
   $\cos\theta = 0.788$
   $\Rightarrow \theta = \cos^{-1}(0.788) = 38°$ (to the nearest degree)

Example 2

Find the coordinates of the point A on the unit circle, shown on the diagram to the right. Give your answer to 2 d.p.

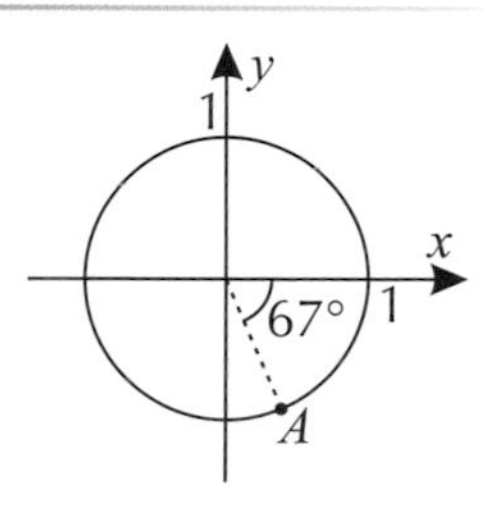

1. You need the angle from the positive x-axis in an **anticlockwise** direction.
   360° – 67° = 293°
2. Work out the x and y-coordinates of A.
   cos 293° = 0.3907...,
   and sin 293° = –0.9205...
   So the coordinates of A are (0.39, –0.92) to 2 d.p.

The trig rules

There are three useful formulas you need to know for working out information about a triangle. There are **two** for finding the **angles** and **sides** (called the **sine rule** and **cosine rule**) and one for finding the **area**. These rules work on **any** triangle, not just right-angled triangles.

To decide which rule to use, look at what you know about the triangle.

The Sine Rule

The sine rule is given by the formula: $\frac{a}{\sin A} = \frac{b}{\sin B} = \frac{c}{\sin C}$

where a, b and c are the lengths of the sides, and A, B and C are the angles opposite the sides with the same letters (so angle C is opposite side c).

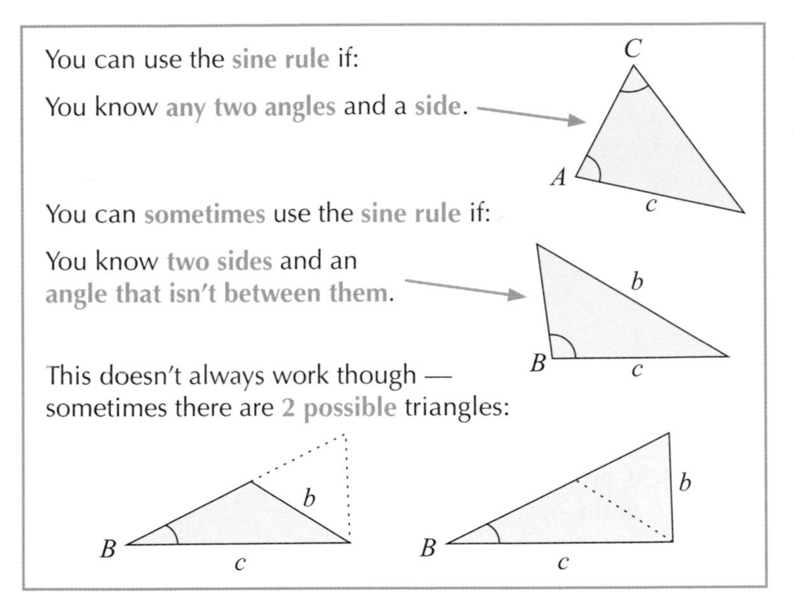

You can use any **two** bits of the sine rule to make a **normal equation** with just one '=' sign.

The sine rule also works if you flip all the fractions **upside down**:

$$\frac{\sin A}{a} = \frac{\sin B}{b} = \frac{\sin C}{c}$$

Tip: Remember, if you know two angles you can work out the third by subtracting them from 180°.

Example 1

Find the missing sides and angles for $\triangle ABC$, in which $A = 40°$, $a = 27$ m and $B = 73°$.

1. Sketch the triangle to help you decide which rule(s) you need.

   You have 2 angles and a side, so can use the **sine rule**.

2. Find angle C (using the fact that the angles in a triangle add up to 180°).

   $\angle C = 180° - 73° - 40° = 67°$

3. Use the sine rule to find the other sides one at a time:

$$\frac{a}{\sin A} = \frac{b}{\sin B} \Rightarrow \frac{27}{\sin 40°} = \frac{b}{\sin 73°}$$

$$\Rightarrow b = \frac{27 \times \sin 73°}{\sin 40°}$$

$$= 40.2 \text{ m (1 d.p.)}$$

$$\frac{c}{\sin C} = \frac{a}{\sin A} \Rightarrow \frac{c}{\sin 67°} = \frac{27}{\sin 40°}$$

$$\Rightarrow c = \frac{27 \times \sin 67°}{\sin 40°}$$

$$= 38.7 \text{ m (1 d.p.)}$$

Exercise 8.1.1

Give all answers to 3 significant figures unless otherwise stated.

Q1 Find the coordinates of points P, Q and R on the unit circle.

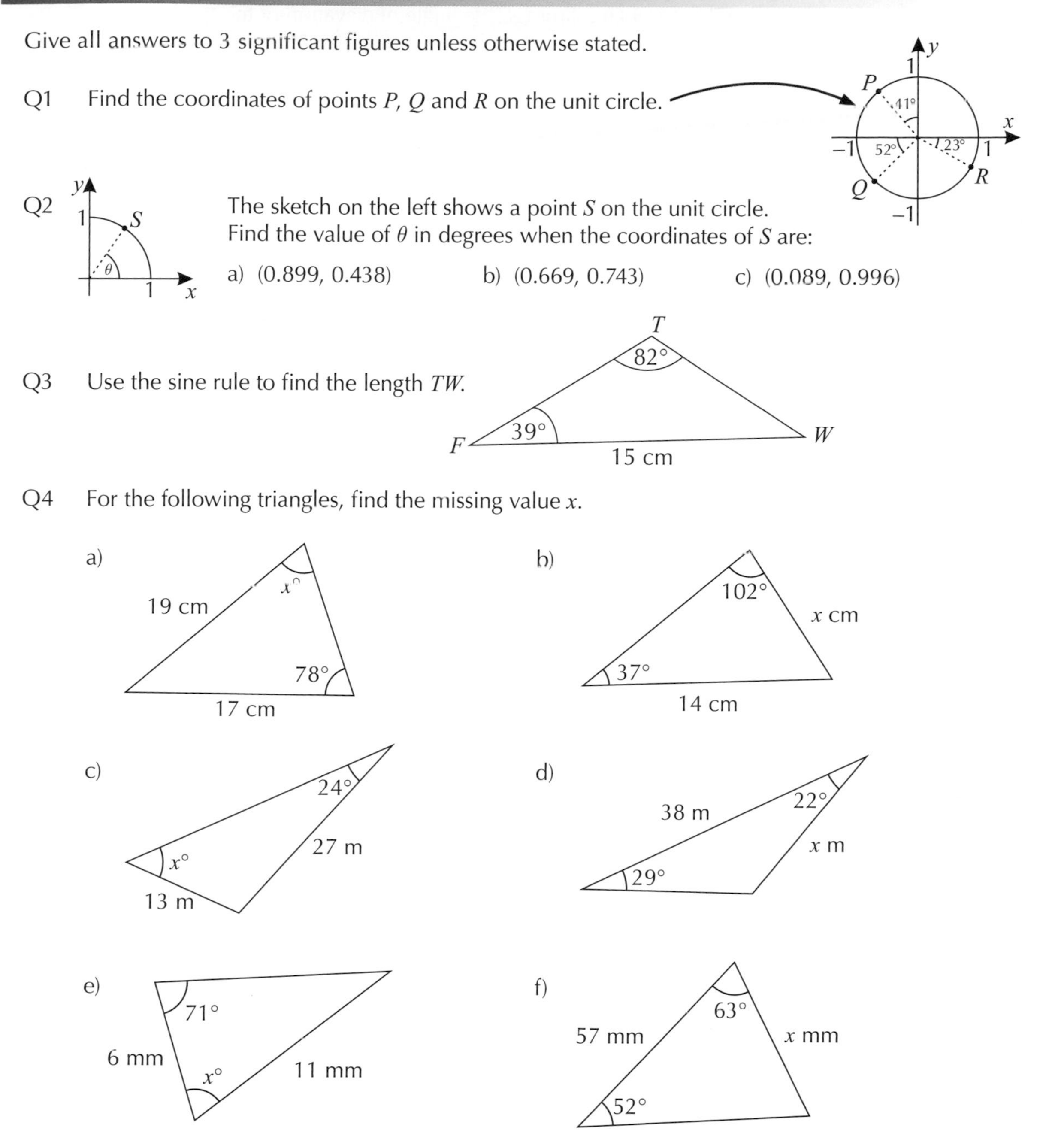

Q2 The sketch on the left shows a point S on the unit circle. Find the value of θ in degrees when the coordinates of S are:

a) (0.899, 0.438) b) (0.669, 0.743) c) (0.089, 0.996)

Q3 Use the sine rule to find the length TW.

Q4 For the following triangles, find the missing value x.

Q5 In triangle PQR: $PR = 48$ m, angle $P = 38°$ and angle $R = 43°$. Find the length PQ.

Q6 In triangle ABC: $AB = 17$ cm, $AC = 14$ cm and angle $C = 67°$. Find the angle A.

PROBLEM SOLVING

Q7 A building has a wall that slopes inward at an angle of 81° to the horizontal ground. A safety inspector calculates that a ladder leaning against the outside of the wall must be at an angle of 78° to the ground. What distance up the wall would a 5 m ladder safely reach?

MODELLING

Q8 Two points, A and B, are both at sea level and on opposite sides of a mountain. The distance between them is 5 km. From A, the angle of elevation of the top of the mountain (M) is 21°, and from B, the angle of elevation is 17°.

a) Find the distance BM.

b) Hence find the height of the mountain to the nearest metre.

The Cosine Rule

The cosine rule is given by the formula: $a^2 = b^2 + c^2 - 2bc \cos A$

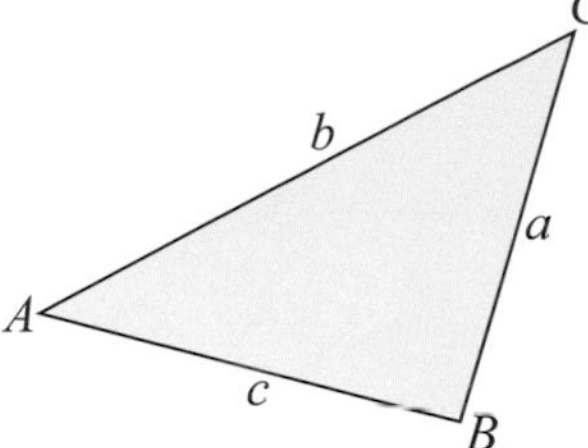

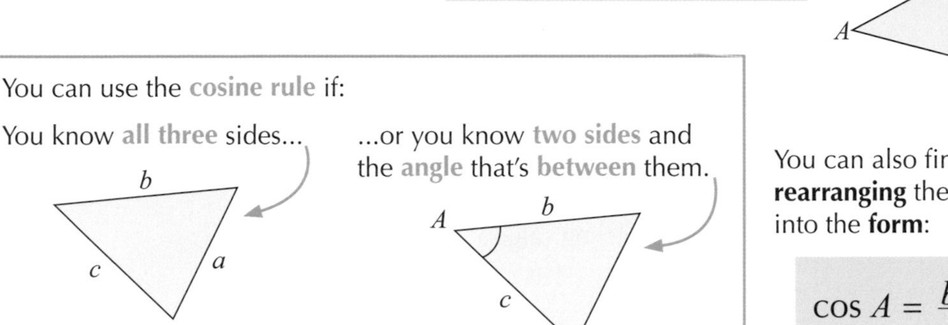

You can also find an angle by **rearranging** the cosine rule into the **form**:

$$\cos A = \frac{b^2 + c^2 - a^2}{2bc}$$

When you're given **three sides**, it's best to find the **largest angle** first, since you don't know whether it'll be **acute** or **obtuse**. The **cosine rule** will always give the right answer for this (the **sine rule** will always give you an **acute** angle).

Example 2

Find the values of X, Y and Z in the triangle on the right.

6.5 cm, 6 cm, 10 cm; angles $Y°$, $Z°$, $X°$

1. You've been given all three sides but none of the angles, so start by using the **cosine rule** to find the largest angle, Y (you'll have to rearrange the formula a bit first).

$$a^2 = b^2 + c^2 - 2bc \cos A$$
$$\Rightarrow \cos A = \frac{b^2 + c^2 - a^2}{2bc}$$
$$\Rightarrow \cos Y = \frac{6^2 + 6.5^2 - 10^2}{2 \times 6 \times 6.5} = -0.278...$$
$$\Rightarrow Y = 106.191... = 106.2° \text{ (1 d.p.)}$$

2. Use the cosine rule **again** to find the value of another angle. It doesn't matter which one you go for (using Z here).

$$\cos Z = \frac{10^2 + 6.5^2 - 6^2}{2 \times 10 \times 6.5} = 0.817...$$
$$\Rightarrow Z = 35.183...° = 35.2° \text{ (1 d.p.)}$$

3. Now that you have two of the angles, you can find the other by subtracting them from 180°:

$$X = 180° - 106.191...° - 35.183...°$$
$$= 38.624...° = 38.6° \text{ (1 d.p.)}$$

Exercise 8.1.2

Give all answers to 3 significant figures unless otherwise stated.

Q1 Use the cosine rule to find the length QR.

Q2 Find the size of the angles D, E and F.

Q3 For the following triangles, find the missing value x.

a)

b)

c)

d)

e)

f)

Q4 In triangle JKL: $JL = 24$ cm, $KL = 29$ cm and angle $L = 62°$. Find the length JK.

Q5 In triangle ABC: $AB = 32$ cm, $AC = 28$ cm and angle $A = 48°$. Find the length BC.

Q6 In triangle DEF: $DE = 8$ cm, $EF = 11$ cm and $DF = 16$ cm. Find the smallest angle.

Q7 In triangle PQR: $PQ = 7.6$ mm, $QR = 6.8$ mm and $PR = 5.9$ mm. Find the largest angle.

Q6-7 Hint: Sketching the triangle will make it easier to see which angle you need to find — the smallest angle is opposite the shortest side and the largest angle is opposite the longest side.

Q8 The vertices of triangle XYZ have coordinates X: (–2, 2), Y: (5, 8) and Z: (3, –2). Find the angle XYZ.

Q9 A triangle ABC has sides $AB = (x - 1)$, $BC = 2x$ and angle $B = 60°$.

a) Given that side $AC = \sqrt{13}$, find the value of x.

b) Find the size of the smallest angle in the triangle ABC.

Area of any triangle

You can use trigonometry to find the area of any triangle, using the formula: $\text{Area} = \frac{1}{2}ab\sin C$

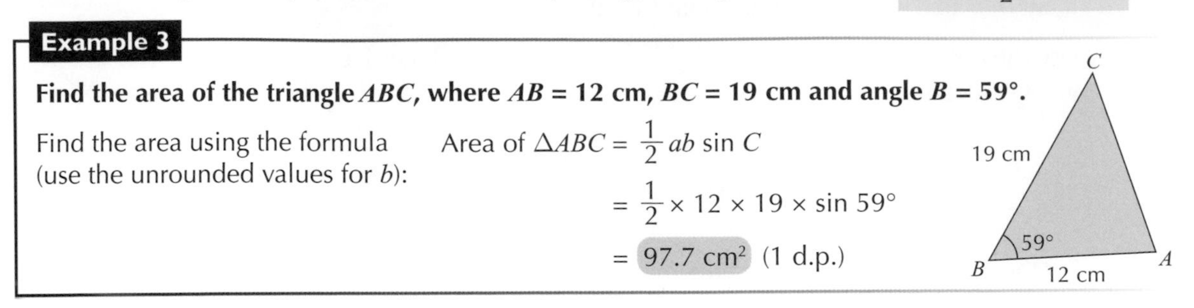

Example 3

Find the area of the triangle ABC, where AB = 12 cm, BC = 19 cm and angle B = 59°.

Find the area using the formula (use the unrounded values for b):

$$\text{Area of } \Delta ABC = \frac{1}{2}ab\sin C$$
$$= \frac{1}{2} \times 12 \times 19 \times \sin 59°$$
$$= 97.7 \text{ cm}^2 \text{ (1 d.p.)}$$

Solving triangle problems

You can find the areas of more **complicated** shapes by turning them into **multiple triangles** stuck together, then using the **sine** and **cosine rules** on each individual triangle.
You might need to use more than one rule, or use a rule multiple times.

This method can be used for working out angles and sides in real-life problems, such as calculating distances travelled or areas covered. Sometimes you'll see a problem that uses **bearings**.

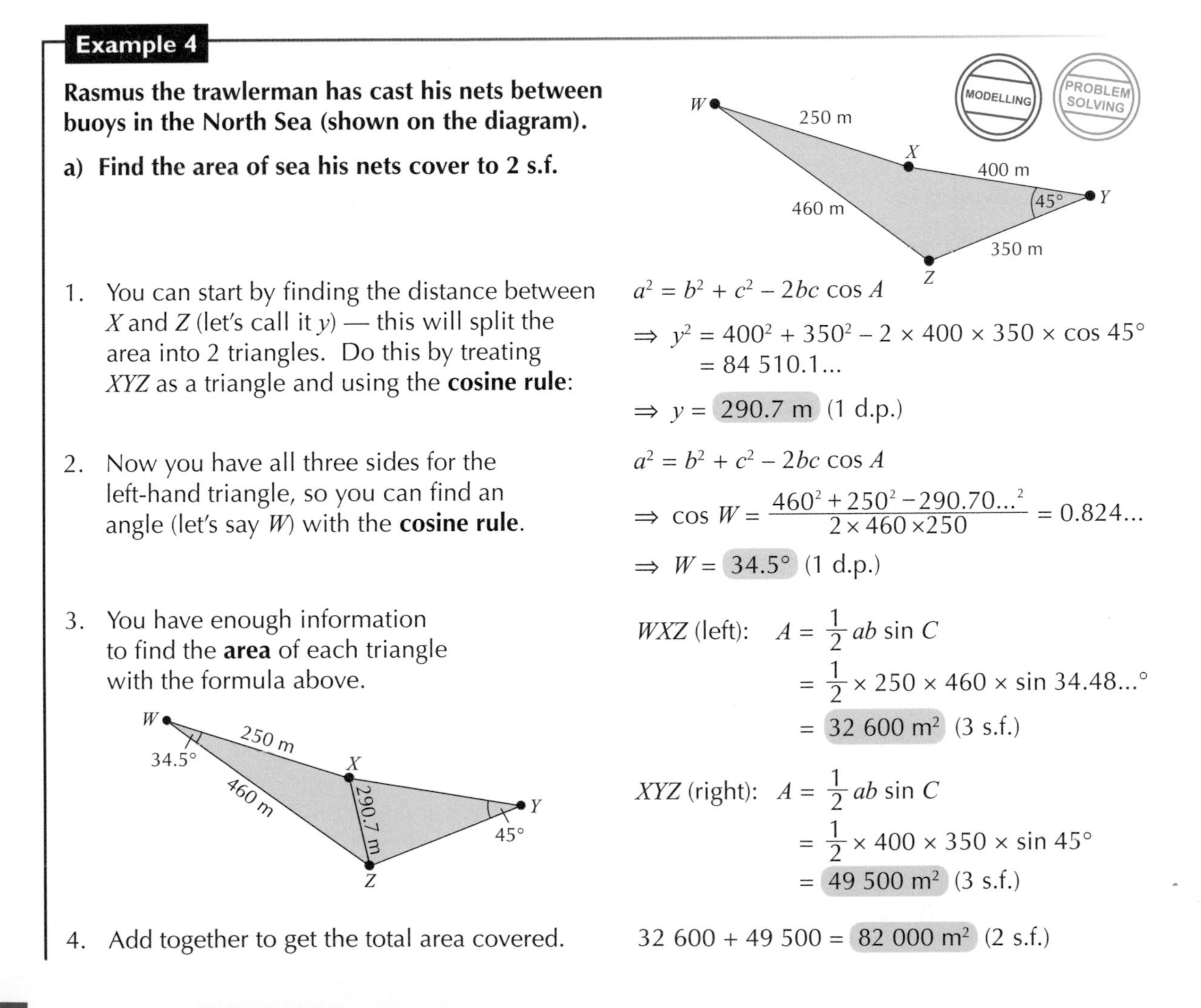

Example 4

Rasmus the trawlerman has cast his nets between buoys in the North Sea (shown on the diagram).

a) Find the area of sea his nets cover to 2 s.f.

1. You can start by finding the distance between X and Z (let's call it y) — this will split the area into 2 triangles. Do this by treating XYZ as a triangle and using the **cosine rule**:

$$a^2 = b^2 + c^2 - 2bc\cos A$$
$$\Rightarrow y^2 = 400^2 + 350^2 - 2 \times 400 \times 350 \times \cos 45° = 84\ 510.1...$$
$$\Rightarrow y = 290.7 \text{ m (1 d.p.)}$$

2. Now you have all three sides for the left-hand triangle, so you can find an angle (let's say W) with the **cosine rule**.

$$a^2 = b^2 + c^2 - 2bc\cos A$$
$$\Rightarrow \cos W = \frac{460^2 + 250^2 - 290.70...^2}{2 \times 460 \times 250} = 0.824...$$
$$\Rightarrow W = 34.5° \text{ (1 d.p.)}$$

3. You have enough information to find the **area** of each triangle with the formula above.

WXZ (left): $A = \frac{1}{2}ab\sin C$
$$= \frac{1}{2} \times 250 \times 460 \times \sin 34.48...°$$
$$= 32\ 600 \text{ m}^2 \text{ (3 s.f.)}$$

XYZ (right): $A = \frac{1}{2}ab\sin C$
$$= \frac{1}{2} \times 400 \times 350 \times \sin 45°$$
$$= 49\ 500 \text{ m}^2 \text{ (3 s.f.)}$$

4. Add together to get the total area covered.

$$32\ 600 + 49\ 500 = 82\ 000 \text{ m}^2 \text{ (2 s.f.)}$$

b) If X is on a bearing of 100° from W, on what bearing does Rasmus have to sail to get from X to Y (to 3 s.f.)?

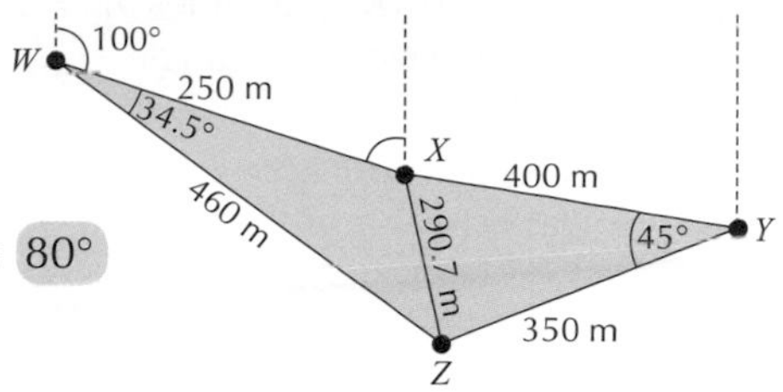

1. To find the bearing, find all the other angles round X. Start with the unknown that is marked.

$180° - 100° = 80°$

2. To find the angle $\angle WXZ$, use the cosine rule on the left-hand triangle:

$a^2 = b^2 + c^2 - 2bc \cos A$

$\Rightarrow \cos \angle WXZ = \dfrac{250^2 + 290.70...^2 - 460^2}{2 \times 250 \times 290.70...}$

$\Rightarrow \angle WXZ = \cos^{-1}(-0.414...) = 116.38°$ (2 d.p.)

3. Then do the same for angle $\angle YXZ$, using the right-hand triangle:

$\Rightarrow \cos \angle YXZ = \dfrac{400^2 + 290.70...^2 - 350^2}{2 \times 400 \times 290.70...}$

$\Rightarrow \cos \angle YXZ = 0.524...$

$\Rightarrow \angle YXZ = 58.36°$ (2 d.p.)

4. Subtract all the angles from 360° to find the bearing Rasmus should sail on to get from X to Y:

$360° - 80° - 116.38° - 58.36° = 105.26°$

$= 105°$ (3 s.f.)

Exercise 8.1.3

Q1 Find the area of the following triangles, giving your answer to 3 significant figures.

a)

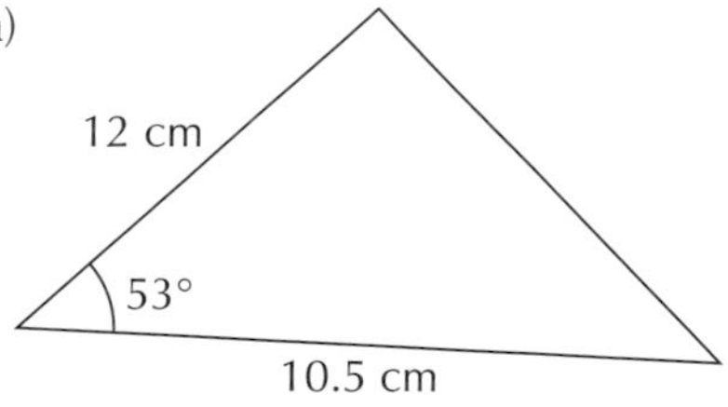

b)

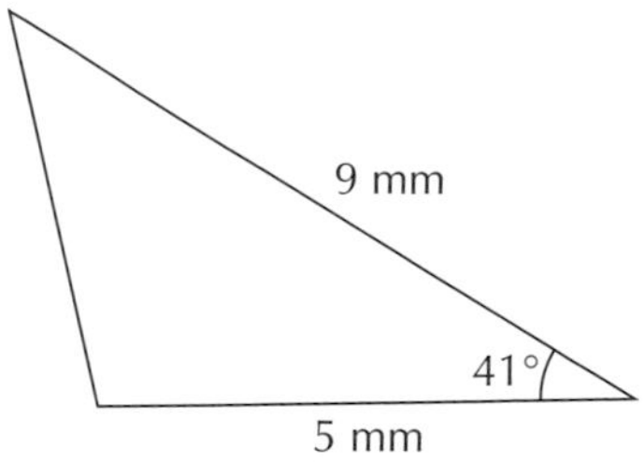

c)

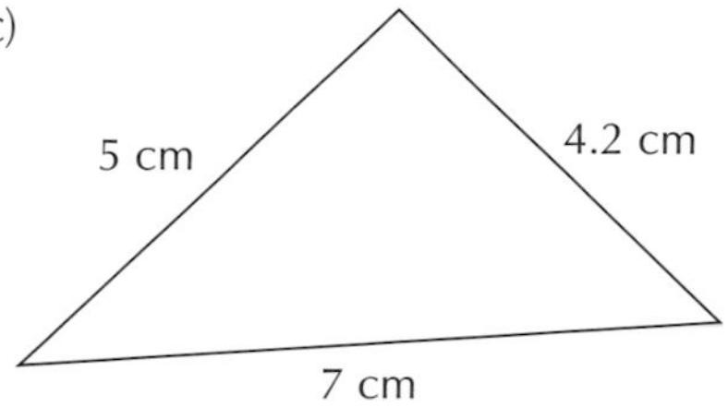

d)

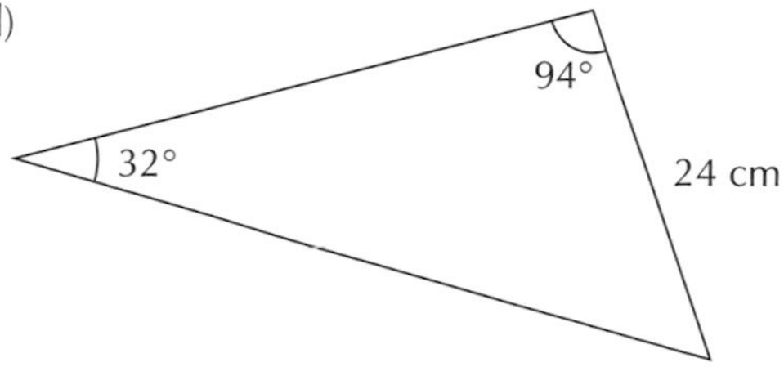

Q2 In triangle PQR: $PQ = 4$ cm, $QR = 7$ cm and angle $Q = 49°$. Find the area of triangle PQR.

Q3 A gardener places three pieces of rope, with lengths 1.9 m, 2.7 m and 2.9 m, end to end to rope off a triangle of land for a vegetable patch. What is the total area of land roped off?

MODELLING

Q4 A triangle JKL has sides $JL = 12$ cm and $KL = 8$ cm, and an area of 30 cm^2. Find the size of the angle JLK, giving your answer to 1 decimal place.

Q5 A triangle ABC has a side $AC = 14$ cm and angles $\angle BAC = 35°$ and $\angle ABC = 52°$. Find the area of the triangle ABC.

Q6 In ΔDEF: $DE = 7$ mm, $DF = 13$ mm and $EF = 16$ mm. Find the area of ΔDEF.

Q7 A ship sails 8 km on a bearing of 070°, and then changes direction to sail 10 km on a bearing of 030°.

a) Draw a diagram to represent the situation.

b) What is the ship's distance from its starting position?

c) On what bearing must it now sail to return to its starting position?

Q7 Hint: Bearings are measured clockwise from the vertical (North).

Q8 Find the area of the quadrilateral $ABCD$ shown on the right.

Q9 a) Find the length of the line QS in the shape below.

b) Given that QST is a straight line, find the total area of the shape.

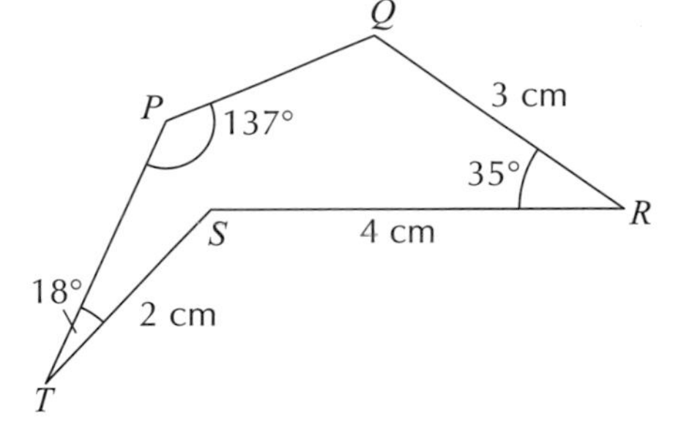

Q10 Find the areas of the following shapes.

a)

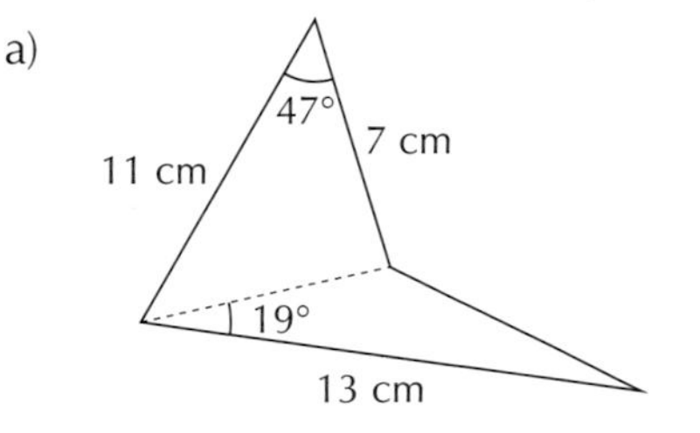

b)

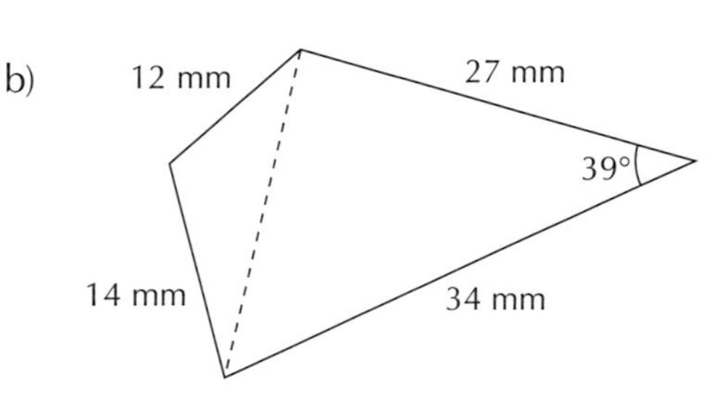

Q11 A hiker walks 3.2 km on a bearing of 110°, then turns and walks p km along a bearing of 67°. Then, they turn again and walk q km on a bearing of 273°, finishing the walk at their starting position.

a) Draw a diagram to represent this situation.

b) Find the three angles in the triangle formed by this walk.

c) Calculate the total length of the walk to 3 significant figures.

Q12 The triangle on the right has an area of 220 cm^2. Find the perimeter of the triangle.

8.2 Trig Identities

Trig identities are really useful for simplifying expressions, and they can make equations easier to solve by replacing one term with another.

Learning Objectives (Spec Ref 5.3):

- Know the trig identities $\tan\theta \equiv \frac{\sin\theta}{\cos\theta}$ and $\sin^2\theta + \cos^2\theta \equiv 1$.
- Use trig identities to prove other relations.
- Use trig identities to find the exact values of an expression.

Prior Knowledge Check: Be familiar with identity notation. See p.6.

Trig identities

There are **two trig identities** you need to know — they're really useful, and also fairly straightforward to find using results you already know.

You know that $\sin\theta = \frac{\text{opp}}{\text{hyp}}$, $\cos\theta = \frac{\text{adj}}{\text{hyp}}$ and $\tan\theta = \frac{\text{opp}}{\text{adj}}$ (see p.134).

So $\sin\theta \div \cos\theta = \frac{\text{opp}}{\text{hyp}} \div \frac{\text{adj}}{\text{hyp}} = \frac{\text{opp}}{\cancel{\text{hyp}}} \times \frac{\cancel{\text{hyp}}}{\text{adj}} = \frac{\text{opp}}{\text{adj}} = \tan\theta$.

So for all values of θ: $\tan\theta \equiv \frac{\sin\theta}{\cos\theta}$

Tip: The '$\equiv$' means that the relation is true for all values of θ.

On page 135 you saw that the coordinates (x, y) of a point on the **unit circle** can be written as $(\cos\theta, \sin\theta)$ — i.e. $x = \cos\theta$ and $y = \sin\theta$.

The equation of the unit circle is $x^2 + y^2 = 1$. (See p.106.)
Substituting in $\cos\theta$ and $\sin\theta$ gives the equation $\cos^2\theta + \sin^2\theta = 1$.

This is true for all values of θ, so you can write it as an **identity**: $\sin^2\theta + \cos^2\theta \equiv 1$

This can be rearranged into $\sin^2\theta \equiv 1 - \cos^2\theta$ or $\cos^2\theta \equiv 1 - \sin^2\theta$.

You can use these identities to **prove** that two expressions are equivalent, although the substitutions aren't always easy to spot. Look out for things like differences of two squares, or 1's that can be replaced by $\sin^2\theta + \cos^2\theta$.

Tip: $\sin^2\theta$ is another way of writing $(\sin\theta)^2$.

Example 1

Show that $\frac{\cos^2\theta}{1+\sin\theta} \equiv 1 - \sin\theta$.

1. Start with the left-hand side of the identity. You know there's a trig identity for $\cos^2\theta$, so start by replacing it in the fraction:

$$\frac{\cos^2\theta}{1+\sin\theta} \equiv \frac{1-\sin^2\theta}{1+\sin\theta}$$

2. The next step isn't quite as obvious, but if you look at the top of the fraction it's a difference of two squares (see p.13).

$$\frac{1-\sin^2\theta}{1+\sin\theta} \equiv \frac{(1+\sin\theta)(1-\sin\theta)}{1+\sin\theta}$$

3. Now you can just cancel the $1 + \sin\theta$ from the top and bottom of the fraction, and you get the answer you were looking for:

$$\frac{(1+\sin\theta)(1-\sin\theta)}{1+\sin\theta} \equiv 1 - \sin\theta$$

This is the right-hand side.

Example 2

Find the exact value of $\sin\theta$ if $\cos\theta = \frac{2}{3}$, given that θ is an acute angle.

1. Rearrange the identity $\sin^2\theta + \cos^2\theta \equiv 1$ to find $\sin\theta$.

$$\sin^2\theta + \cos^2\theta$$
$$\Rightarrow \sin^2\theta \equiv 1 - \cos^2\theta$$
$$\Rightarrow \sin\theta \equiv \sqrt{1-\cos^2\theta}$$

2. Then put in the value of $\cos\theta$ and take the square root of each side.

$$\sin\theta = \sqrt{1-\left(\frac{2}{3}\right)^2} = \sqrt{\frac{5}{9}} = \frac{\sqrt{5}}{3}$$

Tip: θ is acute here, which means that $\sin\theta$ is positive, so you can ignore the negative square root.

Exercise 8.2.1

Q1 Use the identity $\tan\theta \equiv \frac{\sin\theta}{\cos\theta}$ to show that $\frac{\sin\theta}{\tan\theta} - \cos\theta \equiv 0$.

Q2 Use the identity $\sin^2\theta + \cos^2\theta \equiv 1$ to show that $\cos^2\theta \equiv (1 - \sin\theta)(1 + \sin\theta)$.

Q3 Given that x is acute, find the exact value of $\cos x$ if $\sin x = \frac{1}{2}$.

Q4 Given that x is acute, find the exact value of $\tan x$ if $\sin^2 x = \frac{3}{4}$.

Q3-4 Hint: You're told that x is acute, so ignore the negative roots.

Q5 Show that $4\sin^2 x - 3\cos x + 1 \equiv 5 - 3\cos x - 4\cos^2 x$.

Q5 Hint: If you're not told which identity to use, just play around with the ones you know until something works.

Q6 Show that $(\tan x + 1)(\tan x - 1) \equiv \frac{1}{\cos^2 x} - 2$.

Q7 A student is asked to solve the equation $\sin\theta = \frac{1}{2}\tan\theta$, where $0° \leq \theta \leq 90°$. Their working is shown below:

$$\sin\theta = \frac{1}{2}\tan\theta \Rightarrow \sin\theta = \frac{1}{2} \times \frac{\sin\theta}{\cos\theta}$$
$$\Rightarrow \cos\theta\sin\theta = \frac{1}{2}\sin\theta \Rightarrow \cos\theta = \frac{1}{2} \Rightarrow \theta = 60°.$$

Find the error they made and explain how this has resulted in an incomplete solution.

Q8 Show that $\tan x + \frac{1}{\tan x} \equiv \frac{1}{\sin x\cos x}$.

Q8 Hint: $\frac{1}{\tan x} \equiv \frac{\cos x}{\sin x}$

Q9 Show that $2\cos^2 x + 5\sin x + 1 \equiv (3 - \sin x)(2\sin x + 1)$.

Q10 Show that $4 + \sin x - 6\cos^2 x \equiv (2\sin x - 1)(3\sin x + 2)$.

Q11 Show that $\sin^2 x\cos^2 y - \cos^2 x\sin^2 y \equiv \cos^2 y - \cos^2 x$.

Q12 Show that $\frac{\tan^2 x + 1}{\tan^2 x} \equiv \frac{1}{\sin^2 x}$.

Q13 Show that $\frac{\sin^4 x - \cos^4 x}{\sin^2 x - \cos^2 x} \equiv 1$.

PROBLEM SOLVING

Q14 Use the identity $\sin^2\theta + \cos^2\theta \equiv 1$ to prove Pythagoras' theorem.

PROBLEM SOLVING

8.3 Trig Functions

Being able to sketch the graphs of trig functions and their transformations is really useful — it'll come in handy later in the chapter when you have to solve equations within a given interval.

Learning Objectives (Spec Ref 5.2):

- Sketch the graphs of $\sin x$, $\cos x$ and $\tan x$.
- Sketch the common transformations of the graphs of $\sin x$, $\cos x$ and $\tan x$.

Prior Knowledge Check: You should be able to sketch graphs from their key features from GCSE.

Graphs of trig functions

You should be able to draw the graphs of **sin *x***, **cos *x*** and **tan *x*** without looking them up — including all the important points, like where they cross the **axes** and their **maximum** and **minimum** points.

The graphs of $\sin x$ and $\cos x$ are the **same shape** but shifted **90°** along the ***x*-axis**. This makes them easier to remember, but make sure you don't get them mixed up.

sin *x*

- The graph of $y = \sin x$ is **periodic** — it repeats itself every 360°. So $\sin x = \sin (x + 360°) = \sin (x + 720°) = \sin (x + 360n°)$, where n is an integer.
- It bounces between $y = -1$ and $y = 1$, and it can **never** have a value outside this range.
- It goes through the **origin** (as $\sin 0° = 0$) and then crosses the x-axis every **180°**.
- $\sin (-x) = -\sin x$. The graph has **rotational symmetry** about the origin, so you could rotate it 180° about (0, 0) and it would look the same.

The graph of $y = \sin x$ looks like this:

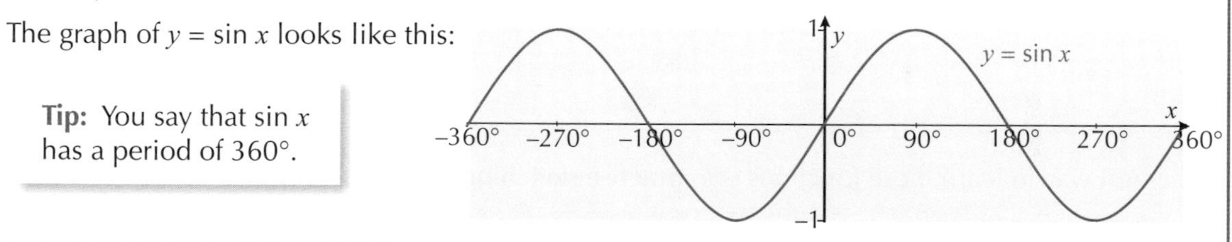

Tip: You say that $\sin x$ has a period of 360°.

cos *x*

- The graph of $y = \cos x$ is also **periodic** with period 360°. $\cos x = \cos (x + 360°) = \cos (x + 720°) = \cos (x + 360n°)$, where n is an integer.
- It also bounces between $y = -1$ and $y = 1$, and it can **never** have a value outside this range.
- It crosses the y-axis at $y = 1$ (as $\cos 0° = 1$) and the x-axis at **±90°**, **±270°** etc.
- $\cos (-x) = \cos x$. The graph is **symmetrical** about the y-axis, so you could reflect it in the ***y*-axis** and it would look the same.

The graph of $y = \cos x$ looks like this:

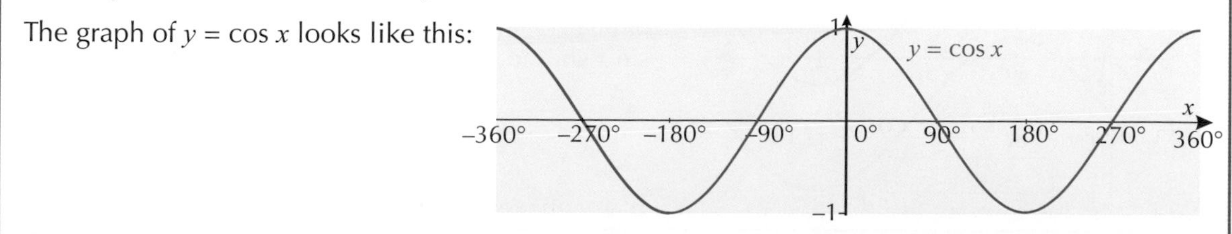

tan x

- The graph of $y = \tan x$ is also **periodic**, but this time it repeats itself every 180°. So $\tan x = \tan (x + 180°) = \tan (x + 360°) = \tan (x + 180n°)$, where n is an integer.
- It takes values between $-\infty$ and ∞ in each **180° interval**.
- It goes through the **origin** (as $\tan 0° = 0$).
- It's **undefined** at ±90°, ±270°, ±450°... — at these points it **jumps** from ∞ to $-\infty$ or vice versa.
- $\tan (-x) = -\tan x$. The graph has **rotational symmetry** about the origin, so you could rotate it 180° about (0, 0) and it would look the same.

The graph of $y = \tan x$ looks like this:

$y = \tan x$ is undefined at certain points because you're **dividing by zero**.

Remember from p.134 that $\tan x = \frac{\sin x}{\cos x}$, so when **$\cos x = 0$**, $\tan x$ is **undefined** — $\cos x = 0$ when $x = 90°$, 270° etc.

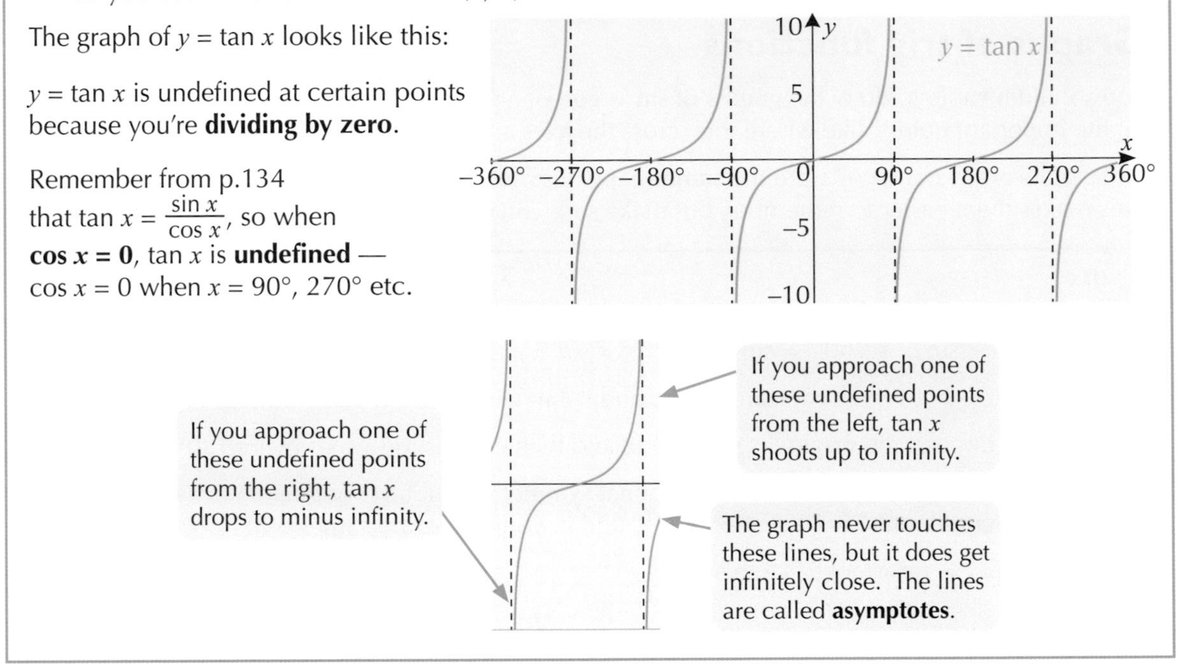

The best way to learn these functions is to **practise sketching them** and marking on the **key points**.

Transformations

You came across different types of **transformations** on pages 99-103.
A **translation** is a horizontal or vertical **shift** that doesn't change the shape of the graph.
A **stretch** is exactly what it says — a horizontal or vertical **stretch** (or **squash**) of the graph.
You need to be able to apply these types of transformation to **trig functions**.

1. A translation along the y-axis: $y = \sin (x) + c$

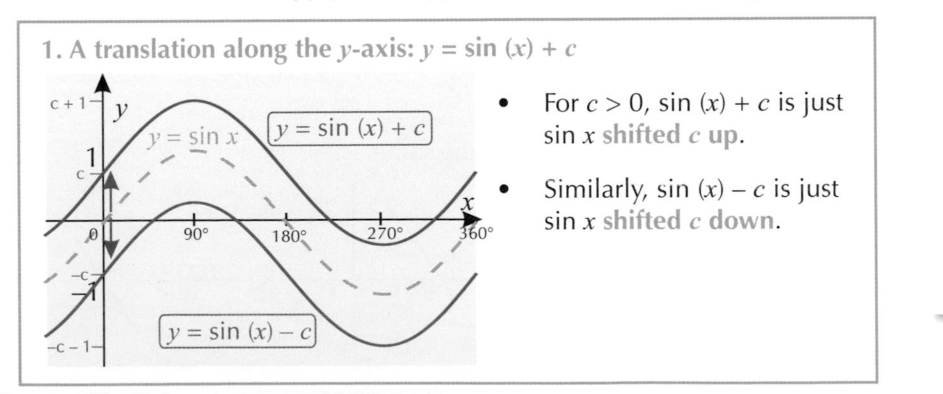

- For $c > 0$, $\sin (x) + c$ is just $\sin x$ **shifted c up**.
- Similarly, $\sin (x) - c$ is just $\sin x$ **shifted c down**.

Tip: The same transformations will apply to the graphs of $y = \cos x$ and $y = \tan x$ as well as $y = \sin x$.

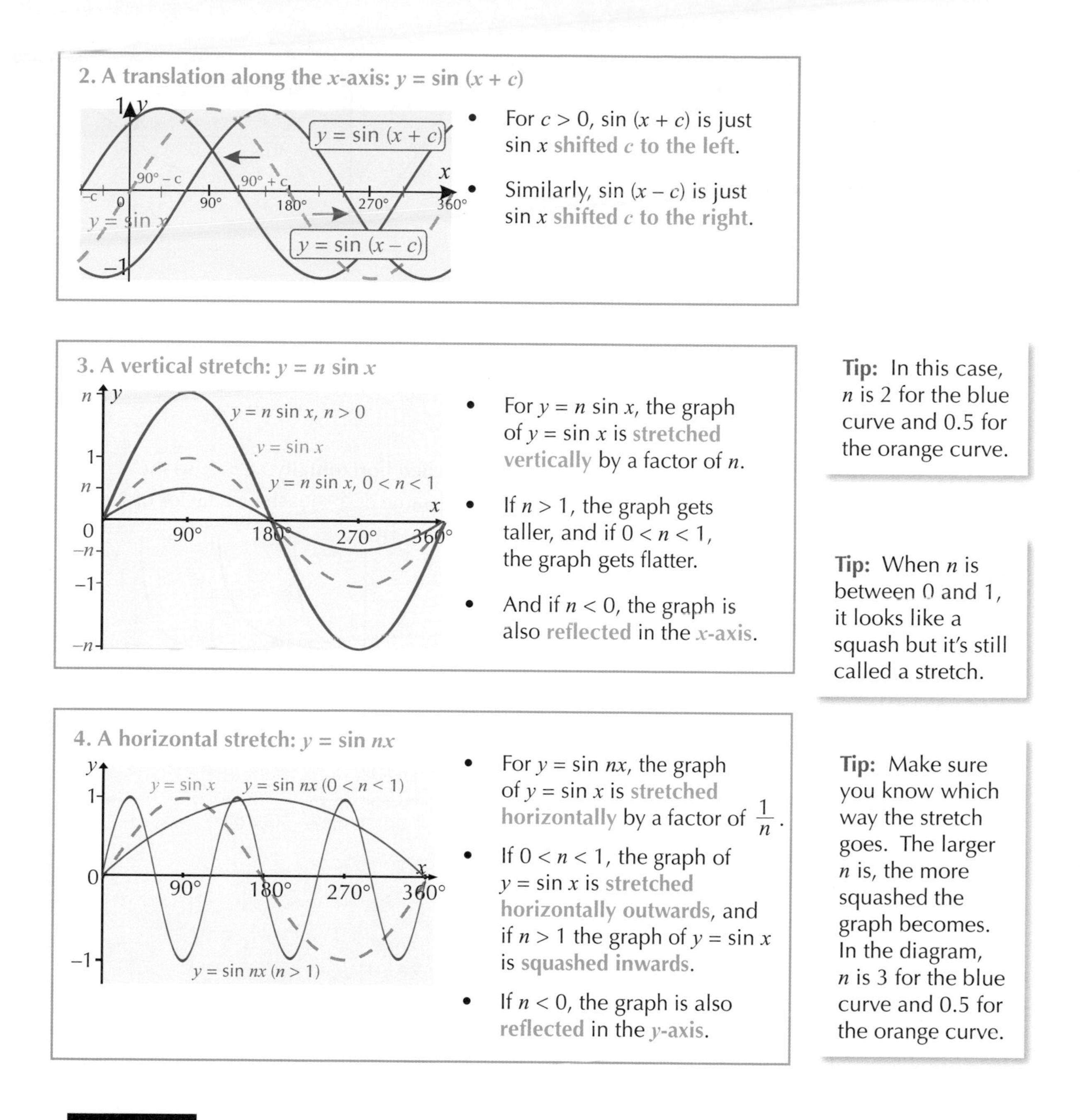

2. A translation along the x-axis: $y = \sin(x + c)$

- For $c > 0$, $\sin(x + c)$ is just $\sin x$ **shifted c to the left.**
- Similarly, $\sin(x - c)$ is just $\sin x$ **shifted c to the right.**

3. A vertical stretch: $y = n \sin x$

- For $y = n \sin x$, the graph of $y = \sin x$ is **stretched vertically** by a factor of n.
- If $n > 1$, the graph gets taller, and if $0 < n < 1$, the graph gets flatter.
- And if $n < 0$, the graph is also **reflected** in the **x-axis**.

Tip: In this case, n is 2 for the blue curve and 0.5 for the orange curve.

Tip: When n is between 0 and 1, it looks like a squash but it's still called a stretch.

4. A horizontal stretch: $y = \sin nx$

- For $y = \sin nx$, the graph of $y = \sin x$ is **stretched horizontally** by a factor of $\frac{1}{n}$.
- If $0 < n < 1$, the graph of $y = \sin x$ is **stretched horizontally outwards**, and if $n > 1$ the graph of $y = \sin x$ is **squashed inwards**.
- If $n < 0$, the graph is also **reflected** in the **y-axis**.

Tip: Make sure you know which way the stretch goes. The larger n is, the more squashed the graph becomes. In the diagram, n is 3 for the blue curve and 0.5 for the orange curve.

Example 1

On the same axes, sketch the graphs of $y = \cos x$ and $y = -2 \cos x$ in the range $-360° \le x \le 360°$.

1. Start by sketching the graph of $\cos x$:

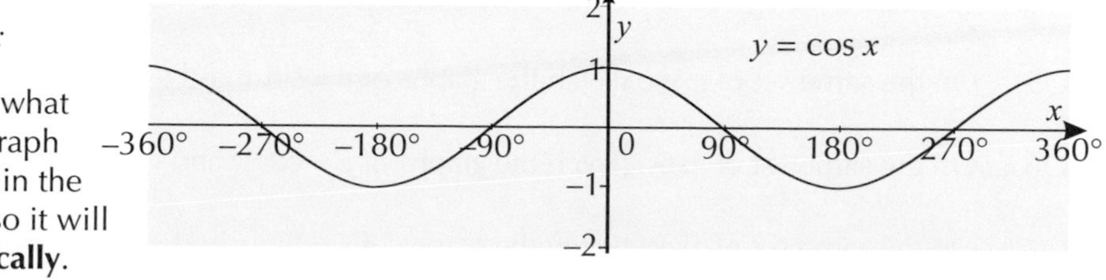

2. Next, think about what the **transformed** graph will look like. It's in the form $y = n \cos x$, so it will be **stretched vertically**.

3. $n = -2$, so it will be stretched by a factor of **2**. As n is negative, it will also be **reflected** in the x-axis. This is all the information you need to be able to sketch the graph.

Example 2

On the same axes, sketch the graphs of $y = \tan x$ and $y = \tan 2x$ in the interval $-180° \leq x \leq 180°$.

1. Again, start by sketching the graph of $y = \tan x$ (see p.146) — this is the dark red line shown below.

2. This time it's in the form $\mathbf{y = \tan nx}$, so it will be **stretched horizontally**. $n > 1$, so the graph will be stretched by a factor of $\frac{1}{2}$, which is the same as a squash by a factor of 2.

3. To make it easier, draw dotted lines for the new asymptotes (divide the x-values of the old ones by 2) then draw the tan shape between them — shown in orange.

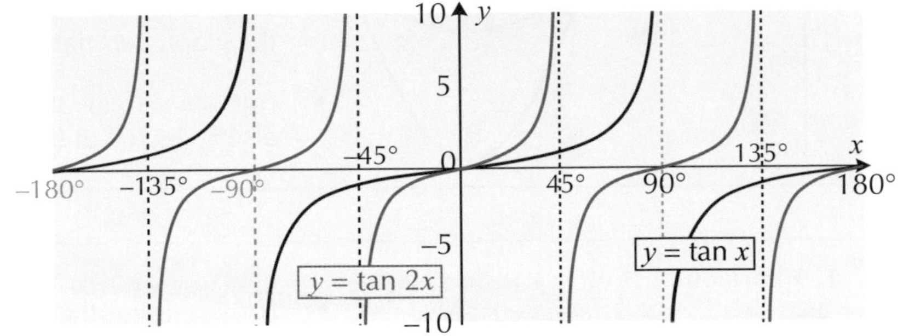

Exercise 8.3.1

Q1 On the same set of axes, sketch the graphs of $y = \cos x$ and $y = \cos x + 3$ in the interval $-360° \leq x \leq 360°$.

Q2 On the same set of axes, sketch the graphs of $y = \cos x$ and $y = \cos (x + 90°)$ in the interval $-180° \leq x \leq 180°$.

Q3 On the same set of axes, sketch the graphs of $y = \tan x$ and $y = \tan (x - 45°)$ in the interval $-180° \leq x \leq 360°$.

Q4 On the same set of axes, sketch the graphs of $y = \sin x$ and $y = \frac{1}{3} \sin x$ in the interval $-180° \leq x \leq 180°$.

Q5 On the same set of axes, sketch the graphs of $y = \sin x$ and $y = \sin 3x$ in the interval $0° \leq x \leq 360°$.

Q6 On the same set of axes, sketch the graphs of $y = \cos x$ and $y = -\cos x$ in the interval $0° \leq x \leq 360°$.

Q7 On the same set of axes, sketch the graphs of $y = \tan x$ and $y = \tan(-x)$ in the interval $0° \leq x \leq 360°$.

Q8 The graph of $y = \sin x$ is translated up by 4 units.
Write down the equation of the transformed graph.

Q9 The graph of $y = \cos x$ is stretched horizontally by a factor of 5.
Write down the equation of the transformed graph.

Q10 The graph of $y = \tan x$ is transformed such that the equation of the transformed graph is $y = \tan 2x$.

a) Find the points where the transformed graph crosses the x-axis in the interval $-180° \leq x \leq 180°$.

b) Write down the equations of the asymptotes of the transformed graph.

Q11 a) Sketch the graph of $f(x) = \tan x$ in the interval $-90° \leq x \leq 270°$.

b) Translate this graph 90° to the left and sketch it on the same set of axes as part a).

c) Write down the equation of the transformed graph.

Q11c) Hint: Look at what's happened to the graph, then think about which type of transformation is needed to achieve it.

Q12 a) Sketch the graph of $y = \sin x$ in the interval $-360° \leq x \leq 360°$.

b) Stretch the graph horizontally by a factor of 2 and sketch it on the same set of axes as part a).

c) Write down the equation of the transformed graph.

Q13 The diagram shows the graph of $y = \sin x$ and a transformed graph.

a) Describe the transformation.

b) Write down the equation of the transformed graph.

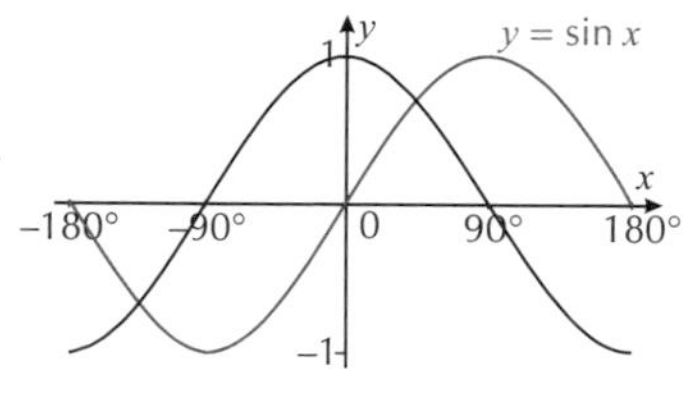

Q14 The diagram shows the graph of $y = \cos x$ and a transformed graph.

a) Describe the transformation.

b) Write down the equation of the transformed graph.

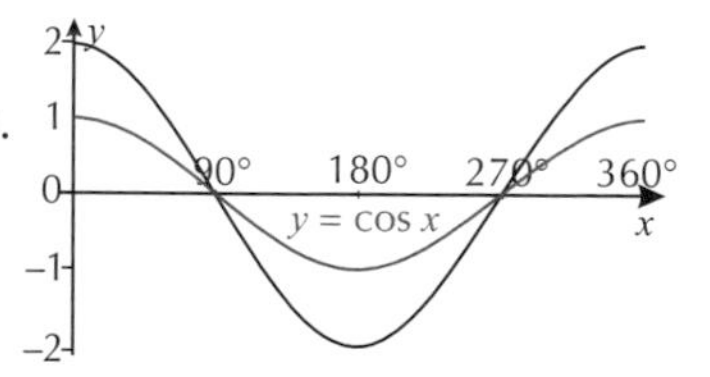

Q15 The diagram shows the graph of $y = \tan x$ and a transformed graph.

a) Describe the transformation.

b) Write down the equation of the transformed graph.

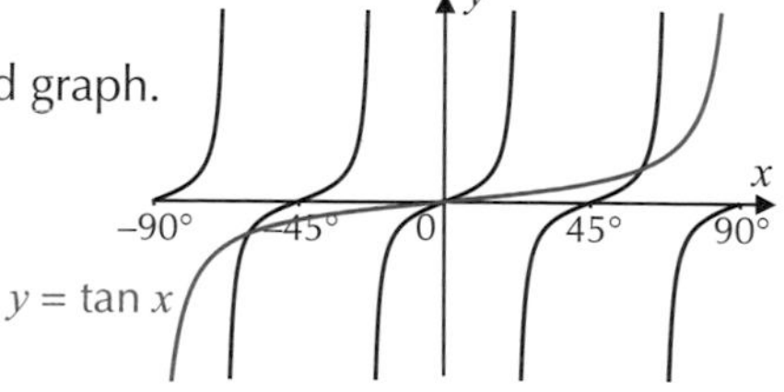

Q16 A transformed graph has the equation $y = \cos px$. Given that the graph crosses the x-axis at the point (180°, 0), find the smallest possible positive value of p.

PROBLEM SOLVING

Q17 A transformed graph has the equation $y = \sin (x + q)$.

PROBLEM SOLVING

a) Given that the graph is at a maximum at the point (60°, 1), find a possible value for q.

b) Explain why there is more than one possible value for q.

8.4 Solving Trig Equations

Once you know how to sketch the graphs of trig functions, you can use them to solve trig equations. Solving a trig equation just means finding the value (or values) of x that satisfies the given equation.

Learning Objectives (Spec Ref 5.2-5.4):
- Solve trig equations by sketching a graph.
- Solve trig equations using a CAST diagram.
- Solve trig equations of the form $\sin kx = n$ and $\sin (x + c) = n$.
- Solve trig equations using trig identities and quadratics.

Sketching a graph

To solve trig equations in a **given interval** you can use one of two methods. The first is drawing a **graph** of the function and reading solutions off the graph. You'll often find that there's **more than one** solution to the equation — in every **360° interval**, there are usually **two** solutions to an equation, and if the interval is bigger (see Example 2 below), there'll be even more solutions.

Example 1

Solve $\cos x = \frac{1}{2}$ for $0° \leq x \leq 360°$.

Tip: This is actually one of the common trig angles from page 134.

1. Start by using your calculator to work out the first value.

   For $\cos x = \frac{1}{2}$, $x =$ 60°

2. Then sketch a graph of $\cos x$ in the interval you're interested in, and draw a horizontal line across for $y = \frac{1}{2}$. The points where the line and curve meet are the solutions of the equation.

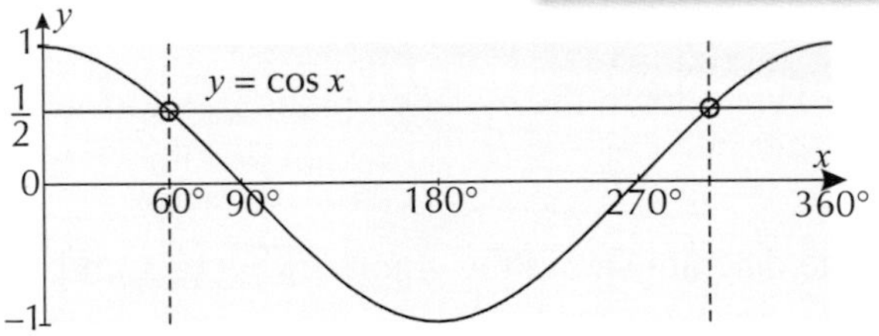

3. Each 360° interval of the graph is symmetrical, so the second solution will be the same distance from 360° as the first is from 0°.

   360° – 60° = 300°

4. You now know all of the solutions in this interval.

   $x =$ 60°, 300°

If you had an interval that was **larger** than **one repetition** of the graph (i.e. 360° for sin and cos, and 180° for tan), you'd just add or subtract **multiples** of 360° (for sin and cos) or 180° (for tan) onto the solutions you've found until you have **all** the solutions in the **given interval**.

Example 2

Solve $\sin x = -0.3$ for $0° \leq x \leq 720°$. Give your answers to 3 s.f.

1. Again, use your calculator to work out the first value.

   For $\sin x = -0.3$, $x = -17.45...°$

2. However, this is outside the given interval for x, so add on 360° to find a solution in the interval:

   $-17.45...° + 360° = 342.54...°$

3. Now sketch a graph of $\sin x$ in the interval you need, and draw a horizontal line across at $y = -0.3$. This time, you'll need to draw the graph between $x = 0°$ and $x = 720°$, so there will be 2 repetitions of the sin wave.

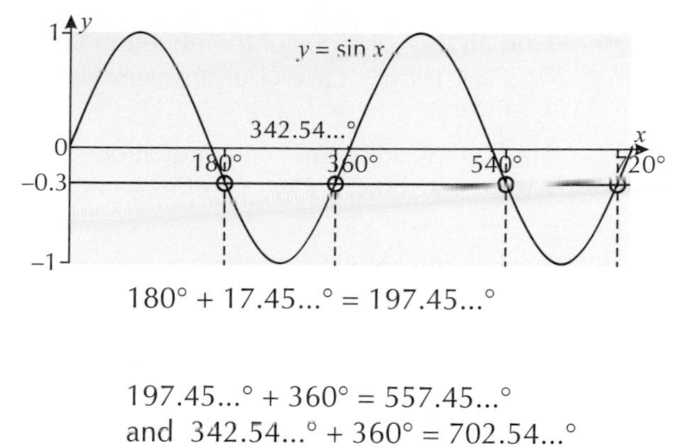

4. Looking at the graph, the other solution between 0° and 360° will be 17.45...° away from 180°.

   $180° + 17.45...° = 197.45...°$

5. For the next two solutions (the ones between 360° and 720°), just add 360° onto the values you've already found:

   $197.45...° + 360° = 557.45...°$
   and $342.54...° + 360° = 702.54...°$

6. So the solutions to $\sin x = -0.3$ for $0 \le x \le 720°$ are: $x =$ **197°, 343°, 557°, 703°** (all to 3 s.f.)

Exercise 8.4.1

Q1 By sketching a graph, find all the solutions to the equations below in the interval $0° \le x \le 360°$. Give your answers to 1 decimal place.

a) $\sin x = 0.75$ b) $\cos x = 0.31$ c) $\tan x = -1.5$

d) $\sin x = -0.42$ e) $\cos x = -0.56$ f) $\tan x = -0.67$

g) $\sin x = 0.32$ h) $\cos x = -0.89$ i) $\tan x = 2.3$

Q2 By sketching a graph, find all the solutions to the equations below in the interval $0° \le x \le 360°$.

a) $\cos x = \frac{1}{\sqrt{2}}$ b) $\tan x = \sqrt{3}$ c) $\sin x = \frac{1}{2}$

d) $\tan x = \frac{1}{\sqrt{3}}$ e) $\tan x = 1$ f) $\cos x = \frac{\sqrt{3}}{2}$

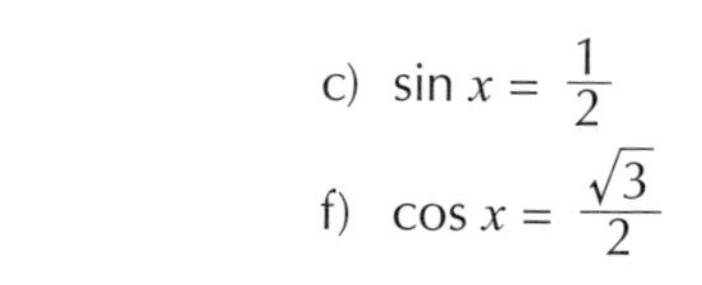

Q3 One solution of $\sin x = 0.7$ is 44.43° (2 d.p.). Use the graph to find all the solutions in the interval $-180° \le x \le 180°$.

Q4 One solution of $\cos x = -0.8$ is 143.1° (1 d.p.). Use the graph to find all the solutions in the interval $0° \le x \le 360°$.

Q5 One solution of $\tan x = 1.9$ is 62.2° (3 s.f.). Use the graph to find all the solutions in the interval $0° \le x \le 540°$.

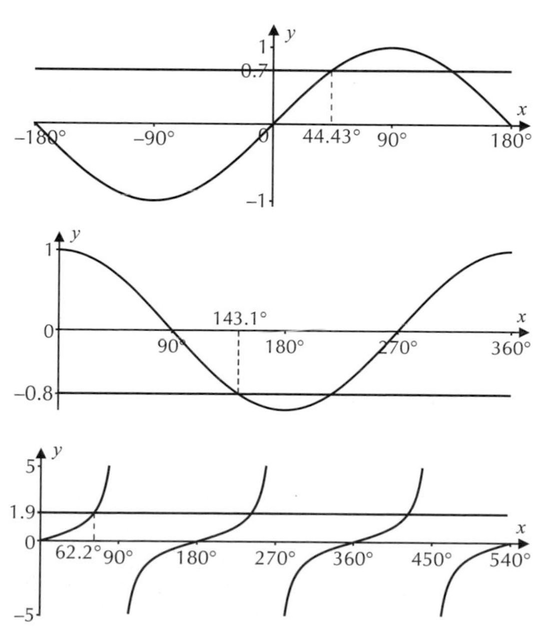

Q6 Find all the solutions of the equation $\tan x = 2.5$ in the interval $0° \leq x \leq 1080°$. Give your answers to 1 decimal place.

Q7 Find all the solutions of the equation $\sin x = 0.81$ in the interval $-360° \leq x \leq 360°$. Give your answers to 3 significant figures.

Q6-7 Hint: Be careful with the intervals here — and remember that tan repeats every 180°.

Q8 Find all the solutions to the equation $\sin x = 0.23$ in the interval $-360° \leq x \leq 540°$, giving your answers to 1 decimal place.

Q9 Find all the solutions to the equation $\cos x = -0.96$ in the interval $-360° \leq x \leq 720°$, giving your answers to 3 significant figures.

Q10 Find all the solutions to the equation $\tan x = -1.75$ in the interval $-360° \leq x \leq 720°$, giving your answers to 1 decimal place.

Q11 Find all the solutions to the equation $5 \sin x - 3 \cos x = 0$ in the interval $0° \leq x \leq 360°$. Give your answers to 3 significant figures.

Q12 Find all the solutions to the equation $3 \cos x + 8 \sin x = 0$ in the interval $0° \leq x \leq 360°$. Give your answers to 2 decimal places.

Q11-12 Hint: Use a trig identity to rearrange the equations so that they're in terms of tan x.

Using a CAST diagram

The second way of finding the solutions to a trig equation is by using a **CAST diagram**. CAST stands for Cos, All, Sin, Tan, and it shows you where each of these functions is **positive** by splitting a 360° period into **quadrants**.

Tip: CAST diagrams are useful because they summarise the information from the graphs of sin, cos and tan without actually having to sketch the graphs.

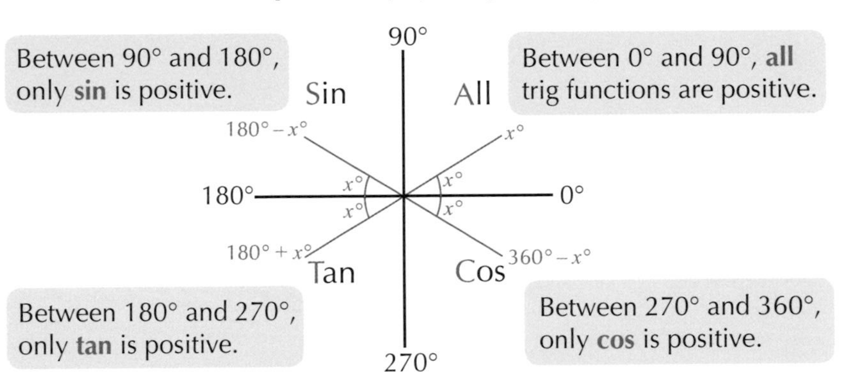

To use a CAST diagram, you need to use your **calculator** to find the first solution of the trig function (or, if it's a common angle, you might just be able to recognise it). The angle you put into the CAST diagram should be **acute** (i.e. between 0° and 90°).

You then make the **same angle** from the **horizontal** in each of the four quadrants (shown in the diagram above), then measure each angle **anticlockwise** from 0°. So if the first solution was 45°, the solution in the 'sin' quadrant would be 135° (180° – 45°) measured anticlockwise from 0°, and so on.

Ignore the ones that give a **negative** result (unless the given value is negative — in which case you want the two quadrants in which the trig function is **negative**).

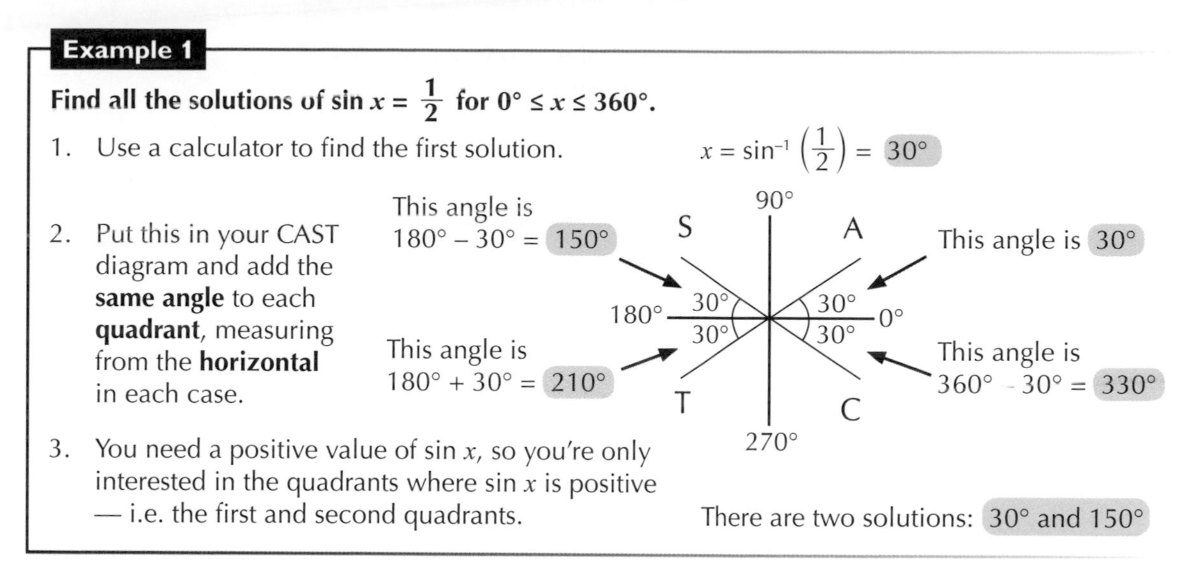

Example 1

Find all the solutions of $\sin x = \frac{1}{2}$ for $0° \leq x \leq 360°$.

1. Use a calculator to find the first solution. $x = \sin^{-1}\left(\frac{1}{2}\right) = 30°$

2. Put this in your CAST diagram and add the **same angle** to each **quadrant**, measuring from the **horizontal** in each case.

This angle is $180° - 30° = 150°$

This angle is $180° + 30° = 210°$

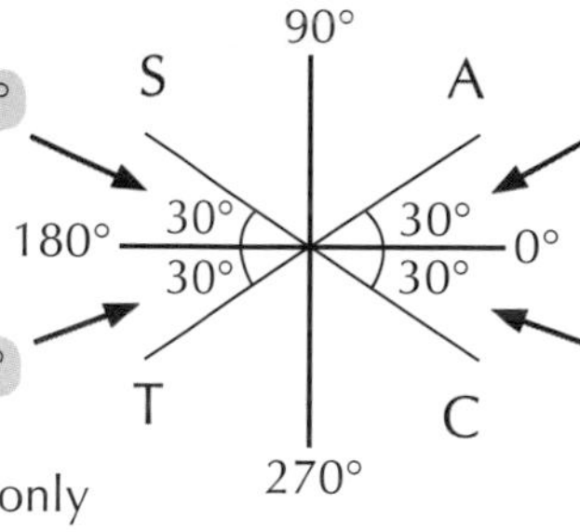

This angle is $30°$

This angle is $360° - 30° = 330°$

3. You need a positive value of $\sin x$, so you're only interested in the quadrants where $\sin x$ is positive — i.e. the first and second quadrants.

There are two solutions: $30°$ and $150°$

For values **outside** the interval $0° \leq x \leq 360°$, just find solutions between 0° and 360° and then **add** or **subtract multiples of 360°** to find solutions in the correct interval (see the example below).

Example 2

Find all the solutions of $\tan x = -6$ for $0° \leq x \leq 720°$. Give your answers to 1 d.p.

1. Find that the first solution using a calculator. $x = \tan^{-1}(-6) = -80.5°$ (1 d.p.)

2. Ignore the negative and just put the value 80.5° into the CAST diagram and add the **same angle** to each **quadrant**, measuring from the **horizontal** in each case.

This angle is $180° - 80.5° = 99.5°$

This angle is $180° + 80.5° = 260.5°$

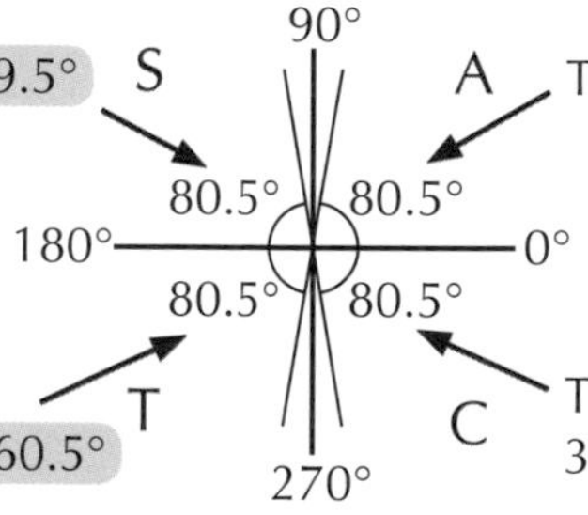

This angle is $80.5°$

This angle is $360° - 80.5° = 279.5°$

3. You need a **negative** value of $\tan x$ (as –6 is negative), so just look at the quadrants where $\tan x$ is negative (i.e. the 'S' and 'C' quadrants).

There are two solutions: 99.5° and 279.5°

> **Tip:** The first value was **negative**, so you could have measured **clockwise** from 0° to find the solution 279.5°.

4. To find the solutions between 360° and 720°, just **add 360°** to the solutions you've already found.

$99.5° + 360° = 459.5°$
$279.5° + 360° = 639.5°$
$x = 99.5°, 279.5°, 459.5°$ and $639.5°$ (1 d.p.)

Exercise 8.4.2

Q1 One solution of $\sin x = 0.45$ is $x = 26.7°$ (1 d.p.).
Use a CAST diagram to find all the solutions in the interval $0° \leq x \leq 360°$.

Q2 The equation $\cos x = 0.68$ has a solution $x = 47.2°$ (1 d.p.).
Use a CAST diagram to find all the solutions in the interval $0° \leq x \leq 360°$.

Q3 Use a CAST diagram to find the solutions of the following equations in the interval $0° \le x \le 360°$. Give your answers to 1 d.p.

a) $\cos x = 0.8$

b) $\tan x = 2.7$

c) $\sin x = -0.15$

d) $\tan x = 0.3$

e) $\tan x = -0.6$

f) $\sin x = -0.29$

Q4 One solution of the equation $\sin x = -0.87$ is $x = -60.5°$ (1 d.p.). Use a CAST diagram to find all the solutions in the interval $-180° \le x \le 360°$.

Q5 Use a CAST diagram to find all the solutions to the following equations in the given interval, giving your answers to 3 s.f.

a) $\tan x = -8.4$ for $0° \le x \le 360°$

b) $\sin x = 0.82$ for $0° \le x \le 720°$

c) $\cos x = 0.72$ for $0° \le x \le 540°$

d) $\tan x = 3.58$ for $-180° \le x \le 180°$

Q6 Use a CAST diagram to find all the solutions to $\sin x = -0.06$ in the interval $0° \le x \le 1080°$. Give your answers to 1 d.p.

Q7 Use a CAST diagram to find all the solutions to $\tan x = 11.8$ in the interval $-360° \le x \le 360°$. Give your answers to 1 d.p.

Q8 Use a CAST diagram to find all the solutions to $5 \cos x - 2 = 0$ in the interval $0° \le x \le 360°$. Give your answers to 1 d.p.

PROBLEM SOLVING

Q9 Use a CAST diagram to find all the solutions to $\frac{1}{4} \tan x = 1.4$ in the interval $0° \le x \le 360°$. Give your answers to 3 s.f.

PROBLEM SOLVING

Q10 Use a CAST diagram to solve the following equations, giving your answers to 1 d.p.

PROBLEM SOLVING

a) $4 \sin x - 3 = 0$ $\quad 0° \le x \le 720°$

b) $2 - 3 \cos x = 0$ $\quad -180° \le x \le 180°$

c) $6 \tan x = -11$ $\quad -180° \le x \le 540°$

d) $8 \sin x = -5$ $\quad -360° \le x \le 720°$

Changing the interval

Sometimes you'll have to solve equations of the form **$\sin kx = n$** or **$\sin (x + c) = n$** (where n, k and c are numbers). It's usually easiest to **change the interval** you're solving for, then **solve as normal** for kx or $x + c$. You'll then need to get the final solutions either by **dividing by k** or **subtracting c** at the end.

Solving equations of the form sin $kx = n$

- First, **multiply** the **interval** you're looking for solutions in by k. E.g. for the equation $\sin 2x = n$ in the interval $0° \le x \le 360°$, you'd look for solutions in the interval $0° \le 2x \le 720°$. Then **solve** the equation over the new interval.
- This gives you solutions for **kx** — then you need to **divide** each solution by **k** to find the values of x.

You can either **sketch the graph** over the new interval (this will show you **how many** solutions there are) or you can use the **CAST method** to find solutions between 0° and 360° then add on multiples of 360° until you have all the solutions in the new interval — use whichever method you prefer.

Example 1

Solve $\cos 4x = 0.6$ for $0° \le x \le 360°$. Give your answers to 1 d.p.

1. First, **change the interval**. The interval is $0° \le x \le 360°$, and the value of k is 4, so **multiply** the interval by 4. $\quad 0° \le 4x \le 1440°$

2. Then **solve** the equation to find the solutions for $4x$. Find the first solution using a calculator. $\quad \cos 4x = 0.6 \Rightarrow 4x = 53.13°$ (2 d.p.)

3. Put this in a CAST diagram (or sketch a graph if you prefer).

4. Use the quadrants where cos is **positive** to find the other solution between 0° and 360°. $\quad 4x = 360° - 53.13° = 306.87°$ (2 d.p.)

5. Now add on **multiples of 360°** to find all the solutions in the interval $0° \le 4x \le 1440°$ (2 d.p.). $\quad$ 53.13°, 306.87°, 413.13°, 666.87°, 773.13°, 1026.87°, 1133.13°, 1386.87°

6. Remember, these are solutions for **$4x$**. To find the solutions for x, **divide** through by 4. So the solutions to $\cos 4x = 0.6$ in the interval $0° \le x \le 360°$ (to 1 d.p.) are: $\quad x =$ 13.3°, 76.7°, 103.3°, 166.7°, 193.3°, 256.7°, 283.3°, 346.7°

7. It's a good idea to check your answers — put your values of x into $\cos 4x$ to check if they give you 0.6. Make sure you have the right number of solutions too — there are 2 solutions to $\cos x = 0.6$ between $0° \le x \le 360°$, so there are 8 solutions to $\cos 4x = 0.6$ in the same interval.

Example 2

Solve $\sin 3x = -\frac{1}{\sqrt{2}}$ for $0° \le x \le 360°$.

Tip: This is one of the common angles from p.134 — so you could have found it without a calculator.

1. Here you've got $3x$ instead of x, so you need to **multiply** the interval by 3. $\quad 0° \le 3x \le 1080°$

2. Use your calculator to find a solution. $\quad 3x = -45°$

3. This is outside the interval for $3x$, so use the pattern of the graph of $\sin x$ to find solutions in the interval. $\quad$ The sine curve repeats every 360°, so a solution is: $-45° + 360° = 315°$

4. Use the **symmetry** of the graph to find the other solution between 0° and 360°. $\quad$ The curve is symmetrical in each interval of 180°, so the other solution is: $180° + 45° = 225°$

5. The graph repeats every 360°, so add on lots of 360° to the answers you've found to find the other solutions in the interval: $3x = 585°, 675°, 945°, 1035°$

6. You have 6 solutions for $3x$, so **divide** them all by 3. $\quad x =$ 75°, 105°, 195°, 225°, 315°, 345°

Example 3

Find all the solutions of $0.9 + 3 \tan 2x = 5.1$ for $0° \le x \le 360°$.
Give your answers to 1 d.p.

PROBLEM SOLVING

1. First, rearrange into a more familiar format. $0.9 + 3 \tan 2x = 5.1 \Rightarrow 3 \tan 2x = 4.2$
   $\Rightarrow \tan 2x = 1.4$
2. It's $2x$, so multiply the interval by 2. $0° \le 2x \le 720°$
3. Use a calculator to work out the first solution. $2x = 54.46°$ (2 d.p.)
4. Then use a CAST diagram to find the other solution between 0° and 360°. You want the quadrants where tan is positive. $2x = 180° + 54.46°$
   $= 234.46°$ (2 d.p.)

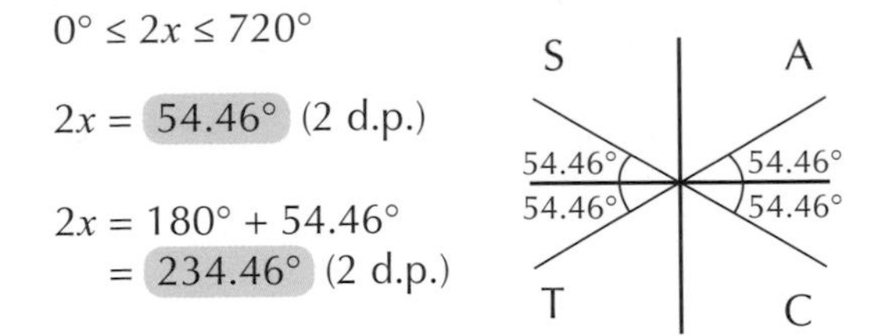

5. Add 360° to the solutions you've found to find the solutions between 360° and 720°. $54.46° + 360° = 414.46°$ (2 d.p.)
   $234.46° + 360° = 594.46°$ (2 d.p.)
6. Finally, to find the solutions for x between 0° and 360°, just divide by 2. $x = 27.2°, 117.2°, 207.2°, 297.2°$ (1 d.p.)

Solving equations of the form sin ($x + c$) = n

The method for solving equations of the form **sin ($x + c$) = n** is similar — but instead of multiplying the interval, you have to add or subtract the value of c.

- **Add** (or **subtract**) the value of c to the **whole interval** — so the interval $0° \le x \le 360°$ becomes $c \le x + c \le 360° + c$ (you add c onto each bit of the interval).
- **Solve** the equation over the **new interval** — you can either sketch a graph or use a CAST diagram.
- Finally, **subtract** (or **add**) c from your solutions to give the values for x.

Example 1

Solve $\cos (x + 60°) = \frac{3}{4}$ for $-360° \le x \le 360°$, giving your answers to 1 d.p.

1. You've got $\cos (x + 60°)$ instead of $\cos x$ — so add 60° to each bit of the interval. The new interval is: $-300° \le x + 60° \le 420°$
2. Use your calculator to get a solution. $\cos (x + 60°) = \frac{3}{4} \Rightarrow x + 60° = 41.4°$ (1 d.p.)
3. Find the other solution between 0° and 360°. The first solution is 41.4° away from 0°, so: $360° - 41.4° = 318.6°$ (1 d.p.).
4. The cos graph repeats every 360°. Find the other solutions by adding and subtracting 360° from the answers you've found, making sure they're still within the interval.

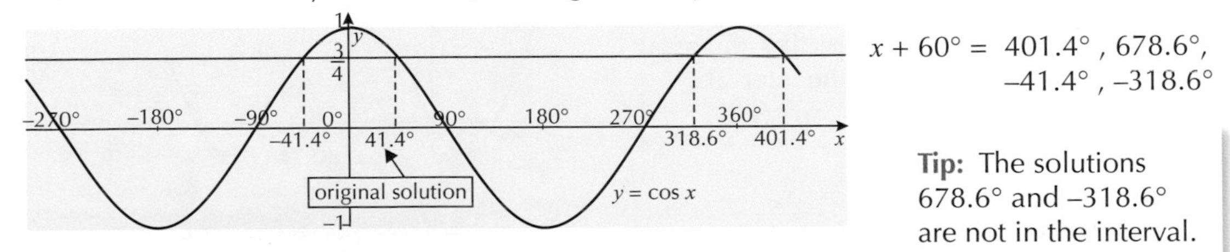

$x + 60° = 401.4°, 678.6°, -41.4°, -318.6°$

Tip: The solutions 678.6° and –318.6° are not in the interval.

5. These solutions are for cos $(x + 60°)$ so you need to subtract 60° from each value to find the solutions for x (to 1 d.p.): $x = -101.4°, -18.6°, 258.6° \text{ and } 341.4°$

6. So there are 4 solutions, and they're all in the required interval $(-360° \le x \le 360°)$. Check your answers by putting them back into cos $(x + 60°)$ and making sure you get $\frac{3}{4}$.

Example 2

Solve $\tan(x - 75°) = 2$ for $0° \le x \le 360°$. Give your answers to 1 d.p.

1. You've got tan $(x - 75°)$ instead of tan x — so subtract 75° from each bit of the interval.
   The new interval is: $-75° \le x - 75° \le 285°$
2. Use your calculator to find a solution.
   $\tan(x - 75°) = 2 \Rightarrow x - 75° = 63.4°$ (1 d.p.)
3. Put 63.4° into a CAST diagram (or use a graph) to find the other solutions in the interval.
   Tan is positive in the first and third quadrants, so the other solution is:
   $63.4° + 180° = 243.4°$ (1 d.p.)

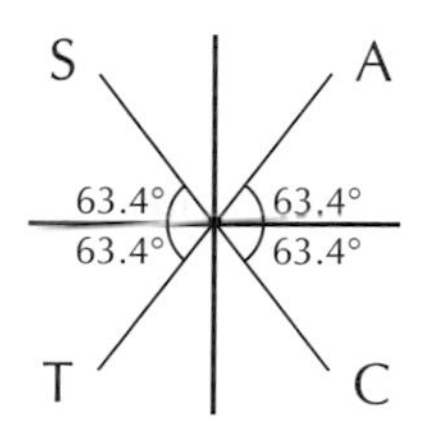

4. Finally, add on 75° to find the solutions in the interval $0° \le x \le 360°$.
   $x = 138.4°, 318.4°$ (1 d.p.)

Example 3

Solve $2\sin(x + 50°) + \sqrt{3} = 0$ for $0° \le x \le 720°$.

1. First, rearrange the equation.
   $2\sin(x + 50°) + \sqrt{3} = 0$
   $\Rightarrow 2\sin(x + 50°) = -\sqrt{3}$
   $\Rightarrow \sin(x + 50°) = -\frac{\sqrt{3}}{2}$
2. Add 50° to each bit of the interval.
   $50° \le x + 50° \le 770°$
3. $\frac{\sqrt{3}}{2}$ is a common trig value — $\sin 60° = \frac{\sqrt{3}}{2}$, so put 60° into the CAST diagram.
4. As you're finding solutions for $-\frac{\sqrt{3}}{2}$, you want the quadrants where sin is **negative** — i.e. the third and fourth quadrants.
   $x + 50° = 180° + 60° = 240°$
   and $x + 50° = 360° - 60° = 300°$
5. To find the other solutions (between 410° and 770°), add 360° to these solutions.
   $x + 50° = 600°, 660°$
6. Finally, subtract 50° from each solution to find the values of x.
   $x = 190°, 250°, 550°, 610°$

Exercise 8.4.3

In this exercise, give all non-integer answers to 1 d.p.

Q1 Solve the following equations in the interval $0° \leq x \leq 360°$:

a) $\sin 2x = 0.6$ b) $\tan 4x = 4.6$ c) $\cos 3x = -0.24$

d) $\sin 3x = 0.94$ e) $\cos 5x = 0.5$ f) $\tan 2x = -6.7$

Q2 Solve $\tan \frac{x}{2} = 2.1$ in the interval $0° \leq x \leq 360°$.

Q2 Hint: Don't let the $\frac{x}{2}$ throw you — just like before, you multiply the interval by $\frac{1}{2}$ (i.e. divide by 2).

Q3 Find all the solutions to $\sin \frac{2x}{3} = 0.52$ in the interval $0° \leq x \leq 270°$.

Q4 Solve $\cos \frac{x}{3} = \frac{\sqrt{3}}{2}$ in the interval $-180° \leq x \leq 180°$.

Q5 Find all the solutions to $\cos (x - 27°) = 0.64$ in the interval $0° \leq x \leq 360°$.

Q6 Solve $\tan (x - 140°) = -0.76$ in the interval $0° \leq x \leq 360°$.

Q7 Find all the solutions to $\tan (x + 73°) = 1.84$ in the interval $0° \leq x \leq 360°$.

Q8 Find all the solutions to $\sin (x - 45°) = -0.25$ in the interval $-180° \leq x \leq 360°$.

Q9 Solve $\cos (x + 22.5°) = 0.13$ in the interval $0° \leq x \leq 360°$.

Q10 Solve $\tan (x - 32°) - 3 = 4.5$ in the interval $-180° \leq x \leq 180°$.

Q11 Solve $4 \cos 3x = 2.8$ in the interval $0° \leq x \leq 360°$.

Q12 Find all the solutions to $\frac{1}{2} \sin 3x - 0.61 = -0.75$ in the interval $0 \leq x \leq 360°$.

Q13 Find all the solutions to $\frac{1}{3} \cos (x - 67°) = 0.23$ in the interval $0° \leq x \leq 540°$.

Q14 Find all the solutions to $2 \sin (x + 19°) + \sqrt{2} = 0$ in the interval $0° \leq x \leq 360°$.

Using trig identities to solve equations

Sometimes you'll be asked to find solutions to an equation that has a **tan** term as well as a sin or cos term in it. In these situations you might need to use the **trig identity** for $\tan x$ (p.143):

$$\tan x \equiv \frac{\sin x}{\cos x}$$

This will **eliminate** the tan term, and you'll be left with an equation just in terms of sin or cos.

Similarly, if you have a $\sin^2 x$ or $\cos^2 x$, you can use the other identity to rewrite one trig function in terms of the other.

$$\sin^2 x + \cos^2 x \equiv 1$$

If you're left with a quadratic equation (e.g. one that contains both $\sin^2 x$ and $\sin x$), you might need to factorise to solve it. To do this, it's usually easiest to make a substitution (e.g. $y = \sin x$ — see p.37).

Example

Solve $6\cos^2 x + \cos x \tan x = 5$ for $0° \leq x \leq 360°$.
Give any non-exact answers to 1 d.p.

1. The equation has both $\cos x$ and $\tan x$ in it, so writing $\tan x$ as $\frac{\sin x}{\cos x}$ is a good place to start.

   $6\cos^2 x + \cos x \tan x = 5$

   $\Rightarrow 6\cos^2 x + \cos x \frac{\sin x}{\cos x} = 5$

2. Now the $\cos x$ terms will cancel.

   $\Rightarrow 6\cos^2 x + \sin x = 5$

3. Now the equation has both $\sin x$ and $\cos x$ in it. So replace the $\cos^2 x$ with $1 - \sin^2 x$.

   $\Rightarrow 6(1 - \sin^2 x) + \sin x = 5$

4. Multiply out the bracket and rearrange it so that you've got zero on one side — you get a quadratic in $\sin x$.

   $\Rightarrow 6 - 6\sin^2 x + \sin x - 5 = 0$

   $\Rightarrow 6\sin^2 x - \sin x - 1 = 0$

5. It's easier to factorise the quadratic if you make the substitution $y = \sin x$.

   $\Rightarrow 6y^2 - y - 1 = 0$

   $\Rightarrow (2y - 1)(3y + 1) = 0$

   $\Rightarrow (2\sin x - 1)(3\sin x + 1) = 0$

6. Now you have two things multiplying together to make zero. That means one of them must be equal to zero.

   $2\sin x - 1 = 0 \Rightarrow \sin x = \frac{1}{2}$

   or $3\sin x + 1 = 0 \Rightarrow \sin x = -\frac{1}{3}$

7. First the first solution for $\sin x = \frac{1}{2}$ — so $x = 30°$.
   Use a CAST diagram to find the other solutions in the interval

   90°
   S A
   180° 30° 30° 0°
   30° 30°
   T C
   270°

   $\sin x$ is positive in the first and second quadrants, so the positive solutions for $\sin x$ are:
   $30°$ and $180° - 30° = 150°$

8. For $\sin x = -\frac{1}{3}$, use your calculator to find $x = -19.47...°$.
   Use a CAST diagram to find the other solutions in the interval.

   90°
   S A
   19.47° 19.47°
   180° 0°
   19.47° 19.47°
   T C
   270°

   $\sin x$ is negative in the third and fourth quadrants, so the negative solutions for $\sin x$ are:
   $180° + 19.47...° = 199.47°$ and
   $360° - 19.47...° = 340.52°$

9. You now have all the solutions.

   $x = 30°, 150°, 199.5°$ and $340.5°$ (to 1 d.p.)

Exercise 8.4.4

In this exercise, give all non-exact answers to 1 d.p.

Q1 Solve each of the following equations for values of x in the interval $0° \leq x \leq 360°$:

a) $(\tan x - 5)(3 \sin x - 1) = 0$

b) $5 \sin x \tan x - 4 \tan x = 0$

c) $\tan^2 x = 9$

d) $4 \cos^2 x = 3 \cos x$

e) $3 \sin x = 5 \cos x$

f) $5 \tan^2 x - 2 \tan x = 0$

g) $6 \cos^2 x - \cos x - 2 = 0$

h) $7 \sin x + 3 \cos x = 0$

Q2 Find the solutions to each of the following equations in the given interval:

a) $\tan x = \sin x \cos x \quad 0° \leq x \leq 360°$

b) $5 \cos^2 x - 9 \sin x = 3 \quad -360° \leq x \leq 720°$

c) $2 \sin^2 x + \sin x - 1 = 0 \quad -360° \leq x \leq 360°$

d) $2 \sin x \tan x = -3 \quad -360° \leq x \leq 360°$

e) $4 - \tan^2 x = 0 \quad -180° \leq x \leq 180°$

Q3 a) Show that the equation $4 \sin^2 x = 3 - 3 \cos x$ can be written as $4 \cos^2 x - 3 \cos x - 1 = 0$.

b) Hence solve the equation $4 \sin^2 x = 3 - 3 \cos x$ in the interval $0° \leq x \leq 360°$.

Q4 Solve the equation $3 \cos x - 2 \sin^2 x = 0$ in the interval $0° \leq x \leq 360°$.

Q5 Solve the equation $2 \sin^2 x + 5 \cos^2 x - 7 \cos x = 0$ in the interval $0° \leq x \leq 360°$.

Q6 Find all the solutions of the equation $9 \sin^2 2x + 3 \cos 2x = 7$ in the interval $0° \leq x \leq 360°$.

Q7 a) Show that $\sin x - \sin x \cos^2 x \equiv \sin^3 x$.

b) Hence solve the equation $3 \sin x - 3 \sin x \cos^2 x - 1 = 0$ in the interval $-180° \leq x \leq 180°$.

Q8 Solve the equation $4 \cos \frac{x}{2} - 3 \sin^2 \frac{x}{2} = -1$ in the interval $0° \leq x \leq 720°$.

Q9 Find all the solutions of the equation $\frac{\cos x}{\tan x} + \sin x = 3$ in the interval $-360° \leq x \leq 360°$.

Q10 Find all the solutions of the equation $\sin^2 x \cos x - \cos x = 0.86$ in the interval $-180° \leq x \leq 180°$.

Q11 Find all the solutions of the equation $4 \cos^2 x \tan x + \sin x = 0$ in the interval $0° \leq x \leq 360°$.

Q12 Find all the solutions of the equation $\frac{\cos^2 x}{\sin x - 1} = -0.25$ in the interval $-360° \leq x \leq 360°$.

Q13 Find the coordinates of all of the points of intersection between the graphs of the functions $f(x) = 2 + 3 \cos^2 x$ and $g(x) = 7 \sin x - 1$ in the interval $0 \leq x \leq 360°$, giving your answers to 3 s.f.

Review Exercise

Q1 Write down the exact values of cos 30°, sin 30°, tan 30°, cos 45°, sin 45°, tan 45°, cos 60°, sin 60° and tan 60°.

Q2 The points below lie on the unit circle. For each point, if a line from the origin to the point makes an angle of θ when measured in an anticlockwise direction from the positive x-axis, find the exact value of θ, where $0 \le \theta \le 180°$.

a) $\left(\frac{1}{2}, \frac{\sqrt{3}}{2}\right)$ b) $\left(\frac{\sqrt{3}}{2}, \frac{1}{2}\right)$ c) $(-1, 0)$

Q3 For triangle ΔABC, in which $A = 30°$, $C = 25°$ and $b = 6$ m:

a) Find all the sides and angles of the triangle.

b) Find the area of the triangle.

Q4 For triangle ΔPQR, in which $p = 13$ km, $q = 23$ km and $R = 20°$:

a) Find all the sides and angles of the triangle.

b) Find the area of the triangle.

Q5 Find all the angles in the triangle below in degrees to 1 d.p.

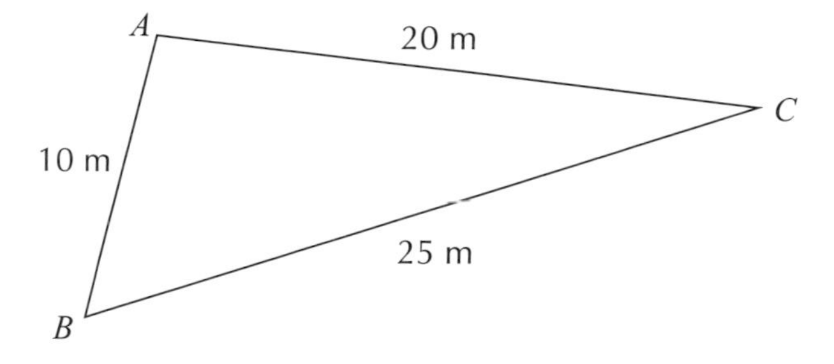

Q6 Find the missing sides and angles for the 2 possible triangles ΔABC which satisfy $b = 5$, $a = 3$, $A = 35°$.

PROBLEM SOLVING

Q7 Show that $\tan x - \sin x \cos x \equiv \sin^2 x \tan x$.

Q8 Show that $\tan^2 x - \cos^2 x + 1 \equiv \tan^2 x(1 + \cos^2 x)$.

Q9 Simplify: $(\sin y + \cos y)^2 + (\cos y - \sin y)^2$.

Q10 Show that $\frac{\sin^4 x + \sin^2 x \cos^2 x}{\cos^2 x - 1} \equiv -1$.

Q11 Sketch the following graphs in the interval $-360° \le x \le 360°$, making sure you label all of the key points.

a) $y = \cos x$ b) $y = \sin x$ c) $y = \tan x$

Review Exercise

Q12 Below is the graph of $y = \cos x$ and a transformation of the graph.
What is the equation of the transformed graph?

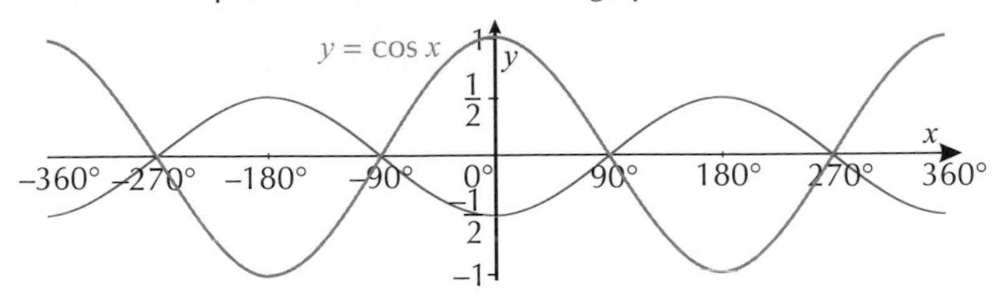

Q13 Below is a graph of $y = \sin x$ and a transformation of the graph.
What is the equation of the transformed graph?

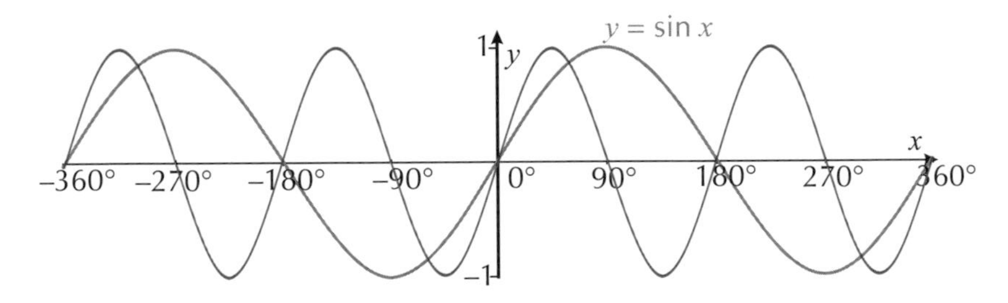

Q14 Sketch the following pairs of graphs on the same axes:

a) $y = \cos x$ and $y = \frac{1}{2} \cos x$ (for $0° \leq x \leq 360°$)

b) $y = \sin x$ and $y = \sin (x + 30°)$ (for $0° \leq x \leq 360°$)

c) $y = \tan x$ and $y = \tan 3x$ (for $0° \leq x \leq 180°$)

Q15 a) Solve each of these equations for $0° \leq \theta \leq 360°$:

(i) $\sin \theta = \frac{\sqrt{3}}{2}$ (ii) $\tan \theta = -1$ (iii) $\cos \theta = -\frac{1}{\sqrt{2}}$

b) Solve each of these equations for $-180° \leq \theta \leq 180°$ (giving your answers to 1 d.p.):

(i) $\cos 4\theta = -\frac{2}{3}$ (ii) $\sin (\theta + 35°) = 0.3$ (iii) $\tan \frac{\theta}{2} = 500$

Q16 Find all the solutions to $6 \sin^2 x = \cos x + 5$ in the interval $0° \leq x \leq 360°$, giving your answers to 1 d.p. where appropriate. PROBLEM SOLVING

Q17 Solve $3 \tan x + 2 \cos x = 0$ for $-90° \leq x \leq 90°$. PROBLEM SOLVING

Q18 Find all the solutions of the equation $8 \sin^2 x + 2 \sin x - 1 = 0$ in the interval $0° \leq x \leq 360°$, giving your answers to 1 d.p. where appropriate. PROBLEM SOLVING

Q19 Find all the solutions of the equation $\tan x - 3 \sin x = 0$ in the interval $0° \leq x \leq 720°$, giving your answers to 1 d.p. PROBLEM SOLVING

Exam-Style Questions

Q1 Prove that $\frac{\cos x}{\tan x} + \sin x \equiv \frac{1}{\sin x}$.

[3 marks]

Q2 The graph below shows the curve with equation $y = p \sin qx$ in the interval $-180° \leq x \leq 180°$.

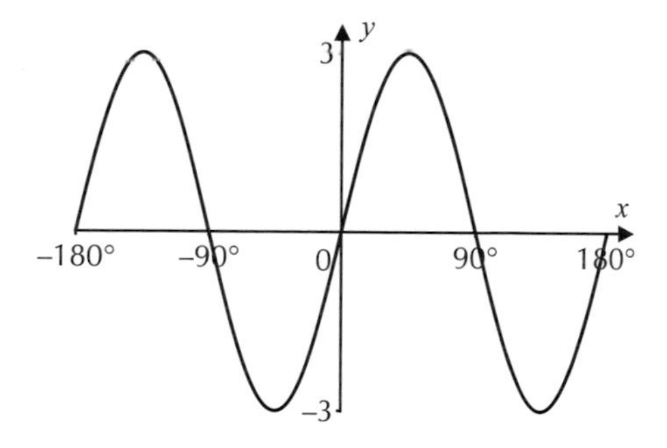

Find the values of p and q.

[3 marks]

Q3 The graph below shows the curve $f(x) = \tan x$ in the interval $-360° \leq x \leq 360°$.

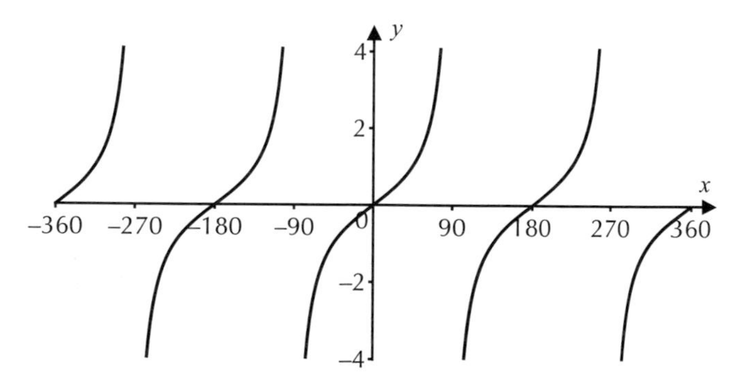

a) (i) Write down the coordinates of any points where the function $g(x) = \tan(x - 30°)$ intersects the x-axis.

[2 marks]

(ii) Write down the equations of the asymptotes of $g(x)$ in the given interval.

[2 marks]

b) Explain why the asymptotes exist in the function $f(x)$.

[1 mark]

c) Given that $\tan(-30°) = -\frac{\sqrt{3}}{3}$, use the graph to determine the value of $\tan 30°$.

[2 marks]

Exam-Style Questions

Q4 a) Given that $3 \sin x - 4 \cos x = 0$, show that $\tan x = \frac{4}{3}$.

[2 marks]

b) Hence, find the values of x in the interval $0° \leq x \leq 360°$ for which $3 \sin x - 4 \cos x = 0$, giving your answers to 1 decimal place.

[3 marks]

Q5 The function $T(t) = 15 + 8 \sin (45t)°$ is used to model how the temperature changes in a room over time, where T represents the temperature in °C and t represents time in hours.

a) What is the temperature after three hours?

[2 marks]

b) Within this model, what are the maximum and minimum possible temperatures?

[2 marks]

c) How many times will the temperature reach 20 °C during the first 8 hours? Show your working.

[6 marks]

Q6 Find the solutions of the equation $2 \sin (x + 30°) = \frac{1}{4}$ for $-360° \leq x \leq 360°$. Give your answers to 1 decimal place.

[5 marks]

Q7 In the triangle PQR, $PQ = 16$ m, $QR = 12$ m and $\angle RPQ = 37°$. Find the possible areas of the triangle, giving your answers to 3 significant figures.

[6 marks]

Q8 Given the functions $f(x) = 3 \cos^2 x$ and $g(x) = 1 - \sin x$, solve $f(x) = g(x)$ in the interval $0° \leq x \leq 360°$. Give your answers to 1 decimal place.

Solutions based only on graphical or numerical methods are not acceptable.

[6 marks]

Chapter 9 Exponentials and Logarithms

9.1 Exponentials

Exponentials are extremely useful functions — they're used in loads of different real-world scenarios, from radioactive decay to rabbit populations. You'll be seeing lots of 'Modelling' stamps in this chapter...

Learning Objectives (Spec Ref 6.1 & 6.2):

- Know the shape of the graph of the exponential function $y = a^x$ and its transformations.
- Understand the significance of the number e, and the function $y = e^x$.
- Know that the gradient of Ae^{kx} is kAe^{kx}.

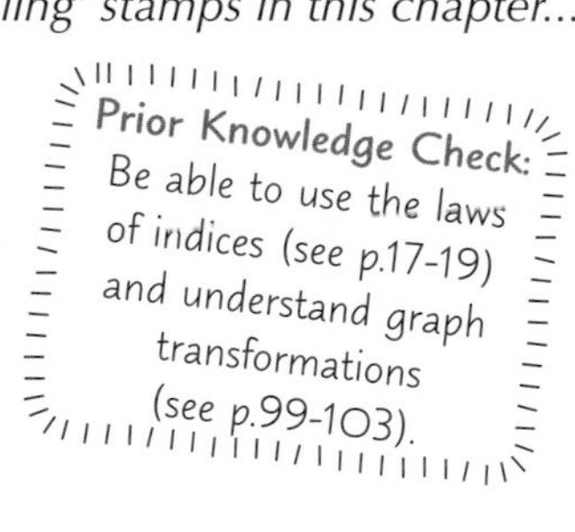

Exponentials

Exponentials are functions of the form $y = a^x$ (or $f(x) = a^x$), where $\boldsymbol{a > 0}$.
All graphs of exponential functions have the **same basic shape**.

$y = a^x$ for $a > 1$

- All the graphs go through 1 at $x = 0$ since $a^0 = 1$ for any a.
- $a > 1$ — so y **increases as x increases**.
- The **bigger** a is, the **quicker** the graph increases (so the curve is **steeper**).
- As x **decreases**, y **decreases** at a **smaller and smaller rate** — y approaches zero, but never actually gets there.
- So as $x \to \infty$, $y \to \infty$ and as $x \to -\infty$, $y \to 0$.

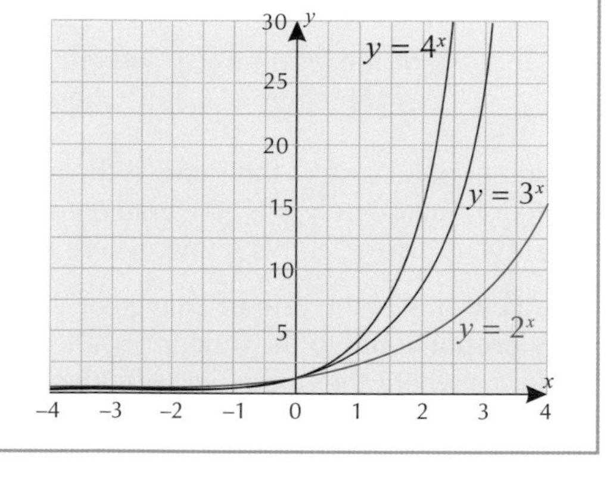

$y = a^x$ for $0 < a < 1$

- All the graphs go through 1 at $x = 0$ since $a^0 = 1$ for any a.
- $0 < a < 1$ — so y **decreases as x increases**.
- The **smaller** a is, the **faster** the graphs decrease (so the curve is **steeper**).
- As x **increases**, y **decreases** at a **smaller and smaller rate** — y will approach zero, but never actually get there.
- So as $x \to \infty$, $y \to 0$ and as $x \to -\infty$, $y \to \infty$.

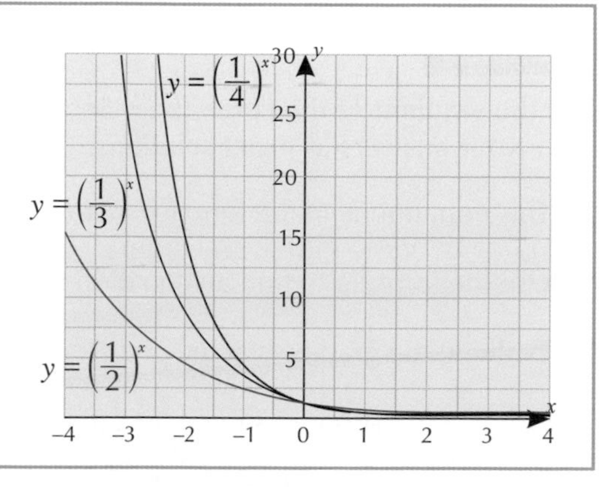

The notation $x \to \infty$ just means 'x tends to ∞' (i.e. x gets bigger and bigger). Similarly $y \to 0$ means 'y tends to 0' (gets smaller and smaller). a^x gets infinitely close to 0 but never reaches it, so $y = 0$ is an asymptote of the graph — see p.97).

The exponential function, e^x

The main feature of **exponential graphs** or functions is that the rate of increase or decrease of the function is **proportional** to the function itself.

You need to know about a value of 'a' for which the **gradient** of $y = a^x$ is **exactly the same** as a^x. That value is known as **e**, an **irrational number** around 2.7183 (it's stored in your calculator just like π). This **special case** of an exponential function, $y = e^x$, is called **'the' exponential function.**

Tip: An irrational number is a real number which can't be written as a fraction $\frac{a}{b}$ (where a and b are both integers and $b \neq 0$).

Because e is just a number, the graph of $y = e^x$ has all the properties of $y = a^x$:

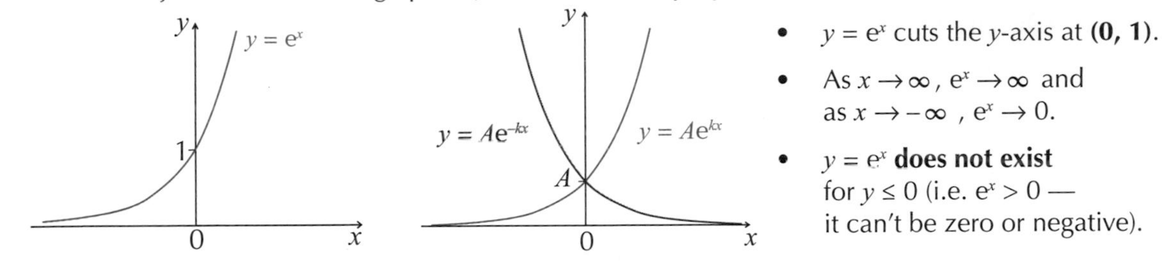

- $y = e^x$ cuts the y-axis at **(0, 1)**.
- As $x \to \infty$, $e^x \to \infty$ and as $x \to -\infty$, $e^x \to 0$.
- $y = e^x$ **does not exist** for $y \leq 0$ (i.e. $e^x > 0$ — it can't be zero or negative).

For **positive constants** A and k, the graph of Ae^{kx} looks similar to the graph of e^x, except that it cuts the y-axis at **(0, A)**. For the graph of Ae^{-kx}, you just reflect the graph of Ae^{kx} in the y-axis. The graphs of Ae^{kx} and Ae^{-kx} come up a lot in modelling (see p.177-179).

Gradient of $y = Ae^{kx}$

As mentioned above, an interesting property of $y = e^x$ is that its gradient is e^x. This can be generalised:

> For the graph $y = e^{kx}$, where k is a constant, its gradient is ke^{kx}.
> For the graph $y = Ae^{kx}$, where A and k are constants, its gradient is kAe^{kx}.

The gradient of a curve is the same as the gradient of the **tangent** to the curve at a point — for example, on the curve $y = e^x$, the gradient of the tangent to the curve at $x = 2$ is e^2 (so the gradient of the curve is also e^2 at this point). There's more on finding the gradient of a curve on p.187-188.

The gradient shows the **rate of change** of the function. So the rate of change of Ae^{kx} is **directly proportional** to the function itself. This means that an **exponential model** is suitable in situations where the rate of increase/decrease of y is proportional to the value of y.

If the **signs** for A and k are **different**, (e.g. $y = -2e^{3x}$ or $y = 4e^{-x}$) then the gradient will be negative — i.e. it's a **decreasing** function (see p.204).

Example 1

Find the gradient of the curve $f(x) = 5e^{8x}$ at the points $x = 0$ and $x = 2$. Leave your answers as exact solutions.

1. This equation is in the form $f(x) = Ae^{kx}$, where $A = 5$ and $k = 8$
2. Use the formula to find an equation for the gradient: $kAe^{kx} = 8 \times 5 \times e^{8x} = 40e^{8x}$
3. Evaluate the gradients at the points:

At $x = 0$, the gradient of $f(x) = 40e^{8 \times 0}$
$= 40e^0 = 40$

At $x = 2$, the gradient of $f(x) = 40e^{8 \times 2}$
$= 40e^{16}$

Tip: If a question asks for an **exact** answer, leave it in terms of e. You could also be asked for a certain number of decimal places or significant figures, e.g. $40e^{16} = 355\ 000\ 000$ to 3 s.f.

Example 2

The number of geese, g, at a nature reserve after t days is modelled using the formula $g = Ae^{kt}$, where A and k are constants. After 1 day, there were 80 geese, and the gradient of the curve plotted from the data collected was 241. Find A and k, giving your answers to 1 s.f.

MODELLING

1. "After 1 day" means $t = 1$.

   When $t = 1$, $g = 80 \Rightarrow 80 = Ae^k$.
   At this point, the gradient is 241 $\Rightarrow 241 = kAe^k$.

2. Solve these simultaneously by substituting $Ae^k = 80$ into $241 = kAe^k$.

   $241 = kAe^k \Rightarrow 241 = 80k \Rightarrow k = \frac{241}{80} = 3.0125$.
   So $k = 3$ (1 s.f.)

3. Put the unrounded value of k back into one of the original equations to find A. In real-life situations, A is called the initial value — and it's usually positive.

   $80 = Ae^k \Rightarrow 80 = Ae^{3.0125} \Rightarrow A = \frac{80}{e^{3.0125}} = 3.933...$
   So $A = 4$ (1 s.f.)

   The equation of the curve is $g = 4e^{3t}$.

Exercise 9.1.1

Q1 On the same axes, sketch the graphs of:

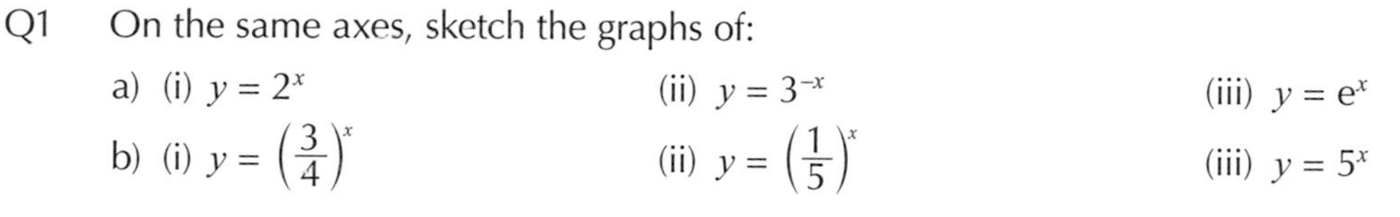

a) (i) $y = 2^x$ (ii) $y = 3^{-x}$ (iii) $y = e^x$

b) (i) $y = \left(\frac{3}{4}\right)^x$ (ii) $y = \left(\frac{1}{5}\right)^x$ (iii) $y = 5^x$

Q2 Match each of the following functions to one of the graphs below:

a) $y = 3e^{3x}$

b) $y = 3^x$

c) $y = 3e^{-x}$

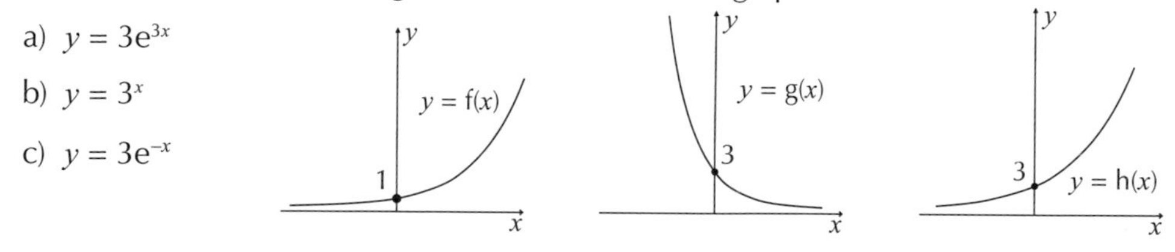

Q3 For the function $f(x) = 1.5^x$, match each of the following equations to one of the graphs below:

a) $y = f(x)$

b) $y = f(x) + 2$

c) $y = f(-x)$

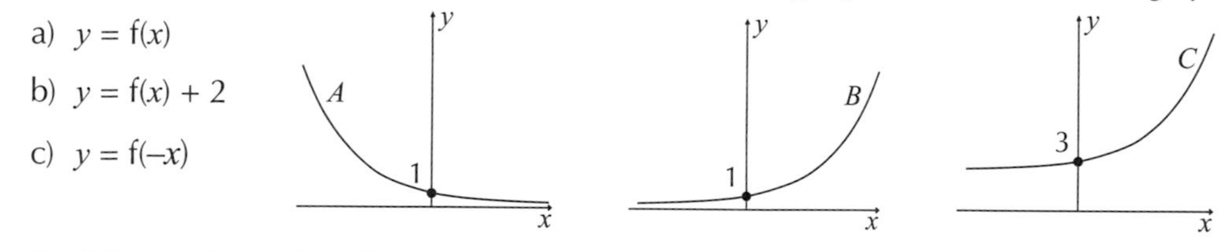

Q4 Find the gradient of each curve at the points given. Leave your answers in exact form.

a) $y = e^{3x}$ at the points $x = 0$ and $x = 1$

b) $y = 5e^{0.5x}$ at the points $x = -2$ and $x = 2$

c) $x = 2.5e^{6t}$ at the points $t = \frac{1}{3}$ and $t = 5$

d) $y = 4e^{-2x}$ at the points $x = -5$ and $x = 0.5$

Q5 The graph of $y = Ae^{kx}$ has gradient $12e^3$ at the point $x = 2$. Find the values of A and k, and hence find the exact y-coordinates when $x = 4$ and $x = 12$.

PROBLEM SOLVING

Q6 The population of rabbits in a wood is modelled using the function $f(t)$, where t is the number of years since the population was first counted. The gradient of the curve of $f(t)$ is always 0.4 times the value of $f(t)$. When first counted, there were 7 rabbits. Find the function $f(t)$ and use it to estimate the number of rabbits in the wood after 5 years.

9.2 Logs

'Logarithm' might sound scary, but it's just a fancy way of describing the power that something has been raised to. Once you know how to use logs, you can solve all sorts of equations that involve powers.

Learning Objectives (Spec Ref 6.3):
- Convert between index and log notation.
- Know and be able to use the natural logarithm, ln.
- Understand the relationship between exponentials and logs.

Prior Knowledge Check: Be able to use the laws of indices. See p.17-19.

Logs

A **logarithm** is just the power that a number needs to be **raised to** to produce a given value.

Before now, you've used **index notation** to represent powers, but sometimes it's easier to work with **log notation**. In **index notation**, 3^5 means that **5** lots of **3** are multiplied together. **3** is known as the **base**. You now need to be able to **switch** from index notation into **log notation**, and vice versa.

Log notation looks like this: $\log_a b = c$ ... which means the **same** as the **index notation**... $a^c = b$

- The little number 'a' after 'log' is the **base**.
- 'c' is the **power** the base is being **raised** to.
- 'b' is the answer you get when a is raised to the power c.
- Log means '**power**', so the log above really just means: "what is the power you need to raise a to if you want to end up with b?"

Tip: If you're struggling with this, try putting in some numbers — e.g. $\log_3 9 = 2$ is the same as $3^2 = 9$.

$\log_a b = c$ is the same as $a^c = b$, so:

$\log_a a = 1$ and $\log_a 1 = 0$

In index notation, this is saying that $a^1 = a$, and $a^0 = 1$.

- The **base** of a log must be a **positive number ≠ 1** (or the log isn't defined for some, or all, values).
- For $a > 1$, if $b > 1$, then c is **positive**. And if $0 < b < 1$, then c is **negative**. For $0 < a < 1$, the **opposite** is true.
- There are two bases that are more common than others: the first is **base 10**. The button marked 'log' on your calculator uses base 10.

Index notation: $10^2 = 100$

or

log notation: $\log_{10} 100 = 2$

So the **logarithm** of 100 to the **base 10** is 2, because 10 raised to the **power** of 2 is 100.

The base goes here but it's usually left out if it's 10.

- The other common base is **base e** — this is known as the **natural logarithm** and is written 'ln':

Index notation: $e^3 = 20.085...$

or

log notation: $\ln 20.085... = 3$

So the **natural logarithm** of 20.085... is 3, because e raised to the **power** of 3 is 20.085...

You don't need to write the base — 'ln' tells you that it's e.

Example 1

a) Write down the value of $\log_2 8$.

1. Compare to $\log_a b = c$.

   Here the base (a) is 2, and the answer (b) is 8.

2. So think about the power (c) that you'll need to raise 2 to to get 8.

   8 is 2 raised to the power of 3, so $2^3 = 8$ and $\log_2 8 = 3$

b) Write down the value of $\log_9 3$.

Work out the power that 9 needs to be raised to to get 3.

3 is the square root of 9, or $9^{\frac{1}{2}} = 3$, so $\log_9 3 = \frac{1}{2}$

c) Write $3^0 = 1$ using log notation.

1. You just need to make sure you get things in the right place.

   0 is the power (c) or logarithm that 3 (a, the base) is raised to to get 1 (b).

2. Substitute the values into $\log_a b = c$:

   So $\log_3 1 = 0$

Inverse Functions

An **inverse function** does the **opposite** to the function. So if the function was '+ 1', then the inverse would be '– 1'. And if the function was '× 2', the inverse would be '÷ 2'. The inverse for a function f(x) is written $\mathbf{f^{-1}(x)}$.

For the function $f(x) = a^x$, its inverse is $f^{-1}(x) = \log_a x$.
And for $g(x) = e^x$, its inverse is $g^{-1}(x) = \ln x$.

Tip: Not every function has an inverse. You'll learn more about which functions have inverses if you carry on to Year 2 of the A Level course.

Doing an inverse function to the original gets you **back to x** on its own. This gives you the following very useful formulas which will help you to solve equations later in the chapter:

$$a^{\log_a x} = x \qquad \log_a (a^x) = x$$

(for any positive $a \neq 1$)

$$e^{\ln x} = x \qquad \ln (e^x) = x$$

Tip: $\log_a a = 1$ and $\ln e = 1$.

The graph of an inverse function is a **reflection** in the line $y = x$:

- $y = \ln x$ is the **reflection** of $y = e^x$ in the line $\mathbf{y = x}$.
- It cuts the x-axis at **(1, 0)** (so **ln 1 = 0**).
- As $x \to \infty$, $\ln x \to \infty$, but it happens very slowly.
- As $x \to 0$, $\ln x \to -\infty$.
- $\ln x$ **does not exist** for $x \leq 0$ (i.e. x can't be zero or negative).

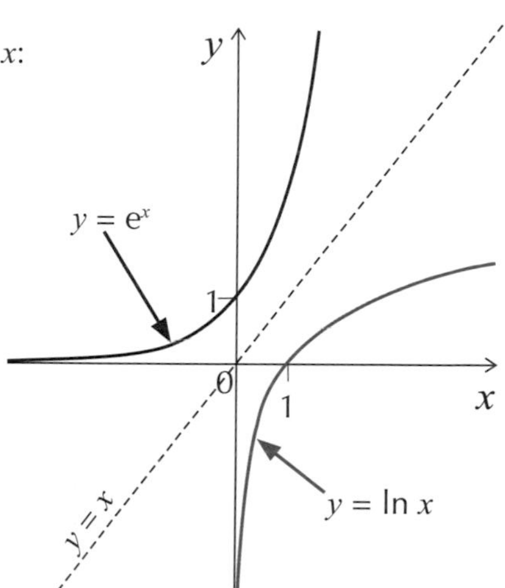

The graphs of $y = a^x$ and $y = \log_a x$ look very similar to the graphs for $y = e^x$ and $y = \ln x$ — they'll just have steeper or gentler curves, depending on the value of a.

Example 2

Solve the following equations, giving your answers as exact solutions.

a) $e^x = 10$

Apply ln to both sides to remove the 'e' function.

$e^x = 10$

$\ln e^x = x$ → $\ln e^x = \ln 10$

$x = \ln 10$

b) $\ln x = 2$

Apply e to both sides to remove the ln function.

$\ln x = 2$

$e^{\ln x} = x$ → $e^{\ln x} = e^2$

$x = e^2$

Exercise 9.2.1

In this exercise, log means $\log_{10}$.

Q1 Hint: Work out a, b and c first, then substitute them into $\log_a b = c$.

Q1 Write the following using log notation:

a) $2^3 = 8$ b) $5^4 = 625$ c) $49^{\frac{1}{2}} = 7$ d) $8^{\frac{2}{3}} = 4$

e) $10^{-2} = \frac{1}{100}$ f) $2^{-3} = 0.125$ g) $4^x = 9$ h) $x^3 = 40$

i) $8^{11} = x$ j) $64^{\frac{3}{2}} = 512$ k) $25^{-\frac{1}{2}} = \frac{1}{5}$ l) $x^{2y} = 4096$

Q2 Write the following using index notation (you don't need to work out any values):

a) $\ln a = 6$ b) $\log_5 t = 0.2$ c) $\log_4 m = 1$ d) $\ln p = 13$

e) $\log k = 5$ f) $\log_x a = m$ g) $\ln k = z$ h) $\log q = r$

Q3 Find the value of the following. Give your answer to 3 d.p. where appropriate.

a) $\log 1000$ b) $\log 0.01$ c) $\log 3$ d) $\log 125$

e) $\ln 2$ f) $\ln 1$ g) $\ln 6$ h) $\ln 20$

Q4 Without using your calculator, find the value of:

a) $\log_2 4$ b) $\log_3 27$ c) $\log_5 0.2$ d) $\log_{0.5} 0.25$

Q5 Find the value of x, where $x \geq 0$, by writing the following in index notation:

a) $\log_x 49 = 2$ b) $\log_x 8 = 3$ c) $\log_x 100\,000 = 5$ d) $\log_x 3125 = 5$

e) $\log_x 3 = \frac{1}{2}$ f) $\log_x 7 = \frac{1}{3}$ g) $\log_x 2 = \frac{1}{5}$ h) $\log_x 9 = \frac{1}{4}$

Q6 Solve these equations, giving (i) an exact solution, and (ii) a solution correct to 3 s.f.

a) $e^x = 5$ b) $\ln x = 8$ c) $e^{3t} = 11$ d) $\ln 10x = 4$

Q7 In each part, use index notation to write y in terms of x, given that: PROBLEM SOLVING

a) $\log_a x = 2$ and $\log_a y = 4$ b) $\log_a x = 3$ and $\log_{2a} y = 3$ c) $\ln x = 5$ and $\ln y = 20$

9.3 Solving Equations

Once you know what logs are, you need to learn what they can do. There are three 'laws of logs' which are really useful for simplifying tricky expressions with lots of logs into expressions with just one log.

Learning Objectives (Spec Ref 6.4 & 6.5):
- Know and be able to use the laws of logs.
- Know and be able to use the change of base formula for logs.
- Solve equations of the form $a^x = b$.

Prior Knowledge Check: Be able to solve quadratic equations involving functions of x. See p.37.

Laws of logs

You'll need to be able to **simplify** expressions containing logs in order to answer trickier questions — e.g. to **add** or **subtract** two logs you can combine them into one. To do this, you'll need to use the **laws of logarithms**. These **only work** if the **base** of each log is the **same**:

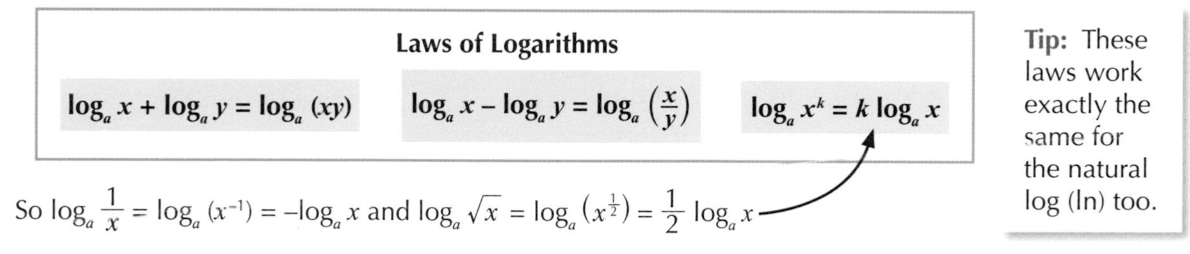

Laws of Logarithms

$\log_a x + \log_a y = \log_a (xy)$ $\quad$ $\log_a x - \log_a y = \log_a \left(\frac{x}{y}\right)$ $\quad$ $\log_a x^k = k \log_a x$

So $\log_a \frac{1}{x} = \log_a (x^{-1}) = -\log_a x$ and $\log_a \sqrt{x} = \log_a (x^{\frac{1}{2}}) = \frac{1}{2} \log_a x$

Tip: These laws work exactly the same for the natural log (ln) too.

Example 1

Simplify the following:

a) $\log_3 4 + \log_3 5$

1. Check that the logs you're **adding** have the same base — both are base 3 so you can combine them.
2. Use the law $\log_a x + \log_a y = \log_a (xy)$. $\quad$ $\log_3 4 + \log_3 5 = \log_3 (4 \times 5) = \log_3 20$

b) $\log_3 4 - \log_3 5$

1. You're **subtracting** one log from another, and the bases are the same.
2. Use the law $\log_a x - \log_a y = \log_a \left(\frac{x}{y}\right)$. $\quad$ $\log_3 4 - \log_3 5 = \log_3 \left(\frac{4}{5}\right)$

c) $3 \ln 2 + 2 \ln 5$

1. Use the law $\ln x^k = k \ln x$ to get rid of the 3 and 2 in front of the logs. $\quad$ $3 \ln 2 = \ln (2^3) = \ln 8$
   $2 \ln 5 = \ln (5^2) = \ln 25$
2. The logs both have the same base (e) so the expression becomes: $\quad$ $\ln 8 + \ln 25 = \ln (8 \times 25) = \ln 200$

Tip: $\ln = \log_e$, so you can use the laws of logs in the same way.

Example 2

a) Write the expression $2 \log_a 6 - \log_a 9$ in the form $\log_a n$, where n is a number to be found.

1. Use $\log_a x^k = k \log_a x$ to simplify $2 \log_a 6$: $2 \log_a 6 = \log_a 6^2 = \log_a 36$

2. Then use $\log_a x - \log_a y = \log_a \left(\frac{x}{y}\right)$: $\log_a 36 - \log_a 9 = \log_a (36 \div 9) = \log_a 4$

b) Write the expression $\ln \frac{(\mathrm{e}x)^2}{y^3}$ in terms of $\ln x$ and $\ln y$.

Use the laws of logs to break up the expression:

1. Use $\log_a \left(\frac{x}{y}\right) = \log_a x - \log_a y$: $\ln \frac{(\mathrm{e}x)^2}{y^3} = \ln (\mathrm{e}x)^2 - \ln y^3$
2. Square the bracket: $= \ln \mathrm{e}^2x^2 - \ln y^3$
3. Use $\log_a (xy) = \log_a x + \log_a y$ on $\ln \mathrm{e}^2x^2$: $= \ln \mathrm{e}^2 + \ln x^2 - \ln y^3$
4. Use $\log_a x^k = k \log_a x$ on each term: $= 2 \ln \mathrm{e} + 2 \ln x - 3 \ln y$
5. Finally, use $\log_a a = 1$, to write $2 \ln \mathrm{e}$ as 2: $= 2 + 2 \ln x - 3 \ln y$

Exercise 9.3.1

Q1 Write each of the following in the form $\log_x n$ (or $\ln n$), where n is a number to be found:

a) $\log_a 2 + \log_a 5$ b) $\ln 8 + \ln 7$ c) $\log_b 8 - \log_b 4$

d) $\log_m 15 - \log_m 5$ e) $3 \log_a 4$ f) $2 \ln 7$

g) $\frac{1}{2} \log_b 16$ h) $\frac{2}{3} \log_a 125$ i) $\frac{1}{5} \ln 4^5$

Q2 Write in the form $a \log b$, where b is the smallest possible positive integer:

a) $\log 0.5$ b) $\log \sqrt{3}$ c) $\log 0.25$ d) $\log \sqrt[3]{5}$

Q3 Write each of the following expressions as a single log:

a) $2 \log_a 5 + \log_a 4$ b) $3 \log_m 2 - \log_m 4$ c) $3 \ln 4 - 2 \ln 8$

d) $\frac{2}{3} \ln 216 - 2 \ln 3$ e) $1 + \log_a 6$ f) $2 - \log_b 5$

Q4 If $\log_a 2 = x$, $\log_a 3 = y$ and $\log_a 5 = z$, write in terms of x, y and z:

a) $\log_a 6$ b) $\log_a 16$ c) $\log_a 60$

Q4 Hint: Rewrite the numbers as products of their prime factors.

Q5 Simplify each of the following as much as possible:

a) $\log_b b^3$ b) $\log_a \sqrt{a}$ c) $\ln 4\mathrm{e} - 2 \ln 2$ d) $\ln 9 + \ln \frac{\mathrm{e}}{3} - \ln 3$

Q6 Show that: a) $\log_2 4^x = 2x$ b) $\frac{\ln 54 - \ln 6}{\ln 3} = 2$

PROBLEM SOLVING

Q7 Find the value of $4 + \log_c \frac{1}{c^2} + \log_c \sqrt{c}$.

Changing the base of a log

As well as the three laws on page 171, there is another rule to learn which lets you **change bases**.

- Your calculator probably has a button to work out any log for you, but some scientific calculators can only work out $\log_{10}$ or **ln** — this makes it trickier to calculate logs with a **different base**.
- You can **change the base** of any log to **any other base** using this formula:

$$\textbf{Change of Base: } \log_a x = \frac{\log_b x}{\log_b a}$$

- So $\log_a x = \frac{\log_{10} x}{\log_{10} a}$ or $\log_a x = \frac{\ln x}{\ln a}$.
- Even if your calculator **can** work out any log, you still need to **learn** this — it's quite useful when it comes to solving equations with logs.

Examples

By converting to $\log_{10}$:

a) Find the value of $\log_7 4$ to 4 d.p.

Tip: You're trying to work out what power you'd need to raise 7 to to get 4.

1. Change the base of the log to 10.
2. Here, $a = 7$ and $x = 4$. And we want $b = 10$. $\log_7 4 = \frac{\log_{10} 4}{\log_{10} 7} = 0.7124$ (4 d.p.)
3. Check this on your calculator by doing $7^{0.7124...} = 4$.

b) Find the value of $\log_3 2$ to 4 d.p.

1. Change the base of the log to 10.
2. Here, $a = 3$ and $x = 2$. And we want $b = 10$. $\log_3 2 = \frac{\log_{10} 2}{\log_{10} 3} = 0.6309$ (4 d.p.)

Exercise 9.3.2

Q1 Write the following in terms of $\log_{10}$:

a) $\log_9 2$ b) $\log_4 8$ c) $\log_{17} 16$ d) $\log_{21} 14$

Q2 Find the value of the following logs to 3 s.f.:

a) $\log_6 3$ b) $\log_9 2$ c) $\log_3 13$ d) $\log_5 4$

Q3 By changing the base, write each of the following expressions as a single log: PROBLEM SOLVING

a) $\frac{\log_{10} 19}{\log_{10} 11}$ b) $\frac{\log_6 2}{\log_6 7}$ c) $\log_3 4 \times \log_4 5$ d) $\ln 2 \times \log_2 10$

Q4 $\log_{10} 3x = \frac{\ln 2}{\ln 10}$. Find the exact value of x without using a calculator. PROBLEM SOLVING

Q5 Write the following in the form $k \ln a$, where a is the smallest possible integer: PROBLEM SOLVING

$$\left(\frac{\log_3 2}{\log_3 e} + \log_e 3\right) - \left(\log_e 6 + \frac{\log_2 9}{\log_2 e}\right)$$

Solving equations

You saw earlier on page 169 that exponentials and logs are the inverses of each other, and that: $a^{\log_a x} = x = \log_a a^x$

You can prove this using the laws of logs: $\log_a a^x = x \log_a a = x$.

Tip: You can also write a^x as $e^{x \ln a}$: $e^{x \ln a} = e^{\ln a^x} = a^x$

So you can use logs to **get rid** of exponentials and vice versa — including using e^x and $\ln x$ to cancel each other out, as you saw on page 169. This is useful for solving equations.

Example 1

Solve $2^{4x} = 3$ to 3 significant figures.

Tip: Here, a log of base 10 has been used, but you could use ln (i.e. base e) instead.

1. To solve the equation, you want x on its own, so take logs of both sides: $\log 2^{4x} = \log 3$
2. Use the log rule: $\log x^k = k \log x$. $4x \log 2 = \log 3$
3. Divide both sides by $4 \log 2$ to get x on its own: $x = \frac{\log 3}{4 \log 2}$
4. $\frac{\log 3}{4 \log 2}$ is just a number you can find using a calculator: $x = 0.396$ (to 3 s.f.)

Example 2

Solve $7 \log_{10} x = 5$ to 3 significant figures.

1. You want x on its own, so divide both sides by 7: $\log_{10} x = \frac{5}{7}$
2. Take exponentials of both sides by doing '10 to the power of' both sides (since the log is to base 10): $10^{\log_{10} x} = 10^{\frac{5}{7}}$
3. Logs and exponentials are inverse functions, so they cancel out: $x = 10^{\frac{5}{7}}$
4. Again, $10^{\frac{5}{7}}$ is just a number you can find using a calculator: $x = 5.18$ (to 3 s.f.)

Problems involving exponentials and logs

In an exam, you might be asked to solve an equation where you have to use a **combination** of the methods covered in this chapter. It can be tricky to work out what's needed, but just remember that you're trying to get ***x* on its own** — and think about which laws will help you do that.

If you're asked to give an **exact** solution, leave your answer in exponential or log form (i.e. don't actually calculate the decimal value of it).

Example 3

Solve $\ln(2x - 1) = 2$, giving your answer as an exact solution.

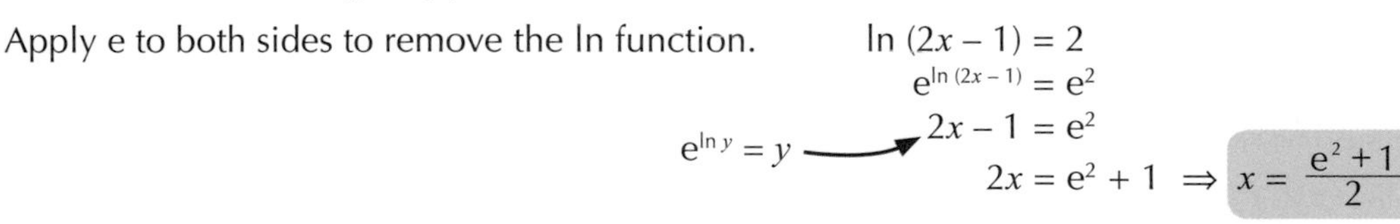

Example 4

Find the two exact solutions of the equation $e^x + 5e^{-x} = 6$.

1. A big clue here is that you're asked for more than one solution — think quadratics... (see page 37)
2. Multiply each part of the equation by e^x to get rid of e^{-x}. Remember basic powers laws: $(e^x)^2 = e^{2x}$ and $e^{-x} \times e^x = e^0 = 1$

$$e^x + 5e^{-x} = 6$$
$$e^{2x} + 5 = 6e^x$$
$$e^{2x} - 6e^x + 5 = 0$$

3. Substitute y for e^x to get a quadratic in y. Since you're asked for exact solutions, it will probably factorise.

$$y^2 - 6y + 5 = 0$$
$$(y - 1)(y - 5) = 0$$
$$y = 1 \text{ and } y = 5$$

4. Put e^x back in and apply the inverse function $\ln x$ to both sides.

$$e^x = 1 \quad \text{and} \quad e^x = 5$$
$$\ln e^x = \ln 1 \quad \text{and} \quad \ln e^x = \ln 5$$
$$x = \ln 1 = 0 \quad \text{and} \quad x = \ln 5$$

$\ln e^x = x$

Example 5

Solve $6^{x-2} = 3^x$, giving your answer to 3 s.f.

1. Start by taking logs of both sides (you can use any base, so use 10).

$$\log 6^{x-2} = \log 3^x$$

2. Now use $\log x^k = k \log x$ on both sides:

$$(x - 2) \log 6 = x \log 3$$

3. Multiply out the brackets and collect all the x terms on one side:

$$x \log 6 - 2 \log 6 = x \log 3$$
$$x \log 6 - x \log 3 = 2 \log 6$$
$$x (\log 6 - \log 3) = 2 \log 6$$

4. Use $\log_a x - \log_a y = \log_a \left(\frac{x}{y}\right)$ on the bracket:

$$x (\log 2) = 2 \log 6$$
$$\Rightarrow x = \frac{2 \log 6}{\log 2} = 5.17 \text{ (3 s.f.)}$$

Example 6

Solve $7^{30x} = 5^{70}$, giving your answer as a single log in the form $a \log_b c$, where a is a number and b and c are integers.

Take logs of both sides.

$$\log_7 7^{30x} = \log_7 5^{70}$$
$$\Rightarrow 30x \log_7 7 = 70 \log_7 5$$
$$\Rightarrow 30x = 70 \log_7 5$$
$$\Rightarrow x = \frac{7}{3} \log_7 5$$

Using $\log_a x^k = k \log_a x$

Tip: By taking logs of base 7, you can replace $\log_7 7$ with 1, which simplifies your working. You could take logs of a different base — but you'd have to use the change of base formula for your final answer.

Exercise 9.3.3

Q1 Solve each of these equations for x, to 3 s.f.

a) $2^x = 3$ b) $7^x = 2$ c) $1.8^x = 0.4$ d) $0.7^x = 3$

e) $3^{5x} = 890$ f) $0.2^{4x} = 0.016$ g) $2^{3x-1} = 5$ h) $0.4^{5x-4} = 2$

Q2 Solve these equations, giving your answers in the form $a \log_b c$, where a, b and c are integers, and c is as small as possible:

a) $2^{4x} = 3^{100}$ b) $11^{6x} = 10^{90}$ c) $6^{50-x} = 2^{50}$ d) $4^{5+x} = 20^5$

Q3 Find the value of x for each case:

a) $\log 5x = 3$ b) $\log_2 (x + 3) = 4$ c) $\log_3 (5 - 2x) = 2.5$

Q4 Solve each of these equations for x:

a) $4^{x+1} = 3^{2x}$ b) $2^{5-x} = 4^{x+3}$ c) $3^{2x-1} = 6^{3-x}$

Q5 Find the value(s) of x which satisfy each of the following equations: PROBLEM SOLVING

a) $\log_6 x = 1 - \log_6 (x + 1)$ b) $\log_2 (2x + 1) = 3 + 2 \log_2 x$

Q6 Solve these equations, giving your answers as exact solutions. PROBLEM SOLVING

a) $5e^{3t} = 11$ b) $e^{(0.5x+3)} = 9$ c) $10 - 3e^{(1-2x)} = 8$

d) $3 \ln (2x) = 7$ e) $\ln (5t - 3) = 4$ f) $6 - \ln (0.5x) = 3$

Q7 Solve these equations, giving your answers in terms of ln 3. PROBLEM SOLVING

a) $e^{3x} = 27$ b) $e^{(6x-1)} = \frac{1}{3}$ c) $\frac{1}{3}e^{(1-x)} - 3 = 0$

Q8 Solve these equations, giving exact answers. PROBLEM SOLVING

a) $\ln 5 + \ln x = 7$ b) $\ln (2x) + \ln (3x) = 15$

c) $2\ln x - \ln (2x) = 2$ d) $\ln (2x - 7) + \ln 4 = -3$

e) $\ln (x^2 - 4) - \ln (2x) = 0$ f) $3 \ln (x^2) + 5 \ln x = 2$

g) $2e^{2x} + e^x = 3$ h) $e^{8x} - e^{4x} - 6 = 0$

Q9 Solve the equations $9^{x-2} = 3^y$ and $\log_3 2x = 1 + \log_3 y$ simultaneously. PROBLEM SOLVING

Q10 Find the exact solutions of $2(10^{2x}) - 7(10^x) + 5 = 0$. PROBLEM SOLVING

Q11 Find, where possible, the solutions of the following equations: PROBLEM SOLVING

a) $2^{2x} - 5(2^x) + 4 = 0$ b) $4^{2x} - 17(4^x) + 16 = 0$ c) $3^{2x+2} - 82(3^x) + 9 = 0$

d) $2^{2x+3} - 9(2^x) + 1 = 0$ e) $e^{4x} + 4e^{2x} + 5 = 0$ f) $3e^{2x} + 10e^x + 3 = 0$

Q12 Find the smallest integer P such that $1.5^P > 1\,000\,000$. PROBLEM SOLVING

9.4 Modelling Exponential Growth and Decay

MODELLING

Modelling growth and decay means using a formula to predict how something will increase or decrease. It's nothing you haven't seen before, but you need to be able to deal with the wordy bits.

Learning Objectives (Spec Ref 6.7):
- Understand and be able to use exponential functions as models for real-world situations, in particular growth and decay.
- Understand the limitations of models using exponential functions.

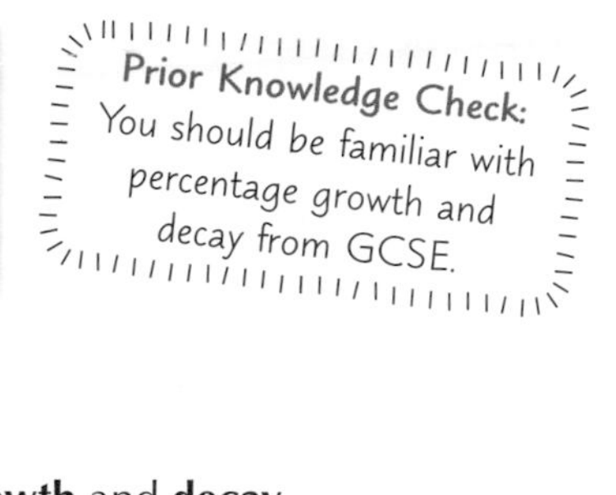

Modelling exponential growth and decay

Logs can be used to solve **real-life** problems involving **exponential growth** and **decay**.

- Exponential **growth** is when the rate of growth **increases** faster as the amount gets bigger.
- Exponential **decay** is just **negative** exponential growth.
  The **rate** of decay gets slower and slower as the amount gets smaller.

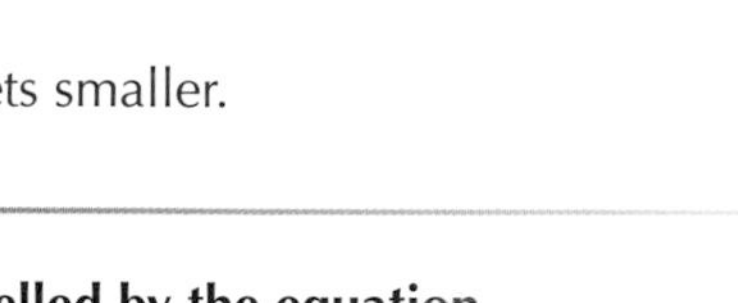

Example 1

The exponential growth of a colony of bacteria can be modelled by the equation $B = 60e^{0.03t}$, where B is the number of bacteria and t is the time in hours from the point at which the colony is first monitored ($t \geq 0$). Use the model to:

a) Work out the initial population of bacteria.

The initial population of bacteria is given by the formula when $t = 0$.

$B = 60e^{0.03t} = 60e^{(0.03 \times 0)} = 60e^0 = 60 \times 1$ ← $e^0 = 1$

$B = 60$

b) Predict the number of bacteria after 4 hours.

1. Substitute $t = 4$ into the equation to find B after 4 hours.
2. Round down because you want to know the number of whole bacteria.

$B = 60 \times e^{(0.03 \times 4)}$
$= 60 \times 1.1274... = 67.6498...$

So $B = 67$ bacteria

c) Predict the time taken for the colony to grow to 1000.

1. You need to find the time, t, when the population is 1000, so substitute in the value of B.

$B = 1000$ so:
$1000 = 60e^{0.03t}$
$e^{0.03t} = 1000 \div 60 = 16.6666...$

2. Take 'ln' of both sides.

$\ln e^{0.03t} = \ln (16.6666...)$
$0.03t = 2.8134...$
$t = 2.8134... \div 0.03$
$= 93.8$ hours (3 s.f.)

Example 2

£350 is initially paid into a bank account that pays 3% interest per year. No further money is deposited or withdrawn from the account. Create a model to show how much money will be in the account after t years. Use this model to calculate how many whole years it will be before there is over £1000 in the account.

1. After 1 year there will be £350 × 1.03, after two years there will be £350 × 1.03^2, etc. So after t years, there will be **£350 × 1.03^t**.
2. Now use this model to find t when $350 \times 1.03^t > 1000$.

$$350 \times 1.03^t > 1000$$
$$\Rightarrow \quad 1.03^t > \frac{20}{7}$$

3. Take logs of both sides.

$$\log 1.03^t > \log \frac{20}{7}$$
$$t \log 1.03 > \log \frac{20}{7}$$

4. log 1.03 > 0, so you can divide both sides of the inequality by it.

$$t > \frac{\log \frac{20}{7}}{\log 1.03}$$
$$\Rightarrow \quad t > 35.516...$$

Tip: If $0 < x < 1$, then $\log x < 0$, so you would need to flip the inequality sign when dividing.

So there'll be over £1000 in the account after **36 years.**

Example 3

The concentration (C) of a drug in the bloodstream, t hours after taking an initial dose, decreases exponentially according to $C = Ae^{-kt}$, where A and k are constants. If the initial concentration is 0.72, and this halves after 5 hours, find the values of A and k and sketch a graph of C against t.

1. 0.72 is the initial concentration — so start by putting this information into the equation and solving for A.

When $t = 0$, $C = 0.72$
So $0.72 = A \times e^0$
$= A \times 1 \Rightarrow A = 0.72$

2. After 5 hours the initial concentration has halved — so you can put in the value of C at $t = 5$, and then solve for k.

When $t = 5$, $C = 0.72 \div 2 = 0.36$

$$0.36 = 0.72 \times e^{-5k}$$
$$0.36 = \frac{0.72}{e^{5k}}$$
$$e^{5k} = \frac{0.72}{0.36} = 2$$

Apply 'ln' to solve for k.

$$\ln e^{5k} = \ln 2$$
$$5k = \ln 2$$
$$k = \ln 2 \div 5 = 0.139 \text{ to 3 s.f.}$$

3. So the equation is **$C = 0.72e^{-0.139t}$**

4. Sketch the graph — the trick to sketching exponential functions is to identify any intercepts and asymptotes.

You only need to draw the graph for $t \geq 0$, since time is always positive.

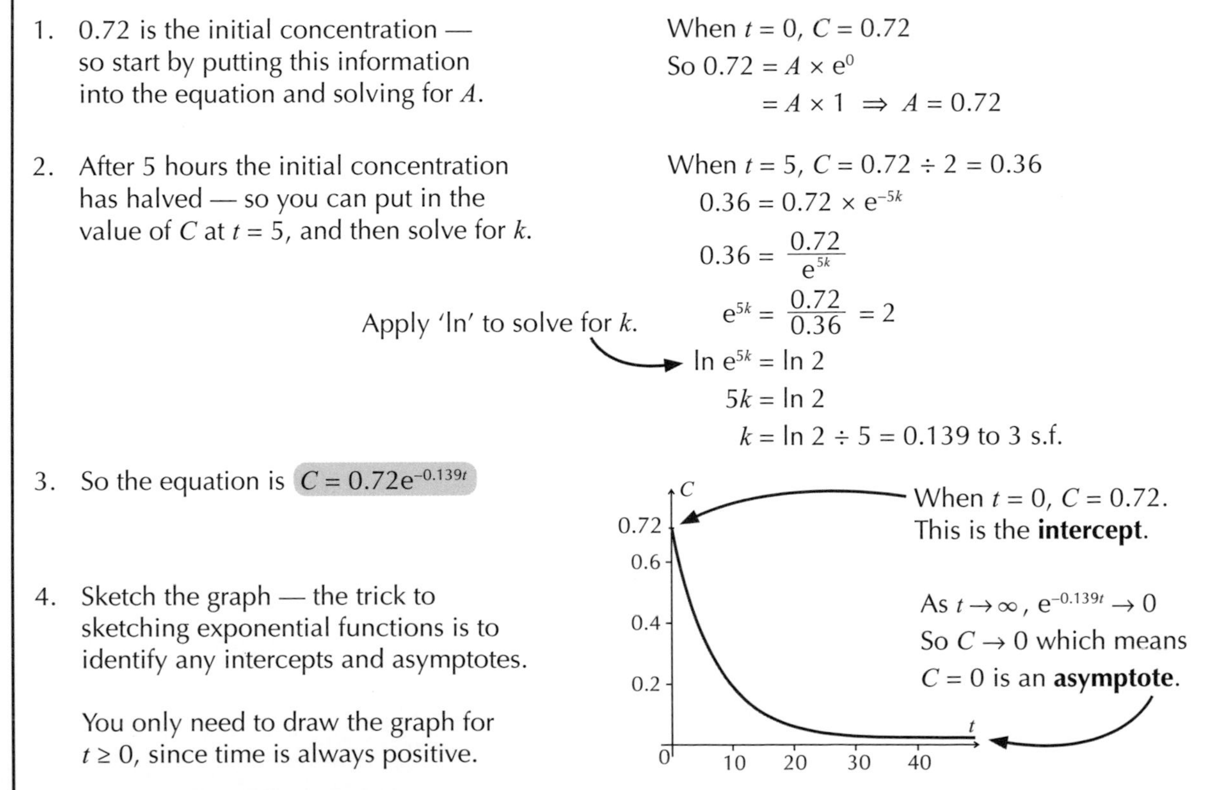

Limitations of modelling

The real world can't always be simplified to a couple of variables — it's usually much more **complicated** than that. Models **ignore** most factors, leaving only the ones you're interested in. This simplification means that models aren't spot on — they have **drawbacks**:

- Often an exponential model will match the real-world scenario for only a **short amount of time** — beyond that the numbers get too big or too small.
- A model may need tweaking for different values to reflect this — for example, there may be an **upper limit** to stop a value increasing to infinity.

You can only use the model to make predictions within the time frame of the model. If it's outside the time frame, then you're **extrapolating**, which is much less accurate.

Example 4

The penguin population of a small island is surveyed. The population, P, can be modelled by the formula $P = 5000e^{0.1t}$, where t is the number of years after the initial survey.

a) What does the '5000' in the formula represent?

5000 is the initial number of penguins (when $t = 0$).

b) Explain why this model may not be appropriate for the long term.

After 60 years ($t = 60$), the penguin population is over 2 million. This seems unrealistic — it's much too large a population. The model doesn't take into account other factors (e.g. predators, food supply) and allows the population to grow infinitely.

Exercise 9.4.1

Give your answers correct to 3 significant figures.

Q1 A radioactive substance has a half-life of 10 years. Its decay is modelled by the equation $A = A_0e^{-kt}$, where A is the activity in Bq (becquerel) after t years and A_0 and k are constants.

a) After how many years will the substance be reduced to a quarter of its original activity?

b) Find the original activity if the activity after 5 years is 200 Bq.

c) Find the activity remaining after 15 years.

Q1 Hint: Half-life is the length of time it takes for the activity of the sample to halve.

Q2 An oven is turned on at 12:00. After t minutes its temperature, T °C, is given by the formula: $T = 225 - 207\,e^{-\frac{t}{8}}$

a) What was the initial temperature of the oven?

b) What was the temperature after 5 minutes?

c) At what time does the oven reach a temperature of 190 °C?

d) Sketch the graph of T against t.

e) Explain how the model restricts the temperature from rising indefinitely.

Q2d) Hint: Make sure you label any significant points, e.g. where the curve meets any axes.

Q3 The value of a motorbike (£V) varies with age (in t years from new) according to $V = 7500e^{-0.2t}$.

a) What is its value after 10 years (to the nearest £)?

b) After how many years will the motorbike's value have fallen below £500?

c) Sketch a graph showing how the value of the motorbike varies with age, labelling any key points.

Q4 A fungus is being grown under controlled conditions in a laboratory. Initially, it covers an area of 4 mm^2. After t hours, its area is F mm^2, where $F = F_0e^{gt}$ (F_0 and g are constants). After 6 hours its area is 10 mm^2.

a) Find the values of F_0 and g.

b) Predict the area of the fungus after 12 hours.

c) How long will it take for the fungus to grow to 15 mm^2?

d) Describe one limitation of the model used.

Q5 A woman is prescribed a medicine, and the concentration of the medicine in her bloodstream is monitored. Initially the concentration in her bloodstream is 3 mg per litre of blood (mg/l). After t hours, the concentration of the drug is N mg/l, where $N = Ae^{-t}$.

a) What is the concentration after 30 minutes?

b) How long does it take for the level to reduce to 0.1 mg/l?

c) Sketch the graph of N against t.

d) What is the gradient of the graph in terms of t?

Q5a) Hint: Read the units carefully — in the question t is in hours but here you're asked about minutes, so you need to convert first.

Q6 The value of a car (£V) t years after purchase can be modelled by the formula: $V = 1500 + 9000\,e^{-\frac{t}{3}}$.

a) Explain the significance of the negative coefficient of t.

b) What was its price when new?

c) What was its value after 5 years?

d) After how many whole years will it be worth less than £2500?

e) Sketch the graph of V against t.

Q7 A forest fire spreads in such a way that the burnt area (H hectares) after t hours is given by the relation $H = 20e^{bt}$. Assume that the fire burns unchecked.

a) Interpret the value 20 used in the model.

b) If $e^b = 1.8$, find b.

c) Find the area burnt after 3 hours.

d) How long would it take to burn an area of 500 hectares?

e) What constant factor is the burnt area multiplied by every hour? What percentage does the burnt area increase by each hour?

f) Describe one limitation of the model.

9.5 Using Logarithmic Graphs

Logarithms can be used to express exponential equations in linear form — they're often much simpler to work with graphically in this format.

Learning Objective (Spec Ref 6.6):

- Plot equations of the form $y = ax^n$ and $y = kb^x$ as straight line graphs, and use these graphs to estimate values.

Prior Knowledge Check: You should remember how to plot scatter graphs with lines of best fit, and find the equations of straight line graphs from GCSE.

Logarithmic graphs in linear form

Equations of the form $\boldsymbol{y = ax^n}$ or $\boldsymbol{y = kb^x}$ can be a bit awkward to use. Fortunately, you can use the **laws of logs** to rewrite them in the form $\boldsymbol{y = mx + c}$, which is much easier to work with. Graphs of $y = ax^n$ or $y = kb^x$ can get **very steep very quickly**, but linear graphs are much more straightforward — and you can use them to read off values, find the gradient of the line, etc.

Logarithmic graphs for $y = ax^n$

To convert $y = ax^n$ to linear form, just **take logs** of both sides and rearrange:

$$y = ax^n \Rightarrow \log y = \log ax^n$$
$$\Rightarrow \log y = \log a + \log x^n$$
$$\Rightarrow \log y = n\log x + \log a$$

This is in the **straight-line** form $y = mx + c$. n is the gradient and $\log a$ is the vertical intercept.

When drawing the graph, plot the values of $\log y$ against $\log x$ and label the axes accordingly.

If you're given values for x and y, you can plot $\log y$ against $\log x$. The line of best fit will have the equation $\log y = n\log x + \log a$, so from this you can work out the equation for y in terms of x.

Example 1

The number of employees, p, working for a company t years after it was founded can be modelled by the equation $p = at^b$. The table below shows the number of employees the company has:

Age of company (t years)	2	5	8	13	25
Number of employees (p)	3	7	10	16	29

MODELLING

Plot a linear graph to represent this data, and use this to find the values a and b.

1. $p = at^b$ can be rearranged into the linear form $\log p = b\log t + \log a$.

   Make a table of the values of $\log t$ and $\log p$, using p and t given above:

$\log t$ (3 d.p.)	0.301	0.699	0.903	1.114	1.398
$\log p$ (3 d.p.)	0.477	0.845	1.000	1.204	1.462

2. Plot a graph of $\log p$ against $\log t$ and draw a line of best fit:

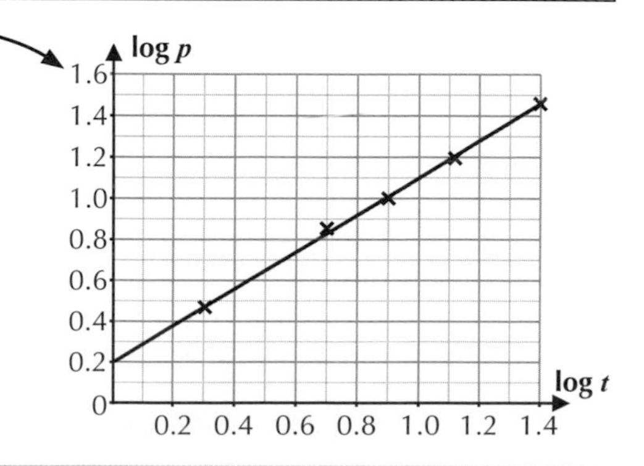

3. The vertical intercept is at 0.2, so:
   $\log a = 0.2 \Rightarrow a = 10^{0.2} = 1.585$ (3 d.p.)

4. Find the gradient by taking two coordinates on the line of best fit, e.g. (0, 0.2) and (1.0, 1.1):
   $$b = \frac{y_2 - y_1}{x_2 - x_1} = \frac{1.1 - 0.2}{1.0 - 0} = 0.9$$

5. So $p = 1.585t^{0.9}$

Logarithmic graphs for $y = kb^x$

Graphs of the form $y = kb^x$ can also be rearranged to be in linear form, but slightly differently:

$$y = kb^x \Rightarrow \log y = \log kb^x$$
$$\Rightarrow \log y = \log k + \log b^x$$
$$\Rightarrow \log y = x \log b + \log k$$

This is in the straight-line form $y = mx + c$ where $\log b$ is the gradient and $\log k$ is the vertical intercept. This time, plot $\log y$ against x.

Example 2

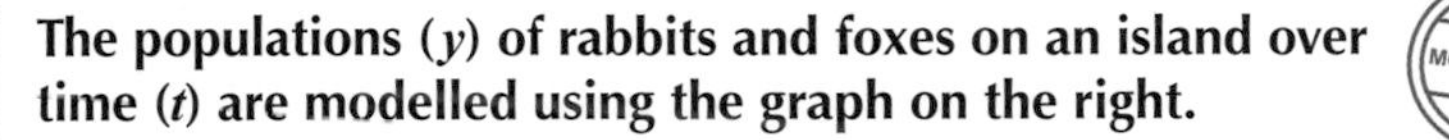

The populations (y) of rabbits and foxes on an island over time (t) are modelled using the graph on the right.

MODELLING

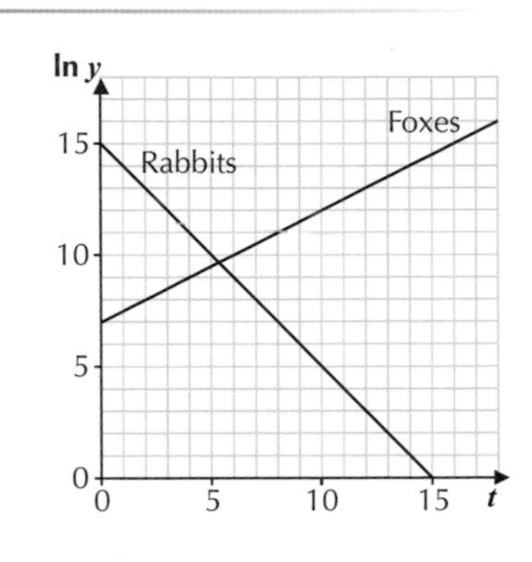

a) Calculate the combined population (to 3 s.f.) when the number of rabbits and foxes are equal.

1. Find the equation for each line.
   Note the vertical axis is $\ln y$, not y.
   Foxes: $\ln y = \frac{1}{2}t + 7$ Rabbits: $\ln y = 15 - t$

2. The populations are equal when the lines on the graph cross:
   $\frac{1}{2}t + 7 = 15 - t \Rightarrow \frac{3}{2}t = 8 \Rightarrow t = \frac{16}{3}$

3. Put this value back into one of the above equations, e.g. rabbits:
   $\ln y = 15 - t \Rightarrow \ln y = 15 - \frac{16}{3} = \frac{29}{3} \Rightarrow y = e^{\frac{29}{3}} = 15\,782.652...$

4. This is the same as the fox population, so the combined population is:
   $2 \times 15\,782.652... = 31\,600$ (3 s.f.)

b) Explain why this model may not be appropriate for an extended timescale.

After 15 years there'll be less than 1 rabbit, and the fox population will continue to grow exponentially — this is clearly unrealistic.

Tip: Think carefully about what the numbers would mean in real life.

Exercise 9.5.1

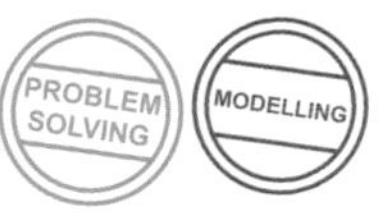

Q1 The value, £V, of a large piece of machinery is modelled by the equation $V = pq^t$, where t is the age of the machinery in years, and p and q are constants.
The line l below plots $\log V$ against t, and the gradient of l is $-\frac{1}{40}$.

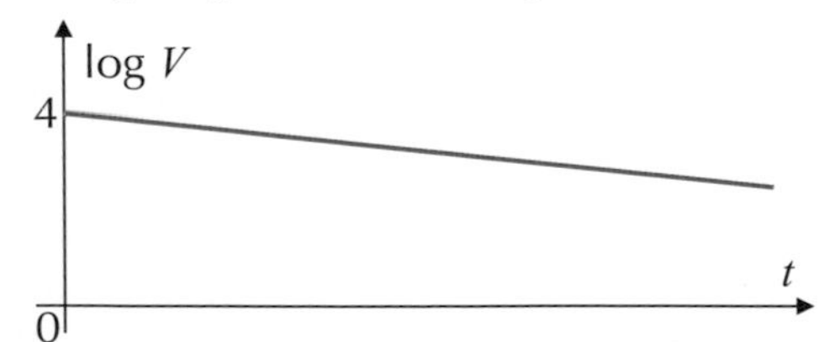

Write an equation for l, and use this to find the value of the machinery when it is 20 years old.

Q2 A blogger believes that the number of subscribers (y) to her blog over time (t days) can be modelled by the formula $y = at^b$. The number of subscribers is shown in the table below:

t (days)	7	14	21	28
y	224	1000	1585	3981

a) Plot a linear graph to show this data.

b) Find the values of a and b.

c) Use the model to predict how many subscribers the blog had after 10 days.

A rival blog is losing subscribers.
The number of subscribers to the rival blog is modelled by the equation $y = 15t^{-1}$.

d) Write this equation in linear form.

e) Plot the linear graph on the same axes as for part a).

f) Using your graphs or otherwise, find the day on which the two blogs had the same number of subscribers.

g) Explain why these models may not be realistic over a long period of time.

Q3 A tank is being filled with water. At time $s = 0$, there is 2 mm of water in the tank, and after s seconds, the height, h mm, of the water in the tank can be modelled by the equation $h = 2 \times 3^s$.

a) (i) Sketch the graph of h against s.

(ii) Sketch the graph of $\log_3 h$ against s.

(iii) Which is the more useful graph for calculations? Explain your reasoning.

b) Explain why this model is unsuitable for long time periods, and suggest an improvement to the model.

Q4 The number of bacteria, p, in a petri dish is observed over a period of time, t. The bacteria population can be modelled by the formula $p = at^b$, where a and b are constants. The results from the observations are shown in the table below.

t (days)	1	3	4	6	9
p (1000s)	2	14	22	44	88

Plot a linear graph to represent this data, and use this to find the values of a and b.

Q5 The activity, x, of a radioactive substance decreases over time, t. The activity follows the formula $x = kb^t$, where k and b are constants. The measurements from an experiment are shown in the table below.

t (days)	5	50	100	200	300
x (Bq)	80.449	32.411	11.803	1.565	0.207

Draw a linear graph of this data, and use a line of best fit to calculate the initial activity of the substance, to the nearest Bq.

Q6 The rules of a sport say that the length, l m, and the width, w m, of the rectangular playing field can be any value, as long as the area is 120 m^2. Use logs to model the relationship between l and w as a straight line, and show this graphically.

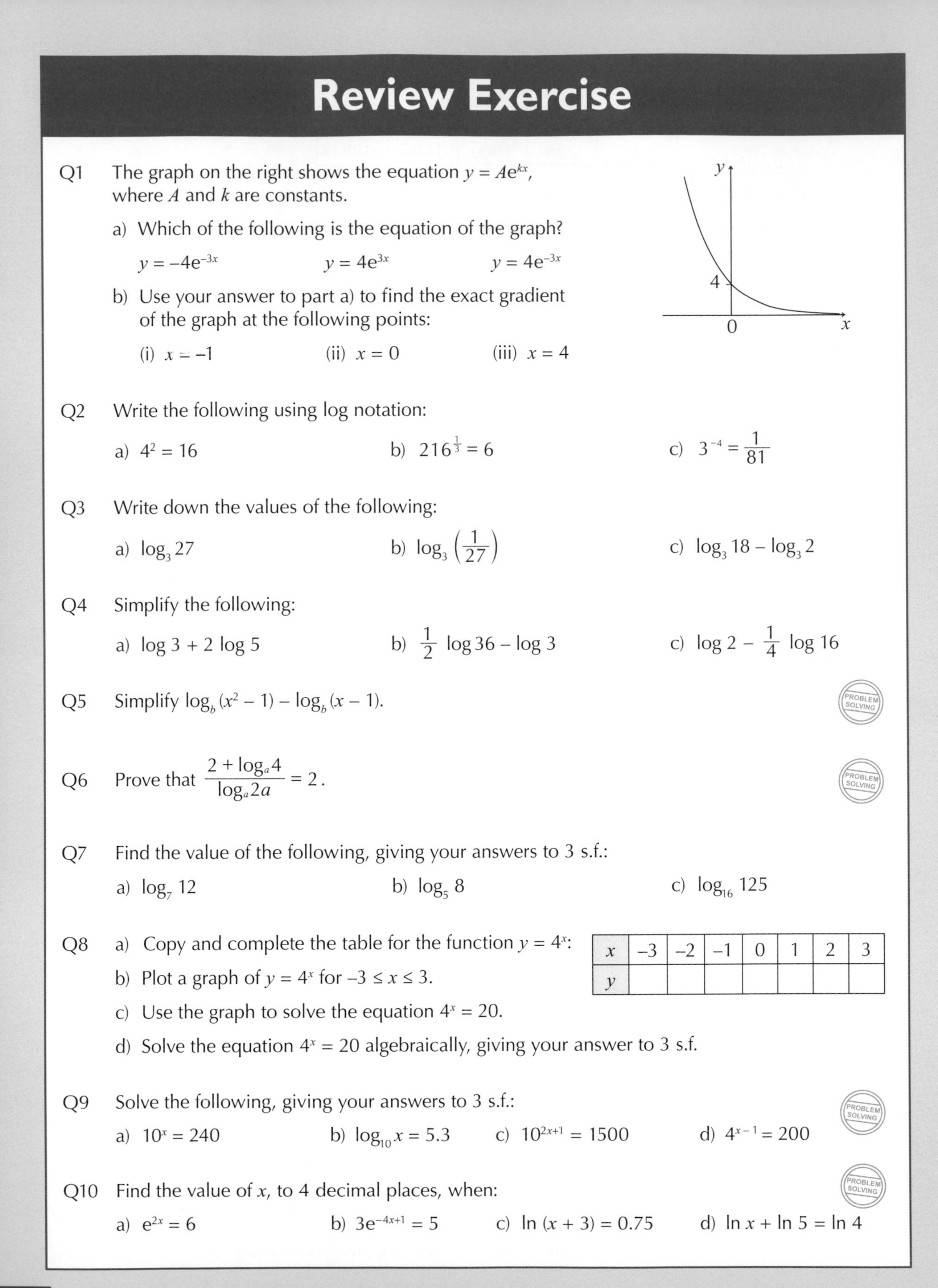

Review Exercise

Q1 The graph on the right shows the equation $y = Ae^{kx}$, where A and k are constants.

a) Which of the following is the equation of the graph?

$y = -4e^{-3x}$ $\quad$ $y = 4e^{3x}$ $\quad$ $y = 4e^{-3x}$

b) Use your answer to part a) to find the exact gradient of the graph at the following points:

(i) $x = -1$ $\quad$ (ii) $x = 0$ $\quad$ (iii) $x = 4$

Q2 Write the following using log notation:

a) $4^2 = 16$ $\quad$ b) $216^{\frac{1}{3}} = 6$ $\quad$ c) $3^{-4} = \frac{1}{81}$

Q3 Write down the values of the following:

a) $\log_3 27$ $\quad$ b) $\log_3 \left(\frac{1}{27}\right)$ $\quad$ c) $\log_3 18 - \log_3 2$

Q4 Simplify the following:

a) $\log 3 + 2 \log 5$ $\quad$ b) $\frac{1}{2} \log 36 - \log 3$ $\quad$ c) $\log 2 - \frac{1}{4} \log 16$

Q5 Simplify $\log_b (x^2 - 1) - \log_b (x - 1)$. PROBLEM SOLVING

Q6 Prove that $\dfrac{2 + \log_a 4}{\log_a 2a} = 2$. PROBLEM SOLVING

Q7 Find the value of the following, giving your answers to 3 s.f.:

a) $\log_7 12$ $\quad$ b) $\log_5 8$ $\quad$ c) $\log_{16} 125$

Q8 a) Copy and complete the table for the function $y = 4^x$:

x	–3	–2	–1	0	1	2	3
y							

b) Plot a graph of $y = 4^x$ for $-3 \leq x \leq 3$.

c) Use the graph to solve the equation $4^x = 20$.

d) Solve the equation $4^x = 20$ algebraically, giving your answer to 3 s.f.

Q9 Solve the following, giving your answers to 3 s.f.: PROBLEM SOLVING

a) $10^x = 240$ $\quad$ b) $\log_{10} x = 5.3$ $\quad$ c) $10^{2x+1} = 1500$ $\quad$ d) $4^{x-1} = 200$

Q10 Find the value of x, to 4 decimal places, when: PROBLEM SOLVING

a) $e^{2x} = 6$ $\quad$ b) $3e^{-4x+1} = 5$ $\quad$ c) $\ln (x + 3) = 0.75$ $\quad$ d) $\ln x + \ln 5 = \ln 4$

Review Exercise

Q11 Scientists are monitoring the population of curly-toed spiders at a secret location. It appears to be dropping at a rate of 25% a year. When the population has dropped below 200, the species will be in danger of extinction. At the moment the population is 2000. In which year will the spiders be in danger of extinction?

MODELLING PROBLEM SOLVING

Q12 A nature reserve has a population of 20 leopards in 2010. The number of leopards in the nature reserve can be modelled by the formula $L = L_0 e^{\frac{t}{12}}$ where L is the number of leopards in the population, L_0 is the initial population size and t is the time in years.

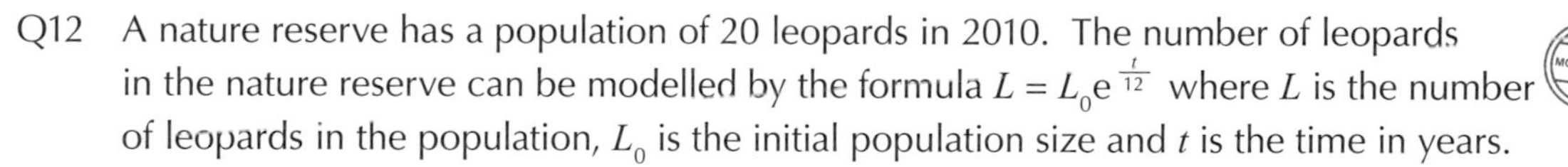

MODELLING

a) How many leopards does the model predict the nature reserve will have after 10 years?

b) The reserve has enough space for 60 leopards.
How long will it be until the reserve runs out of space?

When some of the leopards are released into the wild, the wild population can be modelled by $W = W_0 e^{-\frac{t}{3}}$ where W is the population, t is the time in years and W_0 is the initial population.

c) If the nature reserve releases a population of 15 leopards into the wild, predict how many will be in this population after 5 years in the wild.

Q13 The spread of a zombie apocalypse through a population can be modelled by the formula: $Z = 10 + 20e^t$ where Z is the number of zombies and t is the time in weeks.

MODELLING

a) How many zombies were there initially?

b) Predict how many people will have become zombies after 2 weeks if it spreads according to the model.

c) How many weeks will have passed before there are 60 million zombies?

Q14 The populations (y) of red and grey squirrels in a forest over time (t months) are modelled using the graph below right.
Grey squirrels were introduced to the forest at time $t = 0$.

MODELLING PROBLEM SOLVING

a) Find an exponential equation for the population of red squirrels in the forest.

b) Find an exponential equation for the population of grey squirrels in the forest.

c) The population of red squirrels is considered critical when there are fewer than 20 left in the forest. In which month will the population reach a critical level?

d) Explain why this model may not be suitable to predict the number of red and grey squirrels over a long timescale.

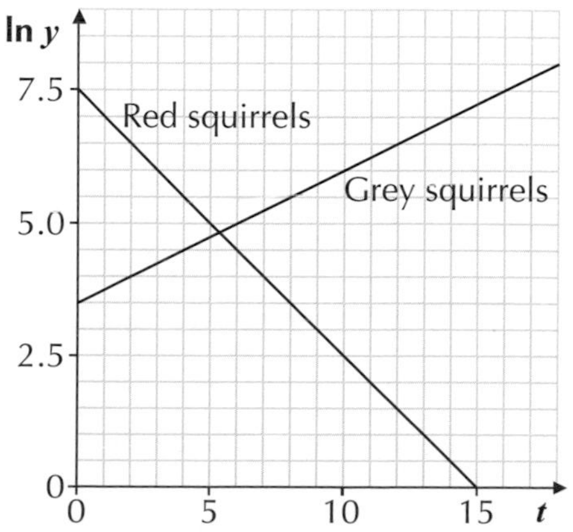

Exam-Style Questions

Q1 Find all the solutions to the equation below in the form $a \ln b$, where b is the smallest possible integer.

$$e^x - \frac{40}{e^x} = 3$$

[3 marks]

Q2 Given that $\log_{10} y = 1 + \log_{10}(x + 2)$:

a) Show that $y = 10x + 20$.

[2 marks]

b) Hence solve the simultaneous equations:
$\log_{10} y = 1 + \log_{10}(x + 2)$ and $10^{15x-y} = 10\,000$

[3 marks]

Q3 Given that $a^x = b^y$ where $a, b > 1$.

a) Show that $x = y \log_a b$.

[1 mark]

b) Hence solve the equation $2^{k-2} = 3^{2k-5}$ giving your answer in the form $\dfrac{p + q\log_2 3}{r + s\log_2 3}$ where p, q, r and s are integers.

[3 marks]

Q4 Show that the tangent line to the curve $y = e^{2x}$ at the point with coordinates $\left(\frac{1}{2}, e\right)$ passes through the origin.

[3 marks]

Q5 A block of ice, initially with mass 100 kg, is exposed to a high temperature and begins to melt. Renata suggests that the mass, M, of the ice t minutes after it begins to melt can be modelled by the equation $M = Ab^t$.
The diagram shows the graph of $\log_{10} M$ against t.

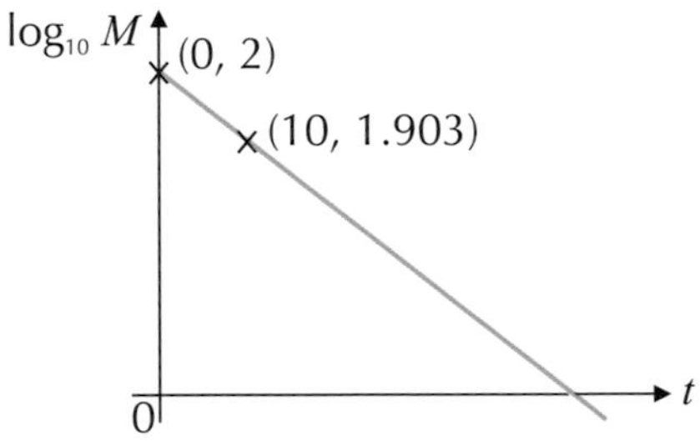

$\log_{10} M$; $(0, 2)$; $(10, 1.903)$; 0; t

a) State the value of A and find the value of b correct to 3 significant figures.

[4 marks]

b) Using the values found in a), calculate how long, to the nearest minute, it will take for 60% of the block of ice to melt according to the model.

[2 marks]

c) Why might Renata's model not be suitable for large values of t?

[1 mark]

Chapter 10 Differentiation

10.1 The Gradient of a Curve

Differentiation is an algebraic process that finds the gradient of a curve.
It is useful for finding out how fast one thing changes with respect to another.

Learning Objectives (Spec Ref 7.1):

- Understand that the derivative of the function f(x) gives the gradient of the tangent to the graph of y = f(x) at the point (x, y).
- Understand that a derivative can be interpreted as a rate of change.
- Understand how the gradient of a tangent to a curve can be found using a limit.
- Find the gradient of a curve by differentiating from first principles.

Prior Knowledge Check: Be able to find gradients of straight lines (see p.82) and expand and factorise equations (see p.27-29).

Finding the gradient of a curve

The **gradient** of a curve is just how **steep** it is. Unlike a straight line, the steepness of a curve **changes** as you move along it — you can only give an **exact value** for the gradient at a **particular point** on the curve.

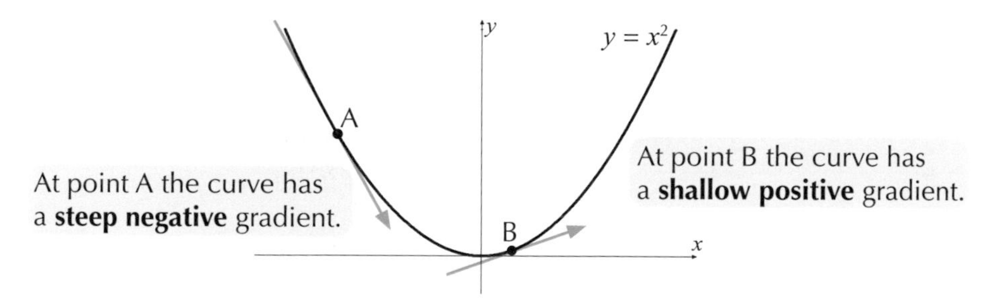

- At a **point**, the gradient of a curve is the same as the gradient of the **tangent line** to the curve at that point.
- The tangent line is a **straight line** which **just touches** the curve at that point, without going through it.
- Sadly, you can't work out the gradient of this tangent using the normal method of picking **two points** and finding the change in y ÷ change in x. This is because you only know **one point** on the line — the point where the tangent **meets the curve**.
- So we need another method to find the gradient of a curve — it's known as **differentiation**.

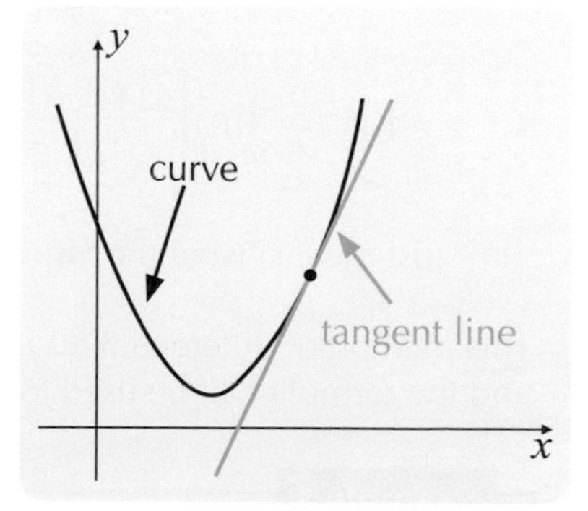

Differentiating produces an **algebraic expression** for the gradient as a **function of x** — its numerical value **changes** as you move along the curve.

Before we get started with differentiation, there's some **notation** to learn:

The function you get from differentiating **y with respect to x** is called the **derivative** of y with respect to x and it's written $\frac{dy}{dx}$.

Tip: $\frac{dy}{dx}$ is **not** a fraction, it's just notation for a derivative.

$\frac{dy}{dx}$ represents the **rate of change** of y with x. In other words, it tells you how quickly y is changing for a given value of x. This is the same as the **gradient** of the curve $y = f(x)$.

The notation **f′(x)** means the derivative of $y = f(x)$ with respect to x. It's sometimes used instead of $\frac{dy}{dx}$.

Differentiating from first principles

To find the derivative of a function you need to find its gradient as a function of x.

You can get **close** to the gradient of the tangent (and so the curve) at a point $(x, f(x))$, by finding the gradient of the line joining $(x, f(x))$ and another point **close to** it on the curve.

- On the diagram, the point $(x + h, f(x + h))$ is a small distance further along the curve from $(x, f(x))$.
- As h gets smaller, the distance between the two points gets smaller.
- The closer the points, the **closer** the line joining them will be **to the tangent line**.

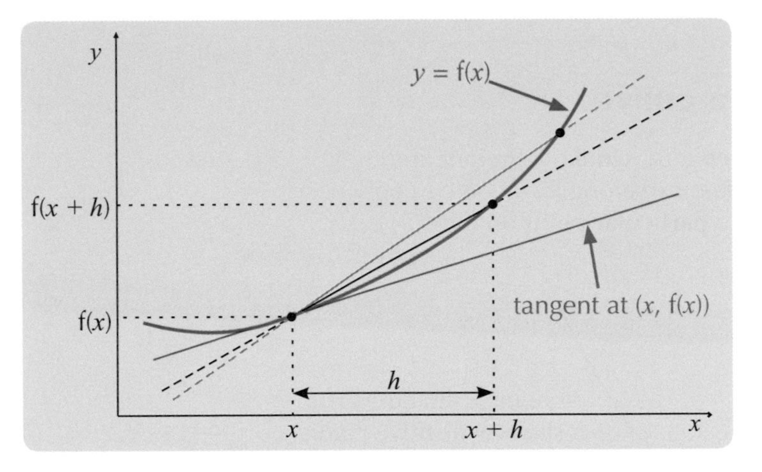

Now you can write an expression for the gradient of the **straight line** joining the two points $(x, f(x))$ and $(x + h, f(x + h))$ like this:

$$\frac{f(x+h)-f(x)}{(x+h)-x}$$

Tip: Remember, the gradient of a line passing through points (x_1, y_1) and (x_2, y_2) is given by: $\frac{y_2-y_1}{x_2-x_1}$.

As h gets **smaller**, the gradient of the straight line gets **closer and closer** to the gradient of the **curve** at $(x, f(x))$. So you can write an expression for the gradient of the curve $y = f(x)$ like this:

$$f'(x) = \lim_{h\to 0}\left[\frac{f(x+h)-f(x)}{(x+h)-x}\right]$$

'$(x + h) - x$' simplifies to 'h'

Tip: You might also see these expressions with 'δx' instead of 'h'.

$\lim_{h\to 0}$ just means 'what the function goes towards as h goes towards zero'.

This method of differentiation is known as differentiating from **first principles** and the formula can be used to find the gradient of a curve as a function of x.

Example 1

Find an expression for the gradient of the function $f(x) = x^2$ by differentiating from first principles.

1. Write down the formula for differentiating from first principles. $f'(x) = \lim_{h\to 0}\left[\frac{f(x+h)-f(x)}{(x+h)-x}\right]$

2. Use the fact that $f(x) = x^2$. $\quad f(x) = \lim_{h \to 0}\left[\frac{(x+h)^2 - x^2}{(x+h) - x}\right]$

3. Multiply out and simplify. $\quad = \lim_{h \to 0}\left[\frac{x^2 + 2xh + h^2 - x^2}{x + h - x}\right] = \lim_{h \to 0}\left[\frac{2xh + h^2}{h}\right] = \lim_{h \to 0}[2x + h]$

4. Decide what happens as h gets close to 0. Here $2x + h$ gets close to $2x$. $\quad = 2x$

Example 2

Find the gradient of the curve $y = 0.5x$ by differentiating from first principles.

1. Use the fact that $y = f(x)$ with $f(x) = 0.5x$ and simplify. $\quad \frac{dy}{dx} = \lim_{h \to 0}\left[\frac{f(x+h) - f(x)}{(x+h) - x}\right] = \lim_{h \to 0}\left[\frac{0.5(x+h) - 0.5x}{(x+h) - x}\right] = \lim_{h \to 0}\left[\frac{0.5h}{h}\right]$

2. Decide what will happen as h gets close to 0. There are no h's so the limit is just 0.5. $\quad = \lim_{h \to 0}[0.5] = 0.5$

Tip: A straight line will always have a constant gradient.

Exercise 10.1.1

Q1 The curve C is given by $y = f(x)$ where $f(x) = x^3$.

a) Find the gradient of the straight line joining the point on the curve where $x = 1$ and the point on the curve where:

(i) $x = 2$ (ii) $x = 1.5$ (iii) $x = 1.1$

b) The gradient of the curve at the point (1, 1) is 3. What do you notice about the gradient of the straight lines in part a) as the value of x moves closer to 1?

Q2 Derive from first principles expressions for the gradients of the following curves:

a) $y = x$ b) $f(x) = x^3$ c) $f(x) = 2x$

d) $f(x) = 2x^2$ e) $f(x) = x - 7$ f) $f(x) = -x^3$

Q3 For the following, find the derivative of y with respect to x by differentiating from first principles.

a) $y = 5x^2 + 1$ b) $y = x - x^2$ c) $y = 3x^3$

d) $y = 2x^3 + 3x$ e) $y = x^3 + x$ f) $y = (2 - x)^2$

Q4 The curve C is given by $y = x^3 + 2x^2 + 3x$.

a) Use differentiation from first principles to show that the gradient of the curve is $\frac{dy}{dx} = 3x^2 + 4x + 3$.

b) Find the gradient of the tangent to the curve where:

(i) $x = -1$ (ii) $x = 2$ (iii) $x = 0$

10.2 Differentiating $y = f(x)$

Differentiating from first principles can take a long time, especially if there are large powers involved. Luckily, there's a formula that will do it quickly for you.

Learning Objectives (Spec Ref 7.2 & 7.3):
- Differentiate powers of x.
- Differentiate sums and differences of powers of x.
- Differentiate more complicated functions containing powers of x.
- Sketch the graph of $y = f'(x)$ by differentiating $y = f(x)$.
- Find tangents and normals to a curve.

Prior Knowledge Check: Be able to use the laws of indices (see p.17), find the equation of straight lines (see p.82) and sketch curves (see p.94).

Differentiating x^n

Expressions are much easier to **differentiate** when they're written using **powers of x** — like writing $\sqrt{x}$ as $x^{\frac{1}{2}}$ or $\frac{3}{x^2}$ as $3x^{-2}$.

When you've done this, you can use this **formula** to differentiate:

$$\text{If } y = x^n, \text{ then } \frac{dy}{dx} = nx^{n-1}$$

The formula comes from differentiating x^n from **first principles**.

Tip: Here, 'differentiate' actually means 'differentiate with respect to x' as it's a function of x you're differentiating.

Example 1

Differentiate each of the following using the formula for powers of x.

a) $y = x^2$

For 'normal' powers, n is just the power of x. Here $n = 2$.

$$\frac{dy}{dx} = nx^{n-1} = 2x^1 = 2x$$

b) $y = \sqrt{x}$

First write the square root as a fractional power of x, then put n into the formula and simplify.

$$y = x^{\frac{1}{2}} \quad \left(n = \frac{1}{2}\right)$$

$$\frac{dy}{dx} = nx^{n-1} = \frac{1}{2}x^{\left(-\frac{1}{2}\right)} = \frac{1}{2\sqrt{x}}$$

c) $y = \frac{1}{x^2}$

Write the fraction as a negative power of x.

$$y = x^{-2} \quad (n = -2)$$

$$\frac{dy}{dx} = nx^{n-1} = -2x^{-3} = -\frac{2}{x^3}$$

d) $y = 4x^3$

This is just a normal power with $n = 3$, but there's a constant (a number) in front of it.

If there's a number in front of the x^n term, multiply the derivative by it. If $y = ax^n$, $\frac{dy}{dx} = anx^{n-1}$.

$$y = 4x^3$$

$$\frac{dy}{dx} = 4(nx^{n-1}) = 4(3x^2) = 12x^2$$

Example 2

Differentiate $y = 5$ using the formula for powers of x.

There are no powers of x in this expression for y so multiply by $x^0 = 1$.

$y = 5x^0, \quad n = 0$

$\frac{dy}{dx} = 5(nx^{n-1}) = 5(0x^{-1}) = 0$

Tip: Differentiating $y = a$ where a is just a constant gives zero, because the line has a gradient of 0.

You could be asked to **use** your gradient function to work out the **numerical value** of the gradient at a **particular point** on the curve.

Example 3

Find the gradient of the curve $y = x^2$ at $x = 1$ and $x = -2$.

1. You need the gradient of the graph, so differentiate the function.

   $y = x^2 \Rightarrow \frac{dy}{dx} = 2x$

2. Use $\frac{dy}{dx} = 2x$ to find the gradient for each x-value.

   When $x = 1$, $\frac{dy}{dx} = 2$. Gradient at $x = 1$ is 2.

   When $x = -2$, $\frac{dy}{dx} = -4$. Gradient at $x = -2$ is -4.

Exercise 10.2.1

Q1 Differentiate to find $\frac{dy}{dx}$ for:

a) $y = x$ b) $y = x^6$ c) $y = x^3$ d) $y = 15x^2$

e) $y = x^{-2}$ f) $y = 3x^2$ g) $y = 7x$ h) $y = \frac{1}{2}x^4$

i) $y = 3$ j) $y = 3\sqrt{x}$ k) $y = 2x^{-1}$ l) $y = \frac{1}{x^3}$

Q2 Differentiate to find $f'(x)$ for:

a) $f(x) = x^5$ b) $f(x) = x^7$ c) $f(x) = x^{-4}$ d) $f(x) = 4x^3$

e) $f(x) = 2x^{-3}$ f) $f(x) = 8\sqrt{x}$ g) $f(x) = 3\sqrt[3]{x}$ h) $f(x) = -7$

i) $f(x) = 4x^{-2}$ j) $f(x) = -3x^3$ k) $f(x) = -5x^{-4}$ l) $f(x) = \frac{4}{x^3}$

Q3 Find the gradient of each of the following functions:

a) $y = 2x^2$ when $x = 4$ b) $y = x^{-1}$ when $x = 2$

c) $y = -4x^5$ when $x = 1$ d) $y = \frac{2}{x}$ when $x = 10$

e) $f(x) = 2\sqrt{x}$ at the point (9, 6) f) $f(x) = x^4$ at the point (–2, 16)

g) $f(x) = -2x^3$ when $f(x) = -250$ h) $f(x) = -3x^{-2}$ when $f(x) = -\frac{3}{4}$

Q4 The line $y = 5x$ and the curves $y = 5x^2$ and $y = 5x^3$ all pass through the point $P(1, 5)$. Show which of these has the steepest gradient at point P. PROBLEM SOLVING

Differentiating functions

Even if there are **loads** of terms in the expression, it doesn't matter. Differentiate each bit **separately** and you'll be fine.

Formally, this means: $$\frac{d}{dx}(x^m + x^n) = \frac{d}{dx}(x^m) + \frac{d}{dx}(x^n)$$

Tip: Remember — if there's a number in front of the function, multiply the derivative by the same number.

Example 1

a) **Differentiate $f(x) = x^4 + 3x^2 - 2$.**

1. Differentiate each bit separately.

$$f(x) = x^4 + 3x^2 - 2$$
$$f'(x) = 4x^3 + 3(2x) - 0$$

2. Put the bits back together using the same signs.

$$= 4x^3 + 6x$$

b) **Find $\frac{d}{dx}\left(6x^2 + \frac{4}{\sqrt[3]{x}} - \frac{2}{x^2} + 1\right)$.**

1. This notation just means the derivative with respect to x of the thing in the brackets. Rewrite the function first to get powers of x.

$$6x^2 + \frac{4}{\sqrt[3]{x}} - \frac{2}{x^2} + 1 = 6x^2 + 4x^{-\frac{1}{3}} - 2x^{-2} + 1$$

2. Then differentiate each bit separately.

$$6x^2 + 4x^{-\frac{1}{3}} - 2x^{-2} + 1$$
$$6(2x) + 4\left(-\frac{1}{3}x^{-\frac{4}{3}}\right) - 2(-2x^{-3}) + 0$$

3. Put the bits back together, being careful with the signs.

$$\frac{d}{dx}\left(6x^2 + \frac{4}{\sqrt[3]{x}} - \frac{2}{x^2} + 1\right) = 12x - \frac{4}{3\sqrt[3]{x^4}} + \frac{4}{x^3}$$

You'll often need to **simplify** a function before you can differentiate it by multiplying out **brackets** or simplifying **fractions**. If you have a fraction to simplify, check first whether the denominator is a **factor** of the numerator, otherwise you'll need to **split it up** into terms.

Example 2

For the function $f(x) = (x + 2)^2(x - 10)$:

a) **Find $f'(x)$.**

1. Multiply out the brackets and simplify.

$$f(x) = (x + 2)^2(x - 10) = (x^2 + 4x + 4)(x - 10)$$
$$= x^3 - 10x^2 + 4x^2 - 40x + 4x - 40$$
$$= x^3 - 6x^2 - 36x - 40$$

2. Differentiate term-by-term.

$$f'(x) = 3x^2 - 12x - 36 - 0$$
$$= 3x^2 - 12x - 36$$

b) Sketch the graph of $y = f'(x)$.

1. $f'(x) = 3x^2 - 12x - 36$ is a quadratic with a positive coefficient of x.

   This means the graph is u-shaped.

2. Find the y-intercept where $x = 0$.

   $f'(0) = 3(0)^2 - 12(0) - 36 = -36$

3. Find the x-intercepts where $f'(x) = 0$.

   $f'(x) = 3x^2 - 12x - 36 = 0$
   $\Rightarrow x^2 - 4x - 12 = 0$
   $\Rightarrow (x + 2)(x - 6) = 0$
   $\Rightarrow x = -2$ or $x = 6$

4. Put all this together to draw the sketch.

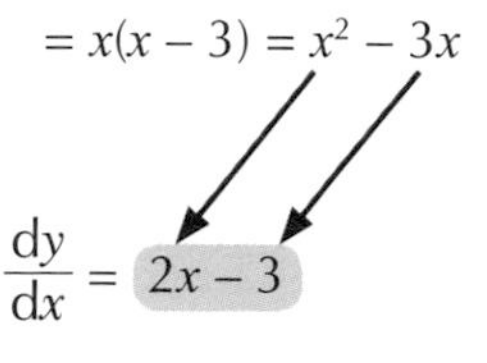

Example 3

a) Differentiate $y = \dfrac{x^3 - 5x^2 + 6x}{x - 2}$.

1. Factorise the numerator.

   $y = \dfrac{x^3 - 5x^2 + 6x}{x - 2} = \dfrac{x(x^2 - 5x + 6)}{x - 2} = \dfrac{x(x-3)(x-2)}{x-2}$

2. Cancel $(x - 2)$ to simplify the fraction and expand the remaining brackets.

   $= x(x - 3) = x^2 - 3x$

3. Differentiate term-by-term.

   $\dfrac{dy}{dx} = 2x - 3$

Tip: If the denominator is an expression instead of just one term, chances are it will cancel with a factor of the numerator.

b) Differentiate the function $f(x) = \dfrac{x^3 + 4x + 1}{2x^2}$.

1. This numerator won't factorise. Instead, split the fraction up into three fractional terms.

   $f(x) = \dfrac{x^3 + 4x + 1}{2x^2} = \dfrac{x^3}{2x^2} + \dfrac{4x}{2x^2} + \dfrac{1}{2x^2} = \dfrac{x}{2} + \dfrac{2}{x} + \dfrac{1}{2x^2}$

2. Write each term as a power of x.

   $= \frac{1}{2}x + 2x^{-1} + \frac{1}{2}x^{-2}$

3. Differentiate term-by-term.

   $f'(x) = \frac{1}{2} + 2(-x^{-2}) + \frac{1}{2}(-2x^{-3})$

   $= \frac{1}{2} - 2x^{-2} - x^{-3} = \dfrac{1}{2} - \dfrac{2}{x^2} - \dfrac{1}{x^3}$

Exercise 10.2.2

Q1 Differentiate these functions:

a) $y = 4x^3 - x^2$

b) $y = x + \dfrac{1}{x}$

c) $y = 3x^2 + \sqrt{x} - 5$

d) $f(x) = -2x^5 + 4x - \dfrac{1}{x^2}$

e) $f(x) = \sqrt{x^3} - x$

f) $f(x) = 5x - \dfrac{2}{x^3} + \sqrt[3]{x}$

Q2 Find:

a) $\frac{d}{dx}(x(x^6 - 1))$ b) $\frac{d}{dx}((x - 3)(x + 4))$ c) $\frac{d}{dx}(x(x - 1)(x - 2))$

d) $\frac{d}{dx}((x - 3)(x + 4)(x - 1))$ e) $\frac{d}{dx}(x^2(x - 4)(3 - x^3))$ f) $\frac{d}{dx}((x - 3)^2(x^2 - 2))$

Q3 Find the gradient of each of the following curves:

a) $y = x^4 - x^2 + 2$ when $x = 3$ b) $y = 2x^5 + \frac{1}{x}$ when $x = -2$

c) $y = x(x - 1)(x - 2)$ when $x = -3$ d) $y = 5(x^2 - 1)(3 - x)$ when $x = 0$

e) $y = \sqrt{x}(x - 1)$ at (4, 6) f) $f(x) = x^3(x^2 - 5)$ at (–1, 4)

g) $f(x) = \frac{1}{x^2}(x^3 - x)$ at $x = 5$ h) $f(x) = \frac{3x^3 + 18x^2 + 24x}{x + 4}$ at (–2, 0)

Q4 For the following graphs, sketch the graph of $f'(x)$ for $0 \le x \le 10$:

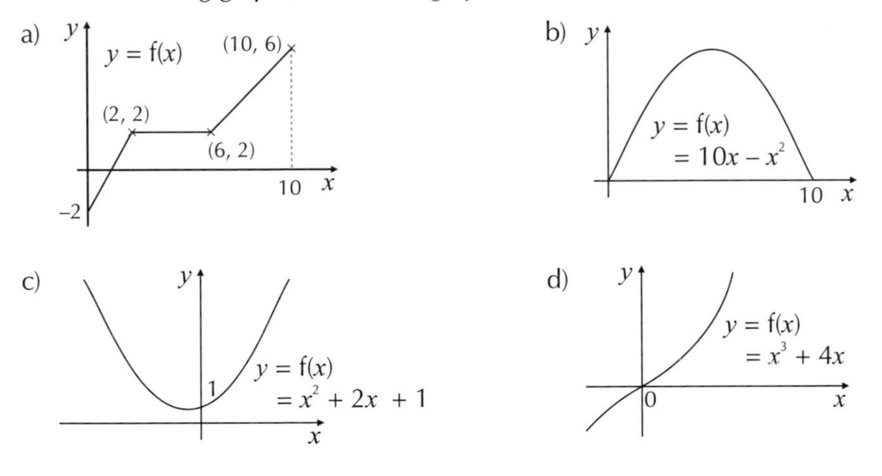

Q5 For each of the following curves, sketch the graph of $y = f'(x)$.

a) $f(x) = (x + 3)(x + 4)$ b) $f(x) = \frac{x^3 - 3x^2 + 2x}{x - 1}$

c) $f(x) = x^4 - 4x^3 + 4x^2 - 9$ d) $f(x) = (x - 1)^2(x + 5)$

Q6 For each of the following functions, find the coordinates of the point or points where the gradient is 0:

PROBLEM SOLVING

a) $y = x^2 - 2x$ b) $y = 3x^2 + 4x$ c) $y = 5x^2 - 3x$

d) $y = 9x - 3x^3$ e) $y = 2x^3 - x^2$ f) $y = 2x^3 + 3x^2 - 12x$

Q7 Differentiate these functions:

a) $y = \frac{x^2 - 3x - 4}{x + 1}$ b) $f(x) = \frac{x^4 - 9}{x^2 + 3}$ c) $f(x) = \frac{x^5 - 16x^3}{x + 4}$

d) $y = \frac{1}{x}(x - 3)(x - 4)$ e) $y = \sqrt{x}(x^3 - \sqrt{x})$ f) $f(x) = \frac{3 - \sqrt{x}}{\sqrt{x}}$

g) $f(x) = \frac{x + 5\sqrt{x}}{\sqrt{x}}$ h) $f(x) = \frac{x - 3\sqrt{x} + 2}{\sqrt{x} - 1}$ i) $y = \frac{4 - x}{2 + \sqrt{x}}$

PROBLEM SOLVING

Finding tangents and normals

Differentiation can be used to find the gradient at a point on a curve.
This makes it easy to find the equation for the **tangent** or **normal** at that point.

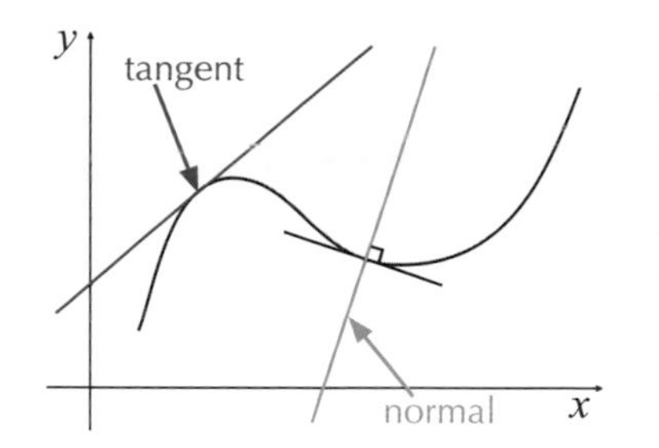

You already know that a **tangent** is a straight line that just **touches** the curve and has the **same gradient** as the curve at that point.

A **normal** is a straight line that is **perpendicular** (at right angles) to the curve at a particular point.

Now, there's one key fact to remember for normals — tangents and normals are perpendicular, and as a result, their **gradients multiply to give –1**:

$$\text{gradient of tangent} \times \text{gradient of normal} = -1$$

$$\text{gradient of normal} = \frac{-1}{\text{gradient of tangent}}$$

Armed with this rule, we can write down a step-by-step method for finding the equation of a tangent or normal to a curve...

To find the equation of the tangent or normal to a curve at a point:

- **Differentiate** the function.
- Find the **gradient** of the curve at that point.
- Use this to deduce the gradient, m, of the tangent or normal:

  gradient of the **tangent** = gradient of the curve

  gradient of the **normal** $= \frac{-1}{\text{gradient of the curve}}$

- Write the **equation** of the tangent or normal in the form $\boldsymbol{y = mx + c}$.
- Work out the **constant value** $\boldsymbol{c}$ in the equation by using the coordinates of the point (which you know lies on the tangent/normal).

Tip: The tangent and normal are always straight lines, so their equations can be written $y = mx + c$. They can also be written in the form $y - y_1 = m(x - x_1)$ or $ax + by + c = 0$ if you prefer.

Example 1

Find the equation of the tangent to the curve $y = (4 - x)(x + 2)$ at the point (2, 8), giving your answer in the form $ax + by + c = 0$, where a, b and c are integers.

1. Write the curve in a form you can differentiate.

   $y = (4 - x)(x + 2) = 4x + 8 - x^2 - 2x$
   $= 8 + 2x - x^2$

2. Differentiate term-by-term.

   $\frac{dy}{dx} = 0 + 2 - 2x = 2 - 2x$

3. Find the gradient of the curve at (2, 8).

   $x = 2 \Rightarrow \frac{dy}{dx} = 2 - (2 \times 2) = 2 - 4 = -2$
   The gradient of the curve is –2 at (2, 8).

4. Gradient of the tangent = gradient of the curve.

   So $m = -2$

5. The equation of the tangent is $y = -2x + c$. Use the point (2, 8) to work out the value of c:

$x = 2, y = 8$
$\Rightarrow 8 = -4 + c \Rightarrow c = 12$

6. Write the equation of the tangent and rearrange into the form $ax + by + c = 0$.

$y = -2x + 12$
$\Rightarrow 2x + y - 12 = 0$

Example 2

Find the equation of the normal to the curve $y = x(x - 3)(x + 2)$ at the point (2, –8), giving your answer in the form $y = mx + c$.

1. Simplify and differentiate.

$$y = x(x - 3)(x + 2) = x^3 - x^2 - 6x$$

$$\frac{dy}{dx} = 3x^2 - 2x - 6$$

2. Find the gradient of the curve at (2, –8).

$$x = 2 \Rightarrow \frac{dy}{dx} = 3(2^2) - 2(2) - 6 = 2$$

The gradient of the curve is 2 at (2, –8).

3. Find the gradient of the normal at (2, –8).

$$m = \frac{-1}{\text{gradient of the curve at } (2, -8)} = -\frac{1}{2}$$

4. The equation of the normal is $y = -\frac{1}{2}x + c$. Use the point (2, –8) to work out the value of c.

$$x = 2, y = -8 \Rightarrow -8 = -1 + c \Rightarrow c = -7$$

5. Write the equation of the normal.

$$y = -\frac{1}{2}x - 7$$

Example 3

Find the equation of the normal to the curve $y = \frac{(x+2)(x+4)}{6\sqrt{x}}$ at the point (4, 4), giving your answer in the form $ax + by + c = 0$, where a, b and c are integers.

1. Simplify the equation. The denominator is one term so it'll probably need splitting up.

$$y = \frac{(x+2)(x+4)}{6\sqrt{x}} = \frac{x^2 + 6x + 8}{6x^{\frac{1}{2}}} = \frac{x^2}{6x^{\frac{1}{2}}} + \frac{6x}{6x^{\frac{1}{2}}} + \frac{8}{6x^{\frac{1}{2}}}$$

$$= \frac{1}{6}x^{\frac{3}{2}} + x^{\frac{1}{2}} + \frac{4}{3}x^{-\frac{1}{2}}$$

2. Differentiate term-by-term.

$$\frac{dy}{dx} = \frac{1}{6}\left(\frac{3}{2}x^{\frac{1}{2}}\right) + \frac{1}{2}x^{-\frac{1}{2}} + \frac{4}{3}\left(-\frac{1}{2}x^{-\frac{3}{2}}\right) = \frac{1}{4}\sqrt{x} + \frac{1}{2\sqrt{x}} - \frac{2}{3\sqrt{x^3}}$$

3. Find the gradient of the curve at (4, 4).

$$x = 4 \Rightarrow \frac{dy}{dx} = \frac{1}{4}\sqrt{4} + \frac{1}{2\sqrt{4}} - \frac{2}{3\sqrt{4^3}} = \frac{1}{2} + \frac{1}{4} - \frac{1}{12} = \frac{2}{3}$$

The gradient of the curve is $\frac{2}{3}$ at (4, 4).

4. Find the gradient of the normal at (4, 4).

$$m = \frac{-1}{\text{gradient of the curve at } (4, 4)} = -\frac{3}{2}$$

5. Use the point (4, 4) to work out c.

$$x = 4, y = 4 \Rightarrow 4 = -\frac{3}{2}(4) + c \Rightarrow c = 10$$

6. Write the equation of the normal in the form $ax + by + c = 0$.

$$y = -\frac{3}{2}x + 10 \Rightarrow 3x + 2y - 20 = 0$$

Exercise 10.2.3

Q1 Find the equation of the tangent to each of these curves at the given point. Give your answer in the form $y = mx + c$.

a) $y = 9x - 2x^2$, (1, 7)

b) $y = x^3 - 2x + 3$, (2, 7)

c) $y = (x + 2)(2x - 3)$, (2, 4)

d) $y = x(x - 1)^2$, (–1, –4)

e) $y = x^2(x + 3) - 10$, (2, 10)

f) $y = x(2x + 4)(x - 3)$, (–1, 8)

Q2 Find the tangent to each of these curves at the given point, giving your answer in the form $ax + by + c = 0$, where a, b and c are integers.

a) $y = \frac{1}{x} + x + 3$, $\left(2, 5\frac{1}{2}\right)$

b) $y = 4x^2 - 3\sqrt{x}$, (1, 1)

c) $y = \frac{3}{x} + 2\sqrt{x}$, $\left(4, 4\frac{3}{4}\right)$

d) $y = \frac{1}{x} + \frac{4}{x^2}$, $\left(2, 1\frac{1}{2}\right)$

e) $y = \frac{1}{3}x^2 - 4\sqrt{x} - \frac{1}{3}$, (4, –3)

f) $y = x - \frac{2}{x} + \frac{3}{x^2}$, (–3, –2)

Q3 Find the normal to each of these curves at the given point, giving your answer in the form $ax + by + c = 0$, where a, b and c are integers.

a) $y = 3x^2 - 4x + 2$, (2, 6)

b) $y = x^2(x + 4) - 5x$, (–1, 8)

c) $y = x(x - 1)(x - 2)$, (3, 6)

d) $y = x(x - 3)(x + 4) - 10$, (–2, 10)

e) $y = \frac{x^3 - 5x^2 - 14x}{x + 2}$, (5, –10)

f) $y = \frac{2x^3 - 32x}{x + 4}$, (3, –6)

Q4 Find the normal to each of these curves at the given point, giving your answer in an appropriate form.

a) $y = \frac{2x^5 - 2x^4}{3x^3}$, (–2, 4)

b) $y = \frac{5x^2 - 2x + 3}{x^2}$, $\left(2, 4\frac{3}{4}\right)$

c) $y = \frac{3x - x^2}{\sqrt{x}}$, (4, –2)

d) $y = \frac{1}{x} - \frac{3}{x^2} - \frac{4}{x^3} + \frac{7}{4}$, (–2, 1)

e) $y = \frac{x^3 - 5x^2 - 4x}{x\sqrt{x}}$, (4, –4)

f) $y = \frac{4x^4 - 2x^2 + 3\sqrt{x}}{\sqrt{x^5}}$, (1, 5)

Q5 Consider the curve with equation $y = f(x)$ where $f(x) = x^3 - 3x^2 + 3$.

PROBLEM SOLVING

a) Find the coordinates of the point where $f'(x) = 9$ and $x > 0$.

b) Find the equation of the tangent to the curve at this point, giving your answer in the form $y = mx + c$.

c) Find the equation of the normal to the curve at this point, giving your answer in the form $ax + by + c = 0$, where a, b and c are integers.

Q6 a) Show that the curve $y = \frac{x^3 + x^2 + x + 5}{x^2}$ passes through the point $\left(-2, -\frac{1}{4}\right)$.

b) Find the equation of the tangent to the curve at this point, giving your answer in the form $ax + by + c = 0$, where a, b and c are integers.

c) Find the equation of the normal to the curve at this point, giving your answer in the form $ax + by + c = 0$, where a, b and c are integers.

10.3 Using Differentiation

Differentiation can be used to find a graph's stationary points and their nature, and to find where a function is increasing or decreasing. But first, you need to know how to find second order derivatives.

Learning Objectives (Spec Ref 7.1 & 7.3):

- Find the second derivative of functions.
- Understand that the second derivative represents the rate of change of the gradient.
- Use differentiation to find all stationary points on a curve.
- Identify the nature of these stationary points.
- Work out where a function is increasing or decreasing.
- Use this information to make an accurate sketch of the graph of a function.

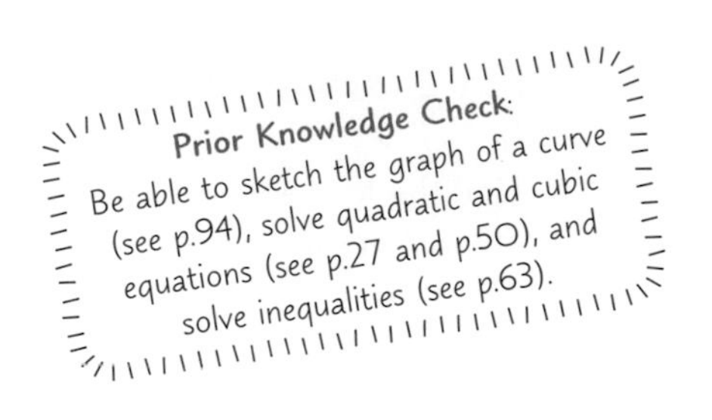

Finding second order derivatives

If you differentiate y with respect to x, you get the derivative $\frac{dy}{dx}$.

If you then differentiate $\frac{dy}{dx}$ with respect to x, you get the **second order derivative**, denoted $\frac{d^2y}{dx^2}$.

The **second derivative** gives the **rate of change** of the **gradient** of the curve with respect to x.
In other words, it tells you how quickly the **gradient** of $y = f(x)$ is changing for any given value of x.
In function notation, the **second derivative** is written $\mathbf{f''(x)}$.

Example

For the function $f(x) = 2x^3 + 4x^2 + x$, find $f'(x)$ and $f''(x)$.

1. Differentiate $f(x)$ term-by-term for $f'(x)$.
2. Differentiate $f'(x)$ to get the second derivative, $f''(x)$.

$f(x) = 2x^3 + 4x^2 + x$

$f'(x) = 2(3x^2) + 4(2x) + 1 = 6x^2 + 8x + 1$

$f''(x) = 6(2x) + 8 = 12x + 8$

Exercise 10.3.1

Q1 Find $\frac{dy}{dx}$ and $\frac{d^2y}{dx^2}$ for each of these functions:

a) $y = x^3$ b) $y = x^5$ c) $y = x^4$ d) $y = x$

e) $y = 2x^2$ f) $y = 4x^3$ g) $y = \frac{1}{2}x^4$ h) $y = 3x^{-2}$

i) $y = \frac{1}{x}$ j) $y = \sqrt{x}$ k) $y = \frac{1}{x^2}$ l) $y = x\sqrt{x}$

Q2 Find the second order derivatives of the following functions:

a) $y = x^2 + x^4$ b) $f(x) = x^3 - x$ c) $y = 2x^6 + 3x^2$

d) $f(x) = -2x^{-2} + 5x$ e) $y = 4x^4 + \sqrt{x}$ f) $f(x) = -x^5 - 4\sqrt{x}$

Q3 Find f′(x) and f″(x) for each of these functions:

a) $f(x) = x(4x^2 - x)$
b) $f(x) = (x^2 - 3)(x - 4)$
c) $f(x) = x^2(3x - x^2)$
d) $f(x) = (x^{-1} + 5)(x^{-1} - 5)$
e) $f(x) = \frac{x^3 + 8}{x}$
f) $f(x) = 3\sqrt{x} + x\sqrt{x}$
g) $f(x) = \frac{4x^5 + 12x^3 - 40x}{4(x^2 + 5)}$
h) $f(x) = \frac{1}{x}(3x^4 - 2x^3)$
i) $f(x) = \frac{x^2 - x\sqrt{x} + 7x}{\sqrt{x}}$

Q4 Find the value of the second derivative at the given value for x.

a) $f(x) = 15x^3,\ x = \frac{1}{3}$
b) $y = \frac{x^4 - 2x^3}{12},\ x = 4$
c) $f(x) = x^3 - x^2,\ x = 3$
d) $y = x\sqrt{x} - \frac{1}{x},\ x = 4$
e) $f(x) = x^2(x - 5)(x^2 + x),\ x = -1$
f) $y = \frac{x^5 + 4x^4 - 12x^3}{x + 6},\ x = 5$
g) $f(x) = \frac{9x^2 + 3x}{3\sqrt{x}},\ x = 1$
h) $y = \left(\frac{1}{x^2} + \frac{1}{x}\right)(5 - x),\ x = -3$

Q5 Find the coordinates of the point(s) with the given value for the second derivative.

PROBLEM SOLVING

a) $y = x^3 + 3x^2,\ \frac{d^2y}{dx^2} = 18$
b) $f(x) = \frac{4}{15}x^{\frac{5}{2}}$ (where $f(x) \geq 0$), $f''(x) = 2$
c) $y = (2x^2 + 4)(2x^2 - 4),\ \frac{d^2y}{dx^2} = 768$
d) $y = \frac{1}{4}x^4 - \frac{5}{3}x^3 + 2x^2 - 20x,\ \frac{d^2y}{dx^2} = 12$

Stationary points

Stationary points occur when the **gradient** of a graph is **zero**. There are three types of stationary point:

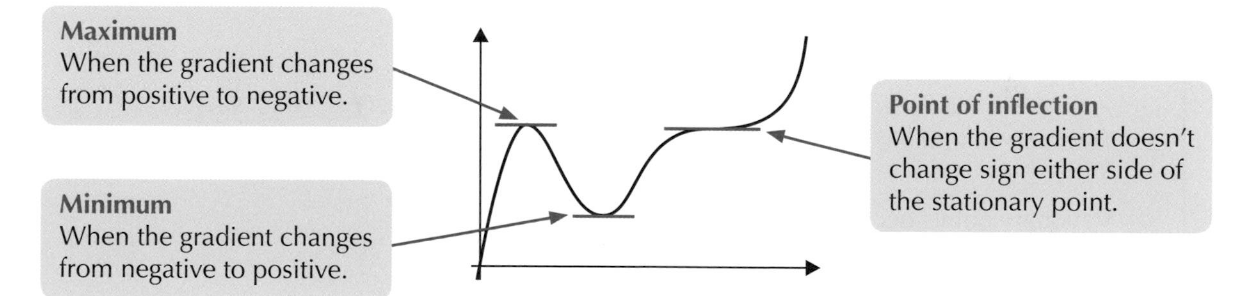

Some stationary points are called **local** maximum or minimum points because the function takes on higher or lower values in other parts of the graph. The ones shown above are both local.

Because stationary points occur when the gradient is zero, you can use **differentiation** to find them:

1. Differentiate $f(x)$.
2. Set $f'(x) = 0$.
3. Solve $f'(x) = 0$ to find the x-values.
4. Put the x-values back into the original equation to find the y-values.

Tip: Don't forget this last step — once you've found x you also need to find y to give the coordinates of each point.

Example 1

Find the stationary points on the curve $y = 2x^3 - 3x^2 - 12x + 5$.

1. You need to find where $\frac{dy}{dx} = 0$, so start by differentiating the function.

$$y = 2x^3 - 3x^2 - 12x + 5 \Rightarrow \frac{dy}{dx} = 6x^2 - 6x - 12$$

2. Then set the derivative equal to zero.

$$6x^2 - 6x - 12 = 0$$

3. Now solve this equation — it's just a normal quadratic.

$$6x^2 - 6x - 12 = 0 \Rightarrow x^2 - x - 2 = 0$$
$$\Rightarrow (x + 1)(x - 2) = 0$$
$$\Rightarrow x = -1 \text{ and } x = 2$$

4. You've found the x-values of the stationary points. To find their coordinates, put these x-values into the original equation.

$$x = -1 \Rightarrow y = 2(-1)^3 - 3(-1)^2 - 12(-1) + 5 = 12$$
$$x = 2 \Rightarrow y = 2(2)^3 - 3(2)^2 - 12(2) + 5 = -15$$

This gives the coordinates $(-1, 12)$ and $(2, -15)$.

Example 2

PROBLEM SOLVING

The sketch to the right is of the graph $y = x^3(x^2 + x - 3)$. One stationary point occurs at $(-1.8, 9.1)$. Show that the other two occur when $x = 0$ and when $x = 1$, and find their coordinates.

y

$y = x^3(x^2 + x - 3)$

x

1. You need to start by **differentiating** the function, but you can't do that in its current form. So first, **multiply out** the brackets:

$$y = x^3(x^2 + x - 3) = x^5 + x^4 - 3x^3$$

2. Then you can differentiate as normal.

$$\frac{dy}{dx} = 5x^4 + 4x^3 - 9x^2$$

3. Stationary points occur when the **gradient** is **equal to zero**, so set $\frac{dy}{dx}$ equal to zero and solve for x.

$$5x^4 + 4x^3 - 9x^2 = 0$$
$$\Rightarrow x^2(5x^2 + 4x - 9) = 0$$
$$\Rightarrow x^2(5x + 9)(x - 1) = 0$$
$$\Rightarrow x = 0,\ x = -\frac{9}{5} = -1.8 \text{ (given above) and } x = 1$$

So the other two stationary points occur at $x = 0$ and $x = 1$.

4. To find the **coordinates** of these points, put the x-values into the original equation.

$$x = 0 \Rightarrow y = x^3(x^2 + x - 3)$$
$$= 0^3(0^2 + 0 - 3) = 0(-3) = 0$$

$$x = 1 \Rightarrow y = x^3(x^2 + x - 3)$$
$$= 1^3(1^2 + 1 - 3) = 1(-1) = -1$$

So coordinates of the stationary points are $(0, 0)$ and $(1, -1)$.

Exercise 10.3.2

Q1 Without doing any calculations, say how many stationary points the graphs below have in the intervals shown.

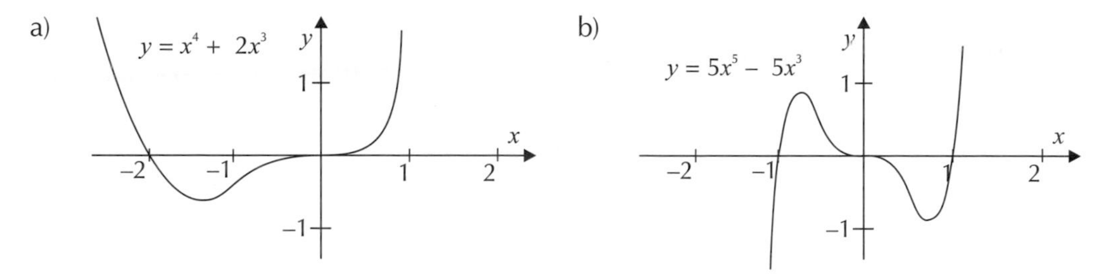

Q2 Find the x-coordinates of the stationary points of the curves with the following equations:

a) $y = x^2 + 3x + 2$

b) $y = (3 - x)(4 + 2x)$

c) $y = x^3 + 4x^2 - 3x$

d) $y = x^4 - 12x^3$

Q3 Find the coordinates of the stationary points of the curves with the following equations:

a) $y = 2x^2 - 5x + 2$

b) $y = -x^2 + 3x - 4$

c) $y = 7 - 6x - 3x^2$

d) $y = (x - 1)(2x + 3)$

Q4 Find the coordinates of the stationary points of the curves with the following equations:

a) $y = x^3 - 3x + 2$

b) $y = 4x^3 + 5$

c) $y = 3x^3 + 6x^2$

d) $4x^3 + 12x^2 + 8$

Q5 Find the coordinates of the stationary points of the graph of $y = x^3 - 6x^2 - 63x + 21$.

Q6 Show that the graph of the function given by $f(x) = x^5 + 3x + 2$ has no stationary points.

Q6 Hint: If there are no stationary points, there are no values of x for which $f'(x) = 0$.

Q7 a) Differentiate $y = x^3 - 7x^2 - 5x + 2$.

b) Hence find the coordinates of the stationary points of the curve with equation $y = x^3 - 7x^2 - 5x + 2$.

Q8 A graph is given by the function $f(x) = x^3 + kx$, where k is a constant. Given that the graph has no stationary points, find the range of possible values for k.

PROBLEM SOLVING

Maximum and minimum points

Once you've found where the stationary points are, you might be asked to decide if each one is a **maximum** or **minimum**. Maximum and minimum points are also known as **turning points**.

To decide whether a stationary point is a maximum or minimum, **differentiate again** to find $\frac{d^2y}{dx^2}$ or $f''(x)$ (see page 198).

Tip: That's what a question means when it asks you to "determine the nature of the turning points".

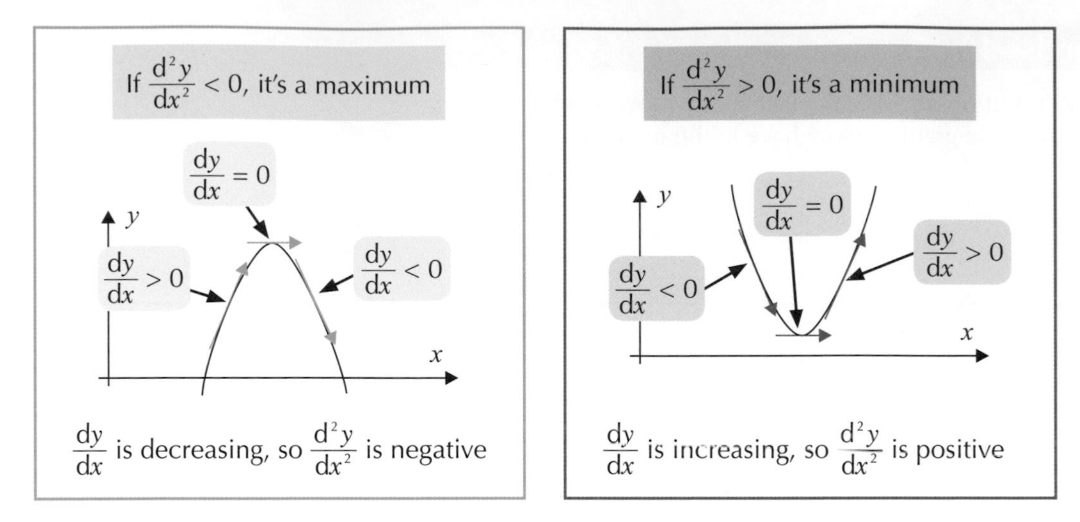

If the second derivative is **equal to zero**, you **can't** tell what type of stationary point it is.

Example

Determine the nature of the stationary points in Example 1 on p.200 ($y = 2x^3 - 3x^2 - 12x + 5$).

1. The first derivative has been found already.
   $\frac{dy}{dx} = 6x^2 - 6x - 12$
2. To determine the nature of the stationary points, differentiate again.
   $\frac{dy}{dx} = 6x^2 - 6x - 12 \Rightarrow \frac{d^2y}{dx^2} = 12x - 6$
3. Then just put in the x-values of the coordinates of the stationary points (you found these on page 200).
   $x = -1, \frac{d^2y}{dx^2} = -18 \qquad x = 2, \frac{d^2y}{dx^2} = 18$
4. Use $\frac{d^2y}{dx^2}$ to determine the nature of the points.
   $\frac{d^2y}{dx^2} < 0$ at $(-1, 12)$ so it's a maximum
   $\frac{d^2y}{dx^2} > 0$ at $(2, -15)$ so it's a minimum
5. Since you know the **turning points** and the fact that it's a **cubic** with a positive coefficient of x^3, you can now **sketch** the graph (though the points of intersection with the x-axis would be difficult to find accurately).

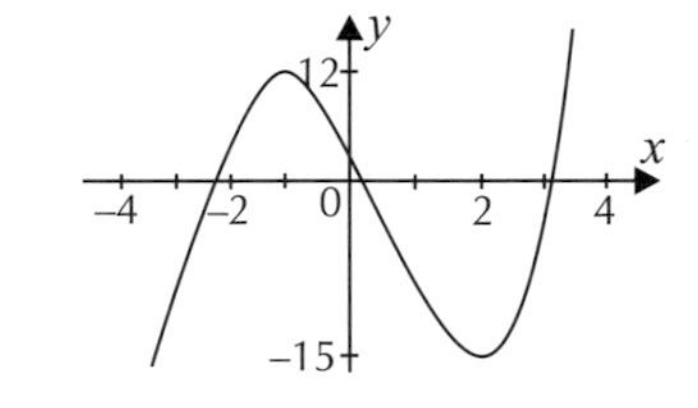

Exercise 10.3.3

Q1 The diagram on the right shows a sketch of the graph of $y = f(x)$. For each turning point, say whether $f''(x)$ would be positive or negative.

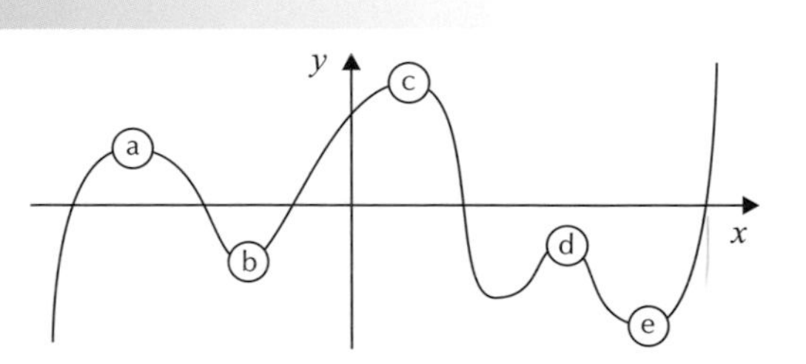

Q2 For each of the following, find the second derivative of the function and say whether the given point is a maximum or a minimum:

a) $y = x^2 + 14x - 3$ at $(-7, -52)$.

b) $y = x^3 - 12x + 4$ at $(2, -12)$.

c) $y = 2x^3 - x^2$ at $\left(\frac{1}{3}, -\frac{1}{27}\right)$.

d) $y = 2x^4 - 16x^3 + 900$ at $(6, 36)$.

e) $y = 4x^5 + 15x^4 - 250$ at $(-3, -7)$.

f) $y = x^5 - 5x^4 + 5x^2 - 40x + 400$ at $(4, 64)$.

Q3 A function $y = f(x)$ is such that $f(1) = 3$, $f'(1) = 0$ and $f''(1) = 7$.

PROBLEM SOLVING

a) Give the coordinates of one of the turning points of $f(x)$.

b) Determine the nature of this turning point, explaining your answer.

Q4 Find the stationary points on the graphs of the following functions and if possible say whether they're maximum or minimum turning points:

a) $y = 5 - x^2$

b) $y = \frac{1}{2}x^2 + 21x + 12$

c) $y = 2x^3 - 6x + 2$

d) $y = x^3 - 3x^2 - 24x + 15$

e) $y = \frac{1}{12}x^4 + \frac{2}{9}x^3$

f) $y = x^4 + 4x^3 + 4x^2 - 10$

Q5 Find the stationary points on the graphs of the following functions and say whether they're maximum or minimum turning points:

a) $f(x) = 8x^3 + 16x^2 + 8x + 1$

b) $f(x) = \frac{1}{3}x^3 - 6x^2 - 45x$

c) $f(x) = \frac{1}{4}x^4 + 2x^3 - 8x^2 + 1$

d) $f(x) = \frac{27}{x^3} + x$

Q6 a) Given that $f(x) = x^3 - 3x^2 + 4$, find $f'(x)$ and $f''(x)$.

b) Hence find the coordinates of any stationary points on the graph $f(x)$ and say whether they're maximum or minimum turning points.

Q7 A function is given by $y = x^2 + \frac{2000}{x}$.

a) Find the value of x at which y is stationary.

b) Is this a minimum or maximum point?

Q8 A curve is given by $f(x) = 2x^3 + 4x^2 + c$, where c is an integer. Find the value of c, given that the y-coordinate of the maximum point is $-\frac{260}{27}$.

PROBLEM SOLVING

Q9 The curve given by $f(x) = x^3 + ax^2 + bx + c$ has a stationary point with coordinates $(3, 10)$. If $f''(x) = 0$ at $(3, 10)$, find a, b and c.

PROBLEM SOLVING

Q10 a) Given that a curve with the equation $y = x^4 + kx^3 + x^2 + 17$ has only one stationary point, show that $k^2 < \frac{32}{9}$.

PROBLEM SOLVING

b) Find the coordinates of the stationary point and say whether it's a maximum or a minimum point.

Increasing and decreasing functions

As differentiation is about finding the gradients of curves, you can use it to find if a function is **increasing** or **decreasing** at a given point. This can help you to sketch the function and determine the nature of turning points.

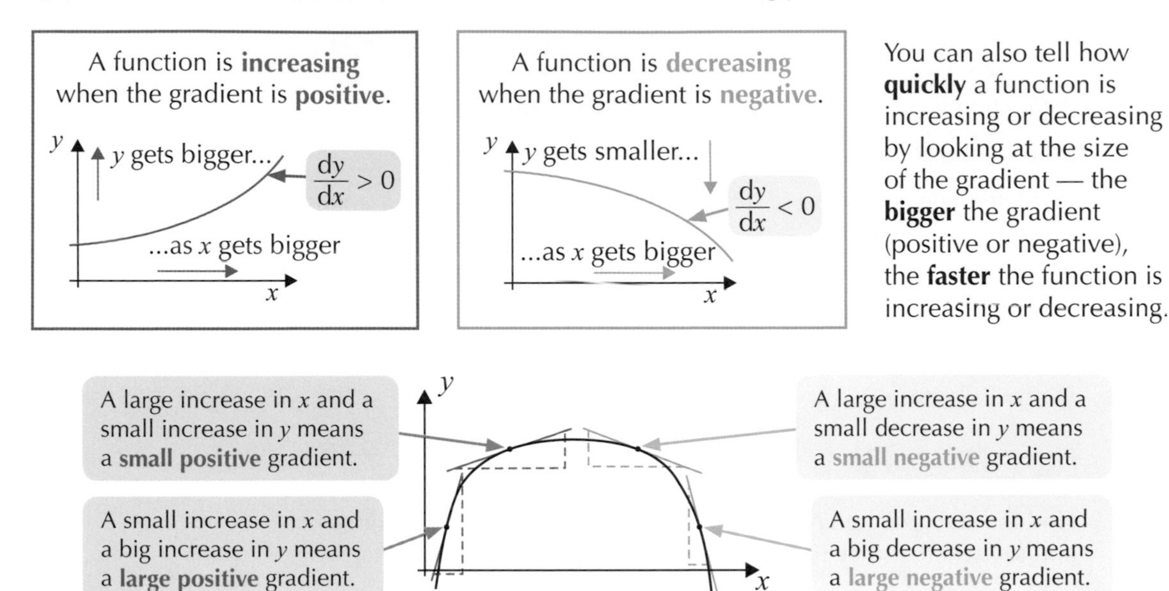

You can also tell how **quickly** a function is increasing or decreasing by looking at the size of the gradient — the **bigger** the gradient (positive or negative), the **faster** the function is increasing or decreasing.

Example

Find the values of x for which the function $y = x^3 - 6x^2 + 9x + 3$, $x > 0$ is increasing.

1. You want to know when y is increasing — so differentiate.

   $y = x^3 - 6x^2 + 9x + 3 \Rightarrow \frac{dy}{dx} = 3x^2 - 12x + 9$

2. It's an increasing function when the derivative is greater than zero, so write it down as an inequality and solve it.

   $\frac{dy}{dx} > 0 \Rightarrow 3x^2 - 12x + 9 > 0$
   $\Rightarrow x^2 - 4x + 3 > 0$
   $\Rightarrow (x - 3)(x - 1) > 0$

3. For the inequality to be true, either both brackets must be positive or both brackets must be negative.

   $x - 1 > 0$ and $x - 3 > 0$
   $\Rightarrow x > 1$ and $x > 3 \Rightarrow x > 3$

   or $x - 1 < 0$ and $x - 3 < 0$
   $\Rightarrow x < 1$ and $x < 3 \Rightarrow x < 1$

4. $x > 0$, so the function is increasing when:

   $0 < x < 1$ and $x > 3$

5. You could also look at the nature of the stationary points — this tells you where the function goes from increasing to decreasing and vice versa. $x = 1$ and $x = 3$ at the stationary points (as $\frac{dy}{dx} = 0$ at these points).

   $\frac{dy}{dx} = 3x^2 - 12x + 9 \Rightarrow \frac{d^2y}{dx^2} = 6x - 12$

   $x = 1$, $\frac{d^2y}{dx^2} = -6$, so it's a maximum

   $x = 3$, $\frac{d^2y}{dx^2} = 6$, so it's a minimum

6. This means the function is increasing as it approaches $x = 1$, is decreasing between $x = 1$ and $x = 3$ and starts increasing after $x = 3$.

   $y = x^3 - 6x^2 + 9x + 3$ is increasing where $0 < x < 1$ and $x > 3$

Exercise 10.3.4

Q1 For each of these functions, calculate the first derivative and use this to find the range of values for which the function is increasing.

a) $y = x^2 + 7x + 5$
b) $y = 5x^2 + 3x - 2$
c) $y = \frac{1}{4}x^2 - 6x + 32$
d) $y = 2 - 9x^2$
e) $y = -3x^2 - 4x + 15$
f) $y = -\left(\frac{3x^2 - 8x + 10}{5}\right)$

Q2 For each of these functions, find $f'(x)$ and find the range of values of x for which $f(x)$ is decreasing.

a) $f(x) = 16 - 3x - 2x^2$
b) $f(x) = (x + 3)(x - 2)$
c) $f(x) = (2x - 1)(x - 7)$
d) $f(x) = (6 - 3x)(6 + 3x)$
e) $f(x) = (1 - 2x)(7 - 3x)$
f) $f(x) = (5x - 2)(2 - 5x)$

Q3 Calculate $\frac{dy}{dx}$ for each of these functions and state the range of values for which the function is increasing.

a) $y = x^3 - 6x^2 - 15x + 25$
b) $y = x^3 + 6x^2 + 12x + 5$
c) $y = x(x^2 - 4x - 16)$
d) $y = \frac{x^3 - 3x^2 - 9x - 1}{9}$

Q4 Find the first derivative of each function and state the range of values for which the function is decreasing.

a) $f(x) = x^3 - 3x^2 - 9x + 1$
b) $f(x) = x^3 - 4x^2 + 4x + 7$
c) $f(x) = 3x^3 + \frac{3}{2}x^2 - 72x + 4$
d) $f(x) = \frac{2x^3 - x^2 - 8x + 3}{4}$

Q5 Use differentiation to explain why $f(x) = x^3 + x$ is an increasing function for all real values of x.

Q5 Hint:
An increasing function is one where $f'(x) > 0$ for all values of x.

Q6 Is the function $f(x) = 3 - 3x - x^3$ an increasing or decreasing function? Explain your answer.

Q7 Use differentiation to find the range of values of x for which each of these functions is decreasing:

a) $y = 2x^4 + x$
b) $y = x^4 - 18x^2 + 7$
c) $y = x^4 - 2x^3 - 5x^2 + 6$
d) $y = 2x^4 + \frac{4}{3}x^3 - 2x^2 + 14$

Q8 Differentiate these functions and find the range of values for which each function is increasing.

a) $y = x^2 + \sqrt{x}, x > 0$
b) $y = x^{-1} + 16x, x > 0$
c) $y = 4x^2 + \frac{1}{x}, x \neq 0$
d) $y = -\left(\frac{3x^{-2} + 18x}{2}\right), x \neq 0$

Q9 The function $y = 5 - 3x - ax^5$ is a decreasing function for all real values of x. Find the range of possible values for a.

PROBLEM SOLVING

Q10 The function $y = x^k + x$, where k is a positive integer, is an increasing function for all real values of x. Find all possible values of k.

PROBLEM SOLVING

Curve sketching

You covered some curve sketching in Chapter 6, so you should know the basic shapes of different types of graph. Now you'll see how differentiation can be used to find out more about the **shape** of the graph and to work out some **key points** like the turning points. Use the following **step-by-step** method to get all the information you need to draw an accurate sketch:

1. **Find where the curve crosses the axes.**

   To find where it crosses the y**-axis**, just put $x = 0$ into the function and find the value of y.

   To find where it crosses the x**-axis**, set the function equal to zero and solve for x (you'll probably have to **factorise** and find the **roots**).

2. **Decide on the shape of the graph.**

   Look at the **highest power** of x and its **coefficient** — this determines the overall **shape** of the graph (have a look back at pages 94-98). The most common ones are **quadratics**, **cubics** and **reciprocals**.

   A **quadratic** with a **positive** coefficient of x^2 will be **u-shaped**, and if the coefficient is **negative**, it'll be **n-shaped**.

   A **cubic** will go from **bottom left** to **top right** if the coefficient of x^3 is **positive**, and **top left** to **bottom right** if the coefficient is **negative**. It'll also have a characteristic '**wiggle**'.

   Reciprocals (e.g. $\frac{1}{x}$) and other **negative powers** have **two separate curves** in **opposite quadrants**, each with **asymptotes**.

3. **Differentiate to find the stationary points.**

   Find the **stationary points** by **differentiating** and setting $f'(x) = 0$.

   Then **differentiate again** to decide whether these points are **maximums** or **minimums**.

Example 1

Sketch the curve of the equation $y = f(x)$, where $f(x) = x^3 - 4x^2 + 4x$.

1. Start by finding where the curve crosses the y-axis.

   When $x = 0$, $y = 0$, so the curve goes through the origin.

2. Find where it crosses the x-axis by solving the equation $f(x) = 0$.

   $$x^3 - 4x^2 + 4x = 0 \Rightarrow x(x^2 - 4x + 4) = 0$$
   $$\Rightarrow x(x - 2)(x - 2) = 0$$
   $$\Rightarrow x = 0 \text{ and } x = 2$$

3. Find $f'(x)$ by differentiating.

   $$f(x) = x^3 - 4x^2 + 4x \Rightarrow f'(x) = 3x^2 - 8x + 4$$

4. Solve $f'(x) = 0$ for x-values at the stationary points.

   $$f'(x) = 0 \Rightarrow 3x^2 - 8x + 4 = 0$$
   $$\Rightarrow (3x - 2)(x - 2) = 0$$
   $$\Rightarrow x = 2 \text{ and } x = \frac{2}{3}$$

5. Put the x-values back into the original equation to find the y-values of the stationary points.

   $$y = (2)^3 - 4(2)^2 + 4(2) = 0 \text{ and}$$
   $$y = \left(\frac{2}{3}\right)^3 - 4\left(\frac{2}{3}\right)^2 + 4\left(\frac{2}{3}\right) = \frac{32}{27}$$

6. Differentiate again to find out if the stationary points are maximums or minimums.

$f''(x) = 6x - 8$
At $x = 2$, $f''(x) = 4$, so this is a **minimum**.
At $x = \frac{2}{3}$, $f''(x) = -4$, so this is a **maximum**.

7. It's a cubic equation with a positive coefficient of x^3, so the graph will go from bottom left to top right.

8. Now you have all the information you need to sketch the graph:

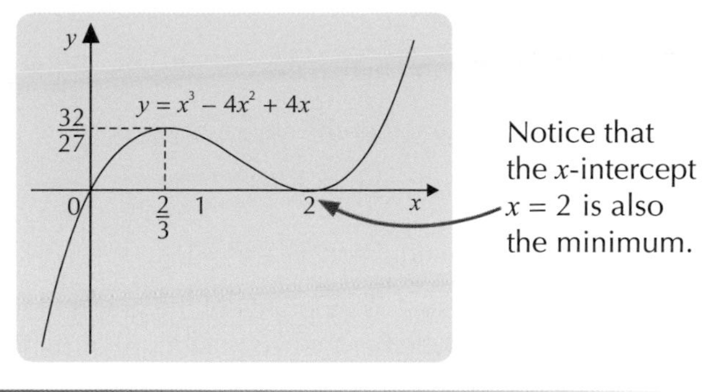

Example 2

Sketch the graph of $f(x) = -8x^3 + 2x$.

1. Start by finding where the curve crosses the axes. When $x = 0$, $f(x) = 0$ so the curve goes through the origin. Now solve $f(x) = 0$.

$-8x^3 + 2x = 0 \Rightarrow 2x(-4x^2 + 1) = 0$
So $x = 0$ or $-4x^2 + 1 = 0 \Rightarrow x^2 = \frac{1}{4} \Rightarrow x = \pm\frac{1}{2}$

2. Next differentiate the function to find the stationary point(s).

$f(x) = -8x^3 + 2x \Rightarrow f'(x) = -24x^2 + 2$
$f'(x) = 0$ so $-24x^2 + 2 = 0$
$\Rightarrow x^2 = \frac{1}{12} \Rightarrow x = \pm\sqrt{\frac{1}{12}} = \pm\frac{1}{2\sqrt{3}}$

$x = \frac{1}{2\sqrt{3}}$ gives $f(x) = \frac{2}{3\sqrt{3}}$ and

$x = -\frac{1}{2\sqrt{3}}$ gives $f(x) = -\frac{2}{3\sqrt{3}}$

3. Differentiate again to see if these points are maximums or minimums.

$f''(x) = -48x$
At $x = \frac{1}{2\sqrt{3}}$, $f''(x) = -\frac{24}{\sqrt{3}}$,
which is negative so it's a **maximum**.
At $x = -\frac{1}{2\sqrt{3}}$, $f''(x) = \frac{24}{\sqrt{3}}$,
which is positive so it's a **minimum**.

4. Finally, think about the overall shape of the graph. The highest power is 3, so it's a cubic and has a negative coefficient, so the graph will go from top left to bottom right.

5. Now you have all the information you need to sketch the graph:

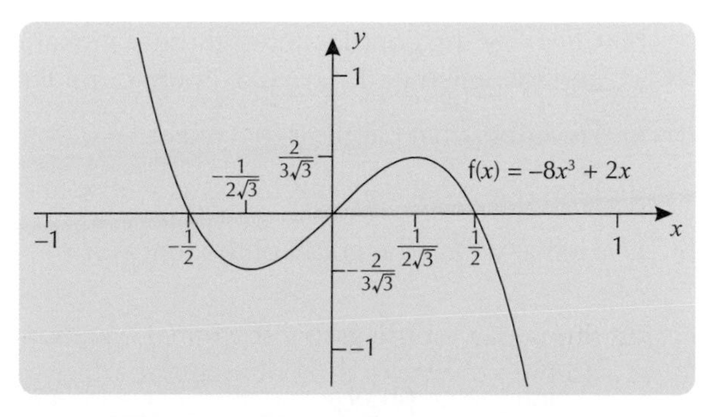

Exercise 10.3.5

Q1 For the graph $y = x^3 - 2x^2$:

a) Find the coordinates of the points at which it crosses each axis.

b) Find $\frac{dy}{dx}$ and hence the coordinates of the points where $\frac{dy}{dx} = 0$.

c) Identify whether the stationary points are maximums or minimums.

d) Sketch the graph of $y = x^3 - 2x^2$.

Q2 a) Solve the equation $x^3 + x^2 = 0$.

b) Find the stationary points of the graph of $f(x) = x^3 + x^2$ and say whether they're maximum or minimum points.

c) Use your answers to parts a) and b) to sketch the graph of $f(x) = x^3 + x^2$, labelling the coordinates of the stationary points and places where the curve meets the axes.

Q3 a) Find the first and second derivatives of the function $f(x) = x^4 - x^3$.

b) Write down the ranges of values of x for which $f(x)$ is increasing and decreasing.

c) Sketch the graph of $y = f(x)$, labelling the coordinates of all stationary points and the points where the curve crosses the axes.

Q4 Sketch the graphs of the equations below, labelling the coordinates of any stationary points and the points where the curves cross the axes.

a) $y = 3x^3 + 3x^2$

b) $y = -x^3 + 9x$

c) $y = x^4 - x^2$

d) $y = x^4 + x^2$

Q5 Given that $x^3 - x^2 - x + 1 = (x + 1)(x - 1)^2$, sketch the graph of $y = x^3 - x^2 - x + 1$, labelling the coordinates of all the stationary points and the points where the curve crosses the axes.

Q6 a) Show that $x^3 - 4x = 0$ when $x = -2$, 0 and 2.

b) Use first and second derivatives to show that the graph of $y = x^3 - 4x$ has a minimum at (1.2, –3.1) and a maximum at (–1.2, 3.1), where all coordinates are given to 1 d.p.

c) Use your answers to parts a) and b) to sketch the graph of $y = x^3 - 4x$, labelling the coordinates of the stationary points and the points at which the curve crosses the axes.

Q7 a) Show that the graph of $f(x) = x + \frac{1}{x}$, $x \neq 0$ has 2 stationary points.

b) Calculate the coordinates of these stationary points and say whether they're maximum or minimum points.

c) Describe what happens to $f(x)$ as $x \to 0$ from both sides.

d) Describe what happens to $f(x)$ as $x \to \infty$ and $x \to -\infty$.

e) Hence sketch the graph of the function $x + \frac{1}{x}$.

Q7 Hint: The '→' symbol means 'tends to' or 'approaches' — you need to say what happens as x gets close to zero and to $\pm\infty$.

Q8 a) Show that for the graph of $y = x^4 + \frac{8}{\sqrt{x}}$, $x > 0$, $\frac{dy}{dx} = 0$ when $x = 1$.

PROBLEM SOLVING

b) Sketch the graph of $y = x^4 + \frac{8}{\sqrt{x}}$, $x > 0$, labelling the coordinates of the stationary point.

10.4 Real-Life Problems

Equations can be used to model real-life contexts. That means you can use differentiation to solve real-life problems involving rates of change or maximum or minimum values.

Learning Objectives (Spec Ref 7.1 & 7.3):
- Solve real-life problems about rates of change.
- Describe a real-life situation in mathematical terms and use differentiation to find maximum and minimum solutions.

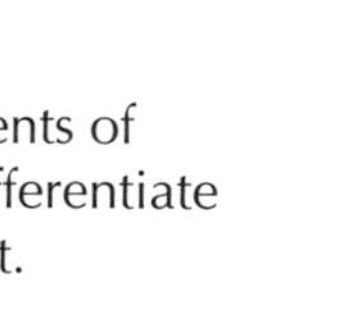

Speed and acceleration problems

Until now, all the examples have been about differentiating functions of x to find gradients of curves. But **real-life** examples often involve a function of time, t, and you'll need to differentiate to find the **rate of change** over time. The maths is the **same**, the **letters** are just different.

The next example looks at the **distance** a car has travelled as a function of **time**.

Example

A car pulls away from a junction and travels x metres in t seconds. For the first 10 seconds, its journey is modelled by the equation $x = 2t^2$.

a) Find the speed of the car after 8 seconds.

1. Speed is the rate of change of distance with respect to time. To work out the speed as a function of t, differentiate x to find $\frac{dx}{dt}$.
   $x = 2t^2$
   $\frac{dx}{dt} = 4t$
2. You've got speed as a function of t, so put t **= 8 seconds** into the expression.
   When $t = 8$, $\frac{dx}{dt} = 32$
3. So after 8 seconds, the car is travelling at... 32 ms^{-1}

b) Find the car's acceleration during this period.

1. Acceleration is the rate of change of speed with respect to time — so differentiate the speed with respect to time to get the second derivative $\frac{d^2x}{dt^2}$.
   $\frac{dx}{dt} = 4t$
   $\frac{d^2x}{dt^2} = 4$
2. This means that the car's acceleration during this period is... 4 ms^{-2}

Exercise 10.4.1

Q1 A particle moves along a path described by the equation $x = 3t^2 - 7t$, where t is the time in seconds and x is the distance in metres.

a) Find the speed, $\frac{dx}{dt}$, of the particle as a function of t.

b) What is the speed of the particle in ms^{-1} at: (i) $t = 2$ seconds? (ii) $t = 5$ seconds?

c) Find the value of t when the speed is 17 ms^{-1}.

d) Find the acceleration $\frac{d^2x}{dt^2}$ of the particle as a function of t.

Q2 A particle moves along a path described by the equation $x = 2t^3 - 4t^2$, $t > 0$, where t is the time in seconds and x is the distance in metres.

a) Find the speed of the particle after t seconds.

b) Find x and t when the speed is 30 ms^{-1}.

c) Find the acceleration of the particle after t seconds.

d) Find the acceleration at $t = 5$ seconds in ms^{-2}.

e) Find the speed when the acceleration is 16 ms^{-2}.

Q2 Hint: For part e), use the information you've been given to work out the value of t and then put the t value into the expression for speed.

Q3 A train moving away from a station travels x metres in t seconds. For the first minute, the train's journey is modelled by the equation $x = \frac{t^2}{4}$.

a) Find the speed of the train after t seconds.

b) Find the acceleration of the train during this period.

c) Find the train's speed at $t = 45$ seconds.

d) How far will the train have travelled when its speed reaches 25 ms^{-1}?

Length, area and volume problems

Because differentiation can be used to find the maximum value of a function, it can be used in **real-life problems** to maximise a quantity subject to certain factors, e.g. maximising the volume of a box that can be made with a set amount of cardboard.

To find the maximum value of something, all you need is an equation **in terms of only one variable** (e.g. x) — then just **differentiate as normal**. Often there'll be too many variables in the question, so you've got to know how to manipulate the information to get rid of the unwanted variables.

Example 1

A farmer wants to build a rectangular sheep pen with length x m and width y m. She has 20 m of fencing in total, and wants the area inside the pen to be as large as possible. How long should each side of the pen be, and what will the area inside the pen be?

1. Start by writing down an expression for the area of the pen:

Area = length × width = xy m^2

(x m by y m rectangle)

2. This has too many variables to work with, so find an expression for y in terms of x. You know how much fencing is available, so find an expression for that in terms of x and y and rearrange it to make y the subject.

Perimeter = 20 m = $2x + 2y$

$\Rightarrow y = \frac{20 - 2x}{2} = 10 - x$

3. Now you can substitute this into the expression you wrote down for the area and use differentiation to maximise it.

$A = xy = x(10 - x) = 10x - x^2$,

so $A = 10x - x^2 \Rightarrow \frac{dA}{dx} = 10 - 2x$

4. Now just find when $\frac{dA}{dx} = 0$.
$\frac{dA}{dx} = 0 \Rightarrow 10 - 2x = 0$, so $x = 5 \Rightarrow y = 10 - x = 5$

5. To check that this value of x gives a maximum for A, differentiate again:
$\frac{d^2A}{dx^2} = -2$, which is negative, so this will give a maximum for A

6. So both x and y should be 5 m, which means the area inside the pen will be...
$5\text{ m} \times 5\text{ m} = 25\text{ m}^2$

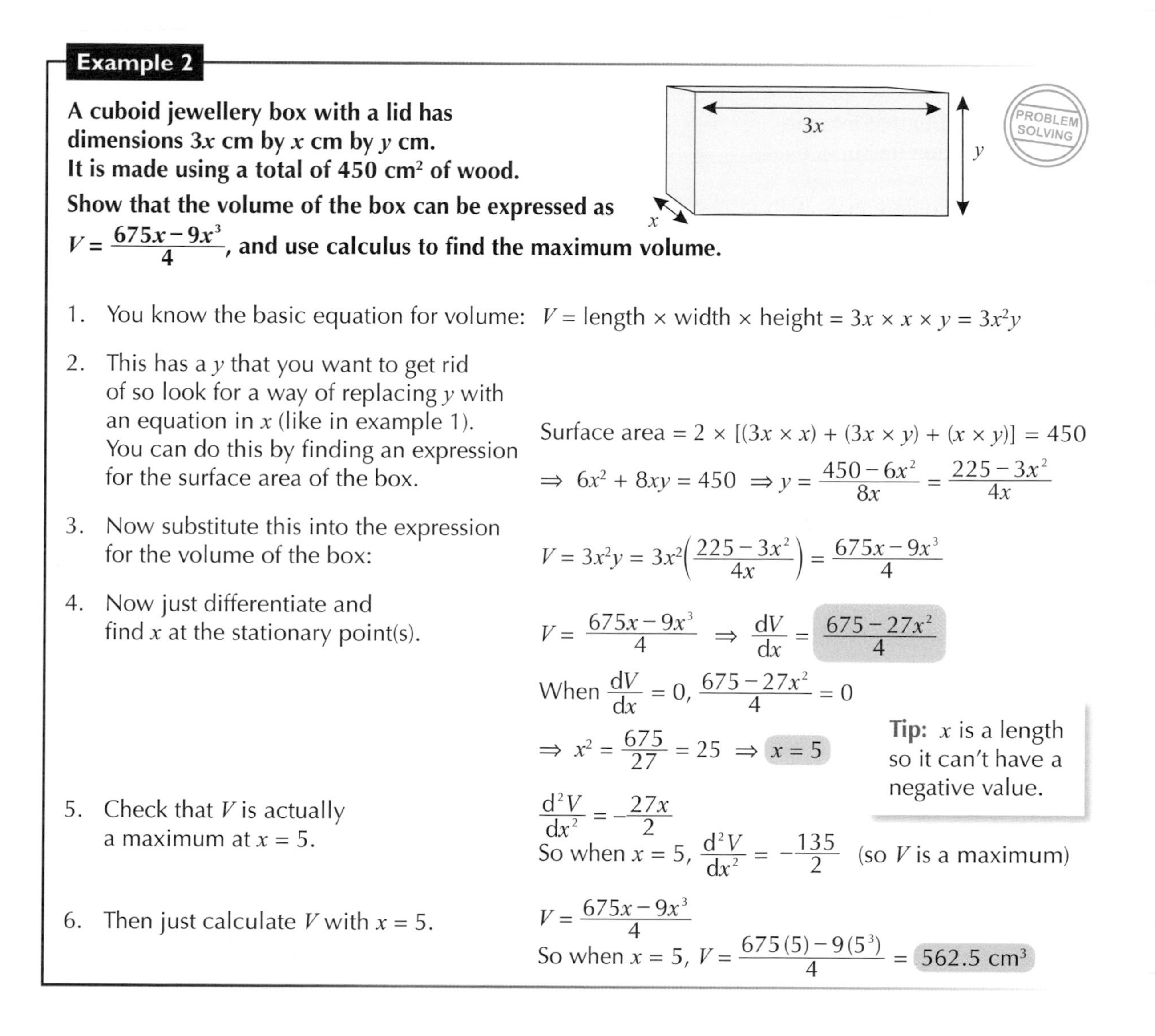

Example 2

A cuboid jewellery box with a lid has dimensions $3x$ cm by x cm by y cm.
It is made using a total of 450 cm² of wood.
Show that the volume of the box can be expressed as $V = \frac{675x - 9x^3}{4}$, and use calculus to find the maximum volume.

1. You know the basic equation for volume: $V = \text{length} \times \text{width} \times \text{height} = 3x \times x \times y = 3x^2y$

2. This has a y that you want to get rid of so look for a way of replacing y with an equation in x (like in example 1). You can do this by finding an expression for the surface area of the box.
Surface area $= 2 \times [(3x \times x) + (3x \times y) + (x \times y)] = 450$
$\Rightarrow 6x^2 + 8xy = 450 \Rightarrow y = \frac{450 - 6x^2}{8x} = \frac{225 - 3x^2}{4x}$

3. Now substitute this into the expression for the volume of the box:
$V = 3x^2y = 3x^2\left(\frac{225 - 3x^2}{4x}\right) = \frac{675x - 9x^3}{4}$

4. Now just differentiate and find x at the stationary point(s).
$V = \frac{675x - 9x^3}{4} \Rightarrow \frac{dV}{dx} = \frac{675 - 27x^2}{4}$
When $\frac{dV}{dx} = 0$, $\frac{675 - 27x^2}{4} = 0$
$\Rightarrow x^2 = \frac{675}{27} = 25 \Rightarrow x = 5$

Tip: x is a length so it can't have a negative value.

5. Check that V is actually a maximum at $x = 5$.
$\frac{d^2V}{dx^2} = -\frac{27x}{2}$
So when $x = 5$, $\frac{d^2V}{dx^2} = -\frac{135}{2}$ (so V is a maximum)

6. Then just calculate V with $x = 5$.
$V = \frac{675x - 9x^3}{4}$
So when $x = 5$, $V = \frac{675(5) - 9(5^3)}{4} = 562.5\text{ cm}^3$

Differentiation isn't just limited to cuboids — it can be used on **any shape** as long as you can describe its (surface) area or volume with variables (i.e x, y).

Example 3

A cylindrical pie tin is t cm high with a diameter of d cm. The volume of the pie tin is 1000 cm³.

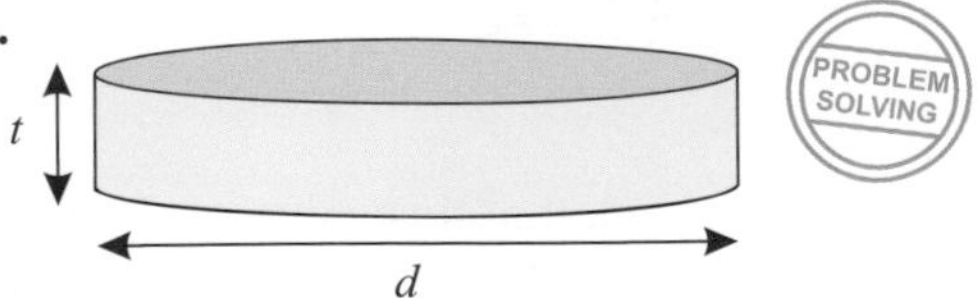

Show that the surface area of the tin is given by $A = \frac{\pi}{4}d^2 + \frac{4000}{d}$ and find the minimum surface area.

1. Write an expression for A.

   A = area of tin's base + area of tin's curved face
   $= \pi\left(\frac{d}{2}\right)^2 + (\pi d \times t) = \frac{\pi d^2}{4} + \pi dt$

2. Find an equation for the volume and rearrange it to make t the subject.

   $V = \pi\left(\frac{d}{2}\right)^2 t = 1000 \Rightarrow t = \frac{1000}{\pi\left(\frac{d}{2}\right)^2} = \frac{4000}{\pi d^2}$

3. Then put that into the equation for surface area.

   $A = \frac{\pi d^2}{4} + \pi dt = \frac{\pi d^2}{4} + (\pi d \times \frac{4000}{\pi d^2}) = \frac{\pi d^2}{4} + \frac{4000}{d}$

4. Next, differentiate with respect to d and find the values of d that make $\frac{dA}{dd} = 0$.

   $\frac{dA}{dd} = \frac{\pi d}{2} - \frac{4000}{d^2}$ so when $\frac{dA}{dd} = 0$,
   $\frac{\pi d}{2} - \frac{4000}{d^2} = 0 \Rightarrow d^3 = \frac{2 \times 4000}{\pi} \Rightarrow d = \frac{20}{\sqrt[3]{\pi}}$

5. Check to see if this value of d gives a minimum for A.

   $\frac{d^2A}{dd^2} = \frac{\pi}{2} + \frac{8000}{d^3} = \frac{\pi}{2} + \frac{8000}{\left(\frac{8000}{\pi}\right)} = \frac{3\pi}{2}$,
   so it's a minimum.

6. Now calculate the surface area for that value of d.

   $A = \frac{\pi}{4}\left(\frac{20}{\sqrt[3]{\pi}}\right)^2 + \frac{4000}{\left(\frac{20}{\sqrt[3]{\pi}}\right)} = 439 \text{ cm}^2$ (to 3 s.f.)

Exercise 10.4.2

Q1 A farmer wants to enclose a rectangular area of 100 m² with a fence. Find the minimum length of fencing he needs to use.

Q2 A ball is catapulted vertically with an initial speed of 30 m/s. After t seconds the height h of the ball, in m, is given by $h = 30t - 4.9t^2$. Use calculus to find the maximum height the ball reaches.

Q3 A rectangular vegetable patch is enclosed by a wall on one side and fencing on three sides as shown in the diagram.

Fencing

Wall

Use calculus to show that the maximum possible area that can be enclosed by 66 m of fencing is 544.5 m².

Q3 Hint: It might look like you're not given enough information here, but just call the length x and the width y and you're on your way.

Q4 A pet food manufacturer designs tins of cat food of capacity 500 cm^3 as shown below. The radius of the tin is r cm and the height is h cm.

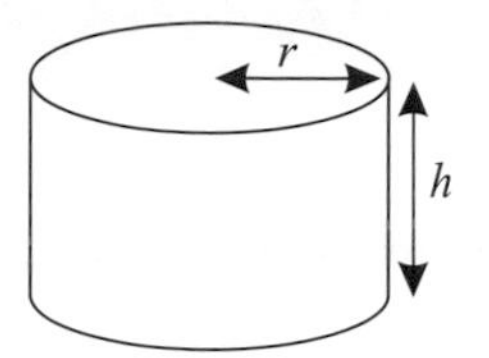

a) Show that the surface area A of the tin is given by $A = 2\pi r^2 + \frac{1000}{r}$.

b) Find the value of r which minimises the surface area (to 3 s.f.).

c) Find the minimum possible surface area for the tin (to 3 s.f.).

Q5 A child makes a box by taking a piece of card measuring 40 × 40 cm and cutting squares with side length x cm, as shown in the diagram. The sides are then folded up to make a box.

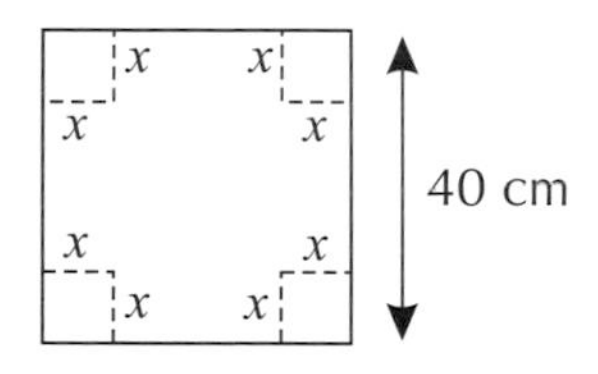

a) Write down a formula for the volume of the box, V.

b) Find the maximum possible volume of the box to 3 s.f.

Q6 A chocolate manufacturer designs a box which is a triangular prism as shown in the diagram. The cross-section of the prism is a right-angled triangle with sides of x cm, x cm and h cm. The length of the prism is l cm and the volume is 300 cm^3.

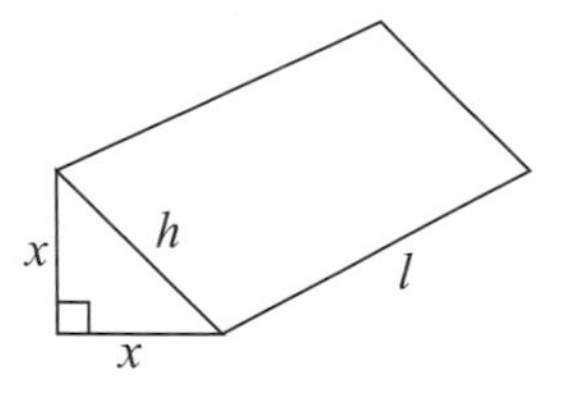

a) Show that the surface area of the prism is given by $A = x^2 + \frac{600(2+\sqrt{2})}{x}$.

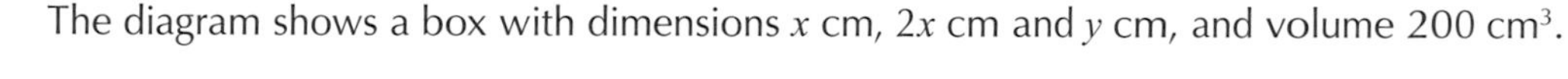

b) Show that the value of x which minimises the surface area of the box is $\sqrt[3]{600 + 300\sqrt{2}}$.

Q7 The diagram shows a box with dimensions x cm, $2x$ cm and y cm, and volume 200 cm^3.

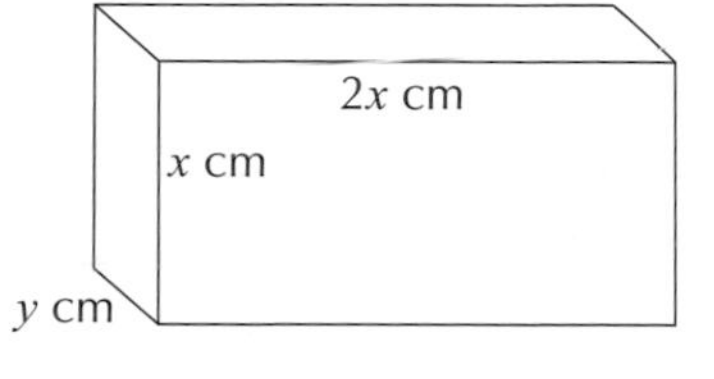

a) Show that the surface area A of the box is given by $A = 4x^2 + \frac{600}{x}$.

b) Use calculus to find the value of x that gives the minimum value of A (to 3 s.f.).

c) Hence find the minimum possible surface area of the box, correct to 3 s.f.

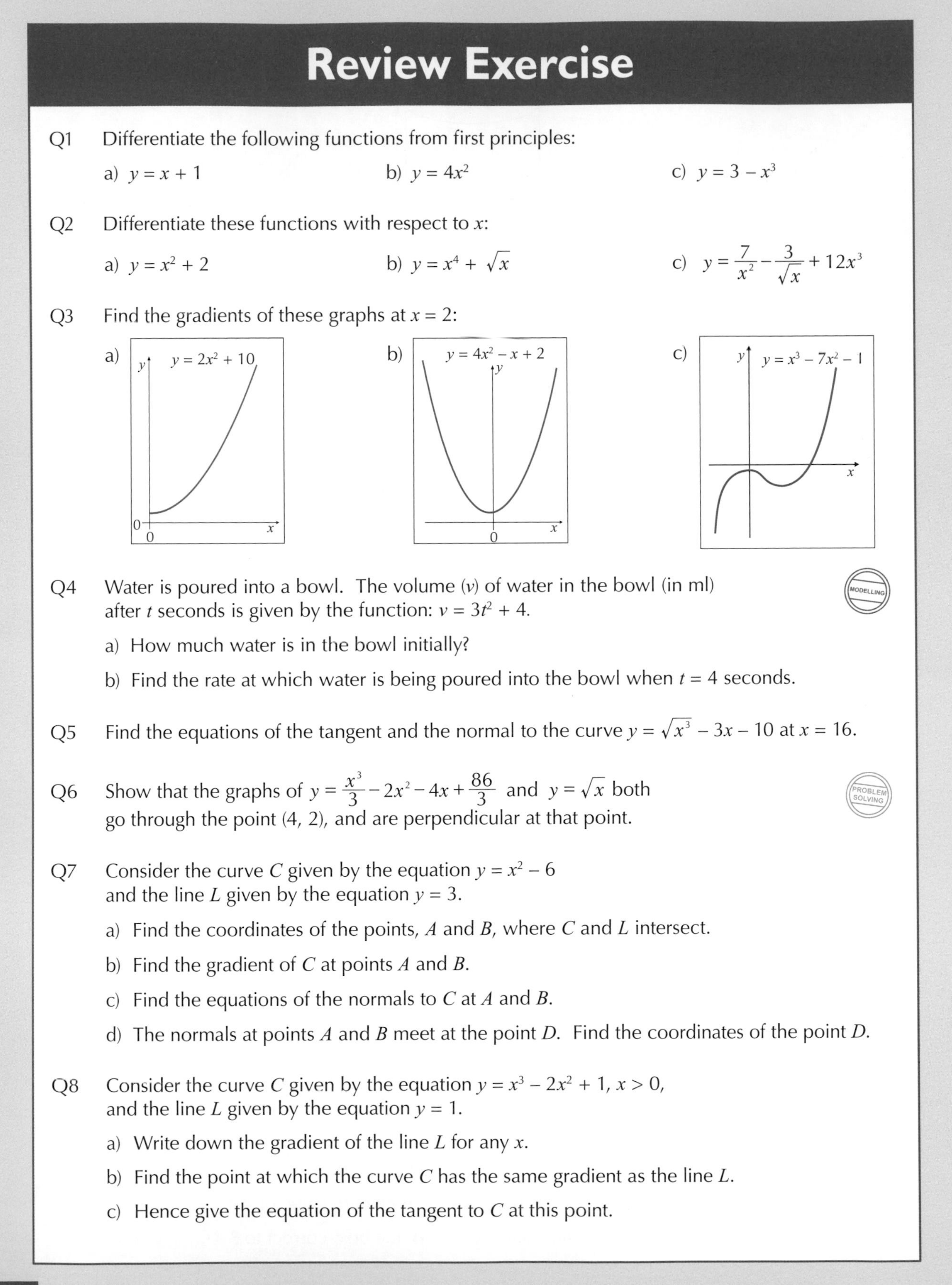

Review Exercise

Q1 Differentiate the following functions from first principles:

a) $y = x + 1$ b) $y = 4x^2$ c) $y = 3 - x^3$

Q2 Differentiate these functions with respect to x:

a) $y = x^2 + 2$ b) $y = x^4 + \sqrt{x}$ c) $y = \frac{7}{x^2} - \frac{3}{\sqrt{x}} + 12x^3$

Q3 Find the gradients of these graphs at $x = 2$:

a)

b)

c)

Q4 Water is poured into a bowl. The volume (v) of water in the bowl (in ml) after t seconds is given by the function: $v = 3t^2 + 4$.

a) How much water is in the bowl initially?

b) Find the rate at which water is being poured into the bowl when $t = 4$ seconds.

Q5 Find the equations of the tangent and the normal to the curve $y = \sqrt{x^3} - 3x - 10$ at $x = 16$.

Q6 Show that the graphs of $y = \frac{x^3}{3} - 2x^2 - 4x + \frac{86}{3}$ and $y = \sqrt{x}$ both go through the point (4, 2), and are perpendicular at that point.

Q7 Consider the curve C given by the equation $y = x^2 - 6$ and the line L given by the equation $y = 3$.

a) Find the coordinates of the points, A and B, where C and L intersect.

b) Find the gradient of C at points A and B.

c) Find the equations of the normals to C at A and B.

d) The normals at points A and B meet at the point D. Find the coordinates of the point D.

Q8 Consider the curve C given by the equation $y = x^3 - 2x^2 + 1$, $x > 0$, and the line L given by the equation $y = 1$.

a) Write down the gradient of the line L for any x.

b) Find the point at which the curve C has the same gradient as the line L.

c) Hence give the equation of the tangent to C at this point.

Review Exercise

Q9 Find the equation of the tangent to the curve $y = x^3 + \frac{4}{x} + 2\sqrt{x}$ at $x = 1$.

Q10 Find the equations of the tangent and the normal to the curve $y = 1 + \sqrt{x^3}$ at $x = 16$.

Q11 Consider the curve with equation $y = f(x)$, where $f(x) = x^3 - 3x$.

a) Work out the gradient of this curve when $x = -1$.

b) Show that $2f''(x) - 3f'(x) + f(x) = x^3 + 9(1 + x - x^2)$.

Q12 Let $f(x) = x^4$. Find $f''(x) + 2f'(x) - 4f(x)$, expressing your answer as the product of factors.

Q13 a) Find the stationary points of the graph of the function $y = x^3 + \frac{3}{x}$.

b) Work out whether each stationary point is a maximum or a minimum.

Q14 Find all the stationary points of the graph of $y = 2x^4 - x^2 + 4$ and determine their nature.

Q15 Find when each of the functions below are increasing and decreasing:

a) $y = 6(x + 2)(x - 3)$

b) $y = \frac{1}{x^2}$

Q16 Sketch the graph of $y = 3x^3 - 16x$, clearly showing the coordinates of any turning points.

Q17 Sketch the graph of $y = -3x^3 + 6x^2$, clearly showing the coordinates of any turning points.

Q18 Given that $xy = 20$ and that both x and y are positive,
find the least possible value of $x^2 + y^2$.

Q19 A particle moves along a path described by the equation $x = t^3 - 8t$, $t > 0$,
where t is the time in seconds and x is the displacement in metres.

a) Find $\frac{dx}{dt}$, the velocity of the particle as a function of t.

b) Find x and t when the velocity is 19 ms^{-1}.

c) Find the acceleration $\frac{d^2x}{dt^2}$ of the particle as a function of t.

d) Find the acceleration, in ms^{-2}, after 2 seconds.

e) Find the velocity, in ms^{-1}, when the acceleration is 18 ms^{-2}.

Q20 The height (h m) a firework can reach is related to the mass (m g) of fuel it carries as shown:

$$h = \frac{m^2}{10} - \frac{m^3}{800}$$

Find the mass of fuel required to achieve the maximum height
and state what the maximum height is to 3 s.f.

Exam-Style Questions

Q1 Differentiate $y = x^3 + 4x$ from first principles to find $\frac{dy}{dx}$.

[5 marks]

Q2 When $y = \frac{x-1}{x^2}$, show that $\frac{d^2y}{dx^2} + \frac{4}{x}\left(\frac{dy}{dx}\right) + \frac{2}{x^2}y = 0$.

[5 marks]

Q3 A cylinder with radius r cm and height $10h$ cm fits perfectly inside a sphere with radius 15 cm.

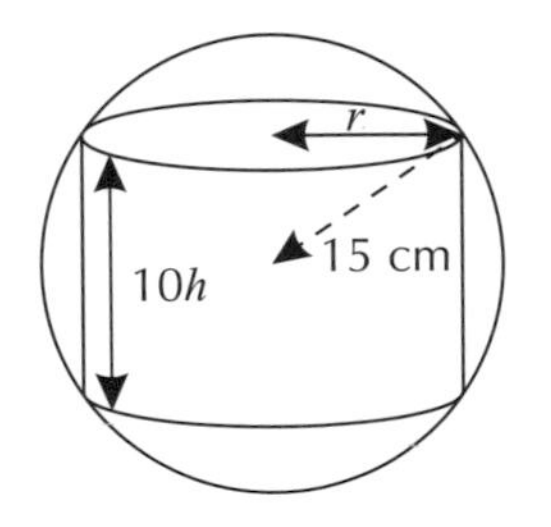

a) Express r in terms of h.

[2 marks]

b) Find the volume of the cylinder, V in terms of h.

[2 marks]

c) Find the maximum volume of the cylinder as h varies to the nearest cm^3.

[5 marks]

Q4 If $y = x^2(x - 3)^2$

a) Find the coordinates of the stationary points of $y = x^2(x - 3)^2$.

[4 marks]

b) Describe the nature of each stationary point.

[3 marks]

c) Sketch $y = x^2(x - 3)^2$.

[3 marks]

Exam-Style Questions

Q5 Determine the x-values for which the function $f(x) = x^4 - 14x^2 + 24x$ is increasing and the x-values for which it is decreasing.

[7 marks]

Q6 A curve has the equation $y = x^2 + 4x - 2$. Find the following:

a) the equation of the tangent, when $x = 2$,

[3 marks]

b) the equation of the normal, when $x = -1$,

[3 marks]

c) the point V, where the tangent at $x = 2$ and the normal at $x = -1$ intersect.

[2 marks]

Q7 $f(x) = 4\sqrt{x} + \frac{a}{x^2} - 6x + b$

If $f(4) = \frac{-97}{8}$ and $f'(1) = -64$ find the values of a and b.

[4 marks]

Q8 A particle moves such that its displacement s metres, at time t seconds is given by:

$$s = 4 + 12t^2 - t^3$$

Find the acceleration when $t = 4.5$ seconds.

[4 marks]

Chapter 11 Integration

11.1 Indefinite Integration

Integration is just the process of getting from $\frac{dy}{dx}$ back to y itself.

Learning Objectives (Spec Ref 8.1 & 8.2):

- Use the integral symbol $\int$.
- Integrate powers of x.
- Integrate more complicated functions containing powers of x.
- Find the equation of a curve, given the gradient and a point on the curve.

Prior Knowledge Check: Be able to use the laws of indices (see p.17), understand differentiation (see Chapter 10) and sketch curves (see p.94).

Integration

Integration is the '**opposite**' of differentiation. When you integrate something, you're trying to find a function that returns to **what you started with** when you differentiate it.
This function is called an **integral**. The integral of a **function** f(x) with respect to x is written:

$\int$ means **the integral of** → $\int f(x)\,dx$ ← dx means **with respect to x**.

For example, 'the integral of $2x$ with respect to x' is written $\int 2x\,dx$.

The answer could be **any function** which differentiates to give $2x$.

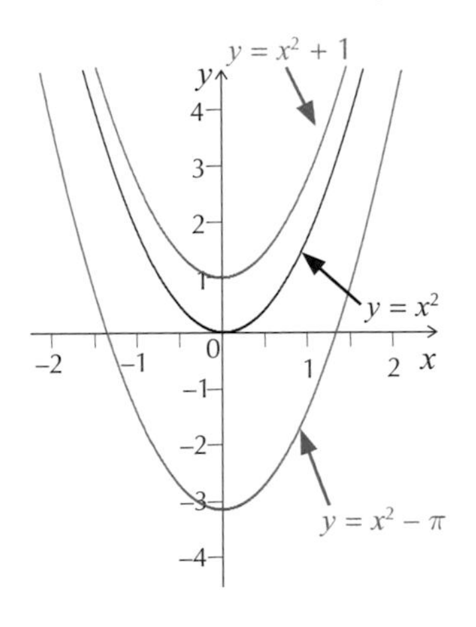

From Chapter 10, we know that:

$$\frac{d}{dx}(x^2) = 2x \qquad \frac{d}{dx}(x^2 + 1) = 2x \qquad \frac{d}{dx}(x^2 - \pi) = 2x$$

If you differentiate any of these functions, you get $2x$ — they're **all possible integrals** because they all have the **same gradient**.

In fact, if you differentiate **any** function which is of the form **x^2 + 'a constant'** you'll get $2x$ because differentiating a constant always gives zero.

So the answer to this integral is actually: $\int 2x\,dx = x^2 + C$

C is a constant representing 'any number' — it's known as the **constant of integration**.

This is an example of **indefinite integration** — a good way to remember this is that C can take an **indefinite number** of values. There are **lots of answers** to an indefinite integral, so you need to add a **constant of integration** to show that it could be **any number**.

Formally, the **Fundamental Theorem of Calculus** states that:

$$\int f(x)\,dx = F(x) + C \Leftrightarrow f(x) = \frac{d}{dx}(F(x))$$

Tip: **Definite** integrals only have one possible answer — you'll learn about these later in this chapter.

Or, more simply, it says that if differentiating takes you from one function to another, then integrating the second function will take you back to the first (with a constant of integration).

Integrating x^n

The formula to the right tells you how to integrate **any power of x** (except x^{-1}):

$$\int x^n \, dx = \frac{x^{n+1}}{n+1} + C$$

This means to integrate a power of x: (i) **Increase the power** by one — then divide by it.
(ii) Add a **constant**.

You can't use this formula for $\frac{1}{x} = x^{-1}$. When you increase the power by 1 and then divide by the power you get $\int x^{-1} \, dx = \frac{x^0}{0}$. This is undefined since you can't divide by 0.

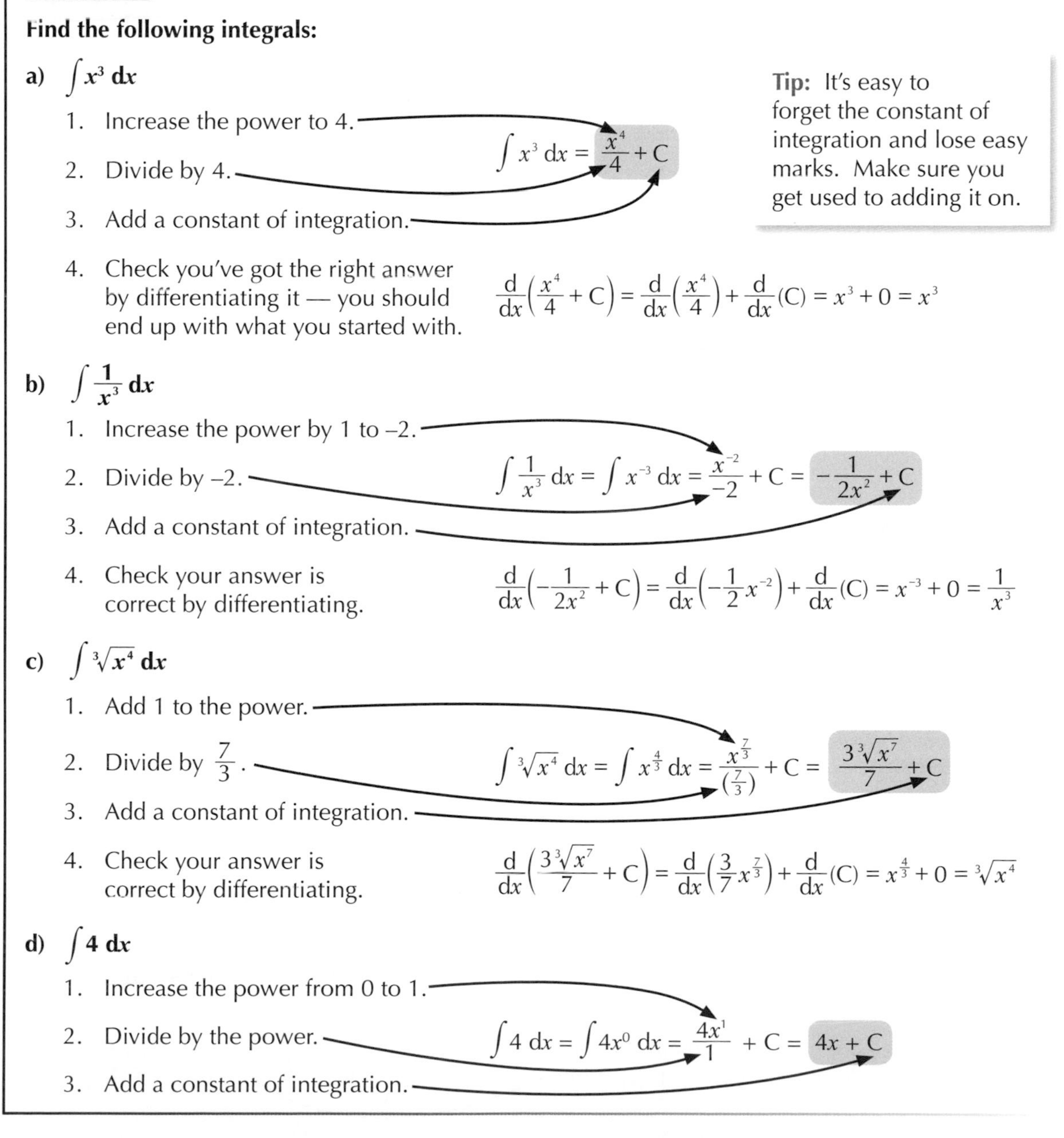

Examples

Find the following integrals:

a) $\int x^3 \, dx$

1. Increase the power to 4.
2. Divide by 4.
3. Add a constant of integration.

$$\int x^3 \, dx = \frac{x^4}{4} + C$$

4. Check you've got the right answer by differentiating it — you should end up with what you started with.

$$\frac{d}{dx}\left(\frac{x^4}{4} + C\right) = \frac{d}{dx}\left(\frac{x^4}{4}\right) + \frac{d}{dx}(C) = x^3 + 0 = x^3$$

Tip: It's easy to forget the constant of integration and lose easy marks. Make sure you get used to adding it on.

b) $\int \frac{1}{x^3} \, dx$

1. Increase the power by 1 to –2.
2. Divide by –2.
3. Add a constant of integration.

$$\int \frac{1}{x^3} \, dx = \int x^{-3} \, dx = \frac{x^{-2}}{-2} + C = -\frac{1}{2x^2} + C$$

4. Check your answer is correct by differentiating.

$$\frac{d}{dx}\left(-\frac{1}{2x^2} + C\right) = \frac{d}{dx}\left(-\frac{1}{2}x^{-2}\right) + \frac{d}{dx}(C) = x^{-3} + 0 = \frac{1}{x^3}$$

c) $\int \sqrt[3]{x^4} \, dx$

1. Add 1 to the power.
2. Divide by $\frac{7}{3}$.
3. Add a constant of integration.

$$\int \sqrt[3]{x^4} \, dx = \int x^{\frac{4}{3}} \, dx = \frac{x^{\frac{7}{3}}}{\left(\frac{7}{3}\right)} + C = \frac{3\sqrt[3]{x^7}}{7} + C$$

4. Check your answer is correct by differentiating.

$$\frac{d}{dx}\left(\frac{3\sqrt[3]{x^7}}{7} + C\right) = \frac{d}{dx}\left(\frac{3}{7}x^{\frac{7}{3}}\right) + \frac{d}{dx}(C) = x^{\frac{4}{3}} + 0 = \sqrt[3]{x^4}$$

d) $\int 4 \, dx$

1. Increase the power from 0 to 1.
2. Divide by the power.
3. Add a constant of integration.

$$\int 4 \, dx = \int 4x^0 \, dx = \frac{4x^1}{1} + C = 4x + C$$

Exercise 11.1.1

Q1 Find an expression for y when $\frac{dy}{dx}$ is the following:

a) x^7 b) $2x^3$ c) $8x$

d) $-5x^4$ e) $200x^{99}$ f) x^{-3}

g) $4x^{-4}$ h) $-6x^{-5}$ i) -12

j) $x^{\frac{1}{2}}$ k) $x^{\frac{1}{3}}$ l) $12\sqrt{x^3}$

Q1 Hint: When the function is multiplied by a constant, the integral of the x^n term is multiplied by the same constant — just like when you differentiate. Look at the formula and examples below if you need to.

Q2 Find f(x) when f′(x) is the following:

a) x^5 b) $-6x$ c) $\frac{4}{3}$ d) $2x^{-2}$

e) $-9x^{-19}$ f) $\frac{x}{5}$ g) $-x^{-6}$ h) $\frac{5}{\sqrt{x}}$

i) $x^{-\frac{4}{3}}$ j) $30x^{-0.8}$ k) $-\frac{2}{3}x^{\frac{7}{3}}$ l) $-\frac{3}{4}x^{-\frac{5}{2}}$

Q3 Find the following:

a) $\int x^{\frac{2}{3}}\,dx$ b) $\int 7x^{\frac{4}{3}}\,dx$ c) $\int x^{-\frac{1}{2}}\,dx$ d) $\int 2x^{-\frac{1}{3}}\,dx$

e) $\int 14x^{0.4}\,dx$ f) $\int -1.2x^{-0.6}\,dx$ g) $\int -2x^{-\frac{5}{4}}\,dx$ h) $\int -\frac{3}{2}x^{-\frac{1}{2}}\,dx$

i) $\int -\frac{4}{3}x^{-\frac{4}{3}}\,dx$ j) $\int \frac{1}{2}\sqrt{x^5}\,dx$ k) $\int \frac{2}{3}x^{\frac{7}{5}}\,dx$ l) $\int 50.5x^{0.01}\,dx$

Q4 Given that a is a constant, find the value of x where $\int 4ax^{a-1}\,dx = \int (a+1)x^a\,dx$. Assume that C = 0 for both integrations.

PROBLEM SOLVING

Integrating functions

Like differentiating, if there are **lots of terms** in an expression, you can just integrate each bit **separately**. If the terms are multiplied by **constants**, take them **outside** the integral like this:

$$\int ax^n\,dx = a\int x^n\,dx$$

When you're doing **lots** of separate integrations, you only need **one constant** of integration for the whole expression — if each integral gives a constant, you can just **add** them up to get a **new constant**.

Examples

a) Find $\int \left(3x^2 - \frac{2}{\sqrt{x}} + \frac{7}{x^2}\right) dx$.

1. Write all the terms as powers of x. $\int \left(3x^2 - \frac{2}{\sqrt{x}} + \frac{7}{x^2}\right) dx = \int (3x^2 - 2x^{-\frac{1}{2}} + 7x^{-2})\,dx$

2. Take the constants outside each integral. $= 3\int x^2\,dx - 2\int x^{-\frac{1}{2}}\,dx + 7\int x^{-2}\,dx$

3. Integrate each term separately and just add one constant of integration. $= \frac{3x^3}{3} - \frac{2x^{\frac{1}{2}}}{\left(\frac{1}{2}\right)} + \frac{7x^{-1}}{-1} + C = x^3 - 4\sqrt{x} - \frac{7}{x} + C$

b) Find y if $\frac{dy}{dx} = \frac{1}{2}x^3 - 4x^{\frac{3}{2}}x$.

1. You need to integrate the derivative of y to get y.

$$y = \int \frac{dy}{dx}\,dx = \int \left(\frac{1}{2}x^3 - 4x^{\frac{3}{2}}x\right) dx$$

2. First take the constants outside each integral.

$$= \int \left(\frac{1}{2}x^3 - 4x^{\frac{5}{2}}\right) dx = \frac{1}{2}\int x^3\,dx - 4\int x^{\frac{5}{2}}\,dx$$

3. Then integrate each term separately and add one constant of integration.

$$= \frac{1}{2} \times \frac{x^4}{4} + (-4) \times \frac{x^{\frac{7}{2}}}{\left(\frac{7}{2}\right)} + C = \frac{x^4}{8} - \frac{8}{7}x^{\frac{7}{2}} + C$$

c) Find $\int \left(\frac{(x-1)^2}{\sqrt{x}}\right) dx$.

1. Expand the bracket.

$$\int \left(\frac{(x-1)^2}{\sqrt{x}}\right) dx = \int \left(\frac{x^2 - 2x + 1}{x^{\frac{1}{2}}}\right) dx$$

2. Split into separate terms.

$$= \int \left(\frac{x^2}{x^{\frac{1}{2}}} - \frac{2x}{x^{\frac{1}{2}}} + \frac{1}{x^{\frac{1}{2}}}\right) dx$$

3. Write all terms as powers of x.

$$= \int (x^{\frac{3}{2}} - 2x^{\frac{1}{2}} + x^{-\frac{1}{2}})\,dx$$

Tip: Some expressions will need simplifying before you integrate.

4. Take the constants outside each integral.

$$= \int x^{\frac{3}{2}}\,dx - 2\int x^{\frac{1}{2}}\,dx + \int x^{-\frac{1}{2}}\,dx$$

5. Integrate each term separately and add one constant of integration.

$$= \frac{x^{\frac{5}{2}}}{\left(\frac{5}{2}\right)} - \frac{2x^{\frac{3}{2}}}{\left(\frac{3}{2}\right)} + \frac{x^{\frac{1}{2}}}{\left(\frac{1}{2}\right)} + C = \frac{2(\sqrt{x})^5}{5} - \frac{4(\sqrt{x})^3}{3} + 2\sqrt{x} + C$$

Exercise 11.1.2

Q1 Find f(x) when f′(x) is given by the following:

a) $5x + 3x^{-4}$ b) $4x(x^2 - 1)$ c) $(x - 3)^2$

d) $x\left(6x + \frac{4}{x^4}\right)$ e) $\left(x + \frac{2}{x}\right)^2$ f) $x\left(3x^{\frac{1}{2}} - \frac{2}{x^{\frac{4}{3}}}\right)$

g) $6\sqrt{x} - \frac{1}{x^2}$ h) $\frac{2}{\sqrt{x}} - 7x^2\sqrt{x}$ i) $5(\sqrt{x})^3 - \frac{3x}{\sqrt{x}}$

j) $\sqrt{x}(1 - x)$ k) $3\sqrt[6]{x} - \frac{\sqrt{x}}{\sqrt[3]{x}}$ l) $\frac{x^3 - 2x^2}{\sqrt{x}}$

Q2 Integrate with respect to x:

a) $\frac{5}{7}x^4 + \frac{2}{3}x + \frac{1}{4}$
b) $\frac{1}{\sqrt{x}} + \sqrt{x}$
c) $\frac{3}{x^2} + \frac{3}{\sqrt[3]{x}}$

d) $x^{\frac{5}{4}} + x^{\frac{4}{5}}$
e) $\frac{1}{2}x^{-\frac{1}{2}} + 3x^{-\frac{2}{3}} + 4x^{-\frac{3}{4}}$
f) $-\frac{4}{x^3} + \frac{1}{\sqrt{x^3}}$

Q3 Find the following integrals:

a) $\int (0.55x^{0.1} - 3x^{-1.5}x)\,dx$
b) $\int \left(8x^3 - \frac{2}{\sqrt{x}} + \frac{5}{x^2}\right) dx$

c) $\int \left((\sqrt{x})^5 + \frac{1}{2\sqrt{x}}\right) dx$
d) $\int \left(\sqrt{x}\left(7x^2 - 1 - \frac{2}{x}\right)\right) dx$

e) $\int (3x - 5\sqrt{x})^2\,dx$
f) $\int \left(\frac{2x^3 - \sqrt{x}}{x}\right) dx$

g) $\int \left(\frac{(5x-3)^2}{\sqrt{x}}\right) dx$
h) $\int (x^{\frac{1}{2}} + 1)(x^{-\frac{1}{2}} - 3)\,dx$

i) $\int x(2x + \sqrt{x})^2\,dx$
j) $\int (\sqrt{x} + 2\sqrt[3]{x})^2\,dx$

Q4 Given that $\frac{dy}{dx} = 1.5x^2 - \frac{4}{x^3}$, find y.

Q5 Given that $f'(x) = \frac{4}{3(x^{\frac{1}{3}})^4} + 5x^{\frac{3}{2}}$, find f($x$).

Q6 Find: a) $\int \left(4x^2 + \frac{3}{\sqrt{x}} - 2\right) dx$
b) $\int (3\sqrt{x} + 3)^2\,dx$

c) $\int \left(\frac{(\sqrt{x}+3)(\sqrt{x}-1)}{\sqrt{x}}\right) dx$
d) $\int \left(\sqrt{x}\left(\sqrt{x} - \frac{1}{\sqrt{x}}\right)^2\right) dx$

e) $\int \left(\frac{(\sqrt{x^3}+6)(\sqrt{x^3}-6)}{\sqrt{x^3}}\right) dx$
f) $\int \left(\frac{3x - \sqrt{x} - 2}{\sqrt{x} - 1}\right) dx$

Q6f) Hint: Factorise the numerator first and then cancel the denominator.

Q7 Integrate $\frac{2n^2x^2 + 3x^3}{nx}$ with respect to x, given that n is a constant.

PROBLEM SOLVING

Integrating to find equations of curves

As you saw in Chapter 10, **differentiating** the equation of a curve gives its **gradient**. **Integrating** the gradient of a curve does the **opposite** — it gives you the **equation** of the curve.

But integrating actually gives you **many** possible curves because of the **constant of integration**, C. C can take any value and each different value represents a different curve (all vertically translated copies of each other).

So to find the equation of a **particular curve** by integration, you need to know the coordinates of **one point** on it, which you can use to find C.

Tip: Have a look at page 99 for more on translations of graphs.

Examples

a) The curve $y = f(x)$ goes through the point (2, 16) and $\frac{dy}{dx} = 2x^3$. Find the equation of the curve.

1. You know the derivative $\frac{dy}{dx} = 2x^3$ and need to find y. So integrating gives:

$$y = \int 2x^3\,dx = \frac{2x^4}{4} + C = \frac{x^4}{2} + C$$

2. Check this is correct by differentiating it and making sure you get what you started with.

$$y = \frac{x^4}{2} + C, \quad \frac{dy}{dx} = \frac{1}{2}(4x^3) + 0 = 2x^3$$

3. So this function has the correct derivative — but you haven't finished yet. You now need to find C — and you do this by using the fact that it goes through the point (2, 16). Put $x = 2$ and $y = 16$ in the equation.

$$y = \frac{x^4}{2} + C$$
$$\Rightarrow 16 = \frac{2^4}{2} + C = 2^3 + C$$
$$\Rightarrow C = 8$$

4. Put the value of C back in the equation.

$$y = \frac{x^4}{2} + 8$$

b) The curve $y = f(x)$ goes through the point (2, 8) and $f'(x) = 6x(x - 1)$. Find $f(x)$.

1. Expand the brackets of $f'(x)$. $\quad f'(x) = 6x(x - 1) = 6x^2 - 6x$

2. Integrate to find $f(x)$.

$$f(x) = \int (6x^2 - 6x)\,dx = \frac{6x^3}{3} - \frac{6x^2}{2} + C = 2x^3 - 3x^2 + C$$

3. Check that this is correct by differentiating...

$$f(x) = 2x^3 - 3x^2 + C$$
$$f'(x) = 2(3x^2) - 3(2x^1) = 6x^2 - 6x$$

4. You now need to find C using the point (2, 8). Put $x = 2$ and $y = 8$ into $f(x) = 2x^3 - 3x^2 + C$.

$$8 = (2 \times 2^3) - (3 \times 2^2) + C$$
$$\Rightarrow 8 = 16 - 12 + C$$
$$\Rightarrow C = 4$$

5. Put C in the equation. $\quad f(x) = 2x^3 - 3x^2 + 4$

Exercise 11.1.3

Q1 For each of the following, the curve $y = f(x)$ passes through the given point. Find $f(x)$.

a) $f'(x) = 4x^3$, (0, 5)

b) $f'(x) = 3x^2 - 4x + 3$, (1, –3)

c) $f'(x) = 6x(x + 2)$, (–1, 1)

d) $f'(x) = \frac{5}{x^2} + 2x$, (5, 4)

e) $f'(x) = 3x^2(x - 4)$, (2, –10)

f) $f'(x) = (3x + 1)(x - 1)$, (3, –3)

g) $f'(x) = x(x + \frac{3}{x^3})$, $(-3, 5)$

h) $f'(x) = \frac{9x^3 + 2x^{-2}}{x}$, $(-1, 2)$

i) $f'(x) = \sqrt{x}\,(3 + x)$, $(1, 4)$

j) $f'(x) = \frac{2 - 3\sqrt[6]{x}}{\sqrt{x}}$, $(64, -33)$

Q2 A curve $y = f(x)$ that passes through the origin has derivative $f'(x) = 6x^2 + 6x - 5$.

a) Find the equation of the curve.

b) Factorise and hence sketch the curve, showing the points where the curve cuts the axes.

Q3 A curve $y = f(x)$ that passes through the point (4, 9) has gradient function $f'(x) = \frac{3}{\sqrt{x}} + 2x$. Find the equation of the curve.

Q3-4 Hint: The gradient function is just the function which tells you the gradient — the derivative.

Q4 The gradient function of a curve is given by $\frac{dy}{dx} = 3\sqrt{x} + \frac{1}{x^2}$
Find the equation of the curve if it passes through the point (1, 7).

Q5 Consider $\frac{dy}{dt} = (\sqrt{t} - 3)^2$. Given that $y = 9$ when $t = 4$, find y as a function of t.

Q6 The curve $y = f(x)$ goes through the point $\left(1, \frac{1}{3}\right)$ and $f'(x) = \sqrt{x}\,(5x - 1)$. Find $f(x)$.

Q7 The curve $y = f(x)$ has derivative $f'(x) = x^2 + \frac{2}{x^{\frac{3}{2}}}$ and passes through the point $\left(1, -\frac{5}{3}\right)$.
Find the equation of the curve.

Q8 The gradient function of a curve is given by $\frac{dy}{dx} = \frac{x - 6}{x^3} + 2$.
Find the equation of the curve if it passes through the point (3, –1).

Q9 The gradient of a curve C is given by $\frac{dy}{dx} = \frac{(x + 2)(x - 2)}{\sqrt{x}}$, $x > 0$.

a) Show that $\frac{dy}{dx}$ can be written in the form $Ax^{\frac{3}{2}} + Bx^{-\frac{1}{2}}$, where A and B are integers.

b) The point $\left(1, \frac{7}{5}\right)$ lies on C. Find the equation of C.

Q10 An object is moving such that its acceleration, in ms^{-2}, is given by the function $\frac{dv}{dt} = 6t - t^2$, for $0 \le t \le 10$, (t = time, measured in seconds). Given that the object comes to rest ($v = 0$ ms^{-1}) at $t = 10$ seconds, find its initial velocity (v, when $t = 0$).

MODELLING

Q11 A function has a gradient given by $f'(x) = 3x^2 + kx - k$, where k is a constant.
Given that $f(x)$ passes through the point (2, 7) with a gradient of 18,
find the value of k, and hence the equation of $f(x)$.

PROBLEM SOLVING

11.2 Definite Integration

Integration can be used to find the area between a graph and the x-axis.
In this section, you'll see how to use a definite integral to work out an area.

Learning Objectives (Spec Ref 8.3):

- Evaluate definite integrals.
- Find the area between a curve and the x-axis using definite integration.

Prior Knowledge Check: Be able to factorise quadratics and cubics (see Chapter 4) and sketch curves (see p.94).

Evaluating definite integrals

Definite integrals have **limits** (little numbers) next to the integral sign. The limits just tell you the ***x*-values** to integrate the function between.

If you're integrating with respect to a different variable, such as t, then the limits tell you the range of that variable instead.

The definite integral of **f(*x*)** with respect to ***x*** between the limits ***x* = *a*** and ***x* = *b*** is written:

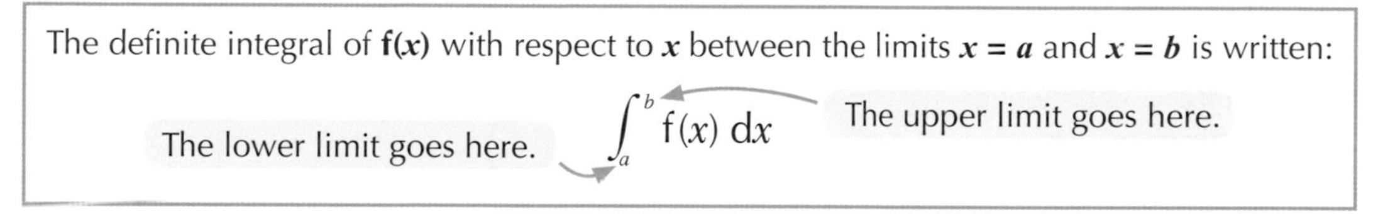

$$\int_a^b f(x)\,dx$$

Finding a definite integral isn't really any harder than an indefinite one — there's just an **extra stage** you have to do. Integrate the function as normal but **don't** add a **constant of integration**. Once you've integrated the function, work out the **value** of the definite integral by **putting in the limits**.

The proper way to write out definite integrals is to use square brackets with the limits to the right as shown below.

Tip: You might be asked to 'evaluate' an integral — this just means find the value. A definite integral always comes out as a number.

If you know that the integral of f(*x*) is $\int f(x)\,dx = g(x) + C$ then:

$$\int_a^b f(x)\,dx = [g(x)]_a^b = g(b) - g(a)$$

Subtract the value of g at the **lower** limit from the value of g at the **upper** limit.

This is the second part of the **Fundamental Theorem of Calculus**.

Example 1

Evaluate $\int_1^3 (x^2 + 2)\,dx$.

1. Find the integral in the normal way — but put the integrated function in square brackets and rewrite the limits on the right-hand side.

$$\int_1^3 (x^2 + 2)\,dx = \left[\frac{x^3}{3} + 2x\right]_1^3$$

Notice that there's no constant of integration.

2. Put the upper limit into the integral and subtract the value of the integral at the lower limit.

$$\left[\frac{x^3}{3} + 2x\right]_1^3 = \left(\frac{3^3}{3} + 6\right) - \left(\frac{1^3}{3} + 2\right)$$

$$= 15 - \frac{7}{3} = \frac{38}{3}$$

The area under a curve

The value of a **definite integral** represents the **area** between the x-axis and the graph of the function you're integrating between the two limits.

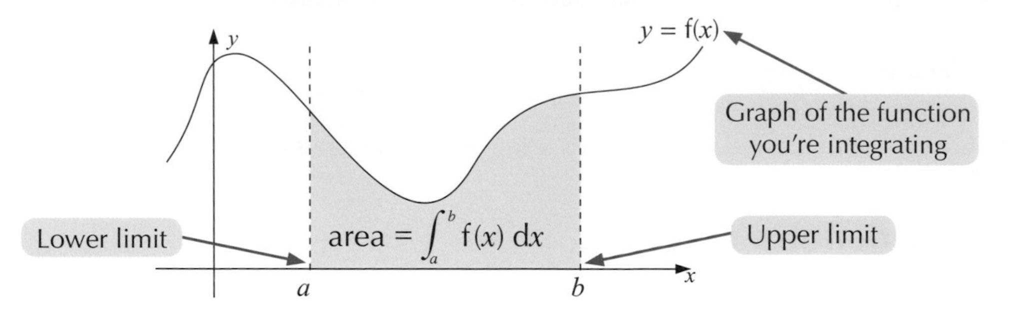

Example 2

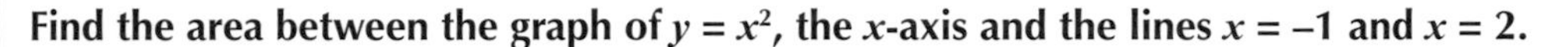

Find the area between the graph of $y = x^2$, the x-axis and the lines $x = -1$ and $x = 2$.

1. You just need to integrate the function $f(x) = x^2$ between -1 and 2 with respect to x.

$f(x) = x^2$, $\int_{-1}^{2} x^2\,dx = \left[\frac{x^3}{3}\right]_{-1}^{2}$

The limits of integration are -1 and 2.

2. Put the upper limit into the integral and subtract the value of the integral at the lower limit.

$= \left(\frac{2^3}{3}\right) - \left(\frac{(-1)^3}{3}\right) = \frac{8}{3} + \frac{1}{3}$

$= \frac{9}{3} = 3$, so the area is 3.

If you integrate a function to find an area that lies **below** the x-axis, it'll give a **negative** value.

If you need to find an area like this, you'll need to make your answer **positive** at the end as you can't have **negative area**.

It's important to note that you're actually finding the area between the curve and the ***x*-axis**, not the area under the curve (the area below a curve that lies under the x-axis will be **infinite**).

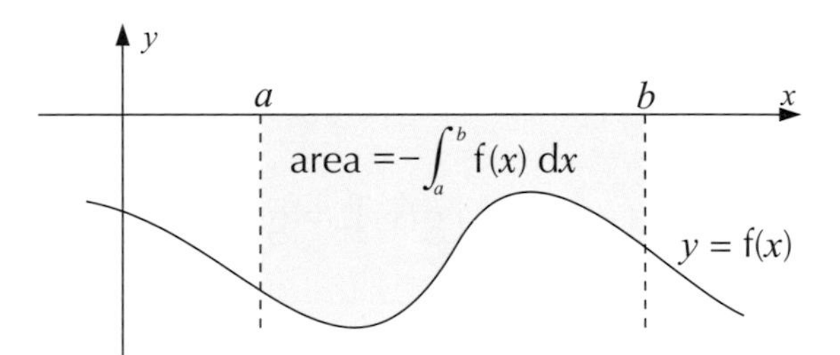

Example 3

Find the area between the graph of $y = 4x - 3x^2 - x^3$ and the x-axis between $x = -4$ and $x = 0$.

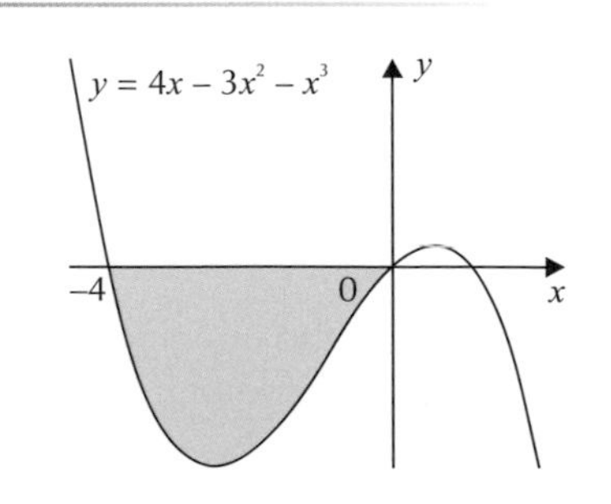

1. You can see from the sketch of the graph that the area you're trying to find lies below the x-axis.

2. Integrate the curve between the given limits.

$\int_{-4}^{0} (4x - 3x^2 - x^3)\,dx = \left[2x^2 - x^3 - \frac{x^4}{4}\right]_{-4}^{0}$

3. Put the limits into the integral and subtract.

$= (0) - \left(2(-4)^2 - (-4)^3 - \frac{(-4)^4}{4}\right)$

$= 0 - (32 + 64 - 64) = -32$

4. Make the area positive.

So the area between the curve and the x-axis between $x = -4$ and $x = 0$ is **32**.

If you need to find the area for a portion of a curve which lies both **above** and **below** the x-axis, you'll need to find the areas above and below **separately** and add them up at the end so that the negative and positive integrals don't **cancel each other out**.

Example 4

a) Evaluate $\int_{-2}^{2} x^3\, dx$.

Integrate the function and put in the limits to find its value.

$$\int_{-2}^{2} x^3\, dx = \left[\frac{x^4}{4}\right]_{-2}^{2} = \left(\frac{2^4}{4}\right) - \left(\frac{(-2)^4}{4}\right) = \frac{16}{4} - \frac{16}{4} = \mathbf{0}$$

b) Find the area between the graph of $y = x^3$, the x-axis and the lines $x = -2$ and $x = 2$.

1. You'd usually just integrate the function between the limits, which gave 0 in part a). But you can see from the diagram that the area is not 0. The 'negative area' below the axis has cancelled out the positive area.

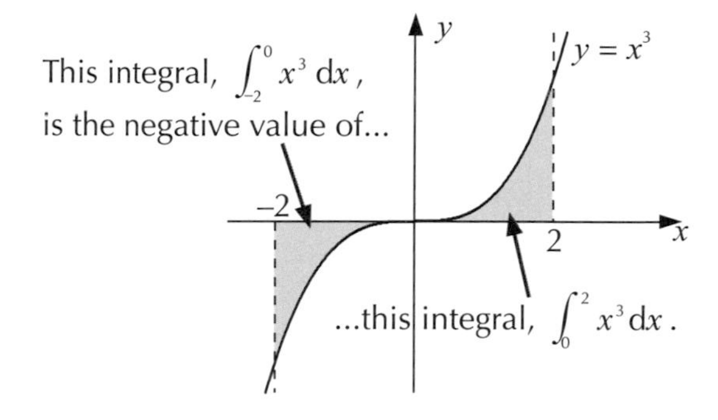

2. So to work out the total area, you need to work out the positive and negative areas separately and then **add** them together.

3. The area **above** the x-axis is between 0 and 2, so integrate the function between these limits:

$$\int_{0}^{2} x^3 = \left[\frac{x^4}{4}\right]_{0}^{2} = \frac{1}{4}[x^4]_{0}^{2} = \frac{1}{4}(2^4 - 0^4) = \frac{16}{4} = 4$$

4. The area **below** the x-axis lies between –2 and 0, so integrate the function between these limits:

$$\int_{-2}^{0} x^3 = \left[\frac{x^4}{4}\right]_{-2}^{0} = \frac{1}{4}[x^4]_{-2}^{0} = \frac{1}{4}(0^4 - (-2)^4) = -\frac{16}{4} = -4$$

5. The area below the x-axis is the positive value of this integral, so is 4. Add these areas together to get the total area.

Area = 4 + 4 = **8**

Some questions might give you an expression for the area in terms of an unknown and ask you to find its value (or possible values).

Example 5

PROBLEM SOLVING

Find the two possible values for A that satisfy: $\int_{1}^{4} \left(\frac{3}{7}x^2 + \frac{2A}{\sqrt{x}}\right) dx = 5A^2$

1. First, you need to evaluate the integral. Treat the A as a constant for now.

$$\int_{1}^{4} \left(\frac{3}{7}x^2 + 2Ax^{-\frac{1}{2}}\right) dx = \left[\frac{1}{7}x^3 + 4Ax^{\frac{1}{2}}\right]_{1}^{4} = \left(\frac{64}{7} + 8A\right) - \left(\frac{1}{7} + 4A\right)$$

$$= \frac{63}{7} + 4A = 9 + 4A$$

2. You know this is equal to $5A^2$, so form a quadratic in A and solve it: $9 + 4A = 5A^2$
$5A^2 - 4A - 9 = 0$
3. Factorise the equation: (you could also use the quadratic formula) $(5A - 9)(A + 1) = 0$
4. So the solutions are: $5A - 9 = 0 \Rightarrow A = \frac{9}{5}$ or $A + 1 = 0 \Rightarrow A = -1$

Exercise 11.2.1

Q1 Find the value of the following, giving exact answers:

a) $\int_{-2}^{0} (4x^3 + 2x)\,dx$ b) $\int_{-2}^{5} (x^3 + x)\,dx$ c) $\int_{-5}^{-2} (x + 1)^2\,dx$

d) $\int_{3}^{4} (6x^{-4} + x^{-2})\,dx$ e) $\int_{1}^{2} \left(x^2 + \frac{1}{x^2}\right)dx$ f) $\int_{1}^{4} (3x^{-4} + \sqrt{x})\,dx$

g) $\int_{0}^{1} ((2x + 3)(x + 2))\,dx$ h) $\int_{1}^{4} \left(\frac{x^2 + 2}{\sqrt{x}}\right)dx$ i) $\int_{4}^{9} \left(\frac{1}{x} + \sqrt{x}\right)^2 dx$

j) $\int_{4}^{16} \left(\frac{\sqrt{x} - 1}{x^2}\right)dx$ k) $\int_{1}^{9} x^{\frac{1}{2}}(5x - x^{-\frac{1}{2}})\,dx$ l) $\int_{1}^{4} (1 + x^2)(1 + \sqrt{x})\,dx$

Q2 Integrate the function $4x - 5x^3 + 7$ between the limits $x = -1$ and $x = 3$.

Q3 Find the integral of $f(x)$ between $x = 0$ and $x = 1$, if $f(x) = 3 - 4\sqrt{x} + \frac{1}{2}x^2$.

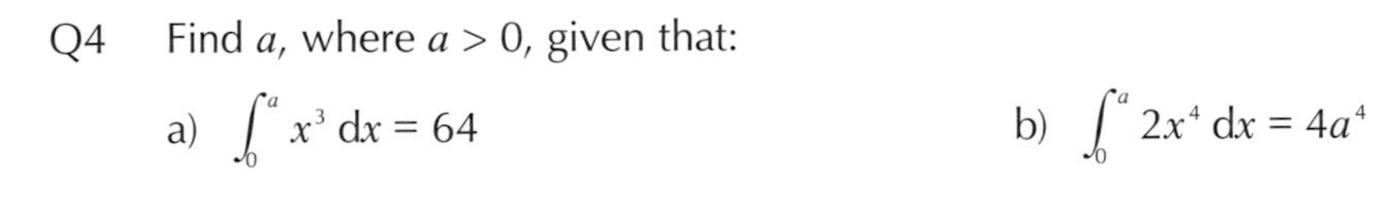

Q4 Find a, where $a > 0$, given that:

a) $\int_{0}^{a} x^3\,dx = 64$ b) $\int_{0}^{a} 2x^4\,dx = 4a^4$

PROBLEM SOLVING

Q5 Calculate the exact shaded area in the following diagrams:

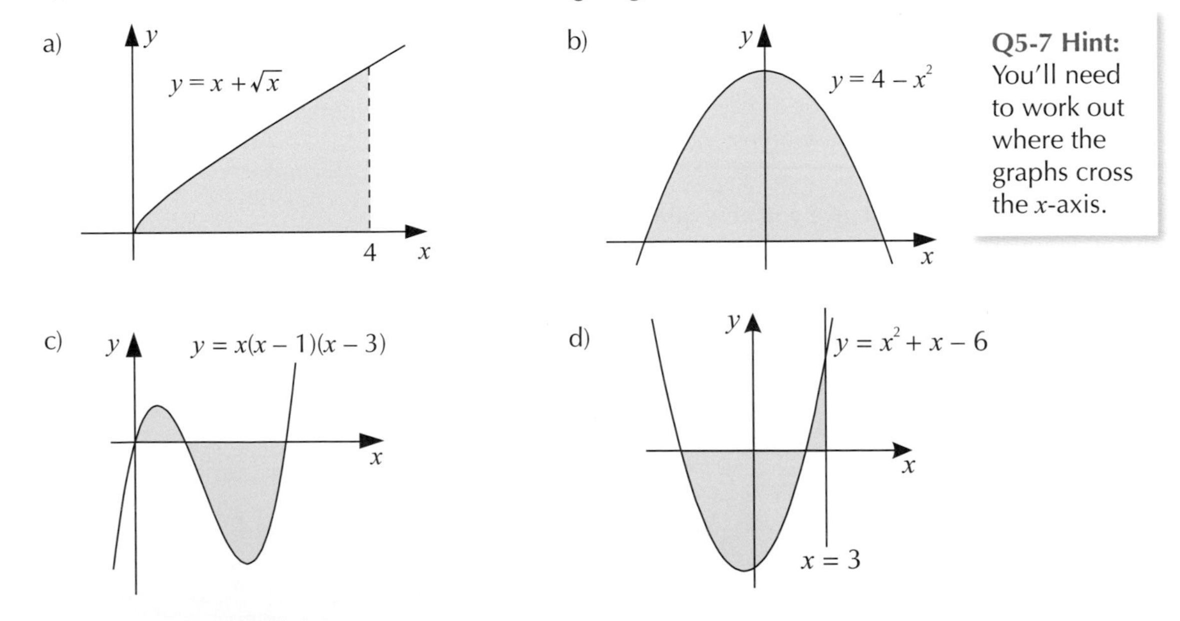

Q5-7 Hint: You'll need to work out where the graphs cross the x-axis.

Q6 Find the area enclosed by the curve with equation $y = (x - 1)(3x + 9)$, the x-axis and the lines:

a) $x = 1$ and $x = 5$ b) $x = -3$ and $x = -4$

c) $x = -2$ and $x = 2$ d) $x = -6$ and $x = 0$

Q6-7 Hint: It may help to sketch the graph so that you can see what's happening.

Q7 Find the area enclosed by the graph of $y = \frac{20}{x^5}$, the x-axis and the lines $x = 1$ and $x = 2$.

Q8 Calculate the area enclosed by the line $y = 3x$, the curve $y = (x - 6)^2$ and the x-axis.

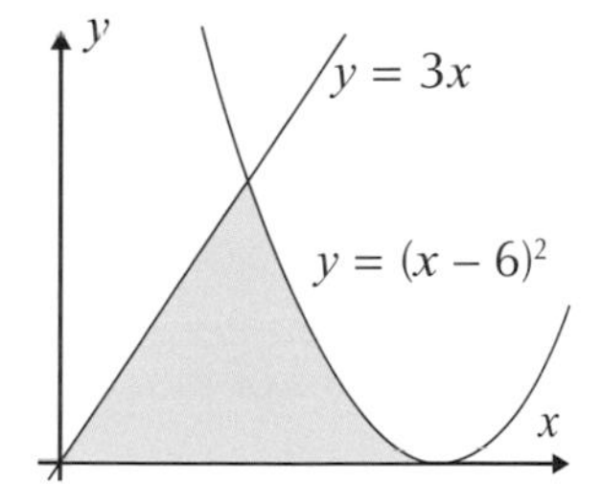

Q9 By first sketching the graph, find the total area between the graph of $y = x^3 + 4x^2 + 3x$ and the x-axis between $x = 0$ and $x = -2$.

Q10 Find the possible values of A that satisfy:

a) $\int_2^3 (1 - 2Ax)\,dx = 6A^2$ b) $\int_{-2}^{2} \left(\frac{21}{8}x^2 + \frac{A}{x^2}\right) dx = 3A^2$

c) $\int_4^5 (2A^2 - 6x^2)\,dx = 120A$ d) $\int_1^4 (A - 2\sqrt{x})^2\,dx = 10 - A^2$

Q11 The area under a velocity-time graph gives the distance travelled. An object's motion is tracked as it speeds up, and then slows to rest. Its velocity, v, at time t is modelled by the function below.

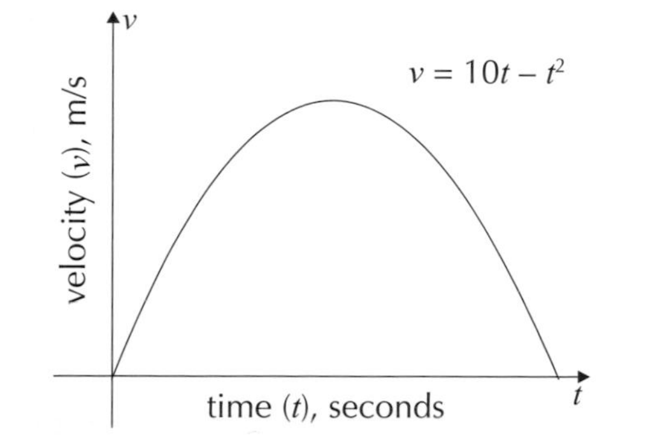

Q11 Hint: All this question is asking you to do is to find the area under the graph between these limits. Remember that you're integrating with respect to t and not x.

How far does the object travel:

a) between $t = 1$ and $t = 3$? b) in total?

Review Exercise

Q1 Find f(x) in each case below. Give each term in its simplest form.

a) $f'(x) = x^{-\frac{1}{2}} + 4 - 5x^3$ b) $f'(x) = 2x + \frac{3}{x^2}$ c) $f'(x) = 6x^2 - \frac{1}{3\sqrt{x}}$

Q2 Work out the equation of the curve that has derivative $\frac{dy}{dx} = 3x^2 - \frac{7}{\sqrt{x}}$ and goes through the point (1, 0).

Q3 The gradient function of a curve is given by $\frac{dy}{dx} = 2(3x - 6.5)$.
The curve passes through the point (1, 2).

a) Find the equation of the curve.

b) Sketch the curve, stating the coordinates of the points where the curve crosses the axes.

Q4 The curve C with equation $y = f(x)$ has derivative $f'(x) = 6x^2 - 12 - \frac{8}{x^2}$, $x > 0$ and passes through the point P with coordinates (–2, 5).
Find the equation of the curve C.

Q5 The curve $y = f(x)$ passes through the point P with coordinates (1, –9).
Given that $f'(x) = \frac{5x^2 + 1}{x^{\frac{1}{2}}} - 10$, $x > 0$, find the equation of the curve.

Q6 Evaluate the following definite integrals:

a) $\int_0^1 (4x^3 + 3x^2 + 2x + 1)\,dx$ b) $\int_1^6 \frac{3}{x^2}\,dx$ c) $\int_1^2 \left(\frac{8}{x^5} + \frac{3}{\sqrt{x}}\right) dx$

d) $\int_4^9 \sqrt{x}(1 - 2\sqrt{x})\,dx$ e) $\int_1^4 \frac{3\sqrt{x} + 2x^4}{x^2}\,dx$ f) $\int_{-1}^{0.5} \frac{2x^5 + 4x^4}{5x^3}\,dx$

Q7 a) Evaluate $\int_{-3}^{3} (9 - x^2)\,dx$.

b) Sketch the area represented by this integral.

Q8 Find the shaded area in the diagrams below:

a) b)

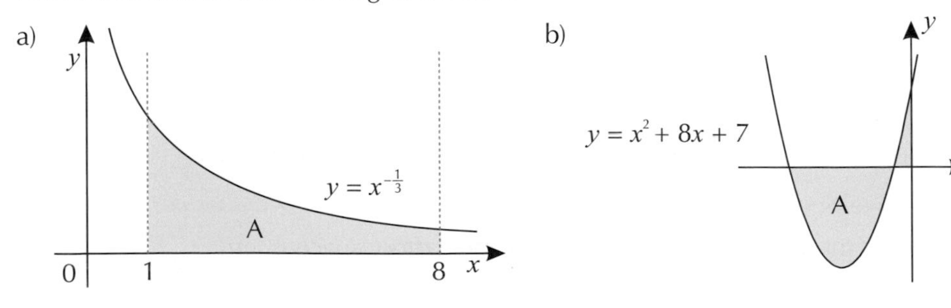

Review Exercise

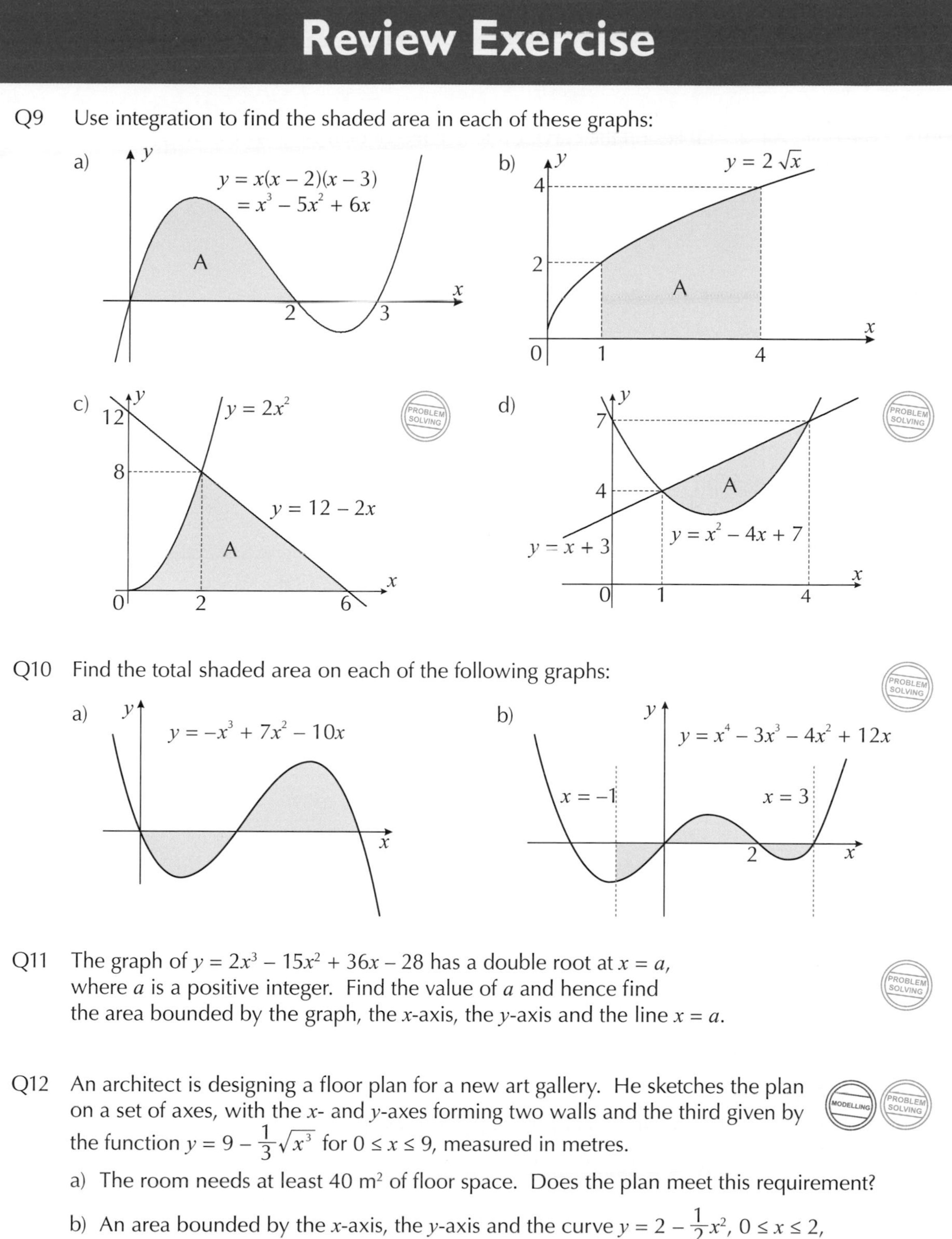

Q9 Use integration to find the shaded area in each of these graphs:

a) $y = x(x-2)(x-3) = x^3 - 5x^2 + 6x$

b) $y = 2\sqrt{x}$

c) $y = 2x^2$, $y = 12 - 2x$

d) $y = x^2 - 4x + 7$, $y = x + 3$

Q10 Find the total shaded area on each of the following graphs:

a) $y = -x^3 + 7x^2 - 10x$

b) $y = x^4 - 3x^3 - 4x^2 + 12x$

Q11 The graph of $y = 2x^3 - 15x^2 + 36x - 28$ has a double root at $x = a$, where a is a positive integer. Find the value of a and hence find the area bounded by the graph, the x-axis, the y-axis and the line $x = a$.

Q12 An architect is designing a floor plan for a new art gallery. He sketches the plan on a set of axes, with the x- and y-axes forming two walls and the third given by the function $y = 9 - \frac{1}{3}\sqrt{x^3}$ for $0 \le x \le 9$, measured in metres.

a) The room needs at least 40 m² of floor space. Does the plan meet this requirement?

b) An area bounded by the x-axis, the y-axis and the curve $y = 2 - \frac{1}{2}x^2$, $0 \le x \le 2$, is reserved for an information kiosk. How much floor space is left?

Exam-Style Questions

Q1 The point A (–2, –13) lies on the curve $y = f(x)$. If the gradient at A is –8 and $\frac{d^2y}{dx^2} = 12x$, use the fact that $\int \frac{d^2y}{dx^2} = \frac{dy}{dx}$ to find the equation of the curve $y = f(x)$.

[7 marks]

Q2 Find $\int \left(\frac{5}{3}x^{\frac{2}{3}} - \frac{4}{\sqrt{x}} + (2x)^3 - \frac{3}{2x^2}\right) dx$.

[5 marks]

Q3 a) Sketch the graph $y = (x + 1)(2x - 1)(x - 3)$, labelling where the curve intersects each axis.

[3 marks]

b) Find the area enclosed between $y = (x + 1)(2x - 1)(x - 3)$ and the x-axis.

[8 marks]

Q4 $I = \int_a^6 (4x - 5)\,dx$. Find the possible values of a, if $I = 24$.

[6 marks]

Q5

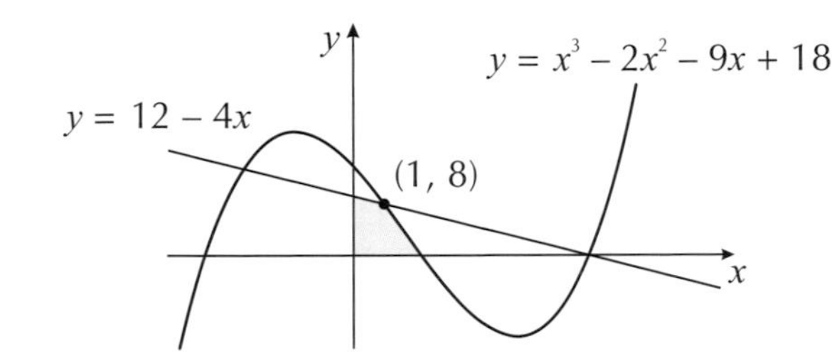

The graph shows the curve $y = x^3 - 2x^2 - 9x + 18$ and the line $y = 12 - 4x$ intersecting at the point (1, 8).

a) Find the x-coordinates where the curve $y = x^3 - 2x^2 - 9x + 18$ crosses the x-axis.

[3 marks]

b) Find the size of the shaded area on the diagram.

[7 marks]

12.1 Vectors

You'll have seen vectors before at GCSE — they've got a size and a direction. In this section you'll see how they work and what you can do with them.

Learning Objectives (Spec Ref 9.1, 9.3 & 9.4):
- Understand what vectors are and how to represent them.
- Add and subtract vectors and multiply them by scalars.
- Show that two vectors are parallel, and that three points are collinear.
- Convert between unit vector form and column vectors.

Prior Knowledge Check: Adding and subtracting vectors — you'll have seen this at GCSE.

Introducing vectors

Scalars are quantities **without a direction** — e.g. a speed of 2 m/s.

Vectors have both **size and direction** — e.g. a velocity of 2 m/s on a bearing of 050°.

Vectors are drawn as **lines** with **arrowheads** on them.

The **length** of the line represents the **magnitude** (size) of the vector.

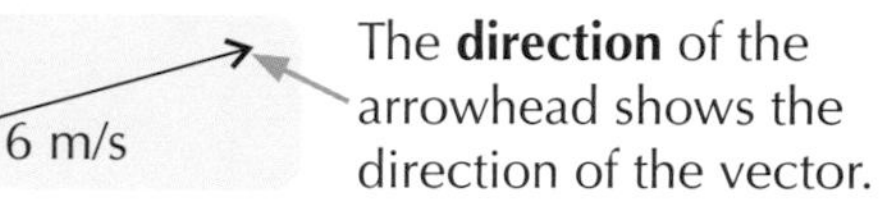

The **direction** of the arrowhead shows the direction of the vector.

Sometimes vectors are drawn to **scale**:

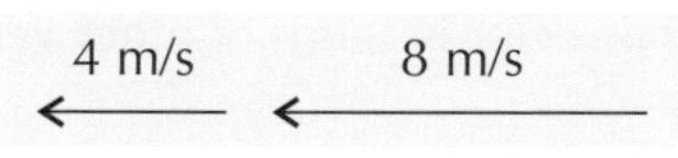

Vectors are usually **written** using either a **lowercase bold** letter or a **lowercase underlined** letter. When the **endpoints** of a vector are labelled, the vector can also be written by putting an **arrow** over the endpoints:

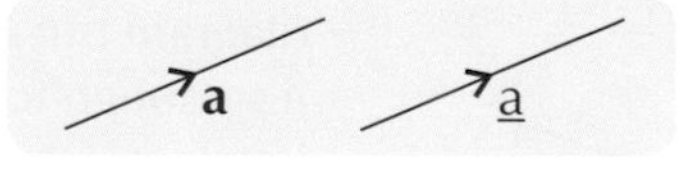

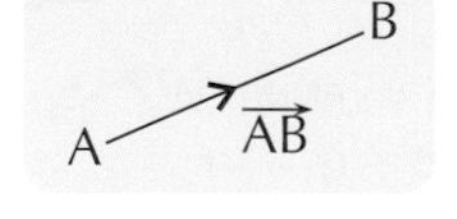

Tip: When a vector is typed it's usually bold, but if you're handwriting a vector you should write it underlined, e.g. a̲.

Adding vectors

You can **add** vectors together by drawing the arrows **nose to tail.**

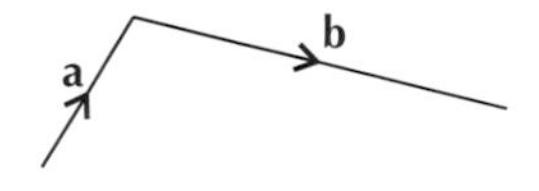

The single vector that goes from the start to the end of the combined vectors is called the **resultant vector**.

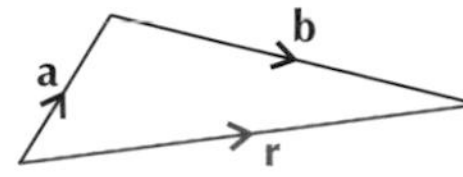

Resultant: $\mathbf{r} = \mathbf{a} + \mathbf{b}$

Tip: You might also see this referred to as the triangle rule.

This method of adding is called the **parallelogram rule** because **a** and **b** form the sides of a parallelogram which has the resultant vector $\mathbf{r} = \mathbf{a} + \mathbf{b}$ as its diagonal.

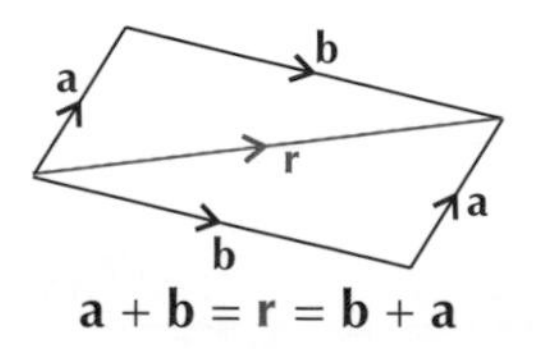

$\mathbf{a} + \mathbf{b} = \mathbf{r} = \mathbf{b} + \mathbf{a}$

When you add two vectors you're really **combining** two **translations**:

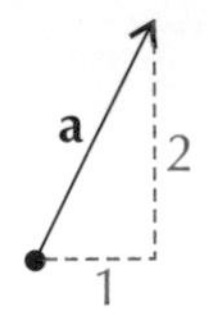

a is a translation of 1 across and 2 up.

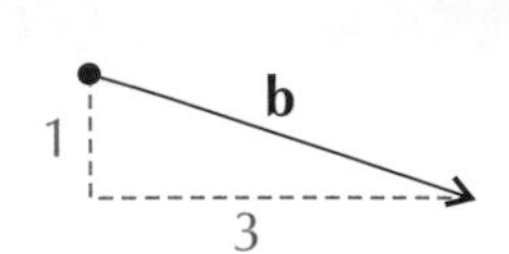

b is a translation of 3 across and 1 down (or –1 up).

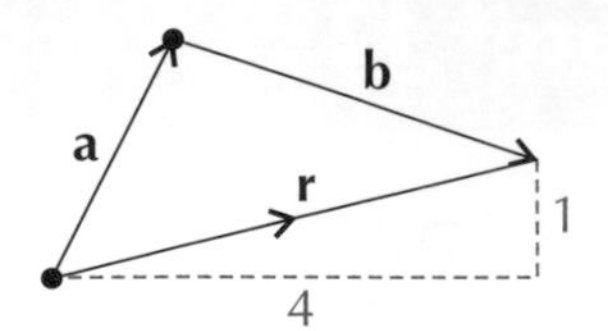

So doing **a** then **b** gives you a combined translation of (1 + 3) = 4 across and (2 – 1) = 1 up.

Subtracting vectors

The vector –**a** points in the opposite direction to the vector **a**, and they're both exactly the **same size**. So **subtracting** a vector is the same as **adding the negative vector**:

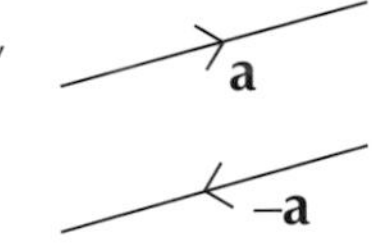

To go from Q to P, you can't just add the vectors **a** and **b** because the arrows don't run from end to end.

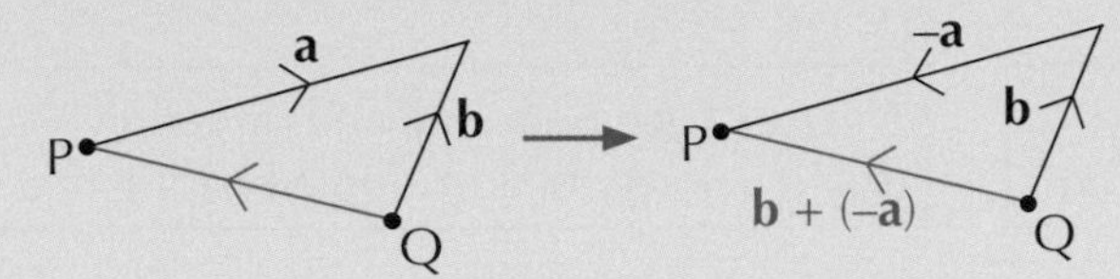

But replace vector **a** with –**a** (which goes in the opposite direction) and now you can add.

So $\overrightarrow{QP}$ = **b** + (–**a**) = **b** – **a**

You can use these rules to find a vector in terms of **other vectors**.

Example 1

Find $\overrightarrow{WZ}$ and $\overrightarrow{ZX}$ in terms of p, q and r.

1. Relabel the vectors on the diagram so that they run from end to end.
2. When you do this you can see that $\overrightarrow{WZ}$ is the resultant vector of –**p** + **q** + (–**r**).
3. So: $\overrightarrow{WZ}$ = –**p** + **q** – **r**
4. Using the addition and subtraction rules in the same way to find $\overrightarrow{ZX}$.
5. So you get: $\overrightarrow{ZX}$ = **r** – **q**

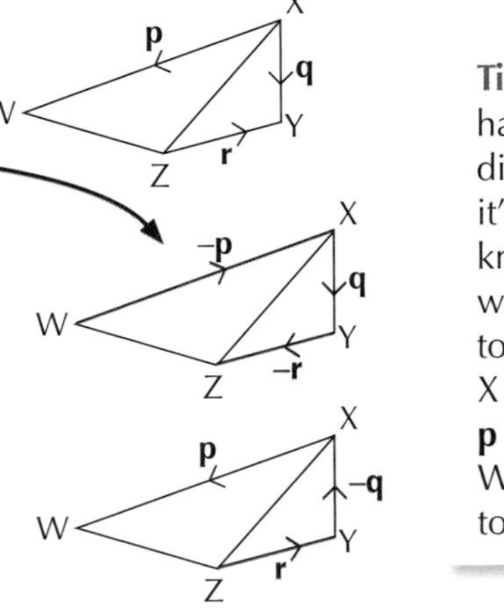

Tip: You don't have to draw this diagram out — it's enough to know that if you want your vector to go from W to X and the vector **p** goes from X to W then you need to use –**p** instead.

Scalar multiplication

You can **multiply** a vector by a **scalar** (just a number).

When you do this the **length changes** but the **direction** stays the **same**...

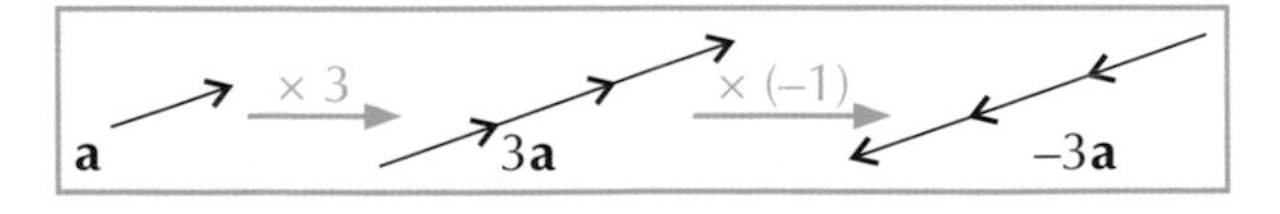

...unless the scalar is **negative**, then the direction's **reversed**.

You can '**divide**' a vector by a scalar as well — you just **multiply** the vector by the **reciprocal** of the scalar. E.g. $\frac{\mathbf{a}}{3} = \frac{1}{3}\mathbf{a}$. **Multiplying** a vector by a **non-zero** scalar always produces a **parallel** vector.

If you're asked to find whether or not two vectors are parallel then you need to check the **scalar factor** for each coefficient. If they're all the **same** then the vectors are parallel, if not then they're not.

Example 2

Show that the vector 9a + 15b is parallel to the vector 6a + 10b.

1. You need to try and write one vector as a scalar multiple of the other. So find the scalar factor that you multiply by.

   $\frac{9\mathbf{a}}{6\mathbf{a}} = 1.5 \quad \frac{15\mathbf{b}}{10\mathbf{b}} = 1.5$

2. The scalar factor is the same for **a** and **b**, so it's possible to write the first vector as a scalar multiple of the second.

   $9\mathbf{a} + 15\mathbf{b} = 1.5(6\mathbf{a} + 10\mathbf{b})$, so the vectors are parallel

All **parallel** vectors are **scalar multiples** of each other, so showing that one vector is a scalar multiple of another is the same as showing they're parallel.

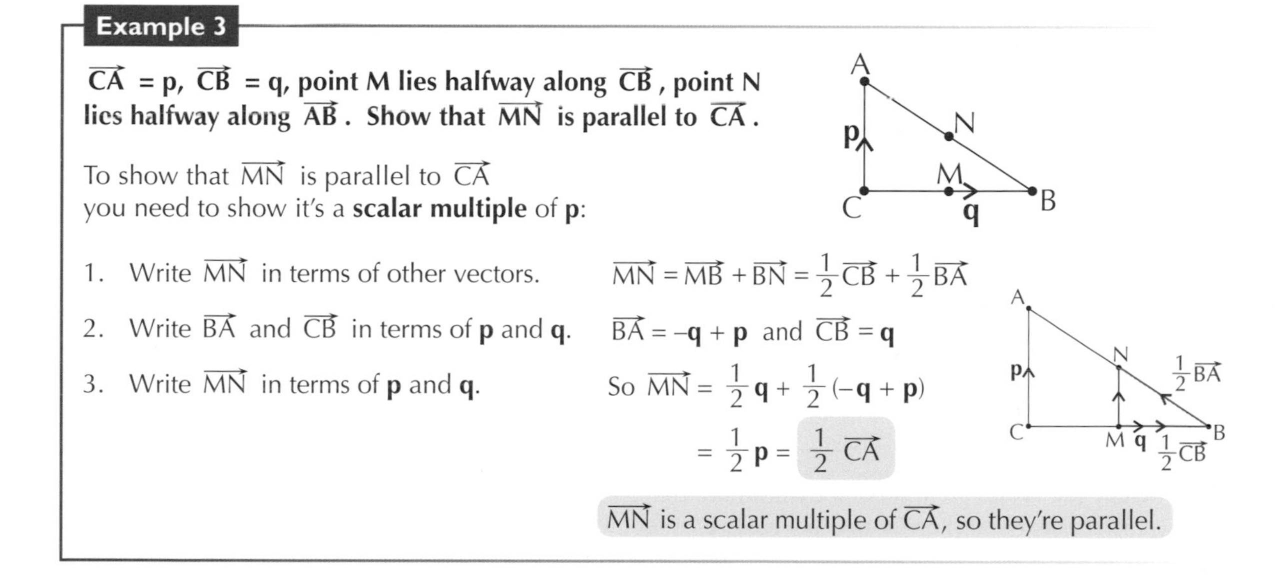

Example 3

$\overrightarrow{CA}$ = p, $\overrightarrow{CB}$ = q, point M lies halfway along $\overrightarrow{CB}$, point N lies halfway along $\overrightarrow{AB}$. Show that $\overrightarrow{MN}$ is parallel to $\overrightarrow{CA}$.

To show that $\overrightarrow{MN}$ is parallel to $\overrightarrow{CA}$ you need to show it's a **scalar multiple** of **p**:

1. Write $\overrightarrow{MN}$ in terms of other vectors. $\quad \overrightarrow{MN} = \overrightarrow{MB} + \overrightarrow{BN} = \frac{1}{2}\overrightarrow{CB} + \frac{1}{2}\overrightarrow{BA}$
2. Write $\overrightarrow{BA}$ and $\overrightarrow{CB}$ in terms of **p** and **q**. $\quad \overrightarrow{BA} = -\mathbf{q} + \mathbf{p}$ and $\overrightarrow{CB} = \mathbf{q}$
3. Write $\overrightarrow{MN}$ in terms of **p** and **q**. $\quad$ So $\overrightarrow{MN} = \frac{1}{2}\mathbf{q} + \frac{1}{2}(-\mathbf{q} + \mathbf{p})$

   $= \frac{1}{2}\mathbf{p} = \frac{1}{2}\overrightarrow{CA}$

$\overrightarrow{MN}$ is a scalar multiple of $\overrightarrow{CA}$, so they're parallel.

A vector can be **anywhere** in **space**.

This means that vectors of the **same size** which are **parallel** and pointing in the **same direction** are the **same**, even if they're not in the same place. Vectors like this are often called **equal factors**.

E.g. in a parallelogram, opposite sides have the **same vector**, so knowing one means you know the other:

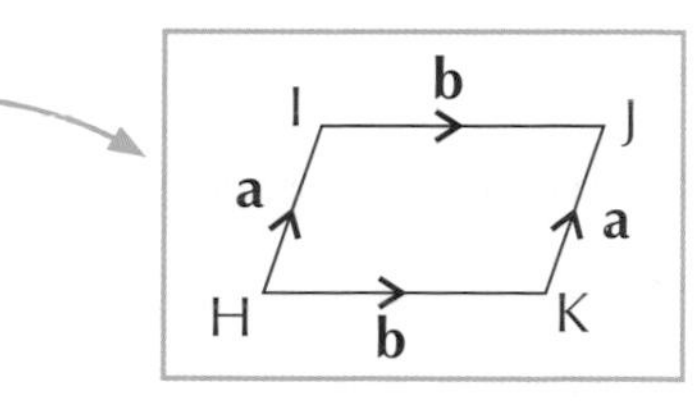

Look out for questions where you have to recognise that two lines are parallel in order to find a vector.

Collinear points

You can also use **vector addition** to show that three points are **collinear** — i.e. they all lie on a **single straight line**. If vectors $\overrightarrow{AB}$ and $\overrightarrow{BC}$ are **parallel**, then the points A, B and C are collinear.

Example 4

$\overrightarrow{AP}$ = m, $\overrightarrow{AQ}$ = m + 2n, $\overrightarrow{AR}$ = m + 6n. Show that P, Q and R are collinear.

1. Find $\overrightarrow{PQ}$ and $\overrightarrow{QR}$.

   $\overrightarrow{PQ} = -\overrightarrow{AP} + \overrightarrow{AQ}$
   $= -\mathbf{m} + \mathbf{m} + 2\mathbf{n} = 2\mathbf{n}$

   $\overrightarrow{QR} = -\overrightarrow{AQ} + \overrightarrow{AR}$
   $= -\mathbf{m} - 2\mathbf{n} + \mathbf{m} + 6\mathbf{n} = 4\mathbf{n}$

2. Now show that $\overrightarrow{QR}$ is a scalar multiple of $\overrightarrow{PQ}$.

   $\overrightarrow{QR} = 4\mathbf{n} = 2(2\mathbf{n}) = 2(\overrightarrow{PQ})$

3. $\overrightarrow{QR}$ is a scalar multiple of $\overrightarrow{PQ}$.

   So the vectors are parallel, meaning points P, Q and R lie on a **straight line** — i.e. they are collinear.

Exercise 12.1.1

Q1 State whether each of these real world examples refers to a scalar quantity, a vector quantity or neither.

MODELLING

a) A pilot flies due south for a distance of 200 kilometres.

b) The time taken to travel from London to Exeter is 3 hours.

c) A force of 20 newtons is required to pull a sledge up the steepest section of a hill — the slope is at an angle of 5° to the horizontal.

Q2 Vectors **a** and **b** are represented by the lines to the right. Draw and label sketches that represent the following vectors:

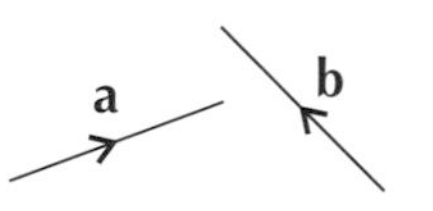

a) $-\mathbf{a}$ b) $2\mathbf{b}$ c) $\mathbf{a} + \mathbf{b}$ d) $\mathbf{a} - \mathbf{b}$

Q3 For the rectangle ABCD shown on the right, write down single vectors that are equivalent to:

a) $\overrightarrow{AB} + \overrightarrow{BC}$ b) $\overrightarrow{BC} + \overrightarrow{CD} + \overrightarrow{DA}$ c) $\overrightarrow{DC} - \overrightarrow{BC}$

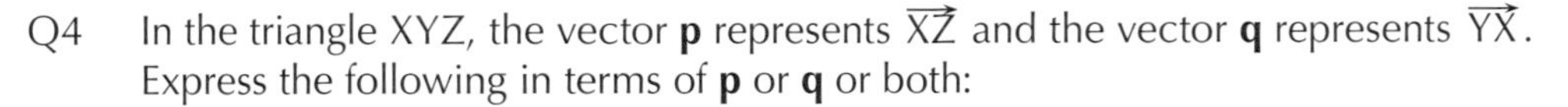

Q4 In the triangle XYZ, the vector **p** represents $\overrightarrow{XZ}$ and the vector **q** represents $\overrightarrow{YX}$. Express the following in terms of **p** or **q** or both:

a) $\overrightarrow{XY}$ b) $\overrightarrow{YZ}$ c) $\overrightarrow{ZY}$

Q5 ABCD is a parallelogram. The vector $\overrightarrow{BA}$ = **n** and $\overrightarrow{BD}$ = **m**. Find $\overrightarrow{BC}$ in terms of **n** and **m**.

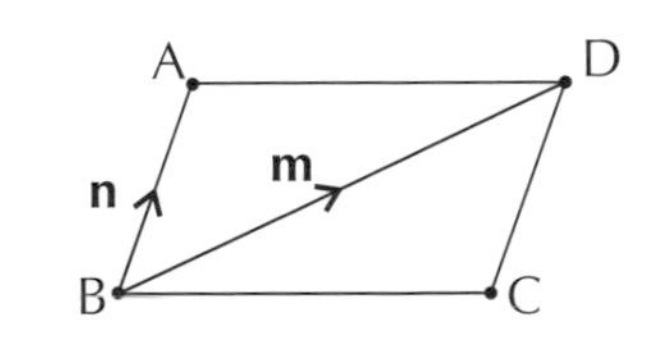

Q6 In the rectangle ABCD, E is the midpoint of AD and F divides DC in the ratio 2 : 1. If $\overrightarrow{AB}$ = **b** and $\overrightarrow{AD}$ = **d**, find the following vectors in terms of **b** and **d**.

a) $\overrightarrow{DF}$ b) $\overrightarrow{BE}$ c) $\overrightarrow{EF}$

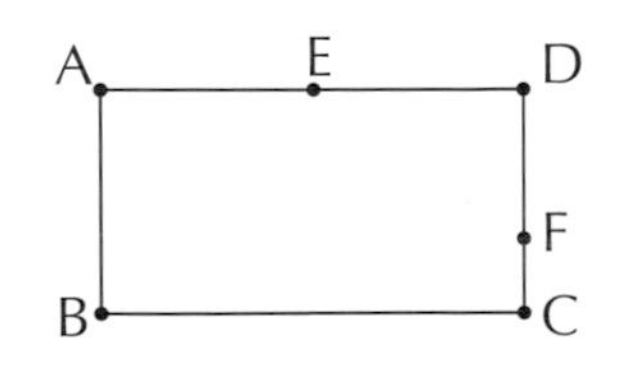

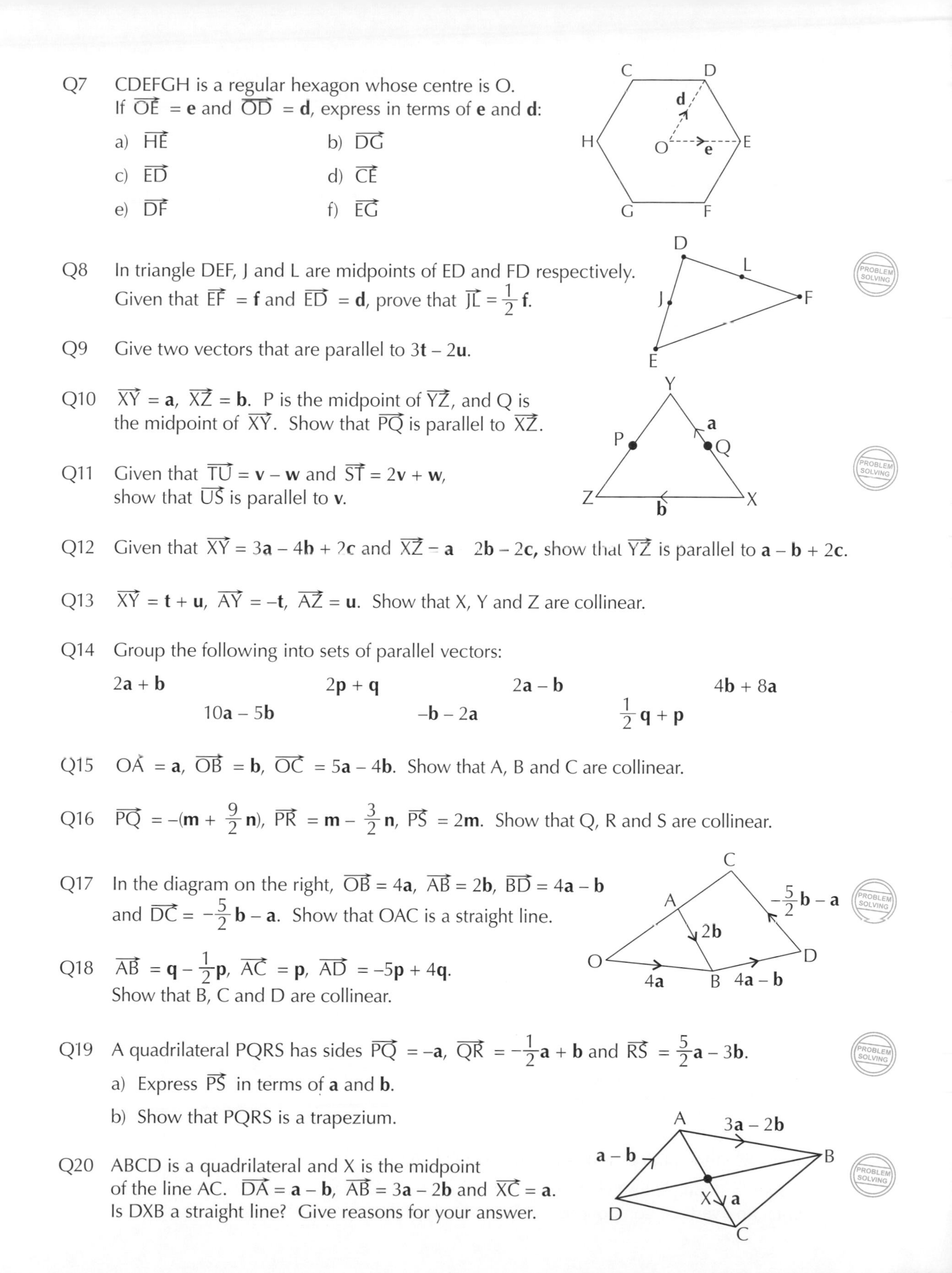

Q7 CDEFGH is a regular hexagon whose centre is O.
If $\overrightarrow{OE} = \mathbf{e}$ and $\overrightarrow{OD} = \mathbf{d}$, express in terms of $\mathbf{e}$ and $\mathbf{d}$:

a) $\overrightarrow{HE}$ b) $\overrightarrow{DG}$

c) $\overrightarrow{ED}$ d) $\overrightarrow{CE}$

e) $\overrightarrow{DF}$ f) $\overrightarrow{EG}$

Q8 In triangle DEF, J and L are midpoints of ED and FD respectively.
Given that $\overrightarrow{EF} = \mathbf{f}$ and $\overrightarrow{ED} = \mathbf{d}$, prove that $\overrightarrow{JL} = \frac{1}{2}\mathbf{f}$.

Q9 Give two vectors that are parallel to $3\mathbf{t} - 2\mathbf{u}$.

Q10 $\overrightarrow{XY} = \mathbf{a}$, $\overrightarrow{XZ} = \mathbf{b}$. P is the midpoint of $\overrightarrow{YZ}$, and Q is the midpoint of $\overrightarrow{XY}$. Show that $\overrightarrow{PQ}$ is parallel to $\overrightarrow{XZ}$.

Q11 Given that $\overrightarrow{TU} = \mathbf{v} - \mathbf{w}$ and $\overrightarrow{ST} = 2\mathbf{v} + \mathbf{w}$,
show that $\overrightarrow{US}$ is parallel to $\mathbf{v}$.

Q12 Given that $\overrightarrow{XY} = 3\mathbf{a} - 4\mathbf{b} + 2\mathbf{c}$ and $\overrightarrow{XZ} = \mathbf{a} - 2\mathbf{b} - 2\mathbf{c}$, show that $\overrightarrow{YZ}$ is parallel to $\mathbf{a} - \mathbf{b} + 2\mathbf{c}$.

Q13 $\overrightarrow{XY} = \mathbf{t} + \mathbf{u}$, $\overrightarrow{AY} = -\mathbf{t}$, $\overrightarrow{AZ} = \mathbf{u}$. Show that X, Y and Z are collinear.

Q14 Group the following into sets of parallel vectors:

$2\mathbf{a} + \mathbf{b}$ $2\mathbf{p} + \mathbf{q}$ $2\mathbf{a} - \mathbf{b}$ $4\mathbf{b} + 8\mathbf{a}$

$10\mathbf{a} - 5\mathbf{b}$ $-\mathbf{b} - 2\mathbf{a}$ $\frac{1}{2}\mathbf{q} + \mathbf{p}$

Q15 $\overrightarrow{OA} = \mathbf{a}$, $\overrightarrow{OB} = \mathbf{b}$, $\overrightarrow{OC} = 5\mathbf{a} - 4\mathbf{b}$. Show that A, B and C are collinear.

Q16 $\overrightarrow{PQ} = -(\mathbf{m} + \frac{9}{2}\mathbf{n})$, $\overrightarrow{PR} = \mathbf{m} - \frac{3}{2}\mathbf{n}$, $\overrightarrow{PS} = 2\mathbf{m}$. Show that Q, R and S are collinear.

Q17 In the diagram on the right, $\overrightarrow{OB} = 4\mathbf{a}$, $\overrightarrow{AB} = 2\mathbf{b}$, $\overrightarrow{BD} = 4\mathbf{a} - \mathbf{b}$ and $\overrightarrow{DC} = -\frac{5}{2}\mathbf{b} - \mathbf{a}$. Show that OAC is a straight line.

Q18 $\overrightarrow{AB} = \mathbf{q} - \frac{1}{2}\mathbf{p}$, $\overrightarrow{AC} = \mathbf{p}$, $\overrightarrow{AD} = -5\mathbf{p} + 4\mathbf{q}$.
Show that B, C and D are collinear.

Q19 A quadrilateral PQRS has sides $\overrightarrow{PQ} = -\mathbf{a}$, $\overrightarrow{QR} = -\frac{1}{2}\mathbf{a} + \mathbf{b}$ and $\overrightarrow{RS} = \frac{5}{2}\mathbf{a} - 3\mathbf{b}$.

a) Express $\overrightarrow{PS}$ in terms of $\mathbf{a}$ and $\mathbf{b}$.

b) Show that PQRS is a trapezium.

Q20 ABCD is a quadrilateral and X is the midpoint of the line AC. $\overrightarrow{DA} = \mathbf{a} - \mathbf{b}$, $\overrightarrow{AB} = 3\mathbf{a} - 2\mathbf{b}$ and $\overrightarrow{XC} = \mathbf{a}$.
Is DXB a straight line? Give reasons for your answer.

Q21 In the diagram on the right,
$\overrightarrow{AB} = \mathbf{b} - \mathbf{a}$, $\overrightarrow{BC} = \mathbf{a}$, $\overrightarrow{AD} = \frac{3}{2}\mathbf{a}$ and $\overrightarrow{AE} = 2\mathbf{a} - \mathbf{b}$.

a) Show that BE is parallel to CD.

b) The point O divides $\overrightarrow{BE}$ in the ratio 1 : 2.
Show that AOC is a straight line.

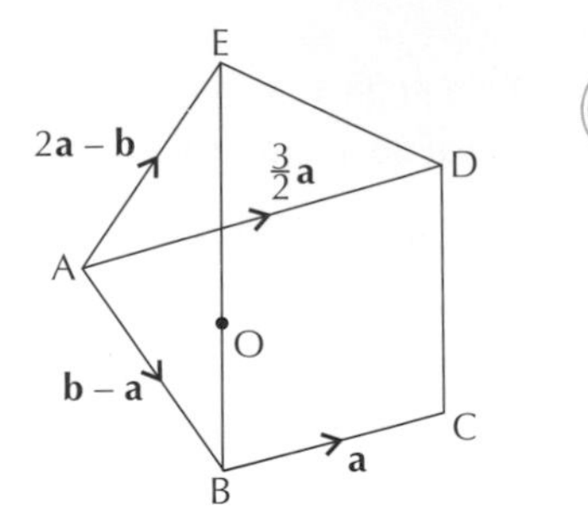

PROBLEM SOLVING

Position vectors

You can use a vector to describe the **position** of a point in relation to the **origin**, O. This vector is called a **position vector** and they always **start** at the origin and finish at the point they're describing.

One way of describing the position of the point P is with its **Cartesian coordinates** which are e.g. (6, 5).

This just tells you how far P is vertically and horizontally from the **origin** O.

Another way of describing how far P is from the origin is using the **position vector** $\overrightarrow{OP} = \mathbf{p}$ which has **horizontal** and **vertical** components.

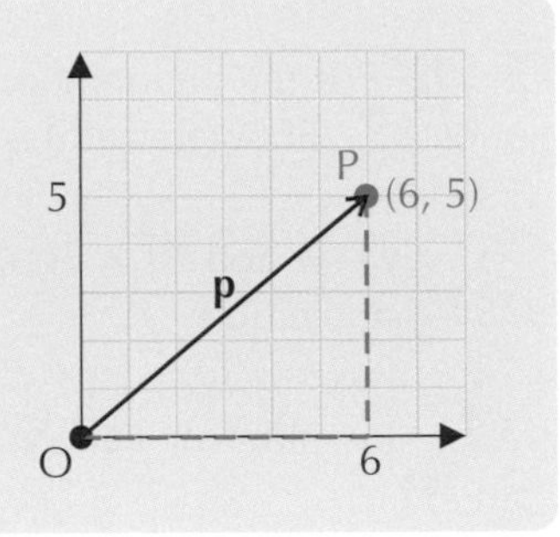

The position vector of any point A is $\overrightarrow{OA}$. It's usually called vector **a**.

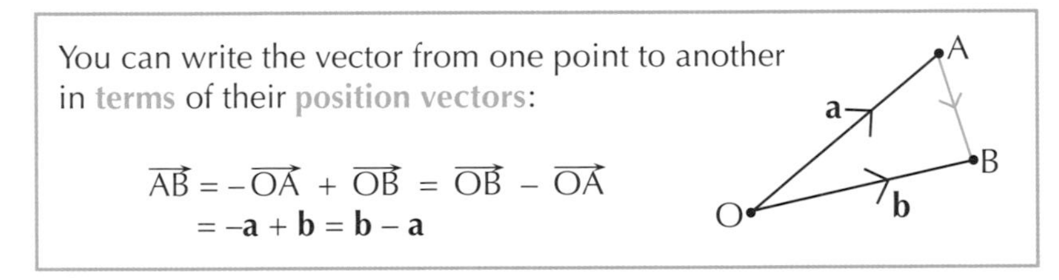

You can write the vector from one point to another in **terms** of their **position vectors**:

$$\overrightarrow{AB} = -\overrightarrow{OA} + \overrightarrow{OB} = \overrightarrow{OB} - \overrightarrow{OA}$$
$$= -\mathbf{a} + \mathbf{b} = \mathbf{b} - \mathbf{a}$$

Tip: This result will be used time after time for finding the vector from one point to another in this chapter.
Make sure you learn it in both its forms:
$\overrightarrow{AB} = \overrightarrow{OB} - \overrightarrow{OA} = \mathbf{b} - \mathbf{a}$

i + j units

A **unit vector** is any vector with a **length of 1 unit**.

The vectors **i** and **j** are **standard unit vectors**, so they each have a length of 1 unit.
i is in the direction of the **positive x-axis**, and **j** is in the direction of the **positive y-axis**.

Every vector in two dimensions is made up of **horizontal** and **vertical components**, so you can express any vector as a **sum** of **i** and **j** unit vectors:

Vector **a** goes from the origin O to the point A.

To get from O to A you move **4** units **to the right** and **3** units **up**.

So **a** is the **resultant** vector when you add a **horizontal vector** that goes **4 units** in the positive x direction and a **vertical vector** that goes **3 units** in the positive y direction.

i and **j** are the **standard** unit vectors we use to express horizontal and vertical components. So $\mathbf{a} = 4\mathbf{i} + 3\mathbf{j}$

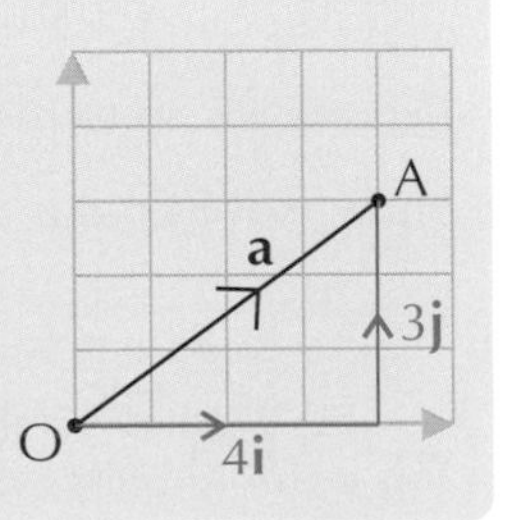

Example

a) Write down the position vectors of A and B in i + j form.

1. Look at the position of point A. — Point A lies 5 units to the right and 7 units above the origin
2. Use this to write A in the form $x\mathbf{i} + y\mathbf{j}$. — $\mathbf{a} = 5\mathbf{i} + 7\mathbf{j}$
3. Repeat for point B. — $\mathbf{b} = 4\mathbf{i} + \mathbf{j}$

b) Hence find $\overrightarrow{AB}$ in terms of i and j.

1. Find $\overrightarrow{AB}$ in terms of $\overrightarrow{OA}$ and $\overrightarrow{OB}$. — $\overrightarrow{AB} = -\overrightarrow{OA} + \overrightarrow{OB} = \overrightarrow{OB} - \overrightarrow{OA}$
2. Write $\overrightarrow{AB}$ in terms of a and b. — $= -\mathbf{a} + \mathbf{b} = \mathbf{b} - \mathbf{a}$
3. Add or subtract the **i** and **j** components separately. — $= (4\mathbf{i} + \mathbf{j}) - (5\mathbf{i} + 7\mathbf{j})$ $= -\mathbf{i} - 6\mathbf{j}$

In the example above, the answer just means that to go from A to B, you **move** 1 unit left and 6 units down. Vectors like this are just like a **translation**.

Column vectors

Column vectors are another way of writing vectors in terms of their **horizontal** and **vertical components**.

You just write the **horizontal** (**i**) **component** on **top** of the **vertical** (**j**) **component** and put a bracket around them:

$$x\mathbf{i} + y\mathbf{j} = \begin{pmatrix} x \\ y \end{pmatrix}$$

Calculating with them is simple. Just add or subtract the **top** row, then add or subtract the **bottom** row **separately**:

Tip: Using column vectors is often quicker and easier than working with sums of **i** and **j** components.

$$\mathbf{a} = 5\mathbf{i} + 7\mathbf{j} = \begin{pmatrix} 5 \\ 7 \end{pmatrix} \qquad \mathbf{b} = 4\mathbf{i} + \mathbf{j} = \begin{pmatrix} 4 \\ 1 \end{pmatrix} \qquad \mathbf{b} - \mathbf{a} = \begin{pmatrix} 4 \\ 1 \end{pmatrix} - \begin{pmatrix} 5 \\ 7 \end{pmatrix} = \begin{pmatrix} 4-5 \\ 1-7 \end{pmatrix} = \begin{pmatrix} -1 \\ -6 \end{pmatrix}$$

When you're **multiplying** a column vector by a **scalar** you multiply **each number** in the column vector by the scalar:

$$2\mathbf{b} - 3\mathbf{a} = 2\begin{pmatrix} 4 \\ 1 \end{pmatrix} - 3\begin{pmatrix} 5 \\ 7 \end{pmatrix} = \begin{pmatrix} 2\times 4 \\ 2\times 1 \end{pmatrix} - \begin{pmatrix} 3\times 5 \\ 3\times 7 \end{pmatrix}$$
$$= \begin{pmatrix} 8 \\ 2 \end{pmatrix} - \begin{pmatrix} 15 \\ 21 \end{pmatrix} = \begin{pmatrix} -7 \\ -19 \end{pmatrix}$$

Exercise 12.1.2

Q1 Given that $\mathbf{a} = \begin{pmatrix} -1 \\ -2 \end{pmatrix}$, $\mathbf{b} = \begin{pmatrix} 3 \\ -2 \end{pmatrix}$ and $\mathbf{c} = \begin{pmatrix} 4 \\ 3 \end{pmatrix}$, calculate the following:

a) $\mathbf{a} + \mathbf{b} + \mathbf{c}$ b) $\mathbf{c} - 2\mathbf{b}$ c) $3\mathbf{a} - \mathbf{b} + 2\mathbf{c}$ d) $5\mathbf{a} - 5\mathbf{c}$

Q2 Given that $\mathbf{p} = \begin{pmatrix} -1 \\ -2 \end{pmatrix}$, $\mathbf{q} = \begin{pmatrix} 3 \\ -2 \end{pmatrix}$ and $\mathbf{r} = \begin{pmatrix} -4 \\ 5 \end{pmatrix}$, show that $\mathbf{p} + 3\mathbf{q} + \mathbf{r}$ is parallel to $12\mathbf{i} - 9\mathbf{j}$.

Q3 On a map, Jack's house has coordinates (2, 3) and his school has coordinates (4, –5). Write down the position vectors of Jack's house and Jack's school, giving your answers as column vectors.

MODELLING

Q4 Give, in **i** and **j** form, the position vectors of the following points:

a) (2, –4) b) (–1, –2) c) (5, 7) d) (3, –7)

Q5 C has position vector $-\mathbf{i} + 2\mathbf{j}$ and D has position vector $4\mathbf{i} - 3\mathbf{j}$.

a) What are the Cartesian coordinates of the points C and D?

b) Write the vectors $\overrightarrow{CD}$ and $\overrightarrow{DC}$ in unit vector form.

Q6 Triangle ABC is shown below. Find the vectors $\overrightarrow{AB}$, $\overrightarrow{BC}$ and $\overrightarrow{CA}$.

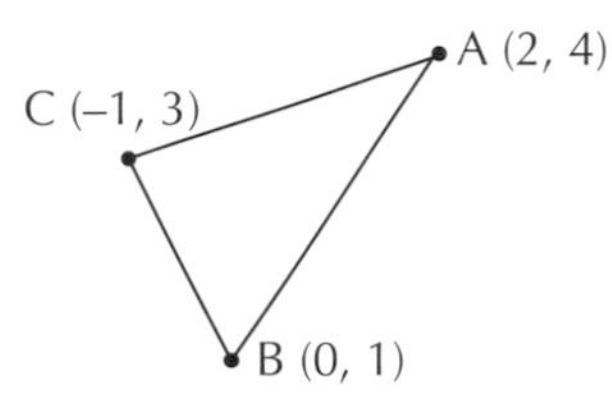

Q7 ABCD is a square, where A has position vector $-4\mathbf{i} + 6\mathbf{j}$ and C has position vector $3\mathbf{i} - \mathbf{j}$. Find the position vectors of B and D.

Q8 M is the midpoint of the line PQ, where P has position vector $-3\mathbf{i} + \mathbf{j}$ and M has position vector $2\mathbf{i} - 5\mathbf{j}$. Find the position vector of Q.

PROBLEM SOLVING

Q9 A point P divides the line AB in the ratio 2 : 1, where $\overrightarrow{OA} = \begin{pmatrix} -2 \\ 4 \end{pmatrix}$ and $\overrightarrow{OB} = \begin{pmatrix} 4 \\ -5 \end{pmatrix}$.

PROBLEM SOLVING

a) Find the position vector of P.

b) Find the vector $\overrightarrow{PB}$.

Q10 A robotic vacuum cleaner models a room using a quadrilateral DEFG, with vertices at the points D (–7, –2), E (–3, –1), F (–1, 5) and G (–3, 10).

MODELLING

a) Give the vectors for the room's walls $\overrightarrow{DE}$, $\overrightarrow{EF}$, $\overrightarrow{FG}$ and $\overrightarrow{GD}$.

b) The vacuum cleaner follows the wall from D to E, then E to F. Give a single vector that the vacuum cleaner could have followed to get from D to F more efficiently.

Q11 PQR is a triangle. Given that $\overrightarrow{OQ} = \begin{pmatrix} 3 \\ 3 \end{pmatrix}$, $\overrightarrow{OP} = \begin{pmatrix} 3 \\ 0 \end{pmatrix}$ and $\overrightarrow{PR} = \begin{pmatrix} 4 \\ 7 \end{pmatrix}$, find the Cartesian coordinates of R and give the vector $\overrightarrow{QR}$.

PROBLEM SOLVING

Q12 The position vectors of the vertices of a quadrilateral ABCD are: $\overrightarrow{OA} = \mathbf{i} - \mathbf{j}$, $\overrightarrow{OB} = -4\mathbf{j}$, $\overrightarrow{OC} = 4\mathbf{i} - 2\mathbf{j}$ and $\overrightarrow{OD} = 3\mathbf{i} + \mathbf{j}$. Find the vectors for the diagonals of the quadrilateral.

PROBLEM SOLVING

12.2 Calculating with Vectors

The magnitude of a vector is a scalar that tells you the vector's length.
In this section you'll see how to calculate it and what you can use it for.

Learning Objectives (Spec Ref 9.2 & 9.4):
- Find the magnitude of any vector in two dimensions.
- Find the unit vector in the direction of any vector in two dimensions.
- Find the distance between two two-dimensional points using vectors.
- Calculate the direction of any vector in two dimensions.
- Use trigonometry to do calculations involving vectors.

Prior Knowledge Check: Be able to use Pythagoras' theorem — seen at GCSE.

Calculating with vectors

Calculating the magnitude of a vector

The **magnitude** of a vector is the **distance** between its start point and end point.
It's sometimes called **modulus** instead of magnitude.

The **magnitude** of a vector **a** is written $|\mathbf{a}|$.

The **magnitude** of a vector $\overrightarrow{AB}$ is written $|\overrightarrow{AB}|$.

Magnitude is a **scalar**, and it's **always positive**. The **i** and **j** components of a vector form a convenient **right-angled triangle**, so you can use **Pythagoras' theorem** to find a vector's magnitude.

Example 1

Find the magnitude of the vector a = 5i + 3j.

1. You know the length of two sides of the right-angled triangle formed by **a** and its **horizontal** and **vertical** components.

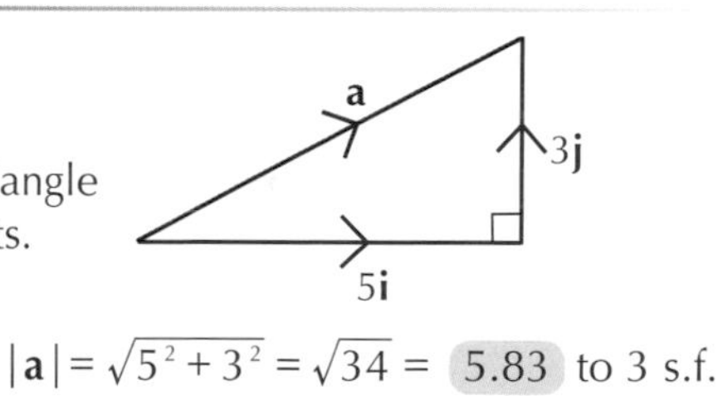

2. The magnitude of **a** is the length of the **hypotenuse** of this triangle. So find $|\mathbf{a}|$ using **Pythagoras**.

$|\mathbf{a}| = \sqrt{5^2 + 3^2} = \sqrt{34} = 5.83$ to 3 s.f.

You can use a vector's magnitude to find the **distance** between two **points**:

Example 2

$\overrightarrow{JK} = \begin{pmatrix} 4 \\ -7 \end{pmatrix}$ **Find the distance between J and K. Give your answer in surd form.**

The distance between J and K is $|\overrightarrow{JK}| = \sqrt{4^2 + (-7)^2} = \sqrt{65}$

You can find **missing components** of the vector using its magnitude:

Example 3

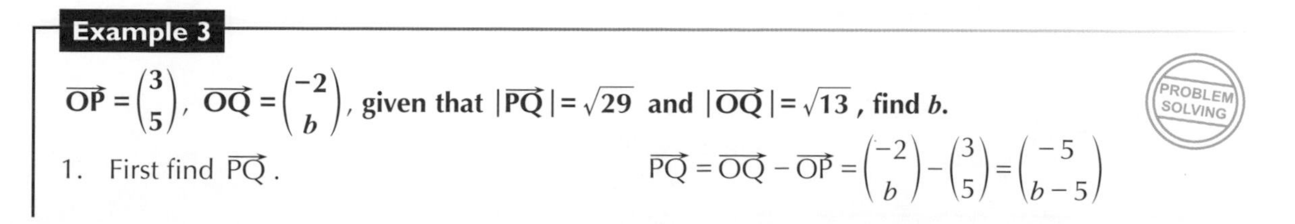

$\overrightarrow{OP} = \begin{pmatrix} 3 \\ 5 \end{pmatrix}$, $\overrightarrow{OQ} = \begin{pmatrix} -2 \\ b \end{pmatrix}$, **given that** $|\overrightarrow{PQ}| = \sqrt{29}$ **and** $|\overrightarrow{OQ}| = \sqrt{13}$, **find** b.

PROBLEM SOLVING

1. First find $\overrightarrow{PQ}$.

$\overrightarrow{PQ} = \overrightarrow{OQ} - \overrightarrow{OP} = \begin{pmatrix} -2 \\ b \end{pmatrix} - \begin{pmatrix} 3 \\ 5 \end{pmatrix} = \begin{pmatrix} -5 \\ b-5 \end{pmatrix}$

2. Now find the magnitude of $\overrightarrow{PQ}$ in terms of b and compare it with the value given.

$|\overrightarrow{PQ}| = \sqrt{25 + (b-5)^2} \Rightarrow 29 = 25 + (b-5)^2$
$\Rightarrow 4 = (b-5)^2 \Rightarrow \pm 2 = b - 5$
$\Rightarrow b = 3$ or $b = 7$

3. Do the same with $\overrightarrow{OQ}$:

$|\overrightarrow{OQ}| = \sqrt{(-2)^2 + b^2} \Rightarrow 13 = 4 + b^2$
$\Rightarrow 9 = b^2 \Rightarrow b = \pm 3$

4. Both the statements '$b = \pm 3$' and '$b = 3$ or $b = 7$' must hold, so $b = 3$

To find a **unit vector** in the direction of a particular vector you just **divide** the vector by its **magnitude** (i.e. multiply by the magnitude's reciprocal).

The **unit vector** in the direction of the vector **a** is: $\frac{1}{|\mathbf{a}|}\mathbf{a} = \frac{\mathbf{a}}{|\mathbf{a}|}$.

A unit vector always has a **magnitude** of $\frac{1}{|\mathbf{a}|} \times |\mathbf{a}| = 1$.

It's a **positive scalar multiple** of **a** (as magnitude is always positive), so it has the **same direction** as **a**.

Example 4

Find the unit vector in the direction of q = 5i – 12j.

1. First find the magnitude of **q**.

$|\mathbf{q}| = \sqrt{5^2 + (-12)^2} = \sqrt{169} = 13$

2. Use the magnitude to work out the unit vector.

$\frac{\mathbf{q}}{|\mathbf{q}|} = \frac{1}{13}(5\mathbf{i} - 12\mathbf{j}) = \frac{5}{13}\mathbf{i} - \frac{12}{13}\mathbf{j}$

Calculating the direction of a vector

The direction of a vector **a** is the **angle** between a line parallel to the x-axis and **a**. It is usually measured **anticlockwise** from the x-axis.

If you know the **i** and **j** components of a vector, then you can find the **direction** of the vector by using **trigonometry**.

Tip: You may see questions asking you to calculate other angles too — like the angle between a vector and the vertical y-axis.

Example 5

The diagram shows the vector a = 2i + 3j.
Find the magnitude and direction of the vector.

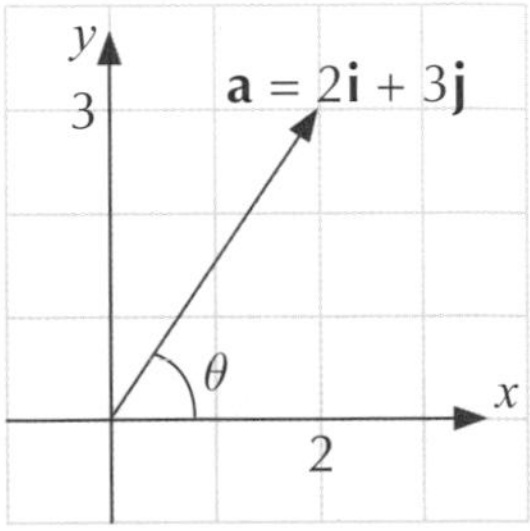

1. The vector forms the hypotenuse of a right-angled triangle of height 3 units and base width 2 units.

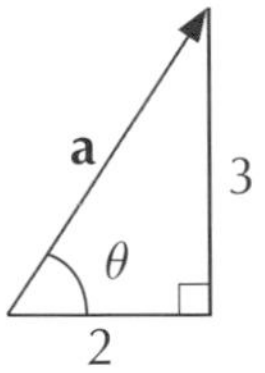

2. Use Pythagoras' theorem as before to find the magnitude.

$|\mathbf{a}| = \sqrt{2^2 + 3^2} = \sqrt{13}$

3. Use trigonometry to find the angle θ.

$\tan\theta = \frac{3}{2} \Rightarrow \theta = \tan^{-1}\left(\frac{3}{2}\right) = 56.3°$ (3 s.f.)

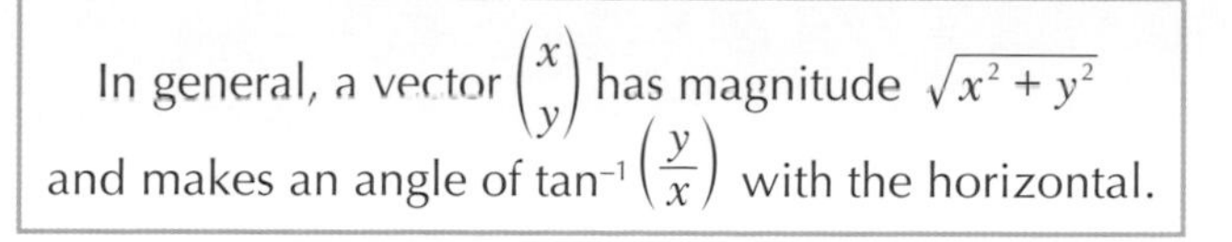

In general, a vector $\begin{pmatrix} x \\ y \end{pmatrix}$ has magnitude $\sqrt{x^2 + y^2}$ and makes an angle of $\tan^{-1}\left(\frac{y}{x}\right)$ with the horizontal.

Similarly, if you know the **magnitude** and **direction** of a vector then you can use **trigonometry** to calculate its horizontal and vertical components.

The direction is measured **anticlockwise** from the positive ***x*-axis**, so the direction isn't always the same as the angle with the **horizontal**. E.g. in the diagram on the right, the direction of the vector is $360° - \theta$.

Example 6

Given a vector v = *a*i + *b*j, with direction 30° and magnitude |v| = 5, calculate *a* and *b*.

1. The vector forms a right-angled triangle, with the magnitude as the length of the hypotenuse.
2. Use trigonometry to find a and b.
   $\cos 30° = \frac{a}{5}$, so $a = 5 \cos 30°$
   $\sin 30° = \frac{b}{5}$, so $b = 5 \sin 30°$
3. The sine and cosine in this example give nice values, so evaluate them (sometimes it's better to leave them unevaluated to give exact results).
   $\mathbf{v} = (5 \cos 30°)\mathbf{i} + (5 \sin 30°)\mathbf{j}$
   $= \frac{5\sqrt{3}}{2}\mathbf{i} + \frac{5}{2}\mathbf{j}$

Calculating the angle between two vectors

The angle between two vectors **a** and **b** can be calculated by constructing a triangle with **a** and **b** as two of its sides.

First calculate the **magnitude** of these vectors, and then use the **cosine rule** to find the angle between them.

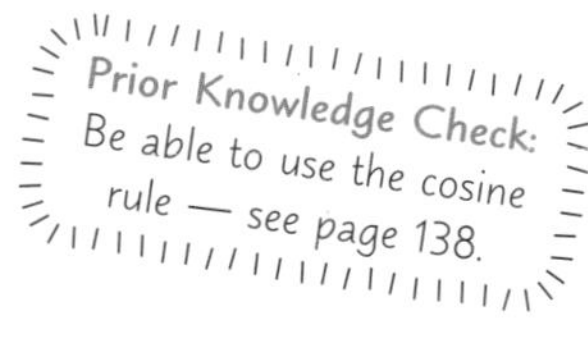

Prior Knowledge Check: Be able to use the cosine rule — see page 138.

Example 7

Find the angle θ between the vectors $\overrightarrow{PQ} = 3\mathbf{i} - \mathbf{j}$ and $\overrightarrow{PR} = -\mathbf{i} + 4\mathbf{j}$.

1. $\overrightarrow{PQ}$ and $\overrightarrow{PR}$ form two sides of a triangle PQR.
2. Work out the lengths of these two sides.
   $|\overrightarrow{PQ}| = \sqrt{3^2 + (-1)^2} = \sqrt{10}$
   and $|\overrightarrow{PR}| = \sqrt{(-1)^2 + 4^2} = \sqrt{17}$
3. Find the vector $\overrightarrow{QR}$ in **i** + **j** form.
   $\overrightarrow{QR} = \overrightarrow{PR} - \overrightarrow{PQ} = -\mathbf{i} + 4\mathbf{j} - (3\mathbf{i} - \mathbf{j}) = -4\mathbf{i} + 5\mathbf{j}$
4. Work out the length of $|\overrightarrow{QR}|$.
   $|\overrightarrow{QR}| = \sqrt{(-4)^2 + 5^2} = \sqrt{41}$
5. Use the cosine rule to find angle θ.
   $$\cos\theta = \frac{(\sqrt{10})^2 + (\sqrt{17})^2 - (\sqrt{41})^2}{2 \times \sqrt{17} \times \sqrt{10}} = \frac{-14}{2\sqrt{170}} = \frac{-7}{\sqrt{170}},$$
   so: $\theta = \cos^{-1}\left(\frac{-7}{\sqrt{170}}\right) = 122.5°$ (1 d.p)

Exercise 12.2.1

Q1 For each of the following vectors, find: (i) the exact magnitude, (ii) the direction (to 2 d.p.)

a) $6\mathbf{i} + 8\mathbf{j}$ b) $12\mathbf{i} - 5\mathbf{j}$ c) $\begin{pmatrix} 2 \\ 4 \end{pmatrix}$ d) $\begin{pmatrix} -3 \\ -1 \end{pmatrix}$

e) $\begin{pmatrix} 24 \\ -7 \end{pmatrix}$ f) $\begin{pmatrix} -\sqrt{13} \\ 6 \end{pmatrix}$ g) $3\mathbf{i} + \sqrt{7}\mathbf{j}$ h) $-7\mathbf{j}$

Q2 S has position vector $10\mathbf{i} + 5\mathbf{j}$. Find the exact length of the line that joins point S to the origin.

Q3 For each of the following pairs of points, find the distance between them using position vectors:

a) (0, 1) and (2, 2) b) (–3, 2) and (4, 3) c) (–1, –1) and (0, 4)

Q4 For each of the pairs of vectors given below, find the exact magnitude of the resultant when the two vectors are added together.

a) $\mathbf{a} = 2\mathbf{i} + \mathbf{j}$ and $\mathbf{b} = 2\mathbf{i} - 4\mathbf{j}$ b) $\mathbf{u} = -5\mathbf{i} + \mathbf{j}$ and $\mathbf{v} = 9\mathbf{i} - 5\mathbf{j}$

c) $\mathbf{f} = \begin{pmatrix} 7 \\ 2 \end{pmatrix}$ and $\mathbf{g} = \begin{pmatrix} 17 \\ -12 \end{pmatrix}$ d) $\mathbf{d} = \begin{pmatrix} 4 \\ -2 \end{pmatrix}$ and $\mathbf{e} = \begin{pmatrix} -1 \\ -4 \end{pmatrix}$

e) $\mathbf{s} = 3\mathbf{i} - 4\mathbf{j}$ and $\mathbf{t} = -(3\mathbf{i} + \mathbf{j})$ f) $\mathbf{w} = \begin{pmatrix} -3 \\ -5 \end{pmatrix}$ and $\mathbf{x} = \begin{pmatrix} -6 \\ 5 \end{pmatrix}$

Q5 For each of the following vectors, give the unit vector in the same direction:

a) $3\mathbf{i}$ b) $\begin{pmatrix} 1 \\ 2 \end{pmatrix}$ c) $5\mathbf{i} - 4\mathbf{j}$ d) $\begin{pmatrix} -2 \\ -6 \end{pmatrix}$

Q6 $\overrightarrow{AB} = 3\mathbf{i} - 2\mathbf{j}$ and $\overrightarrow{BC} = \mathbf{i} + 5\mathbf{j}$. Find the unit vector in the direction of $\overrightarrow{AC}$.

Q7 Point A has position vector $2\mathbf{i} - \mathbf{j}$, and point B has position vector $7\mathbf{i} - 13\mathbf{j}$. Find the unit vector in the direction of $\overrightarrow{BA}$.

Q8 By the finding horizontal and vertical components, express the vectors below in exact **i**, **j** form:

a) **a** has direction 45° and magnitude $\sqrt{2}$ b) **b** has direction 60° and magnitude $\sqrt{7}$

c) **c** has direction 33° and magnitude 3 d) **d** has direction 76° and magnitude 5

Q9 Vector **c** has the same direction as vector **d**. Given that $\mathbf{d} = 8\mathbf{i} - 6\mathbf{j}$ and $|\mathbf{c}| = 70$, find vector **c**.

Q10 The vector $\mathbf{v} = a\mathbf{i} - 4\mathbf{j}$ makes an angle of 51° below the x-axis. Find $|\mathbf{v}|$ to 2 d.p.

PROBLEM SOLVING

Q11 $|\overrightarrow{AB}| = 12$ and $|\overrightarrow{BC}| = 14$ and the angle between vectors $\overrightarrow{AB}$ and $\overrightarrow{BC}$ is 115°. Find $|\overrightarrow{CA}|$ to 2 d.p.

Q12 Two boats set off from a harbour. Each boat's course is modelled by a vector. Boat A's course is given by $\mathbf{a} = \begin{pmatrix} 3 \\ 3 \end{pmatrix}$ and Boat B's by $\mathbf{b} = \begin{pmatrix} -2 \\ 5 \end{pmatrix}$. What is the angle (to 2 d.p.) between the two boats' courses?

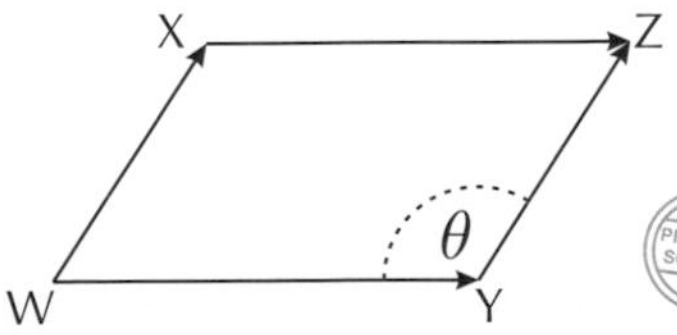

PROBLEM SOLVING MODELLING

Q13 The diagram to the right shows a parallelogram WXYZ. Given $\overrightarrow{WX} = 2\mathbf{i} + 5\mathbf{j}$ and $\overrightarrow{WY} = 8\mathbf{i}$, what is angle θ (to 1 d.p.)?

X Z W Y θ

PROBLEM SOLVING

12.3 Modelling with Vectors

Vectors are really useful and can be used for modelling and solving problems. This section looks at some common uses of vectors in mathematics.

MODELLING

Learning Objectives (Spec Ref 9.5):

- Use vectors to solve problems in different contexts.
- Model forces as vectors.

Prior Knowledge Check: Be able to use bearings to find a direction. See p.140.

Modelling with vectors

An object's **motion** will have a **magnitude** and **direction**, so can be modelled using vectors:

Displacement is the **distance** an object has travelled in a given **direction**.

Velocity is the **speed** of an object with a **direction**.

Acceleration is the rate at which an object's **velocity changes**.

Forces can also be modelled with vectors.

Tip: Be careful — the word acceleration could refer to the vector **a** or its magnitude |**a**|.

Example 1

The acceleration of a particle is given by the vector a = (6i – 2j) ms^{-2}.
Find the magnitude of the acceleration, and the angle this vector makes with the horizontal axis.

1. Start with a diagram — remember, the **j**-component "–2" means "down 2".

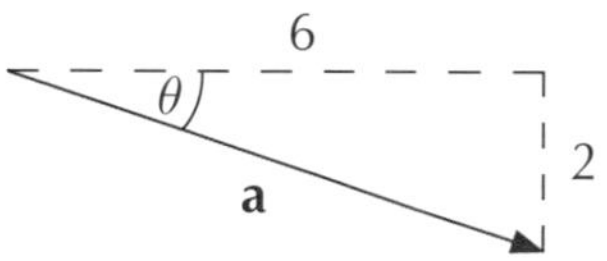

2. Use Pythagoras' theorem to find the magnitude.

$|\mathbf{a}| = \sqrt{6^2 + (-2)^2} = \sqrt{40}$
$= 6.32 \text{ ms}^{-2}$ (3 s.f.)

3. Use trigonometry to find the angle θ.

$\tan\theta = \frac{2}{6}$
$\Rightarrow \theta = \tan^{-1}\left(\frac{2}{6}\right) = 18.4°$ (3 s.f.)

Tip: Make sure you pay attention to whether the components are positive or negative — this tells you which direction the vector acts in.

Example 2

A ball's velocity is modelled by vector v = xi – yj ms^{-1}, with a magnitude of 4 ms^{-1} and direction of 40° below the positive x-axis. Find the x and y components of vector v, giving your answer to 4 s.f.

1. As always, draw a diagram.

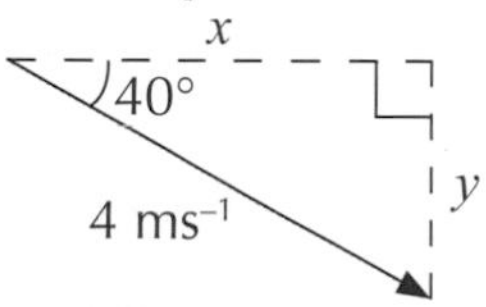

2. Use trigonometry to find x and y.

$\cos 40° = \frac{x}{4} \Rightarrow x = 4\cos 40° = 3.064$ to 4 s.f.

$\sin 40° = \frac{y}{4} \Rightarrow y = 4\sin 40° = 2.571$ to 4 s.f.

3. So the ball's velocity is modelled by: $\mathbf{v} = (3.064\mathbf{i} - 2.571\mathbf{j}) \text{ ms}^{-1}$

The effect of **two forces** working together can also be modelled by vectors.

These two vectors will probably form a triangle **without** a **right angle**, so you will need to use the **sine rule** and **cosine rule** (see p.136-138) for the trigonometry involved.

The size of a force is the magnitude of the vector that represents it.

Example 3

Two tug boats are pulling a ship with an angle of 30° between them. One tug boat exerts a force of 10 kN and is modelled with vector a. The other boat exerts a force of 12 kN and is modelled with vector b.

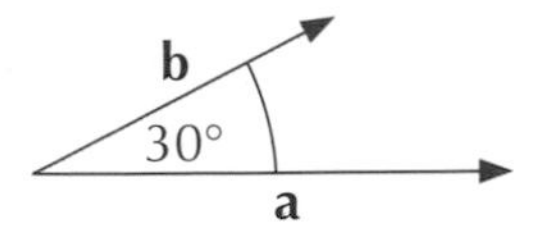

The resultant force on the ship, r, is the resultant vector of these two forces. Calculate the size of the resultant force.

1. Draw the vectors end to end and work out how to find **r**. $\mathbf{r} = \mathbf{a} + \mathbf{b}$
2. Work out the angle opposite **r**. $180° - 30° = 150°$
3. Use the cosine rule to find $|\mathbf{r}|$.

$$|\mathbf{r}|^2 = 10^2 + 12^2 - 2 \times 10 \times 12 \times \cos 150°$$
$$= 244 - 240 \cos 150°$$

So, $|\mathbf{r}| = \sqrt{244 - 240 \cos 150°} =$ **21.3 kN** (to 3 s.f.)

Vectors can also be used to model **lines** and the **sides of polygons** when investigating problems in geometry.

Example 4

The routes from Ayeside to Beesville ($\overrightarrow{AB}$) and to Ceeston ($\overrightarrow{AC}$) are modelled by the vectors p and q respectively. Xander's house lies between Beesville and Ceeston such that its position, X, divides the line BC in the ratio 2:5.

Find the vector $\overrightarrow{AX}$ in terms of p and q.

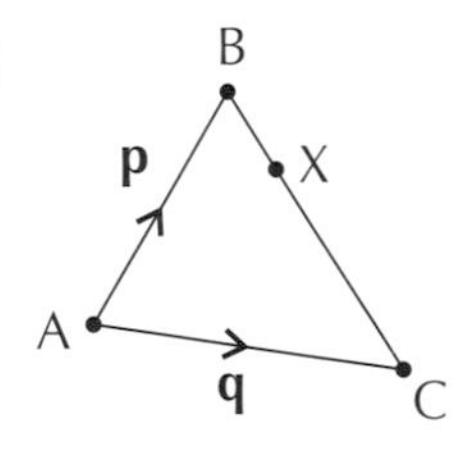

1. Work out how to find $\overrightarrow{AX}$. $\overrightarrow{AX} = \overrightarrow{AB} + \overrightarrow{BX}$
2. Find $\overrightarrow{BC}$ in terms of **p** and **q**. $\overrightarrow{BC} = \overrightarrow{BA} + \overrightarrow{AC} = -\overrightarrow{AB} + \overrightarrow{AC} = -\mathbf{p} + \mathbf{q}$
3. X divides BC in the ratio 2:5, so work out $\overrightarrow{BX}$ in terms of $\overrightarrow{BC}$. BX is $\frac{2}{2+5} = \frac{2}{7}$ of BC, so $\overrightarrow{BX} = \frac{2}{7}\overrightarrow{BC}$
4. You know $\overrightarrow{AB} = \mathbf{p}$, so plug all this into your equation for $\overrightarrow{AX}$.

$$\overrightarrow{AX} = \overrightarrow{AB} + \overrightarrow{BX} = \overrightarrow{AB} + \frac{2}{7}\overrightarrow{BC}$$
$$= \mathbf{p} + \frac{2}{7}(-\mathbf{p} + \mathbf{q}) = \frac{5}{7}\mathbf{p} + \frac{2}{7}\mathbf{q}$$

Tip: 'X divides BC in the ratio 2:5' means X is $\frac{2}{7}$ of the way from B to C. 'X divides CB in the ratio 2:5' would mean X is $\frac{2}{7}$ of the way from C to B.

You might also come across ratio questions **without** contexts — e.g. "P divides the vector $\overrightarrow{QR}$ in the ratio 1 : 3".

Example 5

The position vectors of the vertices of the parallelogram PQRS are:
$\overrightarrow{OP} = 2\mathbf{i} + 3\mathbf{j}$, $\overrightarrow{OQ} = 7\mathbf{i} + 4\mathbf{j}$, $\overrightarrow{OR} = 6(\mathbf{i} + \mathbf{j})$ and $\overrightarrow{OS} = \mathbf{i} + 5\mathbf{j}$.
What are the exact lengths of this parallelogram's diagonals?

1. Find the diagonals in $\mathbf{i} + \mathbf{j}$ form.

$$\overrightarrow{PR} = \overrightarrow{OR} - \overrightarrow{OP} = 6(\mathbf{i} + \mathbf{j}) - (2\mathbf{i} + 3\mathbf{j}) = 4\mathbf{i} + 3\mathbf{j}$$

$$\overrightarrow{SQ} = \overrightarrow{OQ} - \overrightarrow{OS} = (7\mathbf{i} + 4\mathbf{j}) - (\mathbf{i} + 5\mathbf{j}) = 6\mathbf{i} - \mathbf{j}$$

2. Use Pythagoras' theorem to find the lengths of the diagonals.

$$|\overrightarrow{PR}| = \sqrt{4^2 + 3^2} = \sqrt{25} = 5$$

$$|\overrightarrow{SQ}| = \sqrt{6^2 + (-1)^2} = \sqrt{37}$$

Vectors are also really useful for modelling the **direction** something is **travelling** in — like the course of a ship or a plane's flight path.

The **bearing** the vehicle travels on can be used to calculate the vector's **direction**. The **distance** it travels is the **magnitude** of its **displacement vector**, and its **speed** is the magnitude of its **velocity vector**.

Example 6

A ship travels 75 km on a bearing of 140°.
The ship's displacement is modelled by the vector $\mathbf{d} = \begin{pmatrix} x \\ y \end{pmatrix}$. Calculate x and y (to 2 d.p.).

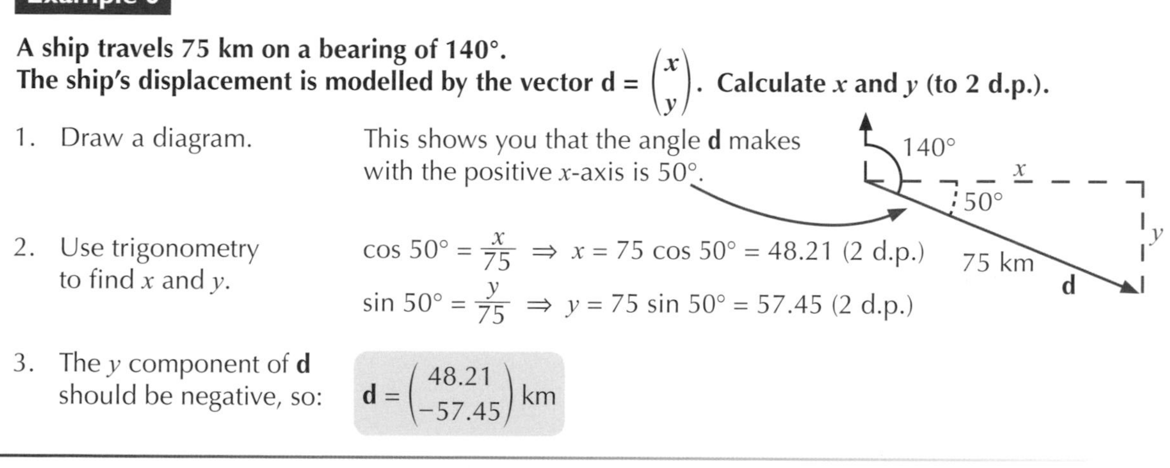

1. Draw a diagram. This shows you that the angle **d** makes with the positive x-axis is 50°.

2. Use trigonometry to find x and y.

$$\cos 50° = \frac{x}{75} \Rightarrow x = 75 \cos 50° = 48.21 \text{ (2 d.p.)}$$

$$\sin 50° = \frac{y}{75} \Rightarrow y = 75 \sin 50° = 57.45 \text{ (2 d.p.)}$$

3. The y component of **d** should be negative, so: $\mathbf{d} = \begin{pmatrix} 48.21 \\ -57.45 \end{pmatrix}$ km

Exercise 12.3.1

MODELLING

Q1 The acceleration of various particles are given by the vectors below. Calculate the exact magnitude of each particle's acceleration.

a) $\mathbf{a} = (\mathbf{i} + 2\mathbf{j})$ ms^{-2} b) $\mathbf{b} = (-\mathbf{i} - \mathbf{j})$ ms^{-2}

c) $\mathbf{c} = (3\mathbf{i} + 2\mathbf{j})$ ms^{-2} d) $\mathbf{d} = (-2\mathbf{i} + 3\mathbf{j})$ ms^{-2}

Q2 The quadrilateral ABCD on the right is used to model a garden. There is a straight path that crosses the garden. Its start point divides $\overrightarrow{AB}$ in the ratio 7:3, and it ends at the midpoint of $\overrightarrow{CD}$.

The path is modelled by vector $\mathbf{p} = \begin{pmatrix} x \\ y \end{pmatrix}$. Calculate x and y.

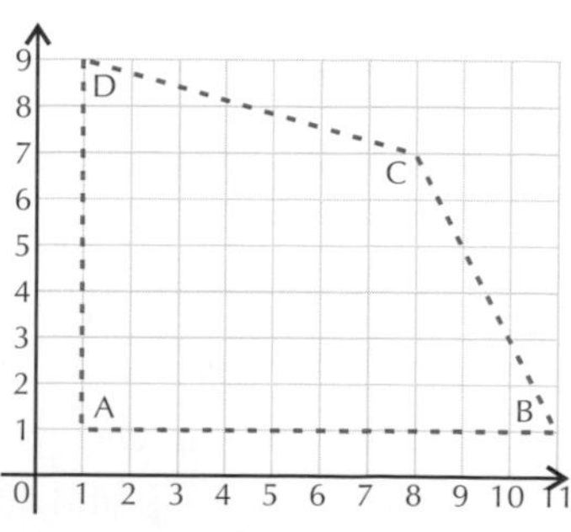

Q3 A simple mathematical model of a ball bouncing off the side of a pool table is constructed using vectors. The ball's velocity has vector $\mathbf{v}_1 = \begin{pmatrix} 3 \\ -2 \end{pmatrix}$ before it hits the side, and vector $\mathbf{v}_2 = \begin{pmatrix} 1 \\ 3 \end{pmatrix}$ afterwards. Find the acute angle θ between $\mathbf{v}_1$ and $\mathbf{v}_2$. Give your answer to 2 d.p.

Q4 The velocity of a fish swimming in the horizontal plane is modelled by a vector $\mathbf{v} = 2t(2\mathbf{i} - \mathbf{j})$ ms^{-1}, where t is time in seconds.

a) State the fish's velocity after 2 seconds.

b) Calculate the fish's exact speed after 3 seconds.

c) Find the direction of the vector $\mathbf{v}$.

Q5 A firework's displacement is modelled by the vector $\mathbf{s} = 2t^2(-\mathbf{i} + 7\mathbf{j})$ m, where t is time in seconds, $\mathbf{i}$ is horizontal and $\mathbf{j}$ is vertical (pointing upwards).

a) Find the firework's displacement after 1.5 seconds.

b) Calculate, to 3 s.f., the direction of the firework above the horizontal.

The firework explodes after 2.5 seconds.

c) What is the firework's vertical distance from the ground when it goes off?

Q6 A roller coaster is accelerating at 30 ms^{-2} at an angle of 35° above the negative horizontal. Express this acceleration as a vector of the form $\mathbf{a} = x\mathbf{i} + y\mathbf{j}$.

Q7 A park is modelled as a quadrilateral EFGH, with sides given by the vectors $\overrightarrow{EF} = 2\mathbf{i} + 3\mathbf{j}$, $\overrightarrow{FG} = \mathbf{i} - \frac{1}{2}\mathbf{j}$, $\overrightarrow{GH} = -\mathbf{i} - \frac{3}{2}\mathbf{j}$ and $\overrightarrow{HE} = -2\mathbf{i} - \mathbf{j}$. Show that the park is a trapezium.

Q8 An aircraft is attempting to fly due north at 600 km/h, but there is a wind from the west at 75 km/h. The aircraft's actual course is modelled by the resultant of these two vectors. Calculate:

a) the actual bearing the plane is flying on.

b) the aircraft's resultant speed in km/h (to 2 d.p.).

Q9 Two lumberjacks are pulling ropes attached to a tree. The angle between the ropes is 100°, and the lumberjacks exert forces of 250 N and 210 N respectively. The resultant force, $\mathbf{f}$, is modelled by the resultant of these two vectors. Calculate the magnitude of $\mathbf{f}$.

Q10 Ray is attempting to paddle his canoe due north at 4 ms^{-1}, but there is a current travelling west at 3 ms^{-1}. Ray's actual course is modelled by the resultant of these two vectors. Calculate:

a) the actual bearing Ray is travelling on.

b) Ray's resultant speed in ms^{-1}.

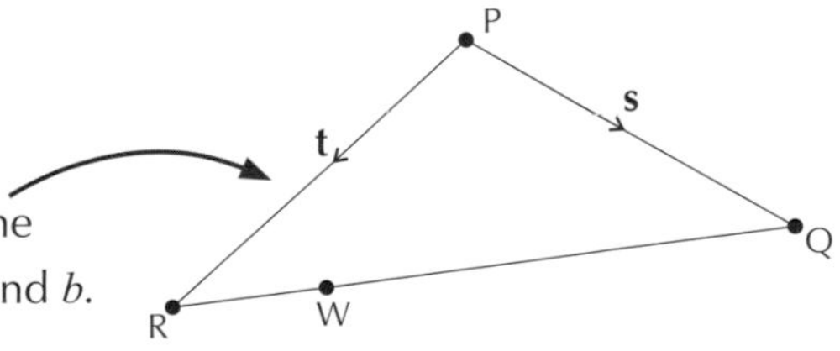

Q11 In the diagram to the right, W divides QR in the ratio $a:b$. Given that $\overrightarrow{PW} = \frac{5}{9}\mathbf{s} + \frac{4}{9}\mathbf{t}$, find a and b.

Q12 The position vectors of the vertices of the parallelogram PQRS are $\overrightarrow{OP} = 2\mathbf{i} + 3\mathbf{j}$, $\overrightarrow{OQ} = 7\mathbf{i} + 4\mathbf{j}$, $\overrightarrow{OR} = 6(\mathbf{i} + \mathbf{j})$ and $\overrightarrow{OS} = \mathbf{i} + 5\mathbf{j}$. What are the exact lengths of this parallelogram's diagonals?

Review Exercise

Q1 Vectors **a** and **b** are represented by the lines to the right. Draw and label sketches that represent the following vectors:

a) $-\mathbf{b}$ b) $3\mathbf{a}$ c) $\mathbf{a} + \mathbf{b}$ d) $2\mathbf{a} - \mathbf{b}$ e) $\mathbf{b} - \mathbf{a}$

Q2 Give two vectors that are parallel to each of the following:

a) $2\mathbf{a}$ b) $3\mathbf{i} + 4\mathbf{j}$ c) $3\mathbf{i} - \mathbf{j}$ d) $\begin{pmatrix} 3 \\ 5 \end{pmatrix}$

Q3 Show that the following sets of vectors are collinear:

a) $\overrightarrow{XY} = 2\mathbf{b} - \mathbf{a}$, $\overrightarrow{OY} = 2\mathbf{b}$, $\overrightarrow{OZ} = 2(\mathbf{a} - \mathbf{b})$

b) $\overrightarrow{OA} = 3\mathbf{p}$, $\overrightarrow{OB} = \mathbf{q}$, $\overrightarrow{OC} = 4\mathbf{q} - 9\mathbf{p}$.

Q4 Using the diagram on the right find these vectors in terms of vectors **a**, **b** and **c**.

a) $\overrightarrow{AB}$ b) $\overrightarrow{BA}$ c) $\overrightarrow{CB}$ d) $\overrightarrow{AC}$

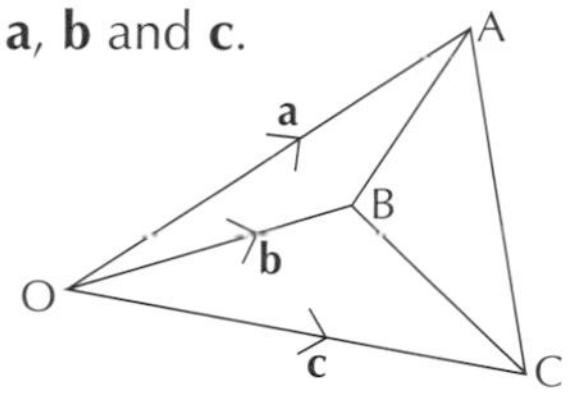

Q5 X is the point (6, –1) and Y is the point (–4, 7). Write the vectors $\overrightarrow{XO}$ and $\overrightarrow{YO}$ in **i** and **j** form and in column vector form.

Q6 Given that $\mathbf{d} = \begin{pmatrix} 3 \\ 2 \end{pmatrix}$, $\mathbf{e} = \begin{pmatrix} -1 \\ -2 \end{pmatrix}$, and $\mathbf{f} = \begin{pmatrix} 3 \\ -1 \end{pmatrix}$, find the following vectors:

a) $2\mathbf{d} - \mathbf{f}$ b) $\mathbf{d} - \mathbf{e} - \mathbf{f}$ c) $\mathbf{f} - 3\mathbf{d} + \mathbf{e}$ d) $-2\mathbf{e} + \mathbf{d} - \mathbf{f}$

Q7 R has position vector $\begin{pmatrix} 3 \\ -1 \end{pmatrix}$ and S has position vector $\begin{pmatrix} -5 \\ -7 \end{pmatrix}$. Find the magnitude of $\overrightarrow{RS}$.

Q8 If $\mathbf{p} = 5\mathbf{i} - 12\mathbf{j}$ and vector **q** is parallel to **p** with magnitude 65, find the vector **q**. PROBLEM SOLVING

Q9 A has position vector $\begin{pmatrix} 2 \\ -3 \end{pmatrix}$ and B has position vector $\begin{pmatrix} x \\ 1 \end{pmatrix}$. PROBLEM SOLVING

Given that $|\overrightarrow{AB}| = 5$, and the direction of $\overrightarrow{OB}$ is less than 90°, find x.

Q10 $\mathbf{v} = 4\mathbf{i} - 6\mathbf{j}$ and $\mathbf{u} = -\mathbf{i} - 3\mathbf{j}$. Find the angle between **v** and **u** to two decimal places.

Q11 A man wants to swim across a river. The current is flowing at 1.8 ms^{-1} parallel to the riverbank. Find the speed and direction that he needs to swim at for his resultant speed to be 1.2 ms^{-1} perpendicular to the riverbank. PROBLEM SOLVING MODELLING

Q12 A girl cycles along a bearing of 171° with speed 16 km/h. Find her velocity in terms of **i** and **j**, where **i** and **j** are the unit vectors directed due east and due north respectively. MODELLING

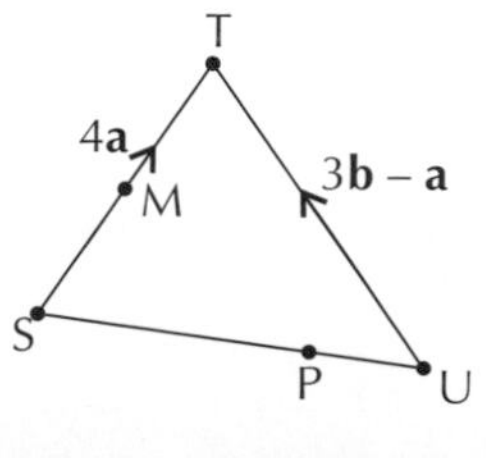

Q13 STU is a triangle, where $\overrightarrow{ST} = 4\mathbf{a}$ and $\overrightarrow{UT} = 3\mathbf{b} - \mathbf{a}$. Point M is the midpoint of line ST and point P divides line SU in the ratio 3 : 1. Find the vector $\overrightarrow{PM}$ in terms of **a** and **b**. PROBLEM SOLVING

Exam-Style Questions

Q1 Given $\mathbf{a} = -3\mathbf{i} + 4\mathbf{j}$ and $\mathbf{b} = \mathbf{i} + 2\mathbf{j}$:

a) Work out $-5\mathbf{a} + 2\mathbf{b}$.

[2 marks]

b) Find the magnitude and direction of **a**, giving your answer to 1 d.p. where necessary.

[4 marks]

Q2 Vector **c** is parallel to vector **d**. Given that $\mathbf{d} = 6\mathbf{i} - 9\mathbf{j}$ and $|\mathbf{c}| = \sqrt{13}$, find **c**.

[4 marks]

Q3 Given $\overrightarrow{OM} = -7\mathbf{i} + 6\mathbf{j}$ and $\overrightarrow{ON} = 3\mathbf{i} + 3\mathbf{j}$, find:

a) $\overrightarrow{MN}$

[2 marks]

b) $|\overrightarrow{MN}|$

[2 marks]

Q4 Two forces are acting on a particle P. The magnitude in newtons (N), and the direction of each force is shown on the diagram below.

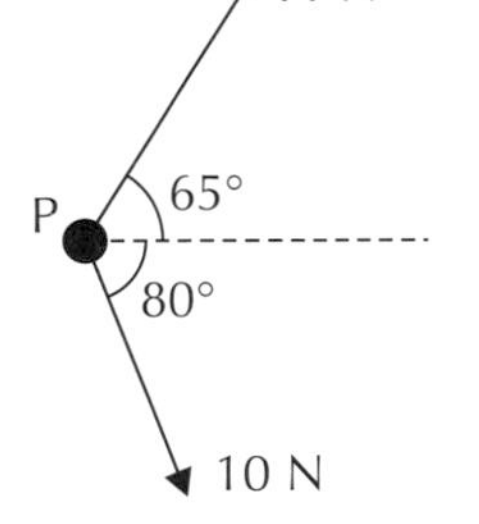

Find the magnitude of the resultant force, R.

[3 marks]

Q5 $PQRS$ is a quadrilateral where $\overrightarrow{OP} = \begin{pmatrix} -4 \\ -5 \end{pmatrix}$, $\overrightarrow{OQ} = \begin{pmatrix} -5 \\ 1 \end{pmatrix}$, $\overrightarrow{OR} = \begin{pmatrix} 1 \\ 5 \end{pmatrix}$ and $\overrightarrow{OS} = \begin{pmatrix} 8 \\ 3 \end{pmatrix}$.

a) Find the exact lengths of the diagonals of $PQRS$.

[3 marks]

b) Show that QR is parallel to PS, and hence show that $PQRS$ is a trapezium.

[4 marks]

Practice Paper

Q1 Given that $\mathbf{p} = \begin{pmatrix} -3 \\ 2 \end{pmatrix}$, $\mathbf{q} = \begin{pmatrix} 6 \\ 10 \end{pmatrix}$ and $\mathbf{r} = \begin{pmatrix} 3 \\ 2 \end{pmatrix}$, show that $3\mathbf{p} - 2\mathbf{q}$ is parallel to $\mathbf{r}$.

[2 marks]

Q2 a) A student says that $\sqrt{x^2} = x$ for all x.
Prove by counter-example that the student's statement is incorrect.

[1 mark]

b) Prove that when n is a positive even number,
$2n^2 + 2n + 6$ is never exactly divisible by 4.

[3 marks]

Q3 Differentiate $f(x) = 4x^2 - 4x + 3$ from first principles.

[4 marks]

Q4 Clearly showing your working, find the exact solution(s) of the equation:

$$2\log_5(2x - 1) = 1 + \log_5(3 - x)$$

[4 marks]

Q5 $p(x) = 2x^3 + ax^2 + bx + 18$ is exactly divisible by $(x - 3)$ and $(x + 2)$,
where a and b are integers.

a) By using the Factor Theorem, find the values of a and b.

[3 marks]

b) Hence, fully factorise $p(x)$.

[1 mark]

Practice Paper

Q6 a) Find the integral $\int \frac{9x^2 - 3x^3}{\sqrt{x^3}}\,dx$.

[3 marks]

The diagram shows part of the graph of $f(x) = \frac{9x^2 - 3x^3}{\sqrt{x^3}}$.

b) Given that $f(x)$ passes through the origin, find the exact value of the area of the shaded region bounded by the curve $f(x)$, the x-axis and the line $x = 2$. Give your answer in the form $p\sqrt{q}$, where p and q are rational numbers.

[3 marks]

Q7 A, B and C are the vertices of a triangle.
The position vectors of A and B are $\begin{pmatrix}1\\1\end{pmatrix}$ and $\begin{pmatrix}3\\7\end{pmatrix}$ respectively.

a) Find $\overrightarrow{AB}$ and $|\overrightarrow{AB}|$, giving your answer as a simplified surd where appropriate.

[3 marks]

b) Given that $|\overrightarrow{AC}| = 2\sqrt{3}$ and $|\overrightarrow{BC}| = 4$, find the angle at vertex A in degrees to one decimal place.

[2 marks]

c) Hence, find the area of the triangle ABC to one decimal place.

[2 marks]

Q8 The straight line l intersects the curve $f(x) = 2 - 6x + 8x^2 - 2x^3$ at two distinct points, P and Q. Given that l is a tangent to $f(x)$ at P and the coordinates of P are $(1, 2)$:

a) find the equation of line l,

[4 marks]

b) find the coordinates of point Q.

[3 marks]

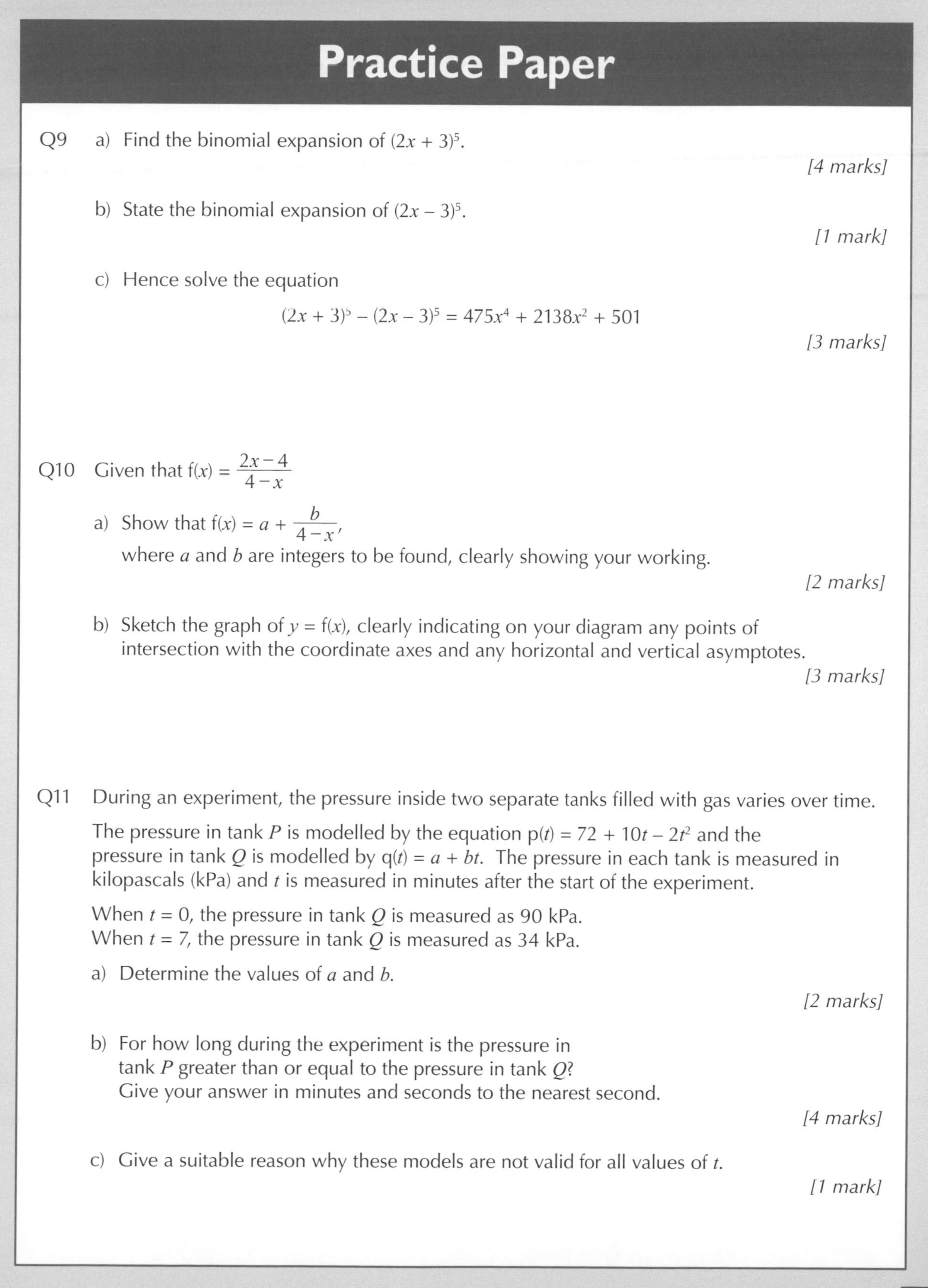

Practice Paper

Q9 a) Find the binomial expansion of $(2x + 3)^5$.

[4 marks]

b) State the binomial expansion of $(2x - 3)^5$.

[1 mark]

c) Hence solve the equation

$$(2x + 3)^5 - (2x - 3)^5 = 475x^4 + 2138x^2 + 501$$

[3 marks]

Q10 Given that $f(x) = \frac{2x-4}{4-x}$

a) Show that $f(x) = a + \frac{b}{4-x}$,
where a and b are integers to be found, clearly showing your working.

[2 marks]

b) Sketch the graph of $y = f(x)$, clearly indicating on your diagram any points of intersection with the coordinate axes and any horizontal and vertical asymptotes.

[3 marks]

Q11 During an experiment, the pressure inside two separate tanks filled with gas varies over time.

The pressure in tank P is modelled by the equation $p(t) = 72 + 10t - 2t^2$ and the pressure in tank Q is modelled by $q(t) = a + bt$. The pressure in each tank is measured in kilopascals (kPa) and t is measured in minutes after the start of the experiment.

When $t = 0$, the pressure in tank Q is measured as 90 kPa.
When $t = 7$, the pressure in tank Q is measured as 34 kPa.

a) Determine the values of a and b.

[2 marks]

b) For how long during the experiment is the pressure in tank P greater than or equal to the pressure in tank Q?
Give your answer in minutes and seconds to the nearest second.

[4 marks]

c) Give a suitable reason why these models are not valid for all values of t.

[1 mark]

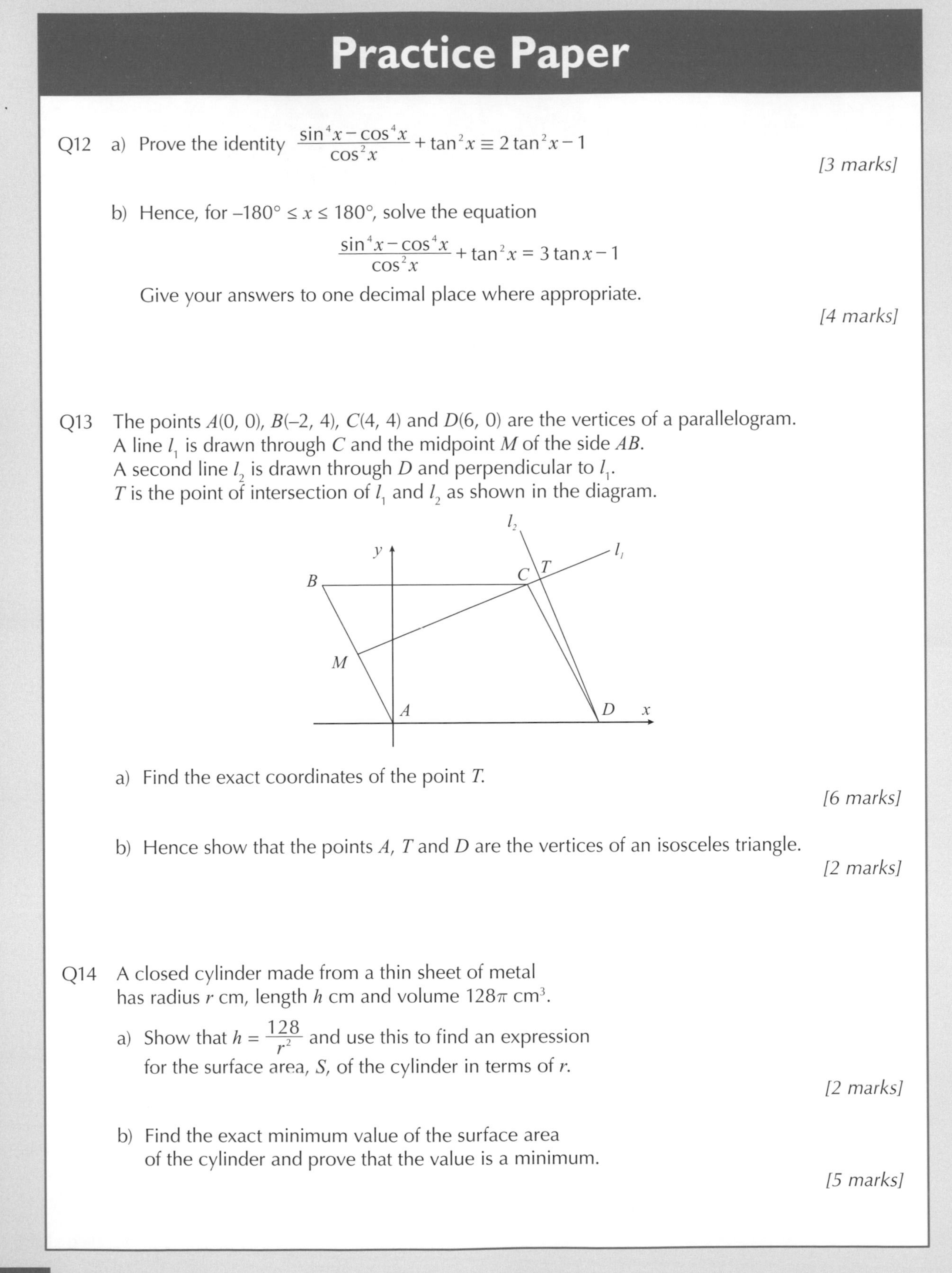

Practice Paper

Q12 a) Prove the identity $\frac{\sin^4 x - \cos^4 x}{\cos^2 x} + \tan^2 x \equiv 2\tan^2 x - 1$

[3 marks]

b) Hence, for $-180° \leq x \leq 180°$, solve the equation

$$\frac{\sin^4 x - \cos^4 x}{\cos^2 x} + \tan^2 x = 3\tan x - 1$$

Give your answers to one decimal place where appropriate.

[4 marks]

Q13 The points $A(0, 0)$, $B(-2, 4)$, $C(4, 4)$ and $D(6, 0)$ are the vertices of a parallelogram.
A line l_1 is drawn through C and the midpoint M of the side AB.
A second line l_2 is drawn through D and perpendicular to l_1.
T is the point of intersection of l_1 and l_2 as shown in the diagram.

a) Find the exact coordinates of the point T.

[6 marks]

b) Hence show that the points A, T and D are the vertices of an isosceles triangle.

[2 marks]

Q14 A closed cylinder made from a thin sheet of metal has radius r cm, length h cm and volume 128π cm^3.

a) Show that $h = \frac{128}{r^2}$ and use this to find an expression for the surface area, S, of the cylinder in terms of r.

[2 marks]

b) Find the exact minimum value of the surface area of the cylinder and prove that the value is a minimum.

[5 marks]

Practice Paper

Q15 The population of a rapidly expanding town is thought to be modelled by the equation $P = ab^t$, where t is measured in years and $t = 0$ corresponds to 1st January 1995. Some data relating to the population of the town is given in the table.

Year (1st January)	1995	2015
$\log_{10} P$	4.762	5.012

Using data from the table:

a) Write down an equation for $\log_{10} P$ in terms of t.

[3 marks]

b) Find the values of a and b, and the population predicted by the model for 1st January 2007 (to the nearest hundred).

[3 marks]

c) In which year did the population reach 100 000 according to the model?

[2 marks]

Q16 a) Explain why points $A(10, 4)$, $B(10, 10)$ and $C(2, 10)$ are the vertices of a right-angled triangle.

[1 mark]

b) Hence find the equation of the circle D_1 which passes through the points A, B and C.

[3 marks]

A second circle D_2 with the centre on the same vertical line as the centre of D_1 has equation $(x - 6)^2 + (y - 12)^2 = 10$. The circles D_1 and D_2 intersect at two distinct points.

c) Show that the point $M(9, 11)$ is a point of intersection of the circles and find the coordinates of the second point of intersection N.

[2 marks]

d) Show that the tangent lines to D_2 at the points M and N intersect on the circumference of the circle D_1.

[6 marks]

Answers

Chapter 2: Proof

2.1 Proof

Exercise 2.1.1 — Proof

Q1 **a)** Take two odd numbers $2l + 1$ and $2m + 1$ (where l and m are integers), then their sum is $2l + 1 + 2m + 1 = 2l + 2m + 2 = 2(l + m + 1) =$ even.

b) Take two even numbers, $2j$ and $2k$ (where j and k are integers), then their product is $2j \times 2k = 4jk = 2(2jk) =$ even.

c) Take one even number, $2l$ and one odd number $2m + 1$ (where l and m are integers), then their product is $2l \times (2m + 1) = 4lm + 2l = 2(2lm + l) =$ even.

Q2 Take one even number $2l$ and one odd number $2m + 1$ (where l and m are integers), then subtracting the odd number from the even number gives $2l - (2m + 1) = 2l - 2m - 1 = 2(l - m) - 1$
$= 2(l - m - 1) + 1$

Q3 E.g. Let $p = 1 \Rightarrow \frac{1}{p^2} = \frac{1}{p}$, so the statement is not true.
You could have also taken p to be a negative integer.

Q4 $(x + 5)^2 + 3(x - 1)^2 = x^2 + 10x + 25 + 3(x^2 - 2x + 1)$
$= x^2 + 10x + 25 + 3x^2 - 6x + 3$
$= 4x^2 + 4x + 28$
$= 4(x^2 + x + 7)$

This has a factor of 4 outside the brackets, so it is always divisible by 4.

Q5 Proof by exhaustion:
Take three consecutive integers $(n - 1)$, n and $(n + 1)$.
Their product is $(n - 1)n(n + 1) = n(n^2 - 1) = n^3 - n$.
Consider the two cases — n even and n odd.
For n even, n^3 is even (as even × even = even) so $n^3 - n$ is also even (as even – even = even). For n odd, n^3 is odd (as odd × odd = odd) so $n^3 - n$ is even (as odd – odd = even). So $n^3 - n$ is even when n is even and when n is odd, and n must be either odd or even, so the product of three consecutive integers is always even.
Another approach to this proof is to take the product of three consecutive integers n(n + 1)(n + 2) and consider n odd and n even. If n is odd: n(n + 1)(n + 2) = (odd × even) × odd = even × odd = even. If n is even: n(n + 1)(n + 2) = (even × odd) × even = even × even = even.

Q6 The simplest way to disprove the statement is to find a counter-example. Try some values of n and see if the statement is true for them:
$n = 3 \Rightarrow n^2 - n - 1 = 3^2 - 3 - 1 = 5$ — prime
$n = 4 \Rightarrow n^2 - n - 1 = 4^2 - 4 - 1 = 11$ — prime
$n = 5 \Rightarrow n^2 - n - 1 = 5^2 - 5 - 1 = 19$ — prime
$n = 6 \Rightarrow n^2 - n - 1 = 6^2 - 6 - 1 = 29$ — prime
$n = 7 \Rightarrow n^2 - n - 1 = 7^2 - 7 - 1 = 41$ — prime
$n = 8 \Rightarrow n^2 - n - 1 = 8^2 - 8 - 1 = 55$ — not prime
$n^2 - n - 1$ is not prime when $n = 8$. So the statement is false.
Sometimes trial and error is the easiest way to find a counter-example. Don't forget, if you've been told to disprove a statement like this, then a counter-example must exist.

Q7 Find a counter-example for which the statement isn't true.
Take $x = -1$ and $y = 2$. Then
$\sqrt{x^2 + y^2} = \sqrt{(-1)^2 + 2^2} = \sqrt{1 + 4} = \sqrt{5} = 2.236...$
and $x + y = -1 + 2 = 1$. $2.236... > 1$,
so the statement is not true.

Q8 Take any two rational numbers a and b. By the definition of rational numbers we know that $a = \frac{p}{q}$ and $b = \frac{r}{s}$ where p, q, r and s are integers, and q and s are non-zero.
So, the sum of a and b is $\frac{p}{q} + \frac{r}{s} = \frac{ps + rq}{qs}$.
ps and rq are the product of integers, so are also integers. This means $ps + rq$ is also an integer. qs is the product of non-zero integers, so must also be a non zero integer. This shows that $a + b$ is the quotient of two integers, and has a non-zero denominator, so by definition $a + b$ is rational.

Q9 Disproof by counter-example:
Zero is a rational number, but dividing by zero does not give a rational result.

Q10 **a)** Proof by exhaustion:
Consider the two cases — n even and n odd. Let n be even.
$n^2 - n = n(n - 1)$.
If n is even, $n - 1$ is odd so $n(n - 1)$ is even (as even × odd = even). This means that $n(n - 1) - 1$ is odd.
Let n be odd. If n is odd, $n - 1$ is even, so $n(n - 1)$ is even (as odd × even = even). This means that $n(n - 1) - 1$ is odd.
As any integer n has to be either odd or even, $n^2 - n - 1$ is odd for any value of n.

b) As $n^2 - n - 1$ is odd, $n^2 - n - 2$ is even.
The product of even numbers is also even, so as $(n^2 - n - 2)^3$ is the product of 3 even numbers, it will always be even.

Exam-Style Questions — Chapter 2

Q1 $f(x) = 2x^2 + 2x + 3 = 2x^2 + 2x + 2 + 1$
$= 2(x^2 + x + 1) + 1$
x is an integer $\Rightarrow x^2 + x + 1$ is an integer
$\Rightarrow 2(x^2 + x + 1)$ is even
$\Rightarrow 2(x^2 + x + 1) + 1$ is odd
So $f(x)$ is odd for all integer values of x.
[2 marks available — 1 mark for rearranging each expression into the form of an odd number, 1 mark for correct interpretation]
You could also do this by expressing x as a general odd or even integer and showing that f(x) is odd in both cases.

Q2 E.g. If $x = -2$, $x^3 = (-2)^3 = -8$
$-8 < -2$, so here $x^3 < x$.
So the statement is not true for all $x \neq 0$.
[2 marks available — 1 mark for using an appropriate x-value, 1 mark for correct interpretation]

Q3 $a^m = a \times a \times a \times ... \times a$ (m lots of a multiplied together)
$a^n = a \times a \times a... \times a$ (n lots of a multiplied together)
$a^m \times a^n = a \times a \times a ... \times a \times a \times a \times ... a$
There are $m + n$ lots of a multiplied together, so $a^m \times a^n = a^{m+n}$.
[3 marks available — 1 mark for expanding a^m and a^n, 1 mark for expanding a^{m+n}, 1 mark for correct reasoning]

Q4 Let the first odd integer be $2n + 1$, then the next consecutive odd integer is $2n + 3$. The sum of the squares of these two numbers is:
$(2n + 1)^2 + (2n + 3)^2 = 4n^2 + 4n + 1 + 4n^2 + 12n + 9$
$= 8n^2 + 16n + 10$
$= 2(4n^2 + 8n + 5)$
The expression in the brackets is an integer, and since it's multiplied by 2 the full expression must be even.
[4 marks available — 1 mark for appropriate expressions for consecutive odd numbers, 1 mark for expanding the sum of squares, 1 mark for simplifying the expression, 1 mark for correct interpretation]

Q5 Take one even number $2l$ and one odd number $2m + 1$ (where l and m are integers), then:
$(2l)^{2m+1} = 2l \times 2l \times 2l \times ... \times 2l$ (this is $2m + 1$ lots of $2l$)
$= 2(l \times 2l \times 2l \times ... \times 2l)$
The expression in the brackets is an integer, and since it's multiplied by 2 the full expression must be even.
[2 marks available — 1 mark for appropriate expressions for even and odd numbers, 1 mark for showing the solution is a multiple of two]

Q6 Take two rational numbers $a = \frac{p}{q}$ and $b = \frac{r}{s}$.
The difference between them is $\frac{p}{q} - \frac{r}{s} = \frac{ps - qr}{qs}$
p, q, r and s are all integers, so $ps - qr$ is an integer.
q and s are non-zero integers, so qs is a non-zero integer.
So $\frac{ps - qr}{qs}$ is a rational number.
[3 marks available — 1 mark for expressing a and b as algebraic fractions, 1 mark for finding an algebraic expression for their difference, 1 mark for correct interpretation]

Q7 $(4n + 1)^2 - (2n - 1) = 16n^2 + 8n + 1 - 2n + 1$
$= 16n^2 + 6n + 2 = 2(8n^2 + 3n + 1)$
The expression in the brackets is an integer, and since it's multiplied by 2 the full expression must be even.
[3 marks available — 1 mark for expanding the brackets, 1 mark for simplifying the expression, 1 mark for correct interpretation]

Q8 Let the odd integer $y = 2n + 1$.
The angles in a triangle add up to 180°, so:
$2y + x = 180 \Rightarrow 2(2n + 1) + x = 180$
$\Rightarrow x = 180 - 4n - 2 = 178 - 4n = 2(89 - 2n)$
The expression in the brackets is an integer, and since it's multiplied by 2 the full expression must be even.
[3 marks available — 1 mark for using a correct expression for an odd number in a suitable equation, 1 mark for rearranging into the form of an even number, 1 mark for correct interpretation]

Q9 E.g. If $x = 0$, $a = 1$ and $b = 2$:
$ax = 1 \times 0 = 0$ and $bx = 2 \times 0 = 0$, so $ax = bx$
But $1 \neq 2$, so $a \neq b$. So the statement is not correct.
[2 marks available — 1 mark for setting x, a and b as appropriate values, 1 mark for correct interpretation]

Chapter 3: Algebra

3.1 Algebraic Expressions

Exercise 3.1.1 — Expanding brackets

Q1
a) $5(x + 4) = 5x + (5 \times 4) = 5x + 20$
b) $a(4 - 2b) = 4a + (a \times -2b) = 4a - 2ab$
c) $-2(x^2 + y) = -2x^2 - 2y$
d) $6mn(m + 1) = 6mnm + 6mn = 6m^2n + 6mn$
e) $-4ht(t^2 - 2ht - 3h^3) = -4ht \times t^2 + (-4ht \times -2ht) + (-4ht \times -3h^3)$
$= -4ht^3 + 8h^2t^2 + 12h^4t$
f) $7z^2(2 + z) = 14z^2 + 7z^2z = 14z^2 + 7z^3$
g) $4(x + 2) + 3(x - 5) = 4x + 8 + 3x - 15 = 7x - 7$
h) $p(3p^2 - 2q) + (q + 4p^3) = (p \times 3p^2) + (p \times -2q) + q + 4p^3$
$= 3p^3 - 2pq + q + 4p^3 = 7p^3 - 2pq + q$
i) $7xy(x^2 + z^2) = (7xy \times x^2) + (7xy \times z^2) = 7x^3y + 7xyz^2$
Don't forget to simplify your answer if possible.

Q2
a) $(x + 5)(x - 3) = x^2 - 3x + 5x - 15 = x^2 + 2x - 15$
b) $(2z + 3)(3z - 2) = 6z^2 - 4z + 9z - 6 = 6z^2 + 5z - 6$
c) $(u + 8)^2 = (u + 8)(u + 8) = u^2 + 8u + 8u + 64 = u^2 + 16u + 64$
d) $(ab + cd)(ac + bd) = abac + abbd + cdac + cdbd$
$= a^2bc + ab^2d + ac^2d + bcd^2$
e) $(10 + f)(2f^2 - 3g) = 20f^2 - 30g + 2f^3 - 3fg$
f) $(7 + q)(7 - q) = 49 - 7q + 7q - q^2 = 49 - q^2$
g) $(2 - 3w)^2 = (2 - 3w)(2 - 3w) = 2^2 - 6w - 6w + 9w^2$
$= 4 - 12w + 9w^2$
h) $(4rs^2 + 3)^2 = (4rs^2 + 3)(4rs^2 + 3) = 16r^2s^4 + 12rs^2 + 12rs^2 + 9$
$= 16r^2s^4 + 24rs^2 + 9$
i) $(5k^2l - 2kn)^2 = (5k^2l - 2kn)(5k^2l - 2kn)$
$= 25k^4l^2 - 10k^3nl - 10k^3nl + 4k^2n^2$
$= 25k^4l^2 - 20k^3ln + 4k^2n^2$
In parts c), g), h) and i), you could get straight to the answer by using $(a + b)^2 = a^2 + 2ab + b^2$.

Q3
a) $(l + 5)(l^2 + 2l + 3) = l(l^2 + 2l + 3) + 5(l^2 + 2l + 3)$
$= l^3 + 2l^2 + 3l + 5l^2 + 10l + 15$
$= l^3 + 7l^2 + 13l + 15$
b) $(2 + q)(3 - q + 4q^2) = 2(3 - q + 4q^2) + q(3 - q + 4q^2)$
$= 6 - 2q + 8q^2 + 3q - q^2 + 4q^3 = 6 + q + 7q^2 + 4q^3$
c) $(m + 1)(m + 2)(m - 4) = (m^2 + 2m + m + 2)(m - 4)$
$= (m^2 + 3m + 2)(m - 4) = m^2(m - 4) + 3m(m - 4) + 2(m - 4)$
$= m^3 - 4m^2 + 3m^2 - 12m + 2m - 8 = m^3 - m^2 - 10m - 8$
d) $(r + s)^3 = (r + s)(r + s)(r + s) = (r^2 + rs + sr + s^2)(r + s)$
$= (r^2 + 2rs + s^2)(r + s) = r^2(r + s) + 2rs(r + s) + s^2(r + s)$
$= r^3 + r^2s + 2r^2s + 2rs^2 + rs^2 + s^3 = r^3 + 3r^2s + 3rs^2 + s^3$
e) $(3x + 2)(x - 4)(2x + 1) = (3x^2 - 12x + 2x - 8)(2x + 1)$
$= (3x^2 - 10x - 8)(2x + 1)$
$= 3x^2(2x + 1) - 10x(2x + 1) - 8(2x + 1)$
$= 6x^3 + 3x^2 - 20x^2 - 10x - 16x - 8$
$= 6x^3 - 17x^2 - 26x - 8$
f) $(4 + x + y)(1 - x - y) = 4(1 - x - y) + x(1 - x - y) + y(1 - x - y)$
$= 4 - 4x - 4y + x - x^2 - xy + y - xy - y^2$
$= 4 - 3x - 3y - 2xy - x^2 - y^2$
g) $(j + 2k - 3)(j^2 + 2j + 1)$
$= j(j^2 + 2j + 1) + 2k(j^2 + 2j + 1) - 3(j^2 + 2j + 1)$
$= j^3 + 2j^2 + j + 2j^2k + 4jk + 2k - 3j^2 - 6j - 3$
$= j^3 - j^2 - 5j + 2j^2k + 4jk + 2k - 3$
h) $(2c^2 - cd + d)(2d - c - 5c^2)$
$= 2c^2(2d - c - 5c^2) - cd(2d - c - 5c^2) + d(2d - c - 5c^2)$
$= 4c^2d - 2c^3 - 10c^4 - 2cd^2 + c^2d + 5c^3d + 2d^2 - dc - 5c^2d$
$= -10c^4 - 2c^3 + 5c^3d - 2cd^2 - cd + 2d^2$
i) $(2f^3 - 4f - 1)(f^2 + 3f + 2)$
$= 2f^3(f^2 + 3f + 2) - 4f(f^2 + 3f + 2) - (f^2 + 3f + 2)$
$= 2f^5 + 6f^4 + 4f^3 - 4f^3 - 12f^2 - 8f - f^2 - 3f - 2$
$= 2f^5 + 6f^4 - 13f^2 - 11f - 2$

Q4 Radius $= 3h - 2$ cm
Volume $= \pi r^2h = \pi(3h - 2)^2h = \pi h(9h^2 - 12h + 4)$
$= 9\pi h^3 - 12\pi h^2 + 4\pi h$ cm³

Q5 Carole's garden: $x \times x = x^2$ m²
Mark's garden: $(x + 3)(2x + 1) = 2x^2 + x + 6x + 3$
$= 2x^2 + 7x + 3$ m²
Difference: $2x^2 + 7x + 3 - x^2 = x^2 + 7x + 3$ m²
For Mark's garden, use the information given to find the length of the sides. Then multiply the brackets and expand.

Exercise 3.1.2 — Factorising

Q1
a) $9k + 15l = (3 \times 3k) + (3 \times 5l) = 3(3k + 5l)$
b) $u^2 - uv = u(u - v)$
c) $2x^2y - 12xy^2 = (2xy \times x) - (2xy \times 6y) = 2xy(x - 6y)$
d) $f^2g^2 - fg = (fg \times fg) - (fg \times 1) = fg(fg - 1)$
e) $p^3 + 3pq^3 + 2p = (p \times p^2) + (p \times 3q^3) + (p \times 2)$
$= p(p^2 + 3q^3 + 2)$
f) $mnp^2 + 7m^2np^3 = (mnp^2 \times 1) + (mnp^2 \times 7mp) = mnp^2(1 + 7mp)$
g) $2ab^4 + 3a^3b^2 - 4ab = (ab \times 2b^3) + (ab \times 3a^2b) - (ab \times 4)$
$= ab(2b^3 + 3a^2b - 4)$
h) $36xyz - 8x^2z^2 + 20y^2z^2 = (4z \times 9xy) - (4z \times 2x^2z) + (4z \times 5y^2z)$
$= 4z(9xy - 2x^2z + 5y^2z)$

Q2
a) $x^2 - y^2 = (x + y)(x - y)$
This is just using the formula for the 'difference of two squares'.
b) $9a^2 - 4b^2 = (3a)^2 - (2b)^2 = (3a + 2b)(3a - 2b)$
c) $25x^2 - 49z^2 = (5x)^2 - (7z)^2 = (5x + 7z)(5x - 7z)$
d) $a^2c - 16b^2c = c(a^2 - 16b^2) = c(a^2 - (4b)^2) = c(a + 4b)(a - 4b)$

e) $y^2 - 2 = y^2 - (\sqrt{2})^2 = (y+\sqrt{2})(y-\sqrt{2})$

f) $m^2 - 11 = m^2 - (\sqrt{11})^2 = (m+\sqrt{11})(m-\sqrt{11})$

g) $4x^2 - 3 = (2x)^2 - (\sqrt{3})^2 = (2x+\sqrt{3})(2x-\sqrt{3})$

h) $7p^2 - 13 = (\sqrt{7}p)^2 - (\sqrt{13})^2 = (\sqrt{7}p+\sqrt{13})(\sqrt{7}p-\sqrt{13})$

Parts e) - h) use the fact that any number can be written as its square root squared.

Q3 a) $(4-z)^2(2-z) + p(2-z) = (2-z)[(4-z)^2 + p]$

b) $(r-d)^3 + 5(r-d)^2 = (r-d)^2[(r-d)+5] = (r-d)^2(r-d+5)$

c) $(b+c)^5(a+b) - (b+c)^5 = (b+c)^5[(a+b)-1]$
$= (b+c)^5(a+b-1)$

d) $l^2m(a-2x) + rp^2(2x-a) = l^2m(a-2x) + rp^2(-(a-2x))$
$= l^2m(a-2x) - rp^2(a-2x)$
$= (a-2x)(l^2m - rp^2)$

You might have factorised this slightly differently and ended up with $(2x-a)(rp^2 - l^2m)$ instead.

Q4 a) $(p+q)^2 + 2q(p+q) = (p+q)[(p+q)+2q] = (p+q)(p+3q)$

b) $2(2x-y)^2 - 6x(2x-y) = 2(2x-y)[(2x-y)-3x]$
$= -2(2x-y)(x+y)$

c) $(l+w+h)^2 - l(l+w+h) = (l+w+h)[(l+w+h)-l]$
$= (l+w+h)(w+h)$

Q5 a) $(m+5)(m^2-5m+25) = m(m^2-5m+25) + 5(m^2-5m+25)$
$= m^3 - 5m^2 + 25m + 5m^2 - 25m + 125 = m^3 + 125$

b) $(p-2q)(p^2+2pq+4q^2)$
$= p(p^2+2pq+4q^2) - 2q(p^2+2pq+4q^2)$
$= p^3 + 2p^2q + 4pq^2 - 2p^2q - 4pq^2 - 8q^3 = p^3 - 8q^3$

Parts a) and b) were likely to need the brackets expanding because the quadratic in the second bracket won't factorise.

c) $(u-v)(u+v) - (u+v)^2 = (u+v)[(u-v)-(u+v)]$
$= (u+v)(-2v) = -2v(u+v)$

d) $(c+d)^3 - c(c+d)^2 - d(c+d)^2 = (c+d)^2[(c+d)-c-d]$
$= (c+d)^2(0) = 0$

Exercise 3.1.3 — Algebraic fractions

Q1 a) The common denominator is 3×4: $\frac{x}{3} + \frac{x}{4} = \frac{4x}{12} + \frac{3x}{12} = \frac{7x}{12}$

b) The common denominator is t^2: $\frac{2}{t} + \frac{13}{t^2} = \frac{2t}{t^2} + \frac{13}{t^2} = \frac{2t+13}{t^2}$

c) The common denominator is $2 \times p \times 5 \times q = 10pq$:

$$\frac{1}{2p} - \frac{1}{5q} = \frac{5q}{10pq} - \frac{2p}{10pq} = \frac{5q-2p}{10pq}$$

d) 12 is the lowest common multiple of 3, 2 and 4, so the common denominator is $12h$.

$$\frac{2}{3h} + \frac{1}{2h} - \frac{3}{4h} = \frac{8}{12h} + \frac{6}{12h} - \frac{9}{12h} = \frac{5}{12h}$$

e) The common denominator is $a \times b \times c$:

$$\frac{ab}{c} + \frac{bc}{a} + \frac{ca}{b} = \frac{abab}{abc} + \frac{bcbc}{abc} + \frac{caca}{abc}$$
$$= \frac{a^2b^2 + b^2c^2 + c^2a^2}{abc}$$

f) The common denominator is mn:

$$\frac{2}{mn} - \frac{3m}{n} + \frac{n^2}{m} = \frac{2}{mn} - \frac{3m^2}{mn} + \frac{n^3}{mn}$$
$$= \frac{2 - 3m^2 + n^3}{mn}$$

g) The common denominator is $a^3 \times b^3 = a^3b^3$:

$$\frac{2}{ab^3} - \frac{9}{a^3b} = \frac{2a^2}{a^3b^3} - \frac{9b^2}{a^3b^3} = \frac{2a^2 - 9b^2}{a^3b^3}$$

h) The common denominator is $x^2 \times y = x^2y$

$$\frac{1}{x} + \frac{2x}{y} + \frac{4}{x^2} = \frac{xy}{x^2y} + \frac{2x^3}{x^2y} + \frac{4y}{x^2y} = \frac{xy + 2x^3 + 4y}{x^2y}$$

i) The common denominator is $a^2 \times b = a^2b$

$$2 + \frac{a^2}{b} - \frac{2b}{a^2} = \frac{2a^2b}{a^2b} + \frac{a^4}{a^2b} - \frac{2b^2}{a^2b} = \frac{2a^2b + a^4 - 2b^2}{a^2b}$$

Q2 a) The common denominator is $(y-1)(y-2)$:

$$\frac{5}{y-1} + \frac{3}{y-2} = \frac{5(y-2)}{(y-1)(y-2)} + \frac{3(y-1)}{(y-1)(y-2)}$$
$$= \frac{5(y-2)+3(y-1)}{(y-1)(y-2)}$$
$$= \frac{5y-10+3y-3}{(y-1)(y-2)}$$
$$= \frac{8y-13}{(y-1)(y-2)}$$

b) The common denominator is $(r-5)(r+3)$:

$$\frac{7}{r-5} - \frac{4}{r+3} = \frac{7(r+3)}{(r-5)(r+3)} - \frac{4(r-5)}{(r-5)(r+3)}$$
$$= \frac{7(r+3)-4(r-5)}{(r-5)(r+3)}$$
$$= \frac{7r+21-4r+20}{(r-5)(r+3)}$$
$$= \frac{3r+41}{(r-5)(r+3)}$$

c) The common denominator is $p(p-3)$:

$$\frac{8}{p} - \frac{1}{p-3} = \frac{8(p-3)}{p(p-3)} - \frac{p}{p(p-3)}$$
$$= \frac{8p-24-p}{p(p-3)} = \frac{7p-24}{p(p-3)}$$

d) The common denominator is $2(w-2)(w-7)$:

$$\frac{w}{2(w-2)} + \frac{3w}{w-7}$$
$$= \frac{w(w-7)}{2(w-2)(w-7)} + \frac{3w \times 2(w-2)}{2(w-2)(w-7)}$$
$$= \frac{w^2-7w}{2(w-2)(w-7)} + \frac{6w(w-2)}{2(w-2)(w-7)}$$
$$= \frac{w^2-7w+6w(w-2)}{2(w-2)(w-7)}$$
$$= \frac{w^2-7w+6w^2-12w}{2(w-2)(w-7)}$$
$$= \frac{7w^2-19w}{2(w-2)(w-7)} = \frac{w(7w-19)}{2(w-2)(w-7)}$$

e) The common denominator is $(z+2)(z+4)$:

$$\frac{z+1}{z+2} - \frac{z+3}{z+4} = \frac{(z+1)(z+4)}{(z+2)(z+4)} - \frac{(z+2)(z+3)}{(z+2)(z+4)}$$
$$= \frac{(z+1)(z+4)-(z+2)(z+3)}{(z+2)(z+4)}$$
$$= \frac{(z^2+5z+4)-(z^2+5z+6)}{(z+2)(z+4)}$$
$$= \frac{-2}{(z+2)(z+4)}$$

f) The common denominator is $(q+1)(q-2)$:

$$\frac{1}{q+1} + \frac{3}{q-2} = \frac{(q-2)}{(q+1)(q-2)} + \frac{3(q+1)}{(q+1)(q-2)}$$
$$= \frac{(q-2)+3(q+1)}{(q+1)(q-2)}$$
$$= \frac{q-2+3q+3}{(q+1)(q-2)}$$
$$= \frac{4q+1}{(q+1)(q-2)}$$

g) The common denominator is $(x+z)(x-z)$:

$$\frac{x}{x+z} + \frac{2z}{x-z} = \frac{x(x-z)}{(x+z)(x-z)} + \frac{2z(x+z)}{(x+z)(x-z)}$$
$$= \frac{x^2 - xz + 2xz + 2z^2}{(x+z)(x-z)} = \frac{x^2 + xz + 2z^2}{(x+z)(x-z)}$$

h) The common denominator is $(2x + 3)(3 - x)$:

$$\frac{y}{2x+3}-\frac{2y}{3-x}=\frac{y(3-x)}{(2x+3)(3-x)}-\frac{2y(2x+3)}{(2x+3)(3-x)}$$

$$=\frac{3y-xy-4xy-6y}{(2x+3)(3-x)}$$

$$=\frac{-3y-5xy}{(2x+3)(3-x)}=\frac{3y+5xy}{(2x+3)(x-3)}$$

In the final step, top and bottom have been multiplied by −1 to make the answer a bit neater.

i) The common denominator is $r(r-4)(r+1)$: $\frac{5}{r-4}+\frac{3}{r}-\frac{r}{r+1}$

$$=\frac{5r(r+1)}{r(r-4)(r+1)}+\frac{3(r-4)(r+1)}{r(r-4)(r+1)}-\frac{r^2(r-4)}{r(r-4)(r+1)}$$

$$=\frac{5r^2+5r+3(r^2+r-4r-4)-(r^3-4r^2)}{r(r-4)(r+1)}$$

$$=\frac{5r^2+5r+3r^2-9r-12-r^3+4r^2}{r(r-4)(r+1)}=\frac{-r^3+12r^2-4r-12}{r(r-4)(r+1)}$$

Q3 a) $\frac{2x+10}{6}=\frac{2(x+5)}{6}=\frac{x+5}{3}$

b) $\frac{6a-12b-15c}{3}=\frac{3(2a-4b-5c)}{3}=2a-4b-5c$

c) $\frac{np^2-2n^2p}{np}=\frac{np(p-2n)}{np}=p-2n$

d) $\frac{4st+6s^2t+9s^3t}{2t}=\frac{st(4+6s+9s^2)}{2t}$

$=\frac{s(4+6s+9s^2)}{2}$

e) $\frac{10yz^3-40y^3z^3+60y^2z^3}{10z^2}=\frac{10yz^3(1-4y^2+6y)}{10z^2}$

$=yz(1-4y^2+6y)$

f) $\frac{12cd-6c^2d+3c^3d^2}{12c^2de}=\frac{3cd(4-2c+c^2d)}{12c^2de}$

$=\frac{4-2c+c^2d}{4ce}$

g) $\frac{2x+x^2y-x^2}{x^2+3x}=\frac{x(2+xy-x)}{x(x+3)}=\frac{2+xy-x}{x+3}$

h) $\frac{2w^3+14w^2}{w^2-49}=\frac{2w^2(w+7)}{(w+7)(w-7)}=\frac{2w^2}{w-7}$

i) $\frac{4g^2-4h^2}{g^2+gh}=\frac{4(g^2-h^2)}{g(g+h)}=\frac{4(g+h)(g-h)}{g(g+h)}=\frac{4(g-h)}{g}$

Q4 Maya's pieces are $\frac{10}{x}$ cm long. Hal's pieces are $\frac{15}{x+3}$ cm long.

So the total length of one of Maya's pieces and one of Hal's pieces is: $\frac{10}{x}+\frac{15}{x+3}=\frac{10(x+3)}{x(x+3)}+\frac{15x}{x(x+3)}$

$=\frac{10x+30+15x}{x(x+3)}=\frac{25x+30}{x(x+3)}=\frac{5(5x+6)}{x(x+3)}$ cm

3.2 Laws of Indices

Exercise 3.2.1 — Laws of indices

Q1 a) $10\times10^4=10^{1+4}=10^5$

b) $y^{-1}\times y^{-2}\times y^7=y^{-1-2+7}=y^4$

c) $5^{\frac12}\times5^3\times5^{-\frac32}=5^{\frac12+3-\frac32}=5^2$

d) $6^5\div6^2=6^{5-2}=6^3$

e) $3^4\div3^{-1}=3^{4-(-1)}=3^{4+1}=3^5$

f) $\frac{6^{11}}{6}=6^{11-1}=6^{10}$

g) $\frac{r^2}{r^6}=r^{2-6}=r^{-4}$

h) $(3^2)^3=3^{2\times3}=3^6$

i) $(k^{-2})^5=k^{(-2)\times5}=k^{-10}$

j) $(z^4)^{-\frac18}=z^{4\times(-\frac18)}=z^{-\frac48}=z^{-\frac12}$

k) $(8^{-6})^{-\frac12}=8^{-6\times-\frac12}=8^{\frac62}=8^3$

l) $\frac{p^5q^4}{p^4q}=(p^{5-4})(q^{4-1})=p^1q^3=pq^3$

m) $\frac{c^{-1}d^{-2}}{c^2d^4}=c^{-1-2}d^{-2-4}=c^{-3}d^{-6}=\frac{1}{c^3d^6}$

n) $(ab^2)^2=(a)^2(b^2)^2=a^2b^{2\times2}=a^2b^4$

o) $\frac{12yz^{-\frac12}}{4yz^{\frac12}}=\left(\frac{12}{4}\right)(y^{1-1})(z^{-\frac12-\frac12})=3y^0z^{-1}=\frac{3}{z}$

p) $(mn^{\frac12})^4=m^4n^{\frac12\times4}=m^4n^2$

Q2 a) $4^{\frac12}\times4^{\frac32}=4^{\frac12+\frac32}=4^2=16$

b) $\frac{2^3\times2}{2^5}=\frac{2^{3+1}}{2^5}=\frac{2^4}{2^5}=2^{4-5}=2^{-1}=\frac12$

c) $\frac{7^5\times7^3}{7^6}=\frac{7^{5+3}}{7^6}=\frac{7^8}{7^6}=7^{8-6}=7^2=49$

d) $\frac{6^4}{6^{\frac54}\times6^{\frac34}}=\frac{6^4}{6^{\frac54+\frac34}}=\frac{6^4}{6^{\frac84}}=\frac{6^4}{6^2}=6^{4-2}=6^2=36$

e) $(3^2)^5\div(3^3)^3=3^{2\times5}\div3^{3\times3}$

$=3^{10}\div3^9=3^{10-9}=3^1=3$

f) $(4^{-\frac12})^2\times(4^{-3})^{-\frac13}=4^{-\frac12\times2}\times4^{(-3)\times(-\frac13)}$

$=4^{-1}\times4^1=4^{-1+1}=4^0=1$

g) $\frac{(2^{\frac12})^6\times(2^{-2})^{-2}}{(2^{-1})^{-1}}=\frac{2^{\frac12\times6}\times2^{(-2)\times(-2)}}{2^{(-1)\times(-1)}}=\frac{2^3\times2^4}{2^1}=\frac{2^{3+4}}{2^1}$

$=\frac{2^7}{2^1}=2^6=64$

h) $1^0=1$

i) $\left(\frac45\right)^0=1$

j) $(-5.726324)^0=1$

k) $8.374936^1=8.374936$

l) $\frac{(3^3)^2}{(9^{\frac12})^4\times(9^{\frac14})^8}=\frac{3^{3\times2}}{9^{\frac12\times4}\times9^{\frac14\times8}}=\frac{3^6}{9^2\times9^2}$

$=\frac{3^6}{9^{2+2}}=\frac{3^6}{9^4}=\frac{(3^2)^3}{9^4}=\frac{9^3}{9^4}=9^{3-4}=9^{-1}=\frac19$

Q3 a) $\frac1p=p^{-1}$

b) $\frac{5}{y^4}=5y^{-4}$

c) $\sqrt{q}=q^{\frac12}$

d) $\sqrt{r^3}=(r^3)^{\frac12}=r^{3\times\frac12}=r^{\frac32}$

e) $\sqrt[4]{s^5}=(s^5)^{\frac14}=s^{5\times\frac14}=s^{\frac54}$

f) $\frac{1}{\sqrt[3]{t}}=\frac{1}{(t^{\frac13})}=t^{-\frac13}$

g) $\left(\frac{1}{\sqrt[3]{x}}\right)^4=\left(\frac{1}{x^{\frac13}}\right)^4=(x^{-\frac13})^4=x^{-\frac13\times4}=x^{-\frac43}$

h) $\frac{\sqrt{z}}{z^3}=\frac{z^{\frac12}}{z^3}=z^{\frac12-3}=z^{-\frac52}$

Q4 a) $9^{\frac12}=\sqrt9=3$

b) $8^{\frac13}=\sqrt[3]{8}=2$

c) $4^{\frac32}=4^{\frac12\times3}=(4^{\frac12})^3=(\sqrt4)^3=(2)^3=8$

d) $27^{-\frac13}=\frac{1}{27^{\frac13}}=\frac{1}{\sqrt[3]{27}}=\frac13$

e) $16^{-\frac34}=\frac{1}{16^{\frac34}}=\frac{1}{(16^{\frac14})^3}=\frac{1}{(\sqrt[4]{16})^3}=\frac{1}{(2)^3}=\frac18$

f) $125^{\frac23}=125^{\frac13\times2}=(125^{\frac13})^2=(\sqrt[3]{125})^2=5^2=25$

g) $81^{\frac14}=\sqrt[4]{81}=3$

h) $64^{\frac12}\times64^{-\frac13}=64^{\frac12}\times\frac{1}{64^{\frac13}}=\sqrt{64}\times\frac{1}{\sqrt[3]{64}}=8\times\frac14=2$

Q5 Taking the cube root gives: $x^{\frac13}$

Raising this to the power 6 gives: $(x^{\frac13})^6=x^{\frac13\times6}=x^2$

Dividing by x gives: $x^2\div x=x$

Q6 a) $p^{\frac12}=\left(\frac1{16}q^2\right)^{\frac12}=\left(\frac1{16}\right)^{\frac12}(q^2)^{\frac12}$

$=\sqrt{\frac1{16}}q^{2\times\frac12}=\frac{1}{\sqrt{16}}q=\frac14q$

b) $2p^{-1}=2\left(\frac1{16}q^2\right)^{-1}=2\left(\frac1{16}\right)^{-1}(q^2)^{-1}$

$=2\times16\times q^{2\times-1}=32q^{-2}=\frac{32}{q^2}$

c) $p^{\frac{1}{2}} \div 2p^{-1} = \frac{1}{4}q \div \left(\frac{32}{q^2}\right) = \frac{1}{4}q \times \frac{q^2}{32}$

$= \frac{q^3}{128} = \frac{1}{128}q^3$

Substitute the answers from a) and b) into the division and use the rule for dividing by fractions. You could also simplify the expression to $\frac{1}{2}p^{\frac{3}{2}}$ first.

d) $p^2q = \left(\frac{1}{16}q^2\right)^2 q = \frac{1}{256}q^4 \times q = \frac{1}{256}q^5$

e) $\frac{4p}{q^3} = \frac{4\left(\frac{1}{16}q^2\right)}{q^3} = \frac{\frac{1}{4}q^2}{q^3} = \frac{1}{4q}$

f) $\frac{q^2}{4p^2} = \frac{q^2}{4\left(\frac{1}{16}q^2\right)^2} = \frac{q^2}{4\left(\frac{1}{256}q^4\right)} = \frac{q^2}{\frac{1}{64}q^4} = \frac{64}{q^2}$

Q7 Start by writing b as a power of 8: $b = 64^3 = (8^2)^3 = 8^6$

Then: $\frac{a^{-5}}{b^{\frac{1}{2}} \times c^4} = \frac{(8^{-3})^{-5}}{(8^6)^{\frac{1}{2}} \times (8^{\frac{1}{2}})^4} = \frac{8^{15}}{8^3 \times 8^2} = \frac{8^{15}}{8^{3+2}}$

$= \frac{8^{15}}{8^5} = 8^{15-5} = 8^{10}$

Q8 a) Write the RHS as 4 to the power 'something':

$\sqrt[3]{16} = (16)^{\frac{1}{3}} = (4^2)^{\frac{1}{3}} = 4^{2\times\frac{1}{3}} = 4^{\frac{2}{3}}$, so $x = \frac{2}{3}$.

b) Write the RHS as 9 to the power 'something':

$\frac{1}{3} = \frac{1}{\sqrt{9}} = \frac{1}{9^{\frac{1}{2}}} = 9^{-\frac{1}{2}}$, so $x = -\frac{1}{2}$.

c) Write the LHS as 5 to the power 'something':

$\sqrt{5} \times 5^x = 5^{\frac{1}{2}} \times 5^x = 5^{x+\frac{1}{2}}$

Write the RHS as 5 to the power 'something':

$\frac{1}{25} = \frac{1}{5^2} = 5^{-2}$

Then equate the two 'something's:

$x + \frac{1}{2} = -2$. So $x = -2 - \frac{1}{2} = -\frac{5}{2}$.

d) Write the LHS as 16 to the power 'something':

$(16^x)^2 = 16^{2x}$

Write the RHS as 16 to the power 'something':

$\frac{1}{4} = \frac{1}{16^{\frac{1}{2}}} = 16^{-\frac{1}{2}}$

Then equate the two 'something's:

$2x = -\frac{1}{2}$, so $x = -\frac{1}{4}$.

e) Write the LHS as 'something' to the power 3:

$x^{-3} = \frac{1}{x^3} = \left(\frac{1}{x}\right)^3$

Write the RHS as 'something' to the power 3:

$-8 = (-2)^3$

Then equate the two 'something's:

$\frac{1}{x} = -2$, so $x = -\frac{1}{2}$.

f) Square both sides:

$\sqrt{100^x} = 0.001 \Rightarrow 100^x = 0.001^2$

$\Rightarrow 100^x = 0.000001$

Write the RHS as a power of 100:

$0.000001 = 10^{-6} = (100^{\frac{1}{2}})^{-6} = 100^{-3}$

Then equate the powers on each side: $x = -3$

In each of the parts of this question, you could have used the laws in a different order to get the same answers.

3.3 Surds

Exercise 3.3.1 — The laws of surds

Q1 a) $\sqrt{8} = \sqrt{4 \times 2} = \sqrt{4}\sqrt{2} = 2\sqrt{2}$

b) $\sqrt{24} = \sqrt{4 \times 6} = \sqrt{4}\sqrt{6} = 2\sqrt{6}$

c) $\sqrt{50} = \sqrt{25 \times 2} = \sqrt{25}\sqrt{2} = 5\sqrt{2}$

d) $\sqrt{63} = \sqrt{9 \times 7} = \sqrt{9}\sqrt{7} = 3\sqrt{7}$

e) $\sqrt{72} = \sqrt{36 \times 2} = \sqrt{36}\sqrt{2} = 6\sqrt{2}$

f) $\sqrt{\frac{5}{4}} = \frac{\sqrt{5}}{\sqrt{4}} = \frac{\sqrt{5}}{2}$

g) $\sqrt{\frac{7}{100}} = \frac{\sqrt{7}}{\sqrt{100}} = \frac{\sqrt{7}}{10}$

h) $\sqrt{\frac{11}{9}} = \frac{\sqrt{11}}{\sqrt{9}} = \frac{\sqrt{11}}{3}$

Q2 a) $2\sqrt{3} \times 4\sqrt{3} = 2 \times 4 \times \sqrt{3} \times \sqrt{3}$

$= 8\sqrt{3}\sqrt{3} = 8 \times 3 = 24$

b) $\sqrt{5} \times 3\sqrt{5} = 3\sqrt{5}\sqrt{5} = 3 \times 5 = 15$

c) $(\sqrt{7})^2 = \sqrt{7}\sqrt{7} = 7$

d) $2\sqrt{2} \times 3\sqrt{5} = 2 \times 3 \times \sqrt{2} \times \sqrt{5} = 6\sqrt{2}\sqrt{5} = 6\sqrt{10}$

e) $(2\sqrt{11})^2 = (2\sqrt{11})(2\sqrt{11})$

$= 4\sqrt{11}\sqrt{11} = 4 \times 11 = 44$

f) $5\sqrt{8} \times 2\sqrt{2} = 5\sqrt{4 \times 2} \times 2\sqrt{2}$

$= 5 \times 2\sqrt{2} \times 2\sqrt{2}$

$= 5 \times 4 \times \sqrt{2} \times \sqrt{2} = 20 \times 2 = 40$

g) $4\sqrt{3} \times 2\sqrt{27} = 4 \times 2 \times \sqrt{3}\sqrt{27}$

$= 8\sqrt{3 \times 27} = 8\sqrt{81} = 8 \times 9 = 72$

h) $2\sqrt{6} \times 5\sqrt{24} = 2 \times 5 \times \sqrt{6} \times \sqrt{24}$

$= 10\sqrt{6 \times 24} = 10\sqrt{144}$

$= 10 \times 12 = 120$

i) $\frac{6}{\sqrt{11}} \times \sqrt{44} = \frac{6}{\sqrt{11}} \times \sqrt{4 \times 11}$

$= \frac{6}{\sqrt{11}} \times \sqrt{4}\sqrt{11} = 6 \times 2 = 12$

j) $2\sqrt{18} \times \frac{5}{\sqrt{8}} = 2\sqrt{9 \times 2} \times \frac{5}{\sqrt{4 \times 2}}$

$= 2\sqrt{9}\sqrt{2} \times \frac{5}{\sqrt{4}\sqrt{2}} = 2 \times 3\sqrt{2} \times \frac{5}{2\sqrt{2}} = 15$

k) $\frac{\sqrt{10}}{6} \times \frac{12}{\sqrt{5}} = \frac{12\sqrt{10}}{6\sqrt{5}}$

$= \frac{12}{6} \times \frac{\sqrt{10}}{\sqrt{5}} = 2 \times \sqrt{\frac{10}{5}} = 2\sqrt{2}$

l) $\frac{\sqrt{12}}{3} \times \frac{2}{\sqrt{27}} = \frac{2\sqrt{12}}{3\sqrt{27}} = \frac{2}{3} \times \frac{\sqrt{12}}{\sqrt{27}}$

$= \frac{2}{3} \times \frac{\sqrt{4 \times 3}}{\sqrt{9 \times 3}} = \frac{2}{3} \times \frac{\sqrt{4}\sqrt{3}}{\sqrt{9}\sqrt{3}}$

$= \frac{2}{3} \times \frac{2\sqrt{3}}{3\sqrt{3}} = \frac{2}{3} \times \frac{2}{3} = \frac{4}{9}$

Q3 a) $\sqrt{20} + \sqrt{5} = \sqrt{4 \times 5} + \sqrt{5} = \sqrt{4}\sqrt{5} + \sqrt{5}$

$= 2\sqrt{5} + \sqrt{5} = 3\sqrt{5}$

b) $\sqrt{32} - \sqrt{8} = \sqrt{16 \times 2} - \sqrt{4 \times 2}$

$= \sqrt{16}\sqrt{2} - \sqrt{4}\sqrt{2} = 4\sqrt{2} - 2\sqrt{2} = 2\sqrt{2}$

c) $\sqrt{27} + 4\sqrt{3} = \sqrt{9 \times 3} + 4\sqrt{3} = \sqrt{9}\sqrt{3} + 4\sqrt{3}$

$= 3\sqrt{3} + 4\sqrt{3} = 7\sqrt{3}$

d) $2\sqrt{8} - 3\sqrt{2} = 2\sqrt{4 \times 2} - 3\sqrt{2}$

$= 2\sqrt{4}\sqrt{2} - 3\sqrt{2} = 4\sqrt{2} - 3\sqrt{2} = \sqrt{2}$

e) $3\sqrt{10} + \sqrt{250} = 3\sqrt{10} + \sqrt{25 \times 10}$

$= 3\sqrt{10} + \sqrt{25}\sqrt{10}$

$= 3\sqrt{10} + 5\sqrt{10} = 8\sqrt{10}$

f) $4\sqrt{27} + 2\sqrt{48} + 5\sqrt{108}$

$= 4\sqrt{9 \times 3} + 2\sqrt{16 \times 3} + 5\sqrt{36 \times 3}$

$= 4\sqrt{9}\sqrt{3} + 2\sqrt{16}\sqrt{3} + 5\sqrt{36}\sqrt{3}$

$= 12\sqrt{3} + 8\sqrt{3} + 30\sqrt{3} = 50\sqrt{3}$

Q4 a) $(1 + \sqrt{2})(2 + \sqrt{2}) = 2 + \sqrt{2} + 2\sqrt{2} + \sqrt{2}\sqrt{2}$

$= 2 + 3\sqrt{2} + 2 = 4 + 3\sqrt{2}$

b) $(3 + 4\sqrt{3})(2 - \sqrt{3}) = 6 - 3\sqrt{3} + 8\sqrt{3} - 4\sqrt{3}\sqrt{3}$

$= 6 + 5\sqrt{3} - 12 = 5\sqrt{3} - 6$

c) By the difference of two squares rule:

$(\sqrt{11} + 2)(\sqrt{11} - 2) = (\sqrt{11})^2 - 2^2 = 11 - 4 = 7$

d) By the difference of two squares rule:
$(9-2\sqrt{5})(9+2\sqrt{5}) = 9^2-(2\sqrt{5})^2 = 81-20 = 61$

e) $(\sqrt{3}+2)^2 = (\sqrt{3}+2)(\sqrt{3}+2)$
$= \sqrt{3}\sqrt{3}+2\sqrt{3}+2\sqrt{3}+4$
$= 3+4\sqrt{3}+4 = 7+4\sqrt{3}$

f) $(3\sqrt{5}-4)^2 = (3\sqrt{5}-4)(3\sqrt{5}-4)$
$= (3\sqrt{5})^2-12\sqrt{5}-12\sqrt{5}+16$
$= 45-24\sqrt{5}+16$
$= 61-24\sqrt{5}$

You could have used the rule $(a+b)^2 = a^2+2ab+b^2$ for parts e and f.

Q5 Area of parallelogram = base length × vertical height
$= 2\sqrt{6}\times\sqrt{3} = 2\sqrt{2\times3}\sqrt{3} = 2\sqrt{2}\sqrt{3}\sqrt{3}$
$= 2\sqrt{2}\times3 = 6\sqrt{2}\text{ cm}^2$

Q6 Perimeter $= 2(3\sqrt{28}+\sqrt{63}) = 2(3\sqrt{4\times7}+\sqrt{9\times7})$
$= 2(3\sqrt{4}\sqrt{7}+\sqrt{9}\sqrt{7}) = 2(3\times2\times\sqrt{7}+3\sqrt{7})$
$= 2(6\sqrt{7}+3\sqrt{7}) = 2(9\sqrt{7}) = 18\sqrt{7}\text{ m}$

Q7 You may want to draw the triangle:
Using Pythagoras:
$(\sqrt{2})^2+(BC)^2 = (5\sqrt{2})^2$
$2+(BC)^2 = 50$
$\Rightarrow (BC)^2 = 48$
$\Rightarrow BC = \sqrt{48} = \sqrt{16\times3} = 4\sqrt{3}\text{ cm}$

A, B, C, $\sqrt{2}$, $5\sqrt{2}$

Exercise 3.3.2 — Rationalising the denominator

Q1 a) $\frac{6}{\sqrt{3}} = \frac{6\sqrt{3}}{\sqrt{3}\sqrt{3}} = \frac{6\sqrt{3}}{3} = 2\sqrt{3}$

b) $\frac{21}{\sqrt{7}} = \frac{21\sqrt{7}}{\sqrt{7}\sqrt{7}} = \frac{21\sqrt{7}}{7} = 3\sqrt{7}$

c) $\frac{30}{\sqrt{5}} = \frac{30\sqrt{5}}{\sqrt{5}\sqrt{5}} = \frac{30\sqrt{5}}{5} = 6\sqrt{5}$

d) $\sqrt{45}+\frac{15}{\sqrt{5}} = \sqrt{45}+\frac{15\sqrt{5}}{\sqrt{5}\sqrt{5}} = \sqrt{45}+\frac{15\sqrt{5}}{5}$
$= \sqrt{9\times5}+3\sqrt{5}$
$= \sqrt{9}\sqrt{5}+3\sqrt{5}$
$= 3\sqrt{5}+3\sqrt{5} = 6\sqrt{5}$

e) $\frac{\sqrt{54}}{3}-\frac{12}{\sqrt{6}} = \frac{\sqrt{9\times6}}{3}-\frac{12\sqrt{6}}{\sqrt{6}\sqrt{6}}$
$= \frac{\sqrt{9}\sqrt{6}}{3}-\frac{12\sqrt{6}}{6}$
$= \frac{3\sqrt{6}}{3}-\frac{12\sqrt{6}}{6} = \sqrt{6}-2\sqrt{6} = -\sqrt{6}$

f) $\frac{\sqrt{300}}{5}+\frac{30}{\sqrt{12}} = \frac{\sqrt{100\times3}}{5}+\frac{30\sqrt{12}}{\sqrt{12}\sqrt{12}}$
$= \frac{\sqrt{100}\sqrt{3}}{5}+\frac{30\sqrt{4\times3}}{12}$
$= \frac{10\sqrt{3}}{5}+\frac{30\sqrt{4}\sqrt{3}}{12}$
$= 2\sqrt{3}+\frac{60\sqrt{3}}{12} = 2\sqrt{3}+5\sqrt{3} = 7\sqrt{3}$

g) $\frac{1}{\sqrt{18}}-\frac{1}{\sqrt{2}} = \frac{1}{\sqrt{9\times2}}-\frac{1}{\sqrt{2}} = \frac{1}{3\sqrt{2}}-\frac{1}{\sqrt{2}}$
$= \frac{1-3}{3\sqrt{2}} = \frac{-2}{3\sqrt{2}} = \frac{-2\sqrt{2}}{3\times2} = -\frac{2\sqrt{2}}{6} = -\frac{1}{3}\sqrt{2}$

h) $\frac{1}{\sqrt{28}}+\frac{3}{\sqrt{7}} = \frac{1}{\sqrt{4\times7}}+\frac{3}{\sqrt{7}} = \frac{1}{2\sqrt{7}}+\frac{3}{\sqrt{7}}$
$= \frac{1+6}{2\sqrt{7}} = \frac{7}{2\sqrt{7}} = \frac{7\sqrt{7}}{2\times7} = \frac{1}{2}\sqrt{7}$

i) $\frac{2}{\sqrt{72}}-\frac{5}{\sqrt{8}} = \frac{2}{\sqrt{9\times8}}-\frac{5}{\sqrt{8}} = \frac{2}{3\sqrt{8}}-\frac{5}{\sqrt{8}} = \frac{2-15}{3\sqrt{8}}$
$= \frac{-13}{3\sqrt{8}} = \frac{-13\sqrt{8}}{3\times8} = -\frac{13}{24}\sqrt{8} = -\frac{13}{24}\times2\sqrt{2} = -\frac{13}{12}\sqrt{2}$

Q2 a) $\frac{4}{1+\sqrt{3}} = \frac{4(1-\sqrt{3})}{(1+\sqrt{3})(1-\sqrt{3})}$
$= \frac{4-4\sqrt{3}}{1-3} = \frac{4-4\sqrt{3}}{-2} = -2+2\sqrt{3}$

The denominator was simplified by using the difference of two squares rule. It will be used in almost every question in the rest of this exercise, so watch out for it and make sure you understand what's going on.

b) $\frac{8}{-1+\sqrt{5}} = \frac{8(-1-\sqrt{5})}{(-1+\sqrt{5})(-1-\sqrt{5})}$
$= \frac{-8-8\sqrt{5}}{1-5} = \frac{-8-8\sqrt{5}}{-4} = 2+2\sqrt{5}$

c) $\frac{18}{\sqrt{10}-4} = \frac{18(\sqrt{10}+4)}{(\sqrt{10}-4)(\sqrt{10}+4)} = \frac{18\sqrt{10}+72}{10-16}$
$= \frac{18\sqrt{10}+72}{-6} = -12-3\sqrt{10}$

d) $\frac{\sqrt{6}}{2-\sqrt{6}} = \frac{\sqrt{6}(2+\sqrt{6})}{(2-\sqrt{6})(2+\sqrt{6})} = \frac{2\sqrt{6}+6}{4-6}$
$= \frac{2\sqrt{6}+6}{-2} = -3-\sqrt{6}$

e) $\frac{3}{5+2\sqrt{7}} = \frac{3(5-2\sqrt{7})}{(5+2\sqrt{7})(5-2\sqrt{7})} = \frac{15-6\sqrt{7}}{25-(2\sqrt{7})^2}$
$= \frac{15-6\sqrt{7}}{25-28} = \frac{15-6\sqrt{7}}{-3} = -5+2\sqrt{7}$

f) $\frac{6}{3\sqrt{2}-4} = \frac{6(3\sqrt{2}+4)}{(3\sqrt{2}-4)(3\sqrt{2}+4)} = \frac{18\sqrt{2}+24}{(3\sqrt{2})^2-16}$
$= \frac{18\sqrt{2}+24}{18-16} = \frac{18\sqrt{2}+24}{2} = 12+9\sqrt{2}$

Q3 a) $\frac{\sqrt{2}+1}{\sqrt{2}-1} = \frac{(\sqrt{2}+1)(\sqrt{2}+1)}{(\sqrt{2}-1)(\sqrt{2}+1)}$
$= \frac{2+\sqrt{2}+\sqrt{2}+1}{2-1} = \frac{2\sqrt{2}+3}{1} = 3+2\sqrt{2}$

b) $\frac{\sqrt{5}+3}{\sqrt{5}-2} = \frac{(\sqrt{5}+3)(\sqrt{5}+2)}{(\sqrt{5}-2)(\sqrt{5}+2)}$
$= \frac{5+2\sqrt{5}+3\sqrt{5}+6}{5-4}$
$= \frac{11+5\sqrt{5}}{1} = 11+5\sqrt{5}$

c) $\frac{3-\sqrt{3}}{4+\sqrt{3}} = \frac{(3-\sqrt{3})(4-\sqrt{3})}{(4+\sqrt{3})(4-\sqrt{3})} = \frac{12-3\sqrt{3}-4\sqrt{3}+3}{16-3}$
$= \frac{15-7\sqrt{3}}{13} = \frac{15}{13}-\frac{7}{13}\sqrt{3}$

d) $\frac{3\sqrt{5}-1}{2\sqrt{5}-3} = \frac{(3\sqrt{5}-1)(2\sqrt{5}+3)}{(2\sqrt{5}-3)(2\sqrt{5}+3)}$
$= \frac{(2\sqrt{5})(3\sqrt{5})+9\sqrt{5}-2\sqrt{5}-3}{(2\sqrt{5})^2-9}$
$= \frac{27+7\sqrt{5}}{(2\sqrt{5})^2-9} = \frac{27+7\sqrt{5}}{11} = \frac{27}{11}+\frac{7}{11}\sqrt{5}$

e) $\frac{\sqrt{2}+\sqrt{3}}{3\sqrt{2}-\sqrt{3}} = \frac{(\sqrt{2}+\sqrt{3})(3\sqrt{2}+\sqrt{3})}{(3\sqrt{2}-\sqrt{3})(3\sqrt{2}+\sqrt{3})}$
$= \frac{3\sqrt{2}\sqrt{2}+\sqrt{2}\sqrt{3}+3\sqrt{3}\sqrt{2}+3}{(3\sqrt{2})^2-(\sqrt{3})^2}$
$= \frac{6+\sqrt{6}+3\sqrt{6}+3}{18-3}$
$= \frac{9+4\sqrt{6}}{15} = \frac{9}{15}+\frac{4}{15}\sqrt{6}$
$= \frac{3}{5}+\frac{4}{15}\sqrt{6}$

f) $\frac{2\sqrt{7}-\sqrt{5}}{\sqrt{7}+2\sqrt{5}} = \frac{(2\sqrt{7}-\sqrt{5})(\sqrt{7}-2\sqrt{5})}{(\sqrt{7}+2\sqrt{5})(\sqrt{7}-2\sqrt{5})}$

$= \frac{2\sqrt{7}\sqrt{7}-4\sqrt{7}\sqrt{5}-\sqrt{5}\sqrt{7}+10}{7-(2\sqrt{5})^2}$

$= \frac{14-4\sqrt{35}-\sqrt{35}+10}{7-20}$

$= \frac{24-5\sqrt{35}}{-13} = -\frac{24}{13}+\frac{5}{13}\sqrt{35}$

g) $\frac{2\sqrt{2}+\sqrt{3}}{\sqrt{3}-\sqrt{12}} = \frac{(2\sqrt{2}+\sqrt{3})(\sqrt{3}+\sqrt{12})}{(\sqrt{3}-\sqrt{12})(\sqrt{3}+\sqrt{12})}$

$= \frac{2\sqrt{2}\sqrt{3}+2\sqrt{2}\sqrt{12}+3+\sqrt{3}\sqrt{12}}{3-12}$

$= \frac{2\sqrt{2}\sqrt{3}+2\sqrt{2}\sqrt{4\times3}+3+\sqrt{3}\sqrt{4\times3}}{-9}$

$= -\frac{2\sqrt{2}\sqrt{3}+4\sqrt{2}\sqrt{3}+3+2\sqrt{3}\sqrt{3}}{9}$

$= -\frac{6\sqrt{2}\sqrt{3}+3+2\times3}{9} = -\frac{6\sqrt{6}+9}{9} = -1-\frac{2}{3}\sqrt{6}$

h) $\frac{6-4\sqrt{2}}{5\sqrt{2}-\sqrt{8}} = \frac{(6-4\sqrt{2})(5\sqrt{2}+\sqrt{8})}{(5\sqrt{2}-\sqrt{8})(5\sqrt{2}+\sqrt{8})}$

$= \frac{30\sqrt{2}+6\sqrt{8}-20\sqrt{2}\sqrt{2}-4\sqrt{2}\sqrt{8}}{(5\sqrt{2})^2-8}$

$= \frac{30\sqrt{2}+6\sqrt{4\times2}-40-4\sqrt{2}\sqrt{4\times2}}{42}$

$= \frac{30\sqrt{2}+12\sqrt{2}-40-4\sqrt{2}\times2\sqrt{2}}{42}$

$= \frac{42\sqrt{2}-40-16}{42} = \frac{42\sqrt{2}-56}{42} = -\frac{4}{3}+\sqrt{2}$

i) $\frac{\sqrt{3}+3}{\sqrt{5}+\sqrt{15}} = \frac{(\sqrt{3}+3)(\sqrt{5}-\sqrt{15})}{(\sqrt{5}+\sqrt{15})(\sqrt{5}-\sqrt{15})}$

$= \frac{\sqrt{3}\sqrt{5}-\sqrt{3}\sqrt{15}+3\sqrt{5}-3\sqrt{15}}{-10}$

$= \frac{\sqrt{15}-\sqrt{3}\sqrt{15}+\sqrt{3}\sqrt{3}\sqrt{5}-3\sqrt{15}}{-10}$

$= \frac{\sqrt{15}-\sqrt{3}\sqrt{15}+\sqrt{3}\sqrt{15}-3\sqrt{15}}{-10} = \frac{-2\sqrt{15}}{-10} = \frac{1}{5}\sqrt{15}$

Q4 a) $\frac{4}{\sqrt{7}-\sqrt{3}} = \frac{4(\sqrt{7}+\sqrt{3})}{(\sqrt{7}-\sqrt{3})(\sqrt{7}+\sqrt{3})}$

$= \frac{4(\sqrt{7}+\sqrt{3})}{7-3} = \frac{4(\sqrt{7}+\sqrt{3})}{4} = \sqrt{7}+\sqrt{3}$

b) $\frac{24}{\sqrt{11}-\sqrt{17}} = \frac{24(\sqrt{11}+\sqrt{17})}{(\sqrt{11}-\sqrt{17})(\sqrt{11}+\sqrt{17})}$

$= \frac{24(\sqrt{11}+\sqrt{17})}{11-17} = \frac{24(\sqrt{11}+\sqrt{17})}{-6}$

$= -4(\sqrt{11}+\sqrt{17})$

c) $\frac{2}{\sqrt{13}+\sqrt{5}} = \frac{2(\sqrt{13}-\sqrt{5})}{(\sqrt{13}+\sqrt{5})(\sqrt{13}-\sqrt{5})}$

$= \frac{2(\sqrt{13}-\sqrt{5})}{13-5} = \frac{2(\sqrt{13}-\sqrt{5})}{8} = \frac{1}{4}(\sqrt{13}-\sqrt{5})$

d) $\frac{\sqrt{5}}{\sqrt{6}+\sqrt{3}} = \frac{\sqrt{5}(\sqrt{6}-\sqrt{3})}{(\sqrt{6}+\sqrt{3})(\sqrt{6}-\sqrt{3})}$

$= \frac{\sqrt{5}\sqrt{6}-\sqrt{5}\sqrt{3}}{6-3} = \frac{1}{3}(\sqrt{30}-\sqrt{15})$

e) $\frac{\sqrt{3}}{\sqrt{21}-3\sqrt{5}} = \frac{\sqrt{3}(\sqrt{21}+3\sqrt{5})}{(\sqrt{21}-3\sqrt{5})(\sqrt{21}+3\sqrt{5})}$

$= \frac{\sqrt{3}\sqrt{21}+3\sqrt{3}\sqrt{5}}{21-(3\sqrt{5})^2} = \frac{\sqrt{3}\sqrt{21}+3\sqrt{3}\sqrt{5}}{21-45}$

$= \frac{\sqrt{3}\sqrt{3\times7}+3\sqrt{3}\sqrt{5}}{-24} = -\frac{3\sqrt{7}+3\sqrt{15}}{24}$

$= -\frac{3}{24}(\sqrt{7}+\sqrt{15}) = -\frac{1}{8}(\sqrt{7}+\sqrt{15})$

f) $\frac{3\sqrt{2}}{2\sqrt{3}+\sqrt{20}} = \frac{3\sqrt{2}(2\sqrt{3}-\sqrt{20})}{(2\sqrt{3}+\sqrt{20})(2\sqrt{3}-\sqrt{20})}$

$= \frac{6\sqrt{2}\sqrt{3}-3\sqrt{2}\sqrt{20}}{(2\sqrt{3})^2-20} = \frac{6\sqrt{6}-3\sqrt{2}\sqrt{4\times5}}{12-20}$

$= \frac{6\sqrt{6}-3\sqrt{2}\times2\sqrt{5}}{-8} = -\frac{6\sqrt{6}-6\sqrt{10}}{8}$

$= -\frac{3}{4}(\sqrt{6}-\sqrt{10}) = \frac{3}{4}(\sqrt{10}-\sqrt{6})$

Q5 $\frac{1}{4\sqrt{3}-2\sqrt{5}} = \frac{4\sqrt{3}+2\sqrt{5}}{(4\sqrt{3}-2\sqrt{5})(4\sqrt{3}+2\sqrt{5})}$

$= \frac{4\sqrt{3}+2\sqrt{5}}{48-20} = \frac{4\sqrt{3}+2\sqrt{5}}{28}$

$= \frac{2\sqrt{3}+\sqrt{5}}{14} = \frac{\sqrt{4}\sqrt{3}+\sqrt{5}}{14} = \frac{1}{14}(\sqrt{12}+\sqrt{5})$

Q6 $12 = \sqrt{3}z \Rightarrow z = \frac{12}{\sqrt{3}} = \frac{12\sqrt{3}}{3} = 4\sqrt{3}$

Q7 $8 = (\sqrt{5}-1)x \Rightarrow x = \frac{8}{(\sqrt{5}-1)} = \frac{8(\sqrt{5}+1)}{(\sqrt{5}-1)(\sqrt{5}+1)}$

$= \frac{8\sqrt{5}+8}{5-1} = \frac{8\sqrt{5}+8}{4} = 2+2\sqrt{5}$

Q8 $5+\sqrt{7} = (3-\sqrt{7})y$

$\Rightarrow y = \frac{5+\sqrt{7}}{3-\sqrt{7}} = \frac{(5+\sqrt{7})(3+\sqrt{7})}{(3-\sqrt{7})(3+\sqrt{7})}$

$= \frac{15+5\sqrt{7}+3\sqrt{7}+7}{9-7} = \frac{22+8\sqrt{7}}{2} = 11+4\sqrt{7}$

Q9 The area of a rectangle is given by area (A) = length (l) × width (w) so:

$(2+\sqrt{2}) = l\times(3\sqrt{2}-4)$

$\Rightarrow l = \frac{(2+\sqrt{2})}{(3\sqrt{2}-4)} = \frac{(2+\sqrt{2})(3\sqrt{2}+4)}{(3\sqrt{2}-4)(3\sqrt{2}+4)} = \frac{6\sqrt{2}+8+6+4\sqrt{2}}{(3\sqrt{2})^2-16}$

$= \frac{14+10\sqrt{2}}{18-16} = \frac{14+10\sqrt{2}}{2} = (7+5\sqrt{2})$ cm

Don't forget the units here.

Review Exercise — Chapter 3

Q1 a) $(a+b)(a-b) = a^2-ab+ba-b^2 = a^2-b^2$

b) $(p+q)(p+q) = p^2+pq+qp+q^2 = p^2+2pq+q^2$

c) $35xy+25y(5y+7x)-100y^2$
$= 35xy+125y^2+175xy-100y^2 = 25y^2+210xy$

d) $(x+3y+2)(3x+y+7)$
$= x(3x+y+7)+3y(3x+y+7)+2(3x+y+7)$
$= 3x^2+xy+7x+9xy+3y^2+21y+6x+2y+14$
$= 3x^2+10xy+3y^2+13x+23y+14$

e) $(c+2d)(c-d)(2d-3c)$
$= (c^2-cd+2dc-2d^2)(2d-3c)$
$= c^2(2d-3c)-cd(2d-3c)+2dc(2d-3c)-2d^2(2d-3c)$
$= 2c^2d-3c^3-2cd^2+3c^2d+4cd^2-6c^2d-4d^3+6cd^2$
$= -c^2d-3c^3+8cd^2-4d^3$

f) $[(s-2t)(s+2t)]^2 = [s^2+2st-2ts-4t^2]^2$
$= [s^2-4t^2]^2 = (s^2-4t^2)(s^2-4t^2)$
$= s^4-4s^2t^2-4t^2s^2+16t^4 = s^4-8s^2t^2+16t^4$

Q2 The side lengths of the box are x cm, $(x-10)$ cm and $(x-8)$ cm. So the volume of one box is $x(x-10)(x-8)$ cm³.
$x(x-10)(x-8) = x(x^2-8x-10x+80)$
$= x^3-18x^2+80x$ cm³
So the volume of 5 boxes is:
$5(x^3-18x^2+80x) = 5x^3-90x^2+400x$ cm³

Q3 a) $2x^2y+axy+2xy^2 = (xy\times2x)+(xy\times a)+(xy\times2y)$
$= xy(2x+a+2y)$

b) $a^2x+a^2b^2x^2 = (a^2x\times1)+(a^2x\times b^2x) = a^2x(1+b^2x)$

c) $16y+8yx+56x = (8\times2y)+(8\times yx)+(8\times7x)$
$= 8(2y+xy+7x)$

d) $24s+60st+15s^2t^2 = (3s\times8)+(3s\times20t)+(3s\times5st^2)$
$= 3s(8+20t+5st^2)$

e) $27c - 9c^3d - 45cd^2 = (9c \times 3) - (9c \times c^2d) - (9c \times 5d^2)$
$= 9c(3 - c^2d - 5d^2)$

f) $x^2y^2z^2 + x^3y^3 - x^2yz = (x^2y \times yz^2) + (x^2y \times xy^2) - (x^2y \times z)$
$= x^2y(yz^2 + xy^2 - z)$

Q4 **a)** $(x + 1) - y(x + 1) = (x + 1)(1 - y)$

b) $z^4 + z^2(3 - z) = z^2(z^2 + 3 - z)$

c) $(x + y)^2 + x(x + y) = (x + y)(x + y + x) = (x + y)(2x + y)$

d) $x(x - 2) + 3(2 - x) = x(x - 2) - 3(x - 2) = (x - 2)(x - 3)$

e) $25 - x^4 = (5 + x^2)(5 - x^2)$

f) $9b^2c^4 - 4c^2d^6 = c^2(9b^2c^2 - 4d^6) = c^2((3bc)^2 - (2d^3)^2)$
$= c^2(3bc + 2d^3)(3bc - 2d^3)$

Q5 **a)** 60 is the lowest common multiple of 3, 12 and 5, so the common denominator is 60:

$$\frac{2x}{3} + \frac{y}{12} + \frac{x}{5} = \frac{40x}{60} + \frac{5y}{60} + \frac{12x}{60} = \frac{52x + 5y}{60}$$

b) The common denominator is $x^2 \times y^2 = x^2y^2$:

$$\frac{5}{xy^2} - \frac{2}{x^2y} = \frac{5x}{x^2y^2} - \frac{2y}{x^2y^2} = \frac{5x - 2y}{x^2y^2}$$

c) The common denominator is $x(x + y)(x - y)$:

$$\frac{1}{x} + \frac{x}{x+y} + \frac{y}{x-y}$$

$$= \frac{(x+y)(x-y)}{x(x+y)(x-y)} + \frac{x^2(x-y)}{x(x+y)(x-y)} + \frac{xy(x+y)}{x(x+y)(x-y)}$$

$$= \frac{x^2 - xy + yx - y^2 + x^3 - x^2y + x^2y + xy^2}{x(x^2 - y^2)}$$

$$= \frac{x^3 + x^2 - y^2 + xy^2}{x(x^2 - y^2)}$$

d) The common denominator is $a^2 \times b = a^2b$:

$$\frac{a}{b} + \frac{4}{a} - \frac{7}{a^2} = \frac{a^3}{a^2b} + \frac{4ab}{a^2b} - \frac{7b}{a^2b} = \frac{a^3 + 4ab - 7b}{a^2b}$$

e) The common denominator is $3xy$:

$$3x - \frac{4}{3xy} = \frac{9x^2y}{3xy} - \frac{4}{3xy} = \frac{9x^2y - 4}{3xy}$$

f) The common denominator is $2 \times s^2 \times t^2 = 2s^2t^2$:

$$\frac{2s}{t^2} + \frac{5}{2t} - \frac{t}{s^2} = \frac{4s^3}{2s^2t^2} + \frac{5s^2t}{2s^2t^2} - \frac{2t^3}{2s^2t^2}$$

$$= \frac{4s^3 + 5s^2t - 2t^3}{2s^2t^2}$$

Q6 **a)** The common denominator is $2b$:

$$\frac{2a}{b} - \frac{a}{2b} = \frac{4a}{2b} - \frac{a}{2b} = \frac{3a}{2b}$$

b) The common denominator is $(p + q)(p - q)$:

$$\frac{2p}{p+q} + \frac{2q}{p-q} = \frac{2p(p-q)}{(p+q)(p-q)} + \frac{2q(p+q)}{(p+q)(p-q)}$$

$$= \frac{2p^2 - 2pq}{p^2 - q^2} + \frac{2qp + 2q^2}{p^2 - q^2} = \frac{2(p^2 + q^2)}{p^2 - q^2}$$

c) The common denominator is $(c - d)^2(c + d)$:

$$\frac{c+d}{(c-d)^2} + \frac{1}{c+d} = \frac{(c+d)^2}{(c-d)^2(c+d)} + \frac{(c-d)^2}{(c-d)^2(c+d)}$$

$$= \frac{c^2 + 2cd + d^2 + c^2 - 2cd + d^2}{(c-d)^2(c+d)} = \frac{2c^2 + 2d^2}{(c-d)^2(c+d)}$$

$$= \frac{2(c^2 + d^2)}{(c-d)^2(c+d)}$$

d) The common denominator is $2x^2(1 + x)$:

$$\frac{1}{1+x} - \frac{1-x}{2x^2} = \frac{2x^2}{2x^2(1+x)} - \frac{(1-x)(1+x)}{2x^2(1+x)}$$

$$= \frac{2x^2 - (1 - x^2)}{2x^2(1+x)} = \frac{3x^2 - 1}{2x^2(1+x)}$$

e) The two denominators in the expression are $k^2 - 1 = (k + 1)(k - 1)$ and $(k - 1)$. So the common denominator is $(k + 1)(k - 1)$.

$$\frac{2k}{k^2 - 1} + \frac{k^2}{k-1} = \frac{2k}{(k+1)(k-1)} + \frac{k^2}{k-1}$$

$$= \frac{2k}{(k+1)(k-1)} + \frac{k^2(k+1)}{(k-1)(k+1)}$$

$$= \frac{2k + k^3 + k^2}{(k+1)(k-1)} = \frac{k(k^2 + k + 2)}{k^2 - 1}$$

f) The common denominator is $(z + 1)(y + z)(y - 1)$.

$$\frac{4}{z+1} + \frac{2}{y+z} - \frac{6}{y-1}$$

$$= \frac{4(y+z)(y-1)}{(z+1)(y+z)(y-1)} + \frac{2(z+1)(y-1)}{(z+1)(y+z)(y-1)} - \frac{6(z+1)(y+z)}{(z+1)(y+z)(y-1)}$$

$$= \frac{4(y^2 - y + zy - z)}{(z+1)(y+z)(y-1)} + \frac{2(zy - z + y - 1)}{(z+1)(y+z)(y-1)} - \frac{6(zy + z^2 + y + z)}{(z+1)(y+z)(y-1)}$$

$$= \frac{4y^2 - 4y + 4zy - 4z}{(z+1)(y+z)(y-1)} + \frac{2zy - 2z + 2y - 2}{(z+1)(y+z)(y-1)} - \frac{6zy + 6z^2 + 6y + 6z}{(z+1)(y+z)(y-1)}$$

$$= \frac{4y^2 - 8y - 12z - 2 - 6z^2}{(z+1)(y+z)(y-1)}$$

Q7 Call the unknown side of the flower bed z.

Then $z = \frac{x^2}{x-3}$ and $y + z = \frac{3x^2}{x+6}$.

So $y = \frac{3x^2}{x+6} - z = \frac{3x^2}{x+6} - \frac{x^2}{x-3} = \frac{3x^2(x-3) - x^2(x+6)}{(x+6)(x-3)}$

$$= \frac{3x^3 - 9x^2 - x^3 - 6x^2}{(x+6)(x-3)} = \frac{2x^3 - 15x^2}{(x+6)(x-3)} = \frac{x^2(2x - 15)}{(x+6)(x-3)}$$

Q8 **a)** $x^3 \cdot x^5 = x^{3+5} = x^8$ **b)** $a^7 \cdot a^8 = a^{7+8} = a^{15}$

c) $\frac{x^8}{x^2} = x^{8-2} = x^6$ **d)** $(a^2)^4 = a^{2 \times 4} = a^8$

e) $(xy^2) \cdot (x^3yz) = x^{1+3}y^{2+1}z = x^4y^3z$

f) $\frac{a^2b^4c^6}{a^3b^2c} = a^{2-3}b^{4-2}c^{6-1} = a^{-1}b^2c^5 = \frac{b^2c^5}{a}$

Q9 **a)** $g^2 \times g^{-5} = g^{2-5} = g^{-3}$

b) $p^4r^2 \div p^5r^{-6} = p^{4-5}r^{2-(-6)} = p^{-1}r^8$

c) $(k^{\frac{1}{3}})^6 = k^{\frac{1}{3} \times 6} = k^2$

d) $(mn^8 \times m^4n^{-11})^{-2} = (m^{1+4}n^{8-11})^{-2} = (m^5n^{-3})^{-2}$
$= m^{5 \times (-2)}n^{(-3) \times (-2)} = m^{-10}n^6$

e) $s^4t^3\left(\frac{1}{s^2t^5}\right)^{-3} = s^4t^3(s^2t^5)^3 = s^4t^3(s^2)^3(t^5)^3 = s^4t^3s^{2 \times 3}t^{5 \times 3}$
$= s^4t^3s^6t^{15} = s^{4+6}t^{3+15} = s^{10}t^{18}$

f) $\frac{a^2}{b^2c} \times \frac{b^6}{a^4c^{-2}} \div \frac{c^2}{a^3b} = \frac{a^2}{b^2c} \times \frac{b^6}{a^4c^{-2}} \times \frac{a^3b}{c^2}$
$= \frac{a^{2+3}b^{6+1}}{a^4b^2c^{1-2+2}} = \frac{a^5b^7}{a^4b^2c} = a^{5-4}b^{7-2}c^{-1} = ab^5c^{-1}$

Q10 **a)** $16^{\frac{1}{2}} = \sqrt{16} = 4$ **b)** $8^{\frac{1}{3}} = \sqrt[3]{8} = 2$

c) $81^{\frac{3}{4}} = (81^{\frac{1}{4}})^3 = 3^3 = 27$ **d)** $x^0 = 1$

e) $49^{-\frac{1}{2}} = \frac{1}{\sqrt{49}} = \frac{1}{7}$

f) $\frac{1}{27^{-\frac{2}{3}}} = 27^{\frac{2}{3}} = (27^{\frac{1}{3}})^2 = 3^2 = 9$

Q11 **a)** $\sqrt{28} = \sqrt{4 \times 7} = \sqrt{4}\sqrt{7} = 2\sqrt{7}$

b) $\sqrt{\frac{5}{36}} = \frac{\sqrt{5}}{\sqrt{36}} = \frac{\sqrt{5}}{6}$

c) $\sqrt{18} = \sqrt{9 \times 2} = \sqrt{9}\sqrt{2} = 3\sqrt{2}$

d) $\sqrt{\frac{9}{16}} = \frac{\sqrt{9}}{\sqrt{16}} = \frac{3}{4}$

Q12 **a)** $\sqrt{3} - \sqrt{12} = \sqrt{3} - \sqrt{4 \times 3} = \sqrt{3} - 2\sqrt{3} = -\sqrt{3}$

b) $3\sqrt{5} + \sqrt{45} = 3\sqrt{5} + \sqrt{9 \times 5} = 3\sqrt{5} + 3\sqrt{5} = 6\sqrt{5}$

c) $\sqrt{7} + \sqrt{448} = \sqrt{7} + \sqrt{64 \times 7} = \sqrt{7} + 8\sqrt{7} = 9\sqrt{7}$

d) $\sqrt{52} + \sqrt{117} = \sqrt{4 \times 13} + \sqrt{9 \times 13}$
$= 2\sqrt{13} + 3\sqrt{13} = 5\sqrt{13}$

e) $4\sqrt{150} + \sqrt{54} - \sqrt{5}\sqrt{120}$
$= 4\sqrt{25 \times 6} + \sqrt{9 \times 6} - \sqrt{5}\sqrt{20 \times 6}$
$= 4 \times 5\sqrt{6} + 3\sqrt{6} - \sqrt{5}\sqrt{4 \times 5}\sqrt{6}$
$= 20\sqrt{6} + 3\sqrt{6} - 2\sqrt{5}\sqrt{5}\sqrt{6}$
$= 20\sqrt{6} + 3\sqrt{6} - 10\sqrt{6} = 13\sqrt{6}$

Q13 The larger square has side length $\sqrt{1920}$ cm and the smaller square has side length $\sqrt{1080}$ cm.
a is the difference between the side lengths of the two squares, so: $a = \sqrt{1920} - \sqrt{1080} = \sqrt{64 \times 30} - \sqrt{36 \times 30}$
$= 8\sqrt{30} - 6\sqrt{30} = 2\sqrt{30}$

Q14 $(6\sqrt{3}+2\sqrt{7})^2$
$= (6\sqrt{3}+2\sqrt{7})(6\sqrt{3}+2\sqrt{7})$
$= 36\sqrt{3}\sqrt{3} + 12\sqrt{3}\sqrt{7} + 12\sqrt{7}\sqrt{3} + 4\sqrt{7}\sqrt{7}$
$= 36 \times 3 + 24\sqrt{3}\sqrt{7} + 4 \times 7$
$= 108 + 24\sqrt{21} + 28$
$= 136 + 24\sqrt{21}$

Q15 **a)** $\frac{8}{\sqrt{2}} = \frac{8}{\sqrt{2}} \times \frac{\sqrt{2}}{\sqrt{2}} = \frac{8\sqrt{2}}{2} = 4\sqrt{2}$

b) $\frac{\sqrt{2}}{2} = \frac{\sqrt{2}}{(\sqrt{2})^2} = \frac{1}{\sqrt{2}}$

Q16 $\frac{2}{3+\sqrt{7}} = \frac{2(3-\sqrt{7})}{(3+\sqrt{7})(3-\sqrt{7})}$
$= \frac{6-2\sqrt{7}}{9-7} = \frac{6-2\sqrt{7}}{2} = 3-\sqrt{7}$

Q17 **a)** $\frac{11+\sqrt{13}}{5-\sqrt{13}} = \frac{(11+\sqrt{13})(5+\sqrt{13})}{(5-\sqrt{13})(5+\sqrt{13})}$
$= \frac{55+11\sqrt{13}+5\sqrt{13}+13}{25-13}$
$= \frac{68+16\sqrt{13}}{12} = \frac{17}{3} + \frac{4}{3}\sqrt{13}$

b) $\frac{2\sqrt{7}+9}{3-\sqrt{7}} = \frac{(2\sqrt{7}+9)(3+\sqrt{7})}{(3-\sqrt{7})(3+\sqrt{7})}$
$= \frac{6\sqrt{7}+14+27+9\sqrt{7}}{9-7}$
$= \frac{15\sqrt{7}+41}{2} = \frac{41}{2} + \frac{15}{2}\sqrt{7}$

c) $\frac{3\sqrt{5}+\sqrt{15}}{\sqrt{60}-\sqrt{20}} = \frac{(3\sqrt{5}+\sqrt{15})(\sqrt{60}+\sqrt{20})}{(\sqrt{60}-\sqrt{20})(\sqrt{60}+\sqrt{20})}$
$= \frac{3\sqrt{5}\sqrt{60}+3\sqrt{5}\sqrt{20}+\sqrt{15}\sqrt{60}+\sqrt{15}\sqrt{20}}{60-20}$
$= \frac{3\sqrt{5}\sqrt{5}\sqrt{3}\sqrt{4}+3\sqrt{5}\sqrt{5}\sqrt{4}}{40}$
$+ \frac{\sqrt{3}\sqrt{5}\sqrt{5}\sqrt{3}\sqrt{4}+\sqrt{3}\sqrt{5}\sqrt{5}\sqrt{4}}{40}$
$= \frac{30\sqrt{3}+30+30+10\sqrt{3}}{40}$
$= \frac{40\sqrt{3}+60}{40} = \frac{3}{2} + \sqrt{3}$

Exam-Style Questions — Chapter 3

Q1 $\frac{(3a^3b^2)^2 \times (2a^2b)^2}{(8a^6b^{-3})^{\frac{1}{3}}} = \frac{9a^6b^4 \times 4a^4b^2}{8^{\frac{1}{3}}a^2b^{-1}} = \frac{36a^{10}b^6}{2a^2b^{-1}} = 18a^8b^7$

[2 marks available — 1 mark for fully simplifying the numerator correctly, 1 mark for the correct answer]

Q2 **a)** Write the RHS as 9 to the power 'something': $3 = \sqrt{9} = 9^{\frac{1}{2}}$
So $x = \frac{1}{2}$ *[1 mark]*

b) $9^{3x} = (3^2)^{3x} = 3^{6x}$
$81 = 9^2 = 3^4$, so $81^{2x-1} = (3^4)^{2x-1} = 3^{8x-4}$
[1 mark for expressing both terms on the left-hand side of the equation as powers of 3]
So $9^{3x} \cdot 81^{2x-1} = 3^{6x} \times 3^{8x-4} = 3^{6x+8x-4} = 3^{14x-4}$
So $9^{3x} \cdot 81^{2x-1} = 27 \Rightarrow 3^{14x-4} = 3^3$
[1 mark for writing both sides of the equation as a single power of 3]
So $14x - 4 = 3 \Rightarrow 14x = 7 \Rightarrow x = \frac{1}{2}$
[1 mark for the correct value of x]

Q3 $6^{\frac{1}{3}} + 6^{\frac{1}{3}} + 6^{\frac{1}{3}} = 3 \times 6^{\frac{1}{3}} = 3 \times (2 \times 3)^{\frac{1}{3}}$
$= 3 \times 2^{\frac{1}{3}} \times 3^{\frac{1}{3}} = 2^{\frac{1}{3}}3^{1+\frac{1}{3}} = 2^{\frac{1}{3}}3^{\frac{4}{3}}$
[2 marks available — 1 mark for writing $6^{\frac{1}{3}}$ as $2^{\frac{1}{3}} \times 3^{\frac{1}{3}}$, 1 mark for the correct answer]

Q4 $(3\sqrt{5}-5\sqrt{3})^2 = (3\sqrt{5}-5\sqrt{3})(3\sqrt{5}-5\sqrt{3})$
$= 9\sqrt{5}\sqrt{5} - 15\sqrt{5}\sqrt{3} - 15\sqrt{3}\sqrt{5} + 25\sqrt{3}\sqrt{3}$
$= 45 - 15\sqrt{15} - 15\sqrt{15} + 75 = 120 - 30\sqrt{15} = 30(4-\sqrt{15})$

[3 marks available — 3 marks for the correct answer, otherwise 2 marks for any two correctly simplified terms (45, $-15\sqrt{15}$ or 75), or 1 mark for attempting to expand the brackets and getting one term correct]

Q5 $\frac{\sqrt{x}}{\sqrt{x}+\sqrt{y}} + \frac{\sqrt{y}}{\sqrt{x}-\sqrt{y}}$
$= \frac{\sqrt{x}(\sqrt{x}-\sqrt{y})}{(\sqrt{x}+\sqrt{y})(\sqrt{x}-\sqrt{y})} + \frac{\sqrt{y}(\sqrt{x}+\sqrt{y})}{(\sqrt{x}+\sqrt{y})(\sqrt{x}-\sqrt{y})}$
$= \frac{\sqrt{x}\sqrt{x}-\sqrt{x}\sqrt{y}+\sqrt{y}\sqrt{x}+\sqrt{y}\sqrt{y}}{x-y} = \frac{x+y}{x-y}$

The common denominator is in the form $(a+b)(a-b) = a^2 - b^2$.

[3 marks available — 1 mark for putting the fractions over a suitable common denominator, 1 mark for at least three correct terms in the expansion of the numerator, 1 mark for the correct answer]

Q6 $3x + 2y = 23 + 5\sqrt{5}$ ①
$\sqrt{5}x + y = 12 + 4\sqrt{5}$ ②
② × 2: $2\sqrt{5}x + 2y = 24 + 8\sqrt{5}$ ③
① − ③:
$(3 - 2\sqrt{5})x = 23 + 5\sqrt{5} - (24 + 8\sqrt{5})$
$(3 - 2\sqrt{5})x = -1 - 3\sqrt{5}$
$x = -\frac{1+3\sqrt{5}}{3-2\sqrt{5}} = -\frac{(1+3\sqrt{5})(3+2\sqrt{5})}{(3-2\sqrt{5})(3+2\sqrt{5})}$
$= -\frac{3+2\sqrt{5}+9\sqrt{5}+6\sqrt{5}\sqrt{5}}{9-(2\sqrt{5})^2}$
$= -\frac{33+11\sqrt{5}}{9-20} = -\frac{33+11\sqrt{5}}{-11} = 3 + \sqrt{5}$

Substitute into ②:
$\sqrt{5}(3+\sqrt{5}) + y = 12 + 4\sqrt{5}$
$\Rightarrow 3\sqrt{5} + 5 + y = 12 + 4\sqrt{5}$
$\Rightarrow y = 7 + \sqrt{5}$

[7 marks available — 1 mark for eliminating y, 1 mark for rearranging to get an expression for x, 1 mark for attempting to rationalise the denominator, 1 mark for at least three correct terms in the numerator, 1 mark for finding the correct value of x, 1 mark for substituting to find the value of y, 1 mark for the correct value of y.]

You could eliminate x instead — then you'd find the value of y first, and substitute it back in to find x.

Q7 **a)** $\frac{(5-2\sqrt{m})+(10+9\sqrt{m})+n\sqrt{m}}{3} = 5 + 4\sqrt{m}$
$\Rightarrow \frac{15+(7+n)\sqrt{m}}{3} = 5 + 4\sqrt{m}$ *[1 mark]*
$\Rightarrow 15 + (7+n)\sqrt{m} = 15 + 12\sqrt{m}$
$\Rightarrow 7 + n = 12$
$\Rightarrow n = 5$ *[1 mark]*

b) $10 + 9\sqrt{m}$ must be the largest number *[1 mark]*, so the range is either:
$10 + 9\sqrt{m} - (5 - 2\sqrt{m}) = 5 + 11\sqrt{m}$
or $10 + 9\sqrt{m} - 5\sqrt{m} = 10 + 4\sqrt{m}$
[1 mark for both expressions]
You're told the range is $5 + 11\sqrt{m}$, so
$5 - 2\sqrt{m}$ must be the smallest number *[1 mark]*.
So $5 - 2\sqrt{m} < 5\sqrt{m} \Rightarrow 5 < 7\sqrt{m} \Rightarrow \frac{5}{7} < \sqrt{m}$ *[1 mark]*
So $m > \frac{25}{49}$ *[1 mark]*

Q8 Let the radius of the pond be p m.
Circumference of pond $= 10\pi = 2\pi p \Rightarrow p = 5$ *[1 mark]*

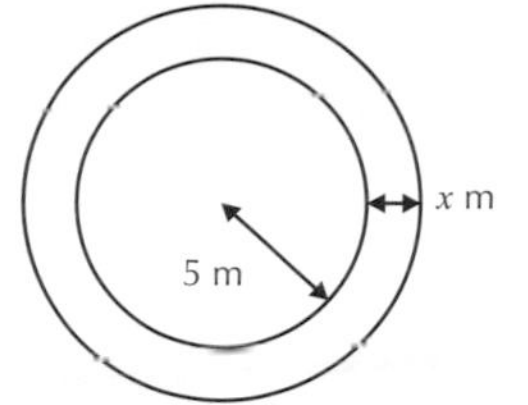

Area of path $= \pi(x + 5)^2 - \pi \times 5^2$
$= \pi((x + 5)^2 - 25)$ m² *[1 mark]*
Maximum area of material $= 20\pi$ m², so the maximum value of x is the solution to: $\pi((x + 5)^2 - 25) = 20\pi$ *[1 mark]*
$\Rightarrow (x + 5)^2 - 25 = 20 \Rightarrow (x + 5)^2 = 45$
$\Rightarrow x + 5 = \pm\sqrt{45} = \pm\sqrt{9 \times 5} = \pm 3\sqrt{5}$ *[1 mark]*
$\Rightarrow x = 3\sqrt{5} - 5$ or $x = -3\sqrt{5} - 5$ *[1 mark]*
But x must be positive, so the second solution isn't possible.
So the maximum value of x is: $x = 3\sqrt{5} - 5$ *[1 mark]*.
The maximum possible width of the path is approximately 1.7 m.

Chapter 4: Quadratics and Cubics

4.1 Quadratic Equations

Exercise 4.1.1 — Factorising a quadratic

Q1 **a)** $x^2 - 6x + 5 = (x - 5)(x - 1)$

b) $x^2 - 3x - 18 = (x - 6)(x + 3)$

c) $x^2 + 22x + 121 = (x + 11)(x + 11) = (x + 11)^2$

d) $x^2 - 12x = x(x - 12)$
Note that if every term contains an x, you can just take a factor of x out of the bracket.

e) $y^2 - 13y + 42 = (y - 6)(y - 7)$

f) $x^2 + 51x + 144 = (x + 48)(x + 3)$

g) $x^2 - 121 = (x + 11)(x - 11)$
If there is no 'b' term, see if the expression is a 'difference of two squares' (chances are it will be).

h) $x^2 - 35x + 66 = (x - 2)(x - 33)$

Q2 **a)** $x^2 - 2x - 8 = 0 \Rightarrow (x - 4)(x + 2) = 0$
$\Rightarrow x - 4 = 0$ or $x + 2 = 0 \Rightarrow x = 4$ or $x = -2$

b) $2x^2 + 2x - 40 = 0 \Rightarrow 2(x^2 + x - 20) = 0$
This is an example of a question where you can simplify the equation before factorising. You can divide through by 2.
$x^2 + x - 20 = 0 \Rightarrow (x + 5)(x - 4) = 0$
$\Rightarrow x + 5 = 0$ or $x - 4 = 0 \Rightarrow x = -5$ or $x = 4$

c) $p^2 + 21p + 38 = 0 \Rightarrow (p + 19)(p + 2) = 0$
$\Rightarrow p + 19 = 0$ or $p + 2 = 0 \Rightarrow p = -19$ or $p = -2$

d) $x^2 - 15x + 54 = 0 \Rightarrow (x - 9)(x - 6) = 0$
$\Rightarrow x - 9 = 0$ or $x - 6 = 0 \Rightarrow x = 9$ or $x = 6$

e) $x^2 + 18x = -65 \Rightarrow x^2 + 18x + 65 = 0$
$(x + 5)(x + 13) = 0 \Rightarrow x + 5 = 0$ or $x + 13 = 0$
$\Rightarrow x = -5$ or $x = -13$

f) $x^2 - x = 42 \Rightarrow x^2 - x - 42 = 0 \Rightarrow (x - 7)(x + 6) = 0$
$\Rightarrow x - 7 = 0$ or $x + 6 = 0 \Rightarrow x = 7$ or $x = -6$

g) $x^2 + 1100x + 100\,000 = 0 \Rightarrow (x + 100)(x + 1000) = 0$
$\Rightarrow x + 100 = 0$ or $x + 1000 = 0 \Rightarrow x = -100$ or $x = -1000$

h) $3x^2 - 3x - 6 = 0 \Rightarrow 3(x^2 - x - 2) = 0$
$\Rightarrow x^2 - x - 2 = 0 \Rightarrow (x - 2)(x + 1) = 0$
$\Rightarrow x - 2 = 0$ or $x + 1 = 0 \Rightarrow x = 2$ or $x = -1$

Q3 **a)** $4x^2 - 4x - 3 = (2x + 1)(2x - 3)$

b) $2x^2 + 23x + 11 = (2x + 1)(x + 11)$

c) $7x^2 - 19x - 6 = (7x + 2)(x - 3)$

d) $-x^2 - 5x + 36 = -(x^2 + 5x - 36) = -(x - 4)(x + 9)$

e) $6x^2 - 7x - 3 = (3x + 1)(2x - 3)$

f) $2x^2 - 2 = 2(x^2 - 1) = 2(x + 1)(x - 1)$

g) $3x^2 - 3 = 3(x^2 - 1) = 3(x + 1)(x - 1)$

h) $-x^2 + 9x - 14 = -(x^2 - 9x + 14) = -(x - 7)(x - 2)$

Q4 **a)** $-5x^2 - 22x + 15 = 0 \Rightarrow 5x^2 + 22x - 15 = 0$
$\Rightarrow (5x - 3)(x + 5) = 0 \Rightarrow x = \frac{3}{5}$ or $x = -5$
If you want to get rid of the minus sign in front of the x^2 just multiply through by −1 — the right-hand side will remain 0 and the left-hand side will change signs.

b) $32x^2 + 60x + 13 = 0 \Rightarrow (4x + 1)(8x + 13) = 0$
$\Rightarrow 4x + 1 = 0$ or $8x + 13 = 0 \Rightarrow x = -\frac{1}{4}$ or $x = -\frac{13}{8}$

c) $5a^2 + 12a = 9 \Rightarrow 5a^2 + 12a - 9 = 0$
$\Rightarrow (5a - 3)(a + 3) = 0$
$\Rightarrow 5a - 3 = 0$ or $a + 3 = 0 \Rightarrow a = \frac{3}{5}$ or $a = -3$

d) $8x^2 + 22x + 15 = 0 \Rightarrow (4x + 5)(2x + 3) = 0$
$\Rightarrow 4x + 5 = 0$ or $2x + 3 = 0 \Rightarrow x = -\frac{5}{4}$ or $x = -\frac{3}{2}$

e) $4q^2 + 6 = 11q \Rightarrow 4q^2 - 11q + 6 = 0$
$\Rightarrow (4q - 3)(q - 2) = 0$
$\Rightarrow 4q - 3 = 0$ or $q - 2 = 0 \Rightarrow q = \frac{3}{4}$ or $q = 2$

f) $24y^2 + 23y - 12 = 0 \Rightarrow (3y + 4)(8y - 3) = 0$
$\Rightarrow 3y + 4 = 0$ or $8y - 3 = 0 \Rightarrow y = -\frac{4}{3}$ or $y = \frac{3}{8}$

Q5 $(x - 1)(x - 2) = 37 - x \Rightarrow x^2 - 3x + 2 = 37 - x$
$\Rightarrow x^2 - 2x - 35 = 0 \Rightarrow (x - 7)(x + 5) = 0$
$\Rightarrow x - 7 = 0$ or $x + 5 = 0 \Rightarrow x = 7$ or $x = -5$

Q6 The function f(x) meets the x-axis when f(x) = 0 so set the expression for f(x) equal to 0.
$-x^2 + 7x + 30 = 0 \Rightarrow x^2 - 7x - 30 = 0$
$\Rightarrow (x - 10)(x + 3) = 0$
$\Rightarrow x - 10 = 0$ or $x + 3 = 0 \Rightarrow x = 10$ or $x = -3$
So the graph of f(x) meets the x-axis when $x = 10$ and $x = -3$.

Q7 The functions intersect when f(x) = g(x), so:
$(x - 8)(x + 10) = (3x + 2)(x - 11) \Rightarrow x^2 + 2x - 80 = 3x^2 - 31x - 22$
$\Rightarrow 0 = 2x^2 - 33x + 58 \Rightarrow 0 = (2x - 29)(x - 2)$
$\Rightarrow 0 = 2x - 29$ or $0 = x - 2 \Rightarrow x = \frac{29}{2} = 14.5$ or $x = 2$
So the functions intersect when $x = 14.5$ and $x = 2$.

Q8 Here, you have to set T equal to 0 and solve the quadratic equation to find the values of h:
$0 = -2h^2 + 13h - 20 \Rightarrow 0 = 2h^2 - 13h + 20$
$0 = (2h - 5)(h - 4) \Rightarrow 0 = 2h - 5$ or $0 = h - 4$
$\Rightarrow h = \frac{5}{2} = 2.5$ or $h = 4$
So the temperature is 0 °C after 2.5 hours and again after 4 hours.

Q9 *This question looks harder because it has y's in it as well as x's — just treat the y as a constant. You'll need two numbers which multiply to give $8y^2$ and add or subtract to give 6y.*
$4y$ and $2y$ multiply to $8y^2$ and add to give $6y$ so these are the numbers you need. $x^2 + 6xy + 8y^2 = (x + 4y)(x + 2y)$

Q10 Width of the square $= \frac{2z + 5}{4}$, so area $= \left(\frac{2z + 5}{4}\right)^2$.
$\left(\frac{2z + 5}{4}\right)^2 = \frac{9z}{2} \Rightarrow \frac{(2z + 5)^2}{16} = \frac{9z}{2}$
$\Rightarrow (2z + 5)^2 = 72z \Rightarrow (2z + 5)(2z + 5) = 72z$
$\Rightarrow 4z^2 + 20z + 25 = 72z \Rightarrow 4z^2 - 52z + 25 = 0$
$\Rightarrow (2z - 1)(2z - 25) = 0 \Rightarrow 2z - 1 = 0$ or $2z - 25 = 0$
$\Rightarrow z = \frac{1}{2} = 0.5$ or $z = \frac{25}{2} = 12.5$

Q11 *First, set up and solve a quadratic equation in x.*
Area of the triangle = $\frac{1}{2}(5x - 3)(x + 2) = (9x + 3)$
$\Rightarrow (5x - 3)(x + 2) = 2(9x + 3) \Rightarrow 5x^2 + 7x - 6 = 18x + 6$
$\Rightarrow 5x^2 - 11x - 12 = 0 \Rightarrow (5x + 4)(x - 3) = 0$
$\Rightarrow 5x + 4 = 0$ or $x - 3 = 0 \Rightarrow x = -\frac{4}{5}$ or $x = 3$
The lengths must be positive, and when $x = -\frac{4}{5}$,
$5x - 3 = -4 - 3 = -7$. So x must be 3, and the lengths are $5(3) - 3 = 12$ m, and $3 + 2 = 5$ m.
Using Pythagoras, hypotenuse = $\sqrt{12^2 + 5^2} = 13$ m.

Exercise 4.1.2 — The quadratic formula

Q1 a) $x^2 - 4x = -2 \Rightarrow x^2 - 4x + 2 = 0 \Rightarrow a = 1, b = -4, c = 2$

$$x = \frac{-b \pm \sqrt{b^2 - 4ac}}{2a} = \frac{-(-4) \pm \sqrt{(-4)^2 - 4 \times 1 \times 2}}{2 \times 1}$$
$$= \frac{4 \pm \sqrt{16 - 8}}{2} = \frac{4 \pm \sqrt{8}}{2} = \frac{4 \pm 2\sqrt{2}}{2} = 2 \pm \sqrt{2}$$

b) $x^2 - 2x - 44 = 0 \Rightarrow a = 1, b = -2, c = -44$

$$x = \frac{-b \pm \sqrt{b^2 - 4ac}}{2a}$$
$$= \frac{-(-2) \pm \sqrt{(-2)^2 - 4 \times 1 \times (-44)}}{2 \times 1}$$
$$= \frac{2 \pm \sqrt{4 + (4 \times 1 \times 44)}}{2}$$
$$= \frac{2 \pm \sqrt{180}}{2} = \frac{2 \pm \sqrt{36 \times 5}}{2}$$
$$= \frac{2 \pm 6\sqrt{5}}{2} = 1 \pm 3\sqrt{5}$$

c) $x^2 + 3x - 12 = 0 \Rightarrow a = 1, b = 3, c = -12$

$$x = \frac{-b \pm \sqrt{b^2 - 4ac}}{2a} = \frac{-3 \pm \sqrt{3^2 - 4 \times 1 \times (-12)}}{2 \times 1}$$
$$= \frac{-3 \pm \sqrt{9 + (4 \times 1 \times 12)}}{2}$$
$$= \frac{-3 \pm \sqrt{57}}{2} = -\frac{3}{2} \pm \frac{1}{2}\sqrt{57}$$

d) $x^2 - 14x + 42 = 0 \Rightarrow a = 1, b = -14, c = 42$

$$x = \frac{-b \pm \sqrt{b^2 - 4ac}}{2a}$$
$$= \frac{-(-14) \pm \sqrt{(-14)^2 - 4 \times 1 \times 42}}{2 \times 1}$$
$$= \frac{14 \pm \sqrt{196 - 168}}{2} = \frac{14 \pm \sqrt{28}}{2} = \frac{14 \pm \sqrt{4 \times 7}}{2}$$
$$= \frac{14 \pm 2\sqrt{7}}{2} = 7 \pm \sqrt{7}$$

e) $4x^2 + 4x - 1 = 0 \Rightarrow a = 4, b = 4, c = -1$

$$x = \frac{-b \pm \sqrt{b^2 - 4ac}}{2a} = \frac{-4 \pm \sqrt{(4)^2 - 4 \times 4 \times (-1)}}{2 \times 4}$$
$$= \frac{-4 \pm \sqrt{16 + 16}}{8} = \frac{-4 \pm \sqrt{32}}{8} = \frac{-4 \pm \sqrt{16 \times 2}}{8}$$
$$= \frac{-4 \pm 4\sqrt{2}}{8} = -\frac{1}{2} \pm \frac{1}{2}\sqrt{2}$$

f) $-x^2 + 4x - 3 = 0 \Rightarrow a = -1, b = 4, c = -3$

$$x = \frac{-b \pm \sqrt{b^2 - 4ac}}{2a} = \frac{-4 \pm \sqrt{4^2 - 4 \times (-1) \times (-3)}}{2 \times (-1)}$$
$$= \frac{-4 \pm \sqrt{16 - (4 \times 1 \times 3)}}{-2}$$
$$= \frac{-4 \pm \sqrt{4}}{-2} = \frac{-4 \pm 2}{-2} = 1 \text{ or } 3$$

g) $x^2 - \frac{5}{6}x + \frac{1}{6} = 0 \Rightarrow 6x^2 - 5x + 1 = 0 \Rightarrow a = 6, b = -5, c = 1$

$$x = \frac{-b \pm \sqrt{b^2 - 4ac}}{2a} = \frac{-(-5) \pm \sqrt{(-5)^2 - 4 \times 6 \times 1}}{2 \times 6}$$
$$= \frac{5 \pm \sqrt{25 - 24}}{12} = \frac{5 \pm 1}{12} = \frac{1}{2} \text{ or } \frac{1}{3}$$

Removing the fractions right at the start here saves you lots of fraction headaches in the working. This one wasn't actually too hard to factorise — you'd get (3x – 1)(2x – 1).

h) $x^2 - 2\sqrt{11}x + 11 = 0 \Rightarrow a = 1, b = -2\sqrt{11}, c = 11$

$$x = \frac{-b \pm \sqrt{b^2 - 4ac}}{2a}$$
$$= \frac{-(-2\sqrt{11}) \pm \sqrt{(-2\sqrt{11})^2 - 4 \times 1 \times 11}}{2 \times 1}$$
$$= \frac{2\sqrt{11} \pm \sqrt{44 - 44}}{2} = \frac{2\sqrt{11} \pm \sqrt{0}}{2}$$
$$= \frac{2\sqrt{11}}{2} = \sqrt{11}$$

There's only one solution to this quadratic — it factorises to give $(x - \sqrt{11})^2 = 0$.

Q2 a) $(x - 2 + \sqrt{5})(x - 2 - \sqrt{5})$
$= x(x - 2 - \sqrt{5}) - 2(x - 2 - \sqrt{5}) + \sqrt{5}(x - 2 - \sqrt{5})$
$= x^2 - 2x - \sqrt{5}x - 2x + 4 + 2\sqrt{5} + \sqrt{5}x - 2\sqrt{5} - 5$
$= x^2 - 4x - 1$
Use the method for multiplying out long brackets from Chapter 3.

b) $x^2 - 4x - 1 = 0 \Rightarrow a = 1, b = -4, c = -1$

$$x = \frac{-b \pm \sqrt{b^2 - 4ac}}{2a}$$
$$= \frac{-(-4) \pm \sqrt{(-4)^2 - 4 \times 1 \times (-1)}}{2 \times 1}$$
$$= \frac{4 \pm \sqrt{16 + 4}}{2} = \frac{4 \pm \sqrt{20}}{2} = \frac{4 \pm 2\sqrt{5}}{2} = 2 \pm \sqrt{5}$$

c) The roots produced by the quadratic formula in part b) are the same as the numbers subtracted from x in the expression from a) — this is because it's just the factorised version of the same quadratic. If you put the factorised version equal to zero and solved the equation, you'd get the same roots.

Q3 $x^2 + 8x + 13 = 0 \Rightarrow a = 1, b = 8, c = 13$

$$x = \frac{-b \pm \sqrt{b^2 - 4ac}}{2a} = \frac{-8 \pm \sqrt{8^2 - 4 \times 1 \times 13}}{2 \times 1}$$
$$= \frac{-8 \pm \sqrt{64 - 52}}{2} = \frac{-8 \pm \sqrt{12}}{2}$$
$$= \frac{-8 \pm 2\sqrt{3}}{2} = -4 \pm \sqrt{3}$$

So A = – 4 and B = 3.

Q4 a) $x^2 + x + \frac{1}{4} = 0 \Rightarrow a = 1, b = 1, c = \frac{1}{4}$

$$x = \frac{-b \pm \sqrt{b^2 - 4ac}}{2a} = \frac{-1 \pm \sqrt{1^2 - 4 \times 1 \times \frac{1}{4}}}{2 \times 1}$$
$$= \frac{-1 \pm \sqrt{1 - 1}}{2} = \frac{-1 \pm 0}{2} = -\frac{1}{2}$$

Multiplying the first equation by 4 gives $4x^2 + 4x + 1 = 0$. This factorises to $(2x + 1)^2 = 0$, giving the same answer.

b) $x^2 - \frac{7}{4}x + \frac{2}{3} = 0 \Rightarrow 12x^2 - 21x + 8 = 0$
$\Rightarrow a = 12, b = -21, c = 8$

$$x = \frac{-b \pm \sqrt{b^2 - 4ac}}{2a}$$
$$= \frac{-(-21) \pm \sqrt{(-21)^2 - 4 \times 12 \times 8}}{2 \times 12}$$
$$= \frac{21 \pm \sqrt{441 - 384}}{24} = \frac{21 \pm \sqrt{57}}{24}$$
$$= \frac{21}{24} \pm \frac{\sqrt{57}}{24} = \frac{7}{8} \pm \frac{1}{24}\sqrt{57}$$

c) $25x^2 - 30x + 7 = 0 \Rightarrow a = 25, b = -30, c = 7$

$$x = \frac{-b \pm \sqrt{b^2 - 4ac}}{2a}$$
$$= \frac{-(-30) \pm \sqrt{(-30)^2 - 4 \times 25 \times 7}}{2 \times 25}$$
$$= \frac{30 \pm \sqrt{900 - 700}}{2 \times 25} = \frac{30 \pm \sqrt{200}}{50}$$
$$= \frac{30 \pm 10\sqrt{2}}{50} = \frac{30}{50} \pm \frac{10}{50}\sqrt{2} = \frac{3}{5} \pm \frac{1}{5}\sqrt{2}$$

d) $60x - 5 = -100x^2 - 3 \Rightarrow 100x^2 + 60x - 2 = 0$
$a = 100,\ b = 60,\ c = -2$

$$x = \frac{-b \pm \sqrt{b^2 - 4ac}}{2a}$$
$$= \frac{-60 \pm \sqrt{60^2 - 4 \times 100 \times (-2)}}{2 \times 100}$$
$$= \frac{-60 \pm \sqrt{3600 + 800}}{200} = \frac{-60 \pm \sqrt{4400}}{200}$$
$$= \frac{-60 \pm \sqrt{44 \times 100}}{200} = \frac{-60 \pm \sqrt{4 \times 11 \times 100}}{200}$$
$$= \frac{-60 \pm \sqrt{4}\sqrt{100}\sqrt{11}}{200} = \frac{-60 \pm 20\sqrt{11}}{200}$$
$$= -\frac{3}{10} \pm \frac{1}{10}\sqrt{11}$$

e) $2x(x - 4) = 7 - 3x \Rightarrow 2x^2 - 8x = 7 - 3x$
$\Rightarrow 2x^2 - 5x - 7 = 0 \Rightarrow a = 2,\ b = -5,\ c = -7$

$$x = \frac{-b \pm \sqrt{b^2 - 4ac}}{2a}$$
$$= \frac{-(-5) \pm \sqrt{(-5)^2 - 4 \times 2 \times (-7)}}{2 \times 2}$$
$$= \frac{5 \pm \sqrt{25 + 56}}{4} = \frac{5 \pm \sqrt{81}}{4} = \frac{5 \pm 9}{4}$$
$$= \frac{5+9}{4} \text{ or } \frac{5-9}{4} = \frac{14}{4} \text{ or } \frac{-4}{4}$$
$$= \frac{7}{2} \text{ or } -1$$

This factorises to $(2x - 7)(x + 1) = 0$

f) $(3x - 5)(x + 2) = 3x - 2 \Rightarrow 3x^2 + x - 10 = 3x - 2$
$\Rightarrow 3x^2 - 2x - 8 = 0 \Rightarrow a = 3,\ b = -2,\ c = -8$

$$x = \frac{-b \pm \sqrt{b^2 - 4ac}}{2a}$$
$$= \frac{-(-2) \pm \sqrt{(-2)^2 - 4 \times 3 \times (-8)}}{2 \times 3}$$
$$= \frac{2 \pm \sqrt{4 + 96}}{6} = \frac{2 \pm \sqrt{100}}{6} = \frac{2 \pm 10}{6}$$
$$= 2 \text{ or } -\frac{4}{3}$$

Q5 Using the quadratic formula:
$kx^2 + 4x - 2 = 0 \Rightarrow a = k,\ b = 4,\ c = -2$:

$$x = \frac{-b \pm \sqrt{b^2 - 4ac}}{2a}$$
$$= \frac{-4 \pm \sqrt{4^2 - 4 \times k \times (-2)}}{2k}$$
$$= \frac{-4 \pm \sqrt{16 + 8k}}{2k} = -\frac{4}{2k} \pm \frac{\sqrt{16 + 8k}}{2k}$$

Comparing with the given root: $-\frac{4}{2k} = -\frac{2}{5} \Rightarrow 4k = 20 \Rightarrow k = 5$

Check that this works for the other part of the root:

If $k = 5$, $\frac{\sqrt{16 + 8k}}{2k} = \frac{\sqrt{16 + 8 \times 5}}{2 \times 5} = \frac{\sqrt{56}}{10}$

$$= \frac{\sqrt{4 \times 14}}{10} = \frac{2\sqrt{14}}{10} = \frac{\sqrt{14}}{5}.$$

So the other root will be $x = -\frac{2}{5} - \frac{\sqrt{14}}{5}$.

Q6 The shopper will arrive home at the first point when $d = 0$, so use the quadratic formula to find the smallest root of the equation $2t^2 - 5t + 1 = 0 \Rightarrow a = 2,\ b = -5,\ c = 1$

$$t = \frac{-b \pm \sqrt{b^2 - 4ac}}{2a}$$
$$= \frac{-(-5) \pm \sqrt{(-5)^2 - 4 \times 2 \times 1}}{2 \times 2}$$
$$= \frac{5 \pm \sqrt{25 - 8}}{4} = \frac{5 \pm \sqrt{17}}{4}$$
$= 0.2192...$ or $2.2807...$

The smallest root is $t = 0.2192...$ hours $= 13.1534...$ minutes.
So they arrive home at 2.13 pm to the nearest minute.

Exercise 4.1.3 — Completing the square

Q1 **a)** Take the square root of both sides to get: $x + 4 = \pm\sqrt{25}$
$\Rightarrow x = -4 \pm \sqrt{25} = -4 \pm 5$. So $x = 1$ or -9

b) Take the square root of both sides to get: $2x + 5 = \pm\sqrt{9}$
$\Rightarrow 2x = -5 \pm \sqrt{9} = -5 \pm 3$. So $2x = -2$ or $-8 \Rightarrow x = -1$ or -4

c) Take the square root of both sides to get: $5x - 3 = \pm\sqrt{21}$
$\Rightarrow 5x = 3 \pm \sqrt{21} \Rightarrow x = \frac{3}{5} \pm \frac{\sqrt{21}}{5}$

Q2 **a)** $x^2 + 6x + 8 = (x + 3)^2 - 9 + 8 = (x + 3)^2 - 1$

b) $x^2 + 8x - 10 = (x + 4)^2 - 16 - 10 = (x + 4)^2 - 26$

c) $x^2 - 3x - 10 = \left(x - \frac{3}{2}\right)^2 - \frac{9}{4} - 10$
$= \left(x - \frac{3}{2}\right)^2 - \frac{9}{4} - \frac{40}{4} = \left(x - \frac{3}{2}\right)^2 - \frac{49}{4}$

d) $x^2 - 20x + 15 = (x - 10)^2 - 100 + 15 = (x - 10)^2 - 85$

e) $x^2 - 2mx + n = (x - m)^2 - m^2 + n = (x - m)^2 + (-m^2 + n)$

f) $x^2 + 6tx + s = (x + 3t)^2 - 9t^2 + s = (x + 3t)^2 + (-9t^2 + s)$

g) $3x^2 - 12x + 7 = 3(x - 2)^2 - 12 + 7 = 3(x - 2)^2 - 5$

h) $2x^2 - 4x - 3 = 2(x - 1)^2 - 2 - 3 = 2(x - 1)^2 - 5$

i) $6x^2 + 30x - 20 = 6\left(x + \frac{5}{2}\right)^2 - \frac{75}{2} - 20 = 6\left(x + \frac{5}{2}\right)^2 - \frac{115}{2}$

j) $-x^2 - 9x + 9 = -\left(x + \frac{9}{2}\right)^2 + \frac{81}{4} + 9 = -\left(x + \frac{9}{2}\right)^2 + \frac{117}{4}$

k) $4x^2 - 22x + 5 = 4\left(x - \frac{11}{4}\right)^2 - \frac{121}{4} + 5 = 4\left(x - \frac{11}{4}\right)^2 - \frac{101}{4}$

l) $-3x^2 + 9x + 1 = -3\left(x - \frac{3}{2}\right)^2 + \frac{27}{4} + 1 = -3\left(x - \frac{3}{2}\right)^2 + \frac{31}{4}$

Q3 **a)** First complete the square of the expression:
$x^2 - 6x - 16 = (x - 3)^2 - 9 - 16 = (x - 3)^2 - 25$
Now set the completed square equal to zero:
$(x - 3)^2 - 25 = 0 \Rightarrow (x - 3)^2 = 25$
$\Rightarrow x - 3 = \pm\sqrt{25} \Rightarrow x = 3 \pm \sqrt{25} = 3 \pm 5 \Rightarrow x = 8$ or -2

b) Write the equation in standard quadratic form:
$p^2 - 10p = 200 \Rightarrow p^2 - 10p - 200 = 0$
Then complete the square of the expression:
$p^2 - 10p - 200 = (p - 5)^2 - 25 - 200 = (p - 5)^2 - 225$
Now set the completed square equal to zero:
$(p - 5)^2 - 225 = 0 \Rightarrow (p - 5)^2 = 225 \Rightarrow p - 5 = \pm\sqrt{225}$
$\Rightarrow p = 5 \pm \sqrt{225} = 5 \pm 15 \Rightarrow p = 20$ or -10

c) First complete the square of the expression:
$x^2 + 2x + k = (x + 1)^2 - 1 + k = (x + 1)^2 + (k - 1)$
Now set the completed square equal to zero:
$(x + 1)^2 + (k - 1) = 0 \Rightarrow (x + 1)^2 = 1 - k$
$\Rightarrow x + 1 = \pm\sqrt{1 - k} \Rightarrow x = -1 \pm \sqrt{1 - k}$

d) First complete the square of the expression:
$x^2 + 4x - 8 = (x + 2)^2 - 4 - 8 = (x + 2)^2 - 12$
Now set the completed square equal to zero:
$(x + 2)^2 - 12 = 0 \Rightarrow (x + 2)^2 = 12$
$\Rightarrow x + 2 = \pm\sqrt{12} \Rightarrow x = \pm\sqrt{12} - 2$. So $x = -2 \pm 2\sqrt{3}$

e) First complete the square of the expression:
$4x^2 + 24x - 13 = 4(x + 3)^2 - 36 - 13 = 4(x + 3)^2 - 49$
Now set the completed square equal to zero:
$4(x + 3)^2 - 49 = 0 \Rightarrow 4(x + 3)^2 = 49 \Rightarrow x + 3 = \pm\sqrt{\frac{49}{4}}$
$\Rightarrow x = -3 \pm \sqrt{\frac{49}{4}} \Rightarrow x = -3 \pm \frac{7}{2}$, so $x = \frac{1}{2}$ or $-\frac{13}{2}$

f) Write the equation in standard quadratic form:
$9x^2 + 18x = 16 \Rightarrow 9x^2 + 18x - 16 = 0$
Then complete the square of the expression:
$9x^2 + 18x - 16 = 9(x + 1)^2 - 9 - 16 = 9(x + 1)^2 - 25$
Now set the completed square equal to zero:
$9(x + 1)^2 - 25 = 0 \Rightarrow 9(x + 1)^2 = 25$
$\Rightarrow (x + 1)^2 = \frac{25}{9} \Rightarrow x + 1 = \pm\sqrt{\frac{25}{9}}$
$\Rightarrow x = -1 \pm \sqrt{\frac{25}{9}} \Rightarrow x = -1 \pm \frac{5}{3}$. So $x = \frac{2}{3}$ or $-\frac{8}{3}$

g) First complete the square of the expression:
$2x^2 - 12x + 9 = 2(x - 3)^2 - 18 + 9 = 2(x - 3)^2 - 9$
Now set the completed square equal to zero:
$2(x - 3)^2 - 9 = 0 \Rightarrow 2(x - 3)^2 = 9 \Rightarrow (x - 3)^2 = \frac{9}{2}$

$\Rightarrow x - 3 = \pm\sqrt{\frac{9}{2}} \Rightarrow x = 3 \pm \sqrt{\frac{9}{2}} \Rightarrow x = 3 \pm \frac{3}{\sqrt{2}} = 3 \pm \frac{3\sqrt{2}}{2}$

Here you should rationalise the denominator by multiplying the top and bottom of the fraction by $\sqrt{2}$.

h) First divide through by 2:
$x^2 - 6x - 27 = (x - 3)^2 - 9 - 27 = (x - 3)^2 - 36$
Now set the completed square equal to zero:
$(x - 3)^2 - 36 = 0 \Rightarrow (x - 3)^2 = 36$
$\Rightarrow x - 3 = \pm 6 \Rightarrow x = 3 \pm 6$. So $x = 9$ or -3

i) Write the equation in standard quadratic form:
$5x^2 + 10x = 1 \Rightarrow 5x^2 + 10x - 1 = 0$

Then complete the square of the expression:
$5x^2 + 10x - 1 = 5(x + 1)^2 - 5 - 1 = 5(x + 1)^2 - 6$
Now set the completed square equal to zero:
$5(x + 1)^2 - 6 = 0 \Rightarrow 5(x + 1)^2 = 6$

$\Rightarrow (x + 1)^2 = \frac{6}{5} \Rightarrow x + 1 = \pm\sqrt{\frac{6}{5}} \Rightarrow x = -1 \pm \sqrt{\frac{6}{5}}$

$\Rightarrow x = -1 \pm \frac{\sqrt{6} \times \sqrt{5}}{\sqrt{5} \times \sqrt{5}}$. So $x = -1 \pm \frac{\sqrt{30}}{5}$

j) First complete the square of the expression:
$-3x^2 - 18x + 2 = -3(x + 3)^2 + 27 + 2 = -3(x + 3)^2 + 29$

Now set the completed square equal to zero:
$-3(x + 3)^2 + 29 = 0 \Rightarrow -3(x + 3)^2 = -29$

$\Rightarrow (x + 3)^2 = \frac{-29}{-3} \Rightarrow x + 3 = \pm\sqrt{\frac{29}{3}}$

$\Rightarrow x = -3 \pm \sqrt{\frac{29}{3}} = -3 \pm \frac{\sqrt{29} \times \sqrt{3}}{\sqrt{3} \times \sqrt{3}} = -3 \pm \frac{\sqrt{87}}{3}$

k) Write the equation in standard quadratic form:
$3x^2 + 2x = \frac{7}{6} \Rightarrow 3x^2 + 2x - \frac{7}{6} = 0$
Then complete the square of the expression:
$3x^2 + 2x - \frac{7}{6} = 3\left(x + \frac{1}{3}\right)^2 - \frac{1}{3} - \frac{7}{6} = 3\left(x + \frac{1}{3}\right)^2 - \frac{3}{2}$
Now set the completed square equal to zero:
$3\left(x + \frac{1}{3}\right)^2 - \frac{3}{2} = 0 \Rightarrow 3\left(x + \frac{1}{3}\right)^2 = \frac{3}{2}$

$\Rightarrow \left(x + \frac{1}{3}\right)^2 = \frac{1}{2} \Rightarrow x + \frac{1}{3} = \pm\sqrt{\frac{1}{2}}$

$\Rightarrow x = -\frac{1}{3} \pm \frac{\sqrt{1} \times \sqrt{2}}{\sqrt{2} \times \sqrt{2}} \Rightarrow x = -\frac{1}{3} \pm \frac{\sqrt{2}}{2}$

l) First complete the square of the expression:

$5x^2 - 3x + \frac{2}{5} = 5\left(x - \frac{3}{10}\right)^2 - \frac{9}{20} + \frac{2}{5}$
$= 5\left(x - \frac{3}{10}\right)^2 + \frac{-9 + 8}{20} = 5\left(x - \frac{3}{10}\right)^2 - \frac{1}{20}$

Now set the completed square equal to zero:

$5\left(x - \frac{3}{10}\right)^2 - \frac{1}{20} = 0 \Rightarrow 5\left(x - \frac{3}{10}\right)^2 = \frac{1}{20}$

$\Rightarrow \left(x - \frac{3}{10}\right)^2 = \frac{1}{100} \Rightarrow x - \frac{3}{10} = \pm\sqrt{\frac{1}{100}}$

$\Rightarrow x - \frac{3}{10} = \pm\frac{1}{10} \Rightarrow x = \frac{3}{10} \pm \frac{1}{10}$. So $x = \frac{2}{5}$ or $\frac{1}{5}$

Q4 Complete the square of the expression:
$3x^2 - 12x + 14 = 3(x - 2)^2 - 12 + 14 = 3(x - 2)^2 + 2$
Since $(x - 2)^2 \geq 0$ for all x, $3(x - 2)^2 + 2 \geq 2$
So $3x^2 - 12x + 14 > 0$ for all x, as required.

Q5 First complete the square of the expression:
$ax^2 + bx + c = a(x^2 + \frac{b}{a}x + \frac{c}{a}) = a\left(x + \frac{b}{2a}\right)^2 - \frac{b^2}{4a} + c$
Now set the completed square equal to zero and rearrange to find the roots:

$a\left(x + \frac{b}{2a}\right)^2 - \frac{b^2}{4a} + c = 0 \Rightarrow \left(x + \frac{b}{2a}\right)^2 = \frac{b^2}{4a^2} - \frac{c}{a}$

$\Rightarrow \left(x + \frac{b}{2a}\right)^2 = \frac{b^2 - 4ac}{4a^2} \Rightarrow x + \frac{b}{2a} = \pm\sqrt{\frac{b^2 - 4ac}{4a^2}}$

$\Rightarrow x = -\frac{b}{2a} \pm \frac{\sqrt{b^2 - 4ac}}{2a} \Rightarrow x = \frac{-b \pm \sqrt{b^2 - 4ac}}{2a}$

This last question was quite tricky, but if you got there you should have noticed something quite special — you've just proved the quadratic formula.

Exercise 4.1.4 — Quadratics involving functions of x

Q1 **a)** $u = x^{\frac{1}{2}} (= \sqrt{x})$

b) $e^x(e^x - 6) = 8 \Rightarrow e^{2x} - 6e^x - 8 = 0$
$\Rightarrow (e^x)^2 - 6(e^x) - 8 = 0$. So $u = e^x$

c) $5^x + 5^{2x} = 4 \Rightarrow 5^{2x} + 5^x - 4 = 0$
$\Rightarrow (5^x)^2 + (5^x) - 4 = 0$. So $u = 5^x$.

d) $2\cos^2 x + 3 = 5\cos x \Rightarrow 2\cos^2 x - 5\cos x + 3 = 0$
$\Rightarrow 2(\cos x)^2 - 5(\cos x) + 3 = 0$. So $u = \cos x$

Q2 **a)** $x^2 + 6x + 7 = (x + 3)^2 - 9 + 7 = (x + 3)^2 - 2$
This is just completing the square.

b) Let $u = (2x + 1)$
So $(2x + 1)^2 + 6(2x + 1) + 7$ becomes
$u^2 + 6u + 7$, which can be written as $(u + 3)^2 - 2$.
$(u + 3)^2 - 2 = 0 \Rightarrow (u + 3)^2 = 2 \Rightarrow u + 3 = \pm\sqrt{2}$
$\Rightarrow u = -3 \pm \sqrt{2}$
Now replace u with $(2x + 1)$:
$(2x + 1) = -3 \pm \sqrt{2} \Rightarrow 2x = -4 \pm \sqrt{2}$
So $x = -2 \pm \frac{\sqrt{2}}{2}$

Q3 Let $u = x^2$
So $x^4 - 17x^2 + 16 = 0$ becomes $u^2 - 17u + 16 = 0$
$(u - 1)(u - 16) = 0 \Rightarrow u = 1$ or $u = 16$
This means $x^2 = 1$ or $x^2 = 16$, so $x = \pm 1$ or $x = \pm 4$
So the four solutions are $x = 1$, $x = -1$, $x = 4$ and $x = -4$.

Q4 **a)** Let $u = \frac{1}{5x + 2}$
So $\frac{3}{(5x + 2)^2} + \frac{1}{5x + 2} = 10$ becomes $3u^2 + u - 10 = 0$

$(3u - 5)(u + 2) = 0 \Rightarrow u = \frac{5}{3}$ or $u = -2$
This means $\frac{1}{5x + 2} = \frac{5}{3} \Rightarrow 3 = 5(5x + 2)$
$\Rightarrow 3 = 25x + 10 \Rightarrow 25x = -7 \Rightarrow x = -\frac{7}{25}$ or $\frac{1}{5x + 2} = -2$
$\Rightarrow 1 = -2(5x + 2) \Rightarrow 1 = -10x - 4 \Rightarrow 10x = -5 \Rightarrow x = -\frac{1}{2}$

b) Let $u = \sqrt{x}$
So $3x + \sqrt{x} = 14$ becomes $3u^2 + u = 14 \Rightarrow 3u^2 + u - 14 = 0$
$(3u + 7)(u - 2) = 0 \Rightarrow u = -\frac{7}{3}$ or $u = 2$
This means $\sqrt{x} = -\frac{7}{3} \Rightarrow x = \left(-\frac{7}{3}\right)^2 = \frac{49}{9}$ or $\sqrt{x} = 2 \Rightarrow x = 4$.

Q5 At the point at which there are 1000 more birds,
$p = 1$, so $5y^3 - y^6 = 1$.
Let $u = y^3$, so $5y^3 - y^6 = 1$ becomes $5u - u^2 = 1$
$\Rightarrow u^2 - 5u + 1 = 0$. This won't factorise, so use the quadratic formula: $a = 1$, $b = -5$, $c = 1$.

$u = \frac{-b \pm \sqrt{b^2 - 4ac}}{2a} = \frac{-(-5) \pm \sqrt{(-5)^2 - 4 \times 1 \times 1}}{2 \times 1}$
$= \frac{5 \pm \sqrt{25 - 4}}{2} = \frac{5 \pm \sqrt{21}}{2}$

You could also complete the square to solve the quadratic instead of using the formula.

So $u = 0.2087...$ or $4.7912...$, which means that
$y^3 = 0.2087... \Rightarrow y = \sqrt[3]{0.2087...} = 0.5931...$
or $y^3 = 4.7912... \Rightarrow y = \sqrt[3]{4.7912...} = 1.6858...$
You need the smallest amount of time, so $y = 0.5931... \times 100 =$ 59 years to the nearest year.
Be careful with the units — p is in 1000s of birds and y is in 100s of years.

4.2 Quadratic Functions and Roots

Exercise 4.2.1 — The roots of a quadratic function

Q1 **a)** 2 real roots. **b)** 1 real root.
c) no real roots. **d)** 2 real roots.

Q2 Completing the square:
$f(x) = x^2 + 6x + 10 = (x + 3)^2 - 9 + 10 = (x + 3)^2 + 1$
The smallest the $(x + 3)^2$ bit can be is 0, and 1 is positive which means that f(x) is always positive and the smallest it can be is 1. So f(x) has no real roots.
$q = 3$, so the graph has a line of symmetry at $x = -3$.

Q3 Comparing $f(x) = -\left(x+\frac{7}{2}\right)^2+\frac{25}{4}$ to $p(x + q)^2 + r$ gives $p = -1$, $r = \frac{25}{4}$
These have different signs, so f(x) has two real roots.
Alternatively, $-\left(x+\frac{7}{2}\right)^2+\frac{25}{4} = 0 \Rightarrow \left(x+\frac{7}{2}\right)^2 = \frac{25}{4}$.
This can be solved by taking the square root (since the RHS is positive) — so it has real roots.

Q4 Completing the square:
$g(x) = 4x^2 - 3x - 5 = 4\left(x-\frac{3}{8}\right)^2-\frac{9}{16}-5$
$= 4\left(x-\frac{3}{8}\right)^2-\frac{89}{16}$, so $p = 4$ and $r = -\frac{89}{16}$.
p and r have different signs, so g(x) has two real roots.
The turning point has coordinates $(-q, r) = \left(\frac{3}{8}, -\frac{89}{16}\right)$

Q5 **a)** Completing the square:
$f(x) = -2x^2 + 3x + k = -2\left(x-\frac{3}{4}\right)^2+\frac{9}{8}+k$.
There is exactly one real root, so $\frac{9}{8} + k = 0 \Rightarrow k = -\frac{9}{8}$.

b) The graph is symmetrical about the single root, i.e. where $-2\left(x-\frac{3}{4}\right)^2 = 0 \Rightarrow x = \frac{3}{4}$ is the line of symmetry.
This is the value of $-q$ when the equation is in the form $p(x + q)^2 + r$.

c) The turning point has coordinates $(-q, r)$ when the equation is in the form $p(x + q)^2 + r$.
$f(x) = -2\left(x-\frac{3}{4}\right)^2$, so the turning point is at $\left(\frac{3}{4}, 0\right)$.
p is negative, so the graph is n-shaped and the turning point is a maximum.

Exercise 4.2.2 — Using the discriminant

Q1 **a)** $a = 1, b = 8, c = 15$.
So $b^2 - 4ac = 8^2 - 4 \times 1 \times 15 = 64 - 60 = 4$.
Discriminant > 0 so the equation has 2 real roots.

b) $a = 1, b = 2\sqrt{3}, c = 3$.
So $b^2 - 4ac = (2\sqrt{3})^2 - 4 \times 1 \times 3 = 12 - 12 = 0$.
Discriminant = 0 so the equation has 1 real root.

c) Write in standard form: $(2x + 1)(5x - 3) = 10x^2 - x - 3$
so $a = 10$, $b = -1$ and $c = -3$.
$b^2 - 4ac = (-1)^2 - 4 \times 10 \times -3 = 1 + 120 = 121$
Discriminant > 0 so the equation has 2 real roots.

d) $a = -3, b = -\frac{11}{5}, c = -\frac{2}{5}$.
So $b^2 - 4ac = \left(-\frac{11}{5}\right)^2 - 4 \times (-3) \times \left(-\frac{2}{5}\right)$
$= \frac{121}{25} - \frac{24}{5} = \frac{121}{25} - \frac{120}{25} = \frac{1}{25}$
Discriminant > 0 so the equation has 2 real roots.

e) $a = 9, b = 20, c = 0$.
So $b^2 - 4ac = 20^2 - 4 \times 9 \times 0 = 400 - 0 = 400$.
Discriminant > 0 so the equation has 2 real roots.

f) $a = \frac{19}{16}, b = 0, c = -4$.
So $b^2 - 4ac = 0^2 - 4 \times \frac{19}{16} \times (-4) = 0 + 19 = 19$.
Discriminant > 0 so the equation has 2 real roots.

Q2 Find the discriminant of the equation by first writing it in standard form: $15x^2 + bx = 2 \Rightarrow 15x^2 + bx - 2 = 0$
$a = 15, b = b, c = -2$.
So $b^2 - 4ac = b^2 - 4 \times 15 \times (-2) = b^2 + 120$.
Now you know that the discriminant is 169 so let $b^2 + 120 = 169$
$\Rightarrow b^2 = 49 \Rightarrow b = \pm 7$.

Q3 First find the discriminant: $a = a, b = 7, c = \frac{1}{4}$.
So $b^2 - 4ac = 7^2 - 4 \times a \times \frac{1}{4} = 49 - a$.
The equation has one real root which means its discriminant must be 0. So $49 - a = 0 \Rightarrow a = 49$.

Q4 **a)** $a = 13, b = 8, c = 2$
so $b^2 - 4ac = 8^2 - 4 \times 13 \times 2 = 64 - 104 = -40$.
The discriminant is negative so the equation has no real roots.

b) $a = \frac{1}{3}, b = \frac{5}{2}, c = 3$
so $b^2 - 4ac = \left(\frac{5}{2}\right)^2 - 4 \times \frac{1}{3} \times 3 = \frac{25}{4} - 4 = \frac{25}{4} - \frac{16}{4} = \frac{9}{4}$
The discriminant is positive so there are two real roots.

c) $a = -\frac{1}{2}, b = -\frac{1}{3}, c = 4$
Don't get caught out by the order of the terms in the equation here.
$b^2 - 4ac = \left(-\frac{1}{3}\right)^2 - 4 \times \left(-\frac{1}{2}\right) \times 4 = \frac{1}{9} + 8 = \frac{73}{9}$
The discriminant is positive so there are two real roots.

Q5 $a = 1, b = -12, c = 27 + p$.
So $b^2 - 4ac = (-12)^2 - 4 \times 1 \times (27 + p)$
$= 144 - (108 + 4p) = 36 - 4p$
If the equation has two distinct real roots, the discriminant must be positive so $36 - 4p > 0 \Rightarrow 36 > 4p \Rightarrow p < 9$.

Q6 $a = 10, b = -10, c = \frac{q}{2}$.
So $b^2 - 4ac = (-10)^2 - 4 \times 10 \times \frac{q}{2} = 100 - 20q$
If the equation has two distinct real roots, the discriminant must be positive so $100 - 20q > 0 \Rightarrow 100 > 20q \Rightarrow q < 5$

Q7 $a = 2, b = 10p + 1, c = 5$
So $b^2 - 4ac = (10p + 1)^2 - 4 \times 2 \times 5$
$= (100p^2 + 20p + 1) - 40 = 100p^2 + 20p - 39$
If the equation has no real roots, the discriminant must be negative so $100p^2 + 20p - 39 < 0 \Rightarrow 100p^2 + 20p < 39$
$\Rightarrow 20p(5p + 1) < 39 \Rightarrow p(5p + 1) < \frac{39}{20}$

Q8 First find the discriminant of the equation. $a = -2, b = -2, c = k$.
So $b^2 - 4ac = (-2)^2 - 4 \times (-2) \times k = 4 + 8k$.

a) If the equation has two distinct real roots, the discriminant must be positive so $4 + 8k > 0 \Rightarrow 8k > -4 \Rightarrow k > -\frac{1}{2}$

b) If the equation has one real root, the discriminant must be zero so $4 + 8k = 0 \Rightarrow k = -\frac{1}{2}$

c) If the equation has no real roots, the discriminant must be negative so $4 + 8k < 0 \Rightarrow k < -\frac{1}{2}$.

Q9 **a)** First work out the discriminant: $a = 1, b = k + 5, c = \frac{k^2}{4}$
So $b^2 - 4ac = (k + 5)^2 - 4 \times 1 \times \frac{k^2}{4}$
$= (k^2 + 10k + 25) - k^2 = 10k + 25$
The equation has no real roots so the discriminant is negative so $10k + 25 < 0$.

b) To find the range of values of k, solve the inequality in part a).
$10k + 25 < 0 \Rightarrow 10k < -25 \Rightarrow k < -\frac{25}{10} = -\frac{5}{2}$. So $k < -\frac{5}{2}$

Q10 **a)** $a = k - \frac{6}{5}, b = \sqrt{k}, c = \frac{5}{4}$
$b^2 - 4ac = (\sqrt{k})^2 - 4 \times \left(k - \frac{6}{5}\right) \times \frac{5}{4} = k - 5\left(k - \frac{6}{5}\right)$
$= k - 5k + 6 = -4k + 6$

b) **(i)** For one real root, discriminant = 0:
$-4k + 6 = 0 \Rightarrow k = \frac{6}{4} = \frac{3}{2}$

(ii) For no real roots, discriminant is negative:
$-4k + 6 < 0$ so $k > \frac{3}{2}$

(iii) For two real roots, discriminant is positive:
$-4k + 6 > 0$ so $k < \frac{3}{2}$

Q11 Find the discriminant: $a = \frac{m}{2}$, $b = \left(\frac{m}{\sqrt{3}} + 1\right)$, $c = \frac{m}{6}$

$b^2 - 4ac = \left(\frac{m}{\sqrt{3}} + 1\right)^2 - 4 \times \frac{m}{2} \times \frac{m}{6}$

$= \frac{m^2}{3} + \frac{2m}{\sqrt{3}} + 1 - \frac{m^2}{3} = \frac{2m}{\sqrt{3}} + 1$

a) For one real root, discriminant = 0:
$\frac{2m}{\sqrt{3}} + 1 = 0 \Rightarrow m = -\frac{\sqrt{3}}{2}$

b) For no real roots, discriminant is negative:
$\frac{2m}{\sqrt{3}} + 1 < 0$ so $m < -\frac{\sqrt{3}}{2}$

c) For two real roots, discriminant is positive:
$\frac{2m}{\sqrt{3}} + 1 > 0$ so $m > -\frac{\sqrt{3}}{2}$

Q12 a) The original rectangle has dimensions of $2w$ by w, so the new rectangle has an area of:
$(2w - 5)(w + 5) = 2w^2 + 5w - 25$.

b) (i) $2w^2 + 5w - 25 = 25$, so $2w^2 + 5w - 50 = 0$.
Find the discriminant: $a = 2$, $b = 5$, $c = -50$.
$b^2 - 4ac = 5^2 - 4 \times 2 \times (-50) = 25 + 400 = 425$
The discriminant is positive, so the equation has two real roots.

(ii) In practice, the value of w needs to be a positive number, and $2w - 5$ needs to be greater than 0.
The discriminant only tells us the number of real roots, not the number of positive roots.

There is only one possible solution in practice: $w = 3.9$ m to 1 d.p.

4.3 Quadratic Graphs

Exercise 4.3.1 — Sketching a quadratic graph

Q1 a) & b)

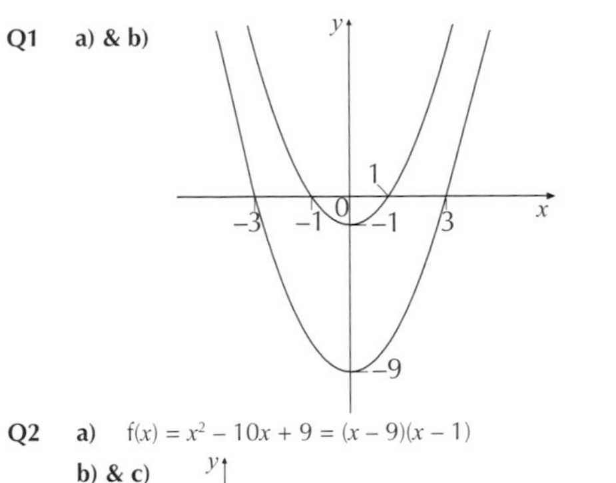

Q2 a) $f(x) = x^2 - 10x + 9 = (x - 9)(x - 1)$

b) & c)

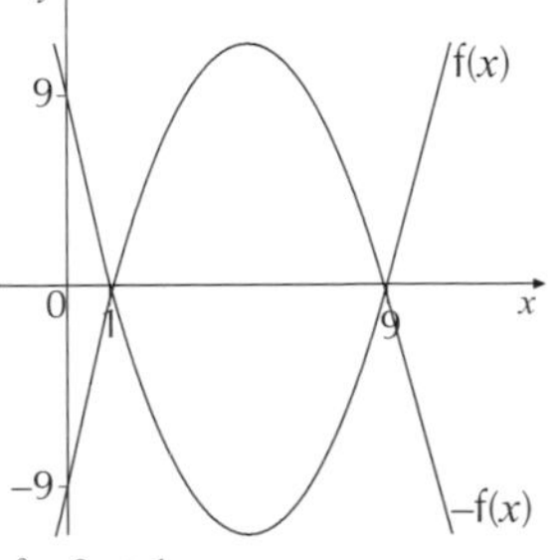

Q3 a) $y = -x^2 + 2x + 1$

(i) The x^2 coefficient is -1 so it is n-shaped.

(ii) Letting $x = 0$, $y = 1$ is the y-intercept.

(iii) Calculate the discriminant to work out the number of roots: $a = -1$, $b = 2$, $c = 1$.
$b^2 - 4ac = 2^2 - 4 \times (-1) \times 1 = 4 + 4 = 8$
The discriminant is positive so there are 2 distinct real roots.

(iv) To find the x-intercepts — find the roots:
$y = -x^2 + 2x + 1 = -(x - 1)^2 + 2$ by completing the square.
Setting this equal to zero: $-(x - 1)^2 + 2 = 0$
$\Rightarrow (x - 1)^2 = 2 \Rightarrow x - 1 = \pm\sqrt{2} \Rightarrow x = 1 \pm \sqrt{2}$

(v) The vertex is a maximum since the graph's n-shaped.
The maximum can be found by looking at the completed square $y = -(x - 1)^2 + 2$.
The highest value $-(x - 1)^2$ can take is 0, so the maximum is at $y = 2$ and $x = 1$ (to make the bracket 0) — i.e. the maximum has coordinates (1, 2).

(vi)

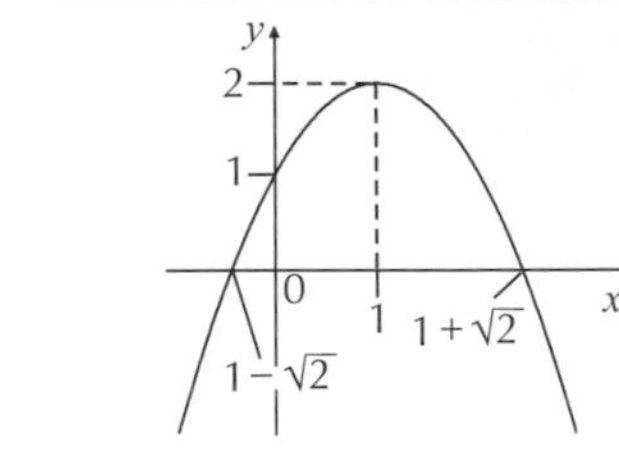

b) $y = x^2 - 7x + 15$

(i) The x^2 coefficient is 1 so the graph's u-shaped.

(ii) Letting $x = 0$, $y = 15$ is the y-intercept.

(iii) Calculate the discriminant to work out the number of roots: $a = 1$, $b = -7$, $c = 15$.
$b^2 - 4ac = (-7)^2 - 4 \times 1 \times 15 = 49 - 60 = -11$
The discriminant is negative so there are no real roots.

(iv) There are no real roots so the graph does not intersect the x-axis.

(v) The vertex is a minimum since the graph's u-shaped.
The minimum can be found by completing the square.
$y = x^2 - 7x + 15 = \left(x - \frac{7}{2}\right)^2 + \frac{11}{4}$

The lowest value $\left(x - \frac{7}{2}\right)^2$ can take is zero — so the minimum is at $y = \frac{11}{4}$ and so $x = \frac{7}{2}$ (to make the bracket 0) — i.e. it has coordinates $\left(\frac{7}{2}, \frac{11}{4}\right)$.

(vi)

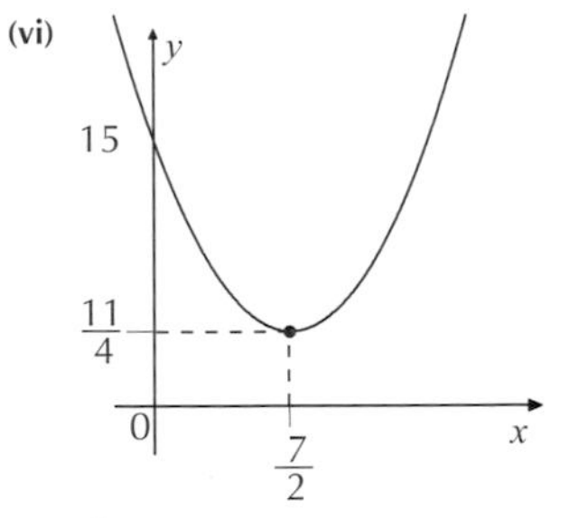

c) $y = 2x^2 + 4x - 9$

(i) The x^2 coefficient is 2 so it is u-shaped.

(ii) Letting $x = 0$, $y = -9$ is the y-intercept.

(iii) Calculate the discriminant to work out the number of roots: $a = 2$, $b = 4$, $c = -9$.
$b^2 - 4ac = 4^2 - 4 \times 2 \times (-9) = 16 + 72 = 88$
The discriminant is positive so there are 2 distinct real roots.

(iv) To find the x-intercepts, find the roots:
$2x^2 + 4x - 9 = 0$
Completing the square gives:
$2(x + 1)^2 - 2 - 9 = 0 \Rightarrow 2(x + 1)^2 - 11 = 0$
Solving: $(x + 1)^2 = \frac{11}{2} \Rightarrow x = -1 \pm \sqrt{\frac{11}{2}}$

(v) The vertex is a minimum since the graph is u-shaped. The minimum occurs when the square is 0, so it's $(-1, -11)$.

(vi)

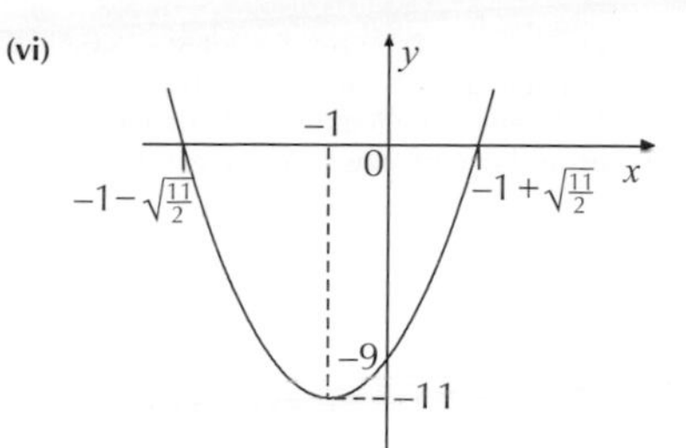

d) $y = -x^2 + 4x - 7$

(i) The x^2 coefficient is –1 so it is n-shaped.

(ii) Letting $x = 0$, $y = -7$ is the y-intercept.

(iii) Calculate the discriminant to work out the number of roots: $a = -1$, $b = 4$, $c = -7$.
$b^2 - 4ac = 4^2 - 4 \times (-1) \times (-7) = 16 - 28 = -12$
The discriminant is negative so there are no real roots.

(iv) There are no real roots so the graph does not intersect the x-axis.

(v) The vertex is a maximum since the graph is n-shaped. The maximum can be found by completing the square.
$y = -x^2 + 4x - 7 = -(x - 2)^2 + 4 - 7 \Rightarrow y = -(x - 2)^2 - 3$
The highest value $-(x - 2)^2$ can take is 0 — so the maximum is at $y = -3$ and $x = 2$ (to make the bracket 0) – i.e. it has coordinates (2, –3).

(vi)

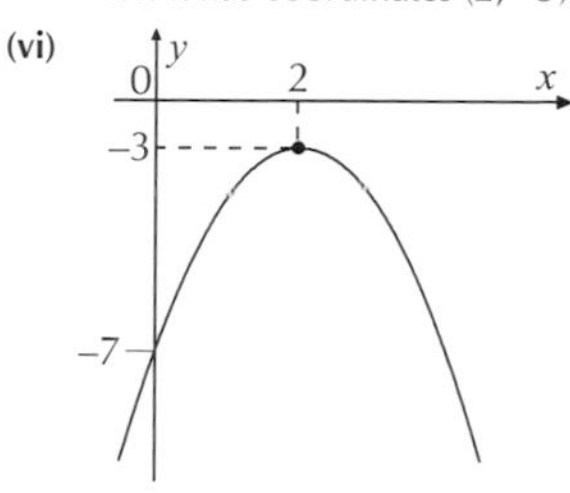

Q4 a) The minimum point is shown on the graph as (–4, 2). The coordinates of the vertex of the function $f(x) = p(x + q)^2 + r$ are $(-q, r)$. In this case $p = 1$ and from the minimum on the graph you can see $-q = -4$, so $q = 4$ and $r = 2$. So you can write the function $f(x) = (x + 4)^2 + 2$.

b) $g(x) = (x + 4)^2$ is in the form $p(x + q)^2 + r$ with $p = 1$ so the graph is u-shaped and the minimum is at $(-q, r) = (-4, 0)$.

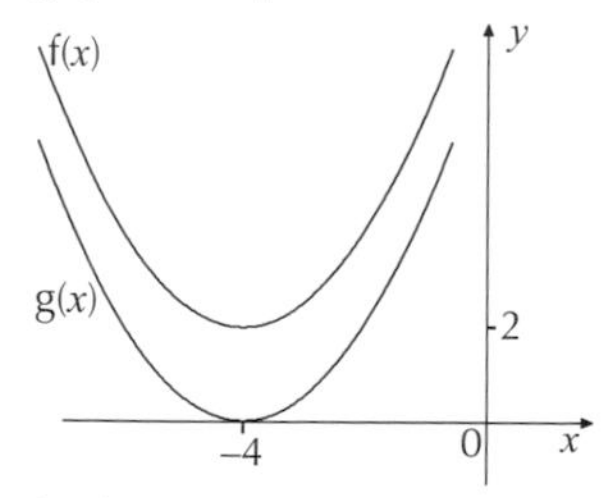

c) f(x) does not have any real roots as its graph does not touch the x-axis. g(x) has one real root as its graph touches the x-axis once (at $x = -4$).

Q5 a) $x^2 - 6x + 5 = (x - 3)^2 - 9 + 5 = (x - 3)^2 - 4$

b) $x^2 - 6x + 5 = 0 \Rightarrow (x - 3)^2 - 4 = 0 \Rightarrow (x - 3)^2 = 4$
$\Rightarrow x - 3 = \pm\sqrt{4} \Rightarrow x = 3 \pm \sqrt{4} = 3 \pm 2 = 5$ or 1

c) The graph is u-shaped. The function has roots $x = 1$ and 5 so these are the x-intercepts. Putting $x = 0$ into the original equation gives $y = 5$, so this is the y-intercept. Completing the square gives the minimum as (3, –4).

Putting all this together gives the following graph:

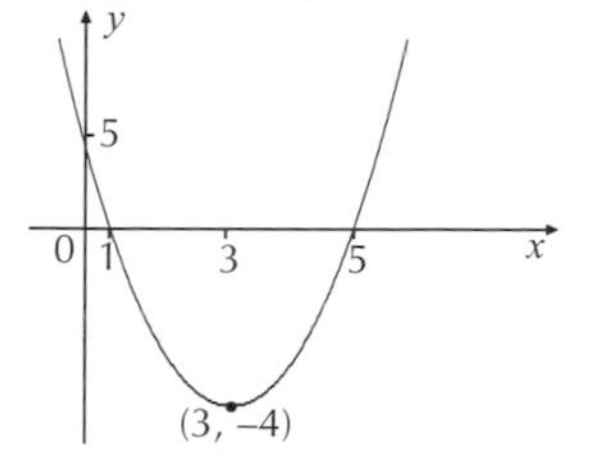

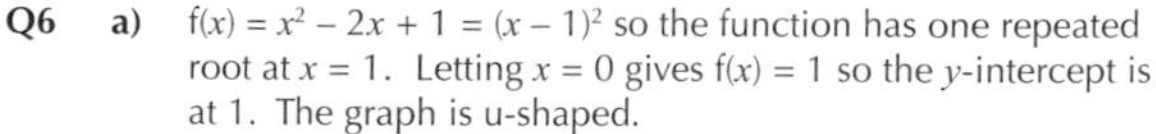

Q6 a) $f(x) = x^2 - 2x + 1 = (x - 1)^2$ so the function has one repeated root at $x = 1$. Letting $x = 0$ gives $f(x) = 1$ so the y-intercept is at 1. The graph is u-shaped.

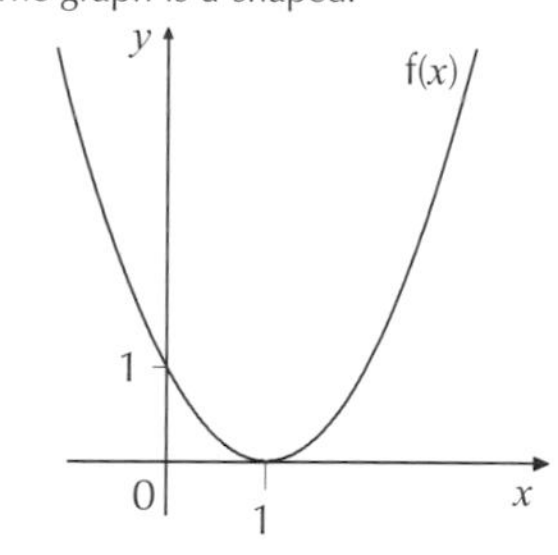

b) $f(x) = x^2 + x - 1 = \left(x + \frac{1}{2}\right)^2 - \frac{5}{4}$ and solving $f(x) = 0$ gives

$x = -\frac{1}{2} \pm \frac{\sqrt{5}}{2}$ as the x-intercepts. Letting $x = 0$ we get $f(x) = -1$ so this is the y-intercept. The graph is u-shaped.

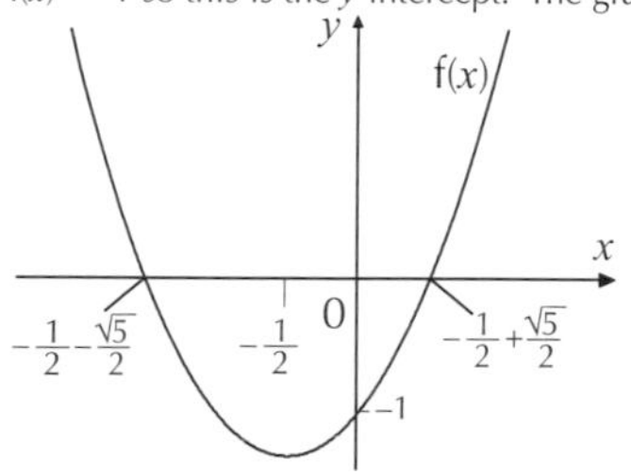

c) $f(x) = x^2 - 8x + 18 = (x - 4)^2 + 2$. Solving $f(x) = 0$ gives $x = 4 \pm \sqrt{-2}$ so there are no x-intercepts as you cannot take the square root of –2.

You could have worked out the discriminant to see that there were no real roots to save you trying to solve the equation.

Letting $x = 0$ gives $f(x) = 18$. The graph is u-shaped but it could be one of two graphs which are u-shaped with a y-intercept of 18. To find out which, work out the vertex. It has a minimum as it is u-shaped and from completing the square, the minimum is at (4, 2).

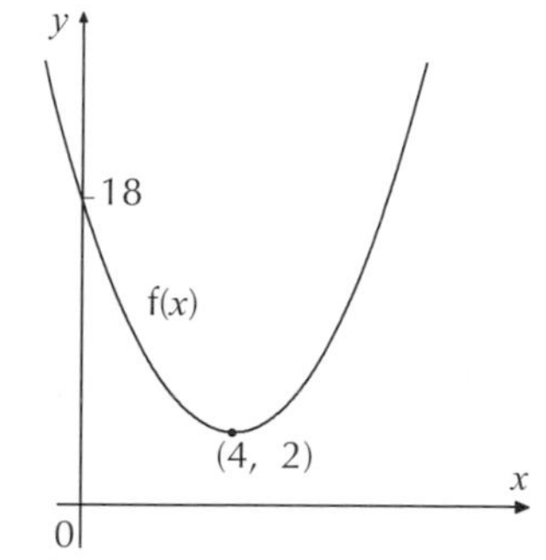

d) $f(x) = -x^2 + 3$ so setting $f(x) = 0$ gives $x = \pm\sqrt{3}$ as the x-intercepts. Letting $x = 0$ gives $f(x) = 3$ so 3 is the y-intercept. The graph is n-shaped.

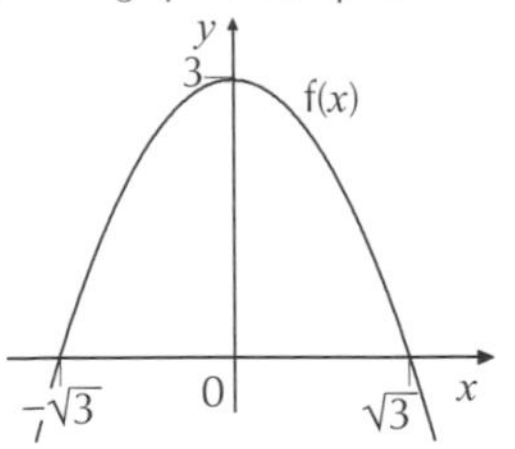

e) $f(x) = 2x^2 + 5x + 2 = 2\left(x + \frac{5}{4}\right)^2 - \frac{9}{8}$ and solving $f(x) = 0$ gives

$x = -\frac{5}{4} \pm \frac{3}{4} = -2$ and $-\frac{1}{2}$ as the x-intercepts.

$f(0) = 2$ so this is the y-intercept. The graph is u-shaped.

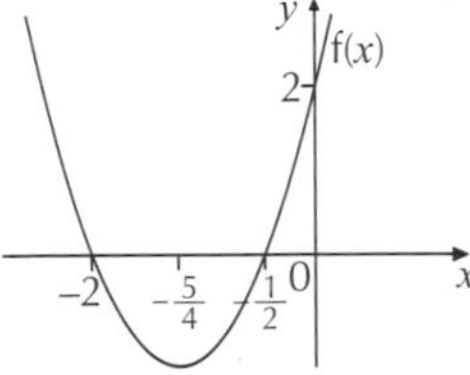

f) $f(x) = 2x^2 - 5x - 1 = 2\left(x - \frac{5}{4}\right)^2 - \frac{33}{8}$ and solving $f(x) = 0$

gives $x = \frac{5}{4} \pm \frac{\sqrt{33}}{4}$ as the x-intercepts. Letting $x = 0$ we get $f(x) = -1$ so this is the y-intercept. The graph is u-shaped.

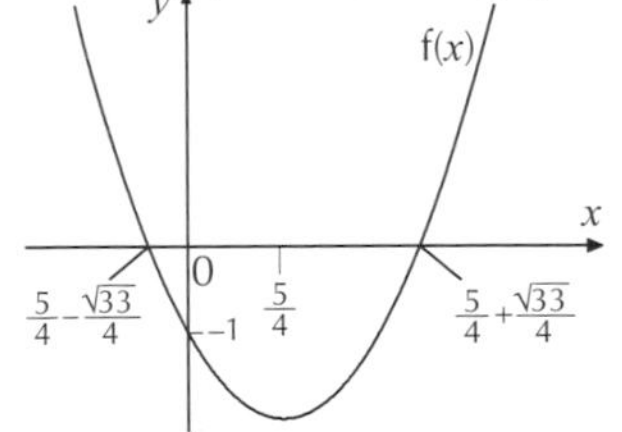

Q7 **a)** The roots of the quadratic function are the values of x where the graph crosses the x-axis. So the roots are $x = -2$ and $x = 1$.

b) One root of the equation is $x = -2$ which means $x + 2$ will be a factor. The other root $x = 1$ means that $x - 1$ will be a factor. So the quadratic function should be of the form $y = a(x + 2)(x - 1)$ for some value of a. But you know the equation has the form $y = -x^2 + px + q$. So a should be -1 to produce the term $-x^2$. So $y = -(x + 2)(x - 1)$ which gives $y = -x^2 - x + 2$, so $p = -1$ and $q = 2$.

The trickiest part of this question is realising you might also need a number factor, a, to form the factorised quadratic. Without it, you'd have got the wrong answer of $x^2 + x - 2$.

Q8 $f(x) = (x + 3)^2$ so setting $f(x) = 0$ gives $x = -3$ as the x-intercept, which means the graph just touches the x-axis at this point. $f(0) = 9$ so 9 is the y-intercept. The graph is u-shaped, so the vertex is a minimum, and occurs when $(x + 3) = 0$, at $(-3, 0)$:

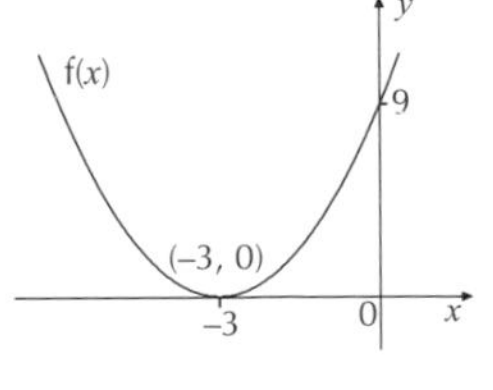

Q9 $g(x) = x^2 - 2x - 15 = (x - 1)^2 - 16$. Setting $g(x) = 0$ gives $x = -3$ and 5 as the x-intercepts. Letting $x = 0$ gives $g(x) = -15$ so -15 is the y-intercept. The graph is u-shaped, so the vertex is a minimum, and occurs when $(x - 1) = 0$, at $(1, -16)$:

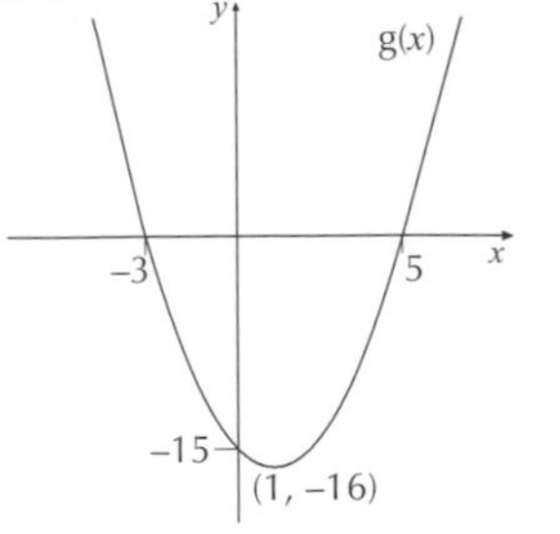

Q10 **a)** $h = 24t - 16t^2 = 8t(3 - 2t)$. Setting $h = 0$ gives $t = 0$ and $\frac{3}{2}$ as the t-intercepts. Letting $t = 0$ gives $h = 0$ so the h-intercept is $(0, 0)$. The coefficient of t^2 is -16 so the graph is n-shaped:

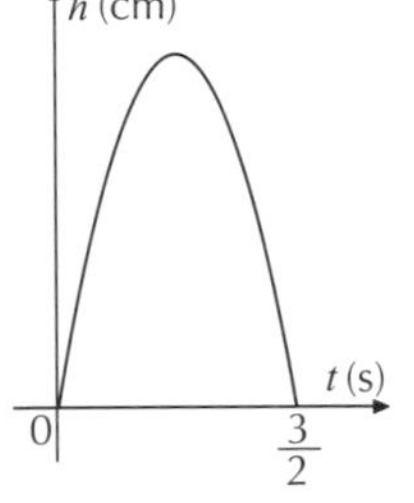

b) The maximum point on the graph occurs halfway between the t-intercepts, when $t = \frac{3}{4}$ s.

So the maximum height $= 24\left(\frac{3}{4}\right) - 16\left(\frac{3}{4}\right)^2 = 9$ cm.

Q11 **a)** Setting $t = 0$ gives the h-intercept at $h = 4$.
Setting $h = 0$ gives $0.25t^2 - 2.5t + 4 = 0$, so
$t^2 - 10t + 16 = 0 \Rightarrow (t - 2)(t - 8) = 0$
so the t-intercepts are at $t = 2$ or $t = 8$.
Setting $t = 10$ gives $h = 25 - 25 + 4 = 4$.
So the graph is u-shaped and looks like this:

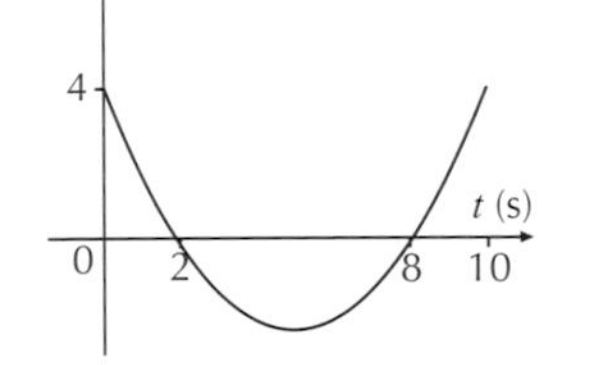

b) **(i)** The height of the raised platform is where the roller coaster sets off from (i.e. at $t = 0$).
From part a), this is 4 m.

(ii) The lowest point of the roller coaster is the value of h at the minimum point of the graph. Using the symmetry of the graph, this point occurs halfway between $t = 2$ and $t = 8$ (i.e. at $t = 5$).
When $t = 5$, $h = 0.25(5)^2 - 2.5(5) + 4 = -2.25$ m.

(iii) The roller coaster is underground when $h < 0$. From the graph, you can see that this is between $t = 2$ and $t = 8$, so it is underground for $8 - 2 = 6$ seconds.

Q12 a) $h = 0.5t^2 - 13t + 100$. Setting $t = 0$ gives the starting height $h = 100$, which is the intercept with the h-axis. To find the time taken to complete the stunt, set $h = 100$ and find the non-zero solution for t: $0.5t^2 - 13t + 100 = 100$
$\Rightarrow 0.5t(t - 26) = 0 \Rightarrow t = 0$ and 26, so the stunt is complete at $t = 26$ seconds. The coefficient of t^2 is 0.5 so the graph is u-shaped. The discriminant
$b^2 - 4ac = (-13)^2 - 4 \times 0.5 \times 100 = -31$, so there are no real roots and the graph will not touch the t-axis:

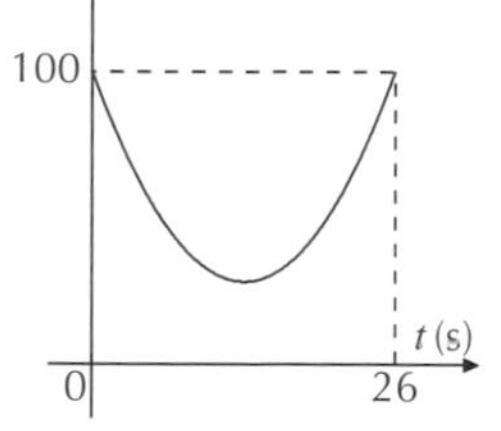

b) The graph is symmetrical so the minimum occurs when $t = 26 \div 2 = 13$ s. This gives a minimum height of $0.5(13)^2 - 13(13) + 100 = 15.5$ m.

4.4 Factorising Cubics

Exercise 4.4.1 — Factorising a cubic (when x is a factor)

Q1 a) $x(x^2 + 5x + 6) = x(x + 2)(x + 3)$

b) $x(x^2 + 6x - 7) = x(x + 7)(x - 1)$

c) $x(x^2 - 18x + 81) = x(x - 9)(x - 9) = x(x - 9)^2$

d) $x(x^2 + 7x + 10) = x(x + 5)(x + 2)$

e) $-x(x^2 - 4x + 3) = -x(x - 3)(x - 1)$

f) $x(x^2 + 4x + 3) = x(x + 1)(x + 3)$

g) $x(x^2 + 2x - 35) = x(x + 7)(x - 5)$

h) $x(x^2 - 6x - 16) = x(x + 2)(x - 8)$

i) $-x(x^2 + 3x - 4) = -x(x + 4)(x - 1)$

j) $x(2x^2 + 15x + 25) = x(2x + 5)(x + 5)$

k) $x(2x^2 - 7x + 6) = x(2x - 3)(x - 2)$

l) $x(4x^2 + 13x - 12) = x(4x - 3)(x + 4)$

m) $x\left(x^2 - \frac{4}{25}\right) = x\left(x + \frac{2}{5}\right)\left(x - \frac{2}{5}\right)$

n) $x(x^2 - 49) = x(x + 7)(x - 7)$

o) $x\left(x^2 - \frac{9}{4}\right) = x\left(x + \frac{3}{2}\right)\left(x - \frac{3}{2}\right)$

Q2 a) $-x^3 + 2x^2 + 24x = 0 \Rightarrow -x(x^2 - 2x - 24) = 0$
$\Rightarrow -x(x - 6)(x + 4) = 0 \Rightarrow -x = 0, x - 6 = 0$ or $x + 4 = 0$.
So $x = 0$, $x = 6$ or $x = -4$.

b) $x^3 - \frac{7}{9}x^2 + \frac{10}{81}x = 0 \Rightarrow x\left(x^2 - \frac{7}{9}x + \frac{10}{81}\right) = 0$
$\Rightarrow x\left(x - \frac{5}{9}\right)\left(x - \frac{2}{9}\right) = 0 \Rightarrow x = 0, x - \frac{5}{9} = 0$ or $x - \frac{2}{9} = 0$.
So $x = 0$, $x = \frac{5}{9}$ or $x = \frac{2}{9}$.

c) $2x^3 + 9x^2 + 4x = 0 \Rightarrow x(2x^2 + 9x + 4) = 0$
$\Rightarrow x(2x + 1)(x + 4) = 0 \Rightarrow x = 0, 2x + 1 = 0$ or $x + 4 = 0$.
So $x = 0$, $x = -\frac{1}{2}$ or $x = -4$.

d) $3x^3 - 3x^2 + 4x = 0 \Rightarrow x(3x^2 - 3x + 4) = 0$
This quadratic won't factorise — so use the quadratic formula: $a = 3$, $b = -3$, $c = 4$

$$x = \frac{-b \pm \sqrt{b^2 - 4ac}}{2a} = \frac{3 \pm \sqrt{(-3)^2 - 4 \times 3 \times 4}}{2 \times 3}$$
$$= \frac{3 \pm \sqrt{9 - 48}}{6} = \frac{3 \pm \sqrt{-39}}{6}$$

These aren't possible solutions as you can't take the square root of a negative number, so the only real solution is $x = 0$.

e) $4x - x^3 = 0 \Rightarrow x^3 - 4x = 0 \Rightarrow x(x^2 - 4) = 0$
$\Rightarrow x(x - 2)(x + 2) = 0 \Rightarrow x = 0, x - 2 = 0$ or $x + 2 = 0$.
So $x = 0$, $x = 2$ or $x - 2$.

f) $5x^3 + 7x^2 - 3x = 0 \Rightarrow x(5x^2 + 7x - 3) = 0$
The quadratic won't factorise — so use the quadratic formula: $a = 5$, $b = 7$, $c = -3$

$$x = \frac{-b \pm \sqrt{b^2 - 4ac}}{2a} = \frac{-7 \pm \sqrt{7^2 - 4 \times 5 \times (-3)}}{2 \times 5}$$
$$= \frac{-7 \pm \sqrt{49 + 60}}{10} = \frac{-7 \pm \sqrt{109}}{10}$$

So $x = 0$ or $x = -\frac{7}{10} \pm \frac{\sqrt{109}}{10}$.

g) $3x^3 + 26x^2 - 9x = 0 \Rightarrow x(3x^2 + 26x - 9) = 0$
$\Rightarrow x(3x - 1)(x + 9) = 0 \Rightarrow x = 0, 3x - 1 = 0$, or $x + 9 = 0$.
So $x = 0$, $x = \frac{1}{3}$ or $x = -9$

h) $x^3 + \frac{2}{3}x^2 - \frac{8}{9}x = 0 \Rightarrow 9x^3 + 6x^2 - 8x = 0$
$\Rightarrow x(9x^2 + 6x - 8) = 0 \Rightarrow x(3x - 2)(3x + 4) = 0$
$\Rightarrow x = 0, 3x - 2 = 0$, or $3x + 4 = 0$.
So $x = 0$, $x = \frac{2}{3}$ or $x = -\frac{4}{3}$

i) $x^2(4x + 3) = x \Rightarrow 4x^3 + 3x^2 = x \Rightarrow 4x^3 + 3x^2 - x = 0$
$\Rightarrow x(4x^2 + 3x - 1) = 0 \Rightarrow x(4x - 1)(x + 1) = 0$
$\Rightarrow x = 0, 4x - 1 = 0$ or $x + 1 = 0$.
So $x = 0$, $x = \frac{1}{4}$ or $x = -1$.

j) $2x^3 + 8x^2 = -3x \quad \Rightarrow 2x^3 + 8x^2 + 3x = 0$
$\Rightarrow x(2x^2 + 8x + 3) = 0$
This quadratic won't factorise, so use the quadratic formula.
Now $a = 2$, $b = 8$ and $c = 3$.

$$x = \frac{-b \pm \sqrt{b^2 - 4ac}}{2a} = \frac{-8 \pm \sqrt{8^2 - 4 \times 2 \times 3}}{2 \times 2}$$
$$= \frac{-8 \pm \sqrt{40}}{4} = -2 \pm \frac{1}{2}\sqrt{10}$$

So $x = 0$ or $x = -2 \pm \frac{1}{2}\sqrt{10}$.

Q3 a) $-x^3 + 36x = -x(x^2 - 36) = -x(x + 6)(x - 6)$

b) $f(x) = 0 \Rightarrow -x(x + 6)(x - 6) = 0$
So the roots are $x = 0$, $x = -6$ and $x = 6$

Q4 $2x^3 + 3x = x^2 \Rightarrow 2x^3 + 3x - x^2 = 0 \Rightarrow x(2x^2 - x + 3) = 0$
So $x = 0$ is one solution.
Find the discriminant of the quadratic factor:
$a = 2$, $b = -1$, $c = 3$. $b^2 - 4ac = (-1)^2 - 4 \times 2 \times 3 = -23$
The discriminant is negative, so the quadratic has no real roots.
So the only real solution to the cubic is $x = 0$.

Exercise 4.4.2 — The Factor Theorem

Q1 a) $a = 1$, find $f(a)$ and show the result is 0:
$f(1) = (1)3 - (1)2 - 3(1) + 3 = 1 - 1 - 3 + 3 = 0$
So by the Factor Theorem, $(x - 1)$ is a factor.
The question asked you to use the Factor Theorem, but you could also show it was a factor by adding the coefficients (1, –1, –3, 3) to get 0. (If the coefficients in a polynomial add up to 0, then (x – 1) is a factor.)

b) $a = -1$, find $f(a)$ and show the result is 0:
$f(-1) = (-1)^3 + 2(-1)^2 + 3(-1) + 2 = -1 + 2 - 3 + 2 = 0$
So by the Factor Theorem, $(x + 1)$ is a factor.

c) $a = -2$, find $f(a)$ and show the result is 0:
$f(-2) = (-2)^3 + 3(-2)^2 - 10(-2) - 24 = -8 + 12 + 20 - 24 = 0$
So by the Factor Theorem, $(x + 2)$ is a factor.

d) $a = 3$, find $f(a)$ and show the result is 0:
$f(3) = (3)^3 + 2(3)^2 - 9(3) - 18 = 0$
So by the Factor Theorem, $(x - 3)$ is a factor.

Q2 a) Substitute $x = \frac{1}{2}$ and show the result is 0:
$$f\left(\frac{1}{2}\right) = 2\left(\frac{1}{2}\right)^3 - \left(\frac{1}{2}\right)^2 - 8\left(\frac{1}{2}\right) + 4 = \frac{2}{8} - \frac{1}{4} - 4 + 4 = 0$$
So by the Factor Theorem, $(2x - 1)$ is a factor.

b) Substitute $x = \frac{2}{3}$ and show the result is 0:

$$f\left(\frac{2}{3}\right) = 3\left(\frac{2}{3}\right)^3 - 5\left(\frac{2}{3}\right)^2 - 16\left(\frac{2}{3}\right) + 12$$
$$= \frac{8}{9} - \frac{20}{9} - \frac{32}{3} + 12$$
$$= \frac{8}{9} - \frac{20}{9} - \frac{96}{9} + 12 = 0$$

So by the Factor Theorem, $(3x - 2)$ is a factor.

Q3 a) Substitute $x = -\frac{1}{5}$ and show the result is 0:

$$f\left(-\frac{1}{5}\right) = 5\left(-\frac{1}{5}\right)^3 - 44\left(-\frac{1}{5}\right)^2 + 61\left(-\frac{1}{5}\right) + 14$$
$$= -\frac{1}{25} - \frac{44}{25} - \frac{61}{5} + 14 = 0$$

So by the Factor Theorem, $(5x + 1)$ is a factor.

b) Substitute $x = \frac{1}{2}$ and show the result is 0:

$$f\left(\frac{1}{2}\right) = -2\left(\frac{1}{2}\right)^3 + 3\left(\frac{1}{2}\right)^2 + 11\left(\frac{1}{2}\right) - 6$$
$$= -\frac{1}{4} + \frac{3}{4} + \frac{11}{2} - 6 = 0$$

So by the Factor Theorem, $(1 - 2x)$ is a factor.

Q4 a) $f(3) = (3)^3 - 2(3)^2 - 5(3) + 6 = 27 - 18 - 15 + 6 = 0$
So by the Factor Theorem, $(x - 3)$ is a factor.

b) $1 - 2 - 5 + 6 = 0$
The coefficients add up to 0, so by the Factor Theorem, $(x - 1)$ is a factor.

Q5 a) $f(-4) = 3(-4)^3 - 5(-4)^2 - 58(-4) + 40$
$= -192 - 80 + 232 + 40 = 0$
So by the Factor Theorem, $(x + 4)$ is a factor.

b) $$f\left(\frac{2}{3}\right) = 3\left(\frac{2}{3}\right)^3 - 5\left(\frac{2}{3}\right)^2 - 58\left(\frac{2}{3}\right) + 40$$
$$= \frac{8}{9} - \frac{20}{9} - \frac{116}{3} + 40 = -\frac{120}{3} + 40 = 0$$

So by the Factor Theorem, $(3x - 2)$ is a factor.

c) $f(5) = 3(5)^3 - 5(5)^2 - 58(5) + 40$
$= 375 - 125 - 290 + 40 = 0$
So by the Factor Theorem, $(x - 5)$ is a factor.

Q6 $(x - 2)$ is a factor of $f(x) = 2x^3 - 7x^2 + px + 20$, so $f(2) = 0$. Using the Factor Theorem in reverse:
$f(2) = 2(2)^3 - 7(2)^2 + p(2) + 20 = 16 - 28 + 2p + 20 = 2p + 8$
$\Rightarrow 2p + 8 = 0 \Rightarrow 2p = -8 \Rightarrow p = -4$
So $f(x) = 2x^3 - 7x^2 - 4x + 20$

Q7 $(x - 3)$ is a factor of $f(x) = qx^3 - 4x^2 - 7qx + 12$, so $f(3) = 0$. Using the Factor Theorem in reverse:
$f(3) = q(3)^3 - 4(3)^2 - 7q(3) + 12 = 27q - 36 - 21q + 12 = 6q - 24$
$\Rightarrow 6q - 24 = 0 \Rightarrow 6q = 24 \Rightarrow q = 4$
So $f(x) = 4x^3 - 4x^2 - 28x + 12$

Q8 $(x - 1)$ and $(x - 2)$ are factors, so using the Factor Theorem in reverse, $f(1) = 0$ and $f(2) = 0$.
$f(1) = (1)^3 + c(1)^2 + d(1) - 2 = 1 + c + d - 2 = 0$
$c + d = 1$ (equation 1)

$f(2) = (2)^3 + c(2)^2 + d(2) - 2 = 8 + 4c + 2d - 2 = 0$
$4c + 2d = -6$ (equation 2)

Rearrange (1) to get $d = 1 - c$, and sub into (2):

$\Rightarrow 4c + 2(1 - c) = -6 \Rightarrow 4c + 2 - 2c = -6 \Rightarrow 2c = -8 \Rightarrow c = -4$

Sub c into rearranged (1): $d = 1 - c = 1 + 4 = 5$
So $f(x) = x^3 - 4x^2 + 5x - 2$

Q9 a) Substitute $x = -\frac{2}{3}$, $x = -\frac{1}{2}$ and $x = -5$ into $V(x)$ and show the result is 0 in each case:

$$V\left(-\frac{2}{3}\right) = 6\left(-\frac{2}{3}\right)^3 + 37\left(-\frac{2}{3}\right)^2 + 37\left(-\frac{2}{3}\right) + 10$$
$$= -\frac{16}{9} + \frac{148}{9} - \frac{74}{3} + 10 = 0$$

$$V\left(-\frac{1}{2}\right) = 6\left(-\frac{1}{2}\right)^3 + 37\left(-\frac{1}{2}\right)^2 + 37\left(-\frac{1}{2}\right) + 10$$
$$= -\frac{6}{8} + \frac{37}{4} - \frac{37}{2} + 10 = 0$$

$V(-5) = 6(-5)^3 + 37(-5)^2 + 37(-5) + 10$
$= -750 + 925 - 185 + 10 = 0$

By the Factor Theorem, $(3x + 2)$, $(2x + 1)$ and $(x + 5)$ are the three factors of $V(x)$. So $V(x) = (3x + 2)(2x + 1)(x + 5)$, and the volume of a cuboid is length × width × height, so the three factors are the length, width and height of the cuboid.
You can see that there are no constant numerical factors of $V(x)$ by considering the product of the x-coefficients of the factors — $3x \times 2x \times x = 6x^3$, which is the required x^3 term in $V(x)$.

b) You're told that x is a positive integer, so the smallest value of x is 1. Substituting $x = 1$ into $V(x)$ gives
$V(x) = (3 + 2)(2 + 1)(1 + 5) = 5 \times 3 \times 6 = 90$ cm³.
You could substitute $x = 1$ into the original equation:
$V(1) = 6(1)^3 + 37(1)^2 + 37(1) + 10 = 6 + 37 + 37 + 10 = 90$ cm³

Exercise 4.4.3 — Factorising a cubic (when x isn't a factor)

Q1 a) x is a common factor, so you get:
$x(x^2 - 3x + 2) = x(x - 1)(x - 2)$

b) Adding the coefficients gives you −12, so $(x - 1)$ is not a factor. Using trial and error, $f(2) = 0$, so $(x - 2)$ is a factor.
Factorise to get: $(x - 2)(2x^2 + 7x + 3) = (x - 2)(x + 3)(2x + 1)$

c) Add the coefficients $(1 - 3 + 3 - 1)$ to get 0, so $(x - 1)$ is a factor. Factorise to get: $(x - 1)(x^2 - 2x + 1) = (x - 1)^3$

d) Adding the coefficients gives you 2, so $(x - 1)$ is not a factor. Using trial and error, $f(2) = 0$, so $(x - 2)$ is a factor. Factorise to get: $(x - 2)(x^2 - x - 2) = (x - 2)(x - 2)(x + 1) = (x - 2)^2(x + 1)$

e) Add the coefficients $(1 - 1 - 7 + 7)$ to get 0, so $(x - 1)$ is a factor. Factorise to get: $(x - 1)(x^2 - 7) = (x - 1)(x + \sqrt{7})(x - \sqrt{7})$

f) Adding the coefficients gives you −8, so $(x - 1)$ is not a factor. Using trial and error, $f(2) = 0$, so $(x - 2)$ is a factor.
Factorise to get: $(x - 2)(x^2 + 4x + 3) = (x - 2)(x + 1)(x + 3)$

Q2 a) Adding the coefficients gives you 0, so $(x - 1)$ is a factor.
Factorise to get: $(x - 1)(x^2 - 2x - 35) = (x - 1)(x + 5)(x - 7)$
So the solutions are $x = 1$, $x = -5$ and $x = 7$.

b) Adding the coefficients gives you 21, so $(x - 1)$ is not a factor. Using trial and error, $f(2) = 0$, so $(x - 2)$ is a factor.
Factorise to get: $(x - 2)(x^2 + 2x - 24) = (x - 2)(x - 4)(x + 6)$
So the solutions are $x = 2$, $x = 4$ and $x = -6$.

Q3 a) Adding the coefficients gives you −3, so $(x - 1)$ is not a factor. Using trial and error, $f(-2) = 0$, so $(x + 2)$ is a factor.
Factorise to get: $(x + 2)(x^2 + 2x - 4)$
You're asked for the product of a linear factor and a quadratic factor, so you don't need to try to factorise the quadratic.

b) There is one solution at $x = -2$. The quadratic doesn't factorise, so use the quadratic formula: $a = 1$, $b = 2$, $c = -4$

$$x = \frac{-b \pm \sqrt{b^2 - 4ac}}{2a} = \frac{-2 \pm \sqrt{2^2 - 4 \times 1 \times (-4)}}{2 \times 1}$$
$$= \frac{-2 \pm \sqrt{4 + 16}}{2} = \frac{-2 \pm \sqrt{20}}{2} = -1 \pm \sqrt{5}$$

So the solutions are $x = -2$, $x = -1 + \sqrt{5}$ and $x = -1 - \sqrt{5}$.

Q4 Add the coefficients $(1 - 2 - 1 + 2)$ to get 0, so $(x - 1)$ is a factor.
Factorise to get: $(x - 1)(x^2 - x - 2) = (x - 1)(x + 1)(x - 2)$.
So the roots are $x = 1$, $x = -1$ and $x = 2$.

Q5 Add the coefficients $(1 - 1 - 3 + 3)$ to get 0, so $(x - 1)$ is a factor.
Factorise to get: $(x - 1)(x^2 - 3)$.
So the roots are $x = 1$ and $x = \pm\sqrt{3}$.

Q6 $(3x - 1)(2x^2 + 13x + 6) = (3x - 1)(2x + 1)(x + 6)$

Q7 a) $(x - 5)$ is a factor, so $f(5) = 0$
$f(5) = (5)^3 - p(5)^2 + 17(5) - 10$
$= 125 - 25p + 85 - 10 = 200 - 25p$
$\Rightarrow 200 - 25p = 0 \Rightarrow 200 = 25p \Rightarrow p = 8$

b) $(x - 5)(x^2 - 3x + 2) = (x - 5)(x - 1)(x - 2)$

c) So the solutions are $x = 5$, 1 and 2.

Q8 a) Adding the coefficients gives you 0, so $(x - 1)$ is a factor.
Factorise to get: $(x - 1)(3x^2 + 5x - 2) = (x - 1)(3x - 1)(x + 2)$

b) Adding the coefficients gives you 0, so $(x - 1)$ is a factor.
Factorise to get: $(x - 1)(5x^2 - 8x - 4) = (x - 1)(5x + 2)(x - 2)$

Q9 Rearrange to give $4x^3 - 7x + 3 = 0$. Adding the coefficients gives you 0, so $(x - 1)$ is a factor. Factorise to get:
$(x - 1)(4x^2 + 4x - 3) = (x - 1)(2x + 3)(2x - 1)$.
$(x - 1)(2x + 3)(2x - 1) = 0$, so $x - 1 = 0$, or $2x + 3 = 0$, or $2x - 1 = 0$. So the solutions are $x = 1$, $-\frac{3}{2}$ and $\frac{1}{2}$.

Q10 If $x = 2$ is a root, $f(2) = 0$:
$f(2) = 2(2)^3 - 2^2 - 2(2) - 8 = 16 - 4 - 4 - 8 = 0$
So $x = 2$ is a root, and $(x - 2)$ is a factor.
Now factorise: $(x - 2)(2x^2 + 3x + 4) = 0$.
The discriminant of the quadratic factor is $b^2 - 4ac$, which is $3^2 - 4 \times 2 \times 4 = 9 - 32 = -23$.
The discriminant is negative so the quadratic has no real roots.
Hence the only real root of the cubic is $x = 2$.
You could have put the values of a, b and c into the quadratic formula here — you'd end up with a negative number inside the square root, which would mean the quadratic has no real roots.

Exercise 4.4.4 — Algebraic division

Q1 a)

$$\begin{array}{r} x^2+x-12 \\ x-3\overline{)x^3-2x^2-15x+36} \\ -\underline{(x^3-3x^2)} \\ x^2-15x \\ -\underline{(x^2-3x)} \\ -12x+36 \\ -\underline{(-12x+36)} \\ 0 \end{array}$$

Factorise the quadratic: $x^2 + x - 12 = (x - 3)(x + 4)$
So $x^3 - 2x^2 - 15x + 36 = (x - 3)(x - 3)(x + 4) = (x - 3)^2(x + 4)$

b)

$$\begin{array}{r} x^2-3x-5 \\ x+2\overline{)x^3-x^2-11x-10} \\ -\underline{(x^3+2x^2)} \\ -3x^2-11x \\ -\underline{(-3x^2-6x)} \\ -5x-10 \\ -\underline{(-5x-10)} \\ 0 \end{array}$$

The quadratic doesn't factorise, so the full factorisation is:
$x^3 - x^2 - 11x - 10 = (x + 2)(x^2 - 3x - 5)$

c)

$$\begin{array}{r} 2x^2+15x+7 \\ x-2\overline{)2x^3+11x^2-23x-14} \\ -\underline{(2x^3-4x^2)} \\ 15x^2-23x \\ -\underline{(15x^2-30x)} \\ 7x-14 \\ -\underline{(7x-14)} \\ 0 \end{array}$$

Factorise the quadratic: $2x^2 + 15x + 7 = (2x + 1)(x + 7)$
So $2x^3 + 11x^2 - 23x - 14 = (x - 2)(2x + 1)(x + 7)$

d)

$$\begin{array}{r} x^2+5x+6 \\ x+5\overline{)x^3+10x^2+31x+30} \\ -\underline{(x^3+5x^2)} \\ 5x^2+31x \\ -\underline{(5x^2+25x)} \\ 6x+30 \\ -\underline{(6x+30)} \\ 0 \end{array}$$

Factorise the quadratic: $x^2 + 5x + 6 = (x + 2)(x + 3)$
So $x^3 + 10x^2 + 31x + 30 = (x + 5)(x + 2)(x + 3)$

Q2 Add the coefficients $(1 - 5 + 4)$ to get 0, so $(x - 1)$ is a factor. Now factorise using long division:

$$\begin{array}{r} x^2+x-4 \\ x-1\overline{)x^3+0x^2-5x+4} \\ -\underline{(x^3-x^2)} \\ x^2-5x \\ -\underline{(x^2-x)} \\ -4x+4 \\ -\underline{(-4x+4)} \\ 0 \end{array}$$

Finally, write the cubic as the product of a linear factor and a quadratic factor: $(x^3 - 5x + 4) = (x - 1)(x^2 + x - 4)$

Q3

$$\begin{array}{r} x^2+4x+1 \\ x-2\overline{)x^3+2x^2-7x-2} \\ -\underline{(x^3-2x^2)} \\ 4x^2-7x \\ -\underline{(4x^2-8x)} \\ x-2 \\ -\underline{(x-2)} \\ 0 \end{array}$$

So $f(x) = (x - 2)(x^2 + 4x + 1)$.

Q4 If $f(-2) = 0$ then $(x + 2)$ is a factor.
Now factorise using long division:

$$\begin{array}{r} x^2-2x-3 \\ x+2\overline{)x^3+0x^2-7x-6} \\ -\underline{(x^3+2x^2)} \\ -2x^2-7x \\ -\underline{(-2x^2-4x)} \\ -3x-6 \\ -\underline{(-3x-6)} \\ 0 \end{array}$$

Then factorise the quadratic: $x^2 - 2x - 3 = (x - 3)(x + 1)$
So $f(x) = (x + 2)(x - 3)(x + 1)$
So the solutions to $f(x) = 0$ are $x = -2$, $x = 3$ and $x = -1$.

Q5 Adding the coefficients gives you -10, so $(x - 1)$ is not a factor.
Using trial and error, $f(2) = 0$, so $(x - 2)$ is a factor.
Factorise using long division:

$$\begin{array}{r} x^2+3x+6 \\ x-2\overline{)x^3+x^2+0x-12} \\ -\underline{(x^3-2x^2)} \\ 3x^2+0x \\ -\underline{(3x^2-6x)} \\ 6x-12 \\ -\underline{(6x-12)} \\ 0 \end{array}$$

So $x^3 + x^2 - 12 = (x - 2)(x^2 + 3x + 6)$

Q6 a)

$$\begin{array}{r} x^2-5x+16 \\ x+3\overline{)x^3-2x^2+x-1} \\ -\underline{(x^3+3x^2)} \\ -5x^2+x \\ -\underline{(-5x^2-15x)} \\ 16x-1 \\ -\underline{(16x+48)} \\ -49 \end{array}$$

So the remainder is -49.

b) $f(x) = (x + 3)(x^2 - 5x + 16) - 49$
If you expand the brackets and simplify, you'll end up with f(x) back in its original form. This is a good way to check your answer.

Q7 a)

$$\begin{array}{r} x^2-6x+8 \\ x-2\overline{)x^3-8x^2+20x-3} \\ -\underline{(x^3-2x^2)} \\ -6x^2+20x \\ -\underline{(-6x^2+12x)} \\ 8x-3 \\ -\underline{(8x-16)} \\ 13 \end{array}$$

So the remainder is 13.

b) From part a), $f(x) = (x - 2)(x^2 - 6x + 8) + 13$.
So if $f(x) - 13 = 0$, then: $(x - 2)(x^2 - 6x + 8) + 13 - 13 = 0$
$\Rightarrow (x - 2)(x^2 - 6x + 8) = 0 \Rightarrow (x - 2)(x - 2)(x - 4) = 0$
$\Rightarrow (x - 2)^2(x - 4) = 0$. So either $x - 2 = 0$ or $x - 4 = 0$, and so the solutions are $x = 2$ and $x = 4$.

Q8 Rearrange to give $x^3 - 15x^2 + 75x - 125 = 0$. $x = 5$ is a solution, so $x - 5$ must be a factor, so divide by $x - 5$ to get the other factors:

$$\begin{array}{r} x^2 - 10x + 25 \\ x-5\overline{)x^3 - 15x^2 + 75x - 125} \\ -(x^3 - 5x^2) \\ -10x^2 + 75x \\ -(-10x^2 + 50x) \\ 25x - 125 \\ -(25x - 125) \\ 0 \end{array}$$

So: $(x - 5)(x^2 - 10x + 25) = 0 \Rightarrow (x - 5)(x - 5)(x - 5) = 0$
$\Rightarrow (x - 5)^3 = 0$. So $x = 5$ is the only solution.

Q9 Use algebraic division to factorise the equation, taking care with the negative coefficient of t^3:

$$\begin{array}{r} -t^2 + 3t + 10 \\ t+1\overline{)-t^3 + 2t^2 + 13t + 10} \\ -(-t^3 - t^2) \\ 3t^2 + 13t \\ -(3t^2 + 3t) \\ 10t + 10 \\ -(10t + 10) \\ 0 \end{array}$$

The car stops when $S = 0$, so solve: $(t + 1)(-t^2 + 3t + 10) = 0$
$\Rightarrow -(t + 1)(t^2 - 3t - 10) = 0 \Rightarrow (t + 1)(t^2 - 3t - 10) = 0$
$\Rightarrow (t + 1)(t - 5)(t + 2) = 0$. So either $t + 1 = 0$, $t - 5 = 0$ or $t + 2 = 0$, and so the solutions are $t = -1$, 5 and -2.
But T must be greater than zero, so $T = 5$ minutes.
Don't forget to link it back to the actual question — you're asked for the value of T, not just the possible solutions.

Review Exercise — Chapter 4

Q1
a) $x^2 + 2x + 1 = (x + 1)(x + 1) = (x + 1)^2$
b) $x^2 - 13x + 30 = (x - 10)(x - 3)$
c) $x^2 - 4 = (x + 2)(x - 2)$
d) $3 + 2x - x^2 = (3 - x)(x + 1)$
e) $2x^2 - 7x - 4 = (2x + 1)(x - 4)$
f) $5x^2 + 7x - 6 = (5x - 3)(x + 2)$

Q2
a) $x^2 - 3x + 2 = 0 \Rightarrow (x - 2)(x - 1) = 0$
$\Rightarrow x - 2 = 0$ or $x - 1 = 0$, so $x = 2$ or 1

b) $x^2 + x - 12 = 0 \Rightarrow (x + 4)(x - 3) = 0$
$\Rightarrow x + 4 = 0$ or $x - 3 = 0$, so $x = -4$ or 3

c) $2 + x - x^2 = 0 \Rightarrow x^2 - x - 2 = 0$
$\Rightarrow (x - 2)(x + 1) = 0$
$\Rightarrow x - 2 = 0$ or $x + 1 = 0$, so $x = 2$ or -1

d) $x^2 + x - 16 = x \Rightarrow x^2 - 16 = 0$
$\Rightarrow (x + 4)(x - 4) = 0 \Rightarrow x + 4 = 0$ or $x - 4 = 0$, so $x = \pm 4$

e) $3x^2 - 15x - 14 = 4x \Rightarrow 3x^2 - 19x - 14 = 0$
$\Rightarrow (3x + 2)(x - 7) = 0$
$\Rightarrow 3x + 2 = 0$ or $x - 7 = 0$, so $x = -\frac{2}{3}$ or 7

f) $4x^2 - 1 = 0 \Rightarrow (2x + 1)(2x - 1) = 0$
$\Rightarrow 2x + 1 = 0$ or $2x - 1 = 0$, so $x = \pm\frac{1}{2}$

g) $6x^2 - 11x + 9 = 2x^2 - x + 3 \Rightarrow 4x^2 - 10x + 6 = 0$
$\Rightarrow 2x^2 - 5x + 3 = 0 \Rightarrow (2x - 3)(x - 1) = 0$
$\Rightarrow 2x - 3 = 0$ or $x - 1 = 0$, so $x = \frac{3}{2}$ or 1

h) $3x^2 + 10x - 8 = 2 - x - 3x^2$
$\Rightarrow 6x^2 + 11x - 10 = 0 \Rightarrow (3x - 2)(2x + 5) = 0$
$\Rightarrow 3x - 2 = 0$ or $2x + 5 = 0$, so $x = \frac{2}{3}$ or $-\frac{5}{2}$

i) $4 - 9x^2 = 0 \Rightarrow (2 + 3x)(2 - 3x) = 0$
$\Rightarrow 2 + 3x = 0$ or $2 - 3x = 0$, so $x = \pm\frac{2}{3}$

Q3 These solutions use the quadratic formula, but you could complete the square to get the same answers.

a) $3x^2 - 7x + 3 = 0$, so $a = 3$, $b = -7$, $c = 3$
$$x = \frac{-b \pm \sqrt{b^2 - 4ac}}{2a} = \frac{-(-7) \pm \sqrt{(-7)^2 - 4 \times 3 \times 3}}{2 \times 3}$$
$$= \frac{7 \pm \sqrt{49 - 36}}{6} = \frac{7 \pm \sqrt{13}}{6}$$

b) $2x^2 - 6x - 2 = 0 \Rightarrow x^2 - 3x - 1 = 0$, so $a = 1$, $b = -3$, $c = -1$
$$x = \frac{-b \pm \sqrt{b^2 - 4ac}}{2a}$$
$$= \frac{-(-3) \pm \sqrt{(-3)^2 - 4 \times 1 \times (-1)}}{2 \times 1}$$
$$= \frac{3 \pm \sqrt{9 + 4}}{2} = \frac{3 \pm \sqrt{13}}{2}$$

c) $x^2 + 4x + 6 = 12 \Rightarrow x^2 + 4x - 6 = 0$, so $a = 1$, $b = 4$, $c = -6$
$$x = \frac{-b \pm \sqrt{b^2 - 4ac}}{2a} = \frac{-4 \pm \sqrt{4^2 - 4 \times 1 \times (-6)}}{2 \times 1}$$
$$= \frac{-4 \pm \sqrt{16 + 24}}{2} = \frac{-4 \pm \sqrt{40}}{2}$$
$$= \frac{-4 \pm 2\sqrt{10}}{2} = -2 \pm \sqrt{10}$$

Q4
a) $x^2 + 6x + 7 = (x + 3)^2 - 9 + 7 = (x + 3)^2 - 2$
b) Let $u = 2x + 1$. From a), $(u + 3)^2 - 2 = 0$
$\Rightarrow (u + 3)^2 = 2 \Rightarrow u + 3 = \pm\sqrt{2} \Rightarrow u = -3 \pm \sqrt{2}$
$\Rightarrow 2x + 1 = -3 \pm \sqrt{2} \Rightarrow x = -2 \pm \frac{\sqrt{2}}{2}$

Q5 Let $u = x^2$: $u^2 - 17u + 16 = 0 \Rightarrow (u - 16)(u - 1) = 0$
So $u - 16 = 0$ or $u - 1 = 0$, so $u = 16$ or $u = 1$.
$x^2 = 16 \Rightarrow x = 4$ or $x = -4$
$x^2 = 1 \Rightarrow x = 1$ or $x = -1$

Q6 $a = 1$, $b = k$, $c = 4$.
So $b^2 - 4ac = k^2 - 4 \times 1 \times 4 = k^2 - 16$
If the equation has two distinct real roots, the discriminant must be positive so $k^2 - 16 > 0 \Rightarrow k^2 > 16 \Rightarrow k > 4$ or $k < -4$.

Q7
a) $x^2 - 4x - 3 = (x - 2)^2 - 4 - 3 = (x - 2)^2 - 7$.
The lowest value $(x - 2)^2$ can take is zero — so the minimum value $= -7$ at $x = 2$ (to make the bracket 0). The graph crosses the x-axis at $(x - 2)^2 - 7 = 0 \Rightarrow x = 2 \pm \sqrt{7}$.

b) $x^2 + 5x + 8 = \left(x + \frac{5}{2}\right)^2 - \frac{25}{4} + 8 = \left(x + \frac{5}{2}\right)^2 + \frac{7}{4}$.
The lowest value $\left(x + \frac{5}{2}\right)^2$ can take is zero — so the minimum value $= \frac{7}{4}$ at $x = -\frac{5}{2}$ (to make the bracket 0), and the graph doesn't cross the x-axis.

c) $3 - 3x - x^2 = -\left(x + \frac{3}{2}\right)^2 + \frac{9}{4} + 3 = \frac{21}{4} - \left(x + \frac{3}{2}\right)^2$.
The lowest value $\left(x + \frac{3}{2}\right)^2$ can take is zero — so the maximum value $= \frac{21}{4}$ at $x = -\frac{3}{2}$ (to make the bracket 0). The graph crosses the x-axis at $\frac{21}{4} - \left(x + \frac{3}{2}\right)^2 = 0 \Rightarrow x = -\frac{3}{2} \pm \frac{\sqrt{21}}{2}$.

d) $2x^2 - 4x + 11 = 2(x - 1)^2 - 2 + 11 = 2(x - 1)^2 + 9$.
The lowest value $(x - 1)^2$ can take is zero — so the minimum value $= 9$ at $x = 1$ (to make the bracket 0), and the graph doesn't cross the x-axis.

e) $4x^2 - 28x + 48 = 4\left(x - \frac{7}{2}\right)^2 - 49 + 48 = 4\left(x - \frac{7}{2}\right)^2 - 1$.
The lowest value $\left(x - \frac{7}{2}\right)^2$ can take is zero — so the minimum value $= -1$ at $x = \frac{7}{2}$ (to make the bracket 0).
The graph crosses the x-axis at $4\left(x - \frac{7}{2}\right)^2 - 1 = 0$
$\Rightarrow x = 4$ or $x = 3$.

f) $-3x^2 + 12x + 14 = -3(x - 2)^2 + 12 + 14 = -3(x - 2)^2 + 26$.
The lowest value $(x - 2)^2$ can take is zero — so the maximum value $= 26$ at $x = 2$ (to make the bracket 0). The graph crosses the x-axis at $-3(x - 2)^2 + 26 = 0 \Rightarrow x = 2 \pm \frac{\sqrt{78}}{3}$.

Q8 **a)** $x^2 - 2x - 3 = 0 \Rightarrow (x-1)^2 - 4 = 0$
$\Rightarrow (x-1)^2 = 4$, so there are 2 roots,
at $x = -2 + 1 = -1$ and $x = 2 + 1 = 3$.
At $x = 0$, $y = -3$. The graph is u-shaped:

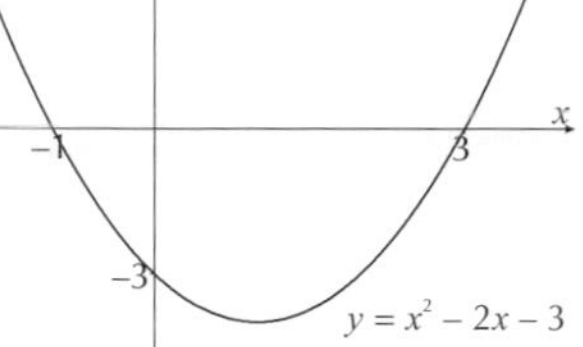

b) $x^2 - 6x + 9 = 0 \Rightarrow (x-3)^2 = 0$, so there is 1 root, at $x = 3$.
Letting $x = 0$ gives 9 as the y-intercept.
The graph is u-shaped:

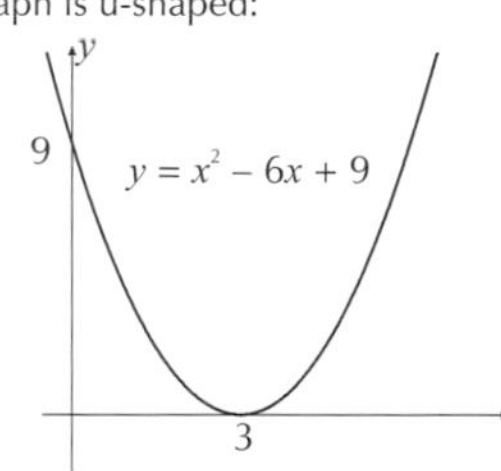

c) $2x^2 + 4x + 3 = 0 \Rightarrow 2(x+1)^2 + 1 = 0$, so there are no real roots. Letting $x = 0$ gives 3 as the y-intercept. The graph is u-shaped, and the lowest value $(x+1)^2$ can take is zero — so the minimum value = 1 at $x = -1$ (to make the bracket 0):

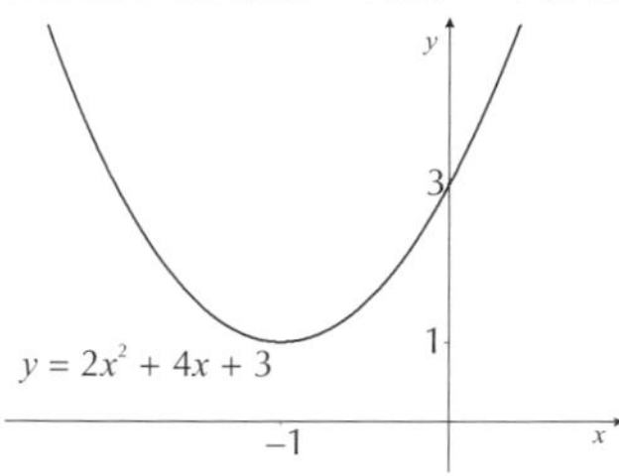

Q9 **a)** $5x^3 - 13x^2 + 6x = -5x^3 + 7x^2 + 6x \Rightarrow 10x^3 - 20x^2 = 0$
$\Rightarrow 10x^2(x-2) = 0 \Rightarrow 10x^2 = 0$ or $x - 2 = 0$, $x = 0$ or $x = 2$.
When $x = 0$, $y = 0 - 0 + 0 = 0$.
When $x = 2$, $y = 40 - 52 + 12 = 0$.
So they intersect at (0, 0) and (2, 0).

b) $f(x) = 5x^3 - 13x^2 + 6x = x(5x^2 - 13x + 6) = x(x-2)(5x-3)$.

Q10 **a)** Set the equations equal to each other:
$x(x-6)^2 = -x(2x-31) \Rightarrow x(x^2 - 12x + 36) = -2x^2 + 31x$
$\Rightarrow x^3 - 12x^2 + 36x = -2x^2 + 31x \Rightarrow x^3 - 10x^2 + 5x = 0$

b) $x^3 - 10x^2 + 5x = 0 \Rightarrow x(x^2 - 10x + 5)$
So the curves meet at $x = 0$ and at
$x^2 - 10x + 5 = 0 \Rightarrow (x-5)^2 - 20 = 0 \Rightarrow x = 5 \pm 2\sqrt{5}$.
You can also use the quadratic formula.

Q11 **a)** When $t = 0$, $h = 6$, so the h-intercept = 6.
When $h = 0$, $0.25t^2 - 2.75t + 6 = 0$
$\Rightarrow t^2 - 11t + 24 = 0 \Rightarrow (t-8)(t-3) = 0$, so there are 2 roots, at $t = 8$ and $t = 3$. The graph is u-shaped:

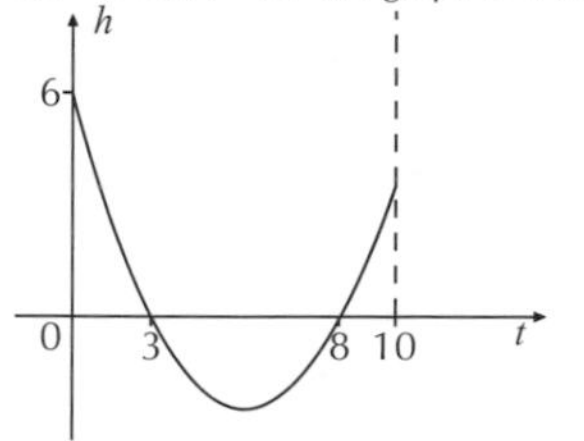

b) **(i)** Initial height when $t = 0$ is 6 m.

(ii) The vertex is halfway between the roots, i.e. when $t = (3 + 8) \div 2 = 5.5$ s, and
$h = 0.25(5.5)^2 - 2.75(5.5) + 6 = -1.5625$ m

(iii) The car is underground between 3 and 8 seconds, so for $8 - 3 = 5$ seconds.

Q12 **a)** $x^3 - 4x^2 = 0 \Rightarrow x^2(x-4) = 0 \Rightarrow x = 0$ or $x = 4$

b) $x^3 + 5x^2 - 6x = 0 \Rightarrow x(x^2 + 5x - 6) = 0$
$\Rightarrow x(x-1)(x+6) = 0 \Rightarrow x = 0, x = 1$ or $x = -6$

c) $x^3 - 6x^2 + 9x = 0 \Rightarrow x(x^2 - 6x + 9) = 0$
$\Rightarrow x(x-3)^2 = 0 \Rightarrow x = 0$ or $x = 3$

d) $2x^3 + 5x^2 + 15x = x^3 - 3x^2$
$\Rightarrow x^3 + 8x^2 + 15x = 0 \Rightarrow x(x^2 + 8x + 15) = 0$
$\Rightarrow x(x+3)(x+5) = 0 \Rightarrow x = 0, x = -3$ or $x = -5$

e) $2x^3 + 20x^2 + 12x = 9x^3 - 20x^2$
$\Rightarrow 7x^3 - 40x^2 - 12x = 0 \Rightarrow x(7x^2 - 40x - 12) = 0$
$\Rightarrow x(x-6)(7x+2) = 0 \Rightarrow x = 0, x = 6$ or $x = -\frac{2}{7}$

f) $6x^3 - 5x^2 - 4x = 0 \Rightarrow x(6x^2 - 5x - 4) = 0$
$\Rightarrow x(3x-4)(2x+1) = 0 \Rightarrow x = 0, x = \frac{4}{3}$ or $x = -\frac{1}{2}$

Q13 **a)** $f(1) = 1^5 - 4(1)^4 + 3(1)^3 + 2(1)^2 - 2 = 0$, so $(x-1)$ is a factor.

b) $f(-1) = (-1)^5 - 4(-1)^4 + 3(-1)^3 + 2(-1)^2 - 2 = -8$, so $(x+1)$ is not a factor.

c) $f(2) = 2^5 - 4(2)^4 + 3(2)^3 + 2(2)^2 - 2 = -2$,
so $(x-2)$ is not a factor.

d) $f\left(\frac{2}{2}\right) = f(1) = 0$ (from a)), so $(2x-2)$ is a factor.

Q14 $(x+2)$ is a factor, so $f(-2) = 0 \Rightarrow (-2+5)(-2-2)(-2-1) + k = 0$
$\Rightarrow 3 \times (-4) \times (-3) + k = 0 \Rightarrow 36 + k = 0 \Rightarrow k = -36$

Q15 **a)**

$$
\begin{array}{r}
x^2 - 4x + 9 \\
x+3\,\overline{)\,x^3 - x^2 - 3x + 3} \\
-\,(x^3 + 3x^2) \\
\hline
-4x^2 - 3x \\
-\,(-4x^2 - 12x) \\
\hline
9x + 3 \\
-\,(9x + 27) \\
\hline
-24
\end{array}
$$

Quotient: $x^2 - 4x + 9$, remainder: -24

b)

$$
\begin{array}{r}
x^2 - x - 7 \\
x-2\,\overline{)\,x^3 - 3x^2 - 5x + 6} \\
-\,(x^3 - 2x^2) \\
\hline
-x^2 - 5x \\
-\,(-x^2 + 2x) \\
\hline
-7x + 6 \\
-\,(-7x + 14) \\
\hline
-8
\end{array}
$$

Quotient: $x^2 - x - 7$, remainder: -8

c)

$$
\begin{array}{r}
x^2 + 0x + 3 \\
x+2\,\overline{)\,x^3 + 2x^2 + 3x + 2} \\
-\,(x^3 + 2x^2) \\
\hline
0x^2 + 3x \\
-\,(0x^2 + 0x) \\
\hline
3x + 2 \\
-\,(3x + 6) \\
\hline
-4
\end{array}
$$

Quotient: $x^2 + 3$, remainder: -4

Q16 **a)**

$$
\begin{array}{r}
3x^2 - 10x + 15 \\
x+2\,\overline{)\,3x^3 - 4x^2 - 5x - 6} \\
-\,(3x^3 + 6x^2) \\
\hline
-10x^2 - 5x \\
-\,(-10x^2 - 20x) \\
\hline
15x - 6 \\
-\,(15x + 30) \\
\hline
-36
\end{array}
$$

So $f(x) = (x+2)(3x^2 - 10x + 15) - 36$

b)
$$\begin{array}{r} x^2 - 0x - 3 \\ x+2\overline{\smash{)}\,x^3 + 2x^2 - 3x + 4} \\ -(x^3 + 2x^2) \\ \hline 0x^2 - 3x \\ -(0x^2 + 0x) \\ \hline -3x + 4 \\ -(-3x - 6) \\ \hline 10 \end{array}$$

So $f(x) = (x + 2)(x^2 - 3) + 10$

c)
$$\begin{array}{r} 2x^2 - 4x + 14 \\ x+2\overline{\smash{)}\,2x^3 + 0x^2 + 6x - 3} \\ -(2x^3 + 4x^2) \\ \hline -4x^2 + 6x \\ -(-4x^2 - 8x) \\ \hline 14x - 3 \\ -(14x + 28) \\ \hline -31 \end{array}$$

So $f(x) = (x + 2)(2x^2 - 4x + 14) - 31$

Q17 $f(2) = 0 \Rightarrow 2(2)^4 + 3(2)^3 + 5(2)^2 + 2c + d = 0$
$2c + d = -76$ — equation 1
$f(-3) = 0 \Rightarrow 2(-3)^4 + 3(-3)^3 + 5(-3)^2 - 3c + d = 0$
$-3c + d = -126$ — equation 2
Subtract equation 2 from equation 1: $5c = 50 \Rightarrow c = 10$.
In equation 1: $2 \times 10 + d = -76 \Rightarrow d = -96$.

Q18 Use algebraic long division or another method to get:
$f(x) = (x - 3)(x^2 - 6x - 11) = 0$
So $x - 3 = 0 \Rightarrow x = 3$, or $x^2 - 6x - 11 = 0$.
Using the quadratic formula, $a = 1$, $b = -6$, $c = -11$:
$$x = \frac{-b \pm \sqrt{b^2 - 4ac}}{2a} = \frac{-(-6) \pm \sqrt{(-6)^2 - 4 \times 1 \times (-11)}}{2 \times 1}$$
$$= \frac{6 \pm \sqrt{36 + 44}}{2} = \frac{6 \pm \sqrt{80}}{2} = \frac{6 \pm 4\sqrt{5}}{2} = 3 \pm 2\sqrt{5}$$
So the solutions are $x = 3$, $x = 3 - 2\sqrt{5}$ and $x = 3 + 2\sqrt{5}$.

Q19 Adding the coefficients gives you 24, so $(x - 1)$ is not a factor. Using trial and error, $f(-1) = 0$, so $(x + 1)$ is a factor. Factorise using algebraic long division or another method to get:
$f(x) = (x + 1)(x^2 + 5x + 6) = (x + 1)(x + 2)(x + 3) = 0$
So the roots are $x = -1$, -2 and -3.

Q20 a)
$$\begin{array}{r} x^2 - 7x + 14 \\ x+2\overline{\smash{)}\,x^3 - 5x^2 + 0x - 2} \\ -(x^3 + 2x^2) \\ \hline -7x^2 + 0x \\ -(-7x^2 - 14x) \\ \hline 14x - 2 \\ -(14x + 28) \\ \hline -30 \end{array}$$

So $f(x) = (x + 2)(x^2 - 7x + 14) - 30$

b) $f(x) + 30 = (x + 2)(x^2 - 7x + 14)$
So when $f(x) + 30 = 0$, $x = -2$ is a solution.
Show that $x^2 - 7x + 14 = 0$ has no real solutions:
$b^2 - 4ac = (-7)^2 - 4 \times 1 \times 14 = -7 < 0$.
Hence $x = -2$ is the only solution to $f(x) + 30 = 0$.

Exam-Style Questions — Chapter 4

Q1 a) (i) $f(x) = x^2 - 2x - 14 = (x - 1)^2 - 1 - 14 = (x - 1)^2 - 15$
[2 marks available — 1 mark for correct a, 1 mark for correct b]

(ii) When $f(x) = 0$, $(x - 1)^2 - 15 = 0 \Rightarrow (x - 1)^2 = 15$
$\Rightarrow x - 1 = \pm\sqrt{15} \Rightarrow x = 1 + \sqrt{15}$ or $x = 1 - \sqrt{15}$
[2 marks available — 1 mark for each correct solution]

b)
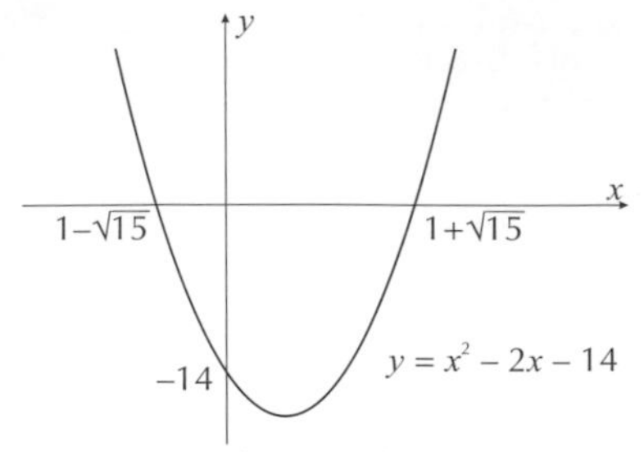

[3 marks available — 1 mark for correct curve shape, 1 mark for both correct x-intercepts, 1 mark for correct y-intercept]

Q2 For a quadratic with repeated roots, the discriminant equals zero,
so $b^2 - 4ac = 0 \Rightarrow 5^2 - 4 \times 2k \times k = 0 \Rightarrow 25 - 8k^2 = 0$
$\Rightarrow 8k^2 = 25 \Rightarrow k^2 = \frac{25}{8} \Rightarrow k = \pm\sqrt{\frac{25}{8}} = \pm\frac{5\sqrt{2}}{4}$
[5 marks available — 1 mark for using the discriminant, 1 mark for correctly substituting in a, b and c, 1 mark for attempting to solve quadratic to find k, 1 mark for unsimplified answer, 1 mark for simplifying and rationalising denominator]

Q3 a) $2x^2 - 5x - 3 = 0 \Rightarrow (2x + 1)(x - 3) = 0 \Rightarrow x = -\frac{1}{2}$ or $x = 3$
[2 marks available — 1 mark for correct factorisation, 1 mark for both correct solutions]

b) Let $u = \sqrt{x}$, then $2x - 5\sqrt{x} - 3 = 2u^2 - 5u - 3$
$2u^2 - 5u - 3 = 0 \Rightarrow u = -\frac{1}{2}$ or $u = 3 \Rightarrow \sqrt{x} = -\frac{1}{2}$ or $\sqrt{x} = 3$
$\sqrt{x} = -\frac{1}{2}$ gives no solution as $\sqrt{x} \geq 0$
$\sqrt{x} = 3 \Rightarrow x = 3^2 = 9$
[3 marks available — 1 mark for substituting to form a quadratic, 1 mark for reason why $\sqrt{x} = -\frac{1}{2}$ gives no solution, 1 mark for correct solution]

Q4 $f(x) = x^3 + x^2 = x^2(x + 1)$
Where $y = f(x)$ intersects the x-axis, $y = f(x) = 0$,
so $x^2(x + 1) = 0 \Rightarrow x = 0$ or $x = -1$
So the curve intersects the x-axis more than once, at (0, 0) and (–1, 0).
[3 marks available — 1 mark for factorising f(x), 1 mark for setting f(x) equal to zero, 1 mark for both correct coordinate pairs]

Q5 a) $x^2 - 1 = (x + 1)(x - 1)$
If $(x^2 - 1)$ is a factor of $f(x)$, both $(x + 1)$ and $(x - 1)$ must be factors of $f(x)$.
Substitute $x = -1$ into $f(x)$ and show the result is 0:
$f(-1) = 2(-1)^3 + 3(-1)^2 - 2(-1) - 3$
$= -2 + 3 + 2 - 3 = 0$
So by the Factor Theorem, $(x + 1)$ is a factor.
Substitute $x = 1$ into $f(x)$ and show the result is 0:
$f(1) = 2(1)^3 + 3(1)^2 - 2(1) - 3 = 2 + 3 - 2 - 3 = 0$
So by the Factor Theorem, $(x - 1)$ is a factor.
So $(x^2 - 1)$ is a factor of $f(x)$.
[3 marks available — 1 mark for factorising $x^2 - 1$, 1 mark for substituting $x = -1$ and $x = 1$ into f(x), 1 mark for correct interpretation]

b) From part a), you know that:
$f(x) = 2x^3 + 3x^2 - 2x - 3 = (x^2 - 1)(mx + n)$
The coefficient of x^3 in $f(x)$ is 2, so $m = 2$.
So $y = f(x) = (x^2 - 1)(2x + n)$
The y-intercept is –3 (i.e. when $x = 0$, $y = -3$), so:
$f(0) = (-1)(n) = -3 \Rightarrow n = 3$
$\Rightarrow f(x) = (x + 1)(x - 1)(2x + 3) = 0$
$\Rightarrow x = -1$ or $x = 1$ or $x = -\frac{3}{2}$

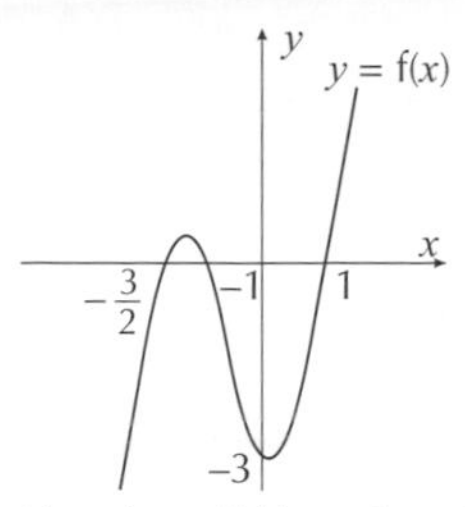

[4 marks available — 1 mark for correct factorisation of f(x), 1 mark for correct curve shape, 1 mark for correct y-intercept, 1 mark for correct x-intercepts]

Q6 **a)** Using the x-intercepts, you know that:
$f(x) = (x + 1)(x - 2)(mx + n)$, where m and n are constants.
You're told that the coefficient of $x^3 = 1$, so $m = 1$.
So $f(x) = (x + 1)(x - 2)(x + n)$.
The y-intercept of the curve is -10
(i.e. when $x = 0$, $y = -10$) so:
$f(0) = (1)(-2)(n) = -10 \Rightarrow n = 5$
So the coordinates of the x-intercept are $(-5, 0)$.
[3 marks available — 1 mark for correct f(x) in terms of m and n (or other constants), 1 mark for showing n = 5, 1 mark for correct coordinate pair]

b) $f(x) = (x + 5)(x + 1)(x - 2) = (x + 5)(x^2 - x - 2)$
$= x^3 - x^2 - 2x + 5x^2 - 5x - 10 = x^3 + 4x^2 - 7x - 10$
[2 marks available — 1 mark for expanding to correct linear factor and quadratic, 1 mark for all correct terms, a maximum of 1 mark if error carried forward from part a) (but correctly expanded)]

Q7 $f(x) = 4x^3 - 6x^2 - 3x + 5$
$f(1) = 4 - 6 - 3 + 5 = 0$, so $(x - 1)$ is a factor.
Use algebraic division to find the quadratic factor:

$$\begin{array}{r} 4x^2 - 2x - 5 \\ x - 1 \overline{)\, 4x^3 - 6x^2 - 3x + 5} \\ -\ \underline{(4x^3 - 4x^2)} \\ -2x^2 - 3x \\ -\ \underline{(-2x^2 + 2x)} \\ -5x + 5 \\ -\ \underline{(-5x + 5)} \\ 0 \end{array}$$

So $f(x) = (x - 1)(4x^2 - 2x - 5)$
Use the quadratic formula to solve $4x^2 - 2x - 5 = 0$:
$$x = \frac{-b \pm \sqrt{b^2 - 4ac}}{2a} = \frac{-(-2) \pm \sqrt{(-2)^2 - 4 \times 4 \times (-5)}}{2 \times 4}$$
$$= \frac{2 \pm \sqrt{4 + 80}}{8} = \frac{1 \pm \sqrt{21}}{4}$$
So $f(x) = 0$ when $x = 1$ and $x = \frac{1 \pm \sqrt{21}}{4}$.
[4 marks available — 1 mark for correct linear factor, 1 mark for correct quadratic factor, 1 mark for correctly substituting into quadratic formula, 1 mark for correct x-values]

Q8 **a)** **(i)** $f(-2) = (-2)^3 + 3(-2)^2 - 4(-2) - 12$
$= -8 + 12 + 8 - 12 = 0$
[2 marks available — 1 mark for correctly substituting in x = −2, 1 mark for correct answer]

(ii) $f(x) = x^3 + 3x^2 - 4x - 12 = (x + 2)(x^2 + nx - 6)$
$\Rightarrow nx^2 + 2x^2 = 3x^2 \Rightarrow n = 1$
So $f(x) = (x + 2)(x^2 + x - 6) = (x + 2)(x + 3)(x - 2)$
[3 marks available — 1 mark for suitable method, 1 mark for one correct factor, 1 mark for other two correct factors]

b)

y
$y = x^3 + 3x^2 - 4x - 12$
x
−3
−2
2
−12

[4 marks available — 1 mark for correct curve shape, 1 mark for correct y-intercept, 1 mark for one correct x-intercept, 1 mark for other two correct x-intercepts]

Q9 $f(x) = 2x^3 - 5x^2 - 4x + 3 = (2x - 1)(x^2 + nx - 3)$
$\Rightarrow 2nx^2 - x^2 = -5x^2 \Rightarrow 2nx^2 = -4x^2 \Rightarrow n = -2$
So $f(x) = 2x^3 - 5x^2 - 4x + 3 = (2x - 1)(x^2 - 2x - 3)$
$= (2x - 1)(x + 1)(x - 3)$
[3 marks available — 1 mark for suitable method, 1 mark each for the correct factors]

Q10 $(3x - 1)(2x + 1) = 6x^2 + x - 1$
$6x^3 - 29x^2 + Px + Q = (6x^2 + x - 1)(ax + b)$
Find a: $6x^3 = 6ax^3 \Rightarrow a = 1$
Find b: $-29x^2 = ax^2 + 6bx^2 = x^2 + 6bx^2$
$\Rightarrow 6bx^2 = -30x^2 \Rightarrow b = -5$
$(6x^2 + x - 1)(x - 5) = 6x^3 - 30x^2 + x^2 - 5x - x + 5$
$= 6x^3 - 29x^2 - 6x + 5$
So $P = -6$ and $Q = 5$.
[4 marks available — 1 mark for suitable method, 1 mark for one correct term of quotient, 1 mark for other correct terms of quotient, 1 mark for correct values of both P and Q]

Chapter 5: Inequalities and Simultaneous Equations

5.1 Inequalities

Exercise 5.1.1 — Linear inequalities

Q1 **a)** $2x - 1 < x + 4 \Rightarrow x < 5$

b) $4 - 3x \geq 10 - 5x \Rightarrow 2x \geq 6 \Rightarrow x \geq 3$

c) $5x + 7 > 3x + 1 \Rightarrow 2x > -6 \Rightarrow x > -3$

d) $3 - 2x \leq 5x - 4 \Rightarrow -7x \leq -7 \Rightarrow x \geq 1$

e) $9 - x \geq 7x + 5 \Rightarrow -8x \geq -4 \Rightarrow x \leq \frac{1}{2}$

f) $12x - 9 \leq 4x + 11 \Rightarrow 8x \leq 20 \Rightarrow x \leq 2.5$

g) $3x - 6 > 6 - 3x \Rightarrow 6x > 12 \Rightarrow x > 2$

h) $-4x < 16 - 7x \Rightarrow 3x < 16 \Rightarrow x < \frac{16}{3}$

Q2 **a)** $2(x + 3) > 3(x + 2) \Rightarrow 2x + 6 > 3x + 6 \Rightarrow -x > 0 \Rightarrow x < 0$
In set notation, this is $\{x : x < 0\}$

b) $5(1 + 3x) \leq 7 \Rightarrow 5 + 15x \leq 7 \Rightarrow 15x \leq 2 \Rightarrow x \leq \frac{2}{15}$
In set notation, this is $\left\{x : x \leq \frac{2}{15}\right\}$

c) $12 \geq 2(5 - 2x) \Rightarrow 6 \geq 5 - 2x \Rightarrow 1 \geq -2x \Rightarrow x \geq -\frac{1}{2}$
In set notation, this is $\left\{x : x \geq -\frac{1}{2}\right\}$

Q3 **a)** $\frac{6 - 5x}{2} < \frac{4 - 8x}{3} \Rightarrow 3(6 - 5x) < 2(4 - 8x)$
$\Rightarrow 18 - 15x < 8 - 16x \Rightarrow x < -10$

b) $\frac{3x - 1}{4} \geq 2x \Rightarrow 3x - 1 \geq 8x \Rightarrow x \leq -\frac{1}{5}$

c) $\frac{x - 2}{2} - \frac{2x + 3}{3} < 7 \Rightarrow 3(x - 2) - 2(2x + 3) < 42$
$\Rightarrow 3x - 6 - 4x - 6 < 42 \Rightarrow -x < 54 \Rightarrow x > -54$

Q4 **a)** $-5 < 2x - 3 < 15 \Rightarrow -2 < 2x < 18 \Rightarrow -1 < x < 9$
In set notation, this is either $\{x : -1 < x < 9\}$
or $\{x : x > -1\} \cap \{x : x < 9\}$

b) $-5 \leq 4 - 3x < 19 \Rightarrow -9 \leq -3x < 15$
$\Rightarrow 3 \geq x > -5 \Rightarrow -5 < x \leq 3$
In set notation, this is either $\{x : -5 < x \leq 3\}$
or $\{x : x > -5\} \cap \{x : x \leq 3\}$

c) $5 \leq 7 + 6x \leq 11 \Rightarrow -2 \leq 6x \leq 4 \Rightarrow -\frac{1}{3} \leq x \leq \frac{2}{3}$
In set notation, this is either $\left\{x : -\frac{1}{3} \leq x \leq \frac{2}{3}\right\}$
or $\left\{x : x \geq -\frac{1}{3}\right\} \cap \left\{x : x \leq \frac{2}{3}\right\}$

Q5 a) $2x \geq 3 - x \Rightarrow 3x \geq 3 \Rightarrow x \geq 1$

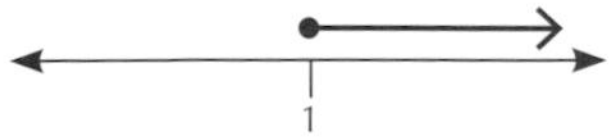

b) $5x - 1 < 3x + 5 \Rightarrow 2x < 6 \Rightarrow x < 3$

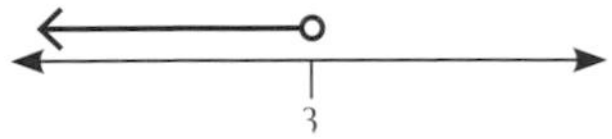

c) $2x + 1 \geq 3x + 2 \Rightarrow -x \geq 1 \Rightarrow x \leq -1$

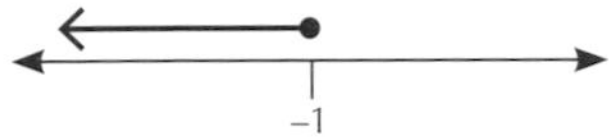

d) $3(x - 3) \leq 5(x - 1) \Rightarrow 3x - 9 \leq 5x - 5$
$\Rightarrow -2x \leq 4 \Rightarrow x \geq -2$

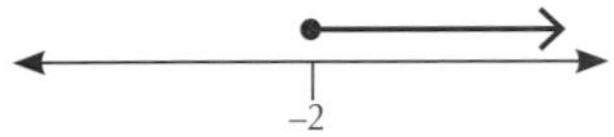

e) $9 - x \leq 3 - 4x \Rightarrow 3x \leq -6 \Rightarrow x \leq -2$

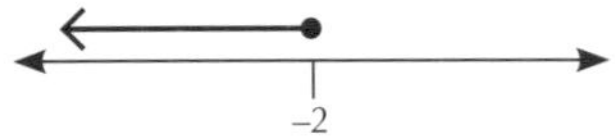

f) $\frac{2(x-3)}{3} + 1 < \frac{2x-1}{2}$
$\Rightarrow 4(x - 3) + 6 < 3(2x - 1)$
$\Rightarrow 4x - 12 + 6 < 6x - 3 \Rightarrow -2x < 3 \Rightarrow x > -\frac{3}{2}$

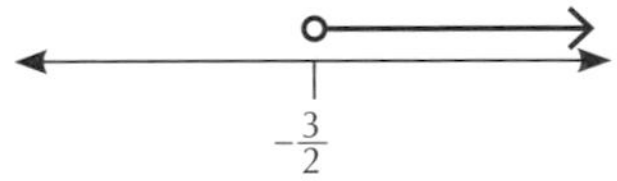

Q6 a) $7 \leq 3x - 2 < 16 \Rightarrow 9 \leq 3x < 18 \Rightarrow 3 \leq x < 6$

b)

Q7 $4 - 2x < 10 \Rightarrow -2x < 6 \Rightarrow x > -3$
$3x - 1 < x + 7 \Rightarrow 2x < 8 \Rightarrow x < 4$

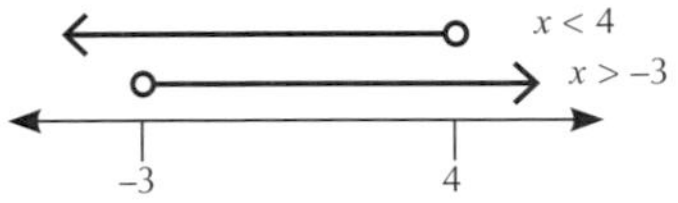

The answer will be the intersection of these solutions, i.e. $\{x : x < 4\} \cap \{x : x > -3\}$

The solutions overlap between –3 and 4, so this is the same as $\{x : -3 < x < 4\}$

Q8 a) $2x \geq 3x - 5 \Rightarrow -x \geq -5 \Rightarrow x \leq 5$
$3x - 2 \geq x - 6 \Rightarrow 2x \geq -4 \Rightarrow x \geq -2$

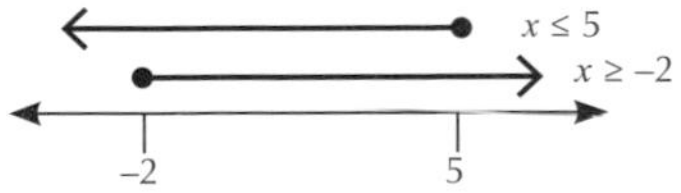

Solution: $-2 \leq x \leq 5$

b) $5x + 1 \leq 11 \Rightarrow 5x \leq 10 \Rightarrow x \leq 2$
$2x - 3 < 5x - 6 \Rightarrow -3x < -3 \Rightarrow x > 1$

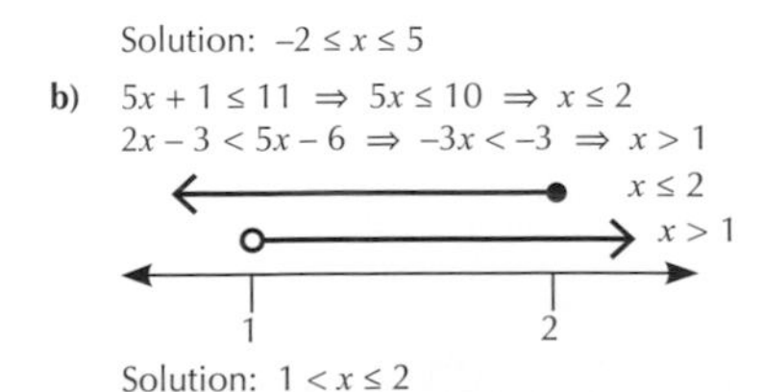

Solution: $1 < x \leq 2$

c) $2x - 1 \leq 3x - 5 \Rightarrow -x \leq -4 \Rightarrow x \geq 4$
$5x - 6 > x + 22 \Rightarrow 4x > 28 \Rightarrow x > 7$

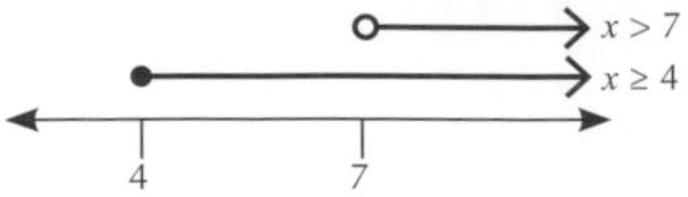

Solution: $x > 7$
Only these values satisfy both of the inequalities.

d) $3x + 5 < x + 1 \Rightarrow 2x < -4 \Rightarrow x < -2$
$6x - 1 \geq 3x + 5 \Rightarrow 3x \geq 6 \Rightarrow x \geq 2$

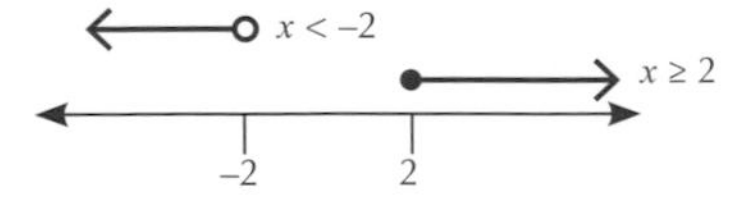

There is no solution that satisfies both inequalities.
The solutions don't overlap on the number line.

Exercise 5.1.2 — Quadratic inequalities

Q1 a) $-3 < x < 1$

b) $x < 0$ or $x > 4$

c) $2x^2 \geq 5 - 9x \Rightarrow 2x^2 + 9x - 5 \geq 0$
$\Rightarrow x \leq -5$ or $x \geq \frac{1}{2}$

d) $x < 1 - \sqrt{6}$ or $x > 1 + \sqrt{6}$

Q2 a) Solve the equation $x^2 = 4$ to find the x-intercepts:
$x^2 = 4 \Rightarrow x^2 - 4 = 0 \Rightarrow (x - 2)(x + 2) = 0$
$x = -2$ or $x = 2$

Solution: $x^2 \leq 4 \Rightarrow x^2 - 4 \leq 0 \Rightarrow -2 \leq x \leq 2$

In set notation, this is either $\{x : -2 \leq x \leq 2\}$
or $\{x : x \geq -2\} \cap \{x : x \leq 2\}$

b) $13x = 3x^2 + 4 \Rightarrow -3x^2 + 13x - 4 = 0 \Rightarrow 3x^2 - 13x + 4 = 0$
$\Rightarrow (3x - 1)(x - 4) = 0 \Rightarrow x = \frac{1}{3}$ or $x = 4$
Solution: $13x < 3x^2 + 4 \Rightarrow -3x^2 + 13x - 4 < 0$
$\Rightarrow x < \frac{1}{3}$ or $x > 4$

In set notation, this is $\left\{x : x < \frac{1}{3}\right\} \cup \{x : x > 4\}$

c) $x^2 + 4 = 6x \Rightarrow x^2 - 6x + 4 = 0$
$\Rightarrow x = \frac{6 \pm \sqrt{36 - 16}}{2} = \frac{6 \pm \sqrt{20}}{2} = \frac{6 \pm 2\sqrt{5}}{2}$
$\Rightarrow x = 3 \pm \sqrt{5}$

Solution: $x^2 + 4 < 6x \Rightarrow x^2 - 6x + 4 < 0$
$\Rightarrow 3 - \sqrt{5} < x < 3 + \sqrt{5}$

In set notation, this is either $\{x : 3 - \sqrt{5} < x < 3 + \sqrt{5}\}$
or $\{x : x > 3 - \sqrt{5}\} \cap \{x : x < 3 + \sqrt{5}\}$

d) $7x = 4 - 2x^2 \Rightarrow -2x^2 - 7x + 4 = 0$
$\Rightarrow (x + 4)(-2x + 1) = 0 \Rightarrow x = -4$ or $x = \frac{1}{2}$
Solution: $7x > 4 - 2x^2 \Rightarrow 0 > -2x^2 - 7x + 4$
$\Rightarrow x < -4$ or $x > \frac{1}{2}$
In set notation, this is $\{x : x < -4\} \cup \left\{x : x > \frac{1}{2}\right\}$

Q3 a) $x^2 + 5x - 6 = 0 \Rightarrow (x + 6)(x - 1) = 0$
$\Rightarrow x = -6$ or $x = 1$

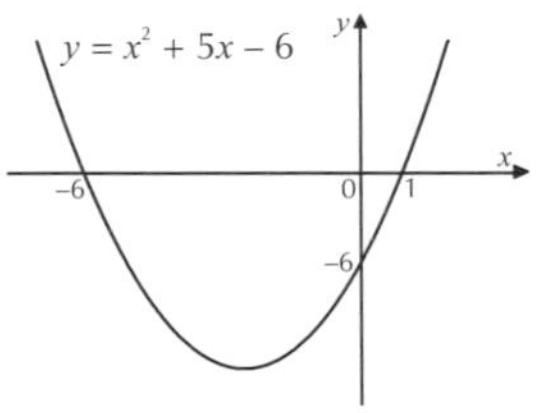

$x^2 + 5x - 6 \geq 0 \Rightarrow x \leq -6$ or $x \geq 1$

b) $x^2 - 3x + 2 = 0 \Rightarrow (x-1)(x-2) = 0 \Rightarrow x = 1$ or $x = 2$

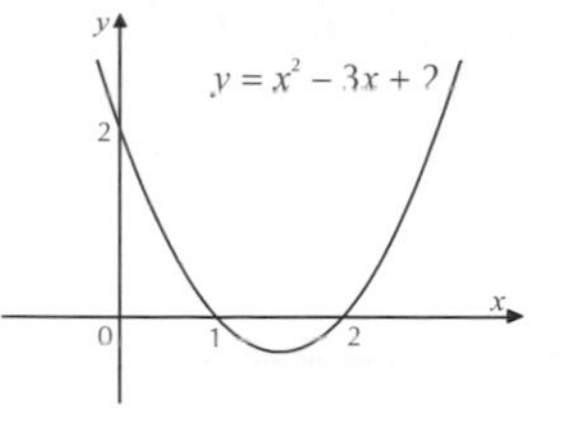

$x^2 - 3x + 2 < 0 \Rightarrow 1 < x < 2$

c) $6 - 5x = 6x^2 \Rightarrow -6x^2 - 5x + 6 = 0 \Rightarrow 6x^2 + 5x - 6 = 0$

$\Rightarrow (3x-2)(2x+3) = 0 \Rightarrow x = \frac{2}{3}$ or $x = -\frac{3}{2}$

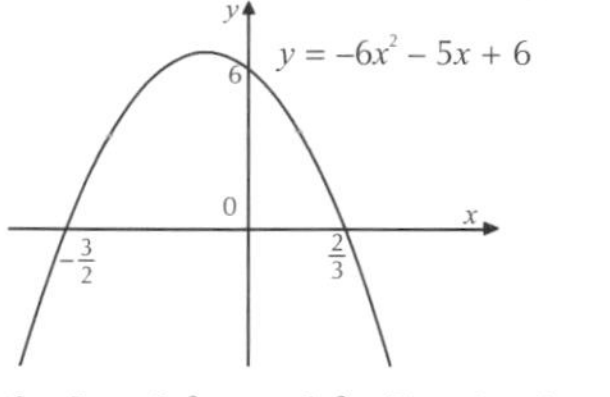

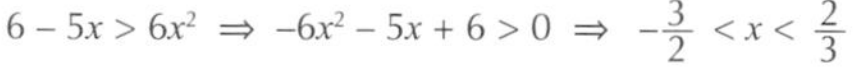

$6 - 5x > 6x^2 \Rightarrow -6x^2 - 5x + 6 > 0 \Rightarrow -\frac{3}{2} < x < \frac{2}{3}$

You could have rearranged the inequality into $6x^2 + 5x - 6 < 0$ and sketched the corresponding graph. You'd get the same final answer, but the graph would be the other way up.

d) $x^2 - 5x + 24 = 5x + 3 \Rightarrow x^2 - 10x + 21 = 0$

$\Rightarrow (x-3)(x-7) = 0 \Rightarrow x = 3$ or $x = 7$

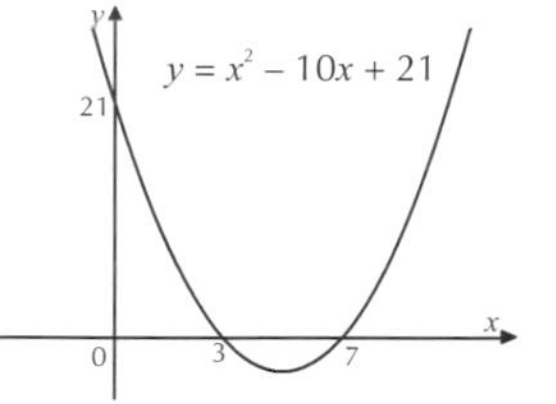

$x^2 - 5x + 24 \le 5x + 3 \Rightarrow x^2 - 10x + 21 \le 0 \Rightarrow 3 \le x \le 7$

e) $36 - 4x^2 = 0 \Rightarrow 9 - x^2 = 0 \Rightarrow (3-x)(3+x) = 0 \Rightarrow x = \pm 3$

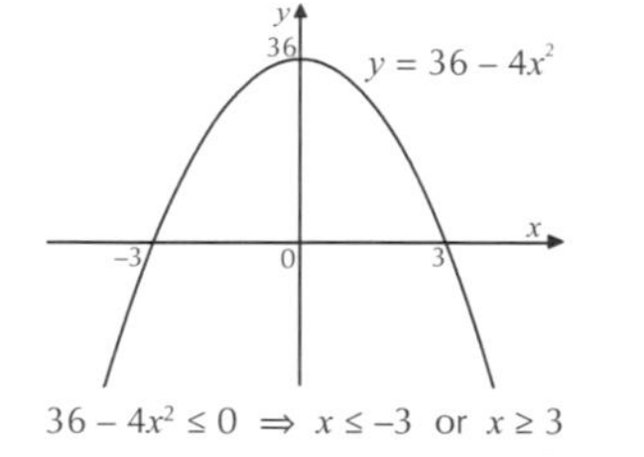

$36 - 4x^2 \le 0 \Rightarrow x \le -3$ or $x \ge 3$

f) $x^2 - 6x + 3 = 0 \Rightarrow x = \frac{6 \pm \sqrt{36 - 12}}{2}$

$\Rightarrow x = \frac{6 \pm \sqrt{24}}{2} = \frac{6 \pm 2\sqrt{6}}{2} = 3 \pm \sqrt{6}$

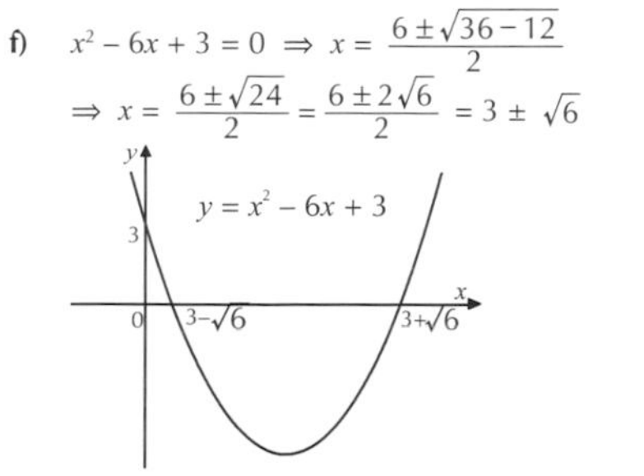

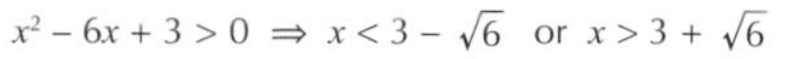

$x^2 - 6x + 3 > 0 \Rightarrow x < 3 - \sqrt{6}$ or $x > 3 + \sqrt{6}$

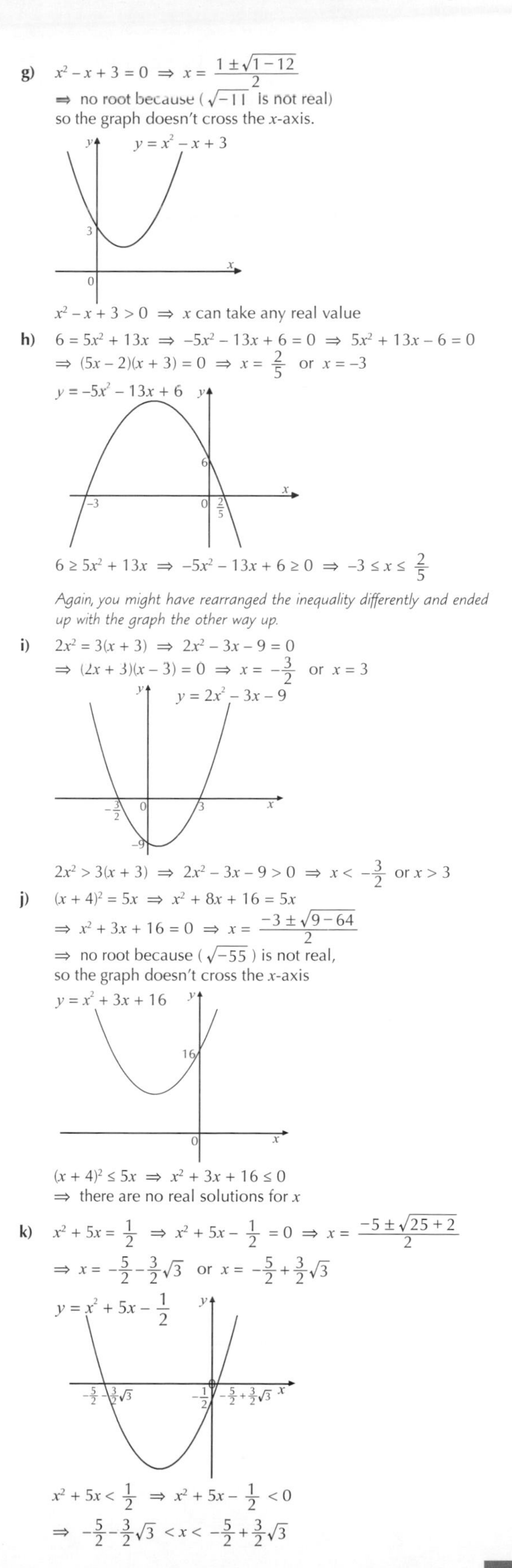

g) $x^2 - x + 3 = 0 \Rightarrow x = \frac{1 \pm \sqrt{1 - 12}}{2}$

$\Rightarrow$ no root because ($\sqrt{-11}$ is not real)
so the graph doesn't cross the x-axis.

$x^2 - x + 3 > 0 \Rightarrow x$ can take any real value

h) $6 = 5x^2 + 13x \Rightarrow -5x^2 - 13x + 6 = 0 \Rightarrow 5x^2 + 13x - 6 = 0$

$\Rightarrow (5x-2)(x+3) = 0 \Rightarrow x = \frac{2}{5}$ or $x = -3$

$6 \ge 5x^2 + 13x \Rightarrow -5x^2 - 13x + 6 \ge 0 \Rightarrow -3 \le x \le \frac{2}{5}$

Again, you might have rearranged the inequality differently and ended up with the graph the other way up.

i) $2x^2 = 3(x+3) \Rightarrow 2x^2 - 3x - 9 = 0$

$\Rightarrow (2x+3)(x-3) = 0 \Rightarrow x = -\frac{3}{2}$ or $x = 3$

$2x^2 > 3(x+3) \Rightarrow 2x^2 - 3x - 9 > 0 \Rightarrow x < -\frac{3}{2}$ or $x > 3$

j) $(x+4)^2 = 5x \Rightarrow x^2 + 8x + 16 = 5x$

$\Rightarrow x^2 + 3x + 16 = 0 \Rightarrow x = \frac{-3 \pm \sqrt{9 - 64}}{2}$

$\Rightarrow$ no root because ($\sqrt{-55}$) is not real,
so the graph doesn't cross the x-axis

$(x+4)^2 \le 5x \Rightarrow x^2 + 3x + 16 \le 0$

$\Rightarrow$ there are no real solutions for x

k) $x^2 + 5x = \frac{1}{2} \Rightarrow x^2 + 5x - \frac{1}{2} = 0 \Rightarrow x = \frac{-5 \pm \sqrt{25 + 2}}{2}$

$\Rightarrow x = -\frac{5}{2} - \frac{3}{2}\sqrt{3}$ or $x = -\frac{5}{2} + \frac{3}{2}\sqrt{3}$

$x^2 + 5x < \frac{1}{2} \Rightarrow x^2 + 5x - \frac{1}{2} < 0$

$\Rightarrow -\frac{5}{2} - \frac{3}{2}\sqrt{3} < x < -\frac{5}{2} + \frac{3}{2}\sqrt{3}$

l) $\frac{3}{4}x^2 = 1 + \frac{1}{4}x \Rightarrow \frac{3}{4}x^2 - \frac{1}{4}x - 1 = 0$
$\Rightarrow (x + 1)(\frac{3}{4}x - 1) = 0 \Rightarrow x = -1$ or $x = \frac{4}{3}$

$\frac{3}{4}x^2 \geq 1 + \frac{1}{4}x \Rightarrow (x + 1)(\frac{3}{4}x - 1) \geq 0 \Rightarrow x \leq 1$ or $x \geq \frac{4}{3}$

Q4 a) $\frac{1}{x} > 5 \Rightarrow x > 5x^2 \Rightarrow 5x^2 - x < 0 \Rightarrow x(5x - 1) < 0$
$\Rightarrow 0 < x < \frac{1}{5}$, or in set notation, $\{x : 0 < x < \frac{1}{5}\}$

b) $7 > \frac{3}{x} \Rightarrow 7x^2 > 3x \Rightarrow 7x^2 - 3x > 0$
$\Rightarrow x(7x - 3) > 0 \Rightarrow x < 0$ or $x > \frac{3}{7}$,
or in set notation, $\{x : x < 0\} \cup \{x : x > \frac{3}{7}\}$

c) $-5 > \frac{2}{x} \Rightarrow -5x^2 > 2x \Rightarrow 0 > 5x^2 + 2x$
$\Rightarrow 0 > x(5x + 2) \Rightarrow -\frac{2}{5} < x < 0$
or in set notation, $\{x : -\frac{2}{5} < x < 0\}$

d) $-\frac{6}{x} > 1 \Rightarrow -6x > x^2 \Rightarrow 0 > x^2 + 6x$
$\Rightarrow 0 > x(x + 6) \Rightarrow -6 < x < 0$
or in set notation, $\{x : -6 < x < 0\}$

Q5 a) $x^2 - 6x - 7 < 0 \Rightarrow -1 < x < 7$. But $x \leq 4$, so the solution is $-1 < x \leq 4$. This is $(-1, 4]$ in interval notation.

b) $-2x^2 + 19x - 30 > 0 \Rightarrow 2 < x < \frac{15}{2}$. But $x > 5$, so the solution is $5 < x < \frac{15}{2}$. This is $(5, \frac{15}{2})$ in interval notation.

c) $x^2 - x \leq 56 \Rightarrow x^2 - x - 56 \leq 0 \Rightarrow -7 \leq x \leq 8$
But $\frac{1}{x} + \frac{1}{x^2} > 0 \Rightarrow x + 1 > 0 \Rightarrow x > -1$, so the solution is $-1 < x \leq 8$. This is $(-1, 8]$ in interval notation.

d) $4x \leq \frac{x^2}{3} \Rightarrow -\frac{x^2}{3} + 4x \leq 0 \Rightarrow x \leq 0$ or $x \geq 12$
$5x - 2 < 4x + 8 \Rightarrow x < 10$
So the solution is $x \leq 0$. This is $(-\infty, 0]$ in interval notation.

Q6 The area of the office will be $(x - 9)(x - 6)$ m², so use this to form an inequality for the necessary floor space:
$(x - 9)(x - 6) \geq 28 \Rightarrow x^2 - 15x + 54 \geq 28 \Rightarrow x^2 - 15x + 26 \geq 0$
Find the x-intercepts of the graph:
$x^2 - 15x + 26 = 0 \Rightarrow (x - 2)(x - 13) = 0 \Rightarrow x = 2$ and $x = 13$

$x^2 - 15x + 26 \geq 0 \Rightarrow x \leq 2$ or $x \geq 13$
But $x \leq 2$ would mean that the sides of the office would have negative lengths, so the only possible values of x are $x \geq 13$ m.

Q7 a) $kx^2 - 6x + k = 0 \Rightarrow a = k, b = -6, c = k$
$b^2 - 4ac = (-6)^2 - (4 \times k \times k) = 36 - 4k^2$
The original equation has two distinct real solutions, so the discriminant must be > 0. So $36 - 4k^2 > 0$.
Factorise the quadratic: $36 - 4k^2 = 4(3 + k)(3 - k)$
So the graph is n-shaped and crosses the k-axis at $k = 3$ and $k = -3$.

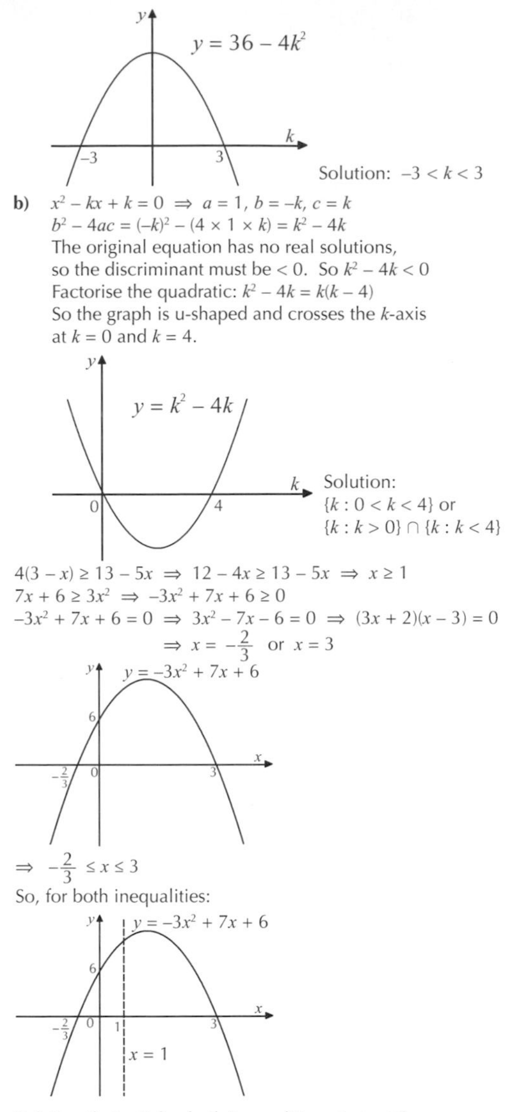

Solution: $-3 < k < 3$

b) $x^2 - kx + k = 0 \Rightarrow a = 1, b = -k, c = k$
$b^2 - 4ac = (-k)^2 - (4 \times 1 \times k) = k^2 - 4k$
The original equation has no real solutions, so the discriminant must be < 0. So $k^2 - 4k < 0$
Factorise the quadratic: $k^2 - 4k = k(k - 4)$
So the graph is u-shaped and crosses the k-axis at $k = 0$ and $k = 4$.

Solution:
$\{k : 0 < k < 4\}$ or
$\{k : k > 0\} \cap \{k : k < 4\}$

Q8 $4(3 - x) \geq 13 - 5x \Rightarrow 12 - 4x \geq 13 - 5x \Rightarrow x \geq 1$
$7x + 6 \geq 3x^2 \Rightarrow -3x^2 + 7x + 6 \geq 0$
$-3x^2 + 7x + 6 = 0 \Rightarrow 3x^2 - 7x - 6 = 0 \Rightarrow (3x + 2)(x - 3) = 0$
$\Rightarrow x = -\frac{2}{3}$ or $x = 3$

$\Rightarrow -\frac{2}{3} \leq x \leq 3$
So, for both inequalities:

Solution that satisfies both inequalities: $1 \leq x \leq 3$
Your graphs might be the other way up if you rearranged the inequality differently at the start.

Exercise 5.1.3 — Graphing inequalities

Q1 a) If (2, 4) lies in the region that satisfies $3y > 8x - 3$ then $3(4) > 8(2) - 3 \Rightarrow 12 > 13$.
This is false so the statement is false.

b) If (–3, –5) lies outside the region that satisfies $4y + x^2 \leq 3$ then it lies within the region that satisfies $4y + x^2 > 3$, so $4(-5) + (-3)^2 > 3 \Rightarrow -20 + 9 > 3 \Rightarrow -11 > 3$.
This is false so the statement is false.

c) If (8, –4) lies in the region that satisfies $y^2 + (x + 6)^2 \geq 68$ then $(-4)^2 + (8 + 6)^2 \geq 68$
$\Rightarrow 16 + 196 \geq 68 \Rightarrow 212 \geq 68$.
This is true so the statement is true.

d) If (1, 3) lies in the region that satisfies $x + 2y > 4$ then $1 + 2(3) > 4 \Rightarrow 7 > 4$.
If (1, 3) is in the region that satisfies $3x^2 > 20 - 4y$ then $3(1)^2 > 20 - 4(3) \Rightarrow 3 > 8$.
The first part of the statement is true, but the second part is false, so the statement is false.
If the point isn't in the region that satisfies all the inequalities in the statement then the statement is false.

e) If $\left(\frac{1}{2}, \frac{3}{2}\right)$ lies outside the region that satisfies $y^2 < 10 - 8x^2$ then it lies within the region that satisfies $y^2 \geq 10 - 8x^2$,

so $\left(\frac{3}{2}\right)^2 \geq 10 - 8\left(\frac{1}{2}\right)^2 \Rightarrow \frac{9}{4} \geq 10 - 2 \Rightarrow \frac{9}{4} \geq 8$.

If $\left(\frac{1}{2}, \frac{3}{2}\right)$ lies outside the region that satisfies $3x + 4y \geq 6$ then it lies within the region that satisfies $3x + 4y < 6$,

so $3\left(\frac{1}{2}\right) + 4\left(\frac{3}{2}\right) < 6 \Rightarrow \frac{3}{2} + \frac{12}{2} < 6 \Rightarrow \frac{15}{2} < 6$.

Both parts of the statement are false, so the statement is false.

It isn't strictly necessary to show that the second part is also false — the statement is shown to be false from the first inequality.

Q2 a) The shaded region is above the solid line $y = -\frac{1}{4}x + 1$ and below the dotted curve $y = 6 - x + x^2$. So the shaded region is defined by $y \geq -\frac{1}{4}x + 1$ and $y < 6 + x - x^2$.

b) The shaded region is above the solid line $y = \frac{3}{2}x + 2$, below the dotted line $y = 14 - \frac{1}{2}x$ and above the curve $y = \frac{1}{2}x^2 - 5x + \frac{9}{2}$. So the shaded region is defined by $y \geq \frac{3}{2}x + 2$, $y < 14 - \frac{1}{2}x$ and $y > \frac{1}{2}x^2 - 5x + \frac{9}{2}$.

Q3 a) Write as equations and rearrange: $y = -x + 5$ (dotted), $y = -2x + 4$ (solid), $y = -\frac{1}{2}x + 3$ (dotted)

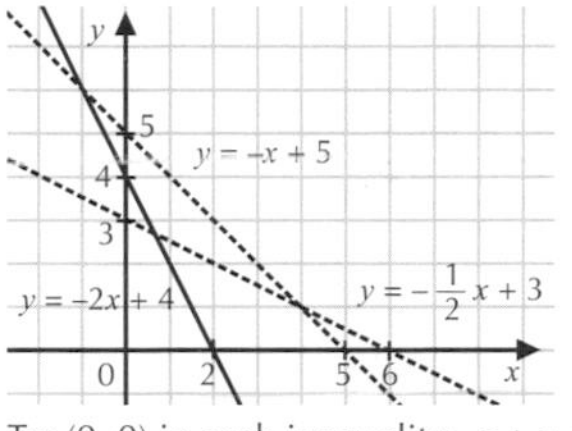

Try (0, 0) in each inequality: $x + y < 5 \Rightarrow 0 < 5$ — this is true so shade this side.
$2x + y \geq 4 \Rightarrow 0 \geq 4$ — this is false so shade the other side.
$x + 2y > 6 \Rightarrow 0 > 6$ — this is false so shade the other side.

So the final region is:

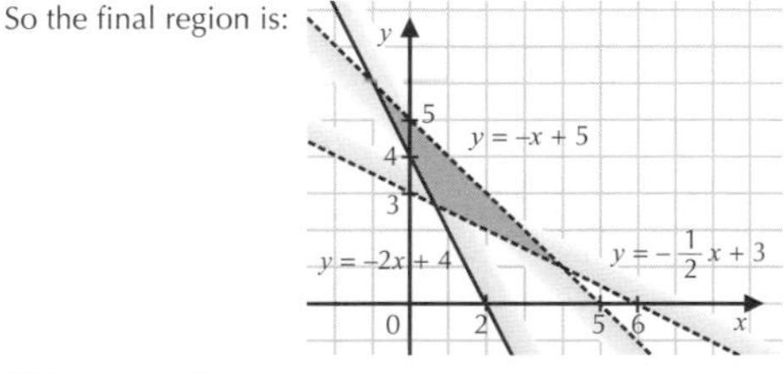

b) Write as equations and rearrange: $x = 4$ (solid), $y = 7$ (solid), $y = -x + 4$ (dotted)

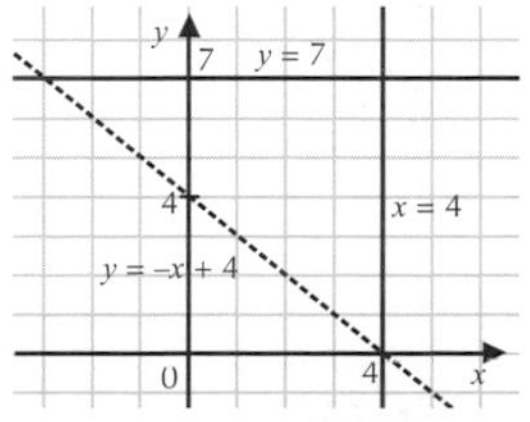

Try (0, 0) in each inequality: $x \leq 4 \Rightarrow 0 \leq 4$ — this is true so shade this side.
$y \leq 7 \Rightarrow 0 \leq 7$ — this is true so shade this side.
$x + y > 4 \Rightarrow 0 > 4$ — this is false so shade the other side.

So the final region is:

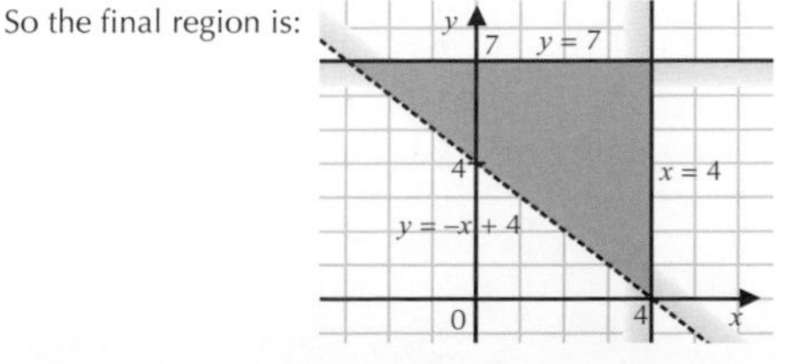

c) Write as equations and rearrange: $y = x^2$ (dotted), $y = x + 3$ (solid)

You can't use (0, 0) as it lies on one of the lines, so try (0, 1) in each inequality: $y > x^2 \Rightarrow 1 > 0$ — this is true so shade this side.
$x - y \geq -3 \Rightarrow -1 \geq -3$ — this is true so shade this side.

So the final region is:

d) Write as equations and rearrange: $y = x^2 + 2$ (solid), $y = 2x^2 - 2$ (dotted)

Try (0, 0) in each inequality: $y - 2 \leq x^2 \Rightarrow -2 \leq 0$ — this is true so shade this side.
$2x^2 - y < 2 \Rightarrow 0 < 2$ — this is true so shade this side.

So the final region is:

e) Write as equations and rearrange: $y = 4x^2 + 5$ (dotted), $y = -\frac{3}{5}x + 8$ (solid)

Try (0, 0) in each inequality: $4x^2 > y - 5 \Rightarrow 0 > -5$ — this is true so shade this side.
$3x + 5y \leq 40 \Rightarrow 0 \leq 40$ — this is true so shade this side.

So the final region is:

f) Write as equations and rearrange:

$y = -2x^2 + 3x + 5$ (dotted), $y = \frac{2}{5}x + 1$ (dotted)

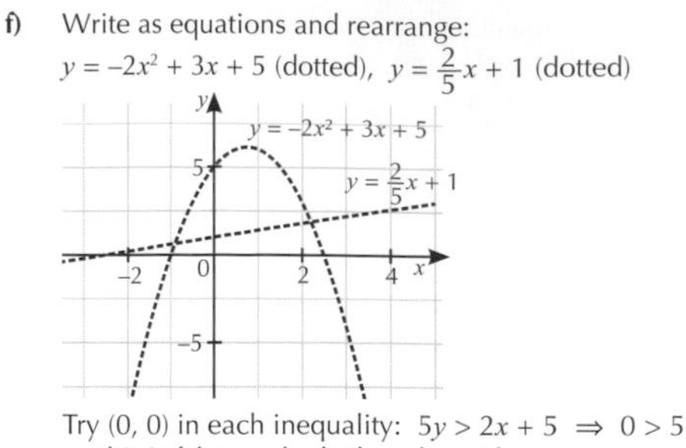

Try (0, 0) in each inequality: $5y > 2x + 5 \Rightarrow 0 > 5$ — this is false so shade the other side.

$2y + 4x^2 < 6x + 10 \Rightarrow 0 < 10$ — this is true so shade this side.

So the final region is:

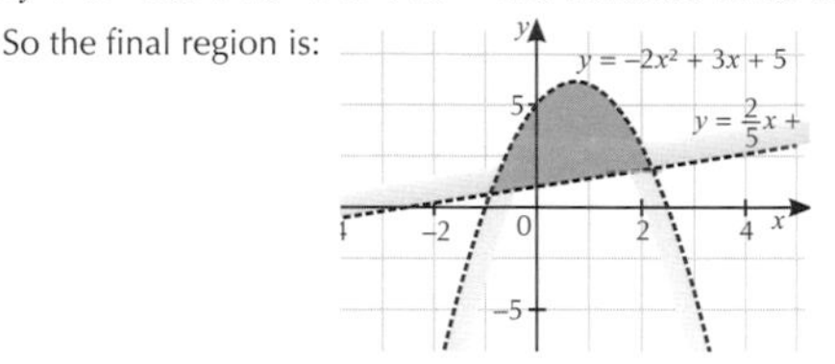

Q4 *A*: Write as equations and rearrange:

$y = 6 - \frac{1}{2}x$, $y = \frac{3}{2}x + 2$, $y = 2$

Then plot these lines (all solid):

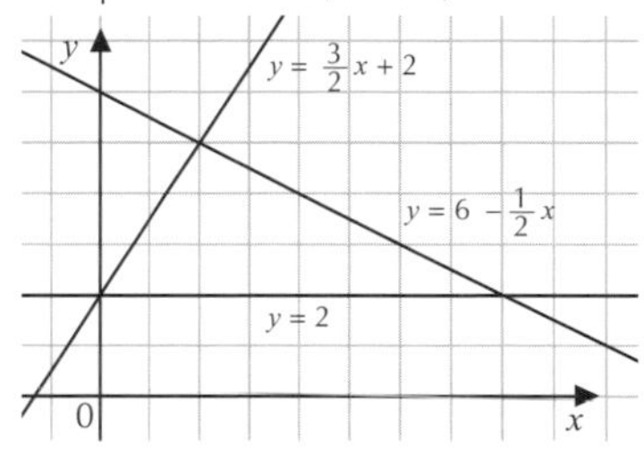

Try (0, 0) in each inequality:

$x + 2y \leq 12 \Rightarrow 0 \leq 12$ — this is true so shade the other side.

$2y - 3x \leq 4 \Rightarrow 0 \leq 4$ — this is true so shade the other side.

$y \geq 2 \Rightarrow 0 \geq 2$ — this is false so shade this side.

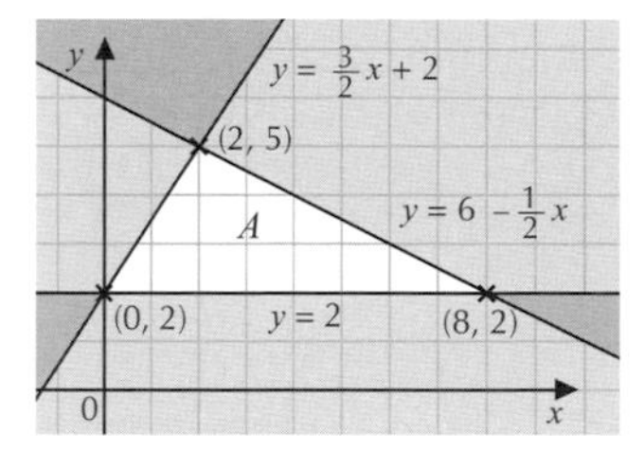

B: Write as equations and rearrange: $x = 3$, $y = 2x - 9$, $y = 5 - \frac{1}{3}x$. Then plot these lines (all solid):

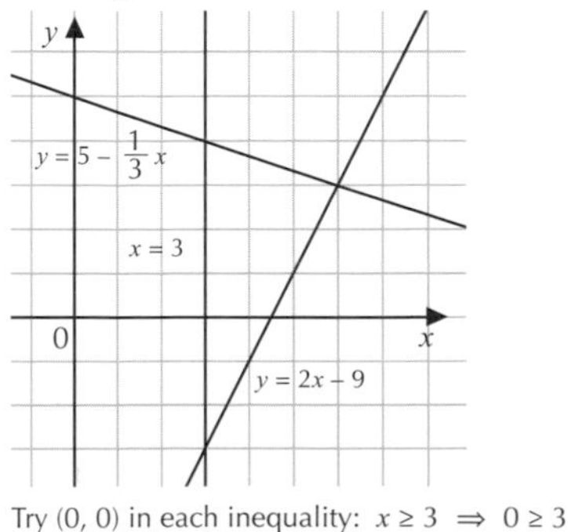

Try (0, 0) in each inequality: $x \geq 3 \Rightarrow 0 \geq 3$

This is false so shade this side.

$2x \leq y + 9 \Rightarrow 0 \leq 9$. This is true so shade the other side.

$x + 3y \leq 15 \Rightarrow 0 \leq 15$. This is true so shade the other side.

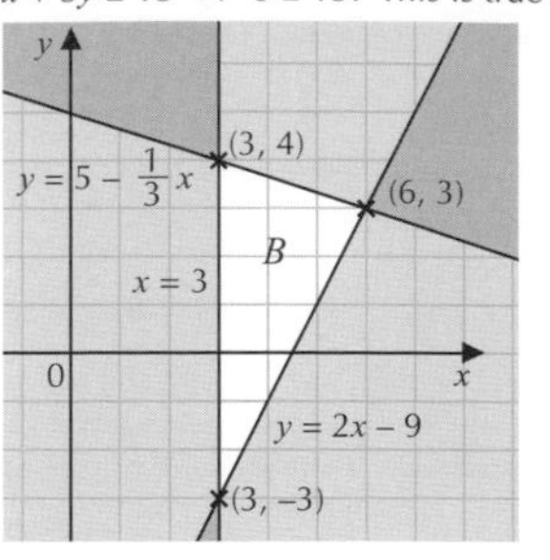

The coordinates of the vertices of triangles *A* and *B* can be read off the graphs. In order to decide which region is larger, calculate the area of each triangle.

You can also find the vertices by equating the lines to find their points of intersection.

A has base $8 - 0 = 8$ and height $5 - 2 = 3$

$\Rightarrow$ Area $A = \frac{1}{2} \times 8 \times 3 = 12$

B has base $4 - (-3) = 7$ and height $6 - 3 = 3$

$\Rightarrow$ Area $B = \frac{1}{2} \times 7 \times 3 = 10.5$

So region *A* is larger.

Q5 **a)** $3x + y \leq 42$

b) Write as equations and rearrange: $y = 12 - \frac{1}{2}x$, $y = 42 - 3x$

Plotting the graphs and testing the point (1, 1) in each inequality gives the region labelled *R*:

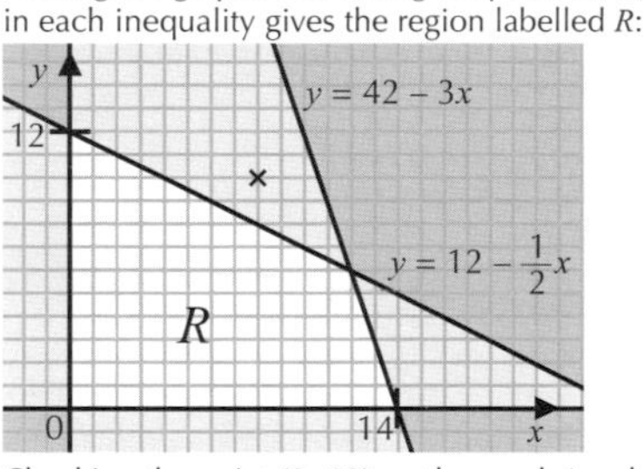

c) Checking the point (8, 10) on the graph (marked in b)) shows that it isn't in the possible region.

Test $x = 8$ and $y = 10$ in both inequalities:

Flour: $x + 2y = 8 + 20 = 28 > 24$

Eggs: $3x + y = 24 + 10 = 34 \leq 42$

So the bakery has enough eggs, but not enough flour to meet the order.

5.2 Simultaneous Equations

Exercise 5.2.1 — Simultaneous equations — both linear

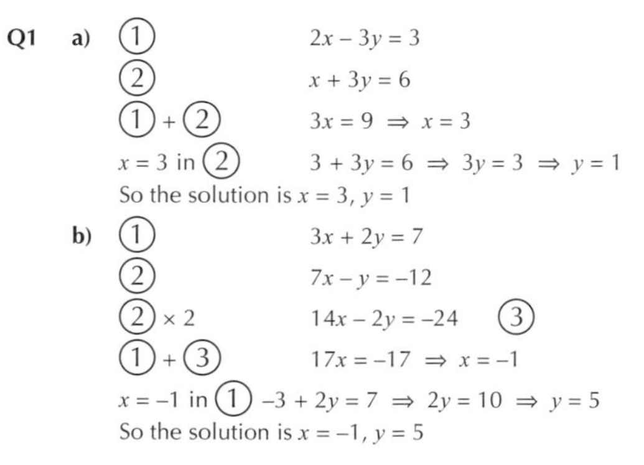

Q1 **a)**

①	$2x - 3y = 3$
②	$x + 3y = 6$
① + ②	$3x = 9 \Rightarrow x = 3$
$x = 3$ in ②	$3 + 3y = 6 \Rightarrow 3y = 3 \Rightarrow y = 1$

So the solution is $x = 3$, $y = 1$

b)

①	$3x + 2y = 7$
②	$7x - y = -12$
② × 2	$14x - 2y = -24$ ③
① + ③	$17x = -17 \Rightarrow x = -1$
$x = -1$ in ①	$-3 + 2y = 7 \Rightarrow 2y = 10 \Rightarrow y = 5$

So the solution is $x = -1$, $y = 5$

c) ① $4x + 3y = -4$
② $6x - 4y = 11$
① × 3 $12x + 9y = -12$ ③
② × 2 $12x - 8y = 22$ ④
③ – ④ $17y = -34 \Rightarrow y = -2$
$y = -2$ in ② $6x + 8 = 11 \Rightarrow 6x = 3 \Rightarrow x = \frac{1}{2}$
So the solution is $x = \frac{1}{2}$, $y = -2$

d) ① $7x - 6y = 4$
② $11x + 9y = -6$
① × 3 $21x - 18y = 12$ ③
② × 2 $22x + 18y = -12$ ④
③ + ④ $43x = 0 \Rightarrow x = 0$
$x = 0$ in ② $0 + 9y = -6 \Rightarrow y = -\frac{6}{9} = -\frac{2}{3}$
So the solution is $x = 0$, $y = -\frac{2}{3}$

e) Rearrange ① $6x + 2y = 8$
Rearrange ② $4x + 3y = -3$
① × 2 $12x + 4y = 16$ ③
② × 3 $12x + 9y = -9$ ④
③ – ④ $-5y = 25 \Rightarrow y = -5$
$y = -5$ in ② $4x - 15 = -3 \Rightarrow 4x = 12 \Rightarrow x = 3$
So the solution is $x = 3$, $y = -5$

f) Rearrange ① $2x + 18y = 21$
Rearrange ② $-3x - 14y = 14$
① × 3 $6x + 54y = 63$ ③
② × 2 $-6x - 28y = 28$ ④
③ + ④ $26y = 91 \Rightarrow y = \frac{7}{2}$
$y = \frac{7}{2}$ in ① $2x + 63 = 21 \Rightarrow 2x = -42$
$\Rightarrow x = -21$
So the solution is $x = -21$, $y = \frac{7}{2}$

g) ① $2x + 16y = 10$
Rearrange ② $3x + 64y = 5$
① × 4 $8x + 64y = 40$ ③
② – ③ $-5x = -35 \Rightarrow x = 7$
$x = 7$ in ① $14 + 16y = 10 \Rightarrow 16y = -4$
$\Rightarrow y = -\frac{1}{4}$
So the solution is $x = 7$, $y = -\frac{1}{4}$

h) ① $4x - 3y = 15$
Rearrange ② $-9x + 5y = 12$
① × 5 $20x - 15y = 75$ ③
② × 3 $-27x + 15y = 36$ ④
③ + ④ $-7x = 111 \Rightarrow x = -\frac{111}{7}$
$x = -\frac{111}{7}$ in ① $-\frac{444}{7} - 3y = 15$
$\Rightarrow 3y = -\frac{549}{7} \Rightarrow y = -\frac{183}{7}$
So the solution is $x = -\frac{111}{7}$, $y = -\frac{183}{7}$

Q2 a) ① $y = 2x - 3$
② $y = \frac{1}{2}x + 3$
② × 4 $4y = 2x + 12$ ③
① – ③ $-3y = -15 \Rightarrow y = 5$
$y = 5$ in ① $5 = 2x - 3 \Rightarrow 8 = 2x \Rightarrow x = 4$
So they intersect at (4, 5)

b) ① $y = -\frac{2}{3}x + 7$
② $y = \frac{1}{2}x + \frac{21}{2}$
① × 3 $3y = -2x + 21$ ③
② × 4 $4y = 2x + 42$ ④
③ + ④ $7y = 63 \Rightarrow y = 9$
$y = 9$ in ① $9 = -\frac{2}{3}x + 7$
$\Rightarrow 2 = -\frac{2}{3}x \Rightarrow x = -3$
So they intersect at (–3, 9)

c) Rearrange ① $x + 2y = -5$
Rearrange ② $3x - 5y = 7$
① × 3 $3x + 6y = -15$ ③
② – ③ $-11y = 22 \Rightarrow y = -2$
$y = -2$ in ① $x - 4 = -5 \Rightarrow x = -1$
So they intersect at (–1, –2)

d) ① $2x - 3y = 7$
② $5x - \frac{15}{2}y = 9$
① × 5 $10x - 15y = 35$ ③
② × 2 $10x - 15y = 18$ ④
③ – ④ $0 = 17$
This is not possible — so these lines do not intersect.
The lines are actually parallel.

e) Rearrange ① $8x + 3y = 10$
Rearrange ② $6x + 9y = 3$
① × 3 $24x + 9y = 30$ ③
② – ③ $-18x = -27 \Rightarrow x = \frac{3}{2}$
$x = \frac{3}{2}$ in ② $9 + 9y = 3$
$\Rightarrow 9y = -6 \Rightarrow y = -\frac{2}{3}$
So they intersect at $\left(\frac{3}{2}, -\frac{2}{3}\right)$

f) ① $7x - 5y = 15$
Rearrange ② $2x - 3y = 9$
① × 2 $14x - 10y = 30$ ③
② × 7 $14x - 21y = 63$ ④
③ – ④ $11y = -33 \Rightarrow y = -3$
$y = -3$ in ② $2x + 9 = 9 \Rightarrow 2x = 0 \Rightarrow x = 0$
So they intersect at (0, –3)

g) ① $6x + 3y = 10$
Rearrange ② $-4x + 8y = -9$
① × 2 $12x + 6y = 20$ ③
② × 3 $-12x + 24y = -27$ ④
③ + ④ $30y = -7 \Rightarrow y = -\frac{7}{30}$
$y = -\frac{7}{30}$ in ① $6x + -\frac{7}{10} = 10 \Rightarrow 6x = \frac{107}{10}$
$\Rightarrow x = \frac{107}{60}$
So they intersect at $\left(\frac{107}{60}, -\frac{7}{30}\right)$

h) Rearrange ① $x + 10y = 3$
Rearrange ② $-6x + 5y = 5$
① × 6 $6x + 60y = 18$ ③
② + ③ $65y = 23 \Rightarrow y = \frac{23}{65}$
y in ① $x + \frac{46}{13} = 3 \Rightarrow x = -\frac{7}{13}$
So they intersect at $\left(-\frac{7}{13}, \frac{23}{65}\right)$

i) Rearrange ① $\frac{7}{3}x - \frac{5}{3}y = 2$

Rearrange ② $-\frac{3}{4}x + y = \frac{1}{3}$

② × $\frac{5}{3}$ $-\frac{5}{4}x + \frac{5}{3}y = \frac{5}{9}$ ③

① + ③ $\frac{13}{12}x = \frac{23}{9} \Rightarrow x = \frac{92}{39}$

x in ② $-\frac{23}{13} + y = \frac{1}{3} \Rightarrow y = \frac{82}{39}$

So they intersect at $\left(\frac{92}{39}, \frac{82}{39}\right)$

j) Rearrange ① $-\frac{9}{5}x + \frac{3}{4}y = -10$

Rearrange ② $\frac{3}{2}x - \frac{3}{5}y = -10$

① × $\frac{3}{5}$ $-\frac{27}{25}x + \frac{9}{20}y = -6$ ③

② × $\frac{3}{4}$ $\frac{9}{8}x - \frac{9}{20}y = -\frac{15}{2}$ ④

③ + ④ $\frac{9}{200}x = -\frac{27}{2} \Rightarrow x = -300$

x in ① $540 + \frac{3}{4}y = -10 \Rightarrow \frac{3}{4}y = -550$
$\Rightarrow y = -\frac{2200}{3}$

So they intersect at $\left(-300, -\frac{2200}{3}\right)$

Q3 There will be a signpost at the point of intersection of each pair of straight lines:

A and B: ① $5x + 2y = -11$

Rearrange ② $2x - y = 1$

② × 2 $4x - 2y = 2$ ③

① + ③ $9x = -9 \Rightarrow x = -1$

$x = -1$ in ② $-2 - y = 1 \Rightarrow y = -3$

So they intersect at $(-1, -3)$

B and C: Rearrange ① $2x - y = 1$

Rearrange ② $-x + 5y = 13$

② × 2 $-2x + 10y = 26$ ③

① + ③ $9y = 27 \Rightarrow y = 3$

$y = 3$ in ① $2x - 3 = 1 \Rightarrow 2x = 4$
$\Rightarrow x = 2$

So they intersect at $(2, 3)$

A and C: ① $5x + 2y = -11$

Rearrange ② $-x + 5y = 13$

② × 5 $-5x + 25y = 65$ ③

① + ③ $27y = 54 \Rightarrow y = 2$

$y = 2$ in ② $-x + 10 = 13$
$\Rightarrow -x = 3 \Rightarrow x = -3$

So they intersect at $(-3, 2)$

So the three signposts are at $(-1, -3)$, $(2, 3)$ and $(-3, 2)$.

Exercise 5.2.2 — Simultaneous equations — if one is not linear

Q1 a) ① $y = 4x + 3$

② $2y - 3x = 1$

Sub ① in ② $2(4x + 3) - 3x = 1$
$8x + 6 - 3x = 1$
$5x = -5 \Rightarrow x = -1$

$x = -1$ in ① $y = 4 \times -1 + 3 = -1$

So the solution is $x = -1$, $y = -1$

b) ① $5x + 2y = 16$

Rearrange ② $x = 2y - 4$

Sub ② in ① $5(2y - 4) + 2y = 16$
$12y - 20 = 16$
$12y = 36 \Rightarrow y = 3$

$y = 3$ in ② $x = 2 \times 3 - 4 = 2$
So the solution is $x = 2$, $y = 3$

Q2 a) ① $y = 2x + 5$

② $y = x^2 - x + 1$

Sub ① in ② $2x + 5 = x^2 - x + 1$
$x^2 - 3x - 4 = 0$
$(x - 4)(x + 1) = 0 \Rightarrow x = 4$ or $x = -1$

From ①, when $x = 4$, $y = 8 + 5 = 13$, and
when $x = -1$, $y = -2 + 5 = 3$

So $x = 4$, $y = 13$ or $x = -1$, $y = 3$

b) ① $y = 2x^2 - 3$

② $y = 3x + 2$

Sub ② in ① $3x + 2 = 2x^2 - 3$
$2x^2 - 3x - 5 = 0$
$(2x - 5)(x + 1) = 0 \Rightarrow x = \frac{5}{2}$ or $x = -1$

From ②, when $x = \frac{5}{2}$, $y = \frac{15}{2} + 2 = \frac{19}{2}$, and
when $x = -1$, $y = -3 + 2 = -1$

So $x = \frac{5}{2}$, $y = \frac{19}{2}$ or $x = -1$, $y = -1$

c) ① $2x^2 - xy = 6$

Rearrange ② $y = 3x - 7$

Sub ② in ① $2x^2 - x(3x - 7) = 6$
$2x^2 - 3x^2 + 7x - 6 = 0$
$-x^2 + 7x - 6 = 0$
$x^2 - 7x + 6 = 0$
$(x - 6)(x - 1) = 0$
$\Rightarrow x = 6$ or $x = 1$

From ②, when $x = 6$, $y = 18 - 7 = 11$, and
when $x = 1$, $y = 3 - 7 = -4$

So $x = 6$, $y = 11$ or $x = 1$, $y = -4$

d) ① $xy = 6$

Rearrange ② $2y + 4 = x$

Sub ② in ① $y(2y + 4) = 6$
$2y^2 + 4y - 6 = 0$
$y^2 + 2y - 3 = 0$
$(y + 3)(y - 1) = 0$
$\Rightarrow y = -3$ or $y = 1$

From ②, when $y = -3$, $x = -6 + 4 = -2$, and
when $y = 1$, $x = 2 + 4 = 6$

So $x = -2$, $y = -3$ or $x = 6$, $y = 1$

e) ① $y = x^2 - 2x - 3$

Rearrange ② $y = -x - 8$

Sub ② in ① $-x - 8 = x^2 - 2x - 3$
$x^2 - x + 5 = 0$

Check discriminant: $b^2 - 4ac = 1 - 20 = -19$,
which is negative so there are no real roots.
So there are no solutions for the simultaneous equations.

f) ① $y = 2x^2 - 3x + 5$

Rearrange ② $5x - 3 = y$

Sub ② in ① $5x - 3 = 2x^2 - 3x + 5$
$2x^2 - 8x + 8 = 0$
$x^2 - 4x + 4 = 0$
$(x - 2)^2 = 0 \Rightarrow x = 2$

From ②, when $x = 2$, $y = 10 - 3 = 7$
So $x = 2$, $y = 7$
There is only one solution here, so the straight line is a tangent to the curve.

g) ① $2x^2 + 3y^2 + 18x = 347$

Rearrange ② $y = -4x + 7$

Sub ② in ① $2x^2 + 3(-4x + 7)^2 + 18x = 347$

$2x^2 + 3(16x^2 - 56x + 49) + 18x - 347 = 0$

$2x^2 + 48x^2 - 168x + 147 + 18x - 347 = 0$

$50x^2 - 150x - 200 = 0$

$x^2 - 3x - 4 = 0$

$(x + 1)(x - 4) = 0 \Rightarrow x = -1$ or $x = 4$

From ②, when $x = -1$, $y = 4 + 7 = 11$, and
when $x = 4$, $y = -16 + 7 = -9$

So $x = -1, y = 11$ or $x = 4, y = -9$

h) ① $2y = 2x^2 + x + 1$

Rearrange ② $y = 2 - 2x$

Sub ② in ① $2(2 - 2x) = 2x^2 + x + 1$

$4 - 4x = 2x^2 + x + 1$

$2x^2 + 5x - 3 = 0$

$(2x - 1)(x + 3) = 0$

$\Rightarrow x = -3$ or $x = \frac{1}{2}$

From ②, when $x = -3$, $y = 2 + 6 = 8$, and
when $x = \frac{1}{2}$, $y = 2 - 1 = 1$

So $x = -3, y = 8$ or $x = \frac{1}{2}, y = 1$

i) ① $x^2 + 4x = 4y + 40$

Rearrange ② $y = -\frac{5}{12}x - \frac{5}{2}$

Sub ② in ① $x^2 + 4x = 4\left(-\frac{5}{12}x - \frac{5}{2}\right) + 40$

$x^2 + 4x = -\frac{5}{3}x - 10 + 40$

$x^2 + \frac{17}{3} - 30 = 0$

$3x^2 + 17x - 90 = 0$

$\Rightarrow (3x - 10)(x + 9) = 0$

$\Rightarrow x = -9$ or $x = \frac{10}{3}$

From ②, when $x = -9$, $y = \frac{45}{12} - \frac{5}{2} = \frac{5}{4}$, and

when $x = \frac{10}{3}$, $y = -\frac{50}{36} - \frac{5}{2} = -\frac{35}{9}$

So $x = -9, y = \frac{5}{4}$ or $x = \frac{10}{3}, y = -\frac{35}{9}$

j) Rearrange ① $y = 2x + 2$

② $\frac{1}{4}y^2 + 25 = 3x^2 + 11x$

Sub ① in ② $\frac{1}{4}(2x + 2)^2 + 25 = 3x^2 + 11x$

$x^2 + 2x + 1 + 25 = 3x^2 + 11x$

$2x^2 + 9x - 26 = 0$

$\Rightarrow (2x + 13)(x - 2) = 0$

$\Rightarrow x = -\frac{13}{2}$ or $x = 2$

From ①, when $x = -\frac{13}{2}$, $y = -13 + 2 = -11$, and

when $x = 2$, $y = 4 + 2 = 6$

So $x = -\frac{13}{2}, y = -11$ or $x = 2, y = 6$

Q3 a) ① $y = x^2 - 5x + 7$

Rearrange ② $y = -2x + 11$

Sub ② in ① $-2x + 11 = x^2 - 5x + 7$

$x^2 - 3x - 4 = 0$

$(x + 1)(x - 4) = 0$

$\Rightarrow x = -1$ or $x = 4$

From ②, when $x = -1$, $y = 2 + 11 = 13$, and
when $x = 4$, $y = -8 + 11 = 3$

So the points of intersection are (–1, 13) and (4, 3)

b) ① $y = -2x^2 + 2x + 12$

Rearrange ② $y - 2x + 4$

Sub ② in ① $2x + 4 = -2x^2 + 2x + 12$

$2x^2 = 8 \Rightarrow x = -2$ or $x = 2$

From ②, when $x = -2$, $y = -4 + 4 = 0$ and
when $x = 2$, $y = 4 + 4 = 8$

So the points of intersection are (–2, 0) and (2, 8)

Q4 a) ① $y = \frac{1}{2}x^2 + 4x - 8$

② $y = 4 + \frac{3}{2}x$

Sub ② in ① $4 + \frac{3}{2}x = \frac{1}{2}x^2 + 4x - 8$

$8 + 3x = x^2 + 8x - 16$

$0 = x^2 + 5x - 24$

$(x + 8)(x - 3) = 0 \Rightarrow x = -8$ or $x = 3$

From ②, when $x = -8$, $y = 4 - 12 = -8$, and
when $x = 3$, $y = 4 + \frac{9}{2} = \frac{17}{2}$

So they intersect at (–8, –8) and $\left(3, \frac{17}{2}\right)$

b) ① $y = 2x^2 + x - 6$

Rearrange ② $y = 5x + 10$

Sub ② in ① $2x^2 + x - 6 = 5x + 10$

$2x^2 - 4x - 16 = 0$

$x^2 - 2x - 8 = 0$

$(x - 4)(x + 2) = 0 \Rightarrow x = 4$ or $x = -2$

From ②, when $x = 4$, $y = 20 + 10 = 30$, and
when $x = -2$, $y = -10 + 10 = 0$

So they intersect at (4, 30) and (–2, 0)

c) ① $x^2 + y^2 = 50$

Rearrange ② $x = -2y + 5$

Sub ② in ① $(-2y + 5)^2 + y^2 = 50$

$4y^2 - 20y + 25 + y^2 - 50 = 0$

$5y^2 - 20y - 25 = 0$

$y^2 - 4y - 5 = 0$

$(y - 5)(y + 1) = 0 \Rightarrow y = 5$ or $y = -1$

From ②, when $y = 5$, $x = -10 + 5 = -5$, and
when $y = -1$, $x = 2 + 5 = 7$

So they intersect at (–5, 5) and (7, –1)

d) ① $2x^2 - y + 3x + 1 = 0$

Rearrange ② $y = x + 5$

Sub ② in ① $2x^2 - (x + 5) + 3x + 1 = 0$

$2x^2 + 2x - 4 = 0$

$x^2 + x - 2 = 0$

$(x + 2)(x - 1) = 0 \Rightarrow x = -2$ or $x = 1$

From ②, when $x = -2$, $y = -2 + 5 = 3$, and
when $x = 1$, $y = 1 + 5 = 6$

So they intersect at (–2, 3) and (1, 6)

e) ① $3x^2 + 9x + 1 = 6y$

Rearrange ② $y = \frac{11}{6} - \frac{2}{3}x$

Sub ② in ① $3x^2 + 9x + 1 = 6\left(\frac{11}{6} - \frac{2}{3}x\right)$

$3x^2 + 9x + 1 = 11 - 4x$

$3x^2 + 13x - 10 = 0$

$(3x - 2)(x + 5) = 0$

$\Rightarrow x = -5$ or $x = \frac{2}{3}$

From ②, when $x = -5$, $y = \frac{11}{6} + \frac{10}{3} = \frac{31}{6}$

when $x = \frac{2}{3}$, $y = \frac{11}{6} - \frac{4}{9} = \frac{25}{18}$

So they intersect at $\left(-5, \frac{31}{6}\right)$ and $\left(\frac{2}{3}, \frac{25}{18}\right)$

f) Rearrange ① $y = 4x + 10$

② $2y - 19 = 4x^2 + 8x$

Sub ① in ② $2(4x + 10) - 19 = 4x^2 + 8x$

$8x + 20 - 19 = 4x^2 + 8x$

$4x^2 = 1 \Rightarrow x = -\frac{1}{2}$ or $x = \frac{1}{2}$

From ①, when $x = -\frac{1}{2}$, $y = -2 + 10 = 8$, and

when $x = \frac{1}{2}$, $y = 2 + 10 = 12$

So they intersect at $\left(-\frac{1}{2}, 8\right)$ and $\left(\frac{1}{2}, 12\right)$

Q5 a) ① $x^2 + y^2 = 10$

Rearrange ② $x = 3y - 10$

Sub ② in ① $(3y - 10)^2 + y^2 = 10$

$9y^2 - 60y + 100 + y^2 - 10 = 0$

$10y^2 - 60y + 90 = 0$

$y^2 - 6y + 9 = 0$

$(y - 3)^2 = 0 \Rightarrow y = 3$

From ②, when $y = 3$, $x = 9 - 10 = -1$

So $x = -1$, $y = 3$.

b) $x^2 + y^2 = 10$ is a circle and $x - 3y + 10 = 0$ is a straight line. Part a) tells us that they intersect at a single point, so the line must actually be a tangent to the circle.

Q6 a) ① $y = x^2 + 6x - 7$

② $y = 2x - 3$

Sub ② in ① $2x - 3 = x^2 + 6x - 7$

$x^2 + 4x - 4 = 0$

So $b^2 - 4ac = 16 + 16 = 32 > 0$

So they will intersect at two points.

b) ① $3x^2 + 4y^2 + 6x = 9$

Rearrange ② $x = 3 - 2y$

Sub ② in ① $3(3 - 2y)^2 + 4y^2 + 6(3 - 2y) = 9$

$27 - 36y + 12y^2 + 4y^2 + 18 - 12y - 9 = 0$

$16y^2 - 48y + 36 = 0$

$4y^2 - 12y + 9 = 0$

Now $b^2 - 4ac = 144 - 144 = 0$

So they will intersect only once — ② is a tangent to the curve ①.

You could have rearranged differently to get an equation in terms of x — you would still get a discriminant of 0.

c) ① $xy + 2x - y = 8$

Rearrange ② $x = 1 - y$

Sub ② in ① $(1 - y)y + 2(1 - y) - y = 8$

$y - y^2 + 2 - 2y - y = 8$

$-y^2 - 2y - 6 = 0$

$y^2 + 2y + 6 = 0$

So $b^2 - 4ac = 4 - 24 = -20 < 0$

So the graphs will not intersect.

Review Exercise — Chapter 5

Q1 a) $7x - 4 > 2x - 42 \Rightarrow 5x > -38 \Rightarrow x > -\frac{38}{5}$

b) $12y - 3 \le 4y + 4 \Rightarrow 8y \le 7 \Rightarrow y \le \frac{7}{8}$

c) $9y - 4 \ge 17y + 2 \Rightarrow -8y \ge 6 \Rightarrow y \le -\frac{3}{4}$

d) $x + 6 < 5x - 4 \Rightarrow -4x < -10 \Rightarrow x > \frac{5}{2}$

e) $4x - 2 > x - 14 \Rightarrow 3x > -12 \Rightarrow x > -4$

f) $7 - x \le 4 - 2x \Rightarrow x \le -3$

g) $11x - 4 < 4 - 11x \Rightarrow 22x < 8 \Rightarrow x < \frac{4}{11}$

h) $1 + 10y \ge 7y - 12 \Rightarrow 3y \ge -13 \Rightarrow y \ge -\frac{13}{3}$

i) $8y - 6 \le 6 - 8y \Rightarrow 16y \le 12 \Rightarrow y \le \frac{3}{4}$

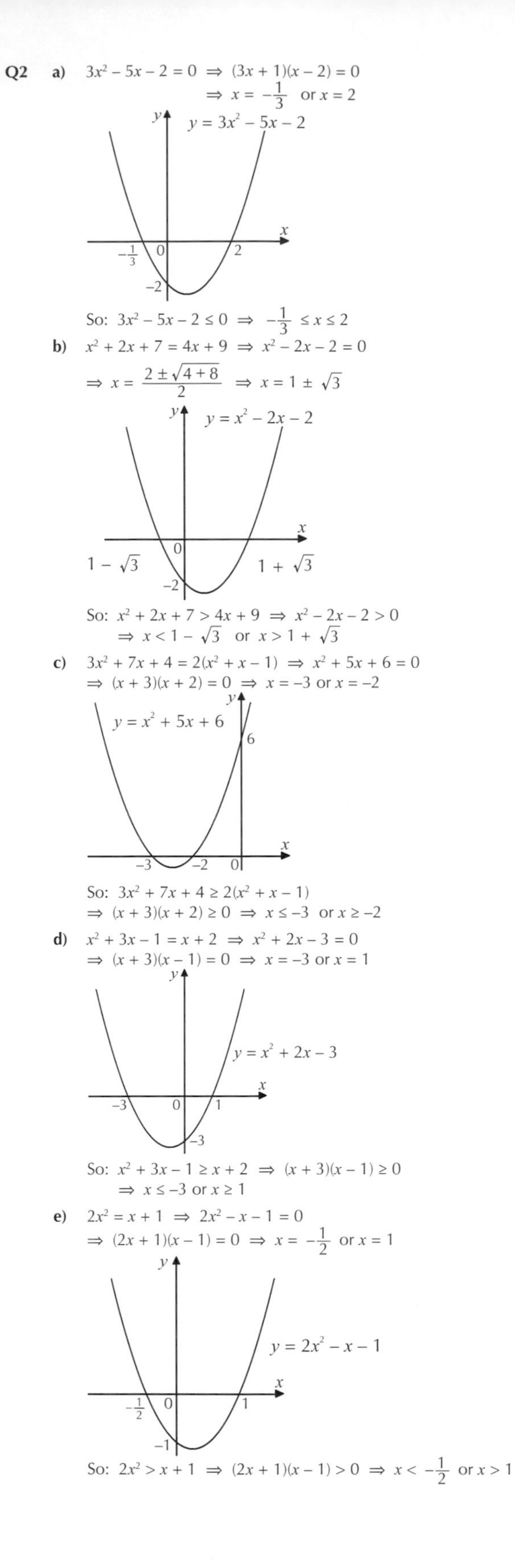

Q2 a) $3x^2 - 5x - 2 = 0 \Rightarrow (3x + 1)(x - 2) = 0$

$\Rightarrow x = -\frac{1}{3}$ or $x = 2$

So: $3x^2 - 5x - 2 \le 0 \Rightarrow -\frac{1}{3} \le x \le 2$

b) $x^2 + 2x + 7 = 4x + 9 \Rightarrow x^2 - 2x - 2 = 0$

$\Rightarrow x = \frac{2 \pm \sqrt{4 + 8}}{2} \Rightarrow x = 1 \pm \sqrt{3}$

So: $x^2 + 2x + 7 > 4x + 9 \Rightarrow x^2 - 2x - 2 > 0$

$\Rightarrow x < 1 - \sqrt{3}$ or $x > 1 + \sqrt{3}$

c) $3x^2 + 7x + 4 = 2(x^2 + x - 1) \Rightarrow x^2 + 5x + 6 = 0$

$\Rightarrow (x + 3)(x + 2) = 0 \Rightarrow x = -3$ or $x = -2$

So: $3x^2 + 7x + 4 \ge 2(x^2 + x - 1)$

$\Rightarrow (x + 3)(x + 2) \ge 0 \Rightarrow x \le -3$ or $x \ge -2$

d) $x^2 + 3x - 1 = x + 2 \Rightarrow x^2 + 2x - 3 = 0$

$\Rightarrow (x + 3)(x - 1) = 0 \Rightarrow x = -3$ or $x = 1$

So: $x^2 + 3x - 1 \ge x + 2 \Rightarrow (x + 3)(x - 1) \ge 0$

$\Rightarrow x \le -3$ or $x \ge 1$

e) $2x^2 = x + 1 \Rightarrow 2x^2 - x - 1 = 0$

$\Rightarrow (2x + 1)(x - 1) = 0 \Rightarrow x = -\frac{1}{2}$ or $x = 1$

So: $2x^2 > x + 1 \Rightarrow (2x + 1)(x - 1) > 0 \Rightarrow x < -\frac{1}{2}$ or $x > 1$

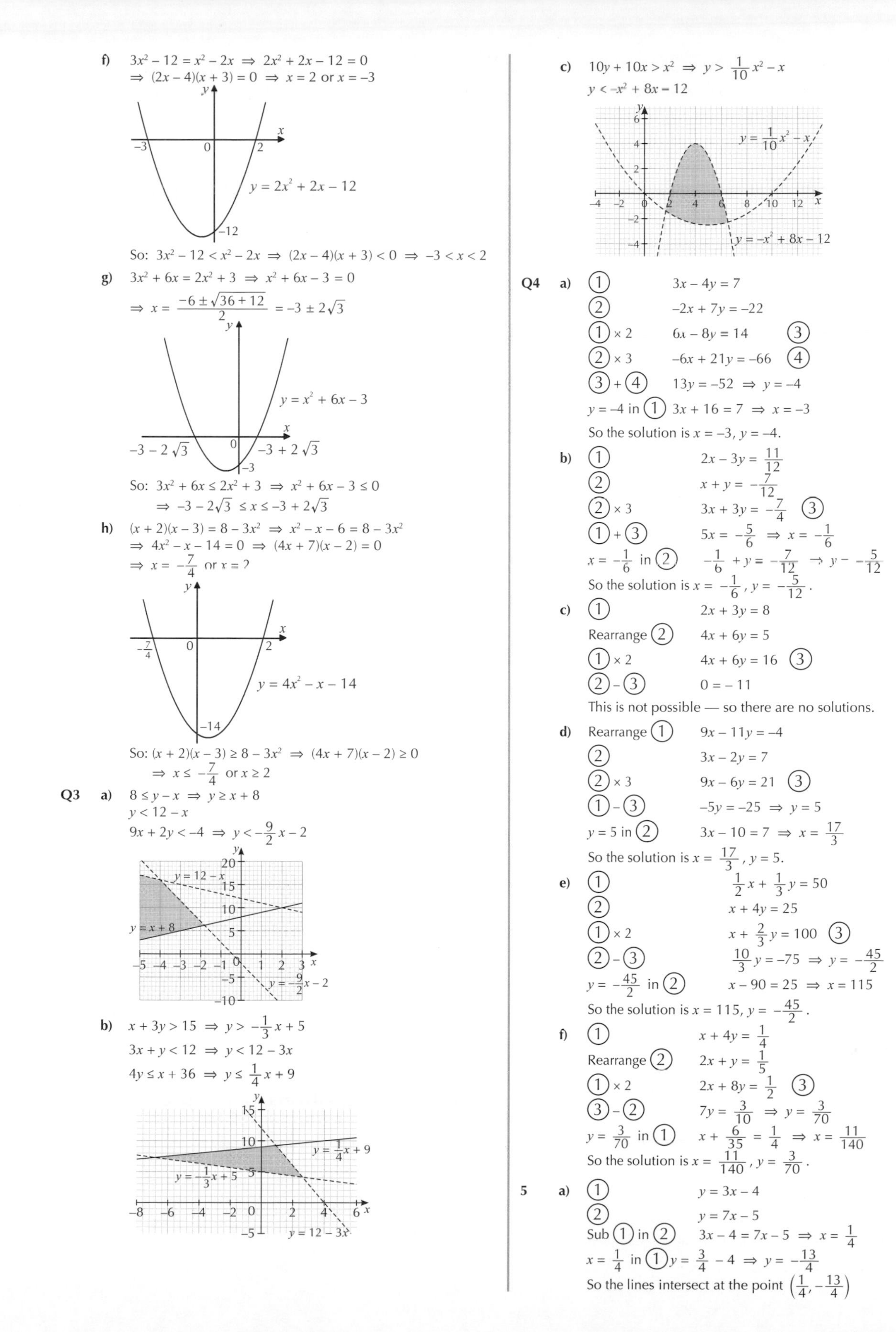

f) $3x^2 - 12 = x^2 - 2x \Rightarrow 2x^2 + 2x - 12 = 0$
$\Rightarrow (2x - 4)(x + 3) = 0 \Rightarrow x = 2$ or $x = -3$

So: $3x^2 - 12 < x^2 - 2x \Rightarrow (2x - 4)(x + 3) < 0 \Rightarrow -3 < x < 2$

g) $3x^2 + 6x = 2x^2 + 3 \Rightarrow x^2 + 6x - 3 = 0$
$\Rightarrow x = \frac{-6 \pm \sqrt{36 + 12}}{2} = -3 \pm 2\sqrt{3}$

So: $3x^2 + 6x \leq 2x^2 + 3 \Rightarrow x^2 + 6x - 3 \leq 0$
$\Rightarrow -3 - 2\sqrt{3} \leq x \leq -3 + 2\sqrt{3}$

h) $(x + 2)(x - 3) = 8 - 3x^2 \Rightarrow x^2 - x - 6 = 8 - 3x^2$
$\Rightarrow 4x^2 - x - 14 = 0 \Rightarrow (4x + 7)(x - 2) = 0$
$\Rightarrow x = -\frac{7}{4}$ or $x = 2$

So: $(x + 2)(x - 3) \geq 8 - 3x^2 \Rightarrow (4x + 7)(x - 2) \geq 0$
$\Rightarrow x \leq -\frac{7}{4}$ or $x \geq 2$

Q3 a) $8 \leq y - x \Rightarrow y \geq x + 8$
$y < 12 - x$
$9x + 2y < -4 \Rightarrow y < -\frac{9}{2}x - 2$

b) $x + 3y > 15 \Rightarrow y > -\frac{1}{3}x + 5$
$3x + y < 12 \Rightarrow y < 12 - 3x$
$4y \leq x + 36 \Rightarrow y \leq \frac{1}{4}x + 9$

c) $10y + 10x > x^2 \Rightarrow y > \frac{1}{10}x^2 - x$
$y < -x^2 + 8x - 12$

Q4 a) ① $3x - 4y = 7$
② $-2x + 7y = -22$
① × 2 $6x - 8y = 14$ ③
② × 3 $-6x + 21y = -66$ ④
③ + ④ $13y = -52 \Rightarrow y = -4$
$y = -4$ in ① $3x + 16 = 7 \Rightarrow x = -3$
So the solution is $x = -3$, $y = -4$.

b) ① $2x - 3y = \frac{11}{12}$
② $x + y = -\frac{7}{12}$
② × 3 $3x + 3y = -\frac{7}{4}$ ③
① + ③ $5x = -\frac{5}{6} \Rightarrow x = -\frac{1}{6}$
$x = -\frac{1}{6}$ in ② $-\frac{1}{6} + y = -\frac{7}{12} \Rightarrow y = -\frac{5}{12}$
So the solution is $x = -\frac{1}{6}$, $y = -\frac{5}{12}$.

c) ① $2x + 3y = 8$
Rearrange ② $4x + 6y = 5$
① × 2 $4x + 6y = 16$ ③
② − ③ $0 = -11$
This is not possible — so there are no solutions.

d) Rearrange ① $9x - 11y = -4$
② $3x - 2y = 7$
② × 3 $9x - 6y = 21$ ③
① − ③ $-5y = -25 \Rightarrow y = 5$
$y = 5$ in ② $3x - 10 = 7 \Rightarrow x = \frac{17}{3}$
So the solution is $x = \frac{17}{3}$, $y = 5$.

e) ① $\frac{1}{2}x + \frac{1}{3}y = 50$
② $x + 4y = 25$
① × 2 $x + \frac{2}{3}y = 100$ ③
② − ③ $\frac{10}{3}y = -75 \Rightarrow y = -\frac{45}{2}$
$y = -\frac{45}{2}$ in ② $x - 90 = 25 \Rightarrow x = 115$
So the solution is $x = 115$, $y = -\frac{45}{2}$.

f) ① $x + 4y = \frac{1}{4}$
Rearrange ② $2x + y = \frac{1}{5}$
① × 2 $2x + 8y = \frac{1}{2}$ ③
③ − ② $7y = \frac{3}{10} \Rightarrow y = \frac{3}{70}$
$y = \frac{3}{70}$ in ① $x + \frac{6}{35} = \frac{1}{4} \Rightarrow x = \frac{11}{140}$
So the solution is $x = \frac{11}{140}$, $y = \frac{3}{70}$.

5 a) ① $y = 3x - 4$
② $y = 7x - 5$
Sub ① in ② $3x - 4 = 7x - 5 \Rightarrow x = \frac{1}{4}$
$x = \frac{1}{4}$ in ① $y = \frac{3}{4} - 4 \Rightarrow y = -\frac{13}{4}$
So the lines intersect at the point $\left(\frac{1}{4}, -\frac{13}{4}\right)$

b) ① $y = 13 - 2x$

② $7x - y - 23 = 0$

Sub ① in ② $7x - (13 - 2x) - 23 = 0$
$\Rightarrow 9x = 36 \Rightarrow x = 4$

$x = 4$ in ① $y = 13 - 8 = 5$

So the lines intersect at the point (4, 5)

c) Rearrange ① $2x - 3y = -4$

Rearrange ② $x - 2y = -1$

② × 2 $2x - 4y = -2$ ③

① – ③ $y = -2$

$y = -2$ in ② $x + 4 = -1 \Rightarrow x = -5$

So the lines intersect at the point (–5, –2)

d) ① $5x - 7y = 22$

Rearrange ② $-4x + 3y = 13$

① × 4 $20x - 28y = 88$ ③

② × 5 $-20x + 15y = 65$ ④

③ + ④ $-13y = 153 \Rightarrow y = -\frac{153}{13}$

$y = -\frac{153}{13}$ in ① $5x + \frac{1071}{13} = 22$
$\Rightarrow x = -\frac{157}{13}$

So the lines intersect at the point $\left(-\frac{157}{13}, -\frac{153}{13}\right)$

e) Rearrange ① $-\frac{2}{3}x - 8y = -9$

② $\frac{1}{3}x + \frac{2}{3}y = 10$

② × 2 $\frac{2}{3}x + \frac{4}{3}y = 20$ ③

① + ③ $-\frac{20}{3}y = 11 \Rightarrow y = -\frac{33}{20}$

$y = -\frac{33}{20}$ in ② $\frac{1}{3}x - \frac{66}{60} = 10 \Rightarrow x = \frac{333}{10}$

So the lines intersect at the point $\left(\frac{333}{10}, -\frac{33}{20}\right)$

f) ① $24x + 15y = 2$

② $18x + 36y = 5$

① × 3 $72x + 45y = 6$ ③

② × 4 $72x + 144y = 20$ ④

④ – ③ $99y = 14 \Rightarrow y = \frac{14}{99}$

$y = \frac{14}{99}$ in ① $24x + \frac{210}{99} = 2 \Rightarrow x = -\frac{1}{198}$

So the lines intersect at the point $\left(-\frac{1}{198}, \frac{14}{99}\right)$

Q6 a) ① $y = x^2 - 7x + 4$

Rearrange ② $y = 2x - 10$

Sub ① in ② $x^2 - 7x + 4 = 2x - 10$
$\Rightarrow x^2 - 9x + 14 = 0$
$\Rightarrow (x - 2)(x - 7) = 0$
$\Rightarrow x = 2$ or $x = 7$

From ②, when $x = 2$, $y = 4 - 10 = -6$, and
when $x = 7$, $y = 14 - 10 = 4$

So the line and the curve meet at the points (2, –6) and (7, 4).

b) Rearrange ① $y = 2x^2 - 6x + 30$

Expand ② $y = 2x + 22$

Sub ① in ② $2x^2 - 6x + 30 = 2x + 22$
$\Rightarrow 2x^2 - 8x + 8 = 0$
$\Rightarrow x^2 - 4x + 4 = 0$
$\Rightarrow (x - 2)^2 = 0 \Rightarrow x = 2$

$x = 2$ in ② $y = 4 + 22 = 26$

So the line is a tangent to the parabola at the point (2, 26).

c) ① $2x^2 + 2y^2 - 3 = 0$

② $y = x + 4$

Sub ② in ① $2x^2 + 2(x + 4)^2 - 3 = 0$
$\Rightarrow 2x^2 + 2x^2 + 16x + 32 - 3 = 0$
$\Rightarrow 4x^2 + 16x + 29 = 0$

Check discriminant: $b^2 - 4ac = 256 - 464 = -208$, which is negative so there are no real roots.
So the line and the curve never meet.

d) Rearrange ① $y = 2 - \frac{3}{4}x$

② $2y - 2x^2 - 4x = 7$

Sub ① in ② $2(2 - \frac{3}{4}x) - 2x^2 - 4x = 7$
$\Rightarrow 4 - \frac{3}{2}x - 2x^2 - 4x = 7$
$\Rightarrow 2x^2 + \frac{11}{2}x + 3 = 0$
$\Rightarrow 4x^2 + 11x + 6 = 0$
$\Rightarrow (4x + 3)(x + 2) = 0$
$\Rightarrow x = -\frac{3}{4}$ or $x = -2$

From ①, when $x = -\frac{3}{4}$, $y = 2 + \frac{9}{16} = \frac{41}{16}$, and
when $x = -2$, $y = 2 + \frac{3}{2} = \frac{7}{2}$

So they meet at the points $\left(-\frac{3}{4}, \frac{41}{16}\right)$ and $\left(-2, \frac{7}{2}\right)$

e) ① $\frac{1}{4}x^2 + 3x + 15 = 4y$

② $2y = 3x + 3$

② × 2 $4y = 6x + 6$ ③

Sub ③ in ① $\frac{1}{4}x^2 + 3x + 15 = 6x + 6$
$\Rightarrow \frac{1}{4}x^2 - 3x + 9 = 0$
$\Rightarrow x^2 - 12x + 36 = 0$
$\Rightarrow (x - 6)^2 = 0 \Rightarrow x = 6$

$x = 6$ in ② $2y = 18 + 3 \Rightarrow y = \frac{21}{2}$

So the line is a tangent to the parabola at the point $\left(6, \frac{21}{2}\right)$.

f) ① $(x - 3)^2 + (y + 4)^2 = 25$

Rearrange ② $x = 7y + 6$

Sub ② in ① $(7y + 3)^2 + (y + 4)^2 = 25$
$\Rightarrow 49y^2 + 42y + 9 + y^2 + 8y + 16 = 25$
$\Rightarrow 50y^2 + 50y = 0$
$\Rightarrow 50y(y + 1) = 0 \Rightarrow y = -1$ or $y = 0$

From ②, when $y = -1$, $x = -7 + 6 = -1$, and
when $y = 0$, $x = 6$

So the line crosses the circle at the points (–1, –1) and (6, 0)

Q7 ① $y = x^2 - 2x - 3$

② $y = 3x + 11$

Sub ① in ② $x^2 - 2x - 3 = 3x + 11$
$\Rightarrow x^2 - 5x - 14 = 0$
$\Rightarrow (x + 2)(x - 7) = 0 \Rightarrow x = -2$ or 7

So the line and curve will intersect at two points.

Exam-Style Questions — Chapter 5

Q1 a) ① $y = 3x^2 + 7x + 15$

② $y = 6x + 25$

Sub ① in ② $3x^2 + 7x + 15 = 6x + 25$
$\Rightarrow 3x^2 + x - 10 = 0$
$\Rightarrow (3x - 5)(x + 2) = 0$
$\Rightarrow x = \frac{5}{3}$ or $x = -2$

From ②, when $x = \frac{5}{3}$, $y = 6\left(\frac{5}{3}\right) + 25 = 35$, and

when $x = -2$, $y = 6(-2) + 25 = 13$.

So they meet at the points $(\frac{5}{3}, 35)$ and $(-2, 13)$.

[5 marks available — 1 mark for setting equations equal to one another, 1 mark for simplifying quadratic equation, 1 mark for solving quadratic equation, 1 mark for one correct coordinate pair, 1 mark for other correct coordinate pair]

b) Using part a):

$y = 3x^2 + x - 10$

So $3x^2 + 7x + 15 > 6x + 25 \Rightarrow 3x^2 + x - 10 > 0$

$\Rightarrow x < -2$ or $x > \frac{5}{3}$

[2 marks available — 1 mark for using root values from part a), 1 mark for correct solution]

Q2

① $3x^2 + 5y = 14$

② $5x - 10y = 38$

① × 2: $3x^2 + 5y = 14 \Rightarrow 10y = 28 - 6x^2$ ③

Rearrange ②: $10y = 5x + 38$ ④

Equating ③ and ④: $28 - 6x^2 = 5x + 38 \Rightarrow 6x^2 + 5x + 10 = 0$

Check discriminant: $b^2 - 4ac = 25 - (4 \times 6 \times 10) = 25 - 240$ which is negative so there are no real roots.

So the line l and curve C never meet.

[4 marks available — 1 mark for rearranging and substituting to eliminate y (or x), 1 mark for simplifying quadratic equation, 1 mark for negative discriminant value, 1 mark for correct explanation]

Q3

① $y = 2x + 7$

② $(x - 9)^2 + (y - 5)^2 = 160$

Sub ① in ② $(x - 9)^2 + (2x + 7 - 5)^2 = 160$

$\Rightarrow x^2 - 18x + 81 + 4x^2 + 8x + 4 = 160$

$\Rightarrow 5x^2 - 10x - 75 = 0$

$\Rightarrow x^2 - 2x - 15 = 0$

$\Rightarrow (x + 3)(x - 5) = 0$

$\Rightarrow x = -3$ or $x = 5$

From ①, when $x = -3$, $y = 2(-3) + 7 = 1$, and when $x = 5$, $y = 2(5) + 7 = 17$.

So the line and the circle meet at the points $(-3, 1)$ and $(5, 17)$.

[5 marks available — 1 mark for forming an equation by substituting x or y, 1 mark for simplifying quadratic equation, 1 mark for solving quadratic equation, 1 mark for one correct coordinate pair, 1 mark for other correct coordinate pair]

Q4 $2x^2 - 5x - 3 = 0 \Rightarrow (2x + 1)(x - 3) = 0$

$\Rightarrow x = -\frac{1}{2}$ or $x = 3$

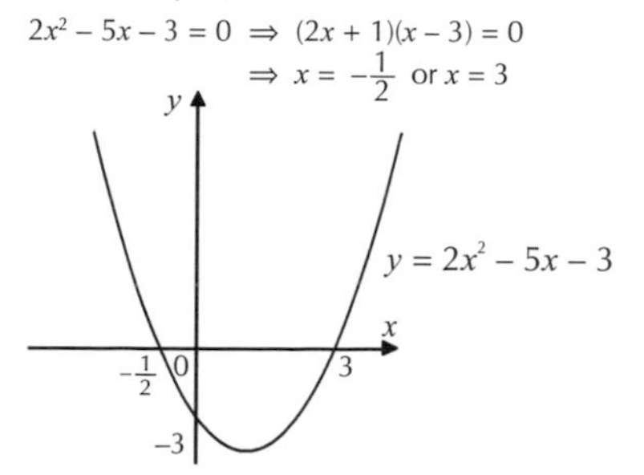

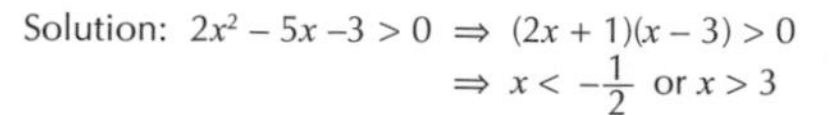

Solution: $2x^2 - 5x - 3 > 0 \Rightarrow (2x + 1)(x - 3) > 0$

$\Rightarrow x < -\frac{1}{2}$ or $x > 3$

In set notation, this is $\left\{x : x < -\frac{1}{2}\right\} \cup \{x : x > 3\}$

[4 marks available — 1 mark for factorising quadratic equation, 1 mark for correct root values, 2 marks for correct solution in set notation, otherwise 1 mark for correct solution not in set notation]

Q5 a)

① $2x + ay = 4$

② $3x - 2y = 1$

① × 3 $6x + 3ay = 12$ ③

② × 2 $6x - 4y = 2$ ④

③ – ④ $3ay + 4y = 10 \Rightarrow y(3a + 4) = 10$

$\Rightarrow y = \frac{10}{3a + 4}$

Sub $y = \frac{10}{3a+4}$ in ②, $3x - 2\left(\frac{10}{3a+4}\right) = 1$

$\Rightarrow 3x - 1 + \frac{20}{3a+4} \Rightarrow 3x = \frac{3a + 4 + 20}{3a + 4} \Rightarrow x = \frac{a + 8}{3a + 4}$

[4 marks available — 1 mark for multiplying both equations to get same coefficient of x or y, 1 mark for adding or subtracting equations to eliminate x or y, 1 mark for correct x or y value, 1 mark for other correct value]

b) The denominator of both x and y is $3a + 4$. This can never be zero, so no solution exists when $3a + 4 = 0 \Rightarrow a = -\frac{4}{3}$

[2 marks available — 1 mark for explanation, 1 mark for correct value]

c) If $a = -\frac{4}{3}$, then ① is $2x - \frac{4}{3}y = 4 \Rightarrow 3x - 2y = 6$.

$3x - 2y = 6$ and ② $3x - 2y = 1$ cannot be simultaneously true as both represent parallel lines.

[2 marks available — 1 mark for substituting a into equation 1 and rearranging, 1 mark for identifying parallel lines]

Q6

① $y = 2x + k$

② $y = x^2 - 4x + 6$

Sub ② in ① $x^2 - 4x + 6 = 2x + k$

$\Rightarrow x^2 - 6x + (6 - k) = 0$

The line is a tangent to the curve so there is only one solution and the discriminant $b^2 - 4ac = 0$. So:

$(-6)^2 - 4 \times 1 \times (6 - k) = 0 \Rightarrow 36 - 24 + 4k = 0 \Rightarrow k = -3$

So $x^2 - 6x + 9 = 0 \Rightarrow (x - 3)^2 = 0 \Rightarrow x = 3$

From ①, $y = 2(3) + (-3) \Rightarrow y = 3$

So the line is a tangent to the curve at $(3, 3)$.

[6 marks available — 1 mark for setting equations 1 and 2 equal to one another, 1 mark for simplifying quadratic equation, 1 mark for using the discriminant, 1 mark for correct value of k, 1 mark for correct value of x, 1 mark for correct value of y]

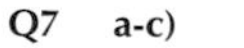

Q7 a-c)

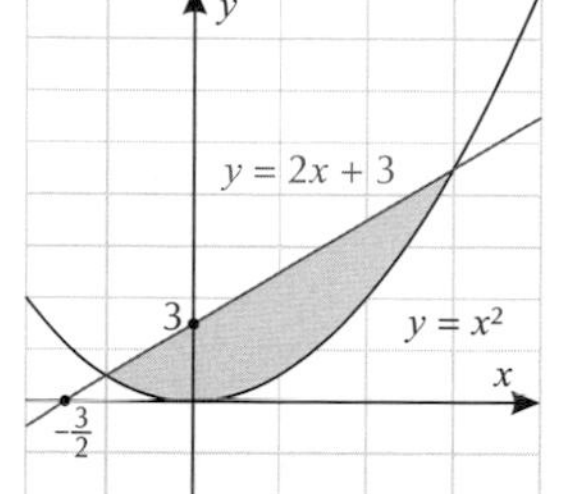

a) *[1 mark for correctly drawn graph of y = x²]*

b) *[2 marks available — 1 mark for straight line with a positive gradient, 1 mark for two correctly labelled axis intercepts]*

c) *[1 mark for correct shading]*

Chapter 6: Coordinate Geometry, Graphs and Circles

6.1 The Equation of a Straight Line

Exercise 6.1.1 — $y - y_1 = m(x - x_1)$ and $y = mx + c$

Q1 **a)** gradient = –4, y-intercept = (0, 11)

b) gradient = –1, y-intercept = (0, 4)

c) gradient = 1.7, y-intercept = (0, –2.3)

Q2 **a)** $y = -3x + 2$

b) $y = 5x - 3$

c) $y = \frac{1}{2}x + 6$

d) $y = 0.8x + 1.2$

e) $y = -0.4x - 7$

f) $y = -\frac{5}{3}x + \frac{1}{2}$

Q3 **a)** $c = 8$

$(x_1, y_1) = (-4, 0)$, $(x_2, y_2) = (0, 8)$

$m = \frac{8-0}{0-(-4)} = \frac{8}{4} = 2 \Rightarrow y = 2x + 8$

b) $c = -5$

$(x_1, y_1) = (-2, 11)$, $(x_2, y_2) = (0, -5)$

$m = \frac{-5-11}{0-(-2)} = \frac{-16}{2} = -8 \Rightarrow y = -8x - 5$

Q4 **a)** $(x_1, y_1) = (2, -2)$, $(x_2, y_2) = (6, 10)$

$m = \frac{10-(-2)}{6-2} = \frac{12}{4} = 3$

(i) $y - (-2) = 3(x - 2) \Rightarrow y + 2 = 3(x - 2)$

(ii) $y = 3x - 8$

b) $(x_1, y_1) = (-1, -6)$, $(x_2, y_2) = (3, 4)$

$m = \frac{4-(-6)}{3-(-1)} = \frac{10}{4} = 2.5$

(i) $y - (-6) = 2.5(x - (-1)) \Rightarrow y + 6 = 2.5(x + 1)$

(ii) $y = 2.5x - 3.5$

c) $(x_1, y_1) = (4, 1)$, $(x_2, y_2) = (0, -3)$

$m = \frac{-3-1}{0-4} = \frac{-4}{-4} = 1$

(i) $y - 1 = 1(x - 4) \Rightarrow y - 1 = x - 4$

(ii) $y = x - 3$

d) $(x_1, y_1) = (12, -3)$, $(x_2, y_2) = (14, 1)$

$m = \frac{1-(-3)}{14-12} = \frac{4}{2} = 2$

(i) $y - (-3) = 2(x - 12) \Rightarrow y + 3 = 2(x - 12)$

(ii) $y = 2x - 27$

e) $(x_1, y_1) = (5, 7)$, $(x_2, y_2) = (-2, 5)$

$m = \frac{5-7}{-2-5} = \frac{-2}{-7} = \frac{2}{7}$

(i) $y - 7 = \frac{2}{7}(x - 5)$

(ii) $y = \frac{2}{7}x + \frac{39}{7}$

f) $(x_1, y_1) = (-3, 6)$, $(x_2, y_2) = (4, -2)$

$m = \frac{-2-6}{4-(-3)} = -\frac{8}{7}$

(i) $y - 6 = -\frac{8}{7}(x - (-3)) \Rightarrow y - 6 = -\frac{8}{7}(x + 3)$

(ii) $y = -\frac{8}{7}x + \frac{18}{7}$

Q5 **a)** $m = \frac{0.65-(-1.85)}{-0.3-0.2} = \frac{2.5}{-0.5} = -5$

b) $y - 0.65 = -5(x + 0.3) \Rightarrow y = -5x - 1.5 + 0.65$
$\Rightarrow y = -5x - 0.85$

Q6 $y = mx + c \Rightarrow -3 = \frac{1}{4} \times (-4) + c \Rightarrow -3 = -1 + c \Rightarrow c = -2$

So $y = \frac{1}{4}x - 2$

Q7 $y - y_1 = m(x - x_1) \Rightarrow y - 2 = -\frac{2}{5}(x - (-8)) \Rightarrow y - 2 = -\frac{2}{5}(x + 8)$

Rearrange into $y = mx + c$ form:

$y - 2 = -\frac{2}{5}(x + 8) \Rightarrow y = -\frac{2}{5}x - \frac{16}{5} + 2 \Rightarrow y = -\frac{2}{5}x - \frac{6}{5}$

Q8 Find the equation of the line first.

$m = 3$, find c using the point (2, –7) on the line:

$y = mx + c \Rightarrow -7 = 3 \times 2 + c \Rightarrow -7 = 6 + c \Rightarrow c = -13$

So $y = 3x - 13$ — the points a), c) and e) lie on the line.

Sub in the x value from each point — if the resulting value for y matches the value of y in the original point, then the point lies on that line.

Q9 Find the equation of the line first: $m = \frac{20-6}{-1-6} = \frac{14}{-7} = -2$

Then $y - 6 = -2(x - 6) \Rightarrow y = -2x + 18$.

So the points a), c), d) and f) lie on the line.

Q10 **a)** The gradient (m) is given as 32. To find c, substitute in the conditions $t = 0$ and $d = 0$:

$d = 32t + c \Rightarrow 0 = 32(0) + c \Rightarrow c = 0$

So the equation is $d = 32t$.

b) Solve the equation where $d = 9.6$:

$9.6 = 32t \Rightarrow t = 9.6 \div 32 = 0.3$ hours

$0.3 \times 60 = 18$ minutes

c) Some possible answers include:

- It is unrealistic that the car would travel at exactly the same speed for any length of time — it would probably vary slightly, which would make the model less accurate.
- In practice, external factors would probably affect the speed of the car during its journey, such as bends in the road, or other vehicles.
- The car wouldn't start at 32 km/h — it would take time for it to reach this speed.

Exercise 6.1.2 — $ax + by + c = 0$

Q1 **a)** $5x - y + 2 = 0$

b) $3y = -\frac{1}{2}x + 3 \Rightarrow \frac{1}{2}x + 3y - 3 = 0 \Rightarrow x + 6y - 6 = 0$

c) $2(x - 1) = 4y - 1 \Rightarrow 2x - 2 = 4y - 1 \Rightarrow 2x - 4y - 1 = 0$

d) $7x - 2y - 9 = 0$

e) $\frac{1}{2}(4x + 3) = 3(y - 2) \Rightarrow 2x + \frac{3}{2} = 3y - 6$

$\Rightarrow 2x - 3y + \frac{15}{2} = 0 \Rightarrow 4x - 6y + 15 = 0$

f) $3(y - 4) = 4(x - 3) \Rightarrow 3y - 12 = 4x - 12 \Rightarrow 4x - 3y = 0$

Q2 **a)** $6x - 2y + 3 = 0 \Rightarrow 2y = 6x + 3 \Rightarrow y = 3x + \frac{3}{2}$

$m = 3$, y-intercept = $\left(0, \frac{3}{2}\right)$

b) $-9x + 3y - 12 = 0 \Rightarrow 3y = 9x + 12 \Rightarrow y = 3x + 4$

$m = 3$, y-intercept = (0, 4)

c) $-x - 4y - 2 = 0 \Rightarrow -4y = x + 2 \Rightarrow y = -\frac{1}{4}x - \frac{1}{2}$

$m = -\frac{1}{4}$, y-intercept = $\left(0, -\frac{1}{2}\right)$

d) $7x + 8y + 11 = 0 \Rightarrow 8y = -7x - 11 \Rightarrow y = -\frac{7}{8}x - \frac{11}{8}$

$m = -\frac{7}{8}$, y-intercept = $\left(0, -\frac{11}{8}\right)$

e) $2x - 14y + 1 = 0 \Rightarrow 14y = 2x + 1 \Rightarrow y = \frac{1}{7}x + \frac{1}{14}$

$m = \frac{1}{7}$, y-intercept = $\left(0, \frac{1}{14}\right)$

f) $-3x + 28y - 16 = 0 \Rightarrow 28y = 3x + 16 \Rightarrow y = \frac{3}{28}x + \frac{4}{7}$

$m = \frac{3}{28}$, y-intercept = $\left(0, \frac{4}{7}\right)$

g) $0.1x + 0.2y + 0.3 = 0 \Rightarrow 0.2y = -0.1x - 0.3$
$\Rightarrow y = -0.5x - 1.5$

$m = -0.5$, y-intercept = (0, –1.5)

h) $-10x + 0.1y + 11 = 0 \Rightarrow 0.1y = 10x - 11 \Rightarrow y = 100x - 110$

$m = 100$, y-intercept = (0, –110)

i) $\frac{6}{7}x - 3y + \frac{3}{4} = 0 \Rightarrow 3y = \frac{6}{7}x + \frac{3}{4} \Rightarrow y = \frac{2}{7}x + \frac{1}{4}$

$m = \frac{2}{7}$, y-intercept $= \left(0, \frac{1}{4}\right)$

Q3 **a)** $(x_1, y_1) = (0, 1), (x_2, y_2) = (-1, -1)$

$m = \frac{-1-1}{-1-0} = \frac{-2}{-1} = 2$

Since you are given the y-intercept (0, 1), you can just use the $y = mx + c$ method before rearranging:

$y = mx + c \Rightarrow y = 2x + 1 \Rightarrow 2x - y + 1 = 0$

b) $(x_1, y_1) = (5, 5), (x_2, y_2) = (0, 0.2)$

$m = \frac{0.2-5}{0-5} = \frac{-4.8}{-5} = \frac{24}{25}$

$y = mx + c \Rightarrow y = \frac{24}{25}x + 0.2 \Rightarrow y = \frac{24}{25}x + \frac{5}{25}$

$\Rightarrow 25y = 24x + 5 \Rightarrow 24x - 25y + 5 = 0$

c) $(x_1, y_1) = (5, 2), (x_2, y_2) = (3, 4)$

$m = \frac{4-2}{3-5} = \frac{2}{-2} = -1$

Here you don't have the y-intercept, so use one of the given points with the $y - y_1 = m(x - x_1)$ method:

$y - y_1 = m(x - x_1) \Rightarrow y - 2 = -1(x - 5)$

$\Rightarrow y - 2 = -x + 5 \Rightarrow x + y - 7 = 0$

d) $(x_1, y_1) = (9, -1), (x_2, y_2) = (7, 2)$

$m = \frac{2-(-1)}{7-9} = -\frac{3}{2}$

$y - y_1 = m(x - x_1)$

$\Rightarrow y - (-1) = -\frac{3}{2}(x - 9) \Rightarrow y + 1 = -\frac{3}{2}x + \frac{27}{2}$

$\Rightarrow \frac{3}{2}x + y - \frac{25}{2} = 0 \Rightarrow 3x + 2y - 25 = 0$

e) $(x_1, y_1) = (-6, 1), (x_2, y_2) = (4, 0)$

$m = \frac{0-1}{4-(-6)} = -\frac{1}{10}$

$y - y_1 = m(x - x_1)$

$\Rightarrow y - 1 = -\frac{1}{10}(x - (-6)) \Rightarrow y - 1 = -\frac{1}{10}x - \frac{6}{10}$

$\Rightarrow \frac{1}{10}x + y - \frac{4}{10} = 0 \Rightarrow x + 10y - 4 = 0$

f) $(x_1, y_1) = (-12, 3), (x_2, y_2) = (5, 7)$

$m = \frac{7-3}{5-(-12)} = \frac{4}{17}$

$y - y_1 = m(x - x_1) \Rightarrow y - 3 = \frac{4}{17}(x - (-12))$

$\Rightarrow y - 3 = \frac{4}{17}x + \frac{48}{17} \Rightarrow -\frac{4}{17}x + y - \frac{99}{17} = 0$

$\Rightarrow -4x + 17y - 99 = 0 \Rightarrow 4x - 17y + 99 = 0$

Q4 **a)** $(x_1, y_1) = (0, -5), (x_2, y_2) = (-5, 0)$

$m = \frac{0-(-5)}{-5-0} = -1$

$y - y_1 = m(x - x_1) \Rightarrow y - (-5) = -1(x - 0)$

$\Rightarrow y + 5 = -x \Rightarrow x + y + 5 = 0$

You can also use the $y = mx + c$ method because you know the y-intercept.

b) $(x_1, y_1) = (0, -2), (x_2, y_2) = (3, 0)$

$m = \frac{0-(-2)}{3-0} = \frac{2}{3}$

$y - y_1 = m(x - x_1)$

$\Rightarrow y - (-2) = \frac{2}{3}(x - 0) \Rightarrow y + 2 = \frac{2}{3}x$

$\Rightarrow -\frac{2}{3}x + y + 2 = 0 \Rightarrow 2x - 3y - 6 = 0$

Q5 $(x_1, y_1) = (-6, 1), (x_2, y_2) = (-2, 7)$

$m = \frac{7-1}{-2-(-6)} = \frac{6}{4} = \frac{3}{2}$

$y - y_1 = m(x - x_1) \Rightarrow y - 7 = \frac{3}{2}(x - (-2)) \Rightarrow y - 7 = \frac{3}{2}x + 3$

$\Rightarrow 2y - 14 = 3x + 6 \Rightarrow 3x - 2y + 20 = 0$

Q6 For diagonal AC: $m_{AC} = \frac{5-2}{6-(-3)} = \frac{3}{9} = \frac{1}{3}$

$y - y_1 = m(x - x_1) \Rightarrow y - 2 = \frac{1}{3}(x - (-3)) \Rightarrow y - 2 = \frac{1}{3}x + 1$

$\Rightarrow 3y - 6 = x + 3 \Rightarrow x - 3y + 9 = 0$

For diagonal BD: $m_{BD} = \frac{2-5}{6-(-3)} = \frac{-3}{9} = -\frac{1}{3}$

$y - y_1 = m(x - x_1) \Rightarrow y - 2 = -\frac{1}{3}(x - 6) \Rightarrow y - 2 = -\frac{1}{3}x + 2$

$\Rightarrow 3y - 6 = -x + 6 \Rightarrow x + 3y - 12 = 0$

Q7 **a)** $3x + 4y = 18 \Rightarrow 3x + 4y - 18 = 0$

b)

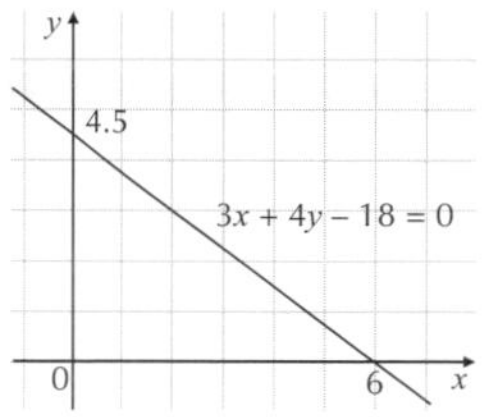

c) $x + 2y = 8 \Rightarrow x + 2y - 8 = 0$

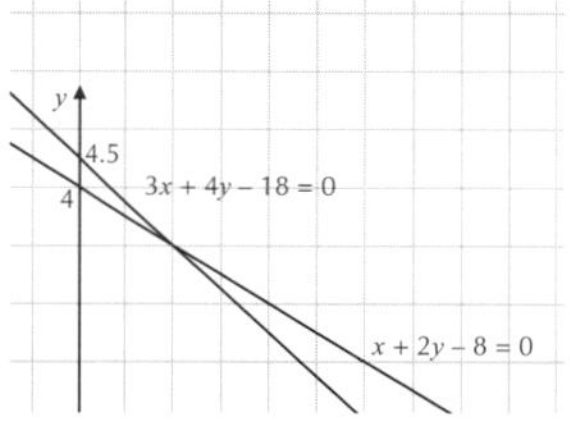

d) Read off the point of intersection from the graph: $x = 2$, $y = 3$ — that is, the cost of a small cup is £2 and the cost of a large cup is £3.

6.2 Parallel and Perpendicular Lines

Exercise 6.2.1 — Parallel lines

Q1 a), c) and e) are parallel.

Rearrange each equation so it's in the form $y = mx + c$ and then compare gradients. If a line is parallel to $y = -3x + c$, it'll have a gradient of -3.

Q2 **a)** $y = 4x + c$

Sub in $x = 3$ and $y = 2 \Rightarrow 2 = 4 \times 3 + c \Rightarrow -10 = c$

So $y = 4x - 10 \Rightarrow 4x - y - 10 = 0$

b) First rearrange the given equation:

$4x - 2y - 1 = 0 \Rightarrow -2y = -4x + 1 \Rightarrow y = 2x - \frac{1}{2}$

So the equation of the line you want is: $y = 2x + c$

Sub in $x = -4, y = -5 \Rightarrow -5 = 2(-4) + c \Rightarrow c = 3$

So $y = 2x + 3 \Rightarrow 2x - y + 3 = 0$

Q3 **a)** no **b)** yes **c)** yes **d)** no

Q4 **a)** Find the gradient of the other line first:

Rearrange $2x - 4y + 3 = 0$

$\Rightarrow 4y = 2x + 3 \Rightarrow y = \frac{1}{2}x + \frac{3}{4}$

So gradient, $m = \frac{1}{2}$

$y = \frac{1}{2}x + c$

Sub in $x = 4$ and $y = 3 \Rightarrow 3 = \frac{1}{2} \times 4 + c \Rightarrow c = 1$

So $y = \frac{1}{2}x + 1$

b) Rearrange $y = \frac{1}{2}x + 1$

$\Rightarrow \frac{1}{2}x - y + 1 = 0 \Rightarrow x - 2y + 2 = 0$

Q5 **a)** Gradient of the given line $m = 2$

$y = mx + c \Rightarrow y = 2x + c$

Sub in $x = 2, y = 1 \Rightarrow 1 = 2(2) + c \Rightarrow c = -3$

So $y = 2x - 3 \Rightarrow 2x - y - 3 = 0$

b) Rearrange the equation: $y = -5x + 11$, so $m = -5$

$y = mx + c \Rightarrow y = -5x + c$

Sub in $x = 3, y = -1 \Rightarrow -1 = -5(3) + c \Rightarrow c = 14$

So $y = -5x + 14 \Rightarrow 5x + y - 14 = 0$

c) Rearrange the equation: $y = \frac{1}{9}x + \frac{2}{3}$, so $m = \frac{1}{9}$

$y = mx + c \Rightarrow y = \frac{1}{9}x + c$

Sub in $x = -6$, $y = 2 \Rightarrow 2 = \frac{1}{9}(-6) + c \Rightarrow c = \frac{8}{3}$

So $y = \frac{1}{9}x + \frac{8}{3} \Rightarrow x - 9y + 24 = 0$

d) Rearrange the equation: $y = 4x + 4$, so $m = 4$

$y = mx + c \Rightarrow y = 4x + c$

Sub in $x = -6$, $y = -5 \Rightarrow -5 = 4(-6) + c \Rightarrow c = 19$

So $y = 4x + 19 \Rightarrow 4x - y + 19 = 0$

e) Rearrange the equation: $y = x - 13$, so $m = 1$

$y = mx + c \Rightarrow y = x + c$

Sub in $x = 0$, $y = 0 \Rightarrow 0 = 1(0) + c \Rightarrow c = 0$

So $y = x \Rightarrow x - y = 0$

f) Rearrange the equation: $y = 100 - \frac{1}{5}x$, so $m = -\frac{1}{5}$

$y = mx + c \Rightarrow y = -\frac{1}{5}x + c$

Sub in $x = 50$, $y = 50 \Rightarrow 50 = -\frac{1}{5}(50) + c \Rightarrow c = 60$

So $y = -\frac{1}{5}x + 60 \Rightarrow x + 5y - 300 = 0$

g) Rearrange the equation:

$y = \frac{60}{11} - \frac{5}{22}x$, so $m = -\frac{5}{22}$

$y = mx + c \Rightarrow y = -\frac{5}{22}x + c$

Sub in $x = 4$, $y = 8 \Rightarrow 8 = -\frac{5}{22}(4) + c \Rightarrow c = \frac{98}{11}$

So $y = -\frac{5}{22}x + \frac{98}{11} \Rightarrow 5x + 22y - 196 = 0$

h) Rearrange the equation:

$3x + 3 - 2y + 2 = 4 \Rightarrow y = \frac{3}{2}x + \frac{1}{2}$, so $m = \frac{3}{2}$

$\Rightarrow y = mx + c \Rightarrow y = \frac{3}{2}x + c$

Sub in $x = -2$, $y = 2 \Rightarrow 2 = \frac{3}{2}(-2) + c \Rightarrow c = 5$

So $y = \frac{3}{2}x + 5 \Rightarrow 3x - 2y + 10 = 0$

i) Rearrange the equation:

$3(y - 3x) = 2(4 + y) \Rightarrow 3y - 9x = 8 + 2y$

$\Rightarrow y = 9x + 8$, so $m = 9$

$\Rightarrow y = mx + c \Rightarrow y = 9x + c$

Sub in $x = 2$, $y = 3 \Rightarrow 3 = 9(2) + c \Rightarrow c = -15$

So $y = 9x - 15 \Rightarrow 9x - y - 15 = 0$

Exercise 6.2.2 — Perpendicular lines

Q1 a) $m = -1 \div 2 = -\frac{1}{2} \Rightarrow y = -\frac{1}{2}x + c$

Sub in $(-2, 5) \Rightarrow 5 = \left(-\frac{1}{2}\right) \times (-2) + c \Rightarrow c = 5 - 1 = 4$

So $y = -\frac{1}{2}x + 4$

b) Rearrange $x - 5y - 30 = 0 \Rightarrow y = \frac{1}{5}x - 6$

$m = -1 \div \frac{1}{5} = -5 \Rightarrow y = -5x + c$

Sub in $(5, 2) \Rightarrow 2 = -5 \times 5 + c \Rightarrow c = 2 + 25 = 27$

So $y = -5x + 27$

Q2 a) $m = -1 \div \frac{1}{4} = -4 \Rightarrow y = -4x + c$

Sub in $(-1, 2) \Rightarrow 2 = (-4) \times (-1) + c \Rightarrow c = 2 - 4 = -2$

So $y = -4x - 2 \Rightarrow 4x + y + 2 = 0$

b) Rearrange to get $y = -\frac{2}{3}x + \frac{1}{3}$

$m = -1 \div -\frac{2}{3} = \frac{3}{2} \Rightarrow y = \frac{3}{2}x + c$

Sub in $(-3, -1) \Rightarrow -1 = \frac{3}{2} \times (-3) + c \Rightarrow c = -1 + \frac{9}{2} = \frac{7}{2}$

So $y = \frac{3}{2}x + \frac{7}{2} \Rightarrow 2y = 3x + 7 \Rightarrow 3x - 2y + 7 = 0$

c) Rearrange to get $y = \frac{1}{2}x + \frac{1}{10}$

$m = -1 \div \frac{1}{2} = -2 \Rightarrow y = -2x + c$

Sub in $(6, -5) \Rightarrow -5 = -2 \times 6 + c \Rightarrow c = -5 + 12 = 7$

So $y = -2x + 7 \Rightarrow 2x + y - 7 = 0$

d) $m = -1 \div \frac{3}{2} = -\frac{2}{3} \Rightarrow y = -\frac{2}{3}x + c$

Sub in $(2, 1) \Rightarrow 1 = -\frac{2}{3} \times 2 + c \Rightarrow c = 1 + \frac{4}{3} = \frac{7}{3}$

So $y = -\frac{2}{3}x + \frac{7}{3} \Rightarrow 3y = -2x + 7 \Rightarrow 2x + 3y - 7 = 0$

e) Rearrange to get $y = \frac{4}{21}x + \frac{2}{21}$

$m = -1 \div \frac{4}{21} = -\frac{21}{4} \Rightarrow y = -\frac{21}{4}x + c$

Sub in $(0.5, 7) \Rightarrow 7 = -\frac{21}{4} \times 0.5 + c \Rightarrow c = 7 + \frac{21}{8} = \frac{77}{8}$

So $y = -\frac{21}{4}x + \frac{77}{8} \Rightarrow 42x + 8y - 77 = 0$

f) Rearrange to get $10 - 5x + 15y = 2 \Rightarrow y = \frac{1}{3}x - \frac{8}{15}$

$m = -1 \div \frac{1}{3} = -3 \Rightarrow y = -3x + c$

Sub in $(-5, -1) \Rightarrow -1 = -3 \times -5 + c \Rightarrow c = -1 - 15 = -16$So

$y = -3x - 16 \Rightarrow 3x + y + 16 = 0$

g) Rearrange to get $56y + 8 = 2x - 3 \Rightarrow y = \frac{1}{28}x - \frac{11}{56}$

$m = -1 \div \frac{1}{28} = -28 \Rightarrow y = -28x + c$

Sub in $(7, 8) \Rightarrow 8 = -28 \times 7 + c \Rightarrow c = 8 + 196 = 204$

So $y = -28x + 204 \Rightarrow 28x + y - 204 = 0$

h) Rearrange to get $y = 4x + 0.2y + 1 \Rightarrow 0.8y = 4x + 1$

$\Rightarrow y = 5x + \frac{5}{4}$

$m = -1 \div 5 = -\frac{1}{5} \Rightarrow y = -\frac{1}{5}x + c$

Sub in $(3, 4.4) \Rightarrow 4.4 = -\frac{1}{5} \times 3 + c \Rightarrow c = 4.4 + \frac{3}{5} = 5$

So $y = -\frac{1}{5}x + 5 \Rightarrow x + 5y - 25 = 0$

Q3 a) Rearrange $3x + 4y - 1 = 0 \Rightarrow y = -\frac{3}{4}x + \frac{1}{4}$

Multiply the gradients of the lines: $\frac{4}{3} \times -\frac{3}{4} = -1$

So the lines are perpendicular.

Remember, if you multiply the gradients of two perpendicular lines you get −1.

b) Rearrange $3x + 2y - 3 = 0 \Rightarrow y = -\frac{3}{2}x + \frac{3}{2}$

Multiply the gradients of the lines: $\frac{3}{2} \times -\frac{3}{2} = -\frac{9}{4}$

So the lines are not perpendicular.

c) Rearrange $4x - y + 3 = 0 \Rightarrow y = 4x + 3$

Rearrange $2x + 8y + 1 = 0 \Rightarrow y = -\frac{1}{4}x - \frac{1}{8}$

Multiply the gradients of the lines: $4 \times -\frac{1}{4} = -1$

So the lines are perpendicular.

d) Rearrange $3x - 5y + 10 = 0 \Rightarrow y = \frac{3}{5}x + 2$

Rearrange $15x + 6y - 4 = 0 \Rightarrow y = \frac{2}{3} - \frac{5}{2}x$

Multiply the gradients of the lines: $\frac{3}{5} \times -\frac{5}{2} = -\frac{3}{2}$

So the lines are not perpendicular.

Q4 PQ: $m = \frac{7-(-2)}{-1-4} = -\frac{9}{5}$

So the gradient of the line perpendicular to PQ is $\frac{5}{9}$

$y = mx + c \Rightarrow y = \frac{5}{9}x + c$

Sub in $(2, 5) \Rightarrow 5 = \frac{5}{9} \times 2 + c \Rightarrow c = 5 - \frac{10}{9} = \frac{35}{9}$

So $y = \frac{5}{9}x + \frac{35}{9} \Rightarrow 5x - 9y + 35 = 0$

Q5 a) AB: $m = \frac{3-2}{4-0} = \frac{1}{4} \Rightarrow y = \frac{1}{4}x + c$

Sub in $(0, 2) \Rightarrow 2 = 0 + c \Rightarrow c = 2$

So $y = \frac{1}{4}x + 2$

BC: $m = \frac{-1-3}{5-4} = -4 \Rightarrow y = -4x + c$

Sub in $(4, 3) \Rightarrow 3 = -4 \times 4 + c \Rightarrow c = 19$

So $y = -4x + 19$

AC: $m = \frac{-1-2}{5-0} = -\frac{3}{5} \Rightarrow y = -\frac{3}{5}x + c$

Sub in $(0, 2) \Rightarrow 2 = 0 + c \Rightarrow c = 2$

So $y = -\frac{3}{5}x + 2$

b) The triangle is right-angled, as AB is perpendicular to BC:

$m_{AB} \times m_{BC} = \frac{1}{4} \times -4 = -1$

Q6 **a)** PR: $m_{PR} = \frac{3-(-1)}{3-1} = \frac{4}{2} = 2 \Rightarrow y = 2x + c$

Sub in $(1, -1) \Rightarrow -1 = 2(1) + c \Rightarrow c = -3$

So $y = 2x - 3$

QS: $m_{QS} = \frac{0-2}{4-0} = \frac{-2}{4} = -\frac{1}{2} \Rightarrow y = -\frac{1}{2}x + c$

Sub in $(0, 2) \Rightarrow 2 = -\frac{1}{2}(0) + c \Rightarrow c = 2$

So $y = -\frac{1}{2}x + 2$

b) From a), the product of the diagonals is $2 \times -\frac{1}{2} = -1$, so they are perpendicular

i.e. PQRS could be a square, a rhombus or a kite.

If PQRS is a square, the diagonals will be the same length.

$PR = \sqrt{(3-1)^2 + (3-(-1))^2} = \sqrt{20} = 2\sqrt{5}$ and

$QS = \sqrt{(4-0)^2 + (0-2)^2} = \sqrt{20} = 2\sqrt{5}$, so PQRS is a square.

Q7 Rearrange the equation into $y = mx + c$:

$-2y = -3x + 6 \Rightarrow y = \frac{3}{2}x - 3$, so the line we want will have gradient $m = -1 \div \frac{3}{2} = -\frac{2}{3}$

Now sub in (a, b) to find c: $y = -\frac{2}{3}x + c$

$$b = -\frac{2}{3}a + c$$

$$c = b + \frac{2}{3}a$$

So the equation of line A is $y = -\frac{2}{3}x + \frac{2}{3}a + b$

You could also have given the line in the form $2x + 3y - 2a - 3b = 0$ as the question didn't tell you which form to use.

Q8 The gradient of the line AB is $\frac{2-4}{5-1} = -\frac{1}{2}$.

So the gradient of the perpendicular bisector is: $m = -1 \div -\frac{1}{2} = 2$

The bisector will pass through the midpoint of the line — this will be at $\left(\frac{1+5}{2}, \frac{4+2}{2}\right) = (3, 3)$.

So $y = 2x + c \Rightarrow 3 = 2(3) + c \Rightarrow c = 3 - 6 = -3$

So the equation of the perpendicular bisector is: $y = 2x - 3$.

6.3 Proportion

Exercise 6.3.1 — Direct proportion

Q1 **a)** $y \propto x \Rightarrow y = kx$

When $x = 8$, $24 = 8k \Rightarrow k = \frac{24}{8} = 3 \Rightarrow y = 3x$

So when $x = 5$, $a = 3 \times 5 = 15$

b) $y \propto x \Rightarrow y = kx$

When $x = 7$, $28 = 7k \Rightarrow k = \frac{28}{7} = 4 \Rightarrow y = 4x$

So when $y = 96$, $96 = 4a \Rightarrow a = \frac{96}{4} = 24$

Q2 **a)** $y \propto \frac{1}{x} \Rightarrow xy = k$

When $x = 6, y = 3 \Rightarrow k = 6 \times 3 = 18$

So $y = \frac{18}{x} = \frac{18}{9} = 2$

b) $y \propto \frac{1}{x} \Rightarrow xy = k$

When $x = 12, y = 12 \Rightarrow k = 12 \times 12 = 144$

So $x = \frac{144}{y} = \frac{144}{36} = 4$

Q3 **a)** The +2 means that the equation cannot be written in the form $y = kx$ (the graph also does not pass through the origin), so y is not directly proportional to x.

b) The equation can be written as $y = (a - b)x$, where the constant of proportionality is equal to $(a - b)$, so y is directly proportional to x.

c) Simplifying the equation gives $y = x - 2$, which cannot be written as $y = kx$, so y is not directly proportional to x.

d) Expanding the brackets gives:

$y = x^2 + 6x + 9 - (x^2 - 6x + 9)$

$\Rightarrow y = x^2 + 6x + 9 - x^2 + 6x - 9$

$\Rightarrow y = 12x$, so they are in direct proportion.

Q4 $y \propto x^2 \Rightarrow y = kx^2$

When $x = 4$, $40 = k \times 4^2 = 16k \Rightarrow k = \frac{5}{2}$

a) $y = \frac{5}{2}x^2 \Rightarrow y = \frac{5}{2} \times 2^2 = 10$

b) $y = \frac{5}{2}x^2 \Rightarrow 45 = \frac{5}{2}x^2 \Rightarrow x^2 = 18 \Rightarrow x = 3\sqrt{2}$

($x > 0$, so ignore the negative root)

Q5 If $y \propto x$ and $y \propto z$, then we can write $y = k_1x$ and $y = k_2z$, where k_1 and k_2 are constants. Equating y's: $k_1x = k_2z \Rightarrow x = \frac{k_2}{k_1}z$

$\frac{k_2}{k_1}$ must be a constant since k_1 and k_2 are constants.

So $x \propto z$ with constant of proportionality $\frac{k_2}{k_1}$.

Q6 $F \propto m$ and $F = 15$ when $m = 12$.

So $F = km \Rightarrow 15 = 12k \Rightarrow k = \frac{15}{12} = 1.25$

So when $m = 18$, $F = 1.25 \times 18 = 22.5$ N

6.4 Curve Sketching

Exercise 6.4.1 — Cubic and quartic functions

Q1 **a)** $y = -1.5x^4$ will be n-shaped and below the x-axis since the power is even and the coefficient is negative, so it must be graph D.

b) $y = 0.5x^3$ has an odd power and a positive coefficient so it will have a bottom-left to top-right curve. It must be graph B.

c) $y = 2x^6$ has an even power and a positive coefficient so it'll be u-shaped and above the x-axis. It must be graph A.

d) $y = -3x^3$ has an odd power of x and a negative coefficient so it must have a top-left to bottom-right curve. It must be graph C.

Q2 a) b) c)

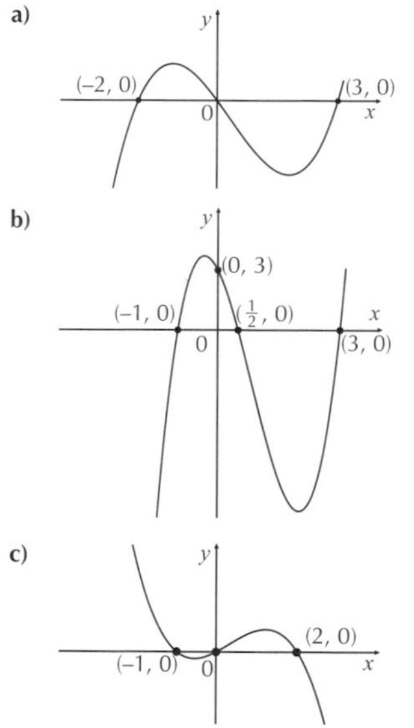

Your graphs don't need to look exactly like these — you don't need to get the size of the 'dips' right, as long as you've got the rough shape and the intercepts with the x-axis.

Q3 a) b)

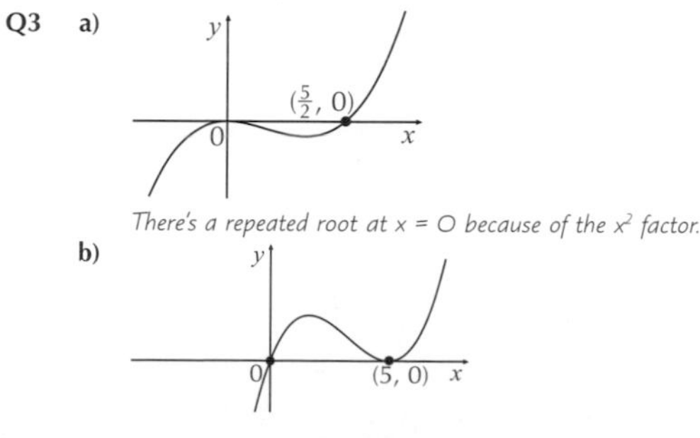

c) 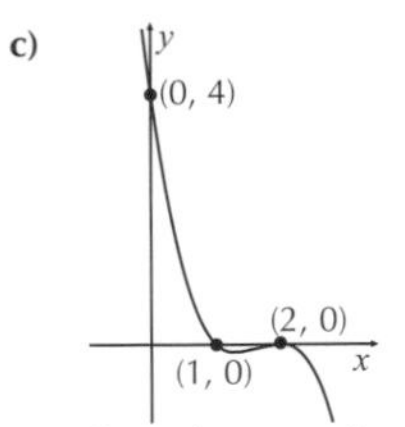

Remember — putting in values for x near the key points can really help you understand the shape of the graphs. E.g. for part c), pop in x = 1.5 and 2.5 to check that both give negative values for y.

Q4 a) b) c)

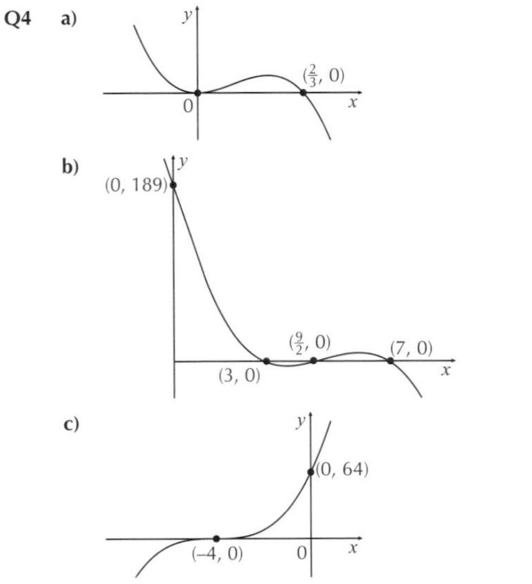

Q5 a) First take out a factor of x: $x^3 - 7x^2 + 12x = x(x^2 - 7x + 12)$
Then factorise the quadratic: $x^2 - 7x + 12 = (x - 3)(x - 4)$
So $x^3 - 7x^2 + 12x = x(x - 3)(x - 4)$.

b) 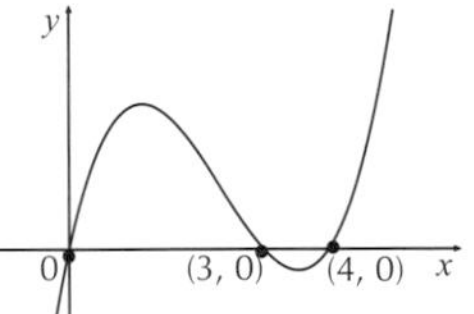

Q6 a) $x^3 - 16x = x(x^2 - 16) = x(x + 4)(x - 4)$
Using this information we can sketch the graph:

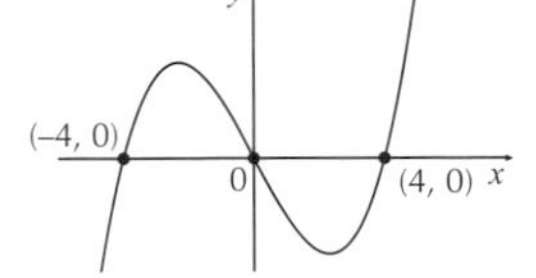

b) $2x^3 - 12x^2 + 18x = 2x(x^2 - 6x + 9) = 2x(x - 3)^2$
Using this information we can sketch the graph:

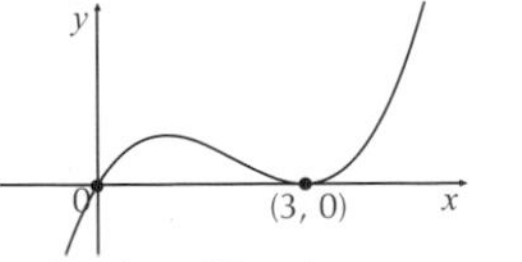

c) $-3x^2 - x^3 = -x^2(3 + x)$
Using this information we can sketch the graph:

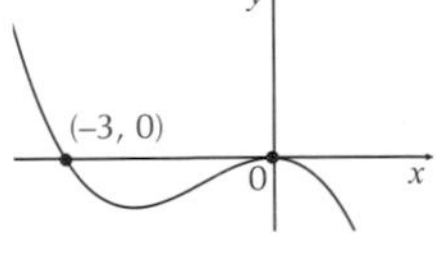

Q7 a) 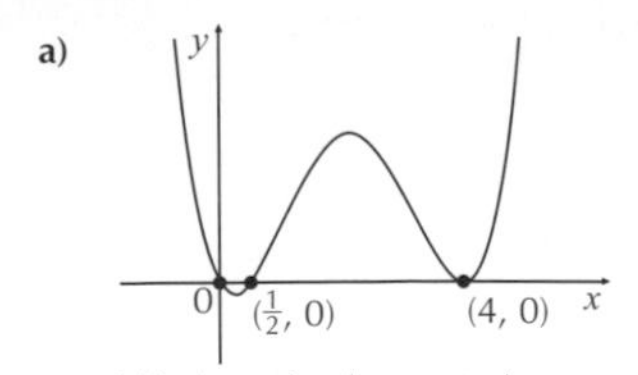

Watch out for the repeated root — the (x – 4) bracket is squared so it only touches the axis at x = 4.

b) 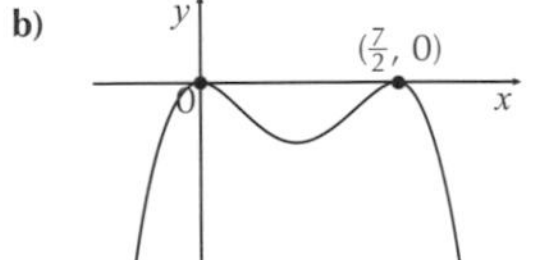

Here there are two double roots, so the graph never crosses the x-axis. Since the coefficient of x^4 is negative, the graph is below the x-axis.

c) 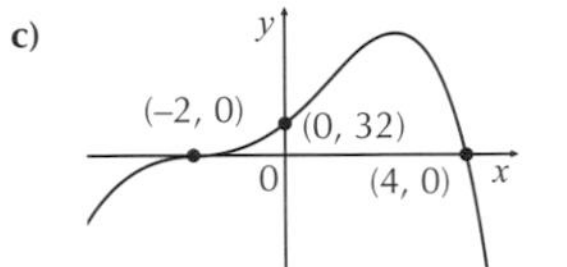

x = –2 is a triple root — the graph at this point looks like the graph of x^3.

Q8 a) $y = x^2(x^2 - 9x + 14) = x^2(x - 2)(x - 7)$

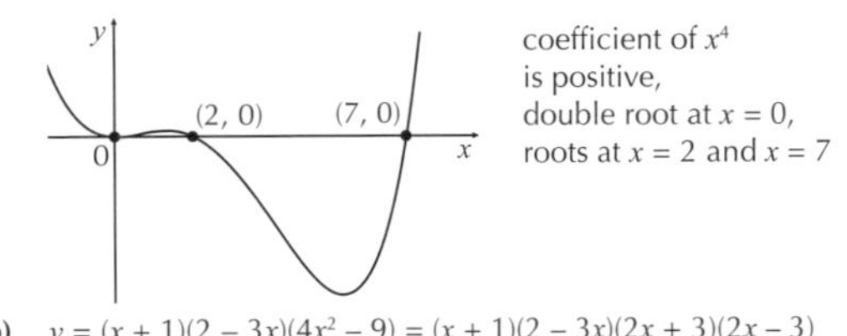

coefficient of x^4 is positive, double root at $x = 0$, roots at $x = 2$ and $x = 7$

b) $y = (x + 1)(2 - 3x)(4x^2 - 9) = (x + 1)(2 - 3x)(2x + 3)(2x - 3)$

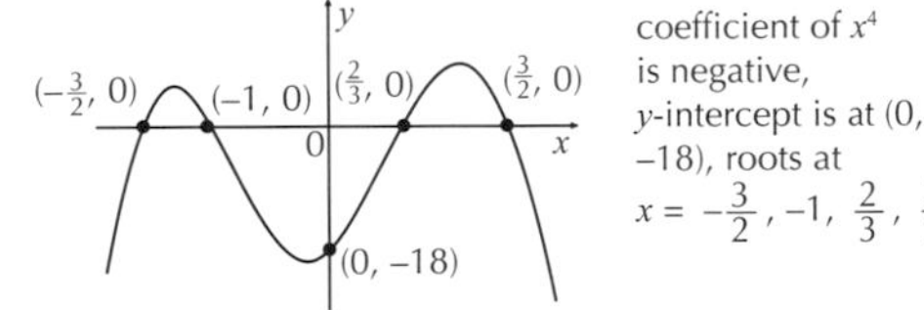

coefficient of x^4 is negative, y-intercept is at (0, –18), roots at $x = -\frac{3}{2}, -1, \frac{2}{3}, \frac{3}{2}$

c) $y = (x - 5)(2x^3 + 5x^2 - 3x)$
$= (x - 5)[x(2x^2 + 5x - 3)] = x(x - 5)(2x - 1)(x + 3)$

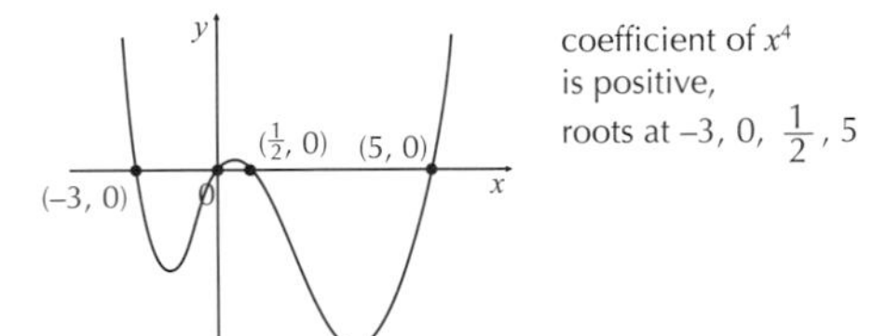

coefficient of x^4 is positive, roots at –3, 0, $\frac{1}{2}$, 5

Q9 The coefficients add up to zero, so $(x - 1)$ must be a factor.
Then use e.g. algebraic division to factorise:
$4x^3 + 4x^2 - 5x - 3 = (x - 1)(4x^2 + 8x + 3) = (x - 1)(2x + 3)(2x + 1)$
Using this information, sketch the graph:

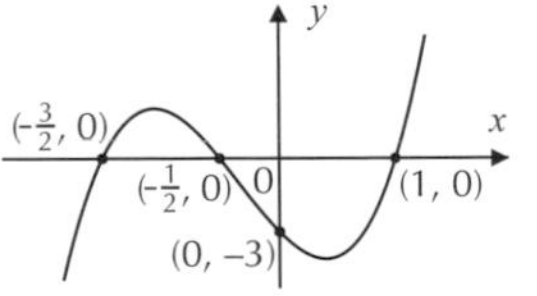

See Chapter 4 for more on factorising cubic and quadratic equations.

Q10 a) $x = 1$ is a root so $(x - 1)$ is a factor.
Using e.g. algebraic division:
$f(x) = (x + 1)(x - 1)(2x^2 - 3x - 2)$
$= (x + 1)(x - 1)(x - 2)(2x + 1)$

b) Using this information, sketch the graph:

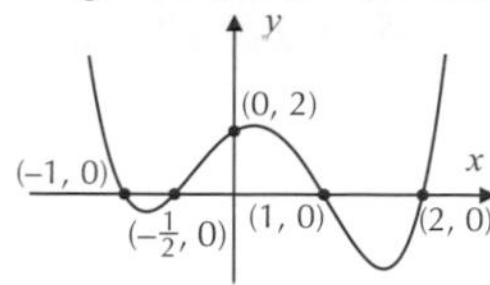

Q11 Fully factorise the quartic — the coefficients in the cubic part add up to zero, so $(x - 1)$ must be a factor.
Using e.g. algebraic division:
$y = (x - 3)(x^3 - 7x^2 + 14x - 8) = (x - 3)(x - 1)(x^2 - 6x + 8)$
$= (x - 3)(x - 1)(x - 2)(x - 4)$

Using this information, you can sketch the graph:

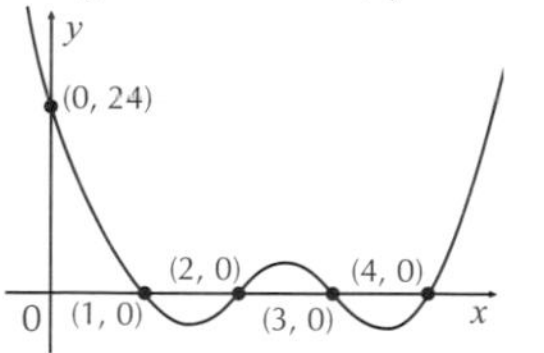

Exercise 6.4.2 — Reciprocal functions and negative powers

Q1 **a)** $y = x^{-2} = \frac{1}{x^2}$.
$n = 2$ is even so you'll get a graph with two bits next to each other. $k = 1$ is positive so the graph will all be above the axis so it must be graph D.

b) $y = -3x^{-3} = -\frac{3}{x^3}$.
$n = 3$ is odd so you'll get a graph with two bits opposite each other. $k = -3$ is negative so the graph will be in the top-left and bottom-right quadrants so it must be graph A.

c) $y = -\frac{3}{x^4}$.
$n = 4$ is even so you'll get a graph with two bits next to each other. $k = -3$ so the graph will all be below the x-axis so it must be graph B.

d) $y = 2x^{-5} = \frac{2}{x^5}$.
$n = 5$ is odd so you'll get a graph with two bits opposite each other. $k = 2$ is positive so the graph will be in the bottom-left and top-right quadrants so it must be graph C.

Q2 **a)**

b)

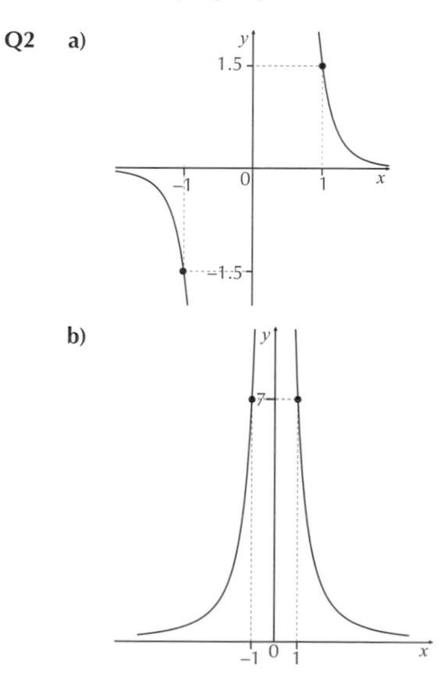

c)

d)

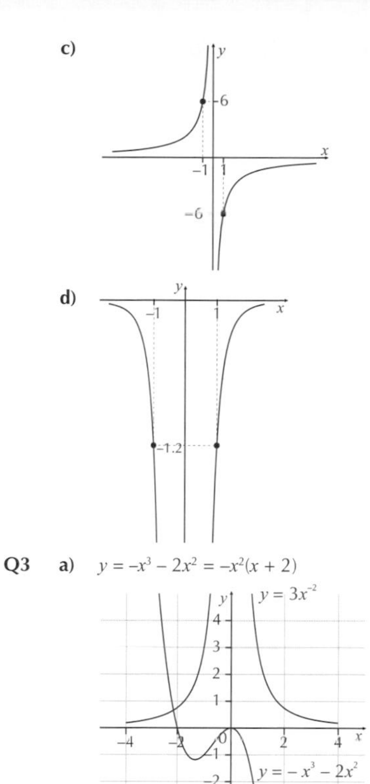

Q3 **a)** $y = -x^3 - 2x^2 = -x^2(x + 2)$

$y = 3x^{-2}$

$y = -x^3 - 2x^2$

b) The number of real roots of the equation $3x^{-2} = -x^3 - 2x^2$ is the number of times the two graphs cross — this equation has 1 real root.

Q4 **a)**

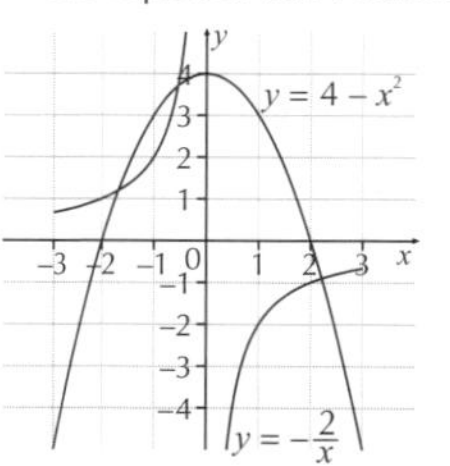

b) The solutions are at the points of intersection on the graph in part a). The actual solutions are $x = -0.54$, -1.68 and 2.21 (to 2 d.p.). Acceptable solutions are: between -0.4 and -0.7, between -1.6 and -1.8 and between 2.1 and 2.3.

6.5 Graph Transformations

Exercise 6.5.1 — Translations

Q1 **a)** **b)**

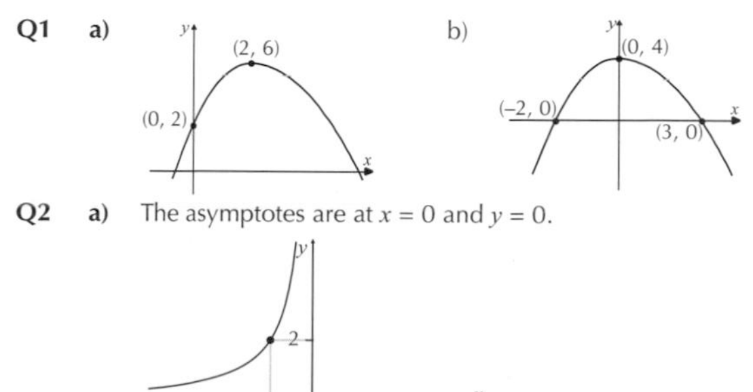

Q2 **a)** The asymptotes are at $x = 0$ and $y = 0$.

b) The asymptotes are at $x = -3$ and $y = 0$.

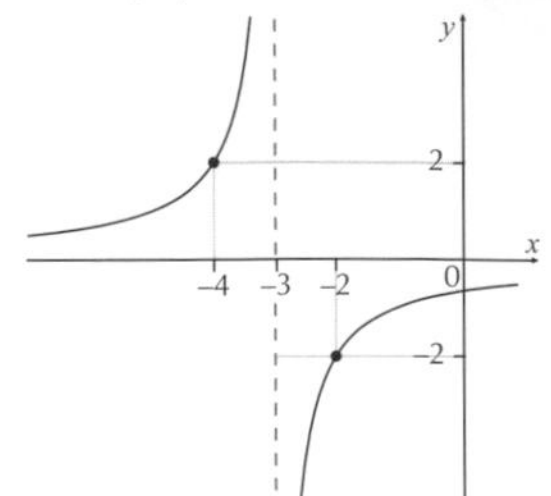

c) The asymptotes are at $x = 0$ and $y = 3$.

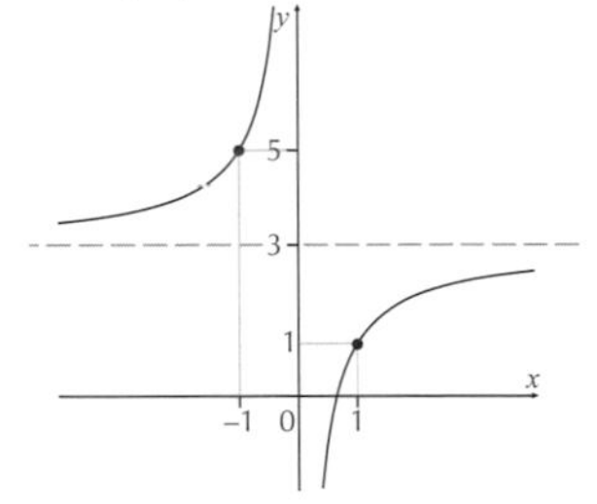

Q3 **a)** If $x^2(x - 4) = f(x)$, then $x^2(x - 4) + 1 = f(x) + 1$. The translation is 1 unit up, i.e. the graph is translated by the vector $\begin{pmatrix} 0 \\ 1 \end{pmatrix}$.

b) If $x^2(x - 4) = f(x)$, then $(x - 2)^2(x - 6) = f(x - 2)$. The translation is 2 units right, i.e. the graph is translated by the vector $\begin{pmatrix} 2 \\ 0 \end{pmatrix}$.

c) If $x^2(x - 4) = f(x)$, then $x(x + 4)^2 = f(x + 4)$. The translation is 4 units left, i.e. the graph is translated by the vector $\begin{pmatrix} -4 \\ 0 \end{pmatrix}$.

Q4 If $x^3 + 3x + 7 = f(x)$, then $x^3 + 3x + 2 = f(x) - 5$. So the translation is 5 units down, i.e. the graph is translated by the vector $\begin{pmatrix} 0 \\ -5 \end{pmatrix}$.

Q5 If $f(x) = x^2 - 3x + 7$, then the translation is
$f(x + 1) = (x + 1)^2 - 3(x + 1) + 7 = x^2 + 2x + 1 - 3x - 3 + 7 = x^2 - x + 5$

Q6 **a) and b)**

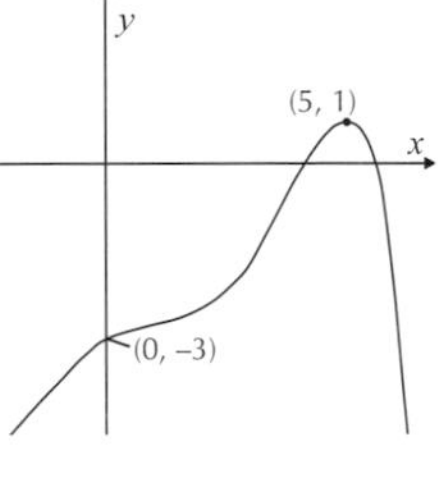

Q7 **a)**

b) $f(x) = (x - 1)(2x - 3)(4 - x)$, so the translation is:
$f(x - 2) = ((x - 2) - 1)(2(x - 2) - 3)(4 - (x - 2))$
$= (x - 3)(2x - 4 - 3)(4 - x + 2) = (x - 3)(2x - 7)(6 - x)$

c)

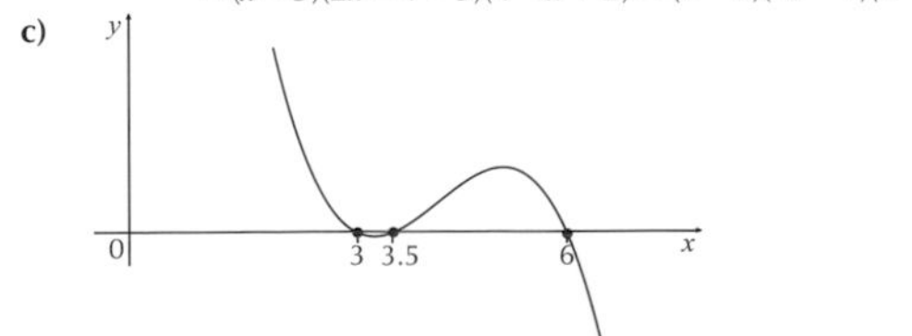

Q8 **a)** $f(x) = x^2 + 5$, so the translation is:
$g(x) = f(x + 3) = (x + 3)^2 + 5 = x^2 + 6x + 9 + 5 = x^2 + 6x + 14$

b) $g(x) = x^2 + 6x + 14$, so the translation is:
$h(x) = g(x) - 4 = x^2 + 6x + 14 - 4 = x^2 + 6x + 10$

Q9 $y = \frac{1}{x}$ has asymptotes at $x = 0$ and $y = 0$. If $f(x) = \frac{1}{x}$:

a) $\frac{1}{x} - 4 = f(x) - 4$, i.e. $f(x)$ has been translated down by 4 units, so the translated asymptotes are $x = 0$ and $y = -4$.

b) $\frac{1}{x+3} = f(x + 3)$, i.e. $f(x)$ has been translated left by 3 units, so the translated asymptotes are $x = -3$ and $y = 0$.

c) $\frac{1}{x-1} + 7 = f(x - 1) + 7$, i.e. $f(x)$ has been translated right by 1 unit and up by 7 units, so the translated asymptotes are $x = 1$ and $y = 7$.

Exercise 6.5.2 — Stretches and reflections

Q1 **a)** **b)** **c)** **d)**

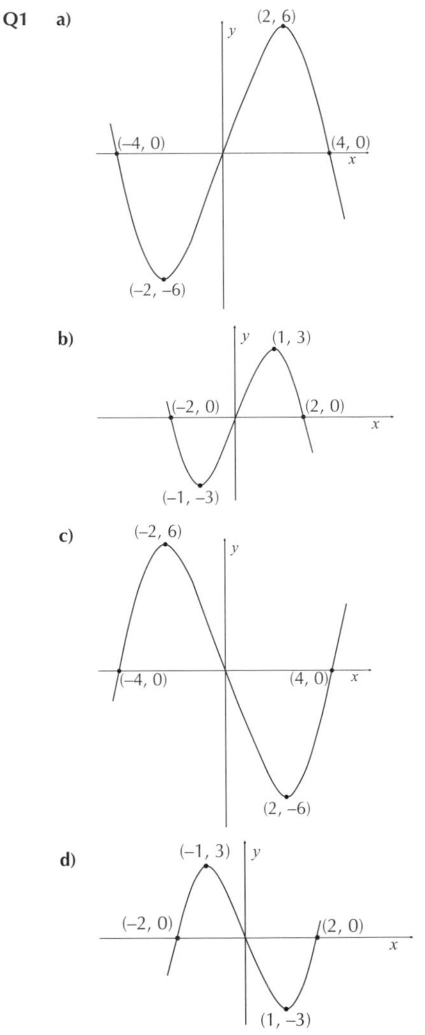

Q2 The graph has been squashed to half its width, so it's a horizontal stretch of scale factor $\frac{1}{2}$, so it must be b).

Q3 The graph has been reflected in the x-axis and stretched vertically by a factor of 3 so it must be b).

Q4 a) $f(x) = x^3 - x = x(x^2 - 1) = x(x + 1)(x - 1)$

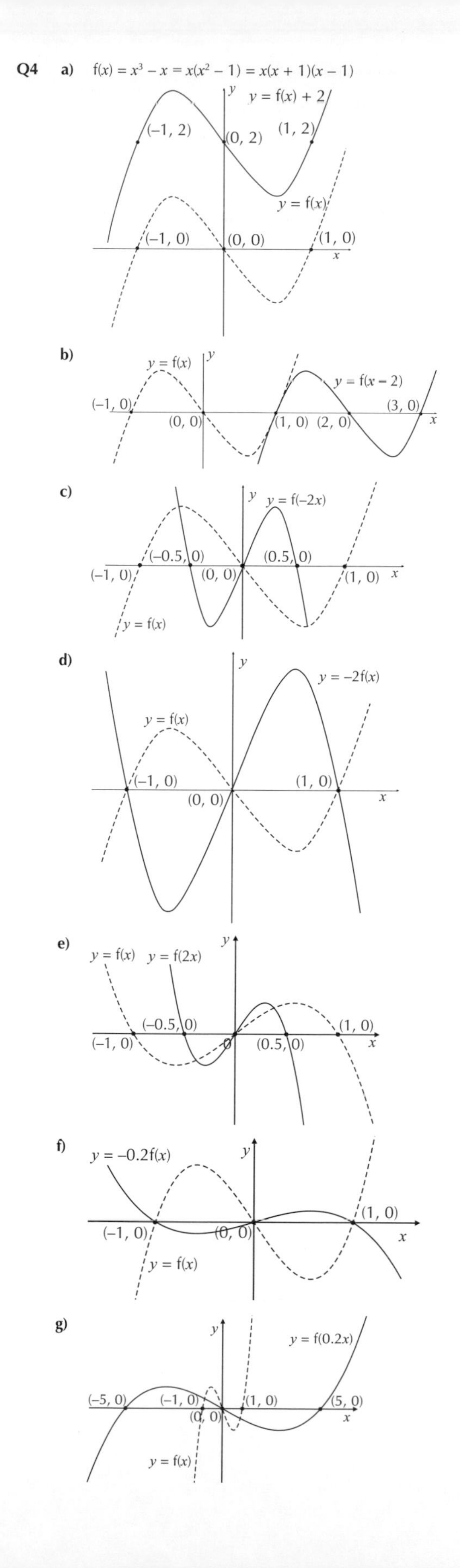

h)

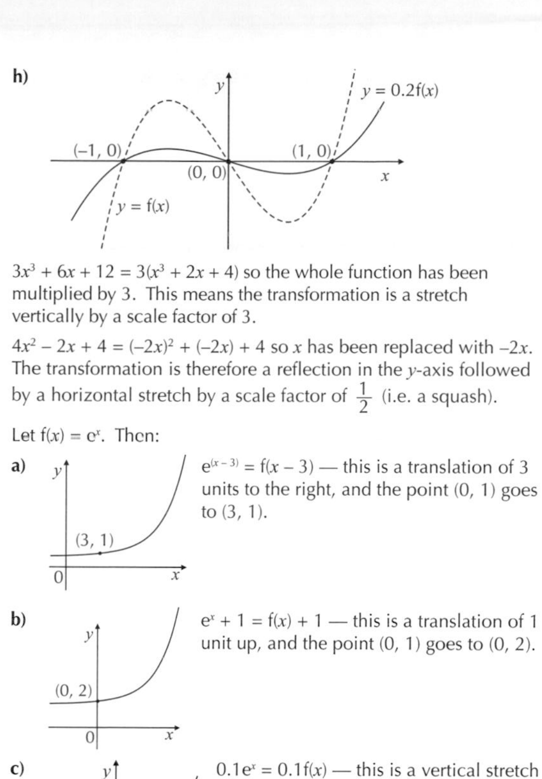

Q5 $3x^3 + 6x + 12 = 3(x^3 + 2x + 4)$ so the whole function has been multiplied by 3. This means the transformation is a stretch vertically by a scale factor of 3.

Q6 $4x^2 - 2x + 4 = (-2x)^2 + (-2x) + 4$ so x has been replaced with $-2x$. The transformation is therefore a reflection in the y-axis followed by a horizontal stretch by a scale factor of $\frac{1}{2}$ (i.e. a squash).

Q7 Let $f(x) = e^x$. Then:

a) $e^{(x-3)} = f(x - 3)$ — this is a translation of 3 units to the right, and the point (0, 1) goes to (3, 1).

(3, 1)

b) $e^x + 1 = f(x) + 1$ — this is a translation of 1 unit up, and the point (0, 1) goes to (0, 2).

(0, 2)

c) $0.1e^x = 0.1f(x)$ — this is a vertical stretch by a scale factor of 0.1 (i.e. a squash), and the point (0, 1) goes to (0, 0.1).

(0, 0.1)

d) $-4e^x = -4f(x)$ — this is a vertical stretch by a scale factor of –4, and the point (0, 1) goes to (0, –4).

(0, –4)

e) $-\frac{1}{2}e^x = -\frac{1}{2}f(x)$ — this is a vertical stretch by a scale factor of $-\frac{1}{2}$ (i.e. a squash)and the point (0, 1) goes to $\left(0, -\frac{1}{2}\right)$.

$(0, -\frac{1}{2})$

f) $e^{7x} = f(7x)$ — this is a horizontal stretch by a scale factor of $\frac{1}{7}$, and the point (0, 1) is unchanged.

(0, 1)

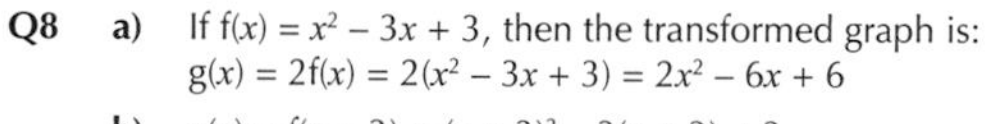

Q8 a) If $f(x) = x^2 - 3x + 3$, then the transformed graph is:
$g(x) = 2f(x) = 2(x^2 - 3x + 3) = 2x^2 - 6x + 6$

b) $g(x) = f(x + 3) = (x + 3)^2 - 3(x + 3) + 3$
$= x^2 + 6x + 9 - 3x - 9 + 3 = x^2 + 3x + 3$

c) $g(x) = f(4x) = (4x)^2 - 3(4x) + 3 = 16x^2 - 12x + 3$

Q9 a) $f(x) = x^2 - 6x - 7 = (x - 3)^2 - 16$, so the minimum point is at (3, –16). Solving $(x - 3)^2 - 16 = 0$ gives $x = -1$ or 7.
So the graph is:

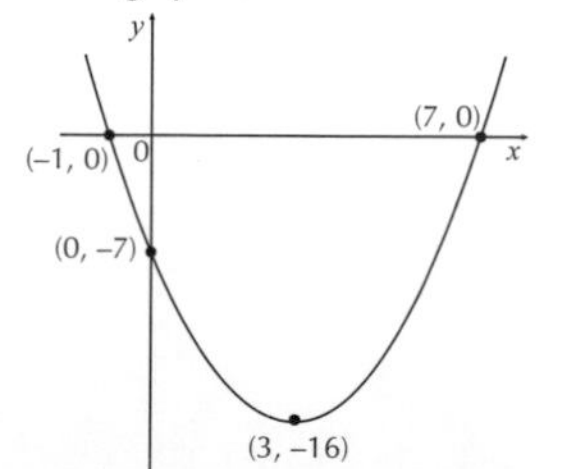

b) $y = -2f(x) = -2(x^2 - 6x - 7) = -2x^2 + 12x + 14$

c)

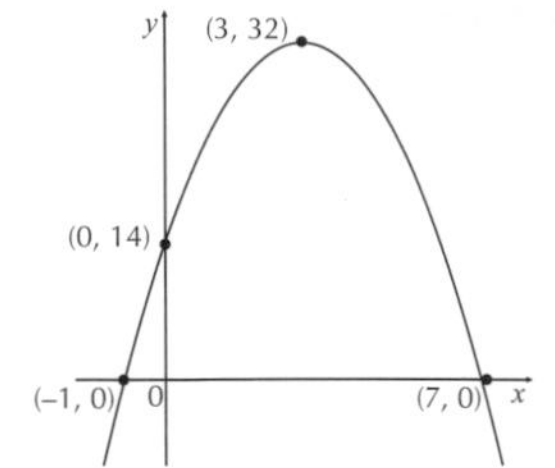

6.6 Circles

Exercise 6.6.1 — The equation of a circle

Q1 **a)** $x^2 + y^2 = 25$

b) $x^2 + y^2 = 49$

c) $x^2 + y^2 = 23$

d) $x^2 + y^2 = 18$

Q2 **a)** $a = 2, b = 5, r = 3 \Rightarrow (x - 2)^2 + (y - 5)^2 = 9$

b) $a = -3, b = 2, r = 5 \Rightarrow (x + 3)^2 + (y - 2)^2 = 25$

c) $a = -2, b = -3, r = 7 \Rightarrow (x + 2)^2 + (y + 3)^2 = 49$

d) $a = 3, b = 0, r = 4 \Rightarrow (x - 3)^2 + y^2 = 16$

e) $a = -1, b = 3, r = 7 \Rightarrow (x + 1)^2 + (y - 3)^2 = 49$

f) $a = 5, b = 4, r = \sqrt{11} \Rightarrow (x - 5)^2 + (y - 4)^2 = 11$

g) $a = -7, b = 5, r = \sqrt{27} \Rightarrow (x + 7)^2 + (y - 5)^2 = 27$

h) $a = -10, b = 7, r = 11 \Rightarrow (x + 10)^2 + (y - 7)^2 = 121$

i) $a = 8, b = 0, r = \sqrt{17} \Rightarrow (x - 8)^2 + y^2 = 17$

Q3 **a)** $a = 3, b = 2, r = 2 \Rightarrow (x - 3)^2 + (y - 2)^2 = 4$

b) $a = 1, b = -2, r = 3 \Rightarrow (x - 1)^2 + (y + 2)^2 = 9$

Q4 **a)** $a = 1, b = 5, r = \sqrt{4}$
So the centre is (1, 5) and the radius is 2.

b) $a = 3, b = 5, r = \sqrt{64}$
So the centre is (3, 5) and the radius is 8.

c) $a = 3, b = -2, r = \sqrt{25}$
So the centre is (3, –2) and the radius is 5.

d) $a = 6, b = 4, r = \sqrt{20}$
So the centre is (6, 4) and the radius is $2\sqrt{5}$.

e) $a = -8, b = -1, r = \sqrt{27}$
So the centre is (–8, –1) and the radius is $3\sqrt{3}$.

f) $a = 0, b = 12, r = \sqrt{147}$
So the centre is (0, 12) and the radius is $7\sqrt{3}$.

Q5 $a = 5, b = 3, r = 8 \Rightarrow (x - 5)^2 + (y - 3)^2 = 64$

Q6 $a = 3, b = 1, r = \sqrt{31} \Rightarrow (x - 3)^2 + (y - 1)^2 = 31$

Q7 $a = 0, b = -3, r = \sqrt{18}$
So the centre is (0, –3) and the radius is $3\sqrt{2}$.

Q8 $a = -3, b = -2, r = \sqrt{5} \Rightarrow (x + 3)^2 + (y + 2)^2 = 5$

Q9 **a)** $a = -9, b = 3$, so the centre is (–9, 3)

b) Radius $= \sqrt{45} = \sqrt{9 \times 5} = 3\sqrt{5}$

Q10 $a = 6, b = 3, r = 4\sqrt{5}$
So the equation of the circle is $(x - 6)^2 + (y - 3)^2 = 80$

a) $(-2 - 6)^2 + (7 - 3)^2 = 64 + 16 = 80$
So (–2, 7) is on the circle.

b) $(10 - 6)^2 + (-3 - 3)^2 = 16 + 36 = 52$
So (10, –3) is not on the circle.

c) $(14 - 6)^2 + (-1 - 3)^2 = 64 + 16 = 80$
So (14, –1) is on the circle.

d) $(2 - 6)^2 + (5 - 3)^2 = 16 + 4 = 20$
So (2, 5) is not on the circle.

Q11 The radius is the distance from the centre of a circle to a point on its circumference, so find the distance from (3, 5) to (10, 4) using Pythagoras:

Radius $= \sqrt{(5-4)^2 + (3-10)^2} = \sqrt{1 + 49} = \sqrt{50}$

So the equation of C is: $(x - 3)^2 + (y - 5)^2 = 50$.

Q12 Radius $= \sqrt{(9-7)^2 + (-2-(-11))^2} = \sqrt{4 + 81} = \sqrt{85}$
So the equation is $(x + 2)^2 + (y - 9)^2 = 85$.

Q13 **a)** Find the distance from A to the origin:
radius $= \sqrt{(1-0)^2 + (1-0)^2} = \sqrt{2}$
So the equation is $(x - 1)^2 + (y - 1)^2 = 2$

b) Radius $= \sqrt{(-7)^2 + 13^2} = \sqrt{218}$
So the equation is $(x + 7)^2 + (y - 13)^2 = 218$

c) Radius $= \sqrt{8^2 + (-6)^2} = 10$
So the equation is $(x - 8)^2 + (y + 6)^2 = 100$

d) Radius $= \sqrt{14^2 + 22^2} = \sqrt{680}$
So the equation is $(x - 14)^2 + (y - 22)^2 = 680$

Q14 **a)** **(i)** Length AB $= \sqrt{(-2-(-10))^2 + (7-(-3))^2}$
$= \sqrt{8^2 + 10^2} = \sqrt{164} = 2\sqrt{41}$

(ii) Midpoint AB $= \left(\frac{-3+7}{2}, \frac{-10+(-2)}{2}\right)$
$= \left(\frac{4}{2}, \frac{-12}{2}\right) = (2, -6)$

b) The radius is half of the length of the diameter AB so radius $= \sqrt{41}$, and the centre of the circle is the midpoint of the diameter (2, –6).
So the equation is $(x - 2)^2 + (y + 6)^2 = 41$.

Exercise 6.6.2 — Rearranging circle equations

Q1 **a)** Complete the square for the x's and y's:
$x^2 + y^2 + 2x - 4y - 3 = 0 \Rightarrow x^2 + 2x + y^2 - 4y - 3 = 0$
$\Rightarrow (x + 1)^2 - 1 + (y - 2)^2 - 4 - 3 = 0 \Rightarrow (x + 1)^2 + (y - 2)^2 = 8$
Centre is (–1, 2).

b) Radius $= \sqrt{8} = \sqrt{2 \times 4} = 2\sqrt{2}$

Q2 **a)** Complete the square for the x's and y's:
$x^2 + y^2 - 3x + 1 = 0 \Rightarrow x^2 - 3x + y^2 + 1 = 0$
$\Rightarrow \left(x - \frac{3}{2}\right)^2 - \frac{9}{4} + y^2 + 1 = 0 \Rightarrow \left(x - \frac{3}{2}\right)^2 + y^2 = \frac{5}{4}$
Centre is $\left(\frac{3}{2}, 0\right)$

b) Radius $= \sqrt{\frac{5}{4}} = \frac{\sqrt{5}}{2}$

Q3 **a)** Complete the square for the x's and y's:
$x^2 + y^2 + 2x - 6y - 6 = 0 \Rightarrow x^2 + 2x + y^2 - 6y - 6 = 0$
$\Rightarrow (x + 1)^2 - 1 + (y - 3)^2 - 9 - 6 = 0 \Rightarrow (x + 1)^2 + (y - 3)^2 = 16$
Radius = 4, centre is (–1, 3).

b) Complete the square for the x's and y's:
$x^2 + y^2 - 2y - 4 = 0 \Rightarrow x^2 + (y - 1)^2 - 1 - 4 = 0$
$\Rightarrow x^2 + (y - 1)^2 = 5$
Radius $= \sqrt{5}$, centre is (0, 1)

c) Complete the square for the x's and y's:
$x^2 + y^2 - 6x - 4y = 12 \Rightarrow x^2 - 6x + y^2 - 4y = 12$
$\Rightarrow (x - 3)^2 - 9 + (y - 2)^2 - 4 = 12 \Rightarrow (x - 3)^2 + (y - 2)^2 = 25$
Radius = 5, centre is (3, 2)

d) Complete the square for the x's and y's:
$x^2 + y^2 - 10x + 6y + 13 = 0 \Rightarrow x^2 - 10x + y^2 + 6y + 13 = 0$
$\Rightarrow (x - 5)^2 - 25 + (y + 3)^2 - 9 + 13 = 0$
$\Rightarrow (x - 5)^2 + (y + 3)^2 = 21$
Radius $= \sqrt{21}$, centre is (5, –3)

e) Complete the square for the x's and y's:
$x^2 + y^2 + 14x - 8y - 1 = 0 \Rightarrow x^2 + 14x + y^2 - 8y - 1 = 0$
$\Rightarrow (x + 7)^2 - 49 + (y - 4)^2 - 16 - 1 = 0$
$\Rightarrow (x + 7)^2 + (y - 4)^2 = 66$
Radius $= \sqrt{66}$, centre is (–7, 4)

f) Complete the square for the x's and y's:
$x^2 + y^2 - 4x + y = 3.75 \Rightarrow x^2 - 4x + y^2 + y = 3.75$
$\Rightarrow (x-2)^2 - 4 + (y+0.5)^2 - 0.25 = 3.75$
$\Rightarrow (x-2)^2 + (y+0.5)^2 = 8$
Radius = $2\sqrt{2}$, centre is $(2, -0.5)$

g) Complete the square for the x's and y's:
$x^2 + y^2 + 2x + 3y - 1.25 = 0 \Rightarrow x^2 + 2x + y^2 + 3y - 1.25 = 0$
$\Rightarrow (x+1)^2 - 1 + (y+1.5)^2 - 2.25 - 1.25 = 0$
$\Rightarrow (x+1)^2 + (y+1.5)^2 = 4.5$
Radius = $\frac{3\sqrt{2}}{2}$, centre is $(-1, -1.5)$

h) Expand the brackets and simplify:
$x^2 - 4x + 4 + y^2 + 2x + 4y - 12 = 0$
$\Rightarrow x^2 - 2x + y^2 + 4y - 8 = 0$
Complete the square for the x's and y's:
$(x-1)^2 - 1 + (y+2)^2 - 4 - 8 = 0 \Rightarrow (x-1)^2 + (y+2)^2 = 13$
Radius = $\sqrt{13}$, centre is $(1, -2)$

Q4 The circle has equation $x^2 + (y+3)^2 = 10$
Expand the brackets: $x^2 + y^2 + 6y + 9 = 10 \Rightarrow x^2 + y^2 + 6y - 1 = 0$
So $f = 0$, $g = 3$ and $c = -1$.

Q5 Find the radius of the circle:
Radius = $\sqrt{(-4)^2 + 2^2} = \sqrt{20} = 2\sqrt{5}$
So the circle has equation $(x+4)^2 + (y-2)^2 = 20$
Expand the brackets: $x^2 + 8x + 16 + y^2 - 4y + 4 = 20$
$\Rightarrow x^2 + y^2 + 8x - 4y = 0$
So $f = 4$, $g = -2$ and $c = 0$.

Q6 a) The circle has centre $(-3, 1)$ and radius 4,
so the equation is $(x+3)^2 + (y-1)^2 = 16$
Expand the brackets: $x^2 + 6x + 9 + y^2 - 2y + 1 = 16$
$\Rightarrow x^2 + y^2 + 6x - 2y - 6 = 0$
So $f = 3$, $g = -1$ and $c = -6$.

b) The circle has centre $(3, -3)$ and goes through $(0, 0)$ so the radius = $\sqrt{3^2 + (-3)^2} = \sqrt{18} = 3\sqrt{2}$
So the equation is $(x-3)^2 + (y+3)^2 = 18$
Expand the brackets:
$x^2 - 6x + 9 + y^2 + 6y + 9 = 18 \Rightarrow x^2 + y^2 - 6x + 6y = 0$
So $f = -3$, $g = 3$ and $c = 0$.

Q7 Divide by 2 to simplify: $x^2 + y^2 + 8x - 4y = 1$
Complete the square for the x's and y's:
$x^2 + 8x + y^2 - 4y - 1 = 0 \Rightarrow (x+4)^2 - 16 + (y-2)^2 - 4 - 1 = 0$
$\Rightarrow (x+4)^2 + (y-2)^2 = 21$
Radius = $\sqrt{21}$, centre is $(-4, 2)$

Q8 a) Complete the square for the x's and y's:

$x^2 + y^2 + 6ax - 7y - \frac{3}{4} = 0 \Rightarrow x^2 + 6ax + y^2 - 7y - \frac{3}{4} = 0$

$\Rightarrow (x+3a)^2 - 9a^2 + (y - \frac{7}{2})^2 - \frac{49}{4} - \frac{3}{4} = 0$

$\Rightarrow (x+3a)^2 + (y - \frac{7}{2})^2 = 13 + 9a^2$

Radius = $\sqrt{13 + 9a^2}$, centre is $(-3a, \frac{7}{2})$

b) The point $\left(3, -\frac{1}{2}\right)$ is on the circle,

so $(3 + 3a)^2 + \left(-\frac{1}{2} - \frac{7}{2}\right)^2 = 13 + 9a^2$

$\Rightarrow 9 + 18a + 9a^2 + 16 = 13 + 9a^2 \Rightarrow 18a = -12 \Rightarrow a = -\frac{2}{3}$

Exercise 6.6.3 — Using circle properties

Q1 a) Centre is $(3, 1)$

b) Gradient of radius = $\frac{1-4}{3-4} = \frac{-3}{-1} = 3$

c) Gradient of the tangent is $-\frac{1}{3}$,
use $y - y_1 = m(x - x_1)$ to find equation of tangent:

$y - 4 = -\frac{1}{3}(x - 4) \Rightarrow 3y - 12 = -x + 4 \Rightarrow x + 3y = 16$

You're asked for the equation in a particular form, so don't forget to rearrange it.

Q2 a) Centre of the circle is $(3, -4)$

Gradient of radius = $\frac{-4-(-10)}{3-2} = 6$

Gradient of the tangent = $-\frac{1}{6}$

Use $y - y_1 = m(x - x_1)$ to find equation of tangent:
$y - (-10) = -\frac{1}{6}(x - 2) \Rightarrow 6y + 60 = -x + 2$
$\Rightarrow x + 6y + 58 = 0$

b) Rearrange $x^2 + y^2 + 8x - 9 = 0$ and complete the square for the x terms to get: $(x+4)^2 + y^2 = 25$
Centre of the circle is $(-4, 0)$
Gradient of radius = $\frac{0-4}{-4-(-7)} = \frac{-4}{3}$

Gradient of the tangent = $\frac{3}{4}$

Use $y - y_1 = m(x - x_1)$ to find equation of tangent:
$y - 4 = \frac{3}{4}(x - (-7)) \Rightarrow y - 4 = \frac{3}{4}x + \frac{21}{4}$
$\Rightarrow 4y - 16 = 3x + 21 \Rightarrow 3x - 4y + 37 = 0$

c) Rearrange $x^2 + y^2 - 6x + 10y = 7$ and complete the square for the x terms: $(x-3)^2 + (y+5)^2 = 41$
Centre of the circle is $(3, -5)$

Gradient of radius = $\frac{-1-(-5)}{8-3} = \frac{4}{5}$

Gradient of the tangent = $-\frac{5}{4}$

Use $y - y_1 = m(x - x_1)$ to find equation of tangent:

$y - (-1) = -\frac{5}{4}(x - 8) \Rightarrow y + 1 = -\frac{5}{4}(x - 8)$

$\Rightarrow 4y + 4 = -5x + 40 \Rightarrow 5x + 4y - 36 = 0$

d) Centre of the circle is $(8, -2)$

Gradient of radius = $\frac{-2-(-5)}{8-6} = \frac{3}{2}$

Gradient of the tangent = $-\frac{2}{3}$

Use $y - y_1 = m(x - x_1)$ to find equation of tangent:
$y - (-5) = -\frac{2}{3}(x - 6) \Rightarrow 3y + 15 = -2x + 12$
$\Rightarrow 2x + 3y + 3 = 0$

e) Centre of the circle is $(4, -2)$

Gradient of radius = $\frac{-2-(-3)}{4-9} = -\frac{1}{5}$

Gradient of the tangent = 5

Use $y - y_1 = m(x - x_1)$ to find equation of tangent:
$y - (-3) = 5(x - 9) \Rightarrow y + 3 = 5x - 45 \Rightarrow 5x - y - 48 = 0$

f) Rearrange $x^2 + y^2 + 16x - 6y + 28 = 0$
and complete the square for the x terms:
$(x+8)^2 + (y-3)^2 = 45$
Centre of the circle is $(-8, 3)$

Gradient of radius = $\frac{9-3}{-11-(-8)} = \frac{6}{-3} = -2$

Gradient of the tangent = $\frac{1}{2}$

Use $y - y_1 = m(x - x_1)$ to find equation of tangent:
$y - 9 = \frac{1}{2}(x - (-11)) \Rightarrow 2y - 18 = x + 11 \Rightarrow x - 2y + 29 = 0$

g) Centre of the circle is $(-3, -1)$

Gradient of radius = $\frac{-4-(-1)}{-5-(-3)} = \frac{-3}{-2} = \frac{3}{2}$

Gradient of the tangent = $-\frac{2}{3}$

Use $y - y_1 = m(x - x_1)$ to find equation of tangent:
$y - (-4) = -\frac{2}{3}(x - (-5)) \Rightarrow 3y + 12 = -2x - 10$
$\Rightarrow 2x + 3y + 22 = 0$

h) Rearrange $x^2 + y^2 - 14x - 12y = 61$ and complete the square for the x terms: $(x - 7)^2 + (y - 6)^2 = 146$
Centre of the circle is $(7, 6)$

Gradient of radius $= \frac{6-1}{7-(-4)} = \frac{5}{11}$

Gradient of the tangent $= -\frac{11}{5}$

Use $y - y_1 = m(x - x_1)$ to find equation of tangent:

$y - 1 = -\frac{11}{5}(x - (-4)) \Rightarrow 5y - 5 = -11x - 44$
$\Rightarrow 11x + 5y + 39 = 0$

Q3 Centre of the circle is $(-1, 2)$
Gradient of radius $= \frac{2-(-1)}{-1-(-3)} = \frac{3}{2}$

Gradient of the tangent $= -\frac{2}{3}$

Use $y - y_1 = m(x - x_1)$ to find equation of tangent:

$y - (-1) = -\frac{2}{3}(x - (-3)) \Rightarrow 3y + 3 = -2(x + 3)$

$\Rightarrow 3y + 3 = -2x - 6 \Rightarrow 2x + 3y + 9 = 0$

Q4 Rearrange $x^2 + y^2 + 2x - 7 = 0$ and complete the square for the x terms to get: $(x + 1)^2 + y^2 = 8$
Centre of the circle is $(-1, 0)$
Gradient of radius $= \frac{0-2}{-1-(-3)} = \frac{-2}{2} = -1$
Gradient of the tangent $= 1$

Use $y - y_1 = m(x - x_1)$ to find equation of tangent:

$y - 2 = 1(x - (-3)) \Rightarrow y - 2 = x + 3 \Rightarrow y = x + 5$
You can give the equation in any form because it's not specified in the question.

Q5 Rearrange $x^2 + y^2 + 2x + 4y = 5$ and complete the square for the x and y terms to get: $(x + 1)^2 + (y + 2)^2 = 10$
Centre of the circle is $(-1, -2)$

Gradient of the radius $= \frac{-2-(-5)}{-1-0} = \frac{3}{-1} = -3$

Gradient of the tangent $= \frac{1}{3}$

Use $y - y_1 = m(x - x_1)$ to find equation of tangent:
$y - (-5) = \frac{1}{3}(x - 0) \Rightarrow 3y + 15 = x \Rightarrow x - 3y = 15$

Q6 Centre of the circle is $(2, 1)$

Gradient of radius $= \frac{7-1}{10-2} = \frac{6}{8} = \frac{3}{4}$

Gradient of the tangent $= -\frac{4}{3}$

Use $y - y_1 = m(x - x_1)$ to find equation of tangent:
$y - 7 = -\frac{4}{3}(x - 10) \Rightarrow 3y - 21 = -4x + 40 \Rightarrow 4x + 3y - 61 = 0$

Q7 The line from the centre of the circle to A has gradient: $\frac{1-4}{n-(-2)} = -\frac{3}{n+2}$
Since the tangent at A is perpendicular to this line,

$m_1 \times m_2 = -1 \Rightarrow -\frac{3}{n+2} \times \frac{5}{3} = -1 \Rightarrow -\frac{5}{n+2} = -1$

$\Rightarrow -5 = -1(n + 2) \Rightarrow n + 2 = 5 \Rightarrow n = 3$

You could also find the equation of this radius in the form $y = mx + c$, then substitute in the point $(n, 1)$ to find n.

Q8 a) Radius $= \sqrt{(4-3)^2 + (3-(-1))^2} = \sqrt{1^2 + 4^2} = \sqrt{17}$
So the equation is $(x - 3)^2 + (y + 1)^2 = 17$.

b) Gradient of radius at $(4, 3) = \frac{3-(-1)}{4-3} = 4$

Gradient of the tangent $= -\frac{1}{4}$

Use $y - y_1 = m(x - x_1)$ to find equation of tangent:
$y - 3 = -\frac{1}{4}(x - 4) \Rightarrow y - 3 = -\frac{1}{4}x + 1 \Rightarrow y = -\frac{1}{4}x + 4$
The tangent crosses the x-axis at $x = 16$ and crosses the y-axis at $y = 4$.
Area of triangle $= \frac{1}{2} \times$ base $\times$ height

$= \frac{1}{2} \times 16 \times 4 = 32$ units^2

Q9 a) The line l is perpendicular to the chord AB.
So find the gradient of AB:

Gradient of AB $= \frac{1-7}{-1-(-3)} = -\frac{6}{2} = -3$

So the gradient of l is $\frac{1}{3}$.
Then sub the gradient of l and point M $(-1, 1)$ into $y - y_1 = m(x - x_1)$ to find the equation:

$y - 1 = \frac{1}{3}(x - (-1)) \Rightarrow 3y - 3 = x + 1 \Rightarrow x - 3y + 4 = 0$

b) The centre is $(2, 2)$, so $a = 2$ and $b = 2$ in the equation $(x - a)^2 + (y - b)^2 = r^2$.
The radius is the length CA, which can be found using Pythagoras:

A (–3, 7)
r
5
5
C (2, 2)

$r^2 = CA^2 = (-3 - 2)^2 + (7 - 2)^2 = (-5)^2 + 5^2 = 50$
So the equation of the circle is: $(x - 2)^2 + (y - 2)^2 = 50$

Q10 a) AB is a diameter so the chords BC and CA must be perpendicular (i.e. $m_{CA} \times m_{BC} = -1$):

$m_{BC} = \frac{-1-1}{4-0} = \frac{-2}{4} = -\frac{1}{2}$

$m_{CA} = \frac{1-a}{0-2} = \frac{a-1}{2}$

$m_{CA} \times m_{BC} = -1 \Rightarrow \frac{a-1}{2} \times -\frac{1}{2} = -1 \Rightarrow -\frac{a-1}{4} = -1$

$\Rightarrow a - 1 = 4 \Rightarrow a = 5$

b) The centre of the circle is the midpoint of the diameter AB, and the radius is half the length AB.

Centre $= \left(\frac{2+4}{2}, \frac{5-1}{2}\right) = (3, 2)$

Radius $= \frac{1}{2}\sqrt{(4-2)^2 + (-1-5)^2}$

$= \frac{1}{2}\sqrt{2^2 + (-6)^2} = \frac{1}{2}\sqrt{40} = \sqrt{10}$
So the equation of the circle is:
$(x - 3)^2 + (y - 2)^2 = 10$

Q11 a) If AC is the diameter, the angle ABC will be 90°.
So find out if AB and BC are perpendicular.

Gradient of AB $= \frac{14-12}{4-(-2)} = \frac{2}{6} = \frac{1}{3}$

Gradient of BC $= \frac{2-14}{8-4} = \frac{-12}{4} = -3$

For perpendicular lines: $m_1 \times m_2 = -1$

$\frac{1}{3} \times -3 = -1$

As AB and BC are perpendicular, the angle ABC must be 90° and so AC must be the diameter of the circle.

b) Since AC is a diameter, the centre of the circle must be the midpoint of A and C.
$\left(\frac{(-2)+8}{2}, \frac{12+2}{2}\right) = \left(\frac{6}{2}, \frac{14}{2}\right) = (3, 7)$
The radius is the distance from the centre to a point on the circle, so choose a point and use Pythagoras' theorem (here using C: $(8 - 3)^2 + (2 - 7)^2 = r^2 \Rightarrow r^2 = 5^2 + 5^2 = 50$
So the equation of the circle is: $(x - 3)^2 + (y - 7)^2 = 50$

Q12 PQ is a chord of the circle, so the line through M and C is the perpendicular bisector of PQ.

Midpoint of PQ = $\left(\frac{-8-3}{2}, \frac{6+7}{2}\right) = \left(\frac{-11}{2}, \frac{13}{2}\right)$

Gradient of PQ = $\frac{7-6}{-3-(-8)} = \frac{1}{5}$

So the gradient of the perpendicular bisector is –5, and it passes through $\left(\frac{-11}{2}, \frac{13}{2}\right)$.

$y - y_1 = m(x - x_1) \Rightarrow y - \frac{13}{2} = -5(x + \frac{11}{2})$

$\Rightarrow y - \frac{13}{2} = -5x - \frac{55}{2} \Rightarrow 5x + y + 21 = 0$

Q13 Find the perpendicular bisectors of two of the line segments AB, BC and CA.

AB: midpoint = $\left(\frac{2+6}{2}, \frac{11+5}{2}\right) = \left(\frac{8}{2}, \frac{16}{2}\right) = (4, 8)$

gradient = $\frac{5-11}{6-2} = \frac{-6}{4} = -\frac{3}{2}$

So the perpendicular bisector passes through (4, 8) and has a gradient of $\left(-1 \div -\frac{3}{2}\right) = \frac{2}{3}$.

$y - y_1 = m(x - x_1) \Rightarrow y - 8 = \frac{2}{3}(x - 4)$

$\Rightarrow y = \frac{2}{3}x - \frac{8}{3} + 8 \Rightarrow y = \frac{2}{3}x + \frac{16}{3}$

BC: midpoint = $\left(\frac{6+(-9)}{2}, \frac{5+0}{2}\right) = \left(-\frac{3}{2}, \frac{5}{2}\right)$

gradient = $\frac{0-5}{(-9)-6} = \frac{-5}{-15} = \frac{1}{3}$

So the perpendicular bisector passes through $\left(-\frac{3}{2}, \frac{5}{2}\right)$ and has a gradient of $\left(-1 \div \frac{1}{3}\right) = -3$.

$y - y_1 = m(x - x_1) \Rightarrow y - \frac{5}{2} = (-3)\left(x - \left(-\frac{3}{2}\right)\right)$

$\Rightarrow y = -3x - \frac{9}{2} + \frac{5}{2} \Rightarrow y = -3x - 2$

Now find the centre of the circle, which is the point where these lines intersect:

$\frac{2}{3}x + \frac{16}{3} = -3x - 2 \Rightarrow 2x + 16 = -9x - 6$
$\Rightarrow 11x = -22 \Rightarrow x = -2$

Substitute $x = -2$ into $y = -3x - 2$:
$y = -3(-2) - 2 = 6 - 2 = 4$

So the circle has its centre at (–2, 4).

You could have used the perpendicular bisector of CA here to find the centre of the circle — it has equation $y = 2 - x$.

The radius is the distance from the centre to one of the points — here we use B:

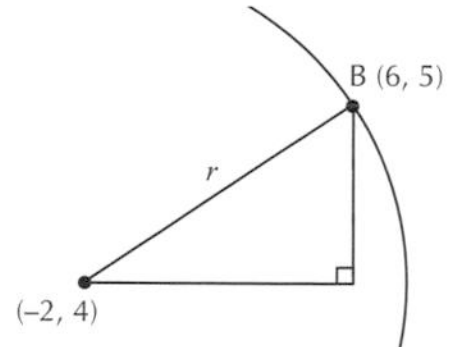

Use Pythagoras' theorem to find r^2:
$(6 - (-2))^2 + (5 - 4)^2 = r^2 \Rightarrow 8^2 + 1^2 = r^2 \Rightarrow r^2 = 65$

So the equation of the circle is: $(x + 2)^2 + (y - 4)^2 = 65$.

Q14 a) Find the perpendicular bisectors of two of the line segments XY, YZ and ZX.

XY: midpoint = $\left(\frac{8-4}{2}, \frac{2-4}{2}\right) = (2, -1)$

gradient = $\frac{-4-2}{-4-8} = \frac{-6}{-12} = \frac{1}{2}$

So the perpendicular bisector passes through (2, –1) and has a gradient of –2.

$y - y_1 = m(x - x_1) \Rightarrow y + 1 = -2(x - 2) \Rightarrow y = -2x + 3$

YZ: midpoint = $\left(\frac{-4+2}{2}, \frac{-4+8}{2}\right) = (-1, 2)$

gradient = $\frac{8-(-4)}{2-(-4)} = \frac{12}{6} = 2$

So the perpendicular bisector passes through (–1, 2) and has a gradient of $-\frac{1}{2}$.

$y - y_1 = m(x - x_1) \Rightarrow y - 2 = -\frac{1}{2}(x + 1) \Rightarrow y = -\frac{1}{2}x + \frac{3}{2}$

Now find the centre of the circle, which is the point where these lines intersect:

$-\frac{1}{2}x + \frac{3}{2} = -2x + 3 \Rightarrow -x + 3 = -4x + 6$
$\Rightarrow 3x = 3 \Rightarrow x = 1$

Substitute $x = 1$ into $y = -2x + 3$: $y = -2(1) + 3 = 1$ so the circle has its centre at (1, 1).

Radius = $\sqrt{(8-1)^2 + (2-1)^2} = \sqrt{49+1} = \sqrt{50}$

So the equation of the circle is: $(x - 1)^2 + (y - 1)^2 = 50$.

b) Find the perpendicular bisectors of two of the line segments XY, YZ and ZX.

XY: midpoint = $\left(\frac{2+5}{2}, \frac{6+9}{2}\right) = \left(\frac{7}{2}, \frac{15}{2}\right)$

gradient = $\frac{9-6}{5-2} = \frac{3}{3} = 1$

So the perpendicular bisector passes through $\left(\frac{7}{2}, \frac{15}{2}\right)$ and has a gradient of –1.

$y - y_1 = m(x - x_1) \Rightarrow y - \frac{15}{2} = -1(x - \frac{7}{2}) \Rightarrow y = -x + 11$

YZ: midpoint = $\left(\frac{5-5}{2}, \frac{9+9}{2}\right) = (0, 9)$

gradient = $\frac{9-9}{-5-5} = 0$

YZ is a horizontal line, so the perpendicular bisector is vertical and passes through (0, 9) — i.e. $x = 0$.

Now find the centre of the circle, which is the point where these lines intersect.
Substitute $x = 0$ into $y = -x + 11 \Rightarrow y = 11$.
So the circle has its centre at (0, 11).

Radius = $\sqrt{(2-0)^2 + (6-11)^2} = \sqrt{2^2 + (-5)^2} = \sqrt{29}$

So the equation of the circle is: $x^2 + (y - 11)^2 = 29$.

c) Find the perpendicular bisectors of two of the line segments XY, YZ and ZX.

XY: midpoint = $\left(\frac{14+22}{2}, \frac{14+6}{2}\right) = (18, 10)$

gradient = $\frac{14-6}{14-22} = \frac{8}{-8} = -1$

So the perpendicular bisector passes through (18, 10) and has a gradient of 1.

$y - y_1 = m(x - x_1) \Rightarrow y - 10 = x - 18 \Rightarrow y = x - 8$

YZ: midpoint = $\left(\frac{22+22}{2}, \frac{6+10}{2}\right) = (22, 8)$

The x-coordinates of Y and Z are the same, so YZ is a vertical line.

So the perpendicular bisector is horizontal and passes through (22, 8) — i.e. $y = 8$.

Now find the centre of the circle, which is the point where these lines intersect.
Substitute $y = 8$ into $y = x - 8 \Rightarrow x = 16$.
So the circle has its centre at (16, 8).

Radius = $\sqrt{(16-22)^2 + (8-6)^2} = \sqrt{(-6)^2 + 2^2} = \sqrt{40}$

So the equation of the circle is: $(x - 16)^2 + (y - 8)^2 = 40$.

d) Find the perpendicular bisectors of two of the line segments XY, YZ and ZX.

XY: midpoint = $\left(\frac{17-7}{2}, \frac{-14+10}{2}\right) = (5, -2)$

gradient = $\frac{-14-10}{17-(-7)} = \frac{-24}{24} = -1$

So the perpendicular bisector passes through (5, –2) and has a gradient of 1.

$y - y_1 = m(x - x_1) \Rightarrow y + 2 = x - 5 \Rightarrow y = x - 7$

YZ: midpoint $= \left(\frac{-7-11}{2}, \frac{10-2}{2}\right) = (-9, 4)$

gradient $= \frac{10-(-2)}{-7-(-11)} = \frac{12}{4} = 3$

So the perpendicular bisector passes through (–9, 4) and has a gradient of $-\frac{1}{3}$.

$y - y_1 = m(x - x_1) \Rightarrow y - 4 = -\frac{1}{3}(x + 9) \Rightarrow y = -\frac{1}{3}x + 1$

Now find the centre of the circle, which is the point where these lines intersect:

$-\frac{1}{3}x + 1 = x - 7 \Rightarrow -x + 3 = 3x - 21$
$\Rightarrow 4x = 24 \Rightarrow x = 6$

Substitute $x = 6$ into $y = x - 7$: $y = 6 - 7 = -1$
so the circle has its centre at (6, –1).

Radius $= \sqrt{(6-(-7))^2 + (-1-10)^2}$
$= \sqrt{13^2 + 11^2} = \sqrt{290}$

So the equation of the circle is: $(x - 6)^2 + (y + 1)^2 = 290$.

Q15 a) Find the perpendicular bisectors of two of the line segments AB, BC and CA.

AB: midpoint $= \left(\frac{2+11}{2}, \frac{-2+1}{2}\right) = \left(\frac{13}{2}, -\frac{1}{2}\right)$

gradient $= \frac{1-(-2)}{11-2} = \frac{3}{9} = \frac{1}{3}$

So the perpendicular bisector passes through $\left(\frac{13}{2}, -\frac{1}{2}\right)$ and has a gradient of –3.

$y - y_1 = m(x - x_1) \Rightarrow y + \frac{1}{2} = -3(x - \frac{13}{2}) \Rightarrow y = -3x + 19$

BC: midpoint $= \left(\frac{11+10}{2}, \frac{1-6}{2}\right) = \left(\frac{21}{2}, -\frac{5}{2}\right)$

gradient $= \frac{1-(-6)}{11-10} = 7$

So the perpendicular bisector passes through $\left(\frac{21}{2}, -\frac{5}{2}\right)$ and has a gradient of $-\frac{1}{7}$.

$y - y_1 = m(x - x_1) \Rightarrow y + \frac{5}{2} = -\frac{1}{7}(x - \frac{21}{2})$
$\Rightarrow y = -\frac{1}{7}x - 1$

Now find the centre of the circle, which is the point where these lines intersect:

$-\frac{1}{7}x - 1 = -3x + 19 \Rightarrow -x - 7 = -21x + 133$
$\Rightarrow 20x = 140 \Rightarrow x = 7$

Substitute $x = 7$ into $y = -3x + 19$: $y = -3(7) + 19 = -2$
so the circle has its centre at (7, –2).

Radius $= \sqrt{(7-2)^2 + (-2-(-2))^2} = \sqrt{5^2} = 5$

So the equation of the circle is: $(x - 7)^2 + (y + 2)^2 = 25$.

b) If BD is a diameter, BA and AD are perpendicular.
Find the gradients of BA and AD:

$m_{BA} = \frac{1}{3}$ from part a).

$m_{AD} = \frac{-2-(-5)}{2-3} = \frac{3}{-1} = -3$

$-3 \times \frac{1}{3} = -1$, so BA and AD are perpendicular.
Therefore BD is a diameter.

You could have used BC and CD instead, or for centre X, you could have shown that $m_{BX} = m_{XD}$

Review Exercise — Chapter 6

Q1 a) (i) gradient $= \frac{-1-(-19)}{2-(-4)} = \frac{18}{6} = 3$

$y - y_1 = m(x - x_1) \Rightarrow y - (-1) = 3(x - 2)$
$\Rightarrow y + 1 = 3(x - 2)$

You could have used (–4, –19) instead — this gives the equation $y + 19 = 3(x + 4)$.

(ii) $y + 1 = 3(x - 2) \Rightarrow y = 3x - 6 - 1 \Rightarrow y = 3x - 7$

(iii) $3x - y - 7 = 0$

b) (i) gradient $= \frac{\frac{2}{3} - \left(-\frac{1}{3}\right)}{5-0} = \frac{1}{5}$

$y - y_1 = m(x - x_1) \Rightarrow y - \left(-\frac{1}{3}\right) = \frac{1}{5}x \Rightarrow y + \frac{1}{3} = \frac{1}{5}x$

(Or $y - \frac{2}{3} = \frac{1}{5}(x - 5)$ using the other point.)

(ii) $y + \frac{1}{3} = \frac{1}{5}x \Rightarrow y = \frac{1}{5}x - \frac{1}{3}$

(iii) $y = \frac{1}{5}x - \frac{1}{3} \Rightarrow 15y = 3x - 5 \Rightarrow 3x - 15y - 5 = 0$

c) (i) gradient $= \frac{-2-7}{-7-8} = \frac{-9}{-15} = \frac{3}{5}$

$y - y_1 = m(x - x_1) \Rightarrow y - 7 = \frac{3}{5}(x - 8)$

(Or $y + 2 = \frac{3}{5}(x + 7)$ using the other point.)

(ii) $y - 7 = \frac{3}{5}(x - 8) \Rightarrow y = \frac{3}{5}x - \frac{24}{5} + 7$
$\Rightarrow y = \frac{3}{5}x + \frac{11}{5}$

(iii) $y = \frac{3}{5}x + \frac{11}{5} \Rightarrow 5y = 3x + 11 \Rightarrow 3x - 5y + 11 = 0$

d) (i) gradient $= \frac{\frac{5}{2} - 5}{2-5} = \frac{-\frac{5}{2}}{-3} = \frac{5}{6}$

$y - y_1 = m(x - x_1) \Rightarrow y - 5 = \frac{5}{6}(x - 5)$

(Or $y - \frac{5}{2} = \frac{5}{6}(x - 2)$ using the other point.)

(ii) $y - 5 = \frac{5}{6}(x - 5) \Rightarrow y = \frac{5}{6}x - \frac{25}{6} + 5$
$\Rightarrow y = \frac{5}{6}x + \frac{5}{6}$

(iii) $y = \frac{5}{6}x + \frac{5}{6} \Rightarrow 6y = 5x + 5 \Rightarrow 5x - 6y + 5 = 0$

e) (i) gradient $= \frac{2-0}{1.3-1.8} = \frac{2}{-0.5} = -4$

$y - y_1 = m(x - x_1) \Rightarrow y = -4(x - 1.8)$

(Or $y - 2 = 4(x - 1.3)$ using the other point.)

(ii) $y = -4(x - 1.8) \Rightarrow y = -4x + 7.2$

(iii) $y = -4x + 7.2 \Rightarrow y = -4x + \frac{36}{5}$
$\Rightarrow 5y = -20x + 36 \Rightarrow 20x + 5y - 36 = 0$

f) (i) gradient $= \frac{-2.3-(-0.3)}{4.6-(-5.4)} = \frac{-2}{10} = -0.2$

$y - y_1 = m(x - x_1) \Rightarrow y - (-0.3) = -0.2(x - (-5.4))$
$\Rightarrow y + 0.3 = -0.2(x + 5.4)$

(Or $y + 2.3 = -0.2(x - 4.6)$ using other point.)

(ii) $y + 0.3 = -0.2(x + 5.4) \Rightarrow y + 0.3 = -0.2x - 1.08$
$\Rightarrow y = -0.2x - 1.38$

(iii) $y = -0.2x - 1.38 \Rightarrow y = -\frac{1}{5}x - \frac{69}{50}$
$\Rightarrow 50y = -10x - 69 \Rightarrow 10x + 50y + 69 = 0$

Q2 a) l_1 is parallel so its gradient is also $\frac{3}{2}$:

$y = mx + c \Rightarrow y = \frac{3}{2}x + c \Rightarrow 2 = \frac{3}{2}(4) + c$
$\Rightarrow c = -4 \Rightarrow y = \frac{3}{2}x - 4$

b) Rearrange the equation to get $y = 2x - 7$.
l_2 is perpendicular so the gradient is $-1 \div 2 = -\frac{1}{2}$

$y = mx + c \Rightarrow y = -\frac{1}{2}x + c \Rightarrow 1 = -\frac{1}{2}(6) + c$
$\Rightarrow c = 4 \Rightarrow y = -\frac{1}{2}x + 4$

Q3 Gradient of RS = $\frac{9-3}{1-10} = \frac{6}{-9} = -\frac{2}{3}$,
so the gradient of the perpendicular is $\frac{3}{2}$.

$y = mx + c \Rightarrow y = \frac{3}{2}x + c \Rightarrow 9 = \frac{3}{2}(1) + c$

$\Rightarrow c = \frac{15}{2} \Rightarrow y = \frac{3}{2}x + \frac{15}{2}$

Q4 y is directly proportional to x, so $y = kx$.
When $y = 3$, $x = 4.5$, so $3 = 4.5k \Rightarrow k = \frac{2}{3}$.

a) When $x = 21$, $y = \frac{2}{3} \times 21 = 14$

b) When $x = -3$, $y = \frac{2}{3} \times -3 = -2$

c) $x = \frac{y}{k}$, so when $y = 58$, $x = 58 \div \frac{2}{3} = 87$

Q5 $s = \frac{k}{t^3} \Rightarrow k = st^3$. When $s = 18$, $t = 6$, so $k = 3888$.

a) When $t = 3$, $s = \frac{3888}{3^3} = 144$

b) When $t = 0.5$, $s = \frac{3888}{0.5^3} = 31\ 104$

c) $t = \sqrt[3]{\frac{3888}{s}}$. When $s = 486$, $t = \sqrt[3]{\frac{3888}{486}} = \sqrt[3]{8} = 2$

Q6 **a)**

(4, 0)

b)

(0, 12) (–2, 0) (3, 0)

c) Factorise y fully: $y = (1 - x)(x - 2)(x - 4)$

(0, 8) (1, 0) (2, 0) (4, 0)

d)

(1, 0) (2, 0) (3, 0) (–6, 0)

e) Factorise y fully: $y = 3x^2(x - 2)$

0 (2, 0)

f) Factorise y fully: $y = x(x^2 - x - 12) = x(x - 4)(x + 3)$

(–3, 0) 0 (4, 0)

Q7 **a)**

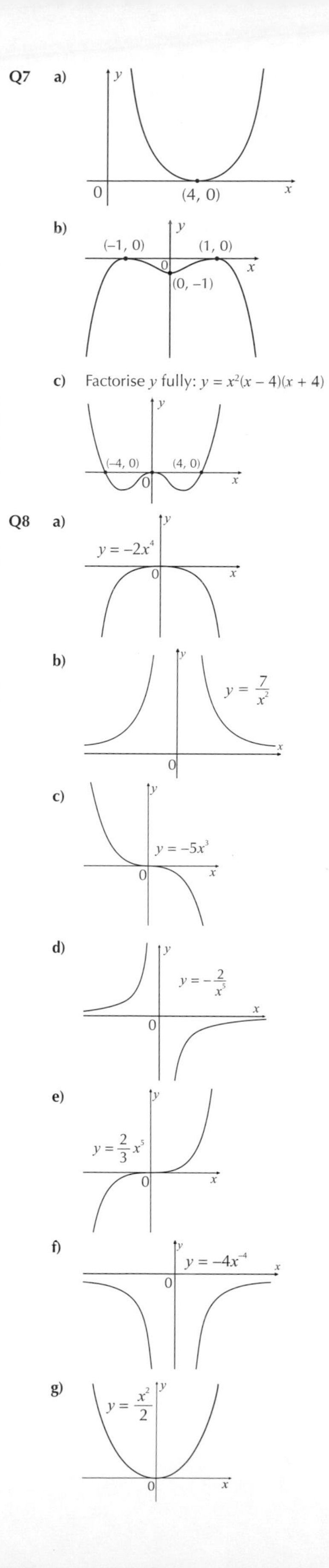

b)

c) Factorise y fully: $y = x^2(x - 4)(x + 4)$

Q8 **a)** **b)** **c)** **d)** **e)** **f)** **g)**

h)

Q9

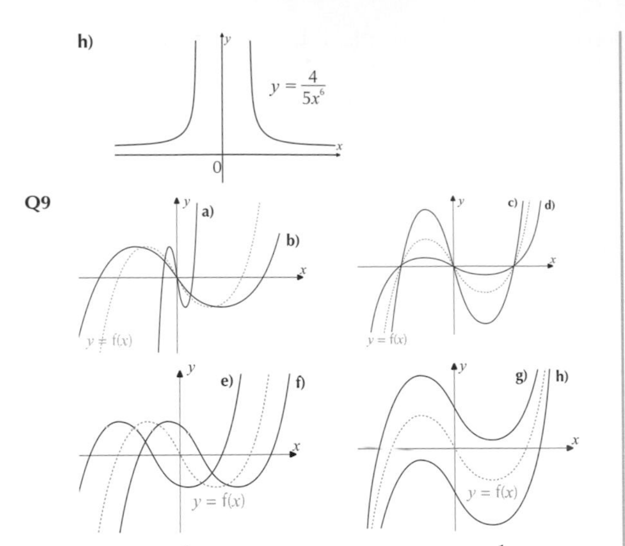

Q10 **a)** $g(x) = -\frac{1}{3}f(x)$ — vertical stretch, scale factor $-\frac{1}{3}$

b) $g(x) = f(\frac{1}{2}x)$ — horizontal stretch, scale factor 2

c) $f(x) = x^2 - 5x + 4$ and $g(x) = x^2 - 5x + 6$

$g(x) = f(x) + 2$ — vertical translation by 2 upwards

Q11 **a)** Since $x = 2$ is a root, $(x - 2)$ is a factor by the Factor Theorem, so use algebraic division (or another suitable method) to factorise $(x - 2)$ out:

$f(x) = x(x^3 - 3x^2 + 4) = x(x - 2)(x^2 - x - 2) = x(x - 2)(x - 2)(x + 1)$

Using this factorisation, sketch the graph:

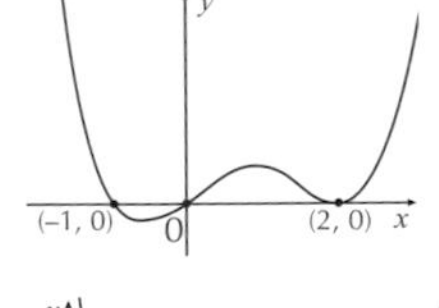

b)

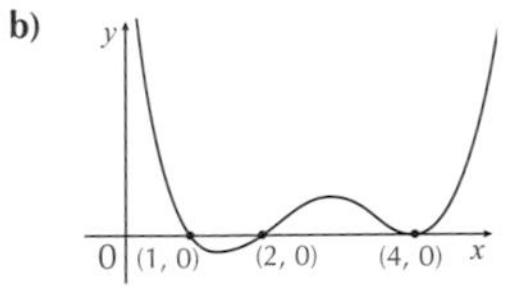

c) $g(x) = (x - 2)^4 - 3(x - 2)^3 + 4(x - 2)$ or

$g(x) = (x - 1)(x - 2)(x - 4)^2$

Substitute (x – 2) in place of x in f(x), or write down g(x) in its factorised form using the roots you know from the graph (or from part a)).

Q12 **a)** **b)** **c)** **d)**

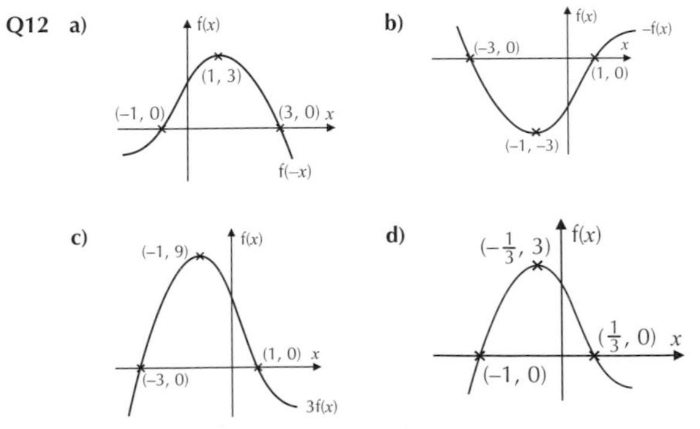

Q13 **a)** $r = 3$, centre (0, 0)

b) $r = 2$, centre (2, –4)

c) Rearrange to get into $(x - a)^2 + (y - b)^2 = r^2$ form:

$x(x + 6) = y(8 - y) \Rightarrow x^2 + 6x + y^2 - 8y = 0$

$\Rightarrow (x + 3)^2 - 9 + (y - 4)^2 - 16 = 0 \Rightarrow (x + 3)^2 + (y - 4)^2 = 25$

So $r = 5$, centre (–3, 4)

Q14 $x^2 + (y + 3)^2 = 14$

Q15 $x^2 + y^2 - 4x + 6y - 68 = 0 \Rightarrow x^2 - 4x + y^2 + 6y - 68 = 0$

$\Rightarrow (x - 2)^2 - 4 + (y + 3)^2 - 9 - 68 = 0 \Rightarrow (x - 2)^2 + (y + 3)^2 = 81$

So the centre is (2, –3) and the radius is 9.

Q16 Rearrange to find the coordinates of the centre:

$x^2 + y^2 - 12x + 2y + 11 = 0 \Rightarrow x^2 - 12x + y^2 + 2y + 11 = 0$

$\Rightarrow (x - 6)^2 - 36 + (y + 1)^2 - 1 + 11 = 0 \Rightarrow (x - 6)^2 + (y + 1)^2 = 26$

So the centre of the circle is (6, –1).

Gradient of the radius: $\frac{-1-(-2)}{6-1} = \frac{1}{5}$

So the gradient of the tangent is –5.

Use $y - y_1 = m(x - x_1)$ to find equation of tangent:

$y - (-2) = -5(x - 1) \Rightarrow y + 2 = -5x + 5 \Rightarrow y = -5x + 3$

Q17

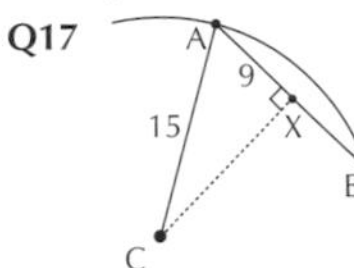

X is the midpoint of AB, so the length of AX is 9. AC is a radius, so the length of AC is 15.

Since CX bisects the chord AB, the angle AXC is 90°, so triangle ACX is a right-angled triangle with hypotenuse AC.

Use Pythagoras' theorem to find CX:

$CX^2 = 15^2 - 9^2 = 144$, so $CX = \sqrt{144} = 12$.

Q18 Find the perpendicular bisectors of two of the line segments AB, BC and CA.

AB: midpoint = $\left(\frac{4.5+2}{2}, \frac{7.5+5}{2}\right) = (3.25, 6.25)$

gradient = $\frac{7.5-5}{4.5-2} = \frac{2.5}{2.5} = 1$

So the perpendicular bisector passes through (3.25, 6.25) and has a gradient of –1.

$y - y_1 = m(x - x_1) \Rightarrow y - 6.25 = -1(x - 3.25) \Rightarrow y = -x + 9.5$

BC: midpoint = $\left(\frac{2+4}{2}, \frac{5+5}{2}\right) = (3, 5)$

gradient = $\frac{5-5}{4-2} = 0$

BC is horizontal so the perpendicular bisector is vertical and passes through (3, 5) — i.e. $x = 3$.

Now find the centre of the circle, which is the point where these lines intersect:

Substitute $x = 3$ into $y = -x + 9.5$: $y = -(3) + 9.5 = 6.5$

so the circle has its centre at (3, 6.5).

Radius = $\sqrt{(3-2)^2+(6.5-5)^2} = \sqrt{1^2+1.5^2} = \sqrt{3.25}$

So the equation of the circle is: $(x - 3)^2 + (y - 6.5)^2 = 3.25$

Q19 **a)** Length = $\sqrt{(-2-3)^2+(4-9)^2}$

$= \sqrt{(-5)^2+(-5)^2} = 5\sqrt{2}$

Midpoint = $\left(\frac{-2+3}{2}, \frac{4+9}{2}\right) = (0.5, 6.5)$

b) Find the perpendicular bisectors of two of the line segments XY, YZ and ZX.

XY: from part a), midpoint = (0.5, 6.5)

gradient = $\frac{9-4}{3-(-2)} = \frac{5}{5} = 1$

So the perpendicular bisector passes through (0.5, 6.5) and has a gradient of –1.

$y - y_1 = m(x - x_1) \Rightarrow y - 6.5 = -(x - 0.5) \Rightarrow y = -x + 7$

YZ: midpoint = $\left(\frac{3+7}{2}, \frac{9+1}{2}\right) = (5, 5)$

gradient = $\frac{9-1}{3-7} = \frac{8}{-4} = -2$

So the perpendicular bisector passes through (5, 5) and has a gradient of $\frac{1}{2}$.

$y - y_1 = m(x - x_1) \Rightarrow y - 5 = \frac{1}{2}(x - 5) \Rightarrow y = \frac{1}{2}x + \frac{5}{2}$

Now find the centre of the circle, which is the point where these lines intersect:

$\frac{1}{2}x + \frac{5}{2} = -x + 7 \Rightarrow x + 5 = -2x + 14$

$\Rightarrow 3x = 9 \Rightarrow x = 3$

Substitute $x = 3$ into $y = -x + 7$: $y = -(3) + 7 = 4$ so the circle has its centre at (3, 4).

Radius = $\sqrt{(3-3)^2 + (9-4)^2} = \sqrt{5^2} = 5$

So the equation of the circle is: $(x - 3)^2 + (y - 4)^2 = 25$

c) CM bisects XY, so it is perpendicular. So CXY has base XY and perpendicular height CM.

Length CM = $\sqrt{(3-0.5)^2 + (4-6.5)^2}$

$= \sqrt{2.5^2 + (-2.5)^2} = \frac{5\sqrt{2}}{2}$

From part a), the length XY = $5\sqrt{2}$.

So the area of CXY = $\frac{1}{2} \times 5\sqrt{2} \times \frac{5\sqrt{2}}{2} = 12.5$ units^2

Alternatively, you might notice that CX and CY are perpendicular, so you can also work out the area of CXY using CX and CY as the base and perpendicular height, both of which have length 5.

Exam-Style Questions — Chapter 6

Q1 a) Length AB = $\sqrt{(4-(-2))^2 + (-10-4)^2}$

$= \sqrt{6^2 + (-14)^2}$

$= \sqrt{232} = 2\sqrt{58}$

[2 marks available — 1 mark for attempting to use Pythagoras' theorem on the x- and y-coordinates, 1 mark for the correct answer in surd form]

b) Gradient $m = \frac{-10-4}{4-(-2)} = \frac{-14}{6} = -\frac{7}{3}$

[2 marks available — 1 mark for change in y over change in x as a fraction, 1 mark for correct answer (or equivalent fraction)]

c) $y - y_1 = m(x - x_1) \Rightarrow y - 4 = -\frac{7}{3}(x - (-2))$

$\Rightarrow y - 4 = -\frac{7}{3}x - \frac{14}{3} \Rightarrow y = -\frac{7}{3}x - \frac{2}{3}$

$\Rightarrow 3y = -7x - 2 \Rightarrow 7x + 3y + 2 = 0$

[3 marks available — 1 mark for substituting a coordinate into an accepted method for the equation of a straight line, 1 mark for a correct line equation in any form, 1 mark for correct final answer as shown with integer coefficients]

Q2 Use the centre of the circle to write the equation in the form $(x - 2)^2 + (y + 3)^2 = r^2$.

Then expand the brackets: $x^2 - 4x + 4 + y^2 + 6y + 9 = r^2$

$\Rightarrow x^2 + y^2 - 4x + 6y + 13 - r^2 = 0$

Comparing this with original equation gives:

$a = -4$, $b = 6$ and $-4 = 13 - r^2 \Rightarrow r^2 = 17 \Rightarrow r = \sqrt{17}$

[4 marks available — 1 mark for correctly substituting into the circle equation, 1 mark for correct value of a, 1 mark for the correct value of b, 1 mark for the correct value of r]

Q3 a) D is the midpoint of AC, i.e. D = (8, 2) = $\left(\frac{1+p}{2}, \frac{7+q}{2}\right)$

So $8 = \frac{1+p}{2} \Rightarrow p = 8 \times 2 - 1 = 15$

and $2 = \frac{7+q}{2} \Rightarrow q = 2 \times 2 - 7 = -3$

[2 marks available — 1 mark for correct value of p, 1 mark for correct value of q]

b) Find the gradient of the line AD: $m_{AD} = \frac{2-7}{8-1} = -\frac{5}{7}$

So the gradient of l is $-1 \div -\frac{5}{7} = \frac{7}{5}$

$y - y_1 = m(x - x_1) \Rightarrow y - 2 = \frac{7}{5}(x - 8) \Rightarrow y - 2 = \frac{7}{5}x - \frac{56}{5}$

$\Rightarrow y = \frac{7}{5}x - \frac{46}{5} \Rightarrow 5y = 7x - 46 \Rightarrow 7x - 5y - 46 = 0$

[4 marks available — 1 mark for finding the gradient of AC, 1 mark for finding the perpendicular gradient, 1 mark for the correct line equation in any form, 1 mark for correct final answer with integer coefficients]

c) The equation of AB is $y = 7$, so substitute this into the equation from b):

$7x - 5(7) - 46 = 0 \Rightarrow 7x = 81 \Rightarrow x = \frac{81}{7}$

[2 marks available — 1 mark for using y = 7, 1 mark for correct answer]

Q4 a) Substitute $(9, p)$ into the circle equation:

$9^2 + p^2 - 10(9) - 7 = 0 \Rightarrow p^2 = 16 \Rightarrow p = 4$

($p > 0$ so ignore the negative root)

[2 marks available — 1 mark for substituting x = 9 into the circle equation, 1 mark for solving the equation for p and taking the positive root]

b) Complete the square to find the centre:

$(x - 5)^2 - 25 + y^2 - 7 = 0$

$\Rightarrow (x - 5)^2 + y^2 = 32$, so the centre is (5, 0).

Gradient of radius: $m = \frac{4-0}{9-5} = 1$

so the gradient of the tangent at A is -1.

$y - y_1 = m(x - x_1) \Rightarrow y - 4 = -1(x - 9)$

$\Rightarrow y = -x + 13 \Rightarrow x + y = 13$

[5 marks available — 1 mark for completing the square on circle equation, 1 mark for finding centre point, 1 mark for finding the gradient of the radius, 1 mark for finding the gradient of the tangent, 1 mark for correct equation of tangent]

c) $x + y = 13$ meets the x-axis at Q (0, 13), so the distance OQ is 13. A is at (9, 4) so the perpendicular height of OAQ is 4.

Area of the triangle = $\frac{1}{2}$ × base × height

$= \frac{1}{2} \times 13 \times 4 = 26$ units^2

[3 marks available — 1 mark for finding coordinates of Q, 1 mark for attempting to use coordinates to find base and height of triangle OAQ, 1 mark for correct area]

Q5 a) $m_{CD} = \frac{6-2}{6-(-4)} = \frac{4}{10} = \frac{2}{5}$

$y - y_1 = m(x - x_1) \Rightarrow y - 6 = \frac{2}{5}(x - 6)$

$\Rightarrow y - 6 = \frac{2}{5}x - \frac{12}{5} \Rightarrow y = \frac{2}{5}x + \frac{18}{5}$

$\Rightarrow 5y = 2x + 18 \Rightarrow 2x - 5y + 18 = 0$

[3 marks available — 1 mark for finding gradient of CD, 1 mark for any correct form of line equation with fractional coefficients, 1 mark for correct final answer with integer coefficients]

b) $y - y_1 = m(x - x_1) \Rightarrow y - 6 = -1(x - 6) \Rightarrow y = -x + 12$

[2 marks available — 1 mark for substituting the gradient and point into a line equation, 1 mark for correct equation of BC in any form]

c) B is the point of intersection between BC and AB. Find an equation for AB:

AB is parallel to CD, so the gradient is $\frac{2}{5}$.

$y - y_1 = m(x - x_1) \Rightarrow y - 7 = \frac{2}{5}(x - (-2))$

$\Rightarrow y - 7 = \frac{2}{5}x + \frac{4}{5} \Rightarrow y = \frac{2}{5}x + \frac{39}{5}$

When AB meets BC:

$-x + 12 = \frac{2}{5}x + \frac{39}{5} \Rightarrow -5x + 60 = 2x + 39$
$\Rightarrow 7x = 21 \Rightarrow x = 3$

Substitute x into an equation to find y:
$y = -(3) + 12 = 9$, so B has coordinates (3, 9).

[4 marks available — 1 mark for finding the equation of AB, 1 mark for attempting to solve the equations simultaneously, 1 mark for correct x-coordinate, 1 mark for correct y-coordinate]

d) Find gradient of DB: $m_{DB} = \frac{9-2}{3-(-4)} = 1$

The product of the gradients of perpendicular lines is –1.
$m_{BC} \times m_{DB} = -1 \times 1 = -1$, so BC and DB are perpendicular — i.e. $\angle DBC = 90°$.

[2 marks available — 1 mark for finding the gradient of DB, 1 mark for using the gradients of DB and BC to show they are perpendicular]

Q6 a) Complete the square of the circle equation:
$x^2 + y^2 - 4x + 12y + 15 = 0 \Rightarrow x^2 - 4x + y^2 + 12y + 15 = 0$
$\Rightarrow (x-2)^2 - 4 + (y+6)^2 - 36 + 15 = 0$
$\Rightarrow (x-2)^2 + (y+6)^2 = 25$
So the centre is (2, –6) and the radius is 5.

[4 marks available — 1 mark for attempting to complete the square on the circle equation, 1 mark for the correct form of the equation as shown, 1 mark for finding the coordinates of the centre, 1 mark for finding the radius]

b) E.g. The y-value of the centre point is –6. Since the radius is 5 and 5 < 6, the entire circle must lie below the x-axis.

[2 marks available — 2 marks for a fully supported argument, otherwise 1 mark for a partially supported argument]

An alternative proof is to show that there are no intersections between the circle and the x-axis — i.e. when $y = 0$, $(x - 2)^2 = -11$, so there are no real roots.
The centre of the circle is below the x-axis, so the entire circle must lie below the x-axis.

c) Substitute $(-1, k)$ into the circle equation:
$(-1-2)^2 + (k+6)^2 = 25 \Rightarrow (k+6)^2 = 16$
$\Rightarrow (k+6) = \pm 4 \Rightarrow k = -2$ or $k = -10$

[3 marks available — 1 mark for substituting in –1 for x into the original equation and attempting to solve a quadratic equation to find k, 1 mark for finding k = –2, 1 mark for k = –10]

Q7 Find the perpendicular bisectors of two of the line segments XY, YZ and ZX.

XY: midpoint = $\left(\frac{7-1}{2}, \frac{6-10}{2}\right) = (3, -2)$

gradient = $\frac{-10-6}{-1-7} = \frac{-16}{-8} = 2$

So the perpendicular bisector passes through (3, –2) and has a gradient of $-\frac{1}{2}$.

$y - y_1 = m(x - x_1) \Rightarrow y - (-2) = -\frac{1}{2}(x - 3) \Rightarrow y = -\frac{1}{2}x - \frac{1}{2}$

YZ: midpoint = $\left(\frac{-1+13}{2}, \frac{-10-12}{2}\right) = (6, -11)$

gradient = $\frac{-12-(-10)}{13-(-1)} = \frac{-2}{14} = -\frac{1}{7}$

So the perpendicular bisector passes through (6, –11) and has a gradient of 7.

$y - y_1 = m(x - x_1) \Rightarrow y - (-11) = 7(x - 6)$
$\Rightarrow y + 11 = 7x - 42 \Rightarrow y = 7x - 53$

Now find the centre of the circle, which is the point where these lines intersect:

$-\frac{1}{2}x - \frac{1}{2} = 7x - 53 \Rightarrow -x - 1 = 14x - 106$
$\Rightarrow 15x = 105 \Rightarrow x = 7$

Substitute $x = 7$ into $y = 7x - 53$: $y = 49 - 53 = -4$
so the circle has its centre at (7, –4).

Radius = $\sqrt{(7-7)^2 + (6-(-4))^2} = \sqrt{10^2} = 10$

So the equation of the circle is: $(x-7)^2 + (y+4)^2 = 100$

[8 marks available — 1 mark for finding gradient of one side of XYZ, 1 mark for equation of that side, 1 mark for gradient of a second side, 1 mark for equation of the second side, 1 mark for solving the two line equations simultaneously for x, 1 mark for substituting for y, 1 mark for using Pythagoras' theorem to calculate the radius, 1 mark for the correct final equation for the circumcircle]

Q8 a) The equation of the circle is $(x-2)^2 + (y-4)^2 = 25$
Substitute $x = 15 - 2y$ into the circle equation to find the y-coordinates of the intersections:
$((15-2y)-2)^2 + (y-4)^2 = 25 \Rightarrow (13-2y)^2 + (y-4)^2 = 25$
$\Rightarrow 169 - 52y + 4y^2 + y^2 - 8y + 16 = 25$
$\Rightarrow 5y^2 - 60y + 160 = 0 \Rightarrow y^2 - 12y + 32 = 0$
$\Rightarrow (y-4)(y-8) = 0$
So A and B have y-coordinates $y = 4$ and $y = 8$.
Substitute the y-values back into one of the original equations to find the x-coordinates:
When $y = 4$, $x = 15 - (2 \times 4) = 7$
When $y = 8$, $x = 15 - (2 \times 8) = -1$
So A (7, 4) and B (–1, 8).
Midpoint of AB, M = $\left(\frac{7-1}{2}, \frac{4+8}{2}\right) = (3, 6)$

So the length CM = $\sqrt{(3-2)^2 + (6-4)^2}$
$= \sqrt{1^2 + 2^2} = \sqrt{5}$ as required.

[7 marks available — 1 mark for finding the circle equation, 1 mark for forming a quadratic equation in x or y by substitution of the line equation, 1 mark for attempting to solve the quadratic, 1 mark for both correct y-values, 1 mark for both correct x-values, 1 mark for finding the coordinates of M, 1 mark for the correct length CM]

You could formulate the quadratic in x instead, then substitute to find the y-values.

b) Since M (3, 6) is the midpoint of the chord AB, CM is perpendicular to AB. So ABC is a triangle with base AB and perpendicular height CM.

Length AB = $\sqrt{(7-(-1))^2 + (4-8)^2}$
$= \sqrt{8^2 + (-4)^2} = \sqrt{80} = 4\sqrt{5}$

So area ABC = $\frac{1}{2} \times \sqrt{5} \times 4\sqrt{5} = 10$ units²

[3 marks available — 1 mark for using the fact that CM is perpendicular to AB, 1 mark for finding the length AB, 1 mark for calculating the final area]

Chapter 7: The Binomial Expansion

7.1 Binomial Expansions

Exercise 7.1.1 — Binomial expansions — $(1 + x)^n$

Q1 Pascal's triangle is

```
        1
      1   1
    1   2   1
  1   3   3   1
1   4   6   4   1
```

The expansion of $(1 + x)^4$ takes its coefficients of each term, in ascending powers of x, from the 5th row:
$(1 + x)^4 = 1 + 4x + 6x^2 + 4x^3 + x^4$

Q2 a) ${}^6C_2 = 15$ **b)** $\binom{12}{5} = {}^{12}C_5 = 792$

c) $\frac{30!}{4!26!} = {}^{30}C_4 = 27\,405$ **d)** ${}^8C_8 = 1$

Q3 a) $\frac{9!}{4!5!} = \frac{9 \times 8 \times 7 \times 6 \times 5 \times 4 \times 3 \times 2 \times 1}{(4 \times 3 \times 2 \times 1)(5 \times 4 \times 3 \times 2 \times 1)}$
$= \frac{9 \times 8 \times 7 \times 6}{4 \times 3 \times 2 \times 1} = 3 \times 7 \times 6 = 126$

b) $^{10}C_3 = \frac{10!}{3!(10-3)!} = \frac{10\times9\times8}{3\times2\times1}$
$= 10\times3\times4 = 120$

c) $\frac{15!}{11!4!} = \frac{15\times14\times13\times12}{4\times3\times2\times1} = 15\times7\times13 = 1365$

d) $\binom{8}{6} = \frac{8!}{6!(8-6)!} = \frac{8\times7}{2\times1} = 4\times7 = 28$

Q4 $(1+x)^{10} = 1 + {}^{10}C_1x + {}^{10}C_2x^2 + {}^{10}C_3x^3 + ...$

You can work out the coefficients ^nC_r using a calculator, or using the method below. Using the notation $\binom{n}{r}$ instead of ^nC_r for the coefficients is fine too.

$^nC_r = \frac{n!}{r!(n-r)!}$,

$^{10}C_1 = \frac{10!}{1!(10-1)!} = \frac{10\times9\times8\times...1}{1\times9\times8\times...1} = 10$

$^{10}C_2 = \frac{10!}{2!(10-2)!} = \frac{10\times9\times8\times...1}{2\times1\times8\times7\times...1}$
$= \frac{10\times9}{2} = 45$

$^{10}C_3 = \frac{10!}{3!(10-3)!} = \frac{10\times9\times8\times7\times...1}{3\times2\times1\times7\times...1}$
$= \frac{10\times9\times8}{3\times2}$
$= 10\times3\times4 = 120$

$(1+x)^{10} = 1 + 10x + 45x^2 + 120x^3 + ...$

For the rest of this exercise you can use one of the methods shown in question 1 or 4 to find the coefficients ^nC_r, or you can use a calculator.

Q5 $(1+x)^6 = 1 + {}^6C_1x + {}^6C_2x^2 + {}^6C_3x^3 + {}^6C_4x^4 + {}^6C_5x^5 + {}^6C_6x^6$
$= 1 + 6x + 15x^2 + 20x^3 + 15x^4 + 6x^5 + x^6$

Q6 $(1+x)^7 = 1 + {}^7C_1x + {}^7C_2x^2 + {}^7C_3x^3 + ... = 1 + 7x + 21x^2 + 35x^3 + ...$

Q7 a) $(1+x)^{11} = 1 + {}^{11}C_1x + {}^{11}C_2x^2 + ... = 1 + 11x + 55x^2 + ...$

b) $(1+x)^{12} = 1 + {}^{12}C_1x + {}^{12}C_2x^2 + ... = 1 + 12x + 66x^2 + ...$

c) $(1+x)^{15} = 1 + {}^{15}C_1x + {}^{15}C_2x^2 + ... = 1 + 15x + 105x^2 + ...$

d) $(1+x)^{30} = 1 + {}^{30}C_1x + {}^{30}C_2x^2 + ... = 1 + 30x + 435x^2 + ...$

Q8 The coefficient of x^{12} is $^{17}C_{12} = 6188$, so $a = 6188$.

Q9 a) First find the term in x in the expansion of $(1+x)^5$:
$(1+x)^5 = 1 + {}^5C_1x + ...$, so the term in x is $5x$. Now add the x-term in the second bracket: $5x + (-3x) = 2x$

b) The coefficient of x^4 is $^5C_4 = 5$.
The second bracket doesn't affect the term in x^4.

Exercise 7.1.2 — Binomial expansions — $(1 + ax)^n$

Q1 a) $(1+3x)^4 = 1 + {}^4C_1(3x) + {}^4C_2(3x)^2 + {}^4C_3(3x)^3 + {}^4C_4(3x)^4$
$= 1 + 4(3x) + 6(9x^2) + 4(27x^3) + 1(81x^4)$
$= 1 + 12x + 54x^2 + 108x^3 + 81x^4$

b) $(1-x)^4 = 1 + {}^4C_1(-x) + {}^4C_2(-x)^2 + {}^4C_3(-x)^3 + {}^4C_4(-x)^4$
$= 1 - 4x + 6x^2 - 4x^3 + x^4$

c) $(1-x)^6 = 1 + {}^6C_1(-x) + {}^6C_2(-x)^2 + {}^6C_3(-x)^3$
$+ {}^6C_4(-x)^4 + {}^6C_5(-x)^5 + {}^6C_6(-x)^6$
$= 1 - 6x + 15x^2 - 20x^3 + 15x^4 - 6x^5 + x^6$

For parts b) and c) you could use the formula for the expansion of $(1-x)^n$: $(1-x)^n = 1 - {}^nC_1x + {}^nC_2x^2 - {}^nC_3x^3 + ...$

d) $(1-2x)^5 = 1 + {}^5C_1(-2x) + {}^5C_2(-2x)^2 + {}^5C_3(-2x)^3$
$+ {}^5C_4(-2x)^4 + {}^5C_5(-2x)^5$
$= 1 + 5(-2x) + 10(4x^2) + 10(-8x^3) + 5(16x^4) + 1(-32x^5)$
$= 1 - 10x + 40x^2 - 80x^3 + 80x^4 - 32x^5$

e) $(1-4x)^3 = 1 + {}^3C_1(-4x) + {}^3C_2(-4x)^2 + {}^3C_3(-4x)^3$
$= 1 + 3(-4x) + 3(16x^2) + 1(-64x^3)$
$= 1 - 12x + 48x^2 - 64x^3$

f) $(1-5x)^5 = 1 + {}^5C_1(-5x) + {}^5C_2(-5x)^2 + {}^5C_3(-5x)^3$
$+ {}^5C_4(-5x)^4 + {}^5C_5(-5x)^5$
$= 1 + 5(-5x) + 10(25x^2) + 10(-125x^3)$
$+ 5(625x^4) + 1(-3125x^5)$
$= 1 - 25x + 250x^2 - 1250x^3 + 3125x^4 - 3125x^5$

g) $(1+2x)^6 = 1 + {}^6C_1(2x) + {}^6C_2(2x)^2 + {}^6C_3(2x)^3$
$+ {}^6C_4(2x)^4 + {}^6C_5(2x)^5 + {}^6C_6(2x)^6$
$= 1 + 6(2x) + 15(4x^2) + 20(8x^3)$
$+ 15(16x^4) + 6(32x^5) + 1(64x^6)$
$= 1 + 12x + 60x^2 + 160x^3 + 240x^4 + 192x^5 + 64x^6$

h) $(1+x)^9 = 1 + {}^9C_1x + {}^9C_2x^2 + {}^9C_3x^3 + {}^9C_4x^4 + {}^9C_5x^5 + ...$
$(1-x)^9 = 1 - {}^9C_1x + {}^9C_2x^2 - {}^9C_3x^3 + {}^9C_4x^4 - {}^9C_5x^5 + ...$
So: $(1+x)^9 - (1-x)^9 = 2({}^9C_1x + {}^9C_3x^3 + {}^9C_5x^5 + ...)$
The even powers cancel out, so only the terms with odd powers appear (and they're doubled because one term comes from each expansion).
$= 2(9x + 84x^3 + 126x^5 + 36x^7 + x^9)$
$= 18x + 168x^3 + 252x^5 + 72x^7 + 2x^9$

Q2 The term in x^4 is: $^{16}C_4(-2x)^4 = 1820(16x^4) = 29\,120x^4$

Q3 The first 3 terms will include 1 and the terms in x and x^2 so expand each bracket up to and including the term in x^2:
$(1+x)^3(1-x)^4 = (1 + 3x + 3x^2 + ...)(1 - 4x + 6x^2 - ...)$
$= 1 - 4x + 6x^2 + ... + 3x - 12x^2$
$+ .... + 3x^2 +$ (*higher power terms*)
$= 1 - x - 3x^2 + ...$

Q4 The expansion of $(1+x)^5(1+y)^7$ is the expansions of $(1+x)^5$ and $(1+y)^7$ multiplied together. We need the x^3 term from $(1+x)^5$ and the y^2 term from $(1+y)^7$. Multiplying the coefficients gives the x^3y^2 coefficient: x^3 coefficient: $\frac{5!}{3!(5-3)!} = \frac{5\times4}{2\times1} = 10$

y^2 coefficient: $\frac{7!}{2!(7-2)!} = \frac{7\times6}{2\times1} = 21$

x^3y^2 coefficient: $10 \times 21 = 210$

Q5 Find all the terms up to x^3 in both expansions:
$(1+4x)^4 = 1 + {}^4C_1(4x) + {}^4C_2(4x)^2 + {}^4C_3(4x)^3 + ...$
$= 1 + 4(4x) + 6(16x^2) + 4(64x^3) + ...$
$= 1 + 16x + 96x^2 + 256x^3 + ...$

$(1-6x)^3 = 1 + {}^3C_1(-6x) + {}^3C_2(-6x)^2 + {}^3C_3(-6x)^3$
$= 1 + 3(-6x) + 3(36x^2) + 1(-216x^3)$
$= 1 - 18x + 108x^2 - 216x^3$

So $(1+4x)^4(1-6x)^3 = (1 + 16x + 96x^2 + 256x^3 + ...)$
$\times (1 - 18x + 108x^2 - 216x^3)$

You're only interested in the term in x^3, so just multiply out the terms which will give you x^3:
Term in $x^3 = (1 \times -216x^3) + (16x \times 108x^2)$
$+ (96x^2 \times -18x) + (256x^3 \times 1)$
$= -216x^3 + 1728x^3 - 1728x^3 + 256x^3 = 40x^3$
So the coefficient of x^3 is 40.

Q6 a) $(1+kx)^8 = 1 + {}^8C_1kx + {}^8C_2(kx)^2 + {}^8C_3(kx)^3 + ...$
$= 1 + 8kx + 28k^2x^2 + 56k^3x^3 + ...$

b) The fourth term is $56k^3x^3$, so $56k^3x^3 = 448x^3 \Rightarrow 56k^3 = 448$
$\Rightarrow k^3 = 8 \Rightarrow k = 2$

Q7 We need the coefficient of x^2 to equal 135:
$^6C_2(-kx)^2 = 15k^2x^2$
So $15k^2 = 135 \Rightarrow k^2 = 9 \Rightarrow k = 3$
The question says k is positive so ignore the negative root.

Q8 $(1-3x)^6 = 1 + {}^6C_1(-3x) + {}^6C_2(-3x)^2 + {}^6C_3(-3x)^3 + ...$
$= 1 - 18x + 135x^2 - 540x^3 + ...$

$(1+x)(1-3x)^6 = (1+x)(1 - 18x + ...)$
$= 1 - 18x + ... + x - 18x^2 + ... \approx 1 - 17x$
You're told you can ignore x^2 and higher terms.

Q9 a) $\left(1+\frac{x}{2}\right)^{12} = 1 + {}^{12}C_1\left(\frac{x}{2}\right) + {}^{12}C_2\left(\frac{x}{2}\right)^2 + {}^{12}C_3\left(\frac{x}{2}\right)^3 + {}^{12}C_4\left(\frac{x}{2}\right)^4 + ...$
$= 1 + 6x + \frac{33}{2}x^2 + \frac{55}{2}x^3 + \frac{495}{16}x^4 + ...$

b) $1 + \left(\frac{x}{2}\right) = 1.005$ when $x = 0.01$.

Substitute this value into the expansion:
$1.005^{12} \approx 1 + 6(0.01) + \frac{33}{2}(0.01)^2 + \frac{55}{2}(0.01)^3 + \frac{495}{16}(0.01)^4$
$1.005^{12} \approx 1.061677809 = 1.0616778$ to 7 d.p.

Q10 Because $x > 1$, higher powers of x will not be small enough to ignore, so only taking the terms up to x^3 will not provide an accurate estimate.

Q11 a) Expansion of $(1 + 7x)^n = 1 + {}^nC_1(7x) + {}^nC_2(7x)^2 + ...$
So term in $x^2 = {}^nC_2 \times 49x^2$

$${}^nC_2 = \frac{n!}{2!(n-2)!} = \frac{n(n-1)(n-2)!}{2(n-2)!} = \frac{n(n-1)}{2}$$

So term in $x^2 = \frac{49n(n-1)}{2}x^2$

The coefficient of the term in x^2 is 490, so:

$$\frac{49n(n-1)}{2} = 490 \Rightarrow \frac{n(n-1)}{2} = 10$$

$\Rightarrow n(n - 1) = 20 \Rightarrow n^2 - n - 20 = 0 \Rightarrow (n + 4)(n - 5) = 0$
So $n = -4$ or $n = 5$. Since n is positive, $n = 5$.

b) Term in $x^3 = {}^5C_3(7x)^3 = 10(343x^3) = 3430x^3$

Exercise 7.1.3 — Binomial expansions — $(a + b)^n$

Q1 Using the formula for the expansion of $(a + b)^n$:

$$(a+b)^n = a^n + \binom{n}{1}a^{n-1}b + \binom{n}{2}a^{n-2}b^2 + ... + b^n$$

In this case $a = 3$ and $b = x$:

$(3 + x)^6 = 3^6 + {}^6C_1 3^5x + {}^6C_2 3^4x^2 + {}^6C_3 3^3x^3 + ...$
$= 729 + 6(243x) + 15(81x^2) + 20(27x^3) + ...$
$= 729 + 1458x + 1215x^2 + 540x^3 + ...$

Q2 a) In this case $a = 2$ and $b = x$:
$(2 + x)^4 = 2^4 + {}^4C_1 2^3x + {}^4C_2 2^2x^2 + {}^4C_3 2x^3 + {}^4C_4x^4$
$= 16 + 4(8x) + 6(4x^2) + 4(2x^3) + x^4$
$= 16 + 32x + 24x^2 + 8x^3 + x^4$

b) In this case $a = 2$ and $b = 2x$:

$(2 + 2x)^4 = 2^4 + {}^4C_1 2^3(2x) + {}^4C_2 2^2(2x)^2 + {}^4C_3 2(2x)^3 + {}^4C_4(2x)^4$
$= 16 + (4 \times 8 \times 2x) + (6 \times 4 \times 4x^2)$
$+ (4 \times 2 \times 8x^3) + 16x^4$
$= 16 + 64x + 96x^2 + 64x^3 + 16x^4$

c) In this case $a = 2$ and $b = -2x$:
$(2 - 2x)^4 = 2^4 + {}^4C_1 2^3(-2x) + {}^4C_2 2^2(-2x)^2$
$+ {}^4C_3 2(-2x)^3 + {}^4C_4(-2x)^4$
$= 16 + (4 \times 8 \times -2x) + (6 \times 4 \times 4x^2)$
$+ (4 \times 2 \times -8x^3) + 16x^4$
$= 16 - 64x + 96x^2 - 64x^3 + 16x^4$

Notice here that you could just use the result from part b) and change the sign of the odd-power terms.

d) In this case $a = \frac{1}{2}$ and $b = \frac{1}{2}x$:

$$\left(\frac{1}{2}+\frac{1}{2}x\right)^4 = \left(\frac{1}{2}\right)^4 + {}^4C_1\left(\frac{1}{2}\right)^3\left(\frac{1}{2}x\right) + {}^4C_2\left(\frac{1}{2}\right)^2\left(\frac{1}{2}x\right)^2 + {}^4C_3\left(\frac{1}{2}\right)\left(\frac{1}{2}x\right)^3 + {}^4C_4\left(\frac{1}{2}x\right)^4$$

$$= \frac{1}{16} + \left(4 \times \frac{1}{8} \times \frac{1}{2}x\right) + \left(6 \times \frac{1}{4} \times \frac{1}{4}x^2\right) + \left(4 \times \frac{1}{2} \times \frac{1}{8}x^3\right) + \frac{1}{16}x^4$$

$$= \frac{1}{16} + \frac{1}{4}x + \frac{3}{8}x^2 + \frac{1}{4}x^3 + \frac{1}{16}x^4$$

Q3 a) The term in x^5 is ${}^8C_5(\lambda x)^5 = 56\lambda^5x^5$
Therefore $56\lambda^5 = 57\,344 \Rightarrow \lambda^5 = 1024 \Rightarrow \lambda = \sqrt[5]{1024} = 4$

b) $(1 + 4x)^8 = 1 + {}^8C_1(4x) + {}^8C_2(4x)^2 + ... = 1 + 32x + 448x^2 + ...$

Q4 a) $(2 + x)^8 = 2^8 + {}^8C_1 2^7x + {}^8C_2 2^6x^2 + {}^8C_3 2^5x^3 + {}^8C_4 2^4x^4 + ...$
$= 256 + 1024x + 1792x^2 + 1792x^3 + 1120x^4 + ...$

b) $2 + x = 2.01$ when $x = 0.01$
Hence: $2.01^8 = 256 + 1024(0.01) + 1792(0.01)^2$
$+ 1792(0.01)^3 + 1120(0.01)^4 + ...$
≈ 266.4210032
An approximation to 2.01^8 is: 266.42100 (to 5 d.p.)

Q5 $(3 + 5x)^7 = 3^7 + {}^7C_1 3^6(5x) + {}^7C_2 3^5(5x)^2 + {}^7C_3 3^4(5x)^3 + ...$
$= 2187 + 25\,515x + 127\,575x^2 + 354\,375x^3 + ...$

Q6 a) $(3 - 2x)^7 = 3^7 + {}^7C_1 3^6(-2x) + {}^7C_2 3^5(-2x)^2 + {}^7C_3 3^4(-2x)^3 + ...$
$= 2187 - 10\,206x + 20\,412x^2 - 22\,680x^3 + ...$
$3 - 2x = 2.998$ when $x = 0.001$
Hence: $2.998^7 = 2187 - 10\,206(0.001) + 20\,412(0.001)^2$
$- 22\,680(0.001)^3 + ...$
$\approx 2176.814... = 2176.81$ (2 d.p.)
For parts b)-d), use the expansion of $(3 - 2x)^7$ above.

b) $3 - 2x = 2.8$ when $x = 0.1$
Hence: $2.8^7 = 2187 - 10\,206(0.1) + 20\,412(0.1)^2$
$- 22\,680(0.1)^3 + ...$
≈ 1347.84

c) $3 - 2x = 2.94$ when $x = 0.03$
Hence: $2.94^7 = 2187 - 10\,206(0.03) + 20\,412(0.03)^2$
$- 22\,680(0.03)^3 + ...$
$\approx 1898.578... = 1898.58$ (2 d.p.)

d) $3 - 2x = 3.002$ when $x = -0.001$
Hence: $3.002^7 = 2187 - 10\,206(-0.001)$
$+ 20\,412(-0.001)^2 - 22\,680(-0.001)^3 + ...$
$\approx 2197.226... = 2197.23$ (2 d.p.)

Q7 Expand the first five terms of $(3 + 2x)^6$ and multiply by $(1 + x)$:
$(3 + 2x)^6 = 3^6 + {}^6C_1 3^5(2x) + {}^6C_2 3^4(2x)^2 + {}^6C_3 3^3(2x)^3 + {}^6C_4 3^2(2x)^4 + ...$
$= 729 + 2916x + 4860x^2 + 4320x^3 + 2160x^4 + ...$

$(1 + x)(3 + 2x)^6 = (3 + 2x)^6 + x(3 + 2x)^6$
$= (729 + 2916x + 4860x^2 + 4320x^3 + 2160x^4 + ...)$
$+ (729x + 2916x^2 + 4860x^3 + 4320x^4 + 2160x^5 + ...)$
$= 729 + 3645x + 7776x^2 + 9180x^3 + 6480x^4 + ...$
(The term in x^5 is the 6th term.)

Q8 a) Term in $x^2 = {}^5C_2 4^3(2x)^2 = (10 \times 64 \times 4)x^2$
So coefficient of $x^2 = 2560$

b) Term in $x^2 = {}^8C_2 2^6(-5x)^2 = (28 \times 64 \times 25)x^2$
So coefficient of $x^2 = 44\,800$

c) Find the term in x in the expansion of $\left(1 + \frac{1}{2}x\right)^7$, then multiply by $2x$ to get the term in x^2.
Term in $x = {}^7C_1 1^6\left(\frac{1}{2}x\right) = \left(7 \times 1 \times \frac{1}{2}\right)x = \frac{7}{2}x$
Multiply by $2x$ to get: $\frac{7}{2}x \times 2x = 7x^2$
So coefficient of $x^2 = 7$

d) Start by expanding $(3x - 1)^9$. You don't need to write out the whole thing, since you're only interested in the terms which will give you the x^2-term in the expansion of $(5 + x)(3x - 1)^9$:
$(3x - 1)^9 = (3x)^9 + {}^9C_1(3x)^8(-1) + {}^9C_2(3x)^7(-1)^2 + ...$
$+ ... + {}^9C_7(3x)^2(-1)^7 + {}^9C_8(3x)(-1)^8 + (-1)^9$

The x^2-term in the expansion of $(5 + x)(3x - 1)^9$ will involve the x^2 and x terms of $(3x - 1)^9$, so work out these two terms:
x^2-term: ${}^9C_7(3x)^2(-1)^7 = (36 \times 9 \times -1)x^2 = -324x^2$
x-term: ${}^9C_8(3x)(-1)^8 = (9 \times 3 \times 1)x = 27x$
So the x^2-term of $(5 + x)(3x - 1)^9$
$= (5 \times -324x^2) + (x \times 27x) = -1620x^2 + 27x^2 = -1593x^2$
So coefficient of $x^2 = -1593$

Q9 a) Expand $(2 - x)^5$ and multiply by $(1 + 3x)$:
$(2 - x)^5 = 2^5 + {}^5C_1 2^4(-x) + {}^5C_2 2^3(-x)^2 + {}^5C_3 2^2(-x)^3$
$+ {}^5C_4 2(-x)^4 + {}^5C_5(-x)^5$
$= 32 - 80x + 80x^2 - 40x^3 + 10x^4 - x^5$
$(1 + 3x)(2 - x)^5 = (2 - x)^5 + 3x(2 - x)^5$
$= (32 - 80x + 80x^2 - 40x^3 + 10x^4 - x^5)$
$+ (96x - 240x^2 + 240x^3 - 120x^4 + 30x^5 - 3x^6)$
$= 32 + 16x - 160x^2 + 200x^3 - 110x^4 + 29x^5 - 3x^6$

b) Expand $(1 + 3x)^5$ and multiply by $(2 - x)$:

$(1 + 3x)^5 = 1 + {}^5C_1(3x) + {}^5C_2(3x)^2 + {}^5C_3(3x)^3 + {}^5C_4(3x)^4 + {}^5C_5(3x)^5$

$= 1 + 15x + 90x^2 + 270x^3 + 405x^4 + 243x^5$

$(1 + 3x)^5(2 - x) = 2(1 + 3x)^5 - x(1 + 3x)^5$

$= (2 + 30x + 180x^2 + 540x^3 + 810x^4 + 486x^5)$

$- (x + 15x^2 + 90x^3 + 270x^4 + 405x^5 + 243x^6)$

$= 2 + 29x + 165x^2 + 450x^3 + 540x^4 + 81x^5 - 243x^6$

c) Expand $(1 + 3x)^2$ and $(2 - x)^4$ and multiply together:

$(1 + 3x)^2 = 1 + {}^2C_1(3x) + {}^2C_2(3x)^2 = 1 + 6x + 9x^2$

$(2 - x)^4 = 2^4 + {}^4C_1 2^3(-x) + {}^4C_2 2^2(-x)^2 + {}^4C_3 2(-x)^3 + {}^4C_4(-x)^4$

$= 16 - 32x + 24x^2 - 8x^3 + x^4$

$(1 + 3x)^2(2 - x)^4 = (1 + 6x + 9x^2) \times (16 - 32x + 24x^2 - 8x^3 + x^4)$

$= (16 - 32x + 24x^2 - 8x^3 + x^4)$

$+ 6x(16 - 32x + 24x^2 - 8x^3 + x^4)$

$+ 9x^2(16 - 32x + 24x^2 - 8x^3 + x^4)$

$= 16 + 64x - 24x^2 - 152x^3 + 169x^4 - 66x^5 + 9x^6$

d) Expand $(1 + 3x)^3$ and $(2 - x)^3$ and multiply together:

$(1 + 3x)^3 = 1 + {}^3C_1(3x) + {}^3C_2(3x)^2 + {}^3C_3(3x)^3$

$= 1 + 9x + 27x^2 + 27x^3$

$(2 - x)^3 = 2^3 + {}^3C_1 2^2(-x) + {}^3C_2 2(-x)^2 + {}^3C_3(-x)^3$

$= 8 - 12x + 6x^2 - x^3$

$(1 + 3x)^3(2 - x)^3 = (1 + 9x + 27x^2 + 27x^3) \times (8 - 12x + 6x^2 - x^3)$

$= (8 - 12x + 6x^2 - x^3) + 9x(8 - 12x + 6x^2 - x^3)$

$+ 27x^2(8 - 12x + 6x^2 - x^3) + 27x^3(8 - 12x + 6x^2 - x^3)$

$= 8 + 60x + 114x^2 - 55x^3 - 171x^4 + 135x^5 - 27x^6$

Q10 a) Expansion of $(1 - 5x)^n = 1 + {}^nC_1(-5x) + {}^nC_2(-5x)^2 + ...$

So term in $x^2 = {}^nC_2 \times 25x^2$

$${}^nC_2 = \frac{n!}{2!(n-2)!} = \frac{n(n-1)(n-2)!}{2(n-2)!} = \frac{n(n-1)}{2}$$

So term in $x^2 = \frac{25n(n-1)}{2}x^2$

The coefficient of the term in x^2 is 150, so:

$\frac{25n(n-1)}{2} = 150$

$\Rightarrow n(n - 1) = 12 \Rightarrow n^2 - n - 12 = 0 \Rightarrow (n + 4)(n - 3) = 0$

So $n = -4$ or $n = 3$. Since n is positive, $n = 3$.

b) Term in $x^3 = {}^3C_3(-5x)^3 = 1 \times -125x^3 = -125x^3$

Q11 The coefficient of x^2 is ${}^8C_2 a^6 3^2$

The coefficient of x^5 is ${}^8C_5 a^3 3^5$

Therefore: $28 \times a^6 \times 3^2 = \frac{32}{27} \times 56 \times a^3 \times 3^5$

$\Rightarrow 28a^3 = \frac{32}{27} \times 56 \times 3^3$

$\Rightarrow a^3 = \frac{32 \times 56 \times 27}{27 \times 28} = 64$

$\Rightarrow a = \sqrt[3]{64} = 4.$

Q12 Expand each bracket up to the term in x^3:

$(1 + 2x)^5 = 1 + 5(2x) + 10(2x)^2 + 10(2x)^3 + ...$

$= 1 + 10x + 40x^2 + 80x^3 + ...$

$(3 - x)^4 = 3^4 + 4(3)^3(-x) + 6(3)^2(-x)^2 + 4(3)(-x)^3 + ...$

$= 81 - 108x + 54x^2 - 12x^3 + ...$

Multiply the terms that will give a result in x^3:

$(1 \times -12x^3) + (10x \times 54x^2) + (40x^2 \times -108x) + (80x^3 \times 81) = 2688x^3$

So the coefficient of x^3 is 2688.

Q13 a) The coefficient of x^3 is: $\frac{n!}{3!(n-3)!} = \frac{n(n-1)(n-2)}{3 \times 2 \times 1}$

The coefficient of x^2 is: $\frac{n!}{2!(n-2)!} = \frac{n(n-1)}{2}$

The coefficient of x^3 is three times the coefficient of x^2, so:

$\frac{n(n-1)(n-2)}{3 \times 2 \times 1} = 3 \times \frac{n(n-1)}{2} \Rightarrow \frac{n-2}{3} = 3$

$\Rightarrow n = 11$

b) $(1 + x)^{11} = 1 + 11x + 55x^2 + ...$

The coefficient of x^2 is $a \times$ (coefficient of x),

so $55 = 11a \Rightarrow a = 5$.

Q14 $(2 + \mu x)^8 = 2^8 + {}^8C_1 2^7(\mu x) + {}^8C_2 2^6(\mu x)^2 + ...$

$= 256 + (8 \times 128)(\mu x) + (28 \times 64)(\mu x)^2 + ...$

$= 256 + 1024\mu x + 1792\mu^2x^2 + ...$

The coefficient of x^2 is:

$87\,808 = 1792\mu^2 \Rightarrow \mu^2 = 49 \Rightarrow \mu = 7$ or -7.

Q15 Start by expanding the outer brackets:

$[(x + 2)^3(x + 3)^2]^2 = (x + 2)^6(x + 3)^4$

Now you can expand both brackets and multiply the expressions together as usual — you can ignore any powers of x greater than 2:

$(x + 2)^6 = x^6 + {}^6C_1 x^5(2) + ... + {}^6C_4 x^2(2)^4 + {}^6C_5 x(2)^5 + {}^6C_6(2)^6$

$= ...+ 240x^2 + 192x + 64$

$(x + 3)^4 = x^4 + {}^4C_1 x^3(3) + {}^4C_2 x^2(3)^2 + {}^4C_3 x(3)^3 + {}^4C_4(3)^4$

$= ...+ 54x^2 + 108x + 81$

$(x + 2)^6(x + 3)^4 = (...+ 240x^2 + 192x + 64) \times (...+ 54x^2 + 108x + 81)$

You're only interested in the term in x^2, so just multiply out the terms which will give you x^2:

Term in $x^2 = (240x^2 \times 81) + (192x \times 108x) + (64 \times 54x^2)$

$= 19\,440x^2 + 20\,736x^2 + 3456x^2$

$= 43\,632x^2$

So the coefficient of x^2 is 43 632.

Review Exercise — Chapter 7

Q1 a) $(1 + x)^{40} = 1 + {}^{40}C_1 x + {}^{40}C_2 x^2 + ... = 1 + 40x + 780x^2 + ...$

b) $(1 - x)^{20} = 1 + {}^{20}C_1(-x) + {}^{20}C_2(-x)^2 + ... = 1 - 20x + 190x^2 + ...$

c) $(1 + 3x)^{20} = 1 + {}^{20}C_1(3x) + {}^{20}C_2(3x)^2 + ...$

$= 1 + 60x + 1710x^2 + ...$

d) $(2 + 3x)^{10} = 2^{10} + {}^{10}C_1 2^9(3x) + {}^{10}C_2 2^8(3x)^2 + ...$

$= 1024 + 15\,360x + 103\,680x^2 + ...$

Q2 a) $\left(1 + \frac{x}{3}\right)^9 = 1 + {}^9C_1\left(\frac{x}{3}\right) + {}^9C_2\left(\frac{x}{3}\right)^2 + {}^9C_3\left(\frac{x}{3}\right)^3 + ...$

$= 1 + \frac{9}{3}x + \frac{36}{9}x^2 + \frac{84}{27}x^3 + ...$

$= 1 + 3x + 4x^2 + \frac{28}{9}x^3 + ...$

b) $1 + \frac{x}{3} = 1.003$ when $x = 0.009$

Hence:

$1.003^9 = 1 + 3(0.009) + 4(0.009)^2 + \frac{28}{9}(0.009)^3 + ...$

$\approx 1.0273262... = 1.027326$ (6 d.p.)

Q3 a) $(1 + ax)^8 = 1 + {}^8C_1 ax + {}^8C_2(ax)^2 + {}^8C_3(ax)^3 + {}^8C_4(ax)^4 + ...$

$= 1 + 8ax + 28a^2x^2 + 56a^3x^3 + 70a^4x^4 + ...$

b) Coefficient of $x^2 = 28a^2$ and coefficient of $x^3 = 56a^3$, so:

$28a^2 = 2(56a^3) \Rightarrow 28a^2 = 112a^3$

$\Rightarrow 1 = 4a$

$\Rightarrow a = \frac{1}{4}$

Coefficient of $x = 8a = 8 \times \frac{1}{4} = 2$

Q4 $(4 - 5x)^7 = 4^7 + {}^7C_1 4^6(-5x) + {}^7C_2 4^5(-5x)^2 + ...$

$= 16\,384 - 143\,360x + 537\,600x^2 + ...$

Q5 x^2 term of the expansion of $(2 + 3x)^5$

$= {}^5C_2 2^3(3x)^2 = (10 \times 8 \times 9)x^2 = 720x^2$

So coefficient of $x^2 = 720$.

Q6 a) $\left(3+\frac{x}{4}\right)^{11} = 3^{11} + {}^{11}C_1 3^{10}\left(\frac{x}{4}\right) + {}^{11}C_2 3^9\left(\frac{x}{4}\right)^2 + {}^{11}C_3 3^8\left(\frac{x}{4}\right)^3 + {}^{11}C_4 3^7\left(\frac{x}{4}\right)^4 + ...$

$= 177\,147 + \frac{649\,539}{4}x + \frac{1\,082\,565}{16}x^2 + \frac{1\,082\,565}{64}x^3 + \frac{360855}{128}x^4 + ...$

b) $3 + \frac{x}{4} = 3.002$ when $x = 0.008$

Hence:

$3.002^{11} = 177\,147 + \frac{649\,539}{4}(0.008) + \frac{1\,082\,565}{16}(0.008)^2 + \frac{1\,082\,565}{64}(0.008)^3 + \frac{360855}{128}(0.008)^4 + ...$

$\approx 178\,450.4169 = 178\,450.417$ (3 d.p.)

Q7 a) $(2 + kx)^{13} = 2^{13} + {}^{13}C_1 2^{12}kx + {}^{13}C_2 2^{11}(kx)^2 + ...$

$= 8192 + 53\,248kx + 159\,744k^2x^2 + ...$

b) Coefficient of x = $53\,248k$ and coefficient of $x^2 = 159\,744k^2$, so:

$6(53\,248k) = 159\,744k^2 \Rightarrow 319\,488k = 159\,744k^2$

$\Rightarrow 159\,744k = 319\,488 \Rightarrow k = 2$

Q8 $(1 - 3x)^8 = 1 + {}^8C_1(-3x) + {}^8C_2(-3x)^2 + {}^8C_3(-3x)^3 + ...$

$= 1 - 24x + 252x^2 - 1512x^3 + ...$

$1 - 3x = 0.97$ when $x = 0.01$

Hence:

$0.97^8 = 1 - 24(0.01) + 252(0.01)^2 - 1512(0.01)^3 + ...$

$\approx 0.7836... = 0.784$ (3 d.p.)

Q9 a) $\left(\frac{x}{3}-2\right)^6 = ... + {}^6C_3\left(\frac{x}{3}\right)^3(-2)^3 + {}^6C_4\left(\frac{x}{3}\right)^2(-2)^4 + {}^6C_5\left(\frac{x}{3}\right)(-2)^5 + {}^6C_6(-2)^6$

You only need the terms up to x^3.

In ascending powers of x, that's:

$64 - 64x + \frac{80}{3}x^2 - \frac{160}{27}x^3 + ...$

b) (i) $4x\left(\frac{x}{3}-2\right)^6 = 4x(64 - 64x + ...) = 256x - 256x^2 + ...$

$\approx 256x - 256x^2$

(ii) $(3 - x)\left(\frac{x}{3}-2\right)^6 = (3 - x)(64 - 64x + \frac{80}{3}x^2 - ...)$

$= 192 - 192x + 80x^2 - ...$

$- 64x + 64x^2 - ...$

$\approx 192 - 256x + 144x^2$

(iii) Expand $(x + 1)^{16}$ — you only need the terms up to the term in x^2:

$(x + 1)^{16} = ... + {}^{16}C_{14}x^2(1)^{14} + {}^{16}C_{15}x(1)^{15} + {}^{16}C_{16}(1)^{16}$

$= ... + 120x^2 + 16x + 1$

$\left(\frac{x}{3}-2\right)^6(x + 1)^{16} = (64 - 64x + \frac{80}{3}x^2 - ...) \times (... + 120x^2 + 16x + 1)$

$= ... + 7680x^2 + 1024x + 64 - ... - 1024x^2 - 64x + ... + \frac{80}{3}x^2$

$\approx 64 + 960x + (6656 + \frac{80}{3})x^2$

$= 64 + 960x + \frac{20\,048}{3}x^2$

Q10 a) $(1 + x)^{15} = 1 + {}^{15}C_1x + {}^{15}C_2x^2 + {}^{15}C_3x^3 + {}^{15}C_4x^4 + ...$

$= 1 + 15x + 105x^2 + 455x^3 + 1365x^4 + ...$

b) To find 2.01^{15}, you would need to substitute $x = 1.01$. Since $x > 1$, x^n is not small enough to ignore for large n, so the estimate would be inaccurate.

Q11 a) (i) $(2 + x)^6 = 2^6 + {}^6C_1 2^5x + {}^6C_2 2^4x^2 + {}^6C_3 2^3x^3 + ...$

$= 64 + 192x + 240x^2 + 160x^3 + ...$

$2 + x = 2.5$ when $x = 0.5$

Hence:

$2.5^6 = 64 + 192(0.5) + 240(0.5)^2 + 160(0.5)^3 + ... \approx 240$

(ii) $(3 - x)^6 = 3^6 + {}^6C_1 3^5(-x) + {}^6C_2 3^4(-x)^2 + {}^6C_3 3^3(-x)^3 + ...$

$= 729 - 1458x + 1215x^2 - 540x^3 + ...$

$3 - x = 2.5$ when $x = 0.5$

Hence:

$2.5^6 = 729 - 1458(0.5) + 1215(0.5)^2 - 540(0.5)^3 + ...$

≈ 236.25

b) Since 244.14 is closer to 240 than to 236.25, $(2 + x)^6$ provides the best approximation.

Exam-Style Questions — Chapter 7

Q1 a) $\binom{37}{33} = \frac{37!}{33!(37-33)!}$, so $\frac{37!}{33!k!} = \frac{37!}{33!(37-33)!}$

Therefore $k = (37 - 33) = 4$ *[1 mark]*

b) $p = \binom{37}{33}(-1)^{33} = -66\,045$

[2 marks available — 1 mark for a correct method to find p, 1 mark for the correct value of p]

Q2 $(n - 4)! = 33! \times 34 \times 35 = 35!$

$n - 4 = 35$, so $n = 39$

[2 marks available — 1 mark for identifying that (n – 4)! = 35!, 1 mark for the correct value of n]

Q3 a) $(2 + 2x)^5 = 2^5 + {}^5C_1 2^4(2x) + {}^5C_2 2^3(2x)^2 + {}^5C_3 2^2(2x)^3 + ...$

$= 2^5 + 5 \times 2^4 \times 2x + 10 \times 2^3 \times (2x)^2 + 10 \times 2^2 \times (2x)^3 + ...$

$= 32 + 160x + 320x^2 + 320x^3 + ...$

[4 marks available — 1 mark for substituting into the binomial formula correctly, 1 mark for one correct simplified term, 1 mark for a second correct simplified term, 1 mark for a fully correct expansion]

b) $2 + 2x = 2.04$ when $x = 0.02$

Hence:

$2.04^5 = 32 + 160(0.02) + 320(0.02)^2 + 320(0.02)^3 + ...$

$\approx 35.3305... = 35.331$ (3 d.p.)

[2 marks available – 1 mark for letting x = 0.02 in the expansion from part a), 1 mark for the correct answer (allow answer correctly calculated from an incorrect expansion in part a)]

Q4 $(x - x^{-1})^3 = x^3 + {}^3C_1x^2(-x^{-1}) + {}^3C_2x(-x^{-1})^2 + (-x^{-1})^3$

$= x^3 + 3x^2(-x^{-1}) + 3x(-x^{-1})^2 + (-x^{-1})^3$

$= x^3 - 3x + 3x^{-1} - x^{-3}$

[4 marks available — 1 mark for substituting into the binomial formula correctly, 1 mark for one correct simplified term, 1 mark for a second correct simplified term, 1 mark for a fully correct expansion]

Q5 The x^2 term is given by $\binom{n}{2}(-3x)^2 = \frac{n(n-1)}{2} \times 9x^2$.

Therefore $\frac{9n(n-1)}{2} = 495$

$\Rightarrow 9n(n - 1) = 990 \Rightarrow n(n - 1) = 110 \Rightarrow n^2 - n - 110 = 0$

$\Rightarrow (n + 10)(n - 11) = 0$

So $n = -10$ or $n = 11$. Since n is positive, $n = 11$.

[4 marks available — 1 mark for the correct x^2 term in terms of n, 1 mark for a correct method leading to a quadratic equation, 1 mark for the correct quadratic equation, 1 mark for the correct answer]

Q6 a) $(5 + 2x)^4 = 5^4 + {}^4C_1 5^3(2x) + {}^4C_2 5^2(2x)^2 + {}^4C_3 5(2x)^3 + (2x)^4$

$= 5^4 + 4 \times 5^3 \times 2x + 6 \times 5^2 \times (2x)^2 + 4 \times 5 \times (2x)^3 + (2x)^4$

$= 625 + 1000x + 600x^2 + 160x^3 + 16x^4$

[4 marks available — 1 mark for substituting into the binomial formula correctly, 1 mark for one correct simplified term, 1 mark for a further two correct simplified terms, 1 mark for a fully correct expansion]

b) $(5 - 2x)^4 = 625 - 1000x + 600x^2 - 160x^3 + 16x^4$

[2 marks available — 1 mark for negating terms involving odd powers of x in the expansion from a), or for a correct method to expand, 1 mark for the correct expansion]

c) $(5 + 2x)^4 + (5 - 2x)^4 = 1250 + 1200x^2 + 32x^4$
Hence:
$(5+2\sqrt{3})^4+(5-2\sqrt{3})^4 = 1250 + 1200(\sqrt{3})^2 + 32(\sqrt{3})^4$
$= 5138$
[3 marks available — 1 mark for adding together the two expansions from a) and b), 1 mark for substituting $x = \sqrt{3}$, 1 mark for the correct answer]

Q7 a) $\left(1-\frac{x}{2}\right)^8 = 1 + {}^8C_1\left(-\frac{x}{2}\right) + {}^8C_2\left(-\frac{x}{2}\right)^2 + ... = 1 - 4x + 7x^2 + ...$
[3 marks available — 1 mark for substituting into the binomial formula correctly, 1 mark for one correct simplified term, 1 mark for a fully correct expansion]

b) $f(x) = \left(1-\frac{x}{2}\right)^8 (p + qx) = (1 - 4x + 7x^2 + ...)(p + qx)$
x term: $(1 \times qx) + (-4x \times p) = qx - 4px = (q - 4p)x$
x^2 term: $(-4x \times qx) + (7x^2 \times p) = -4qx^2 + 7px^2 = (-4q + 7p)x^2$
So coefficient of the x term is $q - 4p$ and coefficient of the x^2 term is $-4q + 7p$, and you can form the following simultaneous equations:
$q - 4p = -26$ (1)
$-4q + 7p = 50$ (2)
(1) × 4:
$4q - 16p = -104$ (3)
Add (2) and (3) to solve for p and q:
$-9p = -54 \Rightarrow p = 6$
$q = -26 + 4 \times 6 = -2$
[6 marks available — 1 mark for a correct method to find either the coefficient of x or x² in the expansion of f(x), 1 mark each for the correct coefficients of x and x² in terms of p and q, 1 mark for the correct simultaneous equations, 1 mark for the correct value of p or for a correct method to solve the simultaneous equations, 1 mark for the correct values of both p and q]

Chapter 8 — Trigonometry

8.1 The Sine and Cosine Rules

Exercise 8.1.1 — The sine rule

Q1 The angle at P measured anticlockwise from the positive x-axis is $41° + 90° = 131°$.
P is on the unit circle, so the coordinates are $(\cos 131°, \sin 131°) = (-0.656, 0.755)$ (3 s.f.)
The angle at Q measured anticlockwise from the positive x-axis is $180° + 52° = 232°$.
Q is on the unit circle, so the coordinates are $(\cos 232°, \sin 232°) = (-0.616, -0.788)$ (3 s.f.)
The angle at R measured anticlockwise from the positive x-axis is $360° - 23° = 337°$.
R is on the unit circle, so the coordinates are $(\cos 337°, \sin 337°) = (0.921, -0.391)$ (3 s.f.)

Q2 a) The coordinates of S are $(\cos\theta, \sin\theta)$ since it's on the unit circle.
$\cos\theta = 0.899 \Rightarrow \theta = \cos^{-1}(0.899) = 26.0°$ (3 s.f.)
You can check the answer using the y-coordinate:
$\sin\theta = 0.438 \Rightarrow \theta = \sin^{-1}(0.438) = 26.0°$ (3 s.f.)

b) $\cos\theta = 0.669 \Rightarrow \theta = \cos^{-1}(0.669) = 48.0°$ (3 s.f.)

c) $\cos\theta = 0.089 \Rightarrow \theta = \cos^{-1}(0.089) = 84.9°$ (3 s.f.)

Q3 Using the sine rule: $\frac{a}{\sin A} = \frac{b}{\sin B}$
$\Rightarrow TW = \frac{FW \times \sin F}{\sin T} = \frac{15 \times \sin 39°}{\sin 82°} = 9.53$ cm (3 s.f.)

Q4 a) Using the sine rule: $\frac{a}{\sin A} = \frac{b}{\sin B}$
$\frac{\sin x}{17} = \frac{\sin 78°}{19} \Rightarrow \sin x = \frac{17 \times \sin 78°}{19}$
$\Rightarrow x = 61.1°$ (3 s.f.)

b) Using the sine rule: $\frac{a}{\sin A} = \frac{b}{\sin B}$
$\frac{x}{\sin 37°} = \frac{14}{\sin 102°} \Rightarrow x = \frac{14 \times \sin 37°}{\sin 102°} = 8.61$ cm (3 s.f.)

c) Using the sine rule: $\frac{a}{\sin A} = \frac{b}{\sin B}$
$\frac{\sin x}{27} = \frac{\sin 24°}{13} \Rightarrow \sin x = \frac{27 \times \sin 24°}{13}$
$\Rightarrow x = 57.6°$ (3 s.f.)

d) The unlabelled angle is $180° - 22° - 29° = 129°$.
Using the sine rule: $\frac{a}{\sin A} = \frac{b}{\sin B}$
$\frac{x}{\sin 29°} = \frac{38}{\sin 129°} \Rightarrow x = \frac{38 \times \sin 29°}{\sin 129°} = 23.7$ m (3 s.f.)

e) You can't find x directly from these values, so find the unlabelled angle instead (call it y) and then use angles in a triangle to find x.
Using the sine rule: $\frac{a}{\sin A} = \frac{b}{\sin B}$
$\frac{\sin y}{6} = \frac{\sin 71°}{11} \Rightarrow \sin y = \frac{6 \times \sin 71°}{11} \Rightarrow y = 31.046...°$
So $x = 180° - 71° - 31.046...° = 78.0°$ (3 s.f.)

f) The unlabelled angle is $180° - 52° - 63° = 65°$
Using the sine rule, $\frac{x}{\sin 52°} = \frac{57}{\sin 65°}$
$\Rightarrow x = \frac{57 \times \sin 52°}{\sin 65°} = 49.6$ mm (3 s.f.)

Q5 Angle $Q = 180° - 38° - 43° = 99°$
Using the sine rule: $\frac{a}{\sin A} = \frac{b}{\sin B}$
$\frac{PQ}{\sin 43°} = \frac{48}{\sin 99°} \Rightarrow PQ = \frac{48 \sin 43°}{\sin 99°} = 33.1$ m (3 s.f.)

Q6 Using the sine rule: $\frac{a}{\sin A} = \frac{b}{\sin B}$
$\frac{\sin B}{14} = \frac{\sin 67°}{17} \Rightarrow \sin B = \frac{14 \times \sin 67°}{17} \Rightarrow B = 49.293...°$
So $A = 180° - 67° - 49.293...° = 63.7°$ (3 s.f.)

Q7 The diagram below models the information given, which is then simplified. You want to find W:

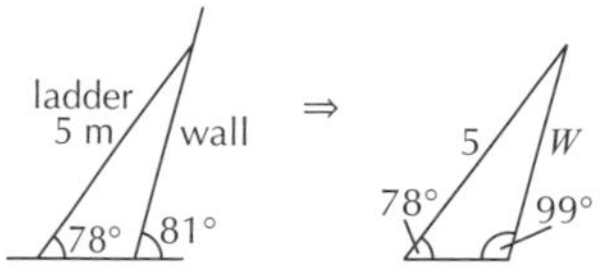

Using the sine rule: $\frac{a}{\sin A} = \frac{b}{\sin B}$
$\frac{W}{\sin 78°} = \frac{5}{\sin 99°} \Rightarrow W = \frac{5 \sin 78°}{\sin 99°} = 4.95$ m (3 s.f.)

Q8 Start by sketching the triangle:

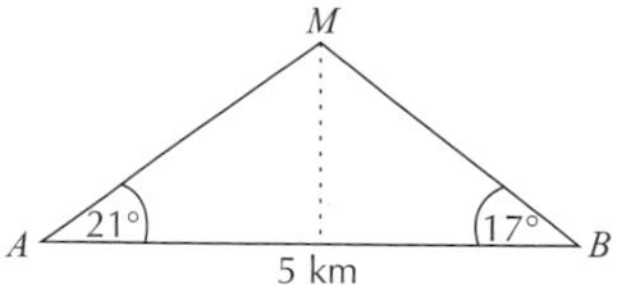

a) Angle $M = 180° - 21° - 17° = 142°$.
Using the sine rule: $\frac{a}{\sin A} = \frac{m}{\sin M}$
$\Rightarrow a = \frac{5 \times \sin 21°}{\sin 142°} = 2.91043... = 2.91$ km (3 s.f.)
Here a is the distance BM and m is the distance AB.

b) To find the height, draw a line through the triangle from M at a right angle to AB (the dotted line shown in the diagram above).
Height $= \sin 17° \times 2.91043... = 0.85092...$ km
$= 851$ m (to the nearest m)
The final step just uses SOHCAHTOA — height is the opposite side and 2.91043... is the hypotenuse.

Exercise 8.1.2 — The cosine rule

Q1 Using the cosine rule: $a^2 = b^2 + c^2 - 2bc\cos A$
$QR^2 = 9^2 + 10^2 - (2 \times 9 \times 10 \times \cos 42°) = 47.2...$
$QR = 6.87$ cm (3 s.f.)

Q2 Using the cosine rule: $a^2 = b^2 + c^2 - 2bc \cos A$

$\Rightarrow \cos A = \dfrac{b^2 + c^2 - a^2}{2bc}$

$\Rightarrow D = \cos^{-1}\left(\dfrac{6^2 + 9^2 - 8^2}{2 \times 6 \times 9}\right) = 60.6°$ (3 s.f.)

$\Rightarrow E = \cos^{-1}\left(\dfrac{6^2 + 8^2 - 9^2}{2 \times 6 \times 8}\right) = 78.6°$ (3 s.f.)

$\Rightarrow F = \cos^{-1}\left(\dfrac{8^2 + 9^2 - 6^2}{2 \times 8 \times 9}\right) = 40.8°$ (3 s.f.)

You could find the third angle by subtracting the other two from 180°.

Q3 **a)** Using the cosine rule: $a^2 = b^2 + c^2 - 2bc \cos A$

$\cos x = \dfrac{15^2 + 12^2 - 9^2}{2 \times 15 \times 12}$

$\Rightarrow x = \cos^{-1}\left(\dfrac{15^2 + 12^2 - 9^2}{2 \times 15 \times 12}\right) = 36.9°$ (3 s.f.)

b) Using the cosine rule: $a^2 = b^2 + c^2 - 2bc \cos A$

$\cos x = \dfrac{14^2 + 12^2 - 21^2}{2 \times 14 \times 12}$

$\Rightarrow x = \cos^{-1}\left(\dfrac{14^2 + 12^2 - 21^2}{2 \times 14 \times 12}\right) = 107°$ (3 s.f.)

c) Using the cosine rule: $a^2 = b^2 + c^2 - 2bc \cos A$

$\cos x = \dfrac{10^2 + 7^2 - 4^2}{2 \times 10 \times 7}$

$\Rightarrow x = \cos^{-1}\left(\dfrac{10^2 + 7^2 - 4^2}{2 \times 10 \times 7}\right) = 18.2°$ (3 s.f.)

d) Using the cosine rule: $a^2 = b^2 + c^2 - 2bc \cos A$

$x^2 = 43^2 + 17^2 - (2 \times 43 \times 17 \times \cos 42°)$

$x = \sqrt{1051.5...} = 32.4$ cm (3 s.f.)

e) Using the cosine rule: $a^2 = b^2 + c^2 - 2bc \cos A$

$x^2 = 56^2 + 32^2 - (2 \times 56 \times 32 \times \cos 27°)$

$x = \sqrt{966.63...} = 31.1$ mm (3 s.f.)

f) Using the cosine rule: $a^2 = b^2 + c^2 - 2bc \cos A$

$x^2 = 7^2 + 13^2 - (2 \times 7 \times 13 \times \cos 54°)$

$x = \sqrt{111.02...} = 10.5$ cm (3 s.f.)

Q4 Using the cosine rule: $a^2 = b^2 + c^2 - 2bc \cos A$

$\Rightarrow (JK)^2 = 24^2 + 29^2 - (2 \times 24 \times 29 \times \cos 62°)$

$\Rightarrow JK = \sqrt{763.4...} = 27.6$ cm (3 s.f.)

Q5 Using the cosine rule: $a^2 = b^2 + c^2 - 2bc \cos A$

$\Rightarrow (BC)^2 = 32^2 + 28^2 - (2 \times 32 \times 28 \times \cos 48°)$

$\Rightarrow BC = \sqrt{608.9...} = 24.7$ cm (3 s.f.)

Q6 The smallest angle is opposite the shortest side, so angle F is the smallest angle.

To be safe you could just work out all 3 angles and then see which is smallest.

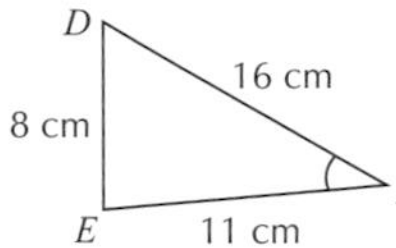

Using the cosine rule: $a^2 = b^2 + c^2 - 2bc \cos A$

$\Rightarrow \cos F = \dfrac{11^2 + 16^2 - 8^2}{2 \times 11 \times 16}$

$\Rightarrow F = \cos^{-1}\left(\dfrac{11^2 + 16^2 - 8^2}{2 \times 11 \times 16}\right) = 27.2°$ (3 s.f.)

Q7 The largest angle is opposite the longest side, so angle R is the largest angle.

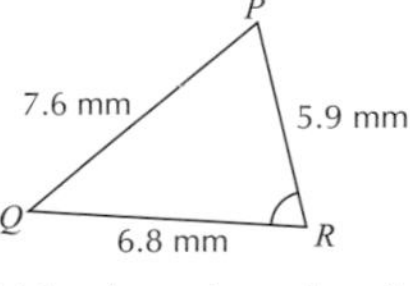

Using the cosine rule: $a^2 = b^2 + c^2 - 2bc \cos A$

$\Rightarrow \cos R = \dfrac{6.8^2 + 5.9^2 - 7.6^2}{2 \times 6.8 \times 5.9}$

$\Rightarrow R = \cos^{-1}\left(\dfrac{6.8^2 + 5.9^2 - 7.6^2}{2 \times 6.8 \times 5.9}\right) = 73.1°$ (3 s.f.)

Q8 Find the lengths XY, YZ and ZX.

$XY = \sqrt{(5 - -2)^2 + (8 - 2)^2} = \sqrt{85}$

$YZ = \sqrt{(5 - 3)^2 + (8 - -2)^2} = \sqrt{104}$

$ZX = \sqrt{(-2 - 3)^2 + (2 - -2)^2} = \sqrt{41}$

Using the cosine rule: $\cos A = \dfrac{b^2 + c^2 - a^2}{2bc}$.

So angle $XYZ = \cos^{-1}\left(\dfrac{104 + 85 - 41}{2 \times \sqrt{104} \times \sqrt{85}}\right) = 38.1°$ (3 s.f.)

It might help to draw a sketch here to make sure you're using the sides in the correct place in the cosine rule.

Q9 **a)**

Using the cosine rule: $a^2 = b^2 + c^2 - 2bc \cos A$

$(AC)^2 = (x - 1)^2 + (2x)^2 - 2(x - 1)(2x) \cos 60°$

$= x^2 - 2x + 1 + 4x^2 - 2(x - 1)(2x) \times \frac{1}{2}$

$= 5x^2 - 2x + 1 - 2x^2 + 2x$

$= 3x^2 + 1$

So $(\sqrt{13})^2 = 13 = 3x^2 + 1 \Rightarrow x^2 = 4 \Rightarrow x = 2$

You can ignore the negative value of x because it is a length.

b) The smallest angle is opposite the shortest side.

$AB = 2 - 1 = 1$, $BC = 2 \times 2 = 4$ and $AC = \sqrt{13}$.

AB is the shortest side, so C is the smallest angle.

Using the cosine rule: $a^2 = b^2 + c^2 - 2bc \cos A$

$\cos C = \dfrac{4^2 + \sqrt{13}^2 - 1^2}{2 \times 4 \times \sqrt{13}}$

$\Rightarrow C = \cos^{-1}\left(\dfrac{4^2 + \sqrt{13}^2 - 1^2}{2 \times 4 \times \sqrt{13}}\right) = 13.9°$ (3 s.f.)

Exercise 8.1.3 — Finding the area of a triangle

Q1 **a)** Area $= \frac{1}{2}ab \sin C = \frac{1}{2} \times 12 \times 10.5 \times \sin 53°$

$= 50.3$ cm² (3 s.f.)

b) Area $= \frac{1}{2}ab \sin C = \frac{1}{2} \times 9 \times 5 \times \sin 41° = 14.8$ mm² (3 s.f.)

c) Start by finding an angle (any angle is fine).

Using the cosine rule: $a^2 = b^2 + c^2 - 2bc \cos A$

$\Rightarrow \cos A = \dfrac{5^2 + 7^2 - 4.2^2}{2 \times 5 \times 7}$

$\Rightarrow A = \cos^{-1}\left(\dfrac{5^2 + 7^2 - 4.2^2}{2 \times 5 \times 7}\right) = 36.3...°$

Now you can find the area:

Area $= \frac{1}{2}ab \sin C = \frac{1}{2} \times 5 \times 7 \times \sin 36.3...°$

$= 10.4$ cm² (3 s.f.)

You could have found any angle to start off, then used the corresponding sides.

d) Find the length of another side (either side is fine):

Using the sine rule: $\dfrac{a}{\sin A} = \dfrac{b}{\sin B}$

$\dfrac{x}{\sin 94°} = \dfrac{24}{\sin 32°} \Rightarrow x = \dfrac{24 \times \sin 94°}{\sin 32°} = 45.17...$ m

The unlabelled angle is $180° - 94° - 32° = 54°$.

Now you can find the area:

Area $= \frac{1}{2}ab \sin C = \frac{1}{2} \times 24 \times 45.17... \times \sin 54°$

$= 439$ cm² (3 s.f.)

Q2 Area $= \frac{1}{2}ab \sin C = \frac{1}{2} \times 4 \times 7 \times \sin 49°$

$= 10.565... = 10.6$ cm² (3 s.f.)

Q3 Using the cosine rule $a^2 = b^2 + c^2 - 2bc \cos A$

$A = \cos^{-1}\left(\dfrac{1.9^2 + 2.7^2 - 2.9^2}{2 \times 1.9 \times 2.7}\right) = 75.954...°$

Then use the area formula:

Area $= \frac{1}{2} \times 1.9 \times 2.7 \times \sin 75.954...°$

$= 2.488... = 2.49$ m² (3 s.f.)

Q4 Area = 30 = $\frac{1}{2}ab\sin C = \frac{1}{2} \times 12 \times 8 \times \sin C$

$\Rightarrow \sin C = \frac{30}{48} \Rightarrow C = \sin^{-1}\left(\frac{30}{48}\right) = 38.7°$ (1 d.p.)

Q5 Find another side using the sine rule: $\frac{a}{\sin A} = \frac{b}{\sin B}$

$\frac{x}{\sin 35°} = \frac{14}{\sin 52°} \Rightarrow x = \frac{14 \times \sin 35°}{\sin 52°} = 10.19...$ cm

The unlabelled angle is 180° – 35° – 52° = 93°.
Now you can find the area:
Area = $\frac{1}{2}ab\sin C = \frac{1}{2} \times 14 \times 10.19... \times \sin 93°$

= 71.2 cm² (3 s.f.)

Q6 Find an angle in the triangle using the cosine rule:
$a^2 = b^2 + c^2 - 2bc\cos A$

$\cos A = \frac{7^2 + 13^2 - 16^2}{2 \times 7 \times 13}$

$\Rightarrow A = \cos^{-1}\left(\frac{7^2 + 13^2 - 16^2}{2 \times 7 \times 13}\right) = 102.05...°$

So the area = $\frac{1}{2}ab\sin C = \frac{1}{2} \times 7 \times 13 \times \sin 102.05...°$
= 44.5 cm² (3 s.f.)

Q7 a)

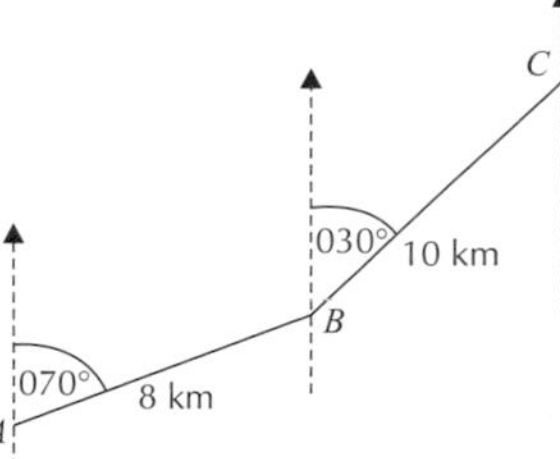

b) The angle anticlockwise from the vertical at B to A is 180° – 70° = 110° (parallel lines), so the angle ABC is 110° + 30° = 140°. Now you can use the cosine rule to find the distance AC:
$a^2 = b^2 + c^2 - 2bc\cos A$

$\Rightarrow AC = \sqrt{(BC)^2 + (AB)^2 - 2(BC)(AB)\cos B}$

$\Rightarrow AC = \sqrt{100 + 64 - 160\cos 140°}$

= 16.928... = 16.9 km (3 s.f.)

c) Find the angle ACB using the cosine rule:
$a^2 = b^2 + c^2 - 2bc\cos A$

$\Rightarrow \cos ACB = \frac{10^2 + 16.928...^2 - 8^2}{2 \times 10 \times 16.928...}$

$\Rightarrow ACB = \cos^{-1}\left(\frac{10^2 + 16.928...^2 - 8^2}{2 \times 10 \times 16.928...}\right) = 17.68°$ (2 d.p.)

The bearing required is therefore
180° + 17.68° + 30° (parallel lines) = 228° (3 s.f.).

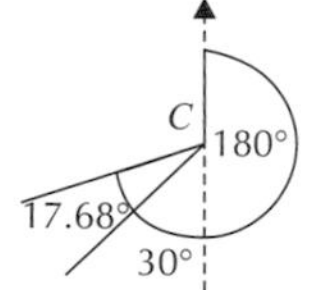

You could also have found angle ACB using the sine rule.

Q8 Using the cosine rule: $a^2 = b^2 + c^2 - 2bc\cos A$

$\Rightarrow AC = \sqrt{9^2 + 11^2 - (2 \times 9 \times 11 \times \cos 148°)}$

= 19.233... = 19.2 m (3 s.f.)

Find the area of each triangle individually.
For the top triangle:
Area = $\frac{1}{2}ab\sin C = \frac{1}{2} \times 9 \times 11 \times \sin 148° = 26.231...$m²
For the bottom triangle:
Area = $\frac{1}{2} \times 8 \times 19.233... \times \sin 79° = 75.519...$ m²
So the area of the quadrilateral is
26.231... + 75.519... = 101.750... = 102 m² (3 s.f.).

Q9 a) Find QS using the cosine rule:
$a^2 = b^2 + c^2 - 2bc\cos A$
$(QS)^2 = 3^2 + 4^2 - (2 \times 3 \times 4 \times \cos 35°)$
$QS = \sqrt{5.3403...} = 2.310... = 2.31$ cm (3 s.f.)

b) Split the shape into two triangles as shown below.

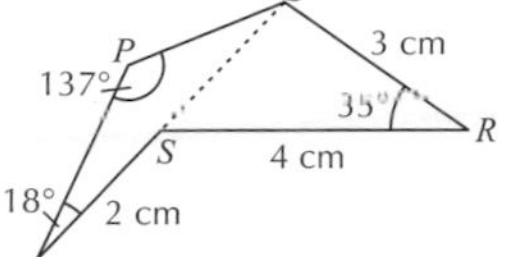

Find the area of each triangle individually. For QRS:
Area = $\frac{1}{2}ab\sin C = \frac{1}{2} \times 3 \times 4 \times \sin 35° = 3.441...$ cm²
For PQT:
Side QT = 2 + 2.310... = 4.310... cm
Find another side using the sine rule: $\frac{a}{\sin A} = \frac{b}{\sin B}$

Here, we find PQ: $\frac{PQ}{\sin 18°} = \frac{4.310...}{\sin 137°}$

$\Rightarrow PQ = \frac{4.310... \times \sin 18°}{\sin 137°} = 1.953...$ m

The missing angle is 180° – 18° – 137° = 25°.
Now you can find the area:
Area = $\frac{1}{2}ab\sin C = \frac{1}{2} \times 4.310... \times 1.952... \times \sin 25°$
= 1.779... cm² (3 s.f.)

So the total area of the shape is:
3.441... + 1.779... = 5.220... = 5.22 cm² (3 s.f.)

Q10 a) Find the area of each triangle individually:
In the top triangle:
Area = $\frac{1}{2}ab\sin C = \frac{1}{2} \times 11 \times 7 \times \sin 47° = 28.157...$ cm²
In the bottom triangle, find the length of the dotted line using the cosine rule: $a^2 = b^2 + c^2 - 2bc\cos A$
$a^2 = 11^2 + 7^2 - 2 \times 11 \times 7 \times \cos 47°$
$a = \sqrt{64.972...} = 8.0605...$ cm
Area = $\frac{1}{2}ab\sin C = \frac{1}{2} \times 13 \times 8.0605... \times \sin 19°$
= 17.057... cm²

So the total area is:
28.157... + 17.057... = 45.214... = 45.2 cm² (3 s.f.)

b) Find the area of each triangle individually.
In the right-hand triangle:

Area = $\frac{1}{2}ab\sin C = \frac{1}{2} \times 27 \times 34 \times \sin 39° = 288.85...$ mm²
In the left-hand triangle, find the length of the dotted line using the cosine rule:
$a^2 = b^2 + c^2 - 2bc\cos A$
$a^2 = 27^2 + 34^2 - 2 \times 27 \times 34 \times \cos 39°$
$a = \sqrt{458.16...} = 21.404...$ mm
Then use the cosine rule again to find an angle:

$\cos A = \frac{12^2 + 14^2 - 458.16...}{2 \times 12 \times 14}$

$\Rightarrow A = \cos^{-1}\left(\frac{12^2 + 14^2 - 458.16...}{2 \times 12 \times 14}\right) = 110.58...°$

Now you can find the area:
Area = $\frac{1}{2}ab\sin C = \frac{1}{2} \times 12 \times 14 \times \sin 110.58...°$
= 78.634... mm²

Total area = 288.85... + 78.634...
= 367.49... = 367 mm² (3 s.f.)

Q11 a)

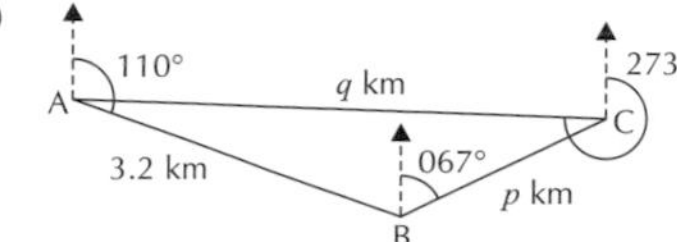

b) Using rules for angles around parallel lines:

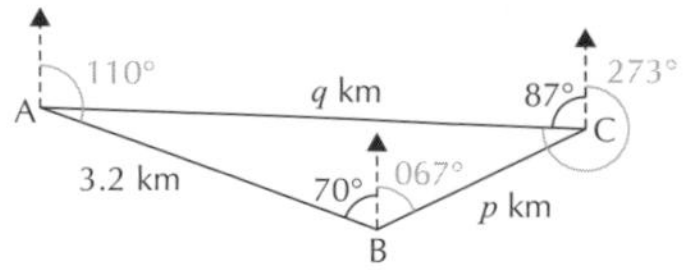

The angle anticlockwise from the bearing B to A is $180° - 110° = 70°$ so $\angle ABC = 70° + 67° = 137°$

The angle anticlockwise from the bearing C to B is $180° - 67° = 113°$ so $\angle ACB = 113° - 87° = 26°$

Then $\angle BAC = 180° - 26° - 137° = 17°$

You could also determine $\angle BAC$ from $110° + 87° - 180° = 17°$.

c) Using the sine rule: $\frac{a}{\sin A} = \frac{b}{\sin B}$

$$\frac{3.2}{\sin 26°} = \frac{p}{\sin 17°} = \frac{q}{\sin 137°},$$

so $p = \frac{3.2 \times \sin 17°}{\sin 26°} = 2.134...$ km

$q = \frac{3.2 \times \sin 137°}{\sin 26°} = 4.978...$ km

So the total length of the walk is
$3.2 + 2.134... + 4.978... = 10.3$ km (3 s.f.)

Q12 Area $= \frac{1}{2}ab \sin C = \frac{1}{2} \times 24 \times b \times \sin 41° = 220$

$\Rightarrow b = \frac{2 \times 220}{24 \times \sin 41°} = 27.944...$ cm

Now you have two sides and an angle, find the third side using the cosine rule: $a^2 = b^2 + c^2 - 2bc \cos A$

$a^2 = (27.944...)^2 + 24^2 - (2 \times 24 \times 27.944... \times \cos 41°)$

$a = \sqrt{344.56...} = 18.562...$ cm

So the total perimeter $= 24 + 27.944... + 18.562...$
$= 70.5$ cm (3 s.f.)

8.2 Trig Identities

Exercise 8.2.1 — Trig identities

Q1 Use $\tan\theta \equiv \frac{\sin\theta}{\cos\theta}$:

$$\frac{\sin\theta}{\tan\theta} - \cos\theta \equiv \frac{\sin\theta}{\left(\frac{\sin\theta}{\cos\theta}\right)} - \cos\theta \equiv \cos\theta - \cos\theta \equiv 0$$

Q2 Use $\sin^2\theta + \cos^2\theta \equiv 1$:
$\cos^2\theta \equiv 1 - \sin^2\theta \equiv (1 - \sin\theta)(1 + \sin\theta)$

Q3 Use $\sin^2 x + \cos^2 x \equiv 1$:
$\cos^2 x \equiv 1 - \sin^2 x$

$$\Rightarrow \cos x = \sqrt{1 - \sin^2 x} = \sqrt{1 - \left(\frac{1}{2}\right)^2} = \sqrt{\frac{3}{4}} = \frac{\sqrt{3}}{2}$$

You could have used your knowledge of common angles for this.

Q4 $\cos^2 x \equiv 1 - \sin^2 x$, $\tan x \equiv \frac{\sin x}{\cos x}$

$$\Rightarrow \tan x \equiv \frac{\sqrt{\sin^2 x}}{\sqrt{1 - \sin^2 x}} \equiv \frac{\frac{\sqrt{3}}{2}}{\frac{1}{2}} = \sqrt{3}$$

Q5 Use $\sin^2 x + \cos^2 x \equiv 1$:

$4\sin^2 x - 3\cos x + 1 \equiv 4(1 - \cos^2 x) - 3\cos x + 1$
$\equiv 4 - 4\cos^2 x - 3\cos x + 1$
$\equiv 5 - 3\cos x - 4\cos^2 x$

Q6 Use $\tan x \equiv \frac{\sin x}{\cos x}$ and $\sin^2 x + \cos^2 x \equiv 1$:

$$(\tan x + 1)(\tan x - 1) \equiv \tan^2 x - 1 \equiv \frac{\sin^2 x}{\cos^2 x} - 1$$

$$\equiv \frac{1 - \cos^2 x}{\cos^2 x} - 1 \equiv \frac{1}{\cos^2 x} - \frac{\cos^2 x}{\cos^2 x} - 1 \equiv \frac{1}{\cos^2 x} - 2$$

Q7 Here the student has divided both sides by $\sin\theta$. But $\sin\theta$ could equal 0, so you shouldn't divide by it. Instead the student should have rearranged:

$\cos\theta \sin\theta = \frac{1}{2}\sin\theta \Rightarrow \cos\theta\sin\theta - \frac{1}{2}\sin\theta = 0$

$\Rightarrow \sin\theta(\cos\theta - \frac{1}{2}) = 0.$

So there's a solution when $\cos\theta = \frac{1}{2} \Rightarrow \theta = 60°$,

as the student found. But there's also a solution when $\sin\theta = 0 \Rightarrow \theta = 0°$. This was not found because the student had cancelled the $\sin\theta$ terms.

Q8 Use $\sin^2 x + \cos^2 x \equiv 1$ and $\tan x \equiv \frac{\sin x}{\cos x}$:

$$\tan x + \frac{1}{\tan x} \equiv \frac{\sin x}{\cos x} + \frac{\cos x}{\sin x} \equiv \frac{\sin^2 x + \cos^2 x}{\sin x \cos x} \equiv \frac{1}{\sin x \cos x}$$

Q9 Use $\sin^2\theta + \cos^2\theta \equiv 1$:

$2\cos^2 x + 5\sin x + 1 \equiv 2(1 - \sin^2 x) + 5\sin x + 1$
$\equiv 2 - 2\sin^2 x + 5\sin x + 1$
$\equiv 3 + 5\sin x - 2\sin^2 x$
$\equiv (3 - \sin x)(2\sin x + 1)$

If you're struggling to factorise, let $y = \sin x$, then the expression becomes $3 + 5y - 2y^2$.

Q10 Use $\sin^2 x + \cos^2 x \equiv 1$:

$4 + \sin x - 6\cos^2 x \equiv 4 + \sin x - 6(1 - \sin^2 x)$
$\equiv -2 + \sin x + 6\sin^2 x$
$\equiv (2\sin x - 1)(3\sin x + 2)$

Q11 Use $\sin^2 x + \cos^2 x \equiv 1$:

$\sin^2 x \cos^2 y - \cos^2 x \sin^2 y$
$\equiv (1 - \cos^2 x)\cos^2 y - \cos^2 x(1 - \cos^2 y)$
$\equiv \cos^2 y - \cos^2 x \cos^2 y - \cos^2 x + \cos^2 x \cos^2 y$
$\equiv \cos^2 y - \cos^2 x$

Q12 Use $\tan x \equiv \frac{\sin x}{\cos x}$ and $\sin^2 x + \cos^2 x \equiv 1$:

$$\frac{\tan^2 x + 1}{\tan^2 x} \equiv \frac{\left(\frac{\sin^2 x}{\cos^2 x}\right) + 1}{\left(\frac{\sin^2 x}{\cos^2 x}\right)} \equiv \frac{\cos^2 x\left(\frac{\sin^2 x}{\cos^2 x}\right) + \cos^2 x}{\cos^2 x\left(\frac{\sin^2 x}{\cos^2 x}\right)}$$

$$\equiv \frac{\sin^2 x + \cos^2 x}{\sin^2 x} \equiv \frac{1}{\sin^2 x}$$

Q13 Use $\sin^2 x + \cos^2 x \equiv 1$:

$$\frac{\sin^4 x - \cos^4 x}{\sin^2 x - \cos^2 x} \equiv \frac{(\sin^2 x)^2 - (\cos^2 x)^2}{\sin^2 x - \cos^2 x}$$

$$\equiv \frac{(\sin^2 x - \cos^2 x)(\sin^2 x + \cos^2 x)}{(\sin^2 x - \cos^2 x)}$$

$\equiv \sin^2 x + \cos^2 x \equiv 1$

$\sin^4 x - \cos^4 x$ can be written as a difference of two squares.

Q14 Look at the right-angled triangle below:

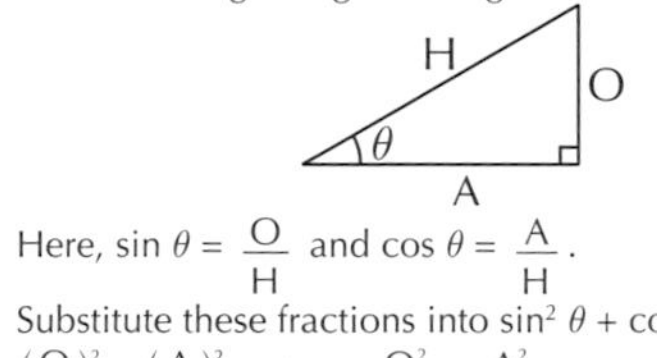

Here, $\sin\theta = \frac{O}{H}$ and $\cos\theta = \frac{A}{H}$.

Substitute these fractions into $\sin^2\theta + \cos^2\theta \equiv 1$:

$$\left(\frac{O}{H}\right)^2 + \left(\frac{A}{H}\right)^2 \equiv 1 \Rightarrow \frac{O^2}{H^2} + \frac{A^2}{H^2} \equiv 1$$

$\Rightarrow O^2 + A^2 \equiv H^2$ — this is Pythagoras' theorem.

8.3 Trig Functions

Exercise 8.3.1 — Graphs of trig functions

Q1

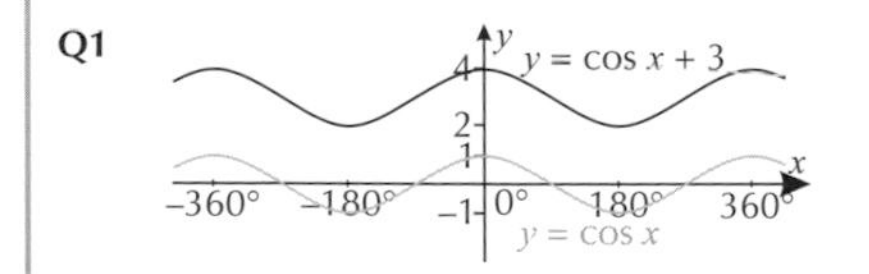

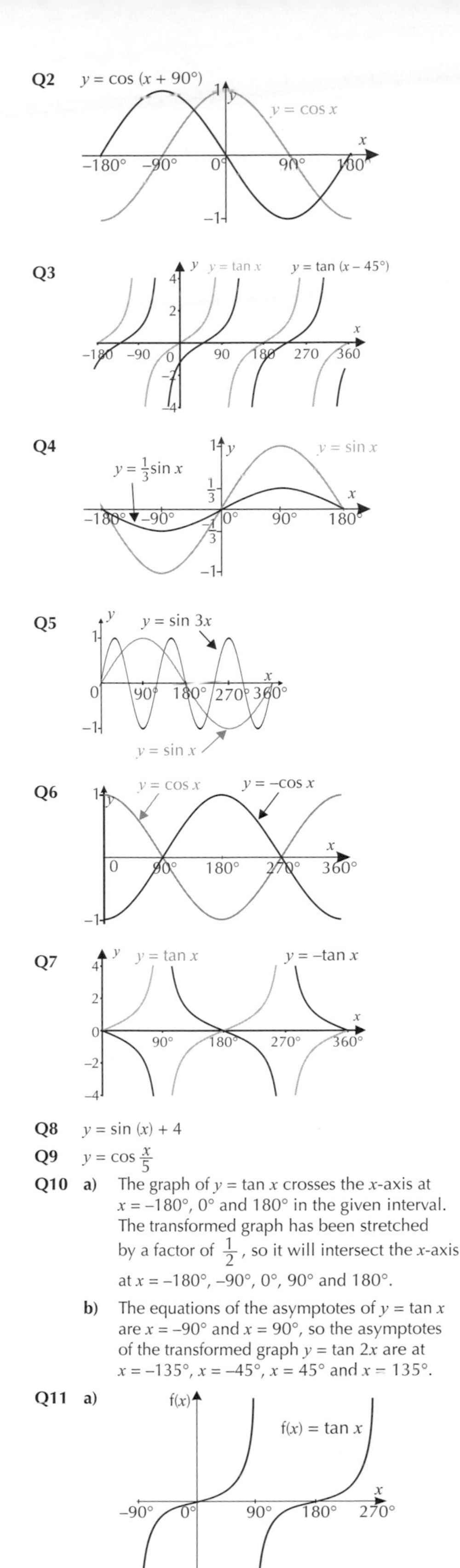

Q8 $y = \sin(x) + 4$

Q9 $y = \cos \frac{x}{5}$

Q10 a) The graph of $y = \tan x$ crosses the x-axis at $x = -180°$, $0°$ and $180°$ in the given interval. The transformed graph has been stretched by a factor of $\frac{1}{2}$, so it will intersect the x-axis at $x = -180°, -90°, 0°, 90°$ and $180°$.

b) The equations of the asymptotes of $y = \tan x$ are $x = -90°$ and $x = 90°$, so the asymptotes of the transformed graph $y = \tan 2x$ are at $x = -135°$, $x = -45°$, $x = 45°$ and $x = 135°$.

Q11 a)

b)

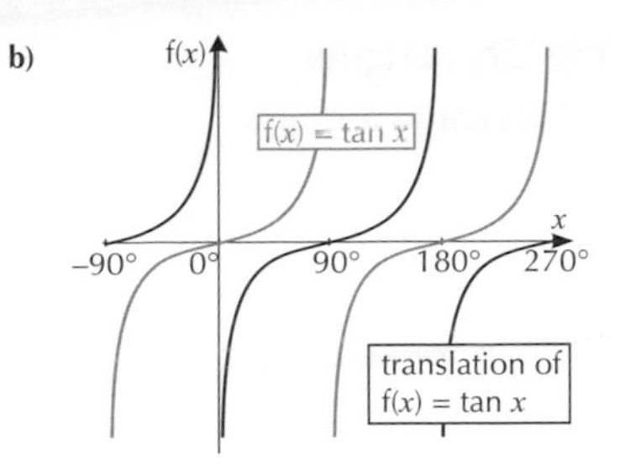

c) $f(x) = \tan(x + 90°)$

Because the graph of tan x repeats every 180°, the transformation could also be f(x) = tan (x – 90°).

Q12 a)

b)

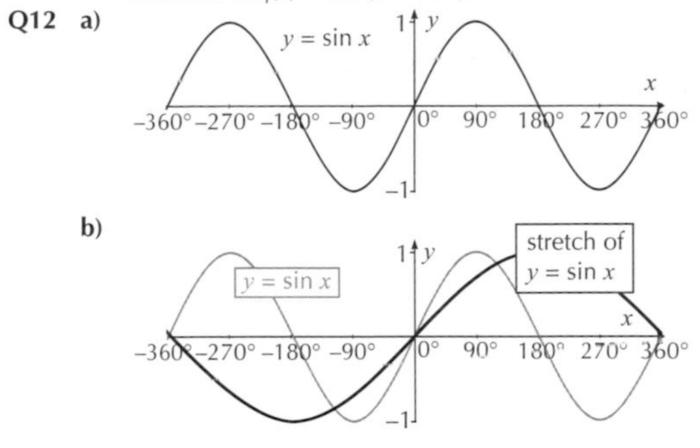

c) $y = \sin \frac{1}{2}x$

Q13 a) The graph has been translated to the left by 90°.

b) $y = \sin(x + 90°)$.

You might have noticed that this graph is exactly the same as the graph of y = cos x.

Q14 a) The graph is stretched vertically by a factor of 2.

b) $y = 2\cos x$

Q15 a) The graph is stretched horizontally by a factor of $\frac{1}{4}$.

b) $y = \tan 4x$

Q16 The graph of $y = \cos x$ crosses the x-axis at $x = 90°$. The transformed graph has the equation $y = \cos px$ (i.e. it's a horizontal stretch), so the graph must have been stretched by a factor of 2. So the equation of the transformed graph is $y = \cos \frac{x}{2}$, i.e. $p = \frac{1}{2}$.

p could be $n + \frac{1}{2}$ *for any integer n, but you are asked for the smallest positive value of p.*

Q17 a) A maximum of $y = \sin x$ occurs at (90°, 1). Since (60°, 1) is a maximum of the transformed graph, the graph of $\sin x$ may have translated 30° to the left. An equation for this transformed graph is $y = \sin(x + 30°)$, i.e. $q = 30°$.

b) $y = \sin x$ is periodic — it repeats itself every 360°. So q could also be $30° + 360° = 390°$,
$30° - 360° = -330°$,
$30° + 720° = 750°$ etc.

8.4 Solving Trig Equations

Exercise 8.4.1 — Sketching a graph

Q1 **a)** Find the first solution using a calculator:
$\sin x = 0.75 \Rightarrow x = 48.6°$ (1 d.p.).
Then sketch a graph:

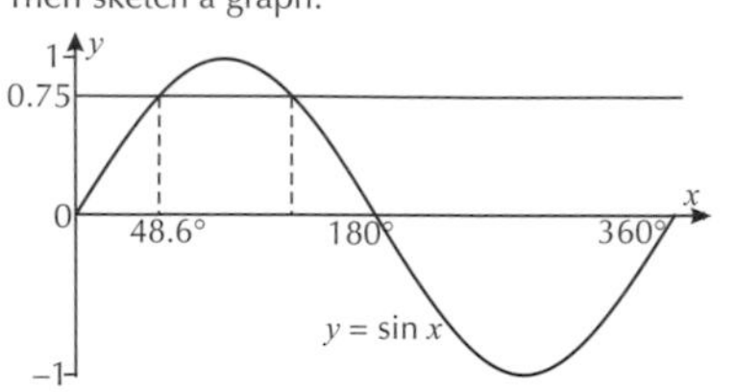

You can see from the graph that there are 2 solutions in the given interval. Using the symmetry of the graph, if one solution is at 48.6°, the other will be at $180° - 48.6° = 131.4°$ (1 d.p.).

b) Find the first solution: $\cos x = 0.31 \Rightarrow x = 71.9°$ (1 d.p.).
Then sketch a graph:

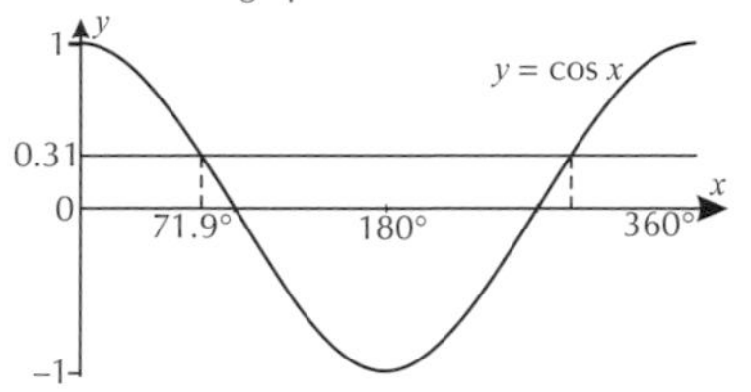

Using the symmetry of the graph to find the second solution: if one solution is at 71.9°, the other will be at $360° - 71.9° = 288.1°$ (1 d.p.).

c) Find the first solution: $\tan x = -1.5 \Rightarrow x = -56.3°$ (1 d.p.). This is outside the given interval, so add on 180° to find the first solution: $-56.3° + 180° = 123.7°$ (1 d.p.).
Then sketch a graph:

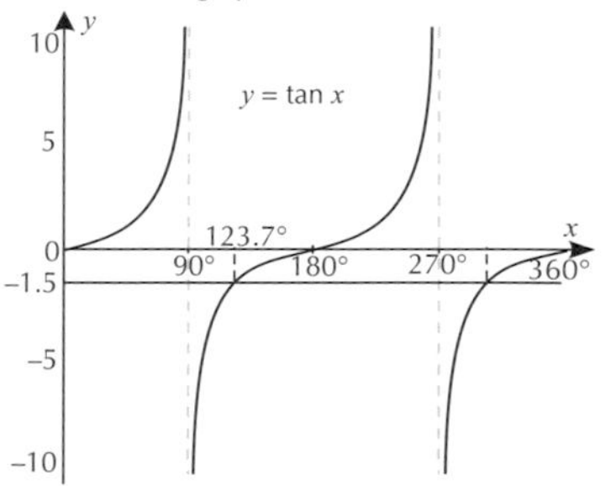

You can see from the graph that the next solution will be at $180° + 123.7° = 303.7°$ (1 d.p.) as $\tan x$ repeats every 180°.

d) Find the first solution: $\sin x = -0.42 \Rightarrow x = -24.8°$ (1 d.p.). This is outside the given interval, so add on 360° to find one solution: $-24.8° + 360° = 335.2°$ (1 d.p.). Then sketch a graph:

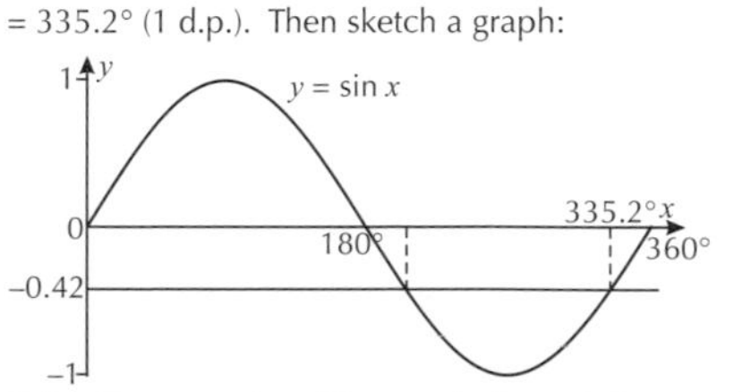

Using the symmetry of the graph, the solution you know is $360° - 335.2° = 24.8°$ away from 360°, so the other solution will be 24.8° away from 180°, i.e. at $180° + 24.8° = 204.8°$ (1 d.p.).

e) Find the first solution: $\cos x = -0.56 \Rightarrow x = 124.1°$ (1 d.p.).
Then sketch a graph:

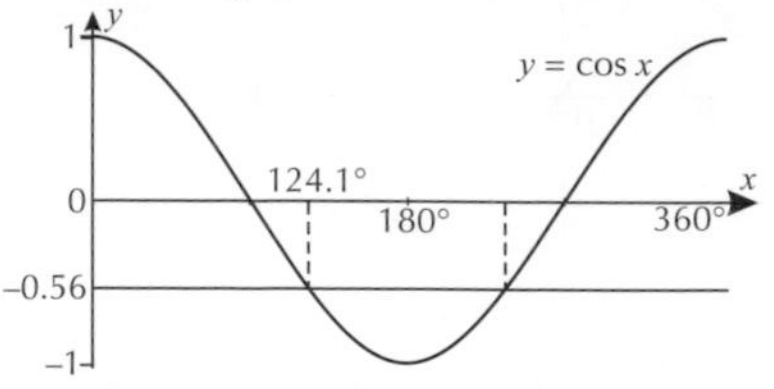

Using the symmetry of the graph to find the second solution: if one solution is at 124.1°, the other will be at $360° - 124.1° = 235.9°$ (1 d.p.).

f) Find the first solution: $\tan x = -0.67 \Rightarrow x = -33.8°$ (1 d.p.). This is outside the interval, so add 180° to find the first solution: $-33.8° + 180° = 146.2°$ (1 d.p.). Then sketch a graph:

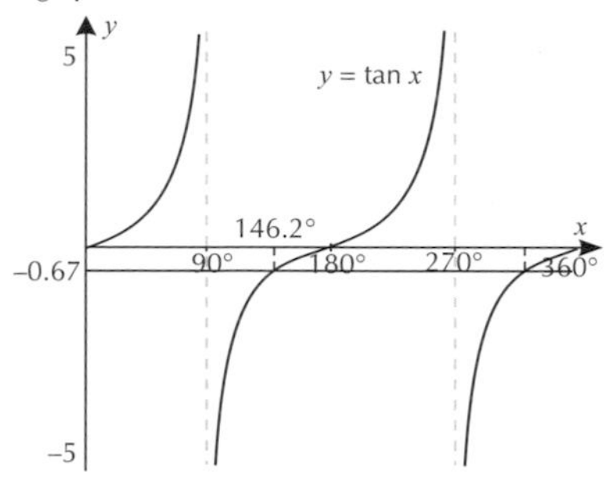

The next solution is at $180° + 146.2° = 326.2°$ (1 d.p.) as $\tan x$ repeats every 180°.

g) Find the first solution: $\sin x = 0.32 \Rightarrow x = 18.7°$ (1 d.p.).
Then sketch a graph:

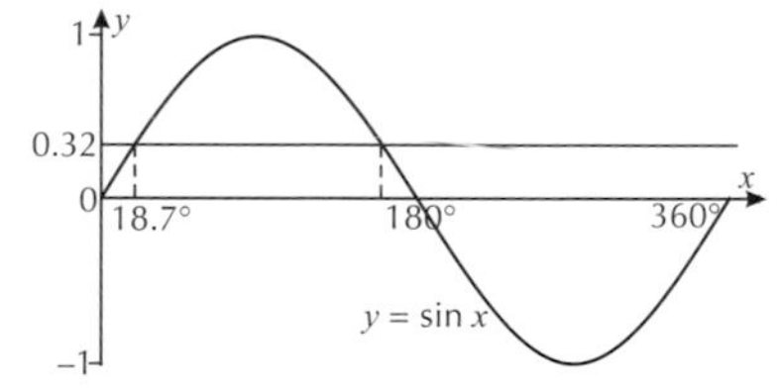

Using the symmetry of the graph, the other solution is $180° - 18.7° = 161.3°$ (1 d.p.).

h) Find the first solution: $\cos x = -0.89 \Rightarrow x = 152.9°$ (1 d.p.).
Then sketch a graph:

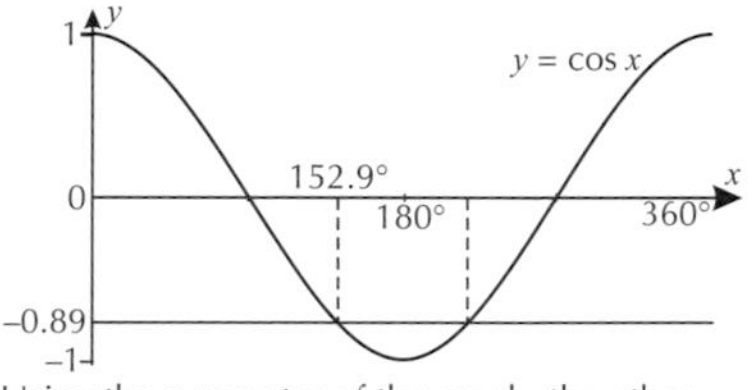

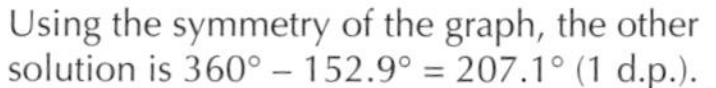

Using the symmetry of the graph, the other solution is $360° - 152.9° = 207.1°$ (1 d.p.).

i) Find the first solution: $\tan x = 2.3 \Rightarrow x = 66.5°$ (1 d.p.). Then sketch a graph:

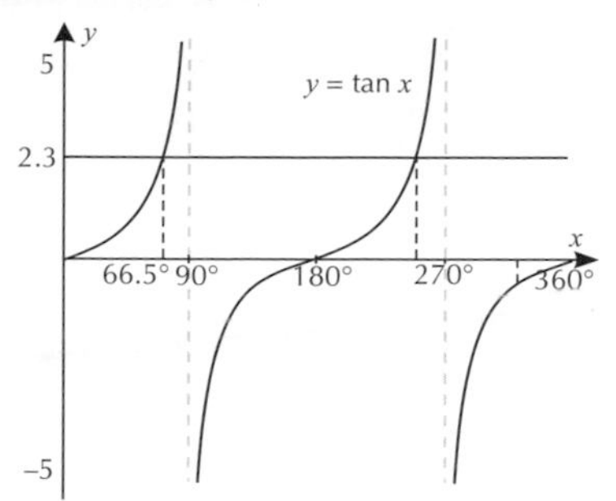

The next solution is at 180° + 66.5° = 246.5° (1 d.p.) as tan x repeats every 180°.

Q2 a) Using your knowledge of common angles, the first solution is at $x = 45°$. Then sketch a graph:

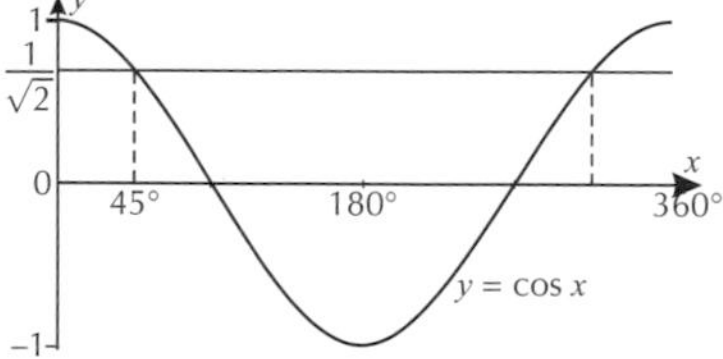

Using the symmetry of the graph, the second solution is at 360° – 45° = 315°.

b) Using common angles, the first solution is at $x = 60°$. Then sketch a graph:

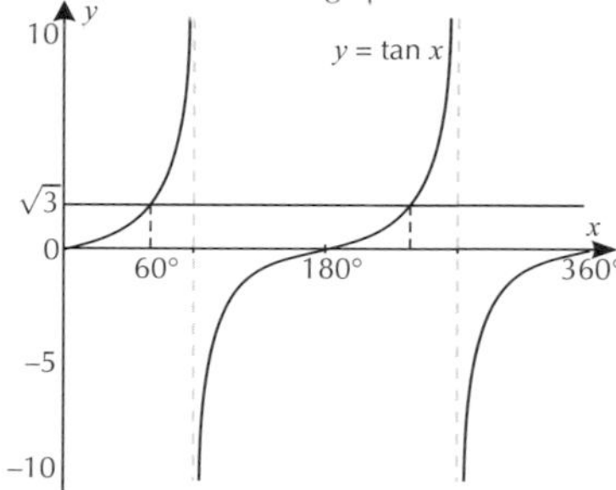

By the symmetry of the graph, the second solution is at 180° + 60° = 240°.

c) Using common angles, the first solution is at $x = 30°$. Then sketch a graph:

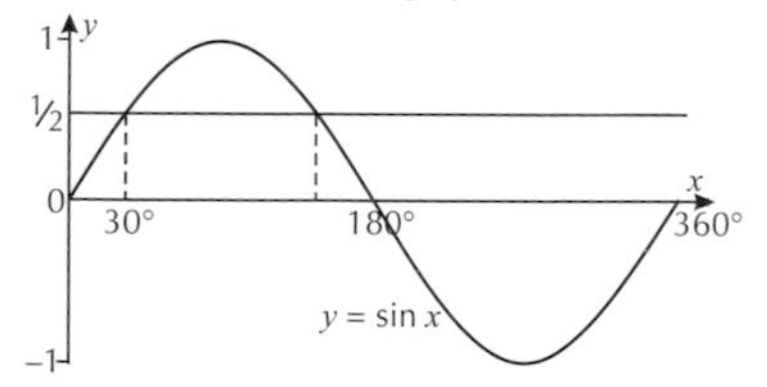

By the symmetry of the graph, the second solution is at 180° – 30° = 150°.

d) Using common angles, the first solution is at $x = 30°$. Then sketch a graph:

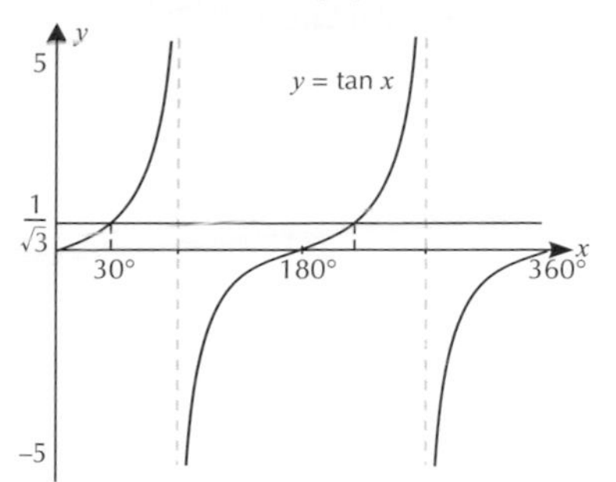

By the symmetry of the graph, the second solution is at 180° + 30° = 210°.

e) Using common angles, the first solution is at $x = 45°$. Then sketch a graph:

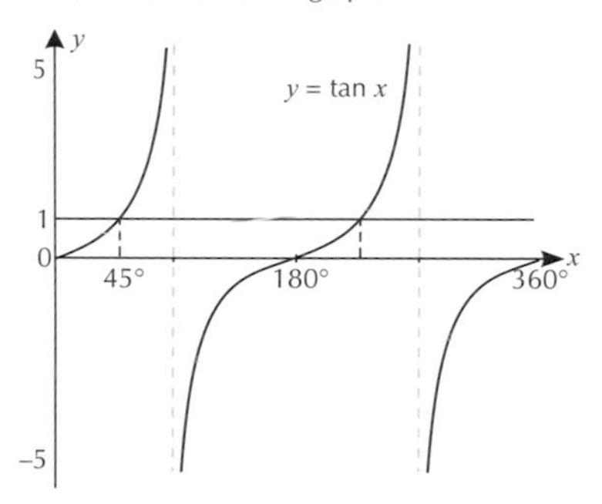

By the symmetry of the graph, the second solution is at 180° + 45° = 225°.

f) Using your knowledge of common angles, the first solution is at $x = 30°$. Then sketch a graph:

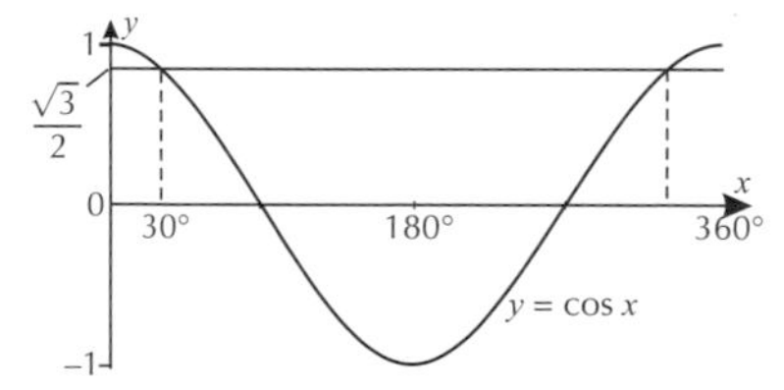

Using the symmetry of the graph, the second solution is at 360° – 30° = 330°.

Q3 You're told that there is a solution at 44.43°. From the graph you can see that there is another solution in the given interval at 180° – 44.43° = 135.57° (2 d.p.).

Q4 You're told that there is a solution at 143.1°, and from the graph you can see that there is another solution in the given interval. The first solution is 180° – 143.1° = 36.9° away from 180°, so the other solution will be at 180° + 36.9° = 216.9° (1 d.p.).
You could have worked this one out by doing 360° – 143.1°.

Q5 You're told that there is a solution at 62.2° (3 s.f.). From the graph you can see that there are two other solutions in the given interval. The solutions are:
180° + 62.2° = 242° (3 s.f.)
360° + 62.2° = 422° (3 s.f.)

Q6 Find the first solution: $\tan x = 2.5 \Rightarrow x = 68.2°$ (1 d.p.). Then sketch a graph — this time the interval is bigger, so you'll need more repetitions of the tan shape:

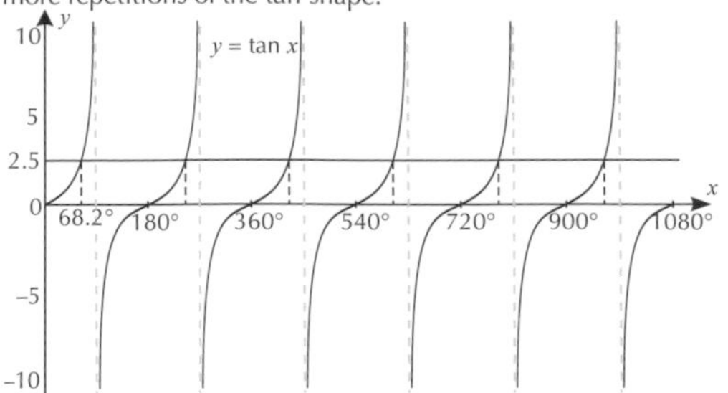

You can see from the graph that there are 6 solutions in the given interval — so just keep adding lots of 180° onto the first solution: $x = 68.2°, 248.2°, 428.2°, 608.2°, 788.2°, 968.2°$ (all to 1 d.p.).

You don't have to draw out the whole graph – sketch the first part to find the first solution, then keep adding on lots of 180° until the solutions are bigger than 1080°.

Q7 Find the first solution: $\sin x = 0.81 \Rightarrow x = 54.1°$ (3 s.f.). Then sketch a graph for the interval $-360° \leq x \leq 360°$:

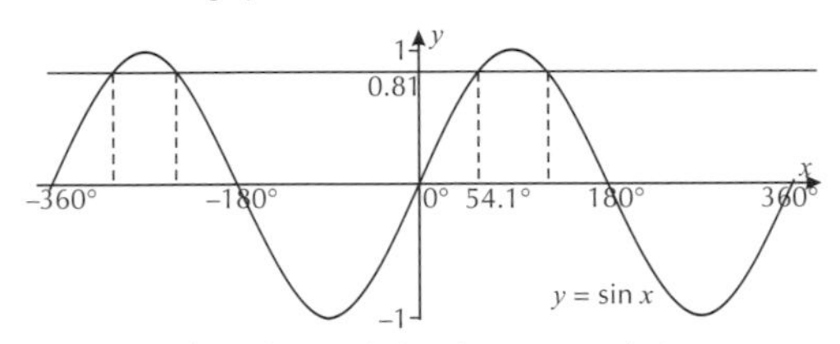

You can see from the graph that there are 4 solutions. Using the symmetry of the graph, there's another solution at $180° - 54.1° = 126°$ (3 s.f.). To find the other 2 solutions, subtract 360° from the values you've just found:
$54.1° - 360° = -306°$ (3 s.f.) and $126° - 360° = -234°$ (3 s.f.).

Q8 Find the first solution: $\sin x = 0.23 \Rightarrow x = 13.3°$ (1 d.p.). Then sketch a graph for the interval $-360° \leq x \leq 540°$:

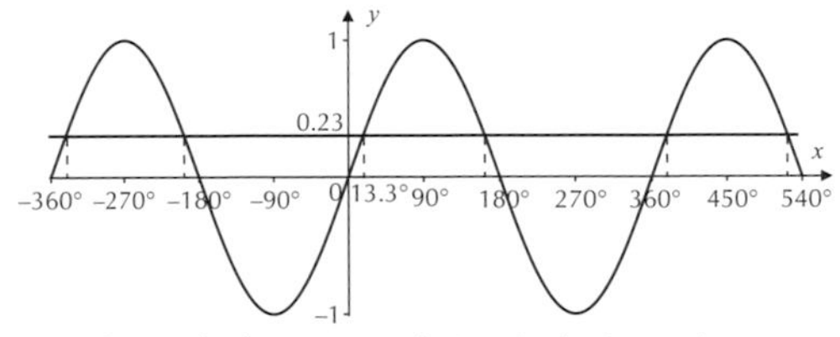

From the graph, there are 6 solutions in the interval. Using the symmetry of the graph, there's another solution at $180° - 13.3° = 166.7°$ (1 d.p.). To find the other solutions, add or subtract 360° to the values you've just found:
$13.3° + 360° = 373.3°$ (1 d.p.)
$166.7° + 360° = 526.7°$ (1 d.p.)
$13.3° - 360° = -346.7°$ (1 d.p.)
$166.7° - 360° = -193.3°$ (1 d.p.)
So the solutions are:
$x = -346.7°, -193.3°, 13.3°, 166.7°, 373.3°$ and $526.7°$

Q9 The first solution: $\cos x = -0.96 \Rightarrow x = 164°$ (3 s.f.). Then sketch a graph for the interval $-360° \leq x \leq 720°$:

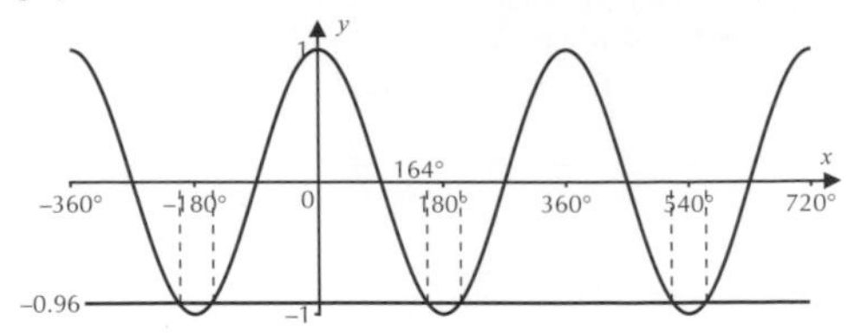

You can see from the graph that there are 6 solutions. Using the symmetry of the graph, the other five are:
$-360° + 164° = -196°$ (3 s.f.)
$0° - 164° = -164°$ (3 s.f.)
$360° - 164° = 196°$ (3 s.f.)
$360° + 164° = 524°$ (3 s.f.)
$720° - 164° = 556°$ (3 s.f.)
So the solutions are:
$x = -196°, -164°, 164°, 196°, 524°, 556°$ (3 s.f.)

Q10 Find the first solution: $\tan x = -1.75 \Rightarrow x = -60.3°$ (1 d.p.). Then sketch a graph for the interval $-360° \leq x \leq 720°$:

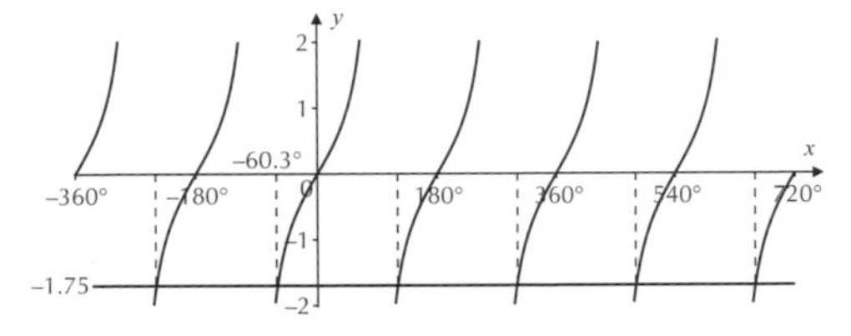

You can see from the graph that there are 6 solutions. Using the symmetry of the graph, the other five can be found by adding or subtracting 180°.
$-60.3° - 180° = -240.3°$ (1 d.p.)
$-60.3° + 180° = 119.7°$ (1 d.p.)
$119.7° + 180° = 299.7°$ (1 d.p.)
$299.7° + 180° = 479.7°$ (1 d.p.)
$479.7° + 180° = 659.7°$ (1 d.p.)
So the solutions are:
$x = -240.3°, -60.3°, 119.7°, 299.7°, 479.7°$ and $659.7°$ (1 d.p.)

Q11 Rearrange the equation:
$5 \sin x = 3 \cos x \Rightarrow \frac{\sin x}{\cos x} = \frac{3}{5} \Rightarrow \tan x = \frac{3}{5}$,
then find the first solution: $x = 30.96...° = 31.0°$ (3 s.f.)
Then sketch a graph for the interval $0° \leq x \leq 360°$:

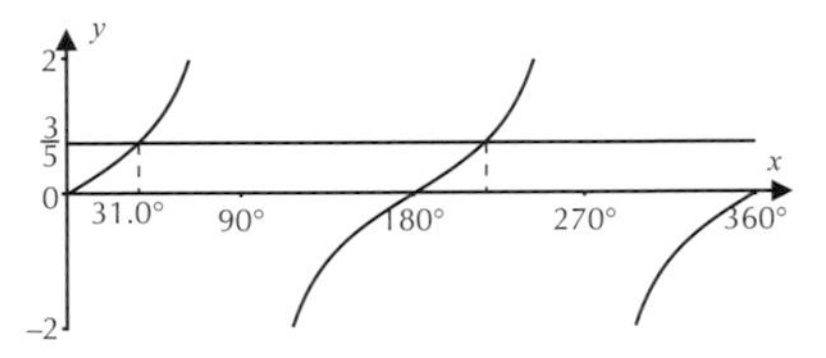

You can see from the graph that there are 2 solutions. Using the symmetry of the graph, the other solution is:
$180° + 30.96...° = 210.96...° = 211°$ (3 s.f.)

Q12 Rearrange the equation:
$3 \cos x = -8 \sin x \Rightarrow \frac{\sin x}{\cos x} = -\frac{3}{8} \Rightarrow \tan x = -\frac{3}{8}$,
then find the first solution: $x = -20.556...° = -20.56°$ (2 d.p.).
This is outside the interval so add 180° to get $x = 159.44°$ (2 d.p.).
Then sketch a graph for the interval $0° \leq x \leq 360°$:

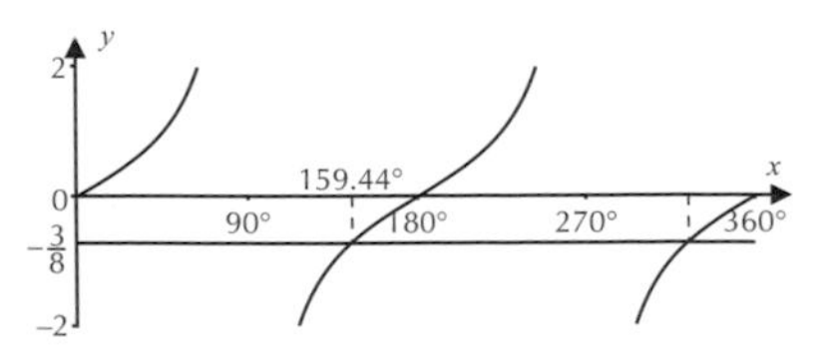

You can see from the graph that there are 2 solutions. Using the symmetry of the graph, the other solution is:
$180° + 159.44° = 339.44°$ (2 d.p.)

Exercise 8.4.2 — Using a CAST diagram

Q1 0.45 is positive, so look at the quadrants where $\sin x$ is positive:

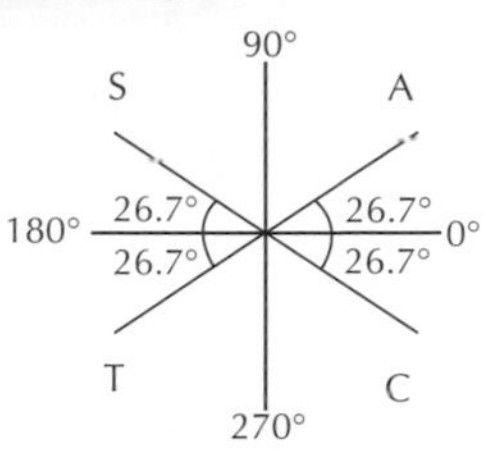

The second quadrant is the other one, so the only other solution is $180° - 26.7° = 153.3°$.

Q2 0.68 is positive, so look at the quadrants where $\cos x$ is positive:

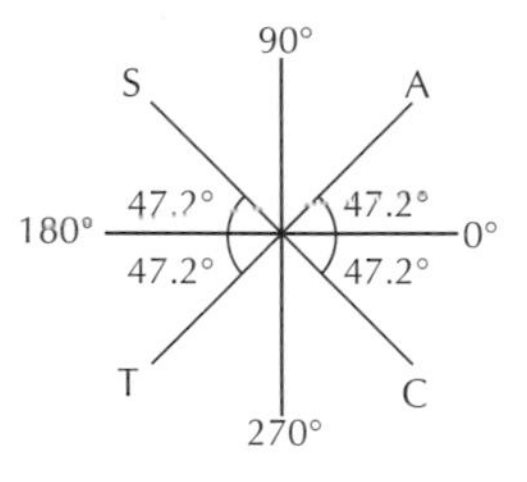

Cos is also positive in the fourth quadrant, so the other solution is $360° - 47.2° = 312.8°$.

Q3 **a)** Use a calculator to find the first solution: $\cos x = 0.8 \Rightarrow x = 36.9°$ (1 d.p.). 0.8 is positive, so look at the quadrants where cos is positive:

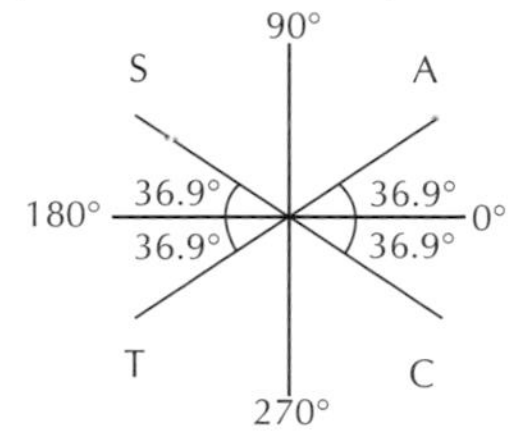

Cos is positive in the 4th quadrant, so the other solution is at $360° - 36.9° = 323.1°$ (1 d.p.).

b) Use a calculator to find the first solution: $\tan x = 2.7 \Rightarrow x = 69.7°$ (1 d.p.). 2.7 is positive, so look at the other quadrants where tan is positive:

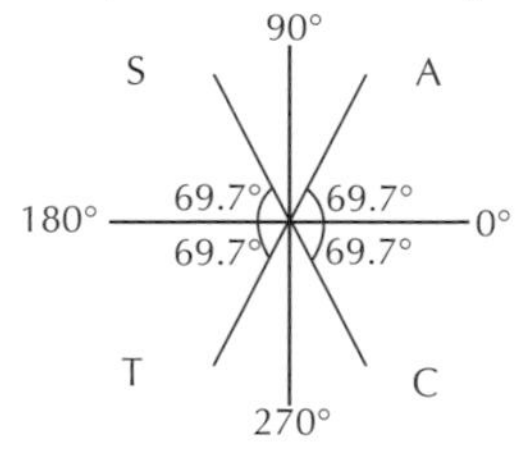

Tan is positive in the 3rd quadrant, so the other solution is at $180° + 69.7° = 249.7°$ (1 d.p.).

c) Use a calculator to find the first solution: $\sin x = -0.15 \Rightarrow x = -8.6°$ (1 d.p.). -0.15 is negative, so look at the quadrants where sin is negative:

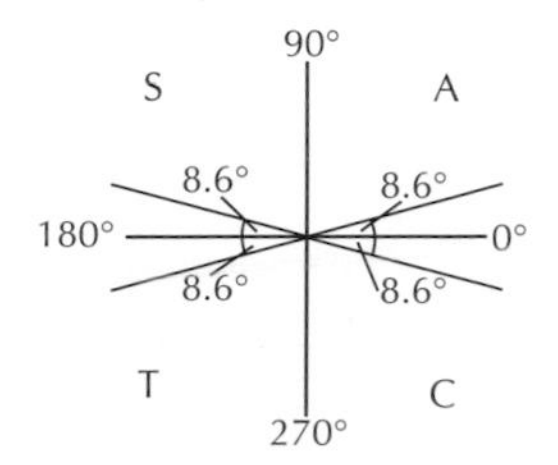

Sin is negative in the 3rd and 4th quadrants, so the solutions are at $180° + 8.6° = 188.6°$ and $360° - 8.6° = 351.4°$ (both to 1 d.p.).

d) Use a calculator to find the first solution: $\tan x = 0.3 \Rightarrow x = 16.7°$ (1 d.p.). 0.3 is positive, so look at the other quadrants where tan is positive:

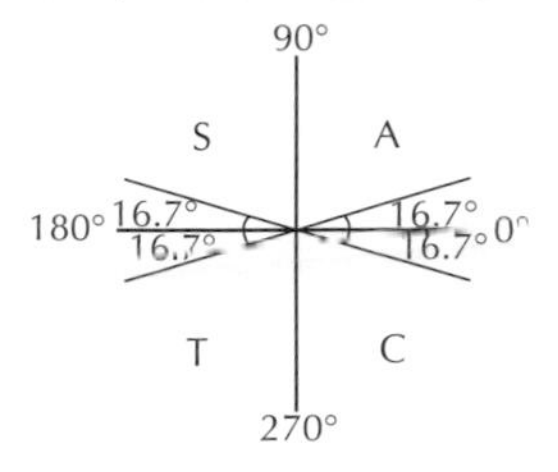

Tan is positive in the 3rd quadrant, so the other solution is at $180° + 16.7° = 196.7°$ (1 d.p.).

e) Use a calculator to find the first solution: $\tan x = -0.6 \Rightarrow x = -31.0°$ (1 d.p.). -0.6 is negative, so look at the quadrants where tan is negative:

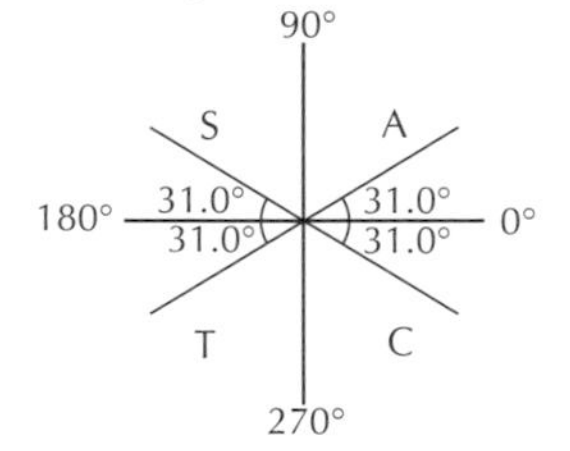

Tan is negative in the 2nd and 4th quadrants So the solutions are at $180° - 31.0° = 149.0°$ and $360° - 31.0° = 329.0°$ (both to 1 d.p.).

f) Use a calculator to find the first solution: $\sin x = -0.29 \Rightarrow x = -16.9°$ (1 d.p.). -0.29 is negative, so look at the quadrants where sin is negative:

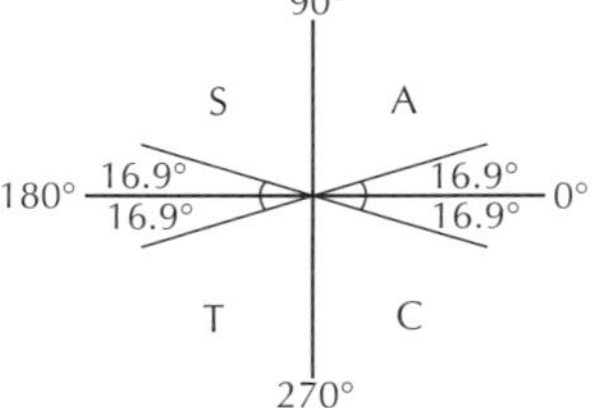

Sin is negative in the 3rd and 4th quadrants, so the solutions are at $180° + 16.9° = 196.9°$ and $360° - 16.9° = 343.1°$ (both to 1 d.p.).

Q4 -0.87 is negative, so look at the quadrants where $\sin x$ is negative:

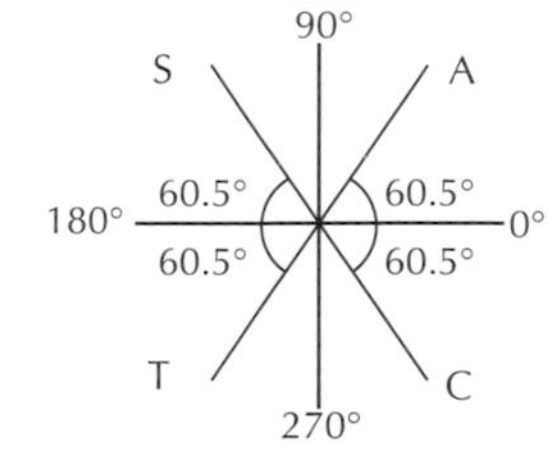

Sin is negative in the third and fourth quadrants, so the first two solutions are $180° + 60.5° = 240.5°$ (1 d.p.), and $360° - 60.5° = 299.5°$ (1 d.p.). To find the other solutions in the interval, subtract 360° from the solutions already found:
$240.5° - 360° = -119.5°$ (1 d.p.) and
$299.5° - 360° = -60.5°$ (1 d.p.)
So the solutions are: $x = -119.5°, -60.5°, 240.5°, 299.5°$ (1 d.p.)
You could have found the negative solutions more directly by reading the CAST diagram in the negative (i.e. clockwise) direction. Reading clockwise from 0°, the angle in the third quadrant is −119.5°.

Q5 a) The first solution is $x = -83.2°$ (3 s.f.).

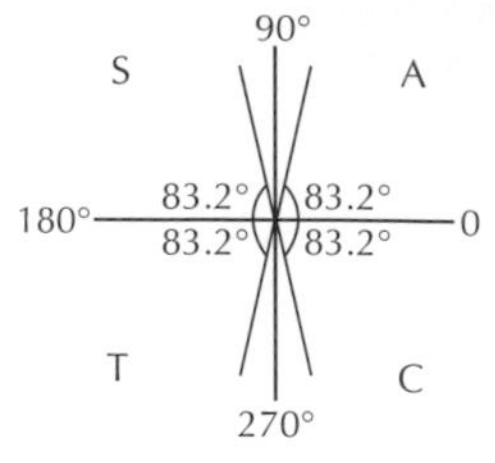

Tan is negative in the 2nd and 4th quadrants, so the solutions are at $180° - 83.2° = 96.8°$ and $360° - 83.2° = 277°$ (both to 3 s.f.).

b) The first solution is $x = 55.1°$ (3 s.f.).

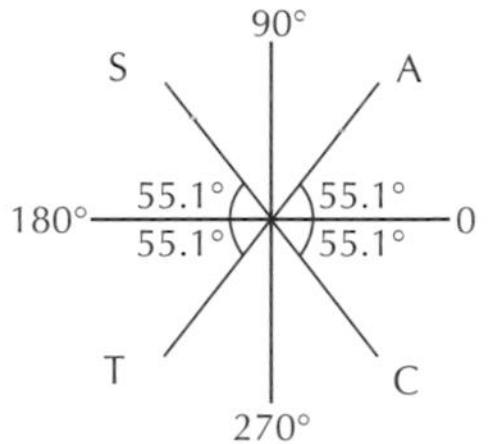

Sin is positive in the second quadrant, so the next solution is $180° - 55.1° = 125°$ (3 s.f.). To find the other solutions in the given interval, add on 360° to the solutions already found: $55.1° + 360° = 415°$ (3 s.f.) and $125° + 360° = 485°$ (3 s.f.).

c) The first solution is $x = 43.9°$ (3 s.f.).

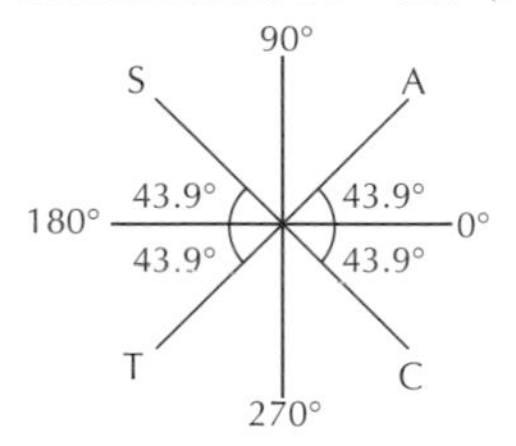

Cos is positive in the fourth quadrant, so the next solution is $360° - 43.9° = 316°$ (3 s.f.). To find the other solution in the interval, add on 360° to the solutions already found: $43.9° + 360° = 404°$ (3 s.f.).

d) The first solution is $x = 74.4°$ (3 s.f.).

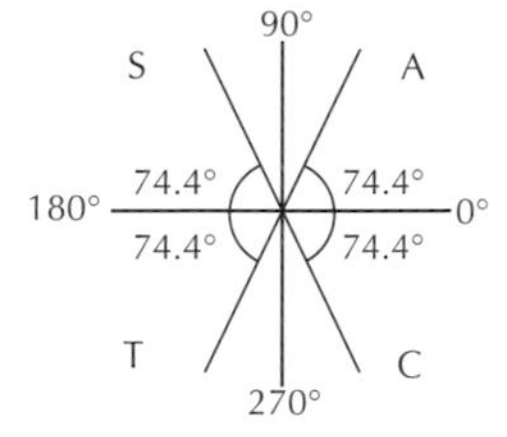

Tan is positive in the third quadrant, so the next solution is $180° + 74.4° = 254°$ (3 s.f.). This is outside the interval, so subtract 360°: $x = -106°$ (3 s.f.).

Q6 The first solution is $x = -3.4°$ (1 d.p.) — note that this is outside the given interval.

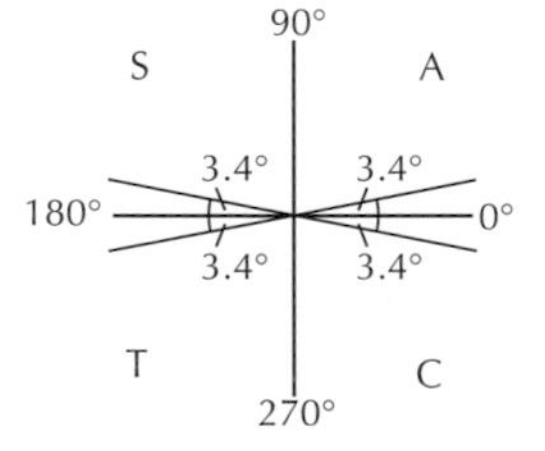

Using a CAST diagram, the solutions of $\sin x$ are negative in the third and fourth quadrants. So the first two solutions are $180° + 3.4° = 183.4°$ (1 d.p.) and $360° - 3.4° = 356.6°$ (1 d.p.). To find the other solutions in the given interval, keep adding 360° to the solutions already found:
$183.4° + 360° = 543.4°$ (1 d.p.)
$356.6° + 360° = 716.6°$ (1 d.p.)
$543.4° + 360° = 903.4°$ (1 d.p.)
$716.6° + 360° = 1076.6°$ (1 d.p.)
So the solutions are:
$x = 183.4°, 356.6°, 543.4°, 716.6°, 903.4°, 1076.6°$ (1 d.p.)

Q7 The first solution is $x = 85.2°$ (1 d.p.).

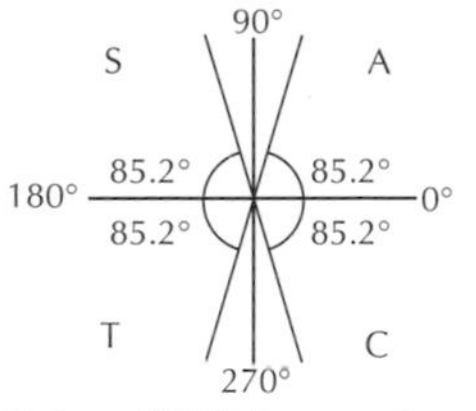

Using a CAST diagram, the solutions of $\tan x$ are also positive in the 3rd quadrant. So the next solution is $180° + 85.2 = 265.2°$. To find the other solutions in the given interval, subtract 360° from the solutions already found: $85.2° - 360° = -274.8°$ (1 d.p.) and $288.1° - 360° = -94.8°$ (1 d.p.)

Q8 Rearrange the equation:
$5 \cos x - 2 = 0 \Rightarrow 5 \cos x = 2$
$\Rightarrow \cos x = \frac{2}{5}$
So the first solution is:
$x = \cos^{-1}\left(\frac{2}{5}\right) = 66.4°$ (1 d.p.)

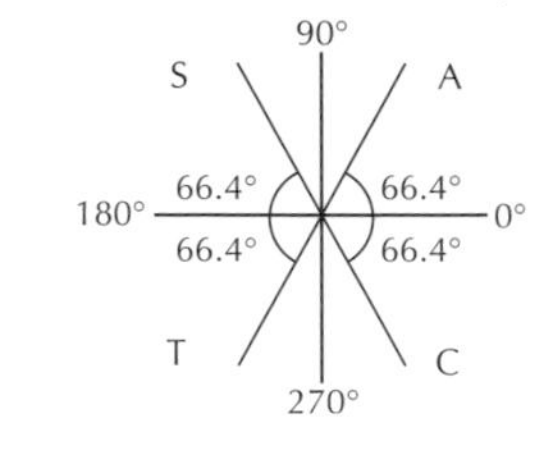

Cos is positive in the fourth quadrant, so the other solution is $360° - 66.4° = 293.6°$ (1 d.p.).

Q9 Rearrange the equation:
$\frac{1}{4} \tan x = 1.4 \Rightarrow \tan x = 5.6$
So the first solution is:
$x = \tan^{-1}(5.6) = 79.9°$ (3 s.f.).

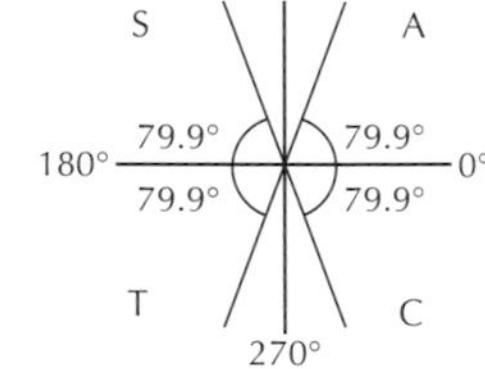

Tan is positive in the third quadrant, so the next solution is $180° + 79.9° = 260°$ (3 s.f.).

Q10 a) Rearrange the equation:
$4 \sin x - 3 = 0 \Rightarrow 4 \sin x = 3 \Rightarrow \sin x = \frac{3}{4}$
The first solution is:
$x = \sin^{-1}\left(\frac{3}{4}\right) = 48.6°$ (1 d.p.)

90°
S
A
180°
48.6°
48.6°
48.6°
48.6°
0°
T
C
270°

Sin is positive in the second quadrant, so the next solution is $180° - 48.6° = 131.4°$ (1 d.p.). To find the other solutions in the interval, add 360° to the solutions you've already found.
$48.6° + 360° = 408.6°$ (1 d.p.) and
$131.4° + 360° = 491.4°$ (1 d.p.)

b) Rearrange the equation:
$2 - 3\cos x = 0 \Rightarrow 3\cos x = 2 \Rightarrow \cos x = \frac{2}{3}$
The first solution is:
$x = \cos^{-1}\left(\frac{2}{3}\right) = 48.2°$ (1 d.p.)

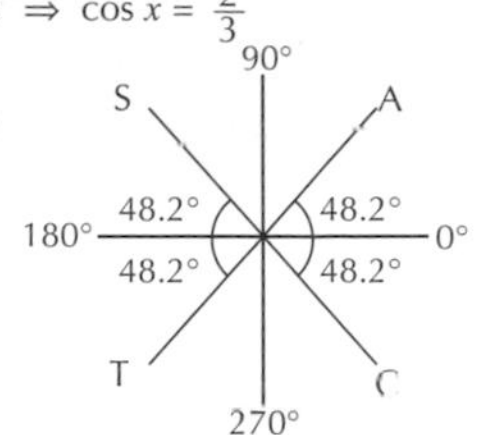

Cos is positive in the fourth quadrant, so the other solution is $0° - 48.2° = -48.2°$ (1 d.p.)

c) Rearrange the equation:
$6\tan x = -11 \Rightarrow \tan x = -\frac{11}{6}$
The first solution is
$x = \tan^{-1}\left(-\frac{11}{6}\right)$
$= -61.4°$ (1 d.p.).

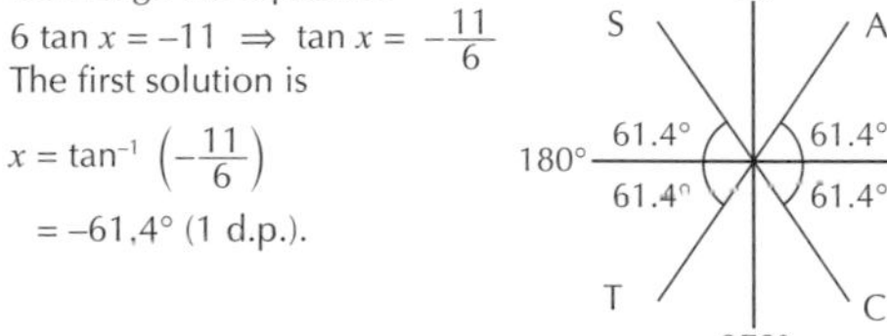

Tan is negative in the 2nd and 4th quadrants so the next solution is $180° - 61.4° = 118.6°$ (1 d.p.).
To find the other solutions in the interval, keep adding 180° to the solutions already found:
$118.6° + 180° = 298.6°$ (1 d.p.) and
$298.6° + 180° = 478.6°$ (1 d.p.)

d) Rearrange the equation:
$8\sin x = -5 \Rightarrow \sin x = -\frac{5}{8}$
The first solution is
$x = \sin^{-1}\left(-\frac{5}{8}\right)$
$= -38.7°$ (1 d.p.).

Sin is negative in the third and fourth quadrant, so the next solution is $180° + 38.7° = 218.7°$ (1 d.p.).
To find other solutions in the interval, add or subtract 360° from the solutions already found:
$218.7° - 360° = -141.3°$ (1 d.p.)
$-38.7° + 360° = 321.3°$ (1 d.p.)
$218.7° + 360° = 578.7°$ (1 d.p.)
$321.3° + 360° = 681.3°$ (1 d.p.)
So the solutions are:
$x = -141.3°, -38.7°, 218.7°, 321.3°, 578.7°, 681.3°$ (1 d.p.)

Exercise 8.4.3 — Changing the interval

For all the questions in this exercise, you can either sketch a graph or use a CAST diagram.

Q1 a) $\sin 2x = 0.6$, so look for solutions in the interval $0° \le 2x \le 720°$. The first solution is $2x = 36.87°$ (2 d.p.). Using a CAST diagram, sin is also positive in the 2nd quadrant, so there's another solution at $2x = 180° - 36.87° = 143.13°$ (2 d.p.). The sin graph repeats every 360°, so add 360° onto the answers already found: $2x = 396.87°, 503.13°$ (2 d.p.).
These are solutions for $2x$, so divide them all by 2:
$x = 18.4°, 71.6°, 198.4°, 251.6°$ (1 d.p.).

b) $\tan 4x = 4.6$, so look for solutions in the interval $0° \le 4x \le 1440°$. The first solution is $4x = 77.74°$ (2 d.p.). From the graph, there will be another solution at $4x = 77.74° + 180° = 257.74°$ (2 d.p.). Then just keep adding on 180° to find the rest of the solutions within the interval: $4x = 437.74°, 617.74°, 797.74°, 977.74°, 1157.74°, 1337.74°$ (2 d.p.)
These are solutions for $4x$, so divide them all by 4:
$x = 19.4°, 64.4°, 109.4°, 154.4°, 199.4°, 244.4°, 289.4°, 334.4°$ (1 d.p.).

c) $\cos 3x = -0.24$, so look for solutions in the interval $0° \le 3x \le 1080°$. The first solution is $3x = 103.89°$ (2 d.p.). Using the symmetry of the graph, there's another solution at $3x = 360° - 103.89° = 256.11°$ (2 d.p.). To find the other solutions within the interval, add on multiples of 360°:
$3x = 463.89°, 616.11°, 823.89°, 976.11°$ (2 d.p.). These are solutions for $3x$, so divide them all by 3: $x = 34.6°, 85.4°, 154.6°, 205.4°, 274.6°, 325.4°$ (1 d.p.).

d) $\sin 3x = 0.94$, so look for solutions in the interval $0° \le 3x \le 1080°$. The first solution is $3x = 70.05°$ (2 d.p.). Using a CAST diagram, sin is also positive in the 2nd quadrant, so there's another solution at $3x = 180° - 70.05° = 109.95°$ (2 d.p.). The sin graph repeats every 360°, so continue adding 360° onto the answers already found:
$3x = 430.05°, 469.95°, 790.05°, 829.95°$ (2 d.p.)
These are solutions for $3x$, so divide them all by 3:
$x = 23.4°, 36.7°, 143.4°, 156.7°, 263.4°, 276.7°$ (1 d.p.).

e) $\cos 5x = 0.5$, so look for solutions in the interval $0° \le 5x \le 1800°$. The first solution is $5x = 60°$. Using the symmetry of the graph, there's another solution at $5x = 360° - 60° = 300°$. To find the other solutions within the interval, add on multiples of 360°:
$5x = 420°, 660°, 780°, 1020°, 1140°, 1380°, 1500°, 1740°$
These are solutions for $5x$, so divide them all by 5:
$x = 12°, 60°, 84°, 132°, 156°, 204°, 228°, 276°, 300°, 348°$

f) $\tan 2x = -6.7$, so look for solutions in the interval $0° \le 2x \le 720°$. The first solution is $2x = -81.51°$ (2 d.p.) — this is outside the given interval, so add 180°: $2x = 98.49°$. From the graph, there will be another solution at $2x = 98.49° + 180° = 278.49°$ (2 d.p.). Then keep adding 180° to find the rest of the solutions within the interval:
$2x = 458.49°, 638.49°$ (2 d.p.)
These are solutions for $2x$, so divide them all by 2:
$x = 49.2°, 139.2°, 229.2°, 319.2°$ (1 d.p.).

Q2 $\tan\frac{x}{2} = 2.1$, so look for solutions in the interval $0° \le \frac{x}{2} \le 180°$. The first solution is $\frac{x}{2} = 64.54°$ (2 d.p.). This is the only solution in the interval, as tan doesn't repeat any values between 0° and 180° (looking at its graph). Multiply by 2 to get the value of x: $x = 129.1°$ (1 d.p.)

Q3 $\sin\frac{2x}{3} = 0.52$, so look for solutions in the interval $0° \le \frac{2x}{3} \le 180°$. The first solution is $\frac{2x}{3} = 31.33°$ (2 d.p.). Using a CAST diagram, sin is also positive in the 2nd quadrant, so there's another solution at
$\frac{2x}{3} = 180° - 31.33° = 148.67°$ (2 d.p.).
These are solutions for $\frac{2x}{3}$, so divide them by $\frac{2}{3}$:
$x = 47.0°, 223.0°$ (1 d.p.)

Q4 $\cos\frac{x}{3} = \frac{\sqrt{3}}{2}$, so look for solutions in the interval $-60° \le \frac{x}{3} \le 60°$. The first solution is $\frac{x}{3} = 30°$ — using the symmetry of the graph, there's another solution at $0 - 30° = -30°$. Multiply by 3 to get the values of x: $x = -90°, 90°$.

Q5 $\cos(x - 27°) = 0.64$, so look for solutions in the interval $-27° \le x - 27° \le 333°$. The first solution is $x - 27° = 50.2°$ (1 d.p.). Using the symmetry of the graph, there's another solution at $x - 27° = 360° - 50.2° = 309.8°$ (1 d.p.).
So the solutions are $x = 77.2°$ and $336.8°$ (1 d.p.).

Q6 $\tan(x - 140°) = -0.76$, so look for solutions in the interval $-140° \le x - 140° \le 220°$. The first solution is $x - 140° = -37.2°$ (1 d.p.). The tan graph repeats every 180°, so there's another solution at $x - 140° = -37.2° + 180° = 142.8°$ (1 d.p.) (if you add on another 180°, the answer is outside the interval).
So the solutions are $x = 102.8°$ and $282.8°$ (1 d.p.).

Q7 $\tan(x + 73°) = 1.84$, so look for solutions in the interval $73° \le x + 73° \le 433°$. The first solution is $x + 73° = 61.5°$ (1 d.p.). This is out of the interval, but use this and the pattern of the graph of $\tan x$ to find the other solutions. $\tan x$ repeats every 180°, so the next two solutions are $x + 73° = 241.5°$ and $x + 73° = 421.5°$ (1 d.p.). So the two solutions are $x = 168.5°$ and $348.5°$ (1 d.p.).

Q8 $\sin(x - 45°) = -0.25$, so look for solutions in the interval $-225° \le x - 45° \le 315°$.
The first solution is $x - 45° = -14.48°$ (2 d.p.).
Using a CAST diagram, the other solutions are at $x - 45° = 180° + 14.48° = 194.48°$ and $x - 45° = -180° + 14.48° = -165.52°$.
So adding 45° to each solution gives $x = -120.5°$, $30.5°$ and $239.5°$ (1 d.p.).
There's no solution at 360° – 14.48° because it's outside the interval.

Q9 $\cos(x + 22.5°) = 0.13$, so look for solutions in the interval $22.5° \le x + 22.5° \le 382.5°$.
The first solution is $x + 22.5° = 82.53°$ (2 d.p.).
Using the symmetry of the graph, there's another solution at $x + 22.5° = 360° - 82.53° = 277.47°$.
So subtracting 22.5° from each solution gives $x = 60.0°$ and $255.0°$ (1 d.p.).

Q10 Rearrange to get $\tan(x - 32°) = 7.5$, so look for solutions in the interval $-212° \le x - 32° \le 148°$. The first solution is $x - 32° = 82.4°$ (1 d.p.). The graph of tan repeats every 180°, so there's another solution at $x - 32° = 82.4° - 180° = -97.6°$ (1 d.p.)
So the solutions are $x = -97.6° + 32° = -65.6°$ (1 d.p) and $x = 82.4° + 32° = 114.4°$ (1 d.p.)

Q11 Rearrange to get $\cos 3x = 0.7$, so look for solutions in the interval $0° \le 3x \le 1080°$. The first solution is $3x = 45.57°$ (2 d.p.).
Using the symmetry of the graph, there's another solution at $3x = 360° - 45.57° = 314.43°$ (2 d.p.).
Add on 360° to find the other solutions in the interval:
$3x = 405.57°, 674.43°, 765.57°, 1034.43°$ (2 d.p.)
These are solutions for $3x$, so divide them all by 3:
$x = 15.2°, 104.8°, 135.2°, 224.8°, 255.2°, 344.8°$ (1 d.p.)

Q12 First, rearrange: $\frac{1}{2}\sin 3x - 0.61 = -0.75$

$\Rightarrow \frac{1}{2}\sin 3x = -0.14 \Rightarrow \sin 3x = -0.28$,

so look for solutions in the interval $0° \le 3x \le 1080°$.
The first solution is $3x = -16.26°$ (2 d.p.). This is outside the interval, but putting 16.26° into a CAST diagram and looking at the quadrants where sin is negative gives
$3x = 180° + 16.26° = 196.26°$
and $3x = 360° - 16.26° = 343.74°$ (2 d.p.).
Add multiples of 360° to find the other solutions in the interval:
$3x = 556.26°, 703.74°, 916.26°, 1063.74°$ (2 d.p.).
These are solutions for $3x$, so divide them all by 3:
$x = 65.4°, 114.6°, 185.4°, 234.6°, 305.4°, 354.6°$ (1 d.p.)

Q13 Rearrange to get $\cos(x - 67°) = 0.69$, so look for solutions in the interval $-67° \le x - 67° \le 473°$.
The first solution is $x - 67° = 46.4°$ (1 d.p.).
Using the symmetry of the graph, there's another solution at $x - 67° = 360° - 46.4° = 313.6°$ (1 d.p.). Add on 360° to find the other solution in the interval:
$x - 67° = 46.4° + 360° = 406.4°$.
So $x = 113.4°, 380.6°, 473.4°$ (1 d.p.)

Q14 Rearrange to get $\sin(x + 19°) = -\frac{\sqrt{2}}{2}$, so look for solutions in the interval $19° \le x + 19° \le 379°$. The first solution is $x + 19° = -45°$.
This is outside the interval, so add 360°: $x + 19° = 315°$.
Using a CAST diagram, sin is also negative in the third quadrant, so there's another solution at $x + 19° = 180° + 45° = 225°$.
So $x = 206°, 296°$

Exercise 8.4.4 — Using trig identities to solve equations

For all the questions in this exercise, you can either sketch a graph or use a CAST diagram.

Q1 **a)** This equation has already been factorised.
Either $\tan x - 5 = 0$ or $3\sin x - 1 = 0$.
$\tan x - 5 = 0 \Rightarrow \tan x = 5 \Rightarrow x = 78.7°$ (1 d.p.)
This is the first solution. tan repeats itself every 180°, so the other solution is 258.7° (1 d.p.).
$3\sin x = 1 \Rightarrow \sin x = \frac{1}{3} \Rightarrow x = 19.5°$ (1 d.p.)
Using the symmetry of the graph, the other solution is $180° - 19.5° = 160.5°$ (1 d.p.).

b) $5\sin x\tan x - 4\tan x = 0 \Rightarrow \tan x(5\sin x - 4) = 0$
So $\tan x = 0$ or $5\sin x - 4 = 0$.
$\tan x = 0 \Rightarrow x = 0°, 180°$ and $360°$ (from the graph of $\tan x$).
$5\sin x - 4 = 0 \Rightarrow \sin x = \frac{4}{5} \Rightarrow x = 53.1°$ (1 d.p.)
Using the symmetry of the graph, the other solution is $180° - 53.1° = 126.9°$ (1 d.p.).

c) $\tan^2 x = 9 \Rightarrow \tan x = 3$ or -3.
$\tan x = 3 \Rightarrow x = 71.6°$ (1 d.p.)
Using the repetition of the tan graph, the other solution is $180° + 71.6° = 251.6°$ (1 d.p.).
$\tan x = -3 \Rightarrow x = -71.6°$ (1 d.p.)
This is outside the interval. Add on 180° until you've found all the solutions within the interval:
$-71.6° + 180° = 108.4°$ (1 d.p.) and
$108.4° + 180° = 288.4°$ (1 d.p.).

d) $4\cos^2 x = 3\cos x \Rightarrow 4\cos^2 x - 3\cos x = 0$
$\Rightarrow \cos x(4\cos x - 3) = 0$

So $\cos x = 0$ or $4\cos x - 3 = 0$
$\cos x = 0 \Rightarrow x = 90°$
Using the cos graph, the other solution is 270°.
$4\cos x - 3 = 0 \Rightarrow \cos x = \frac{3}{4} \Rightarrow x = 41.4°$ (1 d.p.)
Using the symmetry of the graph, the other solution is $360° - 41.4° = 318.6°$ (1 d.p.).

e) $3\sin x = 5\cos x \Rightarrow \tan x = \frac{5}{3}$. The first solution is $x = 59.0°$ (1 d.p.). $\tan x$ repeats every 180°, so the other solution is 239.0° (1 d.p.).

f) $5\tan^2 x - 2\tan x = 0 \Rightarrow \tan x(5\tan x - 2) = 0$.
So either $\tan x = 0$ or $\tan x = 0.4$. If $\tan x = 0$, then the solutions are $x = 0°, 180°$ and $360°$.
If $\tan x = 0.4$, the first solution is 21.8° (1 d.p.).
The graph of $\tan x$ repeats every 180°, so another solution is $x = 201.8°$ (1 d.p.).

g) $6\cos^2 x - \cos x - 2 = 0$
$\Rightarrow (3\cos x - 2)(2\cos x + 1) = 0$
So either $\cos x = \frac{2}{3}$ or $\cos x = -0.5$. If $\cos x = \frac{2}{3}$, the first solution is 48.2° (1 d.p.). Looking at the symmetry of the graph of $\cos x$, the other solution is $x = 360° - 48.2° = 311.8°$ (1 d.p.).
If $\cos x = -0.5$, the first solution is 120°.
Looking at the symmetry of the graph of $\cos x$, the other solution is $360° - 120° = 240°$.

h) $7\sin x + 3\cos x = 0 \Rightarrow 7\sin x = -3\cos x$
$\Rightarrow \tan x = -\frac{3}{7} \Rightarrow x = -23.2°$ (1 d.p.)
This is outside the required interval. Using a CAST diagram, tan is negative in the 2nd and 4th quadrants, so the solutions are $x = 180° - 23.2° = 156.8°$ (1 d.p.) and $x = 360° - 23.2° = 336.8°$ (1 d.p.).

Q2 **a)** $\tan x = \sin x\cos x \Rightarrow \frac{\sin x}{\cos x} - \sin x\cos x = 0$
$\Rightarrow \sin x - \sin x\cos^2 x = 0$
$\Rightarrow \sin x(1 - \cos^2 x) = 0$
$\Rightarrow \sin x(\sin^2 x) = 0$
$\Rightarrow \sin^3 x = 0$
So $\sin x = 0$.
The solutions are $x = 0°, 180°$ and $360°$.

b) $5\cos^2 x - 9\sin x = 3 \Rightarrow 5(1 - \sin^2 x) - 9\sin x = 3$
$\Rightarrow 5\sin^2 x + 9\sin x - 2 = 0$
$\Rightarrow (5\sin x - 1)(\sin x + 2) = 0$
So either $\sin x = 0.2$ or $\sin x = -2$. $\sin x$ can't be -2, so only $\sin x = 0.2$ will give solutions.
The first solution is $x = 11.54°$ (2 d.p.).
The interval covers three intervals of 360°, so there will be 6 solutions.
Looking at the symmetry of the sin graph and adding or subtracting 360°, the other solutions are
$x = -348.5°, -191.5°, 11.5°, 168.5°, 371.5°$ and $528.5°$ (1 d.p.)
If you'd used a CAST diagram here, you'd find 11.5° and 168.5° first, then add or subtract 360°.

c) $2\sin^2 x + \sin x - 1 = 0$
$\Rightarrow (2\sin x - 1)(\sin x + 1) = 0$
So $2\sin x - 1 = 0$ or $\sin x + 1 = 0$.
$2\sin x - 1 = 0 \Rightarrow \sin x = \frac{1}{2} \Rightarrow x = 30°$.
Using the symmetry of the graph, another solution is $180° - 30° = 150°$.
To find the other solutions in the required interval, subtract 360° from each of these:
$30° - 360° = -330°$, and $150° - 360° = -210°$.
$\sin x + 1 = 0 \Rightarrow \sin x = -1$
From the graph, the solutions to this are $x = -90°$ and $x = 270°$.

d) $2\sin x\tan x = -3 \Rightarrow \frac{2\sin^2 x}{\cos x} = -3$
$\Rightarrow 2 - 2\cos^2 x = -3\cos x$
$\Rightarrow 2\cos^2 x - 3\cos x - 2 = 0$
$\Rightarrow (\cos x - 2)(2\cos x + 1) = 0$
$\cos x = 2$ has no solutions.
$\cos x = -\frac{1}{2} \Rightarrow x = 120°$
Using the symmetry of the graph, there is another solution at $360° - 120° = 240°$. Find the other solutions in the interval by subtracting 360°:
$x = -240°, -120°, 120°, 240°$.

e) $4 - \tan^2 x = 0 \Rightarrow (2 - \tan x)(2 + \tan x) = 0$
so either $\tan x = 2$ or $\tan x = -2$.
First, solve $\tan x = 2$: $x = 63.43°$ (2 d.p.)
The graph of tan repeats every 180°, so subtract 180° to find the other solution in the interval:
$x = 63.43° - 180° = -116.57°$.
Now, solve $\tan x = -2$: $x = -63.43°$
Add 180° to find the other solution in the interval:
$x = -63.43° + 180° = 116.57°$
So the solutions to $4 - \tan^2 x = 0$ are:
$x = -116.6°, -63.4°, 63.4°, 116.6°$ (1 d.p.)

Q3 a) $4\sin^2 x = 3 - 3\cos x \Rightarrow 4(1 - \cos^2 x) = 3 - 3\cos x$
$\Rightarrow 4 - 4\cos^2 x = 3 - 3\cos x$
$\Rightarrow 4\cos^2 x - 3\cos x - 1 = 0$
(as required)

b) Solve the equation from a).
$4\cos^2 x - 3\cos x - 1 = 0$
$(4\cos x + 1)(\cos x - 1) = 0$
So $4\cos x + 1 = 0$ or $\cos x - 1 = 0$
$4\cos x + 1 = 0 \Rightarrow \cos x = -\frac{1}{4} \Rightarrow x = 104.5°$ (1 d.p.)
Using the symmetry of the graph, the other solution is $360° - 104.5° = 255.5°$ (1 d.p.)
$\cos x - 1 = 0 \Rightarrow \cos x = 1$
Using the cos graph, the solutions are $x = 0$ and $x = 360°$.

Q4 $3\cos x - 2\sin^2 x = 0 \Rightarrow 3\cos x - 2(1 - \cos^2 x) = 0$
$\Rightarrow 2\cos^2 x + 3\cos x - 2 = 0$
$\Rightarrow (2\cos x - 1)(\cos x + 2) = 0$
$\cos x = -2$ has no solutions and $\cos x = \frac{1}{2}$ has a first solution $x = 60°$. Using the symmetry of cos, there is another solution at $x = 360° - 60° = 300°$.
So the solutions of $3\cos x - 2\sin^2 x = 0$ are: $x = 60°, 300°$

Q5 $2\sin^2 x + 5\cos^2 x - 7\cos x = 0$
$\Rightarrow 2(1 - \cos^2 x) + 5\cos^2 x - 7\cos x = 0$
$\Rightarrow 2 + 3\cos^2 x - 7\cos x = 0$
$\Rightarrow (3\cos x - 1)(\cos x - 2) = 0$
$\cos x = 2$ has no solutions, so you only need to solve $\cos x = \frac{1}{3}$.
The first solution is $x = 70.53°$, and by the symmetry of the cos graph, the other solution in the given interval is $x = 360° - 70.53° = 289.47°$.
So the solutions of $2\sin^2 x + 5\cos^2 x - 7\cos x = 0$ are:
$x = 70.5°, 289.5°$ (1 d.p.)

Q6 $9\sin^2 2x + 3\cos 2x = 7$
$\Rightarrow 9(1 - \cos^2 2x) + 3\cos 2x = 7$
$\Rightarrow 2 - 9\cos^2 2x + 3\cos 2x = 0$
$\Rightarrow 9\cos^2 2x - 3\cos 2x - 2 = 0$
$\Rightarrow (3\cos 2x + 1)(3\cos 2x - 2) = 0$
So either $\cos 2x = -\frac{1}{3}$ or $\cos 2x = \frac{2}{3}$.
For $\cos 2x = -\frac{1}{3}$, look for solutions in the interval $0° \le 2x \le 720°$.
The first solution is $2x = 109.47°$ (2 d.p.). Looking at the symmetry of the graph of $\cos x$, the other solutions are:
$2x = 250.53°, 469.47°$ and $610.53°$ (2 d.p.).
Dividing by 2 gives the solutions:
$x = 54.7°, 125.3°, 234.7°$ and $305.3°$ (1 d.p.).
For $\cos 2x = \frac{2}{3}$, again look for solutions in the interval $0° \le 2x \le 720°$. The first solution is $2x = 48.19°$ (2 d.p.). Looking at the symmetry of the graph of $\cos x$, the other solutions are $2x = 311.81°, 408.19°, 671.81°$ (2 d.p.).
Dividing by 2 gives the solutions:
$x = 24.1°, 155.9°, 204.1°$ and $335.9°$ (1 d.p.).

Q7 a) $\sin x - \sin x\cos^2 x \equiv \sin x(1 - \cos^2 x)$
Then using $\sin^2 x + \cos^2 x \equiv 1$,
$\sin x(1 - \cos^2 x) \equiv \sin x(\sin^2 x)$
$\equiv \sin^3 x$

b) $3\sin x - 3\sin x\cos^2 x - 1 = 0$
$\Rightarrow 3(\sin x - \sin x\cos^2 x) - 1 = 0$
Then using the identity from part a),
$\Rightarrow 3\sin^3 x - 1 = 0$
$\Rightarrow \sin^3 x = \frac{1}{3} \Rightarrow \sin x = \frac{1}{\sqrt[3]{3}}$
Look for solutions in the interval $-180° \le x \le 180°$.
The first solution is $x = 43.9°$ (1 d.p.).
Using the symmetry of the graph of $\sin x$, there is another solution at $x = 180° - 43.9° = 136.1°$ (1 d.p.).
So the solutions are $x = 43.9°$ and $136.1°$ (1 d.p.).

Q8 $4\cos\frac{x}{2} - 3\sin^2\frac{x}{2} - 1 = 0 \Rightarrow 4\cos\frac{x}{2} - 3(1 - \cos^2\frac{x}{2}) - 1 = 0$
$\Rightarrow 3\cos^2\frac{x}{2} + 4\cos\frac{x}{2} - 4 = 0$
$\Rightarrow (3\cos\frac{x}{2} - 2)(\cos\frac{x}{2} + 2) = 0$
$\cos\frac{x}{2} = -2$ has no solutions, so you only need to solve $\cos\frac{x}{2} = \frac{2}{3}$ in the interval $0° \le \frac{x}{2} \le 360°$. The first solution is $\frac{x}{2} = 48.19°$ (2 d.p.). Using the symmetry of the graph, there's another solution at $\frac{x}{2} = 360° - 48.19° = 311.81°$ (2 d.p.).
These are solutions for $\frac{x}{2}$, so multiply them all by 2:
$x = 96.4°, 623.6°$ (1 d.p.)

Q9 $\frac{\cos x}{\tan x} + \sin x = 3 \Rightarrow \frac{\cos x}{\left(\frac{\sin x}{\cos x}\right)} + \sin x = 3$
$\Rightarrow \frac{\cos^2 x}{\sin x} + \sin x = 3 \Rightarrow \cos^2 x + \sin^2 x = 3\sin x$
$\Rightarrow 1 = 3\sin x \Rightarrow \sin x = \frac{1}{3}$
The first solution is $x = 19.5°$ (1 d.p.).
Looking at the symmetry of the graph of $\sin x$, the other solutions are $x = -340.5°, -199.5°$ and $160.5°$ (1 d.p.).

Q10 $\cos x \sin^2 x - \cos x \equiv \cos x (\sin^2 x - 1)$
$\equiv -\cos x (1 - \sin^2 x)$
$\equiv -\cos x (\cos^2 x)$
$\equiv -\cos^3 x$

So solve $\cos^3 x = -0.86 \Rightarrow \cos x = \sqrt[3]{-0.86}$
The first solution is $x = \cos^{-1}(\sqrt[3]{-0.86}) = 162.0°$ (3 s.f.)
Using the symmetry of the graph, the other solution in the given interval is $x = 0° - 162.0° = -162.0°$ (3 s.f.)

Q11 $4\cos^2 x \tan x + \sin x = 0 \Rightarrow 4\cos^2 x \frac{\sin x}{\cos x} + \sin x = 0$
$\Rightarrow 4 \sin x \cos x + \sin x = 0$
$\Rightarrow \sin x (4\cos x + 1) = 0$

Either $\sin x = 0$ or $\cos x = -\frac{1}{4}$, so solve separately:
First solve $\sin x = 0$: $x = 0, 180°, 360°$
Then solve $\cos x = -\frac{1}{4}$:
The first solution is $x = 104.5°$ (1 d.p).
Using the symmetry of cos, there is a solution at $x = 360° - 104.5° = 255.5°$. So the solutions are:
$x = 0, 104.5°, 180°, 255.5°, 360°$ (1 d.p.)

Q12 $\frac{\cos^2 x}{\sin x - 1} \equiv \frac{1 - \sin^2 x}{\sin x - 1} \equiv \frac{(1 - \sin x)(1 + \sin x)}{\sin x - 1}$
$\equiv \frac{-(\sin x - 1)(1 + \sin x)}{\sin x - 1} \equiv -(1 + \sin x)$

So $\frac{\cos^2 x}{\sin x - 1} = -0.25 \Rightarrow 1 + \sin x = 0.25 \Rightarrow \sin x = -0.75$
The first solution is $x = -48.6$ (1 d.p.). Using the symmetry of the graph, there is another solution at $-180° + 48.6° = -131.4°$.
Find the other solutions in the interval by adding 360° to the solutions already found:
$x = -131.4°, -48.6°, 228.6°, 311.4°$ (1 d.p.)

Q13 Points of intersection occur when $f(x) = g(x)$:
$2 + 3\cos^2 x = 7 \sin x - 1 \Rightarrow 3\cos^2 x - 7 \sin x + 3 = 0$
$\Rightarrow 3(1 - \sin^2 x) - 7 \sin x + 3 = 0$
$\Rightarrow 3\sin^2 x + 7 \sin x - 6 = 0$
$\Rightarrow (\sin x + 3)(3 \sin x - 2) = 0$

$\sin x = -3$ has no solutions, so just solve $\sin x = \frac{2}{3}$:
The first solution is $x = 41.8°$ (3 s.f.).
Using the symmetry of the graph of $\sin x$, the other solution is at $x = 180° - 41.8° = 138.2°$ (3 s.f.)
$\sin x = \frac{2}{3}$ for these x-values, so find the y-coordinates:
$g(41.8°) = g(138.2°) = (7 \times \frac{2}{3}) - 1 = 3.67$ (3 s.f.)
So the points are (41.8°, 3.67) and (139.2°, 3.67).

Review Exercise — Chapter 8

Q1 $\cos 30° = \frac{\sqrt{3}}{2}$, $\sin 30° = \frac{1}{2}$, $\tan 30° = \frac{1}{\sqrt{3}}$
$\cos 45° = \frac{1}{\sqrt{2}}$, $\sin 45° = \frac{1}{\sqrt{2}}$, $\tan 45° = 1$
$\cos 60° = \frac{1}{2}$, $\sin 60° = \frac{\sqrt{3}}{2}$, $\tan 60° = \sqrt{3}$

Q2 a) $\cos\theta = \frac{1}{2} \Rightarrow \theta = \cos^{-1}\left(\frac{1}{2}\right) = 60°$

b) $\cos\theta = \frac{\sqrt{3}}{2} \Rightarrow \theta = \cos^{-1}\left(\frac{\sqrt{3}}{2}\right) = 30°$

c) $\cos\theta = -1 \Rightarrow \theta = \cos^{-1}(-1) = 180°$

Q3 a) Angle $B = 180° - 30° - 25° = 125°$.
Using the sine rule: $\frac{a}{\sin A} = \frac{b}{\sin B}$
$\frac{6}{\sin 125°} = \frac{a}{\sin 30°} = \frac{c}{\sin 25°}$ so:
side $c = \frac{6 \sin 25°}{\sin 125°} = 3.095...$ m
side $a = \frac{6 \sin 30°}{\sin 125°} = 3.662...$ m

b) Area $= \frac{1}{2} ab \sin C = \frac{1}{2} \times 3.662 \times 6 \times \sin 25°$
$= 4.643... = 4.64$ m² (3 s.f.)

Q4 a) Using the cosine rule: $a^2 = b^2 + c^2 - 2bc \cos A$
$r^2 = 13^2 + 23^2 - (2 \times 13 \times 23 \times \cos 20°)$
so $r = \sqrt{136.06...} = 11.7$ km (1 d.p.)
$P = \cos^{-1}\left(\frac{23^2 + 136.06... - 13^2}{2 \times 23 \times \sqrt{136.06...}}\right) = 22.4°$ (1 d.p.),
$Q = \cos^{-1}\left(\frac{13^2 + 136.06... - 23^2}{2 \times 13 \times \sqrt{136.06...}}\right) = 137.6°$ (1 d.p.)

b) Area $= \frac{1}{2} ab \sin C = \frac{1}{2} \times 13 \times 23 \times \sin 20°$
$= 51.1$ km² (1 d.p.)

Q5 Using the cosine rule: $\cos A = \frac{b^2 + c^2 - a^2}{2bc}$
$A = \cos^{-1}\left(\frac{10^2 + 20^2 - 25^2}{2 \times 10 \times 20}\right) = 108.2°$ (1 d.p.)
$B = \cos^{-1}\left(\frac{10^2 + 25^2 - 20^2}{2 \times 10 \times 25}\right) = 49.5°$ (1 d.p.)
$C = \cos^{-1}\left(\frac{20^2 + 25^2 - 10^2}{2 \times 20 \times 25}\right) = 22.3°$ (1 d.p.)

Q6 Using the sine rule: $\frac{\sin A}{a} = \frac{\sin B}{b}$
$\frac{\sin 35°}{3} = \frac{\sin B}{5}$ so $B = \sin^{-1}\left(\frac{5 \sin 35°}{3}\right) = 72.93...°$.
By the symmetry of the sine graph, another solution is $B = 180° - 72.93...° = 107.06...°$
When $B = 72.93...°$:
$C = 180° - 35° - 72.93...° = 72.06...°$
Using the sine rule, $\frac{3}{\sin 35°} = \frac{c}{\sin 72.06...°}$
$\Rightarrow c = \frac{3 \sin 72.06...°}{\sin 35°} = 4.976... = 4.98$ cm (3 s.f.)
When $B = 107.06...°$:
$C = 180° - 35° - 107.06...° = 37.93...°$
and $c = \frac{3 \sin 37.93...°}{\sin 35°} = 3.215...$ m $= 3.22$ m (3 s.f.)

Q7 $\tan x - \sin x \cos x \equiv \frac{\sin x}{\cos x} - \sin x \cos x$
$\equiv \frac{\sin x - \sin x \cos^2 x}{\cos x}$
$\equiv \frac{\sin x(1 - \cos^2 x)}{\cos x}$
$\equiv \frac{\sin x(\sin^2 x)}{\cos x}$
$\equiv \sin^2 x \tan x$

Q8 $\tan^2 x - \cos^2 x + 1 \equiv \frac{\sin^2 x}{\cos^2 x} - (1 - \sin^2 x) + 1$
$\equiv \frac{\sin^2 x}{\cos^2 x} + \sin^2 x$
$\equiv \frac{\sin^2 x + \sin^2 x \cos^2 x}{\cos^2 x}$
$\equiv \frac{\sin^2 x(1 + \cos^2 x)}{\cos^2 x}$
$\equiv \tan^2 x(1 + \cos^2 x)$

Q9 $(\sin y + \cos y)^2 + (\cos y - \sin y)^2$
$\equiv (\sin^2 y + 2 \sin y \cos y + \cos^2 y) + (\cos^2 y - 2 \sin y \cos y + \sin^2 y)$
$\equiv 2\sin^2 y + 2 \sin y \cos y - 2 \sin y \cos y + 2\cos^2 y$
$\equiv 2(\sin^2 y + \cos^2 y)$
$\equiv 2$

Q10 $\frac{\sin^4 x + \sin^2 x \cos^2 x}{\cos^2 x - 1} \equiv \frac{\sin^4 x + \sin^2 x(1 - \sin^2 x)}{1 - \sin^2 x - 1}$
$\equiv \frac{\sin^4 x + \sin^2 x - \sin^4 x}{-\sin^2 x}$
$\equiv \frac{\sin^2 x}{-\sin^2 x} \equiv -1$

Q11 a)
y = cos x
y
x
1
-1
−360° −270° −180° −90° 0° 90° 180° 270° 360°

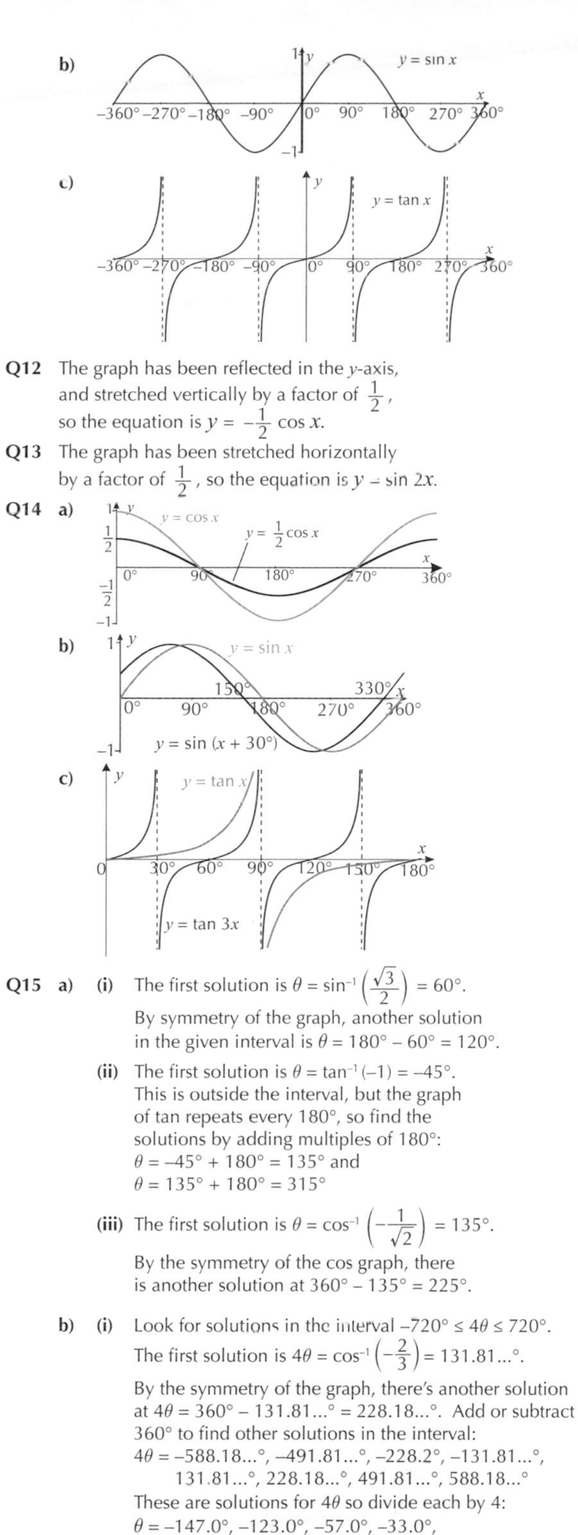

Q12 The graph has been reflected in the y-axis, and stretched vertically by a factor of $\frac{1}{2}$, so the equation is $y = -\frac{1}{2}\cos x$.

Q13 The graph has been stretched horizontally by a factor of $\frac{1}{2}$, so the equation is $y = \sin 2x$.

Q15 a) (i) The first solution is $\theta = \sin^{-1}\left(\frac{\sqrt{3}}{2}\right) = 60°$.
By symmetry of the graph, another solution in the given interval is $\theta = 180° - 60° = 120°$.

(ii) The first solution is $\theta = \tan^{-1}(-1) = -45°$.
This is outside the interval, but the graph of tan repeats every 180°, so find the solutions by adding multiples of 180°:
$\theta = -45° + 180° = 135°$ and
$\theta = 135° + 180° = 315°$

(iii) The first solution is $\theta = \cos^{-1}\left(-\frac{1}{\sqrt{2}}\right) = 135°$.
By the symmetry of the cos graph, there is another solution at $360° - 135° = 225°$.

b) (i) Look for solutions in the interval $-720° \leq 4\theta \leq 720°$.
The first solution is $4\theta = \cos^{-1}\left(-\frac{2}{3}\right) = 131.81...°$.
By the symmetry of the graph, there's another solution at $4\theta = 360° - 131.81...° = 228.18...°$. Add or subtract 360° to find other solutions in the interval:

$4\theta = -588.18...°, -491.81...°, -228.2°, -131.81...°, 131.81...°, 228.18...°, 491.81...°, 588.18...°$
These are solutions for 4θ so divide each by 4:
$\theta = -147.0°, -123.0°, -57.0°, -33.0°, 33.0°, 57.0°, 123.0°. 147.0°$ (1 d.p.)

(ii) Look for solutions in the interval $-145° \leq \theta + 35° \leq 215°$.
The first solution is $\theta + 35° = \sin^{-1}(0.3) = 17.45..°$
By the symmetry of the graph, there's another solution at $\theta + 35° = 180° - 17.45...° = 162.54...°$
So the solutions are $\theta = -17.5°, 127.5°$ (1 d.p.)

(iii) Look for solutions in the interval $-90° \leq \frac{\theta}{2} \leq 90°$.
The first solution is $\frac{\theta}{2} = \tan^{-1}(500) = 89.89...°$
All other solutions are outside the interval.
This is a solution for $\frac{\theta}{2}$, so multiply by 2:
$\theta = 179.8°$ (1 d.p.)

Q16 $6\sin^2 x - \cos x - 5 = 0 \Rightarrow 6(1 - \cos^2 x) - \cos x - 5 = 0$
$\Rightarrow -6\cos^2 x - \cos x + 1 = 0$
$\Rightarrow 6\cos^2 x + \cos x - 1 = 0$
$\Rightarrow (3\cos x - 1)(2\cos x + 1) = 0$

Either $\cos x = \frac{1}{3}$ or $\cos x = -\frac{1}{2}$:
For $\cos x = \frac{1}{3}$, the first solution is $x = 70.52...°$.
By the symmetry of cos, there is another solution at $360° - 70.52...° = 289.47...°$
For $\cos x = -\frac{1}{2}$, the first solution is $x = 120°$. By the symmetry of cos, there is another solution at $360° - 120° = 240°$.
So the solutions to the equation are:
$x = 70.5°$ (1 d.p.), $120°$, $240°$, $289.5°$ (1 d.p.)

Q17 $3\tan x + 2\cos x = 0 \Rightarrow 3\frac{\sin x}{\cos x} + 2\cos x = 0$
$\Rightarrow 3\sin x + 2\cos^2 x = 0$
$\Rightarrow 3\sin x + 2(1 - \sin^2 x) = 0$
$\Rightarrow -2\sin^2 x + 3\sin x + 2 = 0$
$\Rightarrow 2\sin^2 x - 3\sin x - 2 = 0$
$\Rightarrow (\sin x - 2)(2\sin x + 1) = 0$

Either $\sin x = -2$ (this has no solutions) or $\sin x = -\frac{1}{2}$.
The first solution to $\sin x = -\frac{1}{2}$ is $x = -30°$. This is the only solution to the equation in the given interval.

Q18 $8\sin^2 x + 2\sin x - 1 = 0 \Rightarrow (2\sin x + 1)(4\sin x - 1) = 0$
So either $\sin x = -\frac{1}{2}$ or $\sin x = \frac{1}{4}$.
For $\sin x = -\frac{1}{2}$, $x = -30°$. This is outside the interval, so add 360° to get the first solution: $-30° + 360° = 330°$
By the symmetry of $\sin x$, the other solution in the interval is $180° + 30° = 210°$.
For $\sin x = \frac{1}{4}$, the first solution is $x = 14.47...°$
By the symmetry of $\sin x$, the other solution in the interval is $180° - 14.47...° = 165.52...°$
So the solutions are: $x = 14.5°$ (1 d.p.), $165.5°$ (1 d.p.), $210°$, $330°$

Q19 $\tan x - 3\sin x = 0 \Rightarrow \frac{\sin x}{\cos x} - 3\sin x = 0$
$\Rightarrow \sin x - 3\sin x\cos x = 0$
$\Rightarrow \sin x(1 - 3\cos x) = 0$

Either $\sin x = 0$ or $\cos x = \frac{1}{3}$. For $\sin x = 0$:
The solutions are $x = 0°, 180°, 360°, 540°, 720°$.
For $\cos x = \frac{1}{3}$, the first solution is $x = 70.52...°$
By the symmetry of cos, there is another solution at $360° - 70.52...° = 289.47...°$. To find further solutions, add 360° to the solutions found already:
So the solutions are:
$x = 0°$, $70.5°$ (1 d.p.), $180°$, $289.5°$ (1 d.p.), $360°$, $430.5°$ (1 d.p.), $540°$, $649.5°$ (1 d.p.), $720°$

Exam-Style Questions — Chapter 8

Q1 $\frac{\cos x}{\tan x} + \sin x \equiv \cos x \frac{\cos x}{\sin x} + \sin x$
$\equiv \frac{\cos^2 x}{\sin x} + \frac{\sin^2 x}{\sin x}$
$\equiv \frac{\sin^2 x + \cos^2 x}{\sin x}$
$\equiv \frac{1}{\sin x}$

[3 marks available — 1 mark for using $\tan x \equiv \frac{\sin x}{\cos x}$, 1 mark for making a common denominator, 1 mark for using $\sin^2 x + \cos^2 x = 1$]

Q2 Compare this graph with the graph of $y = \sin x$:
$y = \sin x$ has a maximum at 1, and this graph has an maximum at 3, so it has been stretched vertically by a factor of 3, i.e. $p = 3$.
$y = \sin x$ intersects the x-axis at $x = 180°$, and this graph intersects the x-axis at $x = 90°$, so it has been stretched horizontally by a factor of $\frac{1}{2}$, i.e. $q = 2$.

[3 marks available — 1 mark for identifying vertical stretch, 1 mark for identifying horizontal stretch, 1 mark for both correct values of p and q]

Q3 **a)** (i) g(x) intersects the x–axis at:
(–330°, 0), (–150°, 0), (30°, 0), (210°, 0)

[2 marks available — 2 marks for all four coordinates correct, otherwise 1 mark for two coordinates correct]

(ii) The equations of the asymptotes of g(x) are:
$x = -240°$, $x = -60°$, $x = 120°$, $x = 300°$

[2 marks available — 2 marks for all 4 equations correct, otherwise 1 mark for 2 equations correct]

b) As $f(x) = \tan x \equiv \frac{\sin x}{\cos x}$, asymptotes exist for any value of x such that $\cos x = 0$. *[1 mark]*

c) The graph of $\tan x$ has rotational symmetry around the origin, so $\tan(-30°) = -\tan 30°$.
So $\tan 30° = \frac{\sqrt{3}}{3}$.

[2 marks available — 1 mark for correct answer, 1 mark for a valid comment referring to rotational symmetry or odd functions]

Q4 **a)** $3 \sin x - 4 \cos x = 0 \Rightarrow 3 \sin x = 4 \cos x$
$\Rightarrow \frac{\sin x}{\cos x} = \frac{4}{3} \Rightarrow \tan x = \frac{4}{3}$

[2 marks available — 1 mark for rearranging the equation to form $\frac{\sin x}{\cos x}$, 1 mark for using the identity $\tan x \equiv \frac{\sin x}{\cos x}$]

b) The first solution is $x = \tan^{-1}\left(\frac{4}{3}\right) = 53.13...° = 53.1°$ (1 d.p.).
The graph of tan repeats every 180°, so the other solution in the interval is: 53.13...° + 180° = 233.13...° = 233.1° (1 d.p.)

[3 marks available — 1 mark for using the inverse tan function, 1 mark for finding the solution x = 53.1°, 1 mark for both solutions of x correct to 1 d.p.]

Q5 **a)** Evaluate the function when $t = 3$:
$T(3) = 15 + 8 \sin(45 \times 3) = 20.7$ (1 d.p.)
So after 3 hours, the temperature of the room is 20.7 °C.
[2 marks available — 1 mark for substituting t = 3 into the function, 1 mark for the correct answer rounded to an appropriate degree of accuracy]

b) The maximum value that $\sin \theta$ can take is 1 and the minimum value it can take is –1.
So the maximum value is $T(t) = 15 + 8(1) = 23$ °C
and the minimum value is $T(t) = 15 + 8(-1) = 7$ °C.

[2 marks available — 1 mark for the correct maximum and 1 mark for the correct minimum]

c) $T(t) = 15 + 8 \sin(45t) = 20$
$\Rightarrow \sin(45t) = \frac{5}{8} \Rightarrow 45t = \sin^{-1}\left(\frac{5}{8}\right) = 38.682...$
Using the symmetry of sin, there is another solution at $45t = 180 - 38.682... = 141.317...$
Divide the solutions by 45 to find the values of t:
$t = 0.896..., 3.140....$
The next solution is when $45t = 38.682... + 360$, i.e. $t = 398.682... \div 45 = 8.85$ which is outside the interval — the temperature reaches 20 °C twice during the first 8 hours.

[6 marks available — 1 mark for substituting 20 into the function, 1 mark for finding one correct value of 45t, 1 mark for one correct value of t, 1 mark for finding second value of t, 1 mark for showing there are only two solutions in the interval 0 < t < 8, 1 mark for stating the temperature reaches 20 °C twice]

Q6 $2 \sin(x + 30°) = \frac{1}{4} \Rightarrow \sin(x + 30°) = \frac{1}{8}$
Look for solutions in the interval:
$-330° \leq x + 30° \leq 390°$.
The first solution is $(x + 30°) = 7.18...°$.
By the symmetry of sin, there is another solution
$x + 30° = 180° - 7.18...° = 172.82...°$
To find other solutions in the interval, add or subtract 360° from the solutions already found:
$x + 30° = -187.18...°, 7.18...°, 172.82...°, 367.18...°$
So $x = -217.2°, -22.8°, 142.8°, 337.2°$ (1 d.p.)

[5 marks available — 1 mark for using the inverse sin function, 1 mark for adapting the interval given the transformation of sin x to sin (x + 30°), 1 mark for finding the other solutions in the interval, 1 mark for a method to convert the solutions of (x + 30°) to the solutions of x, 1 mark for all four correct values of x]
You could also use a CAST diagram to find the other solutions to the equation.

Q7 Using the sine rule: $\frac{\sin A}{a} = \frac{\sin B}{b}$
$\frac{\sin 37°}{12} = \frac{\sin \angle PRQ}{16}$ so $\angle PRQ = \sin^{-1}\left(\frac{16 \sin 37°}{12}\right) = 53.36...°$
By the symmetry of $\sin x$, there is another solution
$\angle PRQ = 180° - 53.36...° = 126.63...°$
So the possible values of $\angle PQR$ are:
$\angle PQR = 180° - 37° - 53.36° = 89.63...°$ or
$\angle PQR = 180° - 37° - 126.63° = 16.36...°$
So the possible areas of ΔPQR are:
$A_1 = \frac{1}{2} \times 16 \times 12 \times \sin 89.63...° = 95.99... = 96.0 \text{ m}^2$ (3 s.f.)
$A_2 = \frac{1}{2} \times 16 \times 12 \times \sin 16.36...° = 27.04... = 27.0 \text{ m}^2$ (3 s.f.)
[6 marks available — 1 mark for using the sine rule, 1 mark for finding the acute angle PRQ, 1 mark for finding one of the possible values of angle PQR, 1 mark for finding the obtuse angle PRQ, 1 mark for using the area formula, 1 mark for both areas correct to 3 s.f.]

Q8 $f(x) = g(x) \Rightarrow 3 \cos^2 x = 1 - \sin x$
$\Rightarrow 3(1 - \sin^2 x) = 1 - \sin x$
$\Rightarrow 3 \sin^2 x - \sin x - 2 = 0$
$\Rightarrow (3 \sin x + 2)(\sin x - 1) = 0$

So either $\sin x = -\frac{2}{3}$ or $\sin x = 1$.
For $\sin x = -\frac{2}{3}$, the first solution is $x = -41.81...°$
This is outside the interval, so add 360° to get a value for x in the interval: $x = 318.18...°$.
By the symmetry of the sin graph, there is another solution at $180° - (-41.81...°) = 221.81...°$

For $\sin x = 1$, the first solution is $x = 90°$, and there are no other solutions in the given interval.
So the solutions are $x = 90°, 221.8°, 318.2°$ (1 d.p.)

[6 marks available — 1 mark for using the identity $\sin^2 x + \cos^2 x = 1$, 1 mark for constructing a quadratic equation in sin x, 1 mark for factorising to get two values for sin x, 1 mark for using the inverse sin function, 1 mark for at least two correct solutions, 1 mark for all three correct]

Chapter 9: Exponentials and Logarithms

9.1 Exponentials

Exercise 9.1.1 — Exponentials

Q1 **a)**

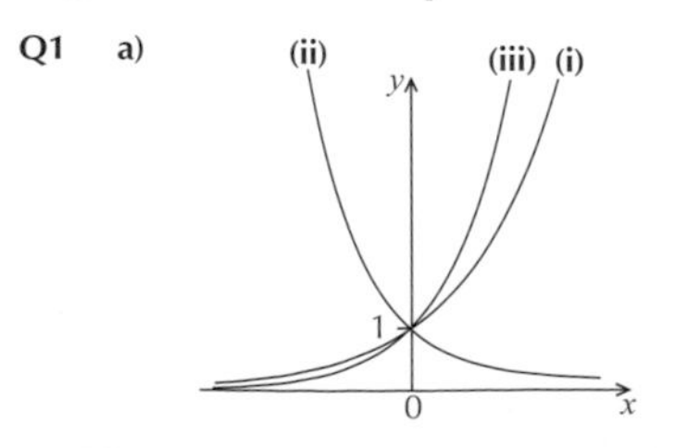

b)

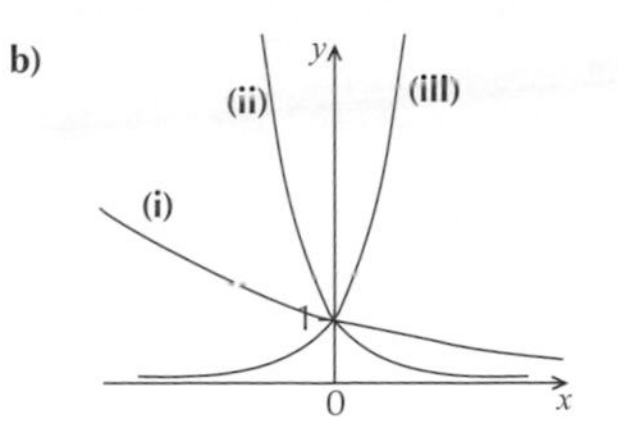

Q2 a) $y = 3e^{3x} = h(x)$ — a positive exponential graph passing through (0, 3) (when $x = 0$, $3e^0 = 3$).

b) $y = 3^x = f(x)$ — a positive exponential graph passing through (0, 1).

c) $y = 3e^{-x} = g(x)$ — a reflection of $y = 3e^x$ in the y-axis.

Q3 a) $y = f(x)$ is graph B — a positive exponential graph passing through (0, 1).

b) $y = f(x) + 2$ is graph C — a vertical translation of f(x) by +2.

c) $y = f(-x)$ is graph A — a reflection of f(x) in the y-axis.

Q4 a) $y = e^{3x}$, so the gradient at any value of x is $3e^{3x}$.
At $x = 0$, gradient $= 3e^{3 \times 0} = 3e^0 = 3 \times 1 = 3$
At $x = 1$, gradient $= 3e^{3 \times 1} = 3e^3$
The question asks for exact answers, so leave them in terms of e.

b) $y = 5e^{0.5x}$, so the gradient at any value of x is $2.5e^{0.5x}$.
At $x = -2$, gradient $= 2.5e^{0.5 \times -2} = 2.5e^{-1} = \frac{2.5}{e}$
At $x = 2$, gradient $= 2.5e^{0.5 \times 2} = 2.5e^1 = 2.5e$

c) $x = 2.5e^{6t}$, so the gradient at any value of t is $15e^{6t}$.
At $t = \frac{1}{3}$, gradient $= 15e^{6 \times \frac{1}{3}} = 15e^2$
At $t = 5$, gradient $= 15e^{6 \times 5} = 15e^{30}$

d) $y = 4e^{-2x}$, so the gradient at any value of x is $-8e^{-2x}$.
At $x = -5$, gradient $= -8e^{-2 \times -5} = -8e^{10}$
At $x = 0.5$, gradient $= -8e^{-2 \times 0.5} = -8e^{-1}$

Q5 $y = Ae^{kx}$, so the gradient at any value of x is kAe^{kx}.
At $x = 2$, gradient $= kAe^{2k} = 12e^3 \Rightarrow 2k = 3 \Rightarrow k = 1.5$,
and $kA = 12 \Rightarrow 1.5A = 12 \Rightarrow A = 8$.
So $y = 8e^{1.5x}$. At $x = 4$, $y = 8e^{1.5 \times 4} = 8e^6$
and at $x = 12$, $y = 8e^{1.5 \times 12} = 8e^{18}$.

Q6 "The gradient of the curve of f(t) is always 0.4 times the value of f(t)" — this tells you that the gradient is directly proportional to the curve, i.e. it's the exponential function where $f(t) = Ae^{kt}$ for constants A and k.
You also know that the gradient of an exponential is kAe^{kt} — so $k = 0.4 \Rightarrow f(t) = Ae^{0.4t}$.
"When first counted, there were 7 rabbits" — this tells you that at $t = 0$, $f(t) = 7$
$\Rightarrow 7 = Ae^{0.4 \times 0} = Ae^0 = A$. So $f(t) = 7e^{0.4t}$.
An estimate for the number of rabbits after 5 years is:
$f(5) = 7 \times e^{0.4 \times 5} = 7 \times e^2 \approx 52$ rabbits.

9.2 Logs

Exercise 9.2.1 — Logs

Q1 a) $\log_2 8 = 3$ b) $\log_5 625 = 4$
c) $\log_{49} 7 = \frac{1}{2}$ d) $\log_8 4 = \frac{2}{3}$
e) $\log_{10} \frac{1}{100} = -2$ f) $\log_2 0.125 = -3$
g) $\log_4 9 = x$ h) $\log_x 40 = 3$
i) $\log_8 x = 11$ j) $\log_{64} 512 = \frac{3}{2}$
k) $\log_{25} \frac{1}{5} = -\frac{1}{2}$ l) $\log_x 4096 = 2y$

Q2 a) $a = e^6$ b) $t = 5^{0.2}$
c) $m = 4^1 = 4$ d) $p = e^{13}$
e) $k = 10^5 = 100\,000$ f) $a = x^m$
g) $k = e^z$ h) $q = 10^r$

Q3 a) 3 b) −2
c) 0.477 (3 d.p.) d) 2.097 (3 d.p.)
e) 0.693 (3 d.p.) f) 0
g) 1.792 (3 d.p.) h) 2.996 (3 d.p.)
Use the 'log' or 'ln' button on your calculator for these.

Q4 a) $a = \log_2 4 \Rightarrow 2^a = 4 = 2^2 \Rightarrow a = 2$
b) $b = \log_3 27 \Rightarrow 3^b = 27 = 3^3 \Rightarrow b = 3$
c) $c = \log_5 0.2 \Rightarrow 5^c = 0.2 = \frac{1}{5} = 5^{-1} \Rightarrow c = -1$
d) $d = \log_{0.5} 0.25 \Rightarrow 0.5^d = 0.25 = 0.5^2 \Rightarrow d = 2$

Q5 a) $x^2 = 49 \Rightarrow x = 7$
b) $x^3 = 8 \Rightarrow x = 2$
c) $x^5 = 100\,000 \Rightarrow x = 10$
d) $x^5 = 3125 \Rightarrow x = 5$
e) $x^{\frac{1}{2}} = 3 \Rightarrow x = 3^2 \Rightarrow x = 9$
f) $x^{\frac{1}{3}} = 7 \Rightarrow x = 7^3 \Rightarrow x = 343$
g) $x^{\frac{1}{5}} = 2 \Rightarrow x = 2^5 \Rightarrow x = 32$
h) $x^{\frac{1}{4}} = 9 \Rightarrow x = 9^4 \Rightarrow x = 6561$

Q6 a) (i) $e^x = 5 \Rightarrow \ln e^x = \ln 5 \Rightarrow x = \ln 5$
(ii) $\ln 5 = 1.61$ (3 s.f.)

b) (i) $\ln x = 8 \Rightarrow e^{\ln x} = e^8 \Rightarrow x = e^8$
(ii) $e^8 = 2980$ (3 s.f.)

c) (i) $e^{3t} = 11 \Rightarrow \ln e^{3t} = \ln 11 \Rightarrow 3t = \ln 11$
$\Rightarrow t = \frac{\ln 11}{3}$
(ii) $\frac{\ln 11}{3} = 0.799$ (3 s.f.)

d) (i) $\ln 10x = 4 \Rightarrow e^{\ln 10x} = e^4 \Rightarrow 10x = e^4 \Rightarrow x = \frac{e^4}{10}$
(ii) $\frac{e^4}{10} = 5.46$ (3 s.f.)

Q7 a) $a^2 = x$ and $a^4 = y$, so $y = x^2$.
b) $a^3 = x$ and $(2a)^3 = y \Rightarrow 8a^3 = y$, so $y = 8x$
c) $e^5 = x$ and $e^{20} = y$, $(e^5)^4 = y$, so $y = x^4$

9.3 Solving Equations

Exercise 9.3.1 — Laws of logs

Q1 a) $\log_a 2 + \log_a 5 = \log_a (2 \times 5) = \log_a 10$
b) $\ln 8 + \ln 7 = \ln (8 \times 7) = \ln 56$
c) $\log_b 8 - \log_b 4 = \log_b (8 \div 4) = \log_b 2$
d) $\log_m 15 - \log_m 5 = \log_m (15 \div 5) = \log_m 3$
e) $3 \log_a 4 = \log_a (4^3) = \log_a 64$
f) $2 \ln 7 = \ln (7^2) = \ln 49$
g) $\frac{1}{2} \log_b 16 = \log_b (16^{\frac{1}{2}}) = \log_b 4$
h) $\frac{2}{3} \log_a 125 = \log_a (125^{\frac{2}{3}}) = \log_a 25$
i) $\frac{1}{5} \ln 4^5 = \ln ((4^5)^{\frac{1}{5}}) = \ln (4^{5 \times \frac{1}{5}}) = \ln 4^1 = \ln 4$

Q2 a) $\log 0.5 = \log \frac{1}{2} = \log (2^{-1}) = -\log 2$
b) $\log \sqrt{3} = \log 3^{\frac{1}{2}} = \frac{1}{2} \log 3$
c) $\log 0.25 = \log \frac{1}{4} = \log (2^{-2}) = -2\log 2$
d) $\log \sqrt[3]{5} = \log 5^{\frac{1}{3}} = \frac{1}{3} \log 5$

Q3 a) $2 \log_a 5 + \log_a 4 = \log_a (5^2) + \log_a 4$
$= \log_a (25 \times 4) = \log_a 100$

b) $3 \log_m 2 - \log_m 4 = \log_m (2^3) - \log_m 4$
$= \log_m (8 \div 4) = \log_m 2$

c) $3 \ln 4 - 2 \ln 8 = \ln (4^3) - \ln (8^2) = \ln (64 \div 64) = \ln 1 = 0$

d) $\frac{2}{3} \ln 216 - 2 \ln 3 = \ln (216^{\frac{2}{3}}) - \ln (3^2) = \ln (36 \div 9) = \ln 4$

e) $1 + \log_a 6 = \log_a a + \log_a 6 = \log_a 6a$

f) $2 - \log_b 5 = 2\log_b b - \log_b 5 = \log_b b^2 - \log_b 5 = \log_b \left(\frac{b^2}{5}\right)$

Q4 a) $\log_a 6 = \log_a (2 \times 3) = \log_a 2 + \log_a 3 = x + y$

b) $\log_a 16 = \log_a 2^4 = 4\log_a 2 = 4x$

c) $\log_a 60 = \log_a (2 \times 2 \times 3 \times 5)$
$= \log_a 2^2 + \log_a 3 + \log_a 5 = 2x + y + z$

Q5 a) $\log_b b^3 = 3\log_b b = 3$

b) $\log_a \sqrt{a} = \log_a a^{\frac{1}{2}} = \frac{1}{2}\log_a a = \frac{1}{2}$

c) $\ln 4e - 2\ln 2 = \ln 4 + \ln e - \ln 2^2 = \ln 4 + 1 - \ln 4 = 1$

d) $\ln 9 + \ln \frac{e}{3} - \ln 3 = \ln 3^2 + \ln e - \ln 3 - \ln 3$
$= 2\ln 3 + \ln e - 2\ln 3 = \ln e = 1$

Q6 a) $\log_2 4^x = x\log_2 4 = x\log_2 2^2 = 2x\log_2 2 = 2x$

b) $\frac{\ln 54 - \ln 6}{\ln 3} = \frac{\ln(54 \div 6)}{\ln 3} = \frac{\ln 9}{\ln 3} = \frac{\ln 3^2}{\ln 3} = \frac{2\ln 3}{\ln 3} = 2$

Q7 $4 + \log_c \frac{1}{c^2} + \log_c \sqrt{c} = 4 + \log_c c^{-2} + \log_c c^{\frac{1}{2}}$
$= 4 - 2\log_c c + \frac{1}{2}\log_c c = 4 - 2 + \frac{1}{2} = 2\frac{1}{2}$

Exercise 9.3.2 — Changing the base of a log

Q1 a) $\frac{\log_{10} 2}{\log_{10} 9}$ b) $\frac{\log_{10} 8}{\log_{10} 4}$

c) $\frac{\log_{10} 16}{\log_{10} 17}$ d) $\frac{\log_{10} 14}{\log_{10} 21}$

Q2 a) 0.613 (3 s.f.) b) 0.315 (3 s.f.)

c) 2.33 (3 s.f.) d) 0.861 (3 s.f.)

If your calculator can't do logs of any base, you'll need to use the change of base formula.

Q3 a) $\log_{11} 19$

b) $\log_7 2$

c) $\log_3 4 \times \log_4 5 = \frac{\log_{10} 4}{\log_{10} 3} \times \frac{\log_{10} 5}{\log_{10} 4} = \frac{\log_{10} 5}{\log_{10} 3} = \log_3 5$

d) $\ln 2 \times \log_2 10 = \log_e 2 \times \log_2 10$
$= \frac{\log_{10} 2}{\log_{10} e} \times \frac{\log_{10} 10}{\log_{10} 2} = \frac{\log_{10} 10}{\log_{10} e} = \log_e 10 = \ln 10$

This question uses the 'change of base' formula in reverse.

Q4 $\log_{10} 3x = \frac{\ln 2}{\ln 10} = \log_{10} 2 \Rightarrow 3x = 2 \Rightarrow x = \frac{2}{3}$

Q5 $\left(\frac{\log_3 2}{\log_3 e} + \log_e 3\right) - \left(\log_e 6 + \frac{\log_2 9}{\log_2 e}\right)$
$= (\log_e 2 + \log_e 3) - (\log_e 6 + \log_e 9)$
$= \ln(2 \times 3) - \ln(6 \times 9)$
$= \ln\frac{6}{54} = \ln\frac{1}{9} = \ln(3^{-2}) = -2\ln 3$

Exercise 9.3.3 — Solving equations

Q1 a) Take logs of both sides: $\log 2^x = \log 3 \Rightarrow x\log 2 = \log 3$
$\Rightarrow x = \frac{\log 3}{\log 2} = 1.584... = 1.58$ (3 s.f.)

b) $7^x = 2 \Rightarrow \log 7^x = \log 2 \Rightarrow x\log 7 = \log 2$
$\Rightarrow x = \frac{\log 2}{\log 7} = 0.3562... = 0.356$ (3 s.f.)

c) $1.8^x = 0.4 \Rightarrow \log 1.8^x = \log 0.4 \Rightarrow x\log 1.8 = \log 0.4$
$\Rightarrow x = \frac{\log 0.4}{\log 1.8} = -1.558... = -1.56$ (3 s.f.)

Notice this solution is negative, because log 0.4 is negative.

d) $0.7^x = 3 \Rightarrow \log 0.7^x = \log 3 \Rightarrow x\log 0.7 = \log 3$
$\Rightarrow x = \frac{\log 3}{\log 0.7} = -3.080... = -3.08$ (3 s.f.)

e) $3^{5x} = 890 \Rightarrow \log 3^{5x} = \log 890$
$\Rightarrow 5x\log 3 = \log 890$
$\Rightarrow x = \frac{\log 890}{5\log 3} = 1.24$ (3 s.f.)

f) $0.2^{4x} = 0.016 \Rightarrow \log 0.2^{4x} = \log 0.016$
$\Rightarrow 4x\log 0.2 = \log 0.016$
$\Rightarrow x = \frac{\log 0.016}{4\log 0.2} = 0.642$ (3 s.f.)

g) $2^{3x-1} = 5 \Rightarrow \log 2^{3x-1} = \log 5 \Rightarrow (3x - 1)\log 2 = \log 5$
$\Rightarrow 3x\log 2 = \log 5 + \log 2 = \log(5 \times 2)$
$\Rightarrow x = \frac{\log 10}{3\log 2} = 1.11$ (3 s.f.)

h) $0.4^{5x-4} = 2 \Rightarrow \log 0.4^{5x-4} = \log 2$
$\Rightarrow (5x - 4)\log 0.4 = \log 2$
$\Rightarrow 5x\log 0.4 = \log 2 + 4\log 0.4$
$\Rightarrow x = \frac{\log 2 + 4\log 0.4}{5\log 0.4} = 0.649$ (3 s.f.)

Q2 a) $2^{4x} = 3^{100} \Rightarrow \log_2 2^{4x} = \log_2 3^{100}$
$\Rightarrow 4x\log_2 2 = 100\log_2 3 \Rightarrow 4x = 100\log_2 3$
$\Rightarrow x = 25\log_2 3$

b) $11^{6x} = 10^{90} \Rightarrow \log_{11} 11^{6x} = \log_{11} 10^{90}$
$\Rightarrow 6x\log_{11} 11 = 90\log_{11} 10 \Rightarrow 6x = 90\log_{11} 10$
$\Rightarrow x = 15\log_{11} 10$

c) $6^{50-x} = 2^{50} \Rightarrow \log_6 6^{50-x} = \log_6 2^{50}$
$\Rightarrow (50 - x)\log_6 6 = 50\log_6 2 \Rightarrow 50 - x = 50\log_6 2$
$\Rightarrow x = 50 - 50\log_6 2 = 50(1 - \log_6 2)$
$= 50(\log_6 6 - \log_6 2) = 50\log_6 3$

d) $4^{5+x} = 20^5 \Rightarrow \log_4 4^{5+x} = \log_4 20^5$
$\Rightarrow (5 + x)\log_4 4 = 5\log_4 20 \Rightarrow 5 + x = 5\log_4 20$
$\Rightarrow x = 5\log_4 20 - 5 = 5(\log_4 20 - 1)$
$= 5(\log_4 20 - \log_4 4) = 5\log_4 5$

Q3 a) Take exponentials of both sides using base 10 (since the logarithm is base 10):
$10^{\log 5x} = 10^3 \Rightarrow 5x = 1000 \Rightarrow x = 200$

b) Take exponentials of both sides (using base 2):
$\Rightarrow 2^{\log_2(x+3)} = 2^4 \Rightarrow x + 3 = 16 \Rightarrow x = 13$

c) Take exponentials of both sides (using base 3):
$\Rightarrow 3^{\log_3(5-2x)} = 3^{2.5} \Rightarrow 5 - 2x = 3^{2.5}$
$\Rightarrow x = \frac{5 - 3^{2.5}}{2} = -5.294... = -5.29$ (3 s.f.)

Q4 a) $4^{x+1} = 3^{2x} \Rightarrow \log 4^{x+1} = \log 3^{2x} \Rightarrow (x + 1)\log 4 = 2x\log 3$
Multiply out the brackets: $x\log 4 + \log 4 = 2x\log 3$
Collect x-terms on one side:
$\log 4 = 2x\log 3 - x\log 4 = x(2\log 3 - \log 4)$
$\Rightarrow x = \frac{\log 4}{2\log 3 - \log 4} = 1.709... = 1.71$ (3 s.f.)

b) $2^{5-x} = 4^{x+3} \Rightarrow \log 2^{5-x} = \log 4^{x+3}$
$\Rightarrow (5 - x)\log 2 = (x + 3)\log 4$
But $\log 4 = \log 2^2 = 2\log 2$
$\Rightarrow (5 - x)\log 2 = 2(x + 3)\log 2$
$\Rightarrow 5 - x = 2(x + 3)$
$\Rightarrow -1 = 3x \Rightarrow x = -\frac{1}{3}$

c) $3^{2x-1} = 6^{3-x} \Rightarrow \log 3^{2x-1} = \log 6^{3-x}$
$\Rightarrow (2x - 1)\log 3 = (3 - x)\log 6$
$\Rightarrow 2x\log 3 - \log 3 = 3\log 6 - x\log 6$
$\Rightarrow 2x\log 3 + x\log 6 = 3\log 6 + \log 3$
$\Rightarrow x(2\log 3 + \log 6) = 3\log 6 + \log 3$
$\Rightarrow x = \frac{3\log 6 + \log 3}{2\log 3 + \log 6} = 1.622... = 1.62$ (3 s.f.)

Q5 **a)** $\log_6 x = 1 - \log_6 (x + 1)$
$\Rightarrow \log_6 x + \log_6 (x + 1) = 1 \Rightarrow \log_6 x(x + 1) = 1$
Take exponentials of base 6 of both sides to get:
$\Rightarrow x(x + 1) = 6^1 \Rightarrow x^2 + x - 6 = 0$
$\Rightarrow (x + 3)(x - 2) = 0 \Rightarrow x = 2$
$x = -3$ is not a solution because logarithms of negative numbers don't exist.

b) $\log_2 (2x + 1) = 3 + 2 \log_2 x$
$\Rightarrow \log_2 (2x + 1) = 3 + \log_2 x^2$
$\Rightarrow \log_2 (2x + 1) - \log_2 x^2 = 3 \Rightarrow \log_2 \frac{2x+1}{x^2} = 3$
Take exponentials of base 2 of both sides to get:
$\Rightarrow \frac{2x+1}{x^2} = 2^3 \Rightarrow 2x + 1 = 8x^2$
$\Rightarrow 8x^2 - 2x - 1 = 0 \Rightarrow (4x + 1)(2x - 1) = 0$
So $x = \frac{1}{2}$

Q6 **a)** $5e^{3t} = 11 \Rightarrow e^{3t} = \frac{11}{5} \Rightarrow \ln e^{3t} = \ln\left(\frac{11}{5}\right)$
$\Rightarrow 3t = \ln\left(\frac{11}{5}\right) \Rightarrow t = \frac{1}{3}\ln\left(\frac{11}{5}\right)$

b) $e^{(0.5x + 3)} = 9 \Rightarrow \ln e^{(0.5x + 3)} = \ln 9 \Rightarrow 0.5x + 3 = \ln 9$
$\Rightarrow 0.5x = \ln 9 - 3 \Rightarrow x = 2 (\ln 9 - 3)$

c) $10 - 3e^{(1-2x)} = 8 \Rightarrow 3e^{(1-2x)} = 2 \Rightarrow e^{(1-2x)} = \frac{2}{3}$
$\Rightarrow \ln e^{(1-2x)} = \ln\frac{2}{3} \Rightarrow 1 - 2x = \ln\frac{2}{3}$
$\Rightarrow 2x = 1 - \ln\frac{2}{3} \Rightarrow x = \frac{1}{2}(1 - \ln\frac{2}{3})$

d) $3 \ln (2x) = 7 \Rightarrow \ln (2x) = \frac{7}{3}$
$\Rightarrow e^{\ln (2x)} = e^{\frac{7}{3}} \Rightarrow 2x = e^{\frac{7}{3}} \Rightarrow x = \frac{1}{2}e^{\frac{7}{3}}$

e) $\ln (5t - 3) = 4 \Rightarrow e^{\ln (5t - 3)} = e^4 \Rightarrow 5t - 3 = e^4$
$\Rightarrow t = \frac{1}{5}(e^4 + 3)$

f) $6 - \ln (0.5x) = 3 \Rightarrow \ln (0.5x) = 3 \Rightarrow e^{\ln (0.5x)} = e^3$
$\Rightarrow 0.5x = e^3 \Rightarrow x = 2e^3$

Q7 **a)** $e^{3x} = 27 \Rightarrow \ln (e^{3x}) = \ln 27 \Rightarrow 3x = \ln 27 = \ln (3^3)$
$\Rightarrow 3x = 3 \ln 3 \Rightarrow x = \ln 3.$
If you're asked to give your answer in the form ln a where a is a number, try and write the number inside the logarithm as a power of a and use the third log law to get it in the form you want.

b) $e^{(6x - 1)} = \frac{1}{3} \Rightarrow \ln e^{(6x - 1)} = \ln\left(\frac{1}{3}\right)$
$\Rightarrow 6x - 1 = \ln (3^{-1}) \Rightarrow 6x = 1 - \ln 3 \Rightarrow x = \frac{1}{6}(1 - \ln 3)$

c) $\frac{1}{3}e^{(1 - x)} - 3 = 0 \Rightarrow e^{(1 - x)} = 9 \Rightarrow \ln e^{(1 - x)} = \ln 9$
$\Rightarrow 1 - x = \ln 9 \Rightarrow x = 1 - \ln (3^2) \Rightarrow x = 1 - 2 \ln 3$

Q8 **a)** $\ln 5 + \ln x = 7 \Rightarrow \ln (5x) = 7 \Rightarrow e^{\ln (5x)} = e^7$
$\Rightarrow 5x = e^7 \Rightarrow x = \frac{e^7}{5}$

b) $\ln (2x) + \ln (3x) = 15 \Rightarrow \ln (2x \times 3x) = 15$
$\Rightarrow \ln (6x^2) = 15 \Rightarrow e^{\ln 6x^2} = e^{15} \Rightarrow 6x^2 = e^{15}$
$\Rightarrow x = \sqrt{\frac{1}{6}e^{15}} = \frac{1}{\sqrt{6}}e^{\frac{15}{2}}$

c) $2 \ln x - \ln 2x = 2 \Rightarrow \ln\left(\frac{x^2}{2x}\right) = 2$
$\Rightarrow \frac{x^2}{2x} = e^2 \Rightarrow x^2 = 2xe^2 \Rightarrow x(x - 2e^2) = 0$
$x \neq 0$ as $\ln x$ is not defined at 0, so $x = 2e^2$
You could also have cancelled the x's from the fraction before using the exponent.

d) $\ln (2x - 7) + \ln 4 = -3 \Rightarrow \ln [4(2x - 7)] = -3$
$\Rightarrow 8x - 28 = e^{-3} \Rightarrow x = \frac{e^{-3} + 28}{8}$ or $\frac{1}{8e^3} + \frac{7}{2}$

e) $\ln (x^2 - 4) - \ln (2x) = 0 \Rightarrow \ln\left(\frac{x^2 - 4}{2x}\right) = 0$
$\Rightarrow \frac{x^2 - 4}{2x} = e^0 = 1 \Rightarrow x^2 - 4 = 2x$
$\Rightarrow x^2 - 2x - 4 = 0 \Rightarrow x = \frac{2 \pm \sqrt{20}}{2} = 1 \pm \sqrt{5}$
But $x > 0$ otherwise $\ln 2x$ would be undefined, so $x = 1 + \sqrt{5}$
You might have spotted that $\ln (x^2 - 4) - \ln (2x) = 0$ $\Rightarrow \ln (x^2 - 4) = \ln (2x) \Rightarrow x^2 - 4 = 2x$, then solved the resulting quadratic to find the answer.

f) $3 \ln (x^2) + 5 \ln x = 2 \Rightarrow 6 \ln x + 5 \ln x = 2$
$\Rightarrow 11 \ln x = 2 \Rightarrow \ln x = \frac{2}{11} \Rightarrow x = e^{\frac{2}{11}}$

g) Let $y = e^x$, then $2e^{2x} + e^x - 3 = 0$ is equivalent to the quadratic equation $2y^2 + y - 3 = 0$
$\Rightarrow (2y + 3)(y - 1) = 0$, so $y = 1$ or $y = -\frac{3}{2}$.
So $e^x = 1$ or $e^x = -\frac{3}{2}$. Negative values of e^x are not possible, so $e^x = 1 \Rightarrow x = \ln 1 = 0$.

h) Let $y = e^{4x}$, then $e^{8x} - e^{4x} - 6 = 0$ is equivalent to the quadratic equation $y^2 - y - 6 = 0$
$\Rightarrow (y - 3)(y + 2) = 0$, so $y = 3$ or $y = -2$.
So $e^{4x} = 3$ or $e^{4x} = -2$. Negative values of e^x are not possible, so $e^{4x} = 3 \Rightarrow x = \frac{1}{4}\ln 3$.

Q9 $9^{x-2} = 3^y \Rightarrow (3^2)^{x-2} = 3^y \Rightarrow 3^{(2(x-2))} = 3^y$ so $2(x - 2) = y$
$\log_3 2x = 1 + \log_3 y \Rightarrow \log_3 2x - \log_3 y = 1$
$\Rightarrow \log_3 \frac{2x}{y} = 1 \Rightarrow \frac{2x}{y} = 3^1 \Rightarrow 2x = 3y$
Solve $2(x - 2) = y$ and $2x = 3y$ simultaneously:
$2x = 3y$ so put this into $2(x - 2) = y$
$\Rightarrow 3y - 4 = y \Rightarrow 2y = 4 \Rightarrow y = 2$ and $x = 3$

Q10 Substitute $y = 10^x \Rightarrow 2y^2 - 7y + 5 = 0$
$\Rightarrow (2y - 5)(y - 1) = 0 \Rightarrow 10^x = \frac{5}{2} = 2.5$ or $10^x = 1$
$\Rightarrow x = \log 2.5$ or $x = 0$

Q11 **a)** Let $y = 2^x$, then $2^{2x} - 5(2^x) + 4 = 0$ is equivalent to the quadratic equation $y^2 - 5y + 4 = 0$
$\Rightarrow (y - 1)(y - 4) = 0$, so $y = 1$ or $y = 4$
So $2^x = 4$ or $2^x = 1 \Rightarrow x = 2$ or $x = 0$.
The y^2 in the quadratic equation comes from $2^{2x} = (2^x)^2$.

b) Let $y = 4^x$, then $4^{2x} - 17(4^x) + 16 = 0$ is equivalent to the quadratic equation $y^2 - 17y + 16 = 0$
$\Rightarrow (y - 1)(y - 16) = 0$, so $y = 1$ or $y = 16$
So $4^x = 1$ or $4^x = 16 \Rightarrow x = 0$ or $x = 2$.

c) Let $y = 3^x$, then $3^{2x + 2} = 3^{2x} \times 3^2 = y^2 \times 9 = 9y^2$.
So $3^{2x + 2} - 82(3^x) + 9 = 0$ is equivalent to the quadratic equation $9y^2 - 82y + 9 = 0$.
$\Rightarrow (9y - 1)(y - 9) = 0$, so $y = \frac{1}{9}$ or $y = 9$.
So $3^x = \frac{1}{9}$ or $3^x = 9 \Rightarrow x = -2$ or $x = 2$.

d) Let $y = 2^x$, then $2^{2x + 3} = 2^{2x} \times 2^3 = y^2 \times 8 = 8y^2$.
So $2^{2x + 3} - 9(2^x) + 1 = 0$ is equivalent to the quadratic equation $8y^2 - 9y + 1 = 0$.
$\Rightarrow (8y - 1)(y - 1) = 0$, so $y = \frac{1}{8}$ or $y = 1$.
So $2^x = \frac{1}{8}$ or $2^x = 1 \Rightarrow x = -3$ or $x = 0$.

e) Substitute $y = e^{2x} \Rightarrow y^2 + 4y + 5 = 0$.
Using the quadratic formula:
$$x = \frac{-4 \pm \sqrt{4^2 - (4 \times 5 \times 1)}}{2} = \frac{-4 \pm \sqrt{16 - 20}}{2} = \frac{-4 \pm \sqrt{-4}}{2}$$
There are no real solutions since there is a negative square root.
The question says "where possible", which implies that there might not be any real solutions.

f) Substitute $y = e^x \Rightarrow 3y^2 + 10y + 3 = 0$
$\Rightarrow (3y + 1)(y + 3) = 0 \Rightarrow e^x = -\frac{1}{3}$ or $e^x = -3$,
both of which are impossible since $e^x > 0$.
There are no solutions.

Q12 Take logs of both sides of the inequality:
$\log 1.5^P > \log 1\,000\,000$
$\Rightarrow P \log 1.5 > \log 1\,000\,000$
$\Rightarrow P > \frac{\log 1\,000\,000}{\log 1.5} \Rightarrow P > 34.0732...$
So the smallest integer $P = 35$.

9.4 Modelling Exponential Growth and Decay

Exercise 9.4.1 — Modelling exponential growth and decay

Q1 When $t = 10$, $A = \frac{A_0}{2}$ so $\frac{A_0}{2} = A_0 e^{-10k} \Rightarrow \frac{1}{2} = e^{-10k}$
$\Rightarrow k = -\frac{1}{10}\ln\left(\frac{1}{2}\right) = 0.0693... = 0.0693$ (3 s.f.)

a) You want to find t when $A = \frac{A_0}{4}$,
so $\frac{A_0}{4} = A_0 e^{(-0.0693... \times t)} \Rightarrow \frac{1}{4} = e^{(-0.0693... \times t)}$
$\Rightarrow t = -\frac{1}{0.0693...}\ln\left(\frac{1}{4}\right) = 20.0$ years (3 s.f.).

So after 20 years the substance will be reduced to a quarter of its original activity.

Note that a much easier way to do this would be to think of it as 'half and half again'. The substance will be a quarter of its activity after two half lives which is 2 × 10 = 20 years.

b) When $t = 5$, $A = 200$, so $200 = A_0 e^{(-0.0693... \times 5)}$,
so $A_0 = 200e^{(0.0693... \times 5)} = 283$ Bq (3 s.f.).

c) If $t = 15$, then $A = 283 \times e^{(-0.0693... \times 15)} = 100$ Bq (3 s.f.)

Q2 a) If $t = 0$, $T = 225 - 207e^0 = 18$ °C.

b) Let $t = 5$, then $T = 225 - 207e^{-\frac{5}{8}} = 114$ °C. (3 s.f.).

c) Let $T = 190$ °C. Then $190 = 225 - 207e^{-\frac{t}{8}}$
$\Rightarrow e^{-\frac{t}{8}} = \frac{190-225}{-207} = \frac{35}{207} \Rightarrow -\frac{t}{8} = \ln\left(\frac{35}{207}\right)$
$\Rightarrow t = -8\ln\left(\frac{35}{207}\right) = 14.2$ min (to 3 s.f.).

So the oven reaches 190 °C just after 12:14.

d)

$T = 225$, $T = 225 - 207e^{\left(-\frac{t}{8}\right)}$, 18, 0, t

e) As $t \to \infty$, $e^{-\frac{t}{8}} \to 0$, so $T \to 225 - 0 = 225$.
So in the model, the asymptote at $T = 225$ acts as a "cap" — even if left on forever, the oven temperature would never rise above 225 °C.

Q3 a) When $t = 10$, $V = 7500e^{-0.2 \times 10} = 7500e^{-2}$
$= 1015.0146... = £1015$ to the nearest £

b) Let $V = 500$, then $500 = 7500e^{-0.2t}$
$\Rightarrow e^{-0.2t} = \frac{500}{7500} = \frac{1}{15}$
$\Rightarrow -0.2t = \ln\left(\frac{1}{15}\right) \Rightarrow t = -5\ln\left(\frac{1}{15}\right) = 13.5$ (3 s.f.).
So the value will have fallen below £500 after 14 years.

c)

V, 7500, 0, t

Q4 a) When $t = 0$, $F = 4$, so $F_0 = 4$.
When $t = 6$, $F = 10$ so $10 = 4e^{6g}$ so
$g = \frac{1}{6}\ln\left(\frac{10}{4}\right) = 0.1527... = 0.153$ (3 s.f.).

b) Let $t = 12$, then $F = 4e^{(12 \times 0.1527...)} = 25$.
So after 12 hours the fungus will be 25 mm^2.

c) Let $F = 15$. Then $15 = 4e^{(0.1527... \times t)}$
$\Rightarrow e^{(0.1527... \times t)} = \frac{15}{4} \Rightarrow t = \frac{1}{0.1527...}\ln\left(\frac{15}{4}\right) = 8.66$ (3 s.f.).

The fungus will take 8.66 hours (or 8 hours 39 minutes) to grow to 15 mm^2.

d) E.g. there's no restriction, so according to the model the fungus could grow infinitely large.

Q5 When $t = 0$, $N = 3$. So $3 = Ae^0 \Rightarrow A = 3$.

a) Let $t = 0.5$ (hours).
Then $N = 3e^{-t} = 3e^{-0.5} = 1.82$ mg/l.

b) Let $N = 0.1$. Then $0.1 = 3e^{-t}$ so
$t = -\ln\left(\frac{0.1}{3}\right) = 3.40$ hours (3 s.f.).

c)

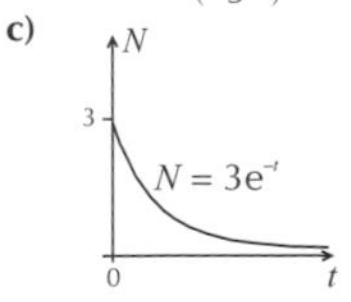

d) The gradient is $-3e^{-t}$.

The gradient of the curve Ae^{kt} is kAe^{kt}. Here, $k = -1$ and $A = 3$.

Q6 a) The negative coefficient means that as t gets larger, V gets smaller — this is exponential decay.
The car will lose value over time.

b) When $t = 0$, $V = 1500 + 9000e^0 = £10\,500$.

c) Let $t = 5$, then $V = 1500 + 9000\,e^{-\frac{5}{3}}$
$= £3200$ (3 s.f.).

d) Let $V = 2500$, then $2500 = 1500 + 9000\,e^{-\frac{t}{3}}$
$\Rightarrow 1000 = 9000\,e^{-\frac{t}{3}} \Rightarrow \frac{1}{9} = e^{-\frac{t}{3}}$
$\Rightarrow t = -3\ln\left(\frac{1}{9}\right) = 6.59$ (3 s.f.). The car will have a value less than £2500 after 7 whole years.

Note: After 6 years the car will still have a value above £2500 so the answer is 7 and not 6.

e)

V, 10500, $V = 1500 + 9000\,e^{\left(-\frac{t}{3}\right)}$, $V = 1500$, 0, t

Q7 a) When $t = 0$, $H = 20\,e^{b \times 0} = 20$. So 20 represents the initial burnt area of forest, in hectares.

b) $b = \ln 1.8 = 0.5877... = 0.588$ (3 s.f.)

c) Let $t = 3$ then $H = 20e^{(0.5877... \times 3)}$
$= 117$ hectares (3 s.f.).

d) If $H = 500$, $500 = 20e^{(0.5877... \times t)} \Rightarrow 25 = e^{(0.5877... \times t)}$
$\Rightarrow \ln 25 = 0.5877... \times t$
$\Rightarrow t = \frac{1}{0.5877...}\ln 25 = 5.48$ hours (3 s.f.)

e) At $t = k$, $H = H_k = 20e^{(\ln 1.8)k}$
At $t = k + 1$, $H = H_{k+1} = 20e^{(\ln 1.8)(k+1)}$
$= 20e^{(\ln 1.8)k + \ln 1.8} = 20e^{(\ln 1.8)k}e^{\ln 1.8} = H_k \times 1.8$
Every hour the burnt area is multiplied by 1.8.
This represents a percentage increase of 80%.

f) E.g. if the fire burnt unchecked, then according to the model $H \to \infty$ as $t \to \infty$, i.e. the area of burnt forest would be infinitely large. This is clearly unrealistic, as the forest will have a finite area.

9.5 Using Logarithmic Graphs

Exercise 9.5.1 — Logarithmic graphs in linear form

Q1 l is a straight line of the form $y = mx + c$, where $y = \log V$, $x = t$, gradient $m = -\frac{1}{40}$ and intercept $c = 4$, so:
$\log V = 4 - \frac{1}{40}t$. When the machine is 20 years old, $t = 20$,
so $\log V = 4 - \frac{1}{40} \times 20 = 4 - \frac{1}{2} = \frac{7}{2}$.
So $V = 10^{\frac{7}{2}} = £3162.2776... = £3200$ (to the nearest £100)

Q2 a) & e)
$y = at^b \Rightarrow \log y = \log a + b\log t$.
To plot the straight-line you need to find logs of both t and y (each to 3 d.p.):

$\log t$	0.845	1.146	1.322	1.447
$\log y$	2.350	3	3.200	3.600

You can now plot $\log y$ against $\log t$:

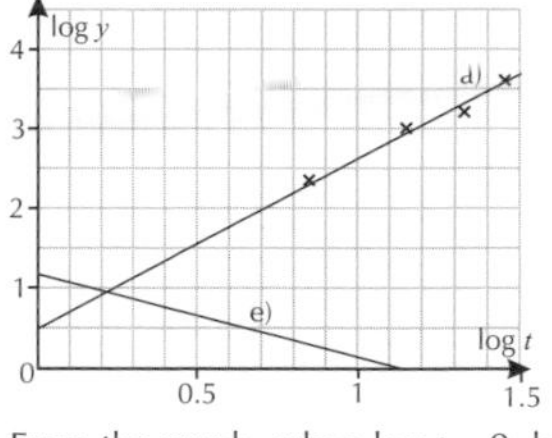

b) From the graph, when $\log t = 0$, $\log a = 0.5$
$\Rightarrow a = 3.16$ (3 s.f.)
b is the gradient of the graph, so take 2 points from the line of best fit — e.g. (0, 0.5) and (1.4, 3.5).
The gradient is: $b = \frac{3.5 - 0.5}{1.4 - 0} = 2.14$ (3 s.f.)
(so $y = 3.16t^{2.14}$)
You might have got slightly different values for a and b depending on your graph — as long as your values are close to the ones given here, that's fine. This will affect your answer to part c) too.

c) $y = 3.16 \times 10^{2.14} = 436.20... = 436$ subscribers

d) $y = 15t^{-1} \Rightarrow \log y = \log(t^{-1} \times 15) = -\log t + \log 15$

e) See part a) for graph.
When $\log t = 0$, $\log y = \log 15 = 1.176$ (3 d.p.).
When $\log y = 0$, $\log t = \log 15 = 1.176$ (3 d.p.).
So plot (0, 1.176) and (1.176, 0) and draw a straight line between them.

f) The graphs intersect at $\log t \approx 0.22$, so $t \approx 1.7$, which is approximately Day 2.

g) E.g. The model predicts that the first blog will continue to grow exponentially for years. In reality, it's likely that the growth will begin to slow over time, e.g. due to competition from other blogs, a decline in their popularity or due to the blog reaching a natural upper limit. The model predicts that the second blog will end up with less than one subscriber — it is unlikely that the blog would continue with so few subscribers, and the blogger would probably try and increase their number of subscribers rather than just watch them disappear.

Q3 a) (i)

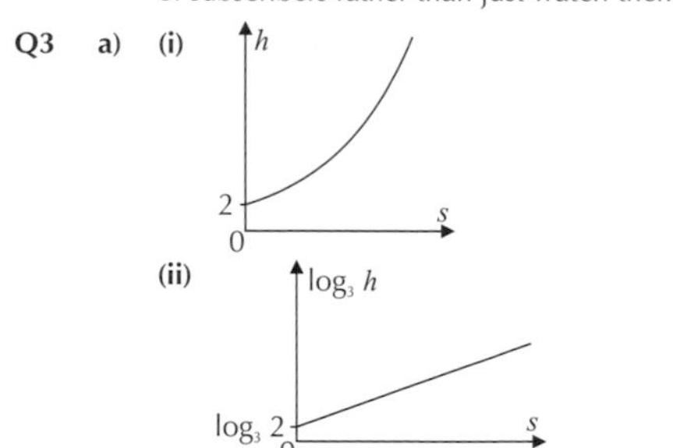

(iii) E.g. the second graph is more useful for calculations — it's easier to calculate the gradient of a straight line than it is a curve.

b) E.g. as $s \to \infty$, $h \to \infty$. This implies that the tank is capable of holding an infinite amount of water, which is obviously unrealistic.
A tank reaches its maximum capacity, H, at time S.
So adjust the model to say that for $s \geq S$, $h = H$.

Q4 The graph is of the form $p = at^b$
$\Rightarrow \log p = \log a + b\log t$. To find the straight-line form you need to find logs of both t and p (each to 3 d.p.):

$\log t$	0	0.477	0.602	0.778	0.954
$\log p$	0.301	1.146	1.342	1.643	1.944

You can now plot $\log p$ against $\log t$:

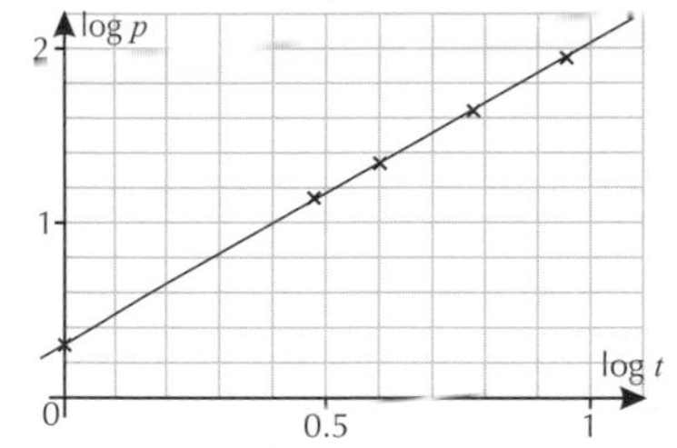

When $\log t = 0$, $\log a = 0.301 = \log 2 \Rightarrow a = 2$
b is the gradient of the graph, so take 2 points — e.g. (0, 0.301) and (0.602, 1.342). The gradient is
$\frac{1.342 - 0.301}{0.602 - 0} = \frac{1.041}{0.602} = 1.72... = 1.7$ (1 d.p.)
So the graph can be approximated using the equation $p = 2t^{1.7}$

Q5 $x = kb^t$, so $\log x = \log k + t\log b$.
So to make a linear graph, you need to find $\log x$ (3 d.p.) for each value of x:

t	5	50	100	200	300
x	80.449	32.411	11.803	1.565	0.207
$\log x$	1.906	1.511	1.072	0.195	−0.684

You can now plot $\log x$ against t:

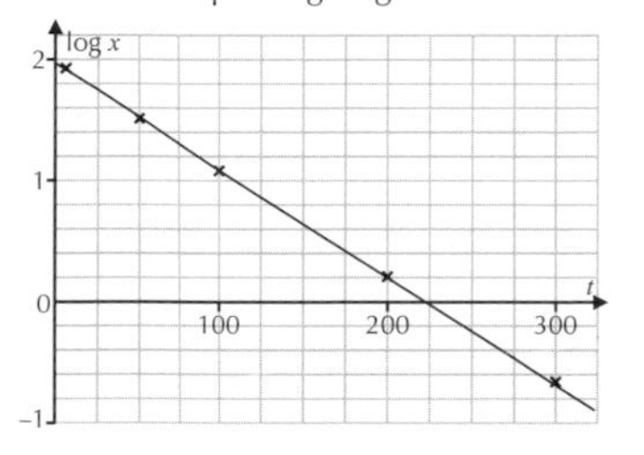

The line of best fit meets the vertical axis at around 1.95 — i.e. when $t = 0$, $\log x = 1.95 \Rightarrow x = 89.12... \approx 89$ Bq.
The line of best fit may vary slightly, so your reading might be a bit different. It should be between 1.92 and 2, so accept any answer between 83 Bq and 100 Bq.

Q6 Area = length × width. The area is always 120 m², so $lw = 120$. Take logs: $\log lw = \log 120$
$\Rightarrow \log l + \log w = \log 120 \Rightarrow \log l = \log 120 - \log w$
This can be plotted on a graph of $\log w$ against $\log l$:

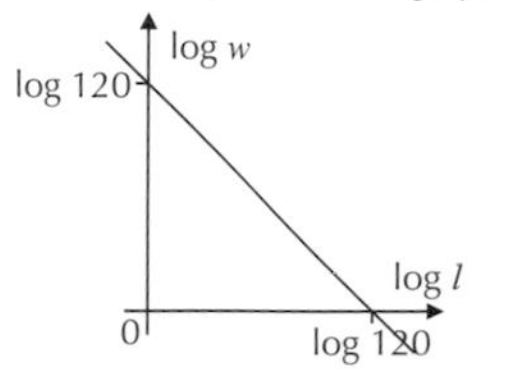

Review Exercise — Chapter 9

Q1 a) $y = 4e^{-3x}$

b) Gradient $= -3 \times 4e^{-3x} = -12e^{-3x}$

(i) When $x = -1$, gradient $= -12e^{(-3 \times -1)} = -12e^3$

(ii) When $x = 0$, gradient $= -12e^{(-3 \times 0)} = -12$

(iii) When $x = 4$, gradient $= -12e^{(-3 \times 4)} = -12e^{-12}$

Q2 a) $\log_4 16 = 2$

b) $\log_{216} 6 = \frac{1}{3}$

c) $\log_3 \frac{1}{81} = -4$

Q3 a) 3 b) -3 c) $\log_3 9 = 2$

Q4 a) $\log 3 + \log 5^2 = \log(3 \times 5^2) = \log 75$

b) $\log 36^{\frac{1}{2}} - \log 3 = \log\left(\frac{\sqrt{36}}{3}\right) = \log 2$

c) $\log 2 - \log 16^{\frac{1}{4}} = \log\left(\frac{2}{\sqrt[4]{16}}\right) = \log 1 = 0$

Q5 $\log_b(x^2 - 1) - \log_b(x - 1) = \log_b\left(\frac{x^2-1}{x-1}\right)$
$= \log_b\left(\frac{(x+1)(x-1)}{x-1}\right) = \log_b(x+1)$

Once you've combined the logs, you have to spot that the numerator is a difference of two squares, and use that to simplify the fraction.

Q6 E.g. $\frac{2 + \log_a 4}{\log_a 2a} = \frac{2 + 2\log_a 2}{\log_a 2 + \log_a a} = \frac{2(1 + \log_a 2)}{\log_a 2 + 1} = 2$

Q7 E.g. using the change of base formula:

a) $\log_7 12 = \frac{\log_{10} 12}{\log_{10} 7} = 1.28$ (3 s.f.)

b) $\log_5 8 = \frac{\log_{10} 8}{\log_{10} 5} = 1.29$ (3 s.f.)

c) $\log_{16} 125 = \frac{\log_{10} 125}{\log_{10} 16} = 1.74$ (3 s.f.)

Q8 a)

x	-3	-2	-1	0	1	2	3
y	0.0156	0.0625	0.25	1	4	16	64

b)

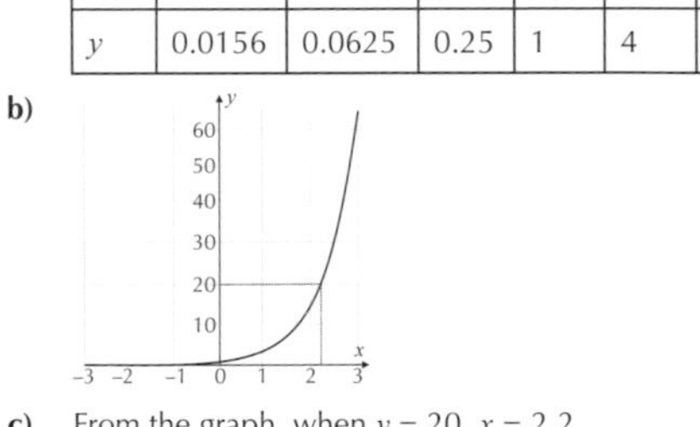

c) From the graph, when $y = 20$, $x = 2.2$

You may get a slightly different answer depending on your graph.

d) $4^x = 20 \Rightarrow \log 4^x = \log 20 \Rightarrow x \log 4 = \log 20$
$\Rightarrow x = \frac{\log 20}{\log 4} = 2.16$ (3 s.f.)

Q9 a) $10^x = 240 \Rightarrow \log 10^x = \log 240 \Rightarrow x \log 10 = \log 240$
$\Rightarrow x = \frac{\log 240}{\log 10} = \log_{10} 240 = 2.38$ (3 s.f.)

Recall that $\log_{10} 10 = 1$.

b) $\log_{10} x = 5.3 \Rightarrow x = 10^{5.3} = 200\,000$ (3 s.f.)

c) $10^{2x+1} = 1500 \Rightarrow \log 10^{2x+1} = \log 1500$
$\Rightarrow (2x + 1)\log 10 = \log 1500$
$\Rightarrow 2x + 1 = \log 1500 \Rightarrow x = \frac{\log 1500 - 1}{2} = 1.09$ (3 s.f.)

d) $4^{x-1} = 200 \Rightarrow \log 4^{x-1} = \log 200 \Rightarrow (x-1)\log 4 = \log 200$
$\Rightarrow x = \frac{\log 200}{\log 4} + 1 = 4.82$ (3 s.f.)

Q10 a) $e^{2x} = 6 \Rightarrow \ln e^{2x} = \ln 6 \Rightarrow 2x \ln e = \ln 6$
$\Rightarrow x = \frac{\ln 6}{2} = 0.8959$ (4 d.p.)

b) $3e^{-4x+1} = 5 \Rightarrow \ln e^{-4x+1} = \ln\frac{5}{3} \Rightarrow -4x + 1 = \ln\frac{5}{3}$
$\Rightarrow x = \frac{\ln\frac{5}{3} - 1}{-4} = 0.1223$ (4 d.p.)

c) $\ln(x + 3) = 0.75 \Rightarrow e^{\ln(x+3)} = e^{0.75}$
$\Rightarrow x + 3 = e^{0.75} \Rightarrow x = e^{0.75} - 3 = -0.8830$ (4 d.p.)

d) $\ln x + \ln 5 = \ln 4 \Rightarrow \ln 5x = \ln 4 \Rightarrow 5x = 4$
$\Rightarrow x = 0.8000$ (4 d.p.)

Q11 The population, P spiders, is linked to the time, t years, by the equation $P = 2000 \times 0.75^t$, so solve:
$2000 \times 0.75^t < 200$
$\Rightarrow 0.75^t < \frac{200}{2000} \Rightarrow \log 0.75^t < \log 0.1$
$\Rightarrow t \log 0.75 < \log 0.1 \Rightarrow t > \frac{\log 0.1}{\log 0.75}$

Watch out — dividing by log 0.75 means dividing by a negative, so you need to flip the inequality sign.

$\Rightarrow t > 8.0039...$ years. So the spiders are in danger of extinction in the 9th year.

Q12 a) Using $L_0 = 20$ and $t = 10$, $L = 20 \times e^{\frac{10}{12}} = 46.01...$
$= 46$ leopards

b) Find t when $L = 20 \times e^{\frac{t}{12}} = 60$
$\Rightarrow e^{\frac{t}{12}} = 3 \Rightarrow t = 12 \ln 3 = 13.18...$
So it will be 13 years until the reserve runs out of space.

c) Using $W_0 = 15$ and $t = 5$, $W = 15 \times e^{-\frac{5}{3}} = 2.833...$
So the model predicts that only 2 leopards will be left in the wild after 5 years.

Q13 a) When $t = 0$, $Z = 10 + 20e^0 = 30$ zombies

b) When $t = 2$, $Z = 10 + 20e^2 = 157.78... = 157$ zombies

You could argue that if you're 0.78... transformed into a zombie, you're probably a zombie, so accept 158 as an answer too.

c) $Z = 60\,000\,000 = 10 + 20e^t$
$\Rightarrow e^t = \frac{60\,000\,000 - 10}{20} = 2\,999\,999.5$
$\Rightarrow t = \ln 2\,999\,999.5 = 14.91... = 15$ weeks

Q14 a) Find an equation in the form $\ln y = mt + c$.
The vertical intercept (c) is 7.5 and the gradient (m) is $-7.5 \div 15 = -0.5$. So $\ln y = -0.5t + 7.5$.
Then take exponentials of both sides to get the exponential equation $y = e^{7.5 - 0.5t}$ (or $y = e^{7.5}e^{-0.5t}$).

b) For this line, the vertical intercept (c) is 3.5 and the gradient (m) is 0.25. So $\ln y = 0.25t + 3.5$.
So the exponential equation is $y = e^{3.5 + 0.25t}$ (or $y = e^{3.5}e^{0.25t}$).

c) When $y = 20$, $\ln y = \ln 20 = 2.9957... \approx 3$.
From the graph, when $\ln y = 3$, $t = 9$, so it will reach a critical level in the 10th month.

You could also solve this algebraically — when $y = 20$, $\ln 20 = 7.5 - 0.5t \Rightarrow 0.5t = 7.5 - \ln 20 \Rightarrow t = 9.008...$

d) E.g. the model predicts that the population of grey squirrels will continue to grow exponentially with time. However, in reality, factors such as the availability of resources will begin to slow their growth after a certain time.

Exam-Style Questions — Chapter 9

Q1 $e^x - \frac{40}{e^x} = 3 \Rightarrow (e^x)^2 - 40 = 3e^x$
Let $y = e^x$: $y^2 - 3y - 40 = 0$ *[1 mark]*
$\Rightarrow (y - 8)(y + 5) = 0 \Rightarrow y = 8, y = -5$, so $e^x = 8$ or $e^x = -5$.
Negative values of e^x are not possible, so $e^x = 8$ *[1 mark]*
$\Rightarrow x = \ln 8 = \ln(2^3) = 3 \ln 2$ *[1 mark]*

Q2 a) $\log_{10} y = 1 + \log_{10}(x + 2)$
$\Rightarrow \log_{10} y = \log_{10} 10 + \log_{10}(x + 2)$ *[1 mark]*
$\Rightarrow \log_{10} y = \log_{10}(10x + 20)$ *[1 mark]*
$\Rightarrow y = 10x + 20$

b) $10^{15x - y} = 10\,000 = 10^4$
$\Rightarrow (15x - y)\log 10 = 4 \log 10 \Rightarrow 15x - y = 4$

So the simultaneous equations are:
$y = 10x + 20$ ①
$15x - y = 4$ ②
E.g. substituting ① into ② gives:
$15x - 10x - 20 = 4 \Rightarrow 5x = 24 \Rightarrow x = 4.8$

Substituting $x = 4.8$ in ①: $y = 10 \times 4.8 + 20 = 68$

[3 marks available — 1 mark for taking logs to form a correct second linear equation, 1 mark for a correct method for solving the simultaneous equations, 1 mark for both answers correct]

Q3 **a)** $a^x = b^y \Rightarrow x = \log_a b^y$ *[1 mark]* $\Rightarrow x = y\log_a b$

b) From part a), if $2^{k-2} = 3^{2k-5}$, then $a = 2$, $b = 3$, $x = k - 2$ and $y = 2k - 5$, so: $k - 2 = (2k - 5)\log_2 3$ *[1 mark]*

$\Rightarrow k - 2 = 2k\log_2 3 - 5\log_2 3$ *[1 mark]*

$\Rightarrow k(1 - 2\log_2 3) = 2 - 5\log_2 3$

$\Rightarrow k = \dfrac{2 - 5\log_2 3}{1 - 2\log_2 3}$ *[1 mark]*

Q4 Gradient of $y = e^{2x}$ at $x = \frac{1}{2}$ is $2 \times e^{2\times\frac{1}{2}} = 2e$.
So the gradient of the tangent to $y = e^{2x}$ at $x = \frac{1}{2}$ $= 2e$ *[1 mark]*.
Equation of tangent at $\left(\frac{1}{2}, e\right)$ is:
$y - e = 2e\left(x - \frac{1}{2}\right) \Rightarrow y = 2ex$ *[1 mark]*
In this equation, when $x = 0$, $y = 0$, so the tangent passes through the origin *[1 mark]*.

Q5 **a)** A = initial mass of ice = 100 kg *[1 mark]*.
$M = Ab^t \Rightarrow \log_{10} M = \log_{10} A + t\log_{10} b$ *[1 mark]*
So the gradient of the line = $\log_{10} b$

$\Rightarrow \log_{10} b = \dfrac{1.903 - 2}{10 - 0}$ *[1 mark]*

$= -0.0097$

$\Rightarrow b = 10^{-0.0097} = 0.9779... = 0.978$ (3 s.f.) *[1 mark]*

b) At the point where 60% of the ice has melted, the remaining mass is 40 kg $\Rightarrow M = 40$.

$\Rightarrow 40 = 100 \times 0.978^t \Rightarrow 0.978^t = 0.4$

$\Rightarrow t\log_{10} 0.978 = \log_{10} 0.4$

$\Rightarrow t = \dfrac{\log 0.4}{\log 0.978} = 41.1897...$

= 41 minutes to the nearest minute.

[2 marks available — 1 mark for a correct method, 1 mark for a correctly rounded final answer]

c) For large values of t the model predicts that some ice will always remain — the mass predicted by the model is always positive. However, in reality, the ice will completely melt and the mass will reduce to zero *[1 mark]*.

Chapter 10: Differentiation

10.1 The Gradient of a Curve

Exercise 10.1.1 — Differentiating from first principles

Q1 *The gradient of the straight line joining points (x_1, y_1) and (x_2, y_2) is given by $\frac{y_2 - y_1}{x_2 - x_1}$.*

a) **(i)** When $x = 1$, $y = 1$ and when $x = 2$, $y = 8$
so the gradient is $\dfrac{8-1}{2-1} = \dfrac{7}{1} = 7$.

(ii) When $x = 1$, $y = 1$ and when $x = 1.5$, $y = 3.375$
so the gradient is $\dfrac{3.375-1}{1.5-1} = \dfrac{2.375}{0.5} = 4.75$.

(iii) When $x = 1$, $y = 1$ and when $x = 1.1$, $y = 1.331$
so the gradient is $\dfrac{1.331-1}{1.1-1} = \dfrac{0.331}{0.1} = 3.31$.

b) The gradients of the straight lines in part a) move closer to 3 as the value of x moves closer to 1.

Q2 **a)** $\dfrac{dy}{dx} = \lim\limits_{h\to0}\left[\dfrac{(x+h)-x}{(x+h)-x}\right] = \lim\limits_{h\to0}[1] = 1$

b) $f'(x) = \lim\limits_{h\to0}\left[\dfrac{(x+h)^3 - x^3}{(x+h)-x}\right]$

$= \lim\limits_{h\to0}\left[\dfrac{x^3 + 3x^2h + 3xh^2 + h^3 - x^3}{(x+h)-x}\right]$

$= \lim\limits_{h\to0}\left[\dfrac{3x^2h + 3xh^2 + h^3}{h}\right]$

$= \lim\limits_{h\to0}[3x^2 + 3xh + h^2] = 3x^2$

c) $f'(x) = \lim\limits_{h\to0}\left[\dfrac{2(x+h) - 2x}{(x+h)-x}\right] = \lim\limits_{h\to0}\left[\dfrac{2x + 2h - 2x}{h}\right]$

$= \lim\limits_{h\to0}\left[\dfrac{2h}{h}\right] = \lim\limits_{h\to0}[2] = 2$

d) $f'(x) = \lim\limits_{h\to0}\left[\dfrac{2(x+h)^2 - 2x^2}{(x+h)-x}\right]$

$= \lim\limits_{h\to0}\left[\dfrac{2(x^2 + 2xh + h^2) - 2x^2}{(x+h)-x}\right]$

$= \lim\limits_{h\to0}\left[\dfrac{2x^2 + 4xh + 2h^2 - 2x^2}{(x+h)-x}\right]$

$= \lim\limits_{h\to0}\left[\dfrac{4xh + 2h^2}{h}\right] = \lim\limits_{h\to0}[4x + 2h] = 4x$

e) $f'(x) = \lim\limits_{h\to0}\left[\dfrac{(x+h) - 7 - (x-7)}{(x+h)-x}\right]$

$= \lim\limits_{h\to0}\left[\dfrac{x + h - 7 - x + 7}{(x+h)-x}\right]$

$= \lim\limits_{h\to0}\left[\dfrac{h}{h}\right] = \lim\limits_{h\to0}[1] = 1$

f) $f'(x) = \lim\limits_{h\to0}\left[\dfrac{-(x+h)^3 - (-x^3)}{(x+h)-x}\right]$

$= \lim\limits_{h\to0}\left[\dfrac{-x^3 - 3x^2h - 3xh^2 - h^3 + x^3}{(x+h)-x}\right]$

$= \lim\limits_{h\to0}\left[\dfrac{-3x^2h - 3xh^2 - h^3}{h}\right]$

$= \lim\limits_{h\to0}[-3x^2 - 3xh - h^2] = -3x^2$

Q3 **a)** $\dfrac{dy}{dx} = \lim\limits_{h\to0}\left[\dfrac{5(x+h)^2 + 1 - (5x^2 + 1)}{(x+h)-x}\right]$

$= \lim\limits_{h\to0}\left[\dfrac{5(x^2 + 2xh + h^2) + 1 - (5x^2 + 1)}{(x+h)-x}\right]$

$= \lim\limits_{h\to0}\left[\dfrac{5x^2 + 10xh + 5h^2 + 1 - 5x^2 - 1}{(x+h)-x}\right]$

$= \lim\limits_{h\to0}\left[\dfrac{10xh + 5h^2}{h}\right] = \lim\limits_{h\to0}[10x + 5h] = 10x$

b) $\dfrac{dy}{dx} = \lim\limits_{h\to0}\left[\dfrac{(x+h) - (x+h)^2 - (x - x^2)}{(x+h)-x}\right]$

$= \lim\limits_{h\to0}\left[\dfrac{(x+h) - (x^2 + 2xh + h^2) - (x - x^2)}{(x+h)-x}\right]$

$= \lim\limits_{h\to0}\left[\dfrac{x + h - x^2 - 2xh - h^2 - x + x^2}{(x+h)-x}\right]$

$= \lim\limits_{h\to0}\left[\dfrac{h - 2xh - h^2}{h}\right] = \lim\limits_{h\to0}[1 - 2x - h] = 1 - 2x$

c) $\dfrac{dy}{dx} = \lim\limits_{h\to0}\left[\dfrac{3(x+h)^3 - 3x^3}{(x+h)-x}\right]$

$= \lim\limits_{h\to0}\left[\dfrac{3(x^3 + 3x^2h + 3xh^2 + h^3) - 3x^3}{(x+h)-x}\right]$

$= \lim\limits_{h\to0}\left[\dfrac{3x^3 + 9x^2h + 9xh^2 + 3h^3 - 3x^3}{(x+h)-x}\right]$

$= \lim\limits_{h\to0}\left[\dfrac{9x^2h + 9xh^2 + 3h^3}{h}\right]$

$= \lim\limits_{h\to0}[9x^2 + 9xh + 3h^2] = 9x^2$

d) $\dfrac{dy}{dx} = \lim\limits_{h\to0}\left[\dfrac{2(x+h)^3 + 3(x+h) - (2x^3 + 3x)}{(x+h)-x}\right]$

$= \lim\limits_{h\to0}\left[\dfrac{2(x^3 + 3x^2h + 3xh^2 + h^3) + 3(x+h) - (2x^3 + 3x)}{(x+h)-x}\right]$

$= \lim\limits_{h\to0}\left[\dfrac{2x^3 + 6x^2h + 6xh^2 + 2h^3 + 3x + 3h - 2x^3 - 3x}{(x+h)-x}\right]$

$= \lim\limits_{h\to0}\left[\dfrac{6x^2h + 6xh^2 + 2h^3 + 3h}{h}\right]$

$= \lim\limits_{h\to0}[6x^2 + 6xh + 2h^2 + 3] = 6x^2 + 3$

e) $\frac{dy}{dx} = \lim_{h\to 0}\left[\frac{(x+h)^3 + (x+h) - (x^3+x)}{(x+h)-x}\right]$

$= \lim_{h\to 0}\left[\frac{x^3 + 3x^2h + 3xh^2 + h^3 + x + h - x^3 - x}{(x+h)-x}\right]$

$= \lim_{h\to 0}\left[\frac{3x^2h + 3xh^2 + h^3 + h}{h}\right]$

$= \lim_{h\to 0}[3x^2 + 3xh + h^2 + 1] = 3x^2 + 1$

f) $f'(x) = \lim_{h\to 0}\left[\frac{(2-(x+h))^2 - (2-x)^2}{(x+h)-x}\right]$

$= \lim_{h\to 0}\left[\frac{4 - 4(x+h) + (x+h)^2 - (4 - 4x + x^2)}{(x+h)-x}\right]$

$= \lim_{h\to 0}\left[\frac{4 - 4x - 4h + x^2 + 2xh + h^2 - 4 + 4x - x^2}{(x+h)-x}\right]$

$= \lim_{h\to 0}\left[\frac{-4h + 2xh + h^2}{h}\right]$

$= \lim_{h\to 0}[-4 + 2x + h] = -4 + 2x$

Q4 a) $\frac{dy}{dx} = \lim_{h\to 0}\left[\frac{(x+h)^3 + 2(x+h)^2 + 3(x+h) - (x^3 + 2x^2 + 3x)}{(x+h)-x}\right]$

$= \lim_{h\to 0}\left[\frac{\left(\begin{array}{l}x^3 + 3x^2h + 3xh^2 + h^3 + 2x^2 + 4xh \\ \quad + 2h^2 + 3x + 3h - x^3 - 2x^2 - 3x\end{array}\right)}{(x+h)-x}\right]$

$= \lim_{h\to 0}\left[\frac{3x^2h + 3xh^2 + h^3 + 4xh + 2h^2 + 3h}{h}\right]$

$= \lim_{h\to 0}[3x^2 + 3xh + h^2 + 4x + 2h + 3]$

$= 3x^2 + 4x + 3$

b) (i) $x = -1$, so $\frac{dy}{dx} = 3(-1)^2 + 4(-1) + 3 = 2$

(ii) $x = 2$, so $\frac{dy}{dx} = 3(2)^2 + 4(2) + 3 = 23$

(iii) $x = 0$, so $\frac{dy}{dx} = 3(0)^2 + 4(0) + 3 = 3$

10.2 Differentiating y = f(x)

Exercise 10.2.1 — Differentiating x^n

Q1 a) $\frac{dy}{dx} = 1$

b) $\frac{dy}{dx} = 6x^5$

c) $\frac{dy}{dx} = 3x^2$

d) $\frac{dy}{dx} = 30x$

e) $\frac{dy}{dx} = -2x^{-3} = -\frac{2}{x^3}$

f) $\frac{dy}{dx} = 6x$

g) $\frac{dy}{dx} = 7$

h) $\frac{dy}{dx} = 2x^3$

i) $\frac{dy}{dx} = 0$

j) $\frac{dy}{dx} = \frac{3}{2}x^{-\frac{1}{2}} = \frac{3}{2\sqrt{x}}$

k) $\frac{dy}{dx} = -2x^{-2} = -\frac{2}{x^2}$

l) $\frac{dy}{dx} = -3x^{-4} = -\frac{3}{x^4}$

Q2 a) $f'(x) = 5x^4$

b) $f'(x) = 7x^6$

c) $f'(x) = -4x^{-5} = -\frac{4}{x^5}$

d) $f'(x) = 12x^2$

e) $f'(x) = -6x^{-4} = -\frac{6}{x^4}$

f) $f'(x) = 4x^{-\frac{1}{2}} = \frac{4}{\sqrt{x}}$

g) $f'(x) = x^{-\frac{2}{3}} = \frac{1}{\sqrt[3]{x^2}}$

h) $f'(x) = 0$

i) $f'(x) = -8x^{-3} = -\frac{8}{x^3}$

j) $f'(x) = -9x^2$

k) $f'(x) = 20x^{-5} = \frac{20}{x^5}$

l) $f'(x) = -12x^{-4} = -\frac{12}{x^4}$

Q3 a) $\frac{dy}{dx} = 4x \Rightarrow$ At $x = 4$, $\frac{dy}{dx} = 16$.

b) $\frac{dy}{dx} = -x^{-2} = -\frac{1}{x^2} \Rightarrow$ At $x = 2$, $\frac{dy}{dx} = -\frac{1}{4}$.

c) $\frac{dy}{dx} = -20x^4 \Rightarrow$ At $x = 1$, $\frac{dy}{dx} = -20$.

d) $y = 2x^{-1} \Rightarrow \frac{dy}{dx} = -2x^{-2} = -\frac{2}{x^2}$

$\Rightarrow$ At $x = 10$, $\frac{dy}{dx} = -\frac{1}{50}$.

e) $f'(x) = x^{-\frac{1}{2}} = \frac{1}{\sqrt{x}} \Rightarrow f'(9) = \frac{1}{3}$

f) $f'(x) = 4x^3 \Rightarrow f'(-2) = -32$

g) $f(x) = -250 \Rightarrow -250 = -2x^3 \Rightarrow 125 = x^3 \Rightarrow x = 5$

$f'(x) = -6x^2 \Rightarrow f'(5) = -150$

h) $f(x) = -\frac{3}{4} \Rightarrow -\frac{3}{4} = -3x^{-2} \Rightarrow \frac{1}{4} = \frac{1}{x^2}$

$\Rightarrow x = -2$ or $x = 2$

$f'(x) = 6x^{-3} = \frac{6}{x^3} \Rightarrow f'(2) = \frac{3}{4}$ and $f'(-2) = -\frac{3}{4}$

Q4 $y = 5x \Rightarrow \frac{dy}{dx} = 5$

$y = 5x^2 \Rightarrow \frac{dy}{dx} = 10x$. When $x = 1$, $\frac{dy}{dx} = 10$.

$y = 5x^3 \Rightarrow \frac{dy}{dx} = 15x^2$. When $x = 1$, $\frac{dy}{dx} = 15$.

So $y = 5x^3$ has the steepest gradient at point P.

Exercise 10.2.2 — Differentiating functions

Q1 a) $\frac{dy}{dx} = 12x^2 - 2x$

b) $\frac{dy}{dx} = 1 + (-x^{-2}) = 1 - \frac{1}{x^2}$

c) $\frac{dy}{dx} = 6x + \frac{1}{2}x^{-\frac{1}{2}} = 6x + \frac{1}{2\sqrt{x}}$

d) $f'(x) = -10x^4 + 4 - (-2x^{-3}) = -10x^4 + 4 + \frac{2}{x^3}$

e) $f'(x) = \frac{3}{2}x^{\frac{1}{2}} - 1 = \frac{3}{2}\sqrt{x} - 1$

f) $f'(x) = 5 - 2(-3x^{-4}) + \frac{1}{3}x^{-\frac{2}{3}} = 5 + \frac{6}{x^4} + \frac{1}{3\sqrt[3]{x^2}}$

Q2 a) $\frac{d}{dx}(x(x^6 - 1)) = \frac{d}{dx}(x^7 - x) = 7x^6 - 1$

b) $\frac{d}{dx}((x-3)(x+4)) = \frac{d}{dx}(x^2 - 3x + 4x - 12)$

$= \frac{d}{dx}(x^2 + x - 12) = 2x + 1$

c) $\frac{d}{dx}(x(x-1)(x-2)) = \frac{d}{dx}(x(x^2 - x - 2x + 2))$

$= \frac{d}{dx}(x(x^2 - 3x + 2))$

$= \frac{d}{dx}(x^3 - 3x^2 + 2x)$

$= 3x^2 - 3(2x) + 2 = 3x^2 - 6x + 2$

d) $\frac{d}{dx}((x-3)(x+4)(x-1)) = \frac{d}{dx}((x-3)(x^2 + 3x - 4))$

$= \frac{d}{dx}(x^3 + 3x^2 - 4x - 3x^2 - 9x + 12)$

$= \frac{d}{dx}(x^3 - 13x + 12) = 3x^2 - 13$

e) $\frac{d}{dx}(x^2(x-4)(3-x^3)) = \frac{d}{dx}(x^2(3x - x^4 - 12 + 4x^3))$

$= \frac{d}{dx}(3x^3 - x^6 - 12x^2 + 4x^5)$

$= 9x^2 - 6x^5 - 24x + 20x^4$

f) $\frac{d}{dx}((x-3)^2(x^2-2)) = \frac{d}{dx}((x^2 - 3x - 3x + 9)(x^2 - 2))$

$= \frac{d}{dx}((x^2 - 6x + 9)(x^2 - 2))$

$= \frac{d}{dx}((x^4 - 6x^3 + 9x^2) + (-2x^2 + 12x - 18))$

$= \frac{d}{dx}(x^4 - 6x^3 + 7x^2 + 12x - 18)$

$= 4x^3 - 18x^2 + 14x + 12$

Q3 a) $\frac{dy}{dx} = 4x^3 - 2x$. At $x = 3$, $\frac{dy}{dx} = 102$.

b) $\frac{dy}{dx} = 10x^4 + (-x^{-2}) = 10x^4 - \frac{1}{x^2}$
At $x = -2$, $\frac{dy}{dx} = 159.75$.

c) $y = x(x-1)(x-2) = x(x^2 - 3x + 2) = x^3 - 3x^2 + 2x$
$\frac{dy}{dx} = 3x^2 - 6x + 2$. At $x = -3$, $\frac{dy}{dx} = 47$.

d) $y = 5(x^2 - 1)(3 - x) = 5(-x^3 + 3x^2 + x - 3)$
$= -5x^3 + 15x^2 + 5x - 15$
$\frac{dy}{dx} = -15x^2 + 30x + 5$. At $x = 0$, $\frac{dy}{dx} = 5$.

e) $y = \sqrt{x}(x-1) = x^{\frac{1}{2}}(x-1) = x^{\frac{3}{2}} - x^{\frac{1}{2}}$
$\frac{dy}{dx} = \frac{3}{2}x^{\frac{1}{2}} - \frac{1}{2}x^{-\frac{1}{2}} = \frac{3}{2}\sqrt{x} - \frac{1}{2\sqrt{x}}$
At $x = 4$, $\frac{dy}{dx} = 2.75$.

f) $f(x) = x^3(x^2 - 5) = x^5 - 5x^3$
$f'(x) = 5x^4 - 15x^2$, $f'(-1) = -10$

g) $f(x) = \frac{1}{x^2}(x^3 - x) = x - x^{-1}$
$f'(x) = 1 + x^{-2} = 1 + \frac{1}{x^2}$, $f'(5) = \frac{26}{25}$

h) $f(x) = \frac{3x^3 + 18x^2 + 24x}{x+4}$
$= \frac{3x(x+4)(x+2)}{x+4} = 3x(x+2) = 3x^2 + 6x$
$f'(x) = 6x + 6$, $f'(-2) = -6$

Q4 a) $y = f'(x)$

Work out the gradient for each bit of the line.
$f'(x) = 2$ for $0 \le x \le 2$ and $f'(x) = 1$ for $6 \le x \le 10$.

b) $y = f'(x) = 10 - 2x$

Differentiate and sketch the graph of the gradient function.

c) $y = f'(x) = 2x + 2$

d) $y = f'(x) = 3x^2 + 4$

Q5 a) $f(x) = (x+3)(x+4) = x^2 + 7x + 12$
$f'(x) = 2x + 7$

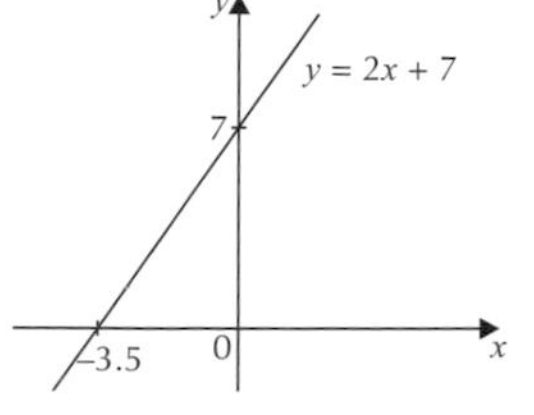

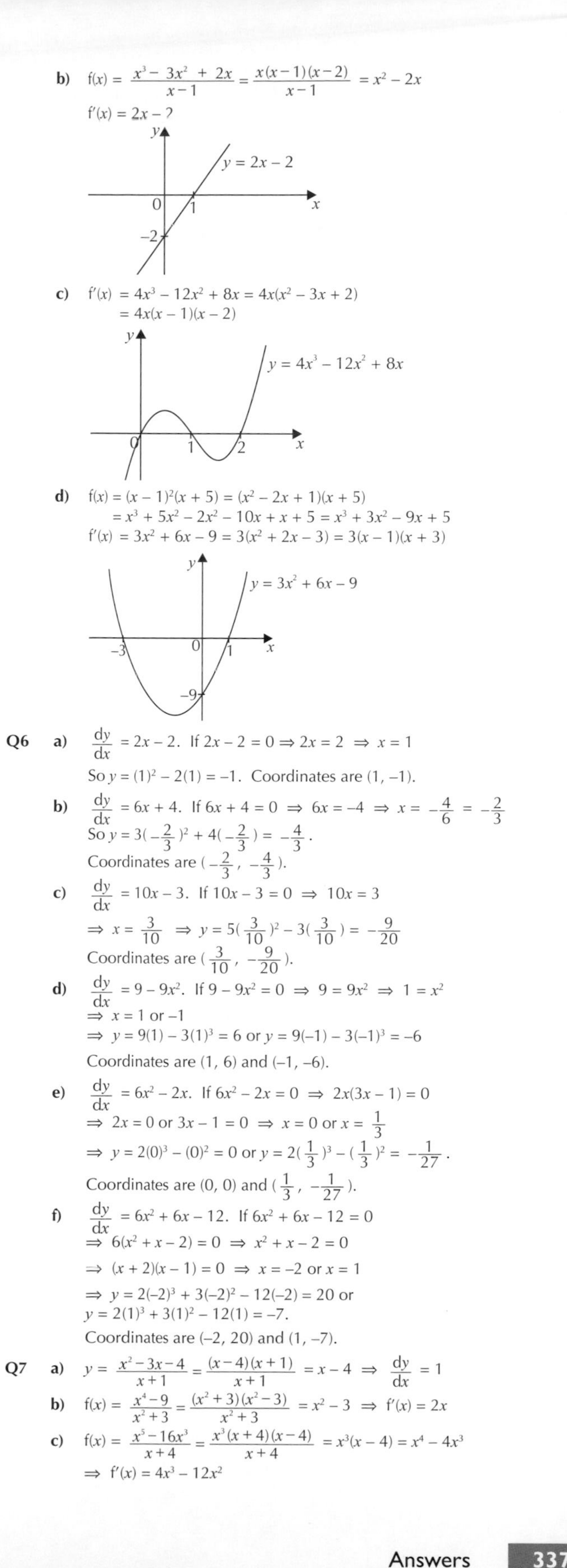

b) $f(x) = \frac{x^3 - 3x^2 + 2x}{x-1} = \frac{x(x-1)(x-2)}{x-1} = x^2 - 2x$
$f'(x) = 2x - 2$

c) $f'(x) = 4x^3 - 12x^2 + 8x = 4x(x^2 - 3x + 2)$
$= 4x(x-1)(x-2)$

d) $f(x) = (x-1)^2(x+5) = (x^2 - 2x + 1)(x+5)$
$= x^3 + 5x^2 - 2x^2 - 10x + x + 5 = x^3 + 3x^2 - 9x + 5$
$f'(x) = 3x^2 + 6x - 9 = 3(x^2 + 2x - 3) = 3(x-1)(x+3)$

Q6 a) $\frac{dy}{dx} = 2x - 2$. If $2x - 2 = 0 \Rightarrow 2x = 2 \Rightarrow x = 1$
So $y = (1)^2 - 2(1) = -1$. Coordinates are $(1, -1)$.

b) $\frac{dy}{dx} = 6x + 4$. If $6x + 4 = 0 \Rightarrow 6x = -4 \Rightarrow x = -\frac{4}{6} = -\frac{2}{3}$
So $y = 3(-\frac{2}{3})^2 + 4(-\frac{2}{3}) = -\frac{4}{3}$.
Coordinates are $(-\frac{2}{3}, -\frac{4}{3})$.

c) $\frac{dy}{dx} = 10x - 3$. If $10x - 3 = 0 \Rightarrow 10x = 3$
$\Rightarrow x = \frac{3}{10} \Rightarrow y = 5(\frac{3}{10})^2 - 3(\frac{3}{10}) = -\frac{9}{20}$
Coordinates are $(\frac{3}{10}, -\frac{9}{20})$.

d) $\frac{dy}{dx} = 9 - 9x^2$. If $9 - 9x^2 = 0 \Rightarrow 9 = 9x^2 \Rightarrow 1 = x^2$
$\Rightarrow x = 1$ or -1
$\Rightarrow y = 9(1) - 3(1)^3 = 6$ or $y = 9(-1) - 3(-1)^3 = -6$
Coordinates are $(1, 6)$ and $(-1, -6)$.

e) $\frac{dy}{dx} = 6x^2 - 2x$. If $6x^2 - 2x = 0 \Rightarrow 2x(3x - 1) = 0$
$\Rightarrow 2x = 0$ or $3x - 1 = 0 \Rightarrow x = 0$ or $x = \frac{1}{3}$
$\Rightarrow y = 2(0)^3 - (0)^2 = 0$ or $y = 2(\frac{1}{3})^3 - (\frac{1}{3})^2 = -\frac{1}{27}$.
Coordinates are $(0, 0)$ and $(\frac{1}{3}, -\frac{1}{27})$.

f) $\frac{dy}{dx} = 6x^2 + 6x - 12$. If $6x^2 + 6x - 12 = 0$
$\Rightarrow 6(x^2 + x - 2) = 0 \Rightarrow x^2 + x - 2 = 0$
$\Rightarrow (x+2)(x-1) = 0 \Rightarrow x = -2$ or $x = 1$
$\Rightarrow y = 2(-2)^3 + 3(-2)^2 - 12(-2) = 20$ or
$y = 2(1)^3 + 3(1)^2 - 12(1) = -7$.
Coordinates are $(-2, 20)$ and $(1, -7)$.

Q7 a) $y = \frac{x^2 - 3x - 4}{x+1} = \frac{(x-4)(x+1)}{x+1} = x - 4 \Rightarrow \frac{dy}{dx} = 1$

b) $f(x) = \frac{x^4 - 9}{x^2 + 3} = \frac{(x^2+3)(x^2-3)}{x^2+3} = x^2 - 3 \Rightarrow f'(x) = 2x$

c) $f(x) = \frac{x^5 - 16x^3}{x+4} = \frac{x^3(x+4)(x-4)}{x+4} = x^3(x-4) = x^4 - 4x^3$
$\Rightarrow f'(x) = 4x^3 - 12x^2$

d) $y = \frac{1}{x}(x-3)(x-4) = \frac{1}{x}(x^2 - 3x - 4x + 12)$
$= \frac{1}{x}(x^2 - 7x + 12) = x - 7 + \frac{12}{x} = x - 7 + 12x^{-1}$
$\Rightarrow \frac{dy}{dx} = 1 - 12x^{-2} = 1 - \frac{12}{x^2}$

e) $y = \sqrt{x}(x^3 - \sqrt{x}) = x^{\frac{1}{2}}(x^3 - x^{\frac{1}{2}}) = x^{\frac{7}{2}} - x$
$\Rightarrow \frac{dy}{dx} = \frac{7}{2}x^{\frac{5}{2}} - 1 = \frac{7}{2}\sqrt{x^5} - 1$

f) $f(x) = \frac{3-\sqrt{x}}{\sqrt{x}} = \frac{3 - x^{\frac{1}{2}}}{x^{\frac{1}{2}}} = x^{-\frac{1}{2}}(3 - x^{\frac{1}{2}})$
$= 3x^{-\frac{1}{2}} - x^0 = 3x^{-\frac{1}{2}} - 1$
$f'(x) = 3(-\frac{1}{2}x^{-\frac{3}{2}}) = -\frac{3}{2}x^{-\frac{3}{2}} = -\frac{3}{2\sqrt{x^3}}$

g) $f(x) = \frac{x + 5\sqrt{x}}{\sqrt{x}} = \frac{x + 5x^{\frac{1}{2}}}{x^{\frac{1}{2}}} = x^{-\frac{1}{2}}(x + 5x^{\frac{1}{2}})$
$= x^{\frac{1}{2}} + 5x^0 = x^{\frac{1}{2}} + 5$
$f'(x) = \frac{1}{2}x^{-\frac{1}{2}} = \frac{1}{2\sqrt{x}}$

h) Factorising the numerator:
$f(x) = \frac{x - 3\sqrt{x} + 2}{\sqrt{x} - 1} = \frac{(\sqrt{x}-2)(\sqrt{x}-1)}{\sqrt{x}-1} = \sqrt{x} - 2 = x^{\frac{1}{2}} - 2$
$f'(x) = \frac{1}{2}x^{-\frac{1}{2}} = \frac{1}{2\sqrt{x}}$

i) Factorising the numerator:
$f(x) = \frac{4-x}{2+\sqrt{x}} = \frac{(2+\sqrt{x})(2-\sqrt{x})}{2+\sqrt{x}} = 2 - \sqrt{x} = 2 - x^{\frac{1}{2}}$
$f'(x) = -\frac{1}{2}x^{-\frac{1}{2}} = -\frac{1}{2\sqrt{x}}$

Exercise 10.2.3 — Finding tangents and normals

Q1 a) $\frac{dy}{dx} = 9 - 4x$. At (1, 7), $\frac{dy}{dx} = 5$
$\Rightarrow$ tangent has a gradient of 5 and has an equation of the form $y = 5x + c$.
Using the point (1, 7), $7 = 5 + c$
$\Rightarrow c = 2$. So the tangent's equation is $y = 5x + 2$.

b) $\frac{dy}{dx} = 3x^2 - 2$. At (2, 7), $\frac{dy}{dx} = 10$
$\Rightarrow$ tangent has a gradient of 10 and has an equation of the form $y = 10x + c$.
Using the point (2, 7), $7 = 20 + c \Rightarrow c = -13$.
So the tangent's equation is $y = 10x - 13$.

c) $y = (x+2)(2x-3) = 2x^2 + x - 6$
$\frac{dy}{dx} = 4x + 1$. At (2, 4), $\frac{dy}{dx} = 9$
$\Rightarrow$ tangent has a gradient of 9 and has an equation of the form $y = 9x + c$.
Using the point (2, 4), $4 = 18 + c \Rightarrow c = -14$.
So the tangent's equation is $y = 9x - 14$.

d) $y = x(x-1)^2 = x(x^2 - 2x + 1) = x^3 - 2x^2 + x$
$\frac{dy}{dx} = 3x^2 - 4x + 1$. At (–1, –4), $\frac{dy}{dx} = 8$
$\Rightarrow$ tangent has a gradient of 8 and has an equation of the form $y = 8x + c$.
Using the point (–1, –4), $-4 = -8 + c \Rightarrow c = 4$.
So the tangent's equation is $y = 8x + 4$.

e) $y = x^2(x+3) - 10 = x^3 + 3x^2 - 10$
$\frac{dy}{dx} = 3x^2 + 6x$. At (2, 10), $\frac{dy}{dx} = 24$
$\Rightarrow$ tangent has a gradient of 24 and has an equation of the form $y = 24x + c$.
Using the point (2, 10), $10 = 48 + c \Rightarrow c = -38$.
So the tangent's equation is $y = 24x - 38$.

f) $y = x(2x^2 - 2x - 12) = 2x^3 - 2x^2 - 12x$
$\frac{dy}{dx} = 6x^2 - 4x - 12$. At (–1, 8), $\frac{dy}{dx} = -2$
$\Rightarrow$ tangent has a gradient of –2 and has an equation of the form $y = -2x + c$.
Using the point (–1, 8), $8 = 2 + c \Rightarrow c = 6$.
So the tangent's equation is $y = -2x + 6$.

Q2 a) $y = x^{-1} + x + 3$
$\frac{dy}{dx} = -x^{-2} + 1$. At (2, $5\frac{1}{2}$), $\frac{dy}{dx} = \frac{3}{4}$
$\Rightarrow$ tangent has a gradient of $\frac{3}{4}$ and has an equation of the form $y = \frac{3}{4}x + c$.
Using the point (2, $5\frac{1}{2}$), $5\frac{1}{2} = 1\frac{1}{2} + c \Rightarrow c = 4$.
So the tangent's equation is $y = \frac{3}{4}x + 4$
$\Rightarrow 4y = 3x + 16 \Rightarrow 3x - 4y + 16 = 0$

b) $y = 4x^2 - 3x^{\frac{1}{2}}$
$\frac{dy}{dx} = 8x - 3(\frac{1}{2}x^{-\frac{1}{2}}) = 8x - \frac{3}{2}x^{-\frac{1}{2}}$.
At (1, 1), $\frac{dy}{dx} = 6\frac{1}{2}$
$\Rightarrow$ tangent has a gradient of $6\frac{1}{2}$ and has an equation of the form $y = 6\frac{1}{2}x + c$.
Using the point (1, 1), $1 = 6\frac{1}{2} + c \Rightarrow c = -5\frac{1}{2}$.
So the tangent's equation is $y = 6\frac{1}{2}x - 5\frac{1}{2}$
$\Rightarrow 2y = 13x - 11 \Rightarrow 13x - 2y - 11 = 0$.

c) $y = 3x^{-1} + 2x^{\frac{1}{2}}$
$\frac{dy}{dx} = 3(-x^{-2}) + 2(\frac{1}{2}x^{-\frac{1}{2}}) = -3x^{-2} + x^{-\frac{1}{2}}$
At (4, $4\frac{3}{4}$), $\frac{dy}{dx} = \frac{5}{16}$
$\Rightarrow$ tangent has a gradient of $\frac{5}{16}$ and has an equation of the form $y = \frac{5}{16}x + c$.
Using the point (4, $4\frac{3}{4}$), $4\frac{3}{4} = \frac{5}{4} + c \Rightarrow c = 3\frac{1}{2}$.
So the tangent's equation is $y = \frac{5}{16}x + 3\frac{1}{2}$
$\Rightarrow 16y = 5x + 56 \Rightarrow 5x - 16y + 56 = 0$.

d) $y = x^{-1} + 4x^{-2}$
$\frac{dy}{dx} = -x^{-2} + 4(-2x^{-3}) = -x^{-2} - 8x^{-3}$
At (2, $1\frac{1}{2}$), $\frac{dy}{dx} = -\frac{5}{4}$
$\Rightarrow$ tangent has a gradient of $-\frac{5}{4}$ and has an equation of the form $y = -\frac{5}{4}x + c$.
Using the point (2, $1\frac{1}{2}$), $1\frac{1}{2} = -\frac{5}{2} + c \Rightarrow c = 4$.
So the tangent's equation is $y = -\frac{5}{4}x + 4$
$\Rightarrow 4y = -5x + 16 \Rightarrow 5x + 4y - 16 = 0$.

e) $y = \frac{1}{3}x^2 - 4x^{\frac{1}{2}} - \frac{1}{3}$
$\frac{dy}{dx} = \frac{1}{3}(2x) - 4(\frac{1}{2}x^{-\frac{1}{2}}) = \frac{2}{3}x - 2x^{-\frac{1}{2}}$.
At (4, –3), $\frac{dy}{dx} = \frac{5}{3}$
$\Rightarrow$ tangent has a gradient of $\frac{5}{3}$ and has an equation of the form $y = \frac{5}{3}x + c$.
Using the point (4, –3), $-3 = \frac{20}{3} + c \Rightarrow c = -\frac{29}{3}$.
So the tangent's equation is $y = \frac{5}{3}x - \frac{29}{3}$
$\Rightarrow 3y = 5x - 29 \Rightarrow 5x - 3y - 29 = 0$.

f) $y = x - 2x^{-1} + 3x^{-2}$
$\frac{dy}{dx} = 1 + 2x^{-2} - 6x^{-3}$. At (–3, –2), $\frac{dy}{dx} = \frac{13}{9}$
$\Rightarrow$ tangent has a gradient of $\frac{13}{9}$ and has an equation of the form $y = \frac{13}{9}x + c$.
Using the point (–3, –2), $-2 = -\frac{13}{3} + c \Rightarrow c = \frac{7}{3}$.
So the tangent's equation is $y = \frac{13}{9}x + \frac{7}{3}$
$\Rightarrow 9y = 13x + 21 \Rightarrow 13x - 9y + 21 = 0$.

Q3 a) $\frac{dy}{dx} = 6x - 4$. At (2, 6), $\frac{dy}{dx} = 8$.
So the normal has a gradient of $-\frac{1}{8}$ and an equation of the form $y = -\frac{1}{8}x + c$.
Don't forget — the gradient of the normal to a curve at a point is $\frac{-1}{\text{Gradient of the curve}}$.
Using the point (2, 6), $6 = -\frac{1}{4} + c \Rightarrow c = 6\frac{1}{4}$
So the normal's equation is $y = -\frac{1}{8}x + 6\frac{1}{4}$
$\Rightarrow 8y = -x + 50 \Rightarrow x + 8y - 50 = 0$.

b) $y = x^3 + 4x^2 - 5x$
$\frac{dy}{dx} = 3x^2 + 8x - 5$. At (–1, 8), $\frac{dy}{dx} = -10$.
So the normal has a gradient of $\frac{1}{10}$ and an equation of the form $y = \frac{1}{10}x + c$.
Using the point (–1, 8), $8 = -\frac{1}{10} + c \Rightarrow c = \frac{81}{10}$
So the normal's equation is $y = \frac{1}{10}x + \frac{81}{10}$
$\Rightarrow 10y = x + 81 \Rightarrow x - 10y + 81 = 0$.

c) $y = x(x^2 - 3x + 2) = x^3 - 3x^2 + 2x$
$\frac{dy}{dx} = 3x^2 - 6x + 2$. At (3, 6), $\frac{dy}{dx} = 11$.
So the normal has a gradient of $-\frac{1}{11}$ and an equation of the form $y = -\frac{1}{11}x + c$.
Using the point (3, 6), $6 = -\frac{3}{11} + c \Rightarrow c = \frac{69}{11}$.
So the normal's equation is $y = -\frac{1}{11}x + \frac{69}{11}$
$\Rightarrow 11y = -x + 69 \Rightarrow x + 11y - 69 = 0$.

d) $y = x(x^2 + x - 12) - 10 = x^3 + x^2 \;\; 12x - 10$
$\frac{dy}{dx} = 3x^2 + 2x - 12$. At (–2, 10), $\frac{dy}{dx} = -4$.
So the normal has a gradient of $\frac{1}{4}$ and an equation of the form $y = \frac{1}{4}x + c$.
Using the point (–2, 10), $10 = -\frac{1}{2} + c \Rightarrow c = \frac{21}{2}$.
So the normal's equation is $y = \frac{1}{4}x + \frac{21}{2}$
$\Rightarrow 4y = x + 42 \Rightarrow x - 4y + 42 = 0$.

e) $y = \frac{(x+2)(x^2-7x)}{x+2} = x^2 - 7x$
$\frac{dy}{dx} = 2x - 7$. At (5, –10), $\frac{dy}{dx} = 3$.
So the normal has a gradient of $-\frac{1}{3}$ and an equation of the form $y = -\frac{1}{3}x + c$.
Using the point (5, –10), $-10 = -\frac{5}{3} + c \Rightarrow c = -\frac{25}{3}$
So the normal's equation is $y = -\frac{1}{3}x - \frac{25}{3}$
$\Rightarrow 3y = -x - 25 \Rightarrow x + 3y + 25 = 0$.

f) $y = \frac{2x(x^2-16)}{x+4} = \frac{2x(x-4)(x+4)}{x+4} = 2x^2 - 8x$
$\frac{dy}{dx} = 4x - 8$. At (3, –6), $\frac{dy}{dx} = 4$.
So the normal has a gradient of $-\frac{1}{4}$ and an equation of the form $y = -\frac{1}{4}x + c$.
Using the point (3, –6), $-6 = -\frac{3}{4} + c \Rightarrow c = -\frac{21}{4}$
So the normal's equation is $y = -\frac{1}{4}x - \frac{21}{4}$
$\Rightarrow 4y = -x - 21 \Rightarrow x + 4y + 21 = 0$.

Q4 a) $y = \frac{2x^5 - 2x^4}{3x^3} = \frac{2}{3}x^2 - \frac{2}{3}x$
Remember — if the denominator is a single term, split the equation up into separate terms.
$\frac{dy}{dx} = \frac{2}{3}(2x) - \frac{2}{3} = \frac{4}{3}x - \frac{2}{3}$
At (–2, 4), $\frac{dy}{dx} = -\frac{10}{3}$.
So the normal has a gradient of $\frac{3}{10}$ and an equation of the form $y = \frac{3}{10}x + c$.
Using the point (–2, 4), $4 = -\frac{6}{10} + c \Rightarrow c = \frac{23}{5}$.
So the normal's equation is $y = \frac{3}{10}x + \frac{23}{5}$
$\Rightarrow 10y = 3x + 46 \Rightarrow 3x - 10y + 46 = 0$.
This question doesn't tell you which form to use to write the equations, so you can choose whichever is easiest for each part.

b) $y = \frac{5x^2 - 2x + 3}{x^2} = 5 - \frac{2}{x} + \frac{3}{x^2}$
$\frac{dy}{dx} = -2(-x^{-2}) + 3(-2x^{-3}) = \frac{2}{x^2} - \frac{6}{x^3}$
At $(2, 4\frac{3}{4})$, $\frac{dy}{dx} = -\frac{1}{4}$.
So the normal has a gradient of 4 and an equation of the form $y = 4x + c$.
Using the point $(2, 4\frac{3}{4})$, $4\frac{3}{4} = 8 + c \Rightarrow c = -\frac{13}{4}$.
So the normal's equation is $y = 4x - \frac{13}{4}$
$\Rightarrow 4y = 16x - 13 \Rightarrow 16x - 4y - 13 = 0$.

c) $y = 3xx^{-\frac{1}{2}} - x^2x^{-\frac{1}{2}} = 3x^{\frac{1}{2}} - x^{\frac{3}{2}}$
$\frac{dy}{dx} = 3(\frac{1}{2}x^{-\frac{1}{2}}) - \frac{3}{2}x^{\frac{1}{2}} = \frac{3}{2\sqrt{x}} - \frac{3}{2}\sqrt{x}$
At (4, –2), $\frac{dy}{dx} = -\frac{9}{4}$.
So the normal has a gradient of $\frac{4}{9}$ and an equation of the form $y = \frac{4}{9}x + c$.
Using the point (4, –2), $-2 = \frac{16}{9} + c \Rightarrow c = -\frac{34}{9}$.
So the normal's equation is
$y = \frac{4}{9}x - \frac{34}{9} \Rightarrow 9y = 4x - 34 \Rightarrow 4x - 9y - 34 = 0$.

d) $y = \frac{1}{x} - \frac{3}{x^2} - \frac{4}{x^3} + \frac{7}{4} = x^{-1} - 3x^{-2} - 4x^{-3} + \frac{7}{4}$
$\frac{dy}{dx} = -x^{-2} - 3(-2x^{-3}) - 4(-3x^{-4}) = -x^{-2} + 6x^{-3} + 12x^{-4}$
At (–2, 1), $\frac{dy}{dx} = -\frac{1}{4}$.
So the normal has a gradient of 4 and an equation of the form $y = 4x + c$.
Using the point (–2, 1), $1 = -8 + c \Rightarrow c = 9$.
So the normal's equation is $y = 4x + 9$.

e) $y = \frac{x^3 - 5x^2 - 4x}{x^{\frac{3}{2}}} = x^{\frac{3}{2}} - 5x^{\frac{1}{2}} - 4x^{-\frac{1}{2}}$
$\frac{dy}{dx} = \frac{3}{2}x^{\frac{1}{2}} - 5(\frac{1}{2}x^{-\frac{1}{2}}) - 4(-\frac{1}{2}x^{-\frac{3}{2}})$
$= \frac{3}{2}\sqrt{x} - \frac{5}{2\sqrt{x}} + \frac{2}{x\sqrt{x}}$
At (4, –4), $\frac{dy}{dx} = 2$.
So the normal has a gradient of $-\frac{1}{2}$ and an equation of the form $y = -\frac{1}{2}x + c$.
Using the point (4, –4), $-4 = -2 + c \Rightarrow c = -2$.
So the equation of the normal is $y = -\frac{1}{2}x - 2$
$\Rightarrow 2y = -x - 4 \Rightarrow x + 2y + 4 = 0$.

f) $y = \frac{4x^4 - 2x^2 + 3\sqrt{x}}{x^{\frac{5}{2}}} = 4x^{\frac{3}{2}} - 2x^{-\frac{1}{2}} + 3x^{-2}$
$\frac{dy}{dx} = 4(\frac{3}{2}x^{\frac{1}{2}}) - 2(-\frac{1}{2}x^{-\frac{3}{2}}) + 3(-2x^{-3}) = 6x^{\frac{1}{2}} + x^{-\frac{3}{2}} - 6x^{-3}$
At (1, 5), $\frac{dy}{dx} = 1$.
So the normal has a gradient of –1 and an equation of the form $y = -x + c$.
Using the point (1, 5), $5 = -1 + c \Rightarrow c = 6$.
So the equation of the normal is $y = 6 - x$.

Q5 **a)** $f'(x) = 3x^2 - 6x$. If $f'(x) = 9$, $3x^2 - 6x = 9 \Rightarrow 3x^2 - 6x - 9 = 0$
$\Rightarrow x^2 - 2x - 3 = 0$
$\Rightarrow (x - 3)(x + 1) = 0 \Rightarrow x = 3$ or $x = -1$.
So $x = 3$ since $x > 0$.
So $y = f(3) = 3^3 - 3(3)^2 + 3 = 3$.
The coordinates are (3, 3).

b) The gradient of the tangent at (3, 3) is 9 from part a). So the equation is of the form $y = 9x + c$. You know the tangent goes through (3, 3) so use this point: $3 = 27 + c \Rightarrow c = -24$.
So the equation is $y = 9x - 24$.

c) The gradient of the normal is $-\frac{1}{9}$ so the equation has the form $y = -\frac{1}{9}x + c$. Again, use the point (3, 3),
so $3 = -\frac{1}{3} + c \Rightarrow c = \frac{10}{3}$.
So the equation is $y = -\frac{1}{9}x + \frac{10}{3}$
$\Rightarrow 9y = -x + 30 \Rightarrow x + 9y - 30 = 0$.

Q6 **a)** Putting $x = -2$ into the equation gives:
$$y = \frac{x^3 + x^2 + x + 5}{x^2}$$
$$= \frac{(-2)^3 + (-2)^2 + (-2) + 5}{(-2)^2}$$
$$= \frac{-8 + 4 - 2 + 5}{4} = -\frac{1}{4}$$
so $(2, -\frac{1}{4})$ is a point on the curve.

b) $y = \frac{x^3 + x^2 + x + 5}{x^2} = x + 1 + \frac{1}{x} + \frac{5}{x^2}$

$\frac{dy}{dx} = 1 + 0 + (-x^{-2}) + 5(-2x^{-3}) = 1 - \frac{1}{x^2} - \frac{10}{x^3}$

At $(-2, -\frac{1}{4})$, $\frac{dy}{dx} = 2$.
So the gradient of the tangent at this point is 2 and it has equation $y = 2x + c$.
Using the point $(-2, -\frac{1}{4})$, $-\frac{1}{4} = -4 + c \Rightarrow c = \frac{15}{4}$.
So the equation of the tangent is $y = 2x + \frac{15}{4}$
$\Rightarrow 4y = 8x + 15 \Rightarrow 8x - 4y + 15 = 0$.

c) The gradient of the normal at $(-2, -\frac{1}{4})$ is $-\frac{1}{2}$ and so it has equation $y = -\frac{1}{2}x + c$.
Using the point $(-2, -\frac{1}{4})$, $-\frac{1}{4} = 1 + c \Rightarrow c = -\frac{5}{4}$.
So the equation of the normal is $y = -\frac{1}{2}x - \frac{5}{4}$
$\Rightarrow 4y = -2x - 5 \Rightarrow 2x + 4y + 5 = 0$.

10.3 Using Differentiation

Exercise 10.3.1 — Finding second order derivatives

Q1 **a)** $\frac{dy}{dx} = 3x^2$ and $\frac{d^2y}{dx^2} = 6x$.

b) $\frac{dy}{dx} = 5x^4$ and $\frac{d^2y}{dx^2} = 20x^3$.

c) $\frac{dy}{dx} = 4x^3$ and $\frac{d^2y}{dx^2} = 12x^2$.

d) $\frac{dy}{dx} = 1$ and $\frac{d^2y}{dx^2} = 0$.

e) $\frac{dy}{dx} = 4x$ and $\frac{d^2y}{dx^2} = 4$.

f) $\frac{dy}{dx} = 12x^2$ and $\frac{d^2y}{dx^2} = 24x$.

g) $\frac{dy}{dx} = 2x^3$ and $\frac{d^2y}{dx^2} = 6x^2$.

h) $\frac{dy}{dx} = -6x^{-3} = -\frac{6}{x^3}$ and $\frac{d^2y}{dx^2} = 18x^{-4} = \frac{18}{x^4}$.

i) $y = x^{-1}$, so $\frac{dy}{dx} = -x^{-2} = -\frac{1}{x^2}$ and $\frac{d^2y}{dx^2} = 2x^{-3} = \frac{2}{x^3}$.

j) $y = x^{\frac{1}{2}}$, so $\frac{dy}{dx} = \frac{1}{2}x^{-\frac{1}{2}} = \frac{1}{2\sqrt{x}}$
and $\frac{d^2y}{dx^2} = -\frac{1}{4}x^{-\frac{3}{2}} = -\frac{1}{4(\sqrt{x})^3}$.

k) $y = x^{-2}$, so $\frac{dy}{dx} = -2x^{-3} = -\frac{2}{x^3}$ and $\frac{d^2y}{dx^2} = 6x^{-4} = \frac{6}{x^4}$.

l) $y = x\sqrt{x} = x^1 x^{\frac{1}{2}} = x^{1+\frac{1}{2}} = x^{\frac{3}{2}}$,
so $\frac{dy}{dx} = \frac{3}{2}x^{\frac{1}{2}} = \frac{3}{2}\sqrt{x}$ and $\frac{d^2y}{dx^2} = \frac{3}{4}x^{-\frac{1}{2}} = \frac{3}{4\sqrt{x}}$.

Q2 **a)** $\frac{dy}{dx} = 2x + 4x^3 \Rightarrow \frac{d^2y}{dx^2} = 2 + 12x^2$.

b) $f'(x) = 3x^2 - 1 \Rightarrow f''(x) = 6x$.

c) $\frac{dy}{dx} = 12x^5 + 6x \Rightarrow \frac{d^2y}{dx^2} = 60x^4 + 6$.

d) $f'(x) = 4x^{-3} + 5 \Rightarrow f''(x) = -12x^{-4} = -\frac{12}{x^4}$.

e) $y = 4x^4 + x^{\frac{1}{2}}$, $\frac{dy}{dx} = 16x^3 + \frac{1}{2}x^{-\frac{1}{2}}$,
$\Rightarrow \frac{d^2y}{dx^2} = 48x^2 - \frac{1}{4}x^{-\frac{3}{2}} = 48x^2 - \frac{1}{4\sqrt{x^3}}$.

f) $f(x) = -x^5 - 4x^{\frac{1}{2}}$, $f'(x) = -5x^4 - 2x^{-\frac{1}{2}}$,
$\Rightarrow f''(x) = -20x^3 + x^{-\frac{3}{2}} = -20x^3 + \frac{1}{\sqrt{x^3}}$.

Q3 **a)** $f(x) = x(4x^2 - x) = 4x^3 - x^2$
$f'(x) = 12x^2 - 2x$, $f''(x) = 24x - 2$

b) $f(x) = (x^2 - 3)(x - 4) = x^3 - 4x^2 - 3x + 12$
$f'(x) = 3x^2 - 8x - 3$, $f''(x) = 6x - 8$

c) $f(x) = x^2(3x - x^2) = 3x^3 - x^4$
$f'(x) = 9x^2 - 4x^3$, $f''(x) = 18x - 12x^2$

d) $f(x) = (x^{-1} + 5)(x^{-1} - 5) = x^{-2} - 25$
$f'(x) = -2x^{-3} = -\frac{2}{x^3}$, $f''(x) = 6x^{-4} = \frac{6}{x^4}$

e) $f(x) = \frac{x^3 + 8}{x} = x^2 + 8x^{-1}$, $f'(x) = 2x - 8x^{-2} = 2x - \frac{8}{x^2}$,
$f''(x) = 2 + 16x^{-3} = 2 + \frac{16}{x^3}$

f) $f(x) = 3x^{\frac{1}{2}} + xx^{\frac{1}{2}} = 3x^{\frac{1}{2}} + x^{\frac{3}{2}}$
$f'(x) = \frac{3}{2}x^{-\frac{1}{2}} + \frac{3}{2}x^{\frac{1}{2}} = \frac{3}{2\sqrt{x}} + \frac{3}{2}\sqrt{x}$
$f''(x) = \frac{3}{2}(-\frac{1}{2}x^{-\frac{3}{2}}) + \frac{3}{2}(\frac{1}{2}x^{-\frac{1}{2}}) = -\frac{3}{4}x^{-\frac{3}{2}} + \frac{3}{4}x^{-\frac{1}{2}}$
$= -\frac{3}{4(\sqrt{x})^3} + \frac{3}{4\sqrt{x}} \quad \left(= -\frac{3}{4x\sqrt{x}} + \frac{3}{4\sqrt{x}}\right)$

g) $f(x) = \frac{4x^5 + 12x^3 - 40x}{4(x^2 + 5)} = \frac{4x(x^4 + 3x^2 - 10)}{4(x^2 + 5)}$
$= \frac{4x(x^2 + 5)(x^2 - 2)}{4(x^2 + 5)} = x(x^2 - 2) = x^3 - 2x$
$f'(x) = 3x^2 - 2$, $f''(x) = 6x$

h) $f(x) = \frac{1}{x}(3x^4 - 2x^3) = 3x^3 - 2x^2$
$f'(x) = 9x^2 - 4x$, $f''(x) = 18x - 4$

i) $f(x) = \frac{x^2 - xx^{\frac{1}{2}} + 7x}{x^{\frac{1}{2}}} = x^2x^{-\frac{1}{2}} - xx^{\frac{1}{2}}x^{-\frac{1}{2}} + 7xx^{-\frac{1}{2}}$
$= x^{\frac{3}{2}} - x + 7x^{\frac{1}{2}}$
$f'(x) = \frac{3}{2}x^{\frac{1}{2}} - 1 + 7(\frac{1}{2}x^{-\frac{1}{2}}) = \frac{3}{2}\sqrt{x} - 1 + \frac{7}{2\sqrt{x}}$
$f''(x) = \frac{3}{2}(\frac{1}{2}x^{-\frac{1}{2}}) + \frac{7}{2}(-\frac{1}{2}x^{-\frac{3}{2}}) = \frac{3}{4\sqrt{x}} - \frac{7}{4(\sqrt{x})^3}$

Q4 **a)** $f'(x) = 45x^2$, so $f''(x) = 90x$ and $f''\left(\frac{1}{3}\right) = 30$.

b) $y = \frac{x^4}{12} - \frac{x^3}{6}$, so $\frac{dy}{dx} = \frac{x^3}{3} - \frac{x^2}{2}$, so $\frac{d^2y}{dx^2} = x^2 - x$
so at $x = 4$, $\frac{d^2y}{dx^2} = 12$.

c) $f'(x) = 3x^2 - 2x$, so $f''(x) = 6x - 2$ and $f''(3) = 16$.

d) $y = xx^{\frac{1}{2}} - x^{-1} = x^{\frac{3}{2}} - x^{-1}$ so $\frac{dy}{dx} = \frac{3}{2}x^{\frac{1}{2}} + x^{-2}$
so $\frac{d^2y}{dx^2} = \frac{3}{2}(\frac{1}{2}x^{-\frac{1}{2}}) - 2x^{-3} = \frac{3}{4\sqrt{x}} - \frac{2}{x^3}$
so at $x = 4$, $\frac{d^2y}{dx^2} = \frac{11}{32}$.

e) $f(x) = x^2(x^3 - 4x^2 - 5x) = x^5 - 4x^4 - 5x^3$
so $f'(x) = 5x^4 - 16x^3 - 15x^2$
and $f''(x) = 20x^3 - 48x^2 - 30x$.
$f''(-1) = -38$.

f) $y = \frac{x^3(x+6)(x-2)}{(x+6)} = x^3(x-2) = x^4 - 2x^3$
so $\frac{dy}{dx} = 4x^3 - 6x^2$, $\frac{d^2y}{dx^2} = 12x^2 - 12x$.
At $x = 5$, $\frac{d^2y}{dx^2} = 240$.

g) $f(x) = \frac{9x^2 + 3x}{3\sqrt{x}} = 3x^{\frac{3}{2}} + x^{\frac{1}{2}}$ so
$f'(x) = 3(\frac{3}{2}x^{\frac{1}{2}}) + \frac{1}{2}x^{-\frac{1}{2}} = \frac{9}{2}\sqrt{x} + \frac{1}{2\sqrt{x}}$ and so
$f''(x) = \frac{9}{2}(\frac{1}{2}x^{-\frac{1}{2}}) + \frac{1}{2}(-\frac{1}{2}x^{-\frac{3}{2}}) = \frac{9}{4\sqrt{x}} - \frac{1}{4(\sqrt{x})^3}$
$f''(1) = 2$.

h) $y = (x^{-2} + x^{-1})(5 - x) = 5x^{-2} - x^{-2}x + 5x^{-1} - xx^{-1}$
$= 5x^{-2} - x^{-1} + 5x^{-1} - 1 = 5x^{-2} + 4x^{-1} - 1$
$\frac{dy}{dx} = 5(-2x^{-3}) + 4(-x^{-2}) = -10x^{-3} - 4x^{-2}$
so $\frac{d^2y}{dx^2} = 30x^{-4} + 8x^{-3} = \frac{30}{x^4} + \frac{8}{x^3}$.
At $x = -3$, $\frac{d^2y}{dx^2} = \frac{2}{27}$.

Q5 a) $y = x^3 + 3x^2 \Rightarrow \frac{dy}{dx} = 3x^2 + 6x \Rightarrow \frac{d^2y}{dx^2} = 6x + 6$
$\frac{d^2y}{dx^2} = 18 \Rightarrow 6x + 6 = 18 \Rightarrow x = 2$
$\Rightarrow y = 2^3 + 3(2)^2 = 20$
So the coordinates of the point are (2, 20).

b) $f(x) = \frac{4}{15}x^{\frac{5}{2}} \Rightarrow f'(x) = \frac{2}{3}x^{\frac{3}{2}} \Rightarrow f''(x) = x^{\frac{1}{2}} = \sqrt{x}$
$f''(x) = 2 \Rightarrow \sqrt{x} = 2 \Rightarrow x = 4 \Rightarrow y = \frac{4}{15}(4)^{\frac{5}{2}} = \frac{128}{15}$
So the coordinates of the point are $(4, \frac{128}{15})$.

c) $y = (2x^2 + 4)(2x^2 - 4) = 4x^4 - 16 \Rightarrow \frac{dy}{dx} = 16x^3$
$\Rightarrow \frac{d^2y}{dx^2} = 48x^2$
$\frac{d^2y}{dx^2} = 768 \Rightarrow 48x^2 = 768 \Rightarrow x = -4$ or $x = 4$
$y = 4(\pm 4)^4 - 16 = 1008$
So the coordinates of the points are (–4, 1008) and (4, 1008).

d) $y = \frac{1}{4}x^4 - \frac{5}{3}x^3 + 2x^2 - 20x \Rightarrow \frac{dy}{dx} = x^3 - 5x^2 + 4x - 20$
$\Rightarrow \frac{d^2y}{dx^2} = 3x^2 - 10x + 4$
$\frac{d^2y}{dx^2} = 12 \Rightarrow 3x^2 - 10x + 4 = 12$
$\Rightarrow 3x^2 - 10x - 8 = 0$
$\Rightarrow (3x + 2)(x - 4) = 0$
$\Rightarrow x = -\frac{2}{3}$ or $x = 4$
When $x = -\frac{2}{3}$, $y = \frac{1}{4}\left(-\frac{2}{3}\right)^4 - \frac{5}{3}\left(-\frac{2}{3}\right)^3 + 2\left(-\frac{2}{3}\right)^2 - 20\left(-\frac{2}{3}\right)$
$= \frac{1196}{81}$
When $x = 4$, $y = \frac{1}{4}(4)^4 - \frac{5}{3}(4)^3 + 2(4)^2 - 20(4) = -\frac{272}{3}$
So the coordinates of the points are:
$\left(-\frac{2}{3}, \frac{1196}{81}\right)$ and $(4, -\frac{272}{3})$.

Exercise 10.3.2 — Stationary points

Q1 a) The graph has 2 stationary points — a minimum and a point of inflection.

b) The graph has 3 stationary points — a maximum, a minimum and a point of inflection.

Q2 a) $\frac{dy}{dx} = 2x + 3$. When $\frac{dy}{dx} = 0$, $2x + 3 = 0 \Rightarrow x = -\frac{3}{2}$

b) $y = (3 - x)(4 + 2x) = 12 + 2x - 2x^2$
$\frac{dy}{dx} = 2 - 4x$. When $\frac{dy}{dx} = 0$, $2 - 4x = 0 \Rightarrow x = \frac{1}{2}$

c) $\frac{dy}{dx} = 3x^2 + 8x - 3$. When $\frac{dy}{dx} = 0$, $3x^2 + 8x - 3 = 0$
$\Rightarrow (3x - 1)(x + 3) \Rightarrow x = \frac{1}{3}$ and $x = -3$

d) $\frac{dy}{dx} = 4x^3 - 36x^2$. When $\frac{dy}{dx} = 0$, $4x^3 - 36x^2 = 0$
$\Rightarrow 4x^2(x - 9) \Rightarrow x = 0$ and $x = 9$

Q3 a) $\frac{dy}{dx} = 4x - 5$. When $\frac{dy}{dx} = 0$, $4x - 5 = 0 \Rightarrow x = \frac{5}{4}$
When $x = \frac{5}{4}$, $y = 2\left(\frac{5}{4}\right)^2 - 5\left(\frac{5}{4}\right) + 2 = -\frac{9}{8}$
So the coordinates are $\left(\frac{5}{4}, -\frac{9}{8}\right)$.

b) $\frac{dy}{dx} = -2x + 3$. When $\frac{dy}{dx} = 0$, $-2x + 3 = 0 \Rightarrow x = \frac{3}{2}$
When $x = \frac{3}{2}$, $y = -\left(\frac{3}{2}\right)2 + 3\left(\frac{3}{2}\right) - 4 = -\frac{7}{4}$.
So the coordinates are $(\frac{3}{2}, -\frac{7}{4})$.

c) $\frac{dy}{dx} = -6 - 6x$.
When $\frac{dy}{dx} = 0$, $-6 - 6x = 0 \Rightarrow x = -1$
When $x = -1$, $y = 7 - 6(-1) - 3(-1)^2 = 10$.
So the coordinates are (–1, 10).

d) $y = (x - 1)(2x + 3) = 2x^2 + x - 3$
$\frac{dy}{dx} = 4x + 1$. When $\frac{dy}{dx} = 0$, $4x + 1 = 0 \Rightarrow x = -\frac{1}{4}$
When $x = -\frac{1}{4}$, $y = (-\frac{1}{4} - 1)(2(-\frac{1}{4}) + 3) = -\frac{25}{8}$.
So the coordinates are $(-\frac{1}{4}, -\frac{25}{8})$.

Q4 a) $\frac{dy}{dx} = 3x^2 - 3$.
When $\frac{dy}{dx} = 0$, $3x^2 - 3 = 0 \Rightarrow x = \pm 1$.
When $x = 1$, $y = 1^3 - 3(1) + 2 = 0$.
When $x = -1$, $y = (-1)^3 - 3(-1) + 2 = 4$.
So the coordinates are (1, 0) and (–1, 4).

b) $\frac{dy}{dx} = 12x^2$. When $\frac{dy}{dx} = 0$, $12x^2 = 0 \Rightarrow x = 0$
When $x = 0$, $y = 4(0)^3 + 5 = 5$.
So the coordinates are (0, 5).

c) $\frac{dy}{dx} = 9x^2 + 12x$.
When $\frac{dy}{dx} = 0$, $9x^2 + 12x = 0 \Rightarrow 3x(3x + 4) = 0$
$\Rightarrow x = 0$ and $x = -\frac{4}{3}$
When $x = 0$, $y = 3(0)^3 + 6(0)^2 = 0$.
When $x = -\frac{4}{3}$, $= 3\left(-\frac{4}{3}\right)^3 + 6\left(-\frac{4}{3}\right)^2 = \frac{32}{9}$.
So the coordinates are (0, 0) and $\left(-\frac{4}{3}, \frac{32}{9}\right)$.

d) $\frac{dy}{dx} = 12x^2 + 24x$.
When $\frac{dy}{dx} = 0$, $12x^2 + 24x = 0 \Rightarrow 12x(x + 2)$
$\Rightarrow x = 0$ and $x = -2$
When $x = 0$, $y = 4(0)^3 + 12(0)^2 + 8 = 8$.
When $x = -2$, $y = 4(-2)^3 + 12(-2)^2 + 8 = 24$.
So the coordinates are (0, 8) and (–2, 24).

Q5 $\frac{dy}{dx} = 3x^2 - 12x - 63$.
When $\frac{dy}{dx} = 0$, $3x^2 - 12x - 63 = 0 \Rightarrow x^2 - 4x - 21$
$\Rightarrow (x + 3)(x - 7) = 0$
$\Rightarrow x = -3$ and $x = 7$
When $x = -3$, $y = (-3)^3 - 6(-3)^2 - 63(-3) + 21 = 129$
When $x = 7$, $y = 7^3 - 6(7)^2 - 63(7) + 21 = -371$
So the coordinates are (–3, 129) and (7, –371).

Q6 $f'(x) = 5x^4 + 3$. When $f'(x) = 0$, $5x^4 + 3 = 0 \Rightarrow x^4 = -\frac{3}{5}$.
Finding a solution would involve finding the fourth root of a negative number. But $x^4 = (x^2)^2$, so x^4 is always positive and so there are no stationary points.

Q7 **a)** $\frac{dy}{dx} = 3x^2 - 14x - 5$

b) When $\frac{dy}{dx} = 0$, $3x^2 - 14x - 5 = 0$
$\Rightarrow (3x + 1)(x - 5) = 0$, so $x = -\frac{1}{3}$ and $x = 5$.
When $x = -\frac{1}{3}$, $y = \left(-\frac{1}{3}\right)^3 - 7\left(-\frac{1}{3}\right)^2 - 5\left(-\frac{1}{3}\right) + 2$
$= \frac{77}{27}$.
When $x = 5$, $y = 5^3 - 7(5)^2 - 5(5) + 2 = -73$.
So the coordinates are $(-\frac{1}{3}, \frac{77}{27})$ and $(5, -73)$.

Q8 For stationary points to occur, $f'(x)$ must equal zero, so
$f'(x) = 3x^2 + k = 0 \Rightarrow -\frac{k}{3} = x^2$.
For this equation to have a solution, k can't be positive (or it would be taking the square root of a negative number), so $k \leq 0$.
Therefore, if the graph has no stationary points, $k > 0$.

Exercise 10.3.3 — Maximum and minimum points

Q1 **a)** negative **b)** positive **c)** negative
d) negative **e)** positive

Q2 **a)** $\frac{dy}{dx} = 2x + 14$ $\frac{d^2y}{dx^2} = 2$
At $(-7, -52)$, $\frac{d^2y}{dx^2} = 2 > 0$, so $(-7, -52)$ is a minimum.
The curve is a quadratic so has only one stationary point and you know it'll be a minimum because it's a positive u-shaped quadratic.

b) $\frac{dy}{dx} = 3x^2 - 12$ $\frac{d^2y}{dx^2} = 6x$
At $(2, -12)$, $\frac{d^2y}{dx^2} = 6 \times 2 = 12 > 0$, so $(2, -12)$ is a minimum.

c) $\frac{dy}{dx} = 6x^2 - 2x$ $\frac{d^2y}{dx^2} = 12x - 2$
At $\left(\frac{1}{3}, -\frac{1}{27}\right)$, $\frac{d^2y}{dx^2} = 12\left(\frac{1}{3}\right) - 2 = 2 > 0$,
so $\left(\frac{1}{3}, -\frac{1}{27}\right)$ is a minimum.

d) $\frac{dy}{dx} = 8x^3 - 48x^2$ $\frac{d^2y}{dx^2} = 24x^2 - 96x$
At $(6, 36)$, $\frac{d^2y}{dx^2} = 24 \times 6^2 - 96 \times 6 = 288 > 0$,
so $(6, 36)$ is a minimum.

e) $\frac{dy}{dx} = 20x^4 + 60x^3$ $\frac{d^2y}{dx^2} = 80x^3 + 180x^2$
At $(-3, -7)$, $\frac{d^2y}{dx^2} = 80 \times (-3)^3 + 180 \times (-3)^2$
$= -540 < 0$, so $(-3, -7)$ is a maximum.

f) $\frac{dy}{dx} = 5x^4 - 20x^3 + 10x - 40$ $\frac{d^2y}{dx^2} = 20x^3 - 60x^2 + 10$
At $(4, 64)$, $\frac{d^2y}{dx^2} = 20 \times 4^3 - 60 \times 4^2 + 10 = 330 > 0$, so
$(4, 64)$ is a minimum.

Q3 **a)** $(1, 3)$
All the clues are in the question — the derivative when $x = 1$ is zero so you know it's a stationary point, and the y-value when $x = 1$ is 3.

b) The second derivative at $x = 1$ is positive, so it's a minimum.

Q4 **a)** $\frac{dy}{dx} = -2x$. When $\frac{dy}{dx} = 0$, $x = 0$. When $x = 0$,
$y = 5 - 0 = 5$. So the coordinates are $(0, 5)$.
$\frac{d^2y}{dx^2} = -2$, so it's a maximum turning point.

b) $\frac{dy}{dx} = x + 21$.
When $\frac{dy}{dx} = 0$, $x + 21 = 0 \Rightarrow x = -21$
When $x = -21$, $y = \frac{441}{2} - 441 + 12 = -\frac{417}{2}$
So the coordinates are $(-21, -\frac{417}{2})$.
$\frac{d^2y}{dx^2} = 1$, so it's a minimum turning point.

c) $\frac{dy}{dx} = 6x^2 - 6$. When $\frac{dy}{dx} = 0$, $6x^2 = 6 \Rightarrow x = \pm 1$
When $x = 1$, $y = 2 - 6 + 2 = -2$.
When $x = -1$, $y = -2 + 6 + 2 = 6$. So the coordinates are $(1, -2)$ and $(-1, 6)$. $\frac{d^2y}{dx^2} = 12x$.
At $(1, -2)$, $\frac{d^2y}{dx^2} = 12$, so it's a minimum.
At $(-1, 6)$, $\frac{d^2y}{dx^2} = -12$ so it's a maximum.

d) $\frac{dy}{dx} = 3x^2 - 6x - 24$. When $\frac{dy}{dx} = 0$, $x^2 - 2x - 8 = 0$
$\Rightarrow (x - 4)(x + 2) = 0 \Rightarrow x = 4$ and -2.
When $x = 4$, $y = 64 - 48 - 96 + 15 = -65$.
When $x = -2$, $y = -8 - 12 + 48 + 15 = 43$.
So the coordinates are $(4, -65)$ and $(-2, 43)$.
$\frac{d^2y}{dx^2} = 6x - 6$. At $(4, -65)$, $\frac{d^2y}{dx^2} = 24 - 6 = 18$,
so it's a minimum.
At $(-2, 43)$, $\frac{d^2y}{dx^2} = -12 - 6 = -18$, so it's a maximum.

e) $\frac{dy}{dx} = \frac{1}{3}x^3 + \frac{2}{3}x^2$. When $\frac{dy}{dx} = 0$, $\frac{1}{3}x^3 + \frac{2}{3}x^2 = 0$
$\Rightarrow \frac{1}{3}x^2(x + 2)$, so $x = 0$ and $x = -2$.
When $x = 0$, $y = 0 + 0 = 0$.
When $x = -2$, $y = \frac{4}{3} - \frac{16}{9} = -\frac{4}{9}$.
So the stationary points are $(0, 0)$ and $(-2, -\frac{4}{9})$.
$\frac{d^2y}{dx^2} = x^2 + \frac{4}{3}x$. At $(0, 0)$, $\frac{d^2y}{dx^2} = 0 + 0 = 0$, so you can't tell whether it's a maximum or miminum.
At $(-2, -\frac{4}{9})$, $\frac{d^2y}{dx^2} = 4 - \frac{8}{3} = \frac{4}{3}$, so it's a minimum.

f) $\frac{dy}{dx} = 4x^3 + 12x^2 + 8x$.
When $\frac{dy}{dx} = 0$, $x^3 + 3x^2 + 2x = 0$
$\Rightarrow x(x + 2)(x + 1) = 0$, so $x = 0$, -1 and -2.
When $x = 0$, $y = 0 + 0 + 0 - 10 = -10$.
When $x = -1$, $y = 1 - 4 + 4 - 10 = -9$.
When $x = -2$, $y = 16 - 32 + 16 - 10 = -10$.
So the stationary points are $(0, -10)$, $(-1, -9)$ and $(-2, -10)$.
$\frac{d^2y}{dx^2} = 12x^2 + 24x + 8$.
At $(0, -10)$, $\frac{d^2y}{dx^2} = 0 + 0 + 8 = 8$, so it's a minimum.
At $(-1, -9)$, $\frac{d^2y}{dx^2} = 12 - 24 + 8 = -4$, so it's a maximum.
At $(-2, -10)$, $\frac{d^2y}{dx^2} = 48 - 48 + 8 = 8$, so it's a minimum.

Q5 **a)** $f'(x) = 24x^2 + 32x + 8$. When $f'(x) = 0$, $3x^2 + 4x + 1 = 0$
$\Rightarrow (3x + 1)(x + 1) = 0$, so $x = -1$ and $-\frac{1}{3}$.
When $x = -1$, $f(x) = -8 + 16 - 8 + 1 = 1$.
When $x = -\frac{1}{3}$, $f(x) = -\frac{8}{27} + \frac{16}{9} - \frac{8}{3} + 1 = -\frac{5}{27}$.
So the coordinates are $(-1, 1)$ and $(-\frac{1}{3}, -\frac{5}{27})$.
$f''(x) = 48x + 32$. At $(-1, 1)$ $f'(x) = -48 + 32 = -16$,
so it's a maximum.
At $(-\frac{1}{3}, -\frac{5}{27})$, $f''(x) = -\frac{48}{3} + 32 = 16$, so it's a minimum.

b) $f'(x) = x^2 - 12x - 45$. When $f'(x) = 0$, $x^2 - 12x - 45 = 0$
$\Rightarrow (x + 3)(x - 15) = 0 \Rightarrow x = -3$ and 15.
When $x = -3$, $y = -9 - 54 + 135 = 72$.
When $x = 15$, $y = 1125 - 1350 - 675 = -900$.
So the coordinates are $(-3, 72)$ and $(15, -900)$.
$f''(x) = 2x - 12$.
At $(-3, 72)$, $f''(x) = -6 - 12 = -18$, so it's a maximum.
At $(15, -900)$, $f''(x) = 30 - 12 = 18$, so it's a minimum.

c) $f'(x) = x^3 + 6x^2 - 16x$. When $f'(x) = 0$, $x^3 + 6x^2 - 16x = 0$
$\Rightarrow x(x^2 + 6x - 16) = 0 \Rightarrow x(x + 8)(x - 2)$, so $x = -8$, 0 and 2.
When $x = -8$, $f(x) = 1024 - 1024 - 512 + 1 = -511$.
When $x = 0$, $f(x) = 0 + 0 + 0 + 1 = 1$.
When $x = 2$, $f(x) = 4 + 16 - 32 + 1 = -11$.
So the coordinates are $(-8, -511)$, $(0, 1)$ and $(2, -11)$.
$f''(x) = 3x^2 + 12x - 16$.
At $(-8, -511)$, $f''(x) = 192 - 96 - 16 = 80$, so it's a minimum.
At $(0, 1)$, $f''(x) = -16$, so it's a maximum.
At $(2, -11)$, $f''(x) = 12 + 24 - 16 = 20$, so it's a minimum.

d) $f(x) = \frac{27}{x^3} + x = 27x^{-3} + x \Rightarrow f'(x) = -81x^{-4} + 1$.
When $f'(x) = 0$, $x^4 = 81 \Rightarrow x = \pm 3$.
When $x = 3$, $f(x) = \frac{27}{27} + 3 = 4$.
When $x = -3$, $f(x) = -\frac{27}{27} - 3 = -4$.
So the coordinates are $(3, 4)$ and $(-3, -4)$.
$f''(x) = 324x^{-5}$. At $(3, 4)$ $f''(x) = \frac{4}{3}$, so it's a minimum.
At $(-3, -4)$ $f''(x) = -\frac{4}{3}$, so it's a maximum.

Q6 **a)** $f'(x) = 3x^2 - 6x$. $f''(x) = 6x - 6$.

b) When $f'(x) = 0$, $3x^2 - 6x = 0 \Rightarrow x(x - 2) = 0$,
so $x = 0$ and $x = 2$.
When $x = 0$, $f(x) = 0 - 0 + 4 = 4$.
When $x = 2$, $f(x) = 8 - 12 + 4 = 0$.
So the coordinates are $(0, 4)$ and $(2, 0)$.
At $(0, 4)$ $f''(x) = 0 - 6 = -6$, so it's a maximum.
At $(2, 0)$ $f''(x) = 12 - 6 = 6$, so it's a minimum.

Q7 **a)** $y = x^2 + \frac{2000}{x} = x^2 + 2000x^{-1} \Rightarrow \frac{dy}{dx} = 2x - \frac{2000}{x^2}$
When $\frac{dy}{dx} = 0$, $2x = \frac{2000}{x^2} \Rightarrow x^3 = 1000 \Rightarrow x = 10$

b) $\frac{d^2y}{dx^2} = 2 + \frac{4000}{x^3}$. When $x = 10$,
$\frac{d^2y}{dx^2} = 2 + 4 = 6$, so it's a minimum.

Q8 c is an integer so differentiates to 0.
$f(x) = 2x^3 + 4x^2 + c \Rightarrow f'(x) = 6x^2 + 8x \Rightarrow f''(x) = 12x + 8$
At stationary points $f'(x) = 0$, $6x^2 + 8x = 0 \Rightarrow 2x(3x + 4) = 0$
$\Rightarrow x = 0$ and $x = -\frac{4}{3}$
When $x = 0$, $f''(x) = 12(0) + 8 = 8 > 0$, so the point is a minimum.
When $x = -\frac{4}{3}$, $f''(x) = 12\left(-\frac{4}{3}\right) + 8 = -8 < 0$,
so the point is a maximum.
When $x = -\frac{4}{3}$, $y = -\frac{260}{27}$, so $-\frac{260}{27} = 2\left(-\frac{4}{3}\right)^3 + 4\left(-\frac{4}{3}\right)^2 + c$
$\Rightarrow c = -\frac{260}{27} - 2\left(-\frac{4}{3}\right)^3 - 4\left(-\frac{4}{3}\right)^2 = -12$

Q9 $f(x) = x^3 + ax^2 + bx + c \Rightarrow f'(x) = 3x^2 + 2ax + b$.
$\Rightarrow f''(x) = 6x + 2a$. At the point $(3, 10)$:
$10 = 3^3 + a(3^2) + b(3) + c \Rightarrow 10 = 27 + 9a + 3b + c$
As $(3, 10)$ is a stationary point, $0 = 3(3^2) + 2a(3) + b$
$\Rightarrow 0 = 27 + 6a + b$. We know that $f''(3) = 0$,
so $0 = 6(3) + 2a \Rightarrow 0 = 18 + 2a \Rightarrow a = -9$.
Then $0 = 27 + 6a + b = 27 + 6(-9) + b \Rightarrow b = 27$
And $10 = 27 + 9a + 3b + c = 27 + 9(-9) + 3(27) + c$
$\Rightarrow c = -17$. So $f(x) = x^3 - 9x^2 + 27x - 17$.

Q10 **a)** $\frac{dy}{dx} = 4x^3 + 3kx^2 + 2x$.
Stationary points occur when $\frac{dy}{dx} = 0$,
so $4x^3 + 3kx^2 + 2x = 0 \Rightarrow x(4x^2 + 3kx + 2) = 0$
so $x = 0$ or $4x^2 + 3kx + 2 = 0$.
As you know the only stationary point occurs at $x = 0$, the part in brackets can't have any solutions. This gives you information about the discriminant of the quadratic equation:
$b^2 - 4ac < 0 \Rightarrow 9k^2 < 32 \Rightarrow k^2 < \frac{32}{9}$.

b) When $x = 0$, $y = 0 + 0 + 0 + 17 = 17$,
so the coordinates are $(0, 17)$.
$\frac{d^2y}{dx^2} = 12x^2 + 6kx + 2$.
When $x = 0$, $\frac{d^2y}{dx^2} = 2$, so it's a minimum.

Exercise 10.3.4 — Increasing and decreasing functions

Q1 **a)** $\frac{dy}{dx} = 2x + 7$. If the function is increasing, $\frac{dy}{dx} > 0$
$\Rightarrow 2x > -7 \Rightarrow x > -\frac{7}{2}$.

b) $\frac{dy}{dx} = 10x + 3$. If the function is increasing,
$\frac{dy}{dx} > 0 \Rightarrow 10x > -3 \Rightarrow x > -\frac{3}{10}$.

c) $\frac{dy}{dx} = \frac{1}{2}x - 6$. If the function is increasing, $\frac{dy}{dx} > 0$
$\Rightarrow \frac{1}{2}x - 6 > 0 \Rightarrow \frac{1}{2}x > 6 \Rightarrow x > 12$.

d) $\frac{dy}{dx} = -18x$. If the function is increasing, $\frac{dy}{dx} > 0$
$\Rightarrow -18x > 0 \Rightarrow x < 0$.
Be careful with the direction of the inequality sign if you're dividing by a negative number.

e) $\frac{dy}{dx} = -6x - 4$. If the function is increasing, $\frac{dy}{dx} > 0$
$\Rightarrow -6x - 4 > 0 \Rightarrow -6x > 4 \Rightarrow x < -\frac{2}{3}$.

f) $y = -\frac{3}{5}x^2 + \frac{8}{5}x - 2 \Rightarrow \frac{dy}{dx} = -\frac{6}{5}x + \frac{8}{5}$.
If the function is increasing, $\frac{dy}{dx} > 0$
$\Rightarrow -\frac{6}{5}x + \frac{8}{5} > 0 \Rightarrow -\frac{6}{5}x > -\frac{8}{5} \Rightarrow x < \frac{4}{3}$

Q2 **a)** $f'(x) = -3 - 4x$. If the function is decreasing,
$f'(x) < 0 \Rightarrow -4x < 3 \Rightarrow x > -\frac{3}{4}$.

b) $f(x) = (x + 3)(x - 2) = x^2 + x - 6 \Rightarrow f'(x) = 2x + 1$.
If the function is decreasing, $f'(x) < 0$
$\Rightarrow 2x + 1 < 0 \Rightarrow 2x < -1 \Rightarrow x < -\frac{1}{2}$.

c) $f(x) = (2x - 1)(x - 7) = 2x^2 - 15x + 7$
$\Rightarrow f'(x) = 4x - 15$. If the function is decreasing,
$f'(x) < 0 \Rightarrow 4x - 15 < 0 \Rightarrow x < \frac{15}{4}$.

d) $f(x) = (6 - 3x)(6 + 3x) = 36 - 9x^2 \Rightarrow f'(x) = -18x$.
If the function is decreasing, $f'(x) < 0$
$\Rightarrow -18x < 0 \Rightarrow x > 0$.

e) $f(x) = (1 - 2x)(7 - 3x) = 7 - 17x + 6x^2$
$f'(x) = -17 + 12x$. If the function is decreasing,
$f'(x) < 0 \Rightarrow 12x < 17 \Rightarrow x < \frac{17}{12}$.

f) $f(x) = (5x - 2)(2 - 5x) = -25x^2 + 20x - 4$
$\Rightarrow f'(x) = -50x + 20$. If the function is decreasing, $f'(x) < 0$
$\Rightarrow -50x + 20 < 0 \Rightarrow -50x < -20 \Rightarrow x > \frac{2}{5}$.

Q3 **a)** $\frac{dy}{dx} = 3x^2 - 12x - 15$. If the function is increasing,
$\frac{dy}{dx} > 0 \Rightarrow 3x^2 - 12x - 15 > 0$
$\Rightarrow x^2 - 4x - 5 > 0 \Rightarrow (x - 5)(x + 1) > 0$
For this expression to be > 0, both brackets must be positive or both brackets must be negative.
So either $x > 5$ and $x > -1$ or $x < 5$ and $x < -1$.
So the function is increasing when $x < -1$ and when $x > 5$.
Remember that you can use a different method, e.g. sketching the quadratic, to solve the inequality if you prefer.

b) $\frac{dy}{dx} = 3x^2 + 12x + 12$.
If the function is increasing, $\frac{dy}{dx} > 0$
$\Rightarrow 3x^2 + 12x + 12 > 0 \Rightarrow x^2 + 4x + 4 > 0$
$\Rightarrow (x + 2)(x + 2) > 0 \Rightarrow (x + 2)^2 > 0$
So x can be any real value except $x = -2$, which means that the function is increasing for all values of x except $x = -2$.

c) $y = x(x^2 - 4x - 16) = x^3 - 4x^2 - 16x$
$\frac{dy}{dx} = 3x^2 - 8x - 16$.
If the function is increasing, $\frac{dy}{dx} > 0$
$\Rightarrow 3x^2 - 8x - 16 > 0 \Rightarrow (3x + 4)(x - 4) > 0$
For this expression to be > 0, both brackets must be positive or both brackets must be negative.
So either $x > -\frac{4}{3}$ and $x > 4$ or $x < -\frac{4}{3}$ and $x < 4$.
So the function is increasing when $x < -\frac{4}{3}$ and when $x > 4$.

d) $y = \frac{x^3 + 3x^2 - 9x - 1}{9} = \frac{x^3}{9} - \frac{x^2}{3} - x - \frac{1}{9}$
$\frac{dy}{dx} = \frac{x^2}{3} - \frac{2x}{3} - 1$
If the function is increasing, $\frac{dy}{dx} > 0$
$\Rightarrow \frac{x^2}{3} - \frac{2x}{3} - 1 > 0 \Rightarrow x^2 - 2x - 3 > 0 \Rightarrow (x + 1)(x - 3) > 0$
For this expression to be > 0, both brackets must be positive or both brackets must be negative.
So either $x > -1$ and $x > 3$ or $x < -1$ and $x < 3$.
So the function is increasing when $x < -1$ and $x > 3$.

Q4 a) $f'(x) = 3x^2 - 6x - 9$. If the function is decreasing, $f'(x) < 0 \Rightarrow 3x^2 - 6x - 9 < 0 \Rightarrow x^2 - 2x - 3 < 0$
$\Rightarrow (x - 3)(x + 1) < 0$. For the expression to be < 0, one bracket must be positive and one negative.
So either $x < 3$ and $x > -1$ or $x > 3$ and $x < -1$.
The second situation is impossible, so $-1 < x < 3$.

b) $f'(x) = 3x^2 - 8x + 4$. If the function is decreasing, $f'(x) < 0 \Rightarrow 3x^2 - 8x + 4 < 0 \Rightarrow (3x - 2)(x - 2) < 0$.
For the expression to be < 0, either $x < \frac{2}{3}$ and $x > 2$ or $x > \frac{2}{3}$ and $x < 2$. The first situation is impossible, so $\frac{2}{3} < x < 2$.

c) $f'(x) = 9x^2 + 3x - 72$. If the function is decreasing, $f'(x) < 0$
$\Rightarrow 9x^2 + 3x - 72 < 0 \Rightarrow 3x^2 + x - 24 < 0$
$\Rightarrow (x + 3)(3x - 8) < 0$.
For the expression to be < 0, one bracket must be positive and one negative. So either $x > -3$ and $x < \frac{8}{3}$ or $x < -3$ and $x > \frac{8}{3}$. The second situation is impossible, so $-3 < x < \frac{8}{3}$.

d) $f(x) = \frac{2x^3 - x^2 - 8x + 3}{4} = \frac{1}{2}x^3 - \frac{1}{4}x^2 - 2x + \frac{3}{4}$
$f'(x) = \frac{3}{2}x^2 - \frac{1}{2}x - 2$. If the function is decreasing
$f'(x) < 0 \Rightarrow \frac{3}{2}x^2 - \frac{1}{2}x - 2 < 0 \Rightarrow 3x^2 - x - 4 < 0$
$\Rightarrow (3x - 4)(x + 1) < 0$. For the expression to be < 0, one bracket must be positive and one negative.
So either $x < \frac{4}{3}$ and $x > -1$ or $x > \frac{4}{3}$ and $x < -1$.
The second situation is impossible, so $-1 < x < \frac{4}{3}$.

Q5 $f'(x) = 3x^2 + 1$. x^2 can't be negative ($x^2 \geq 0$), so $f'(x)$ must always be positive and so $f(x)$ is an increasing function for all real values of x.

Q6 $f'(x) = -3 - 3x^2$. x^2 can't be negative ($x^2 \geq 0$), so $f'(x)$ is always ≤ -3 (so negative), so $f(x)$ is a decreasing function.

Q7 a) $\frac{dy}{dx} = 8x^3 + 1$. If the function is decreasing,
$\frac{dy}{dx} < 0 \Rightarrow 8x^3 + 1 < 0 \Rightarrow x^3 < -\frac{1}{8} \Rightarrow x < -\frac{1}{2}$

b) $\frac{dy}{dx} = 4x^3 - 36x$. If the function is decreasing,
$\frac{dy}{dx} < 0 \Rightarrow 4x^3 - 36x < 0 \Rightarrow 4x(x + 3)(x - 3) < 0$
For this to be true, there are 4 possibilities — all are less than zero, or one is less than zero and the other two are not:
Either $x < 0$ and $x < -3$ and $x < 3$, so $x < -3$
Remember, if x must be smaller than 0, −3 and 3, you can dismiss the two higher numbers and simplify it to x being smaller than −3.
Or $x < 0$ and $x > -3$ and $x > 3$ (impossible)
Or $x > 0$ and $x < -3$ and $x > 3$ (impossible)
Or $x > 0$ and $x > -3$ and $x < 3$
This gives ranges $x < -3$ and $0 < x < 3$.
You could also consider the graph of 4x(x + 3)(x − 3) and see where this takes negative values.

c) $\frac{dy}{dx} = 4x^3 - 6x^2 - 10x$. If the function is decreasing, $\frac{dy}{dx} < 0 \Rightarrow 4x^3 - 6x^2 - 10x < 0$
$\Rightarrow x(2x - 5)(x + 1) < 0$. For this to be true, there are 4 possibilities — all are less than zero, or one is less than zero and the other two are not:
Either $x < 0$ and $x < \frac{5}{2}$ and $x < -1$, so $x < -1$
Or $x < 0$ and $x > \frac{5}{2}$ and $x > -1$ (impossible)
Or $x > 0$ and $x < \frac{5}{2}$ and $x > -1$, so $0 < x < \frac{5}{2}$
Or $x > 0$ and $x > \frac{5}{2}$ and $x < -1$ (impossible)
This gives the ranges $x < -1$ and $0 < x < \frac{5}{2}$.

d) $\frac{dy}{dx} = 8x^3 + 4x^2 - 4x$. If the function is decreasing,
$\frac{dy}{dx} < 0 \Rightarrow 8x^3 + 4x^2 - 4x < 0 \Rightarrow 2x^3 + x^2 - x < 0$
$\Rightarrow x(2x - 1)(x + 1) < 0$
For this to be true, there are 4 possibilities — all are less than zero, or one is less than zero and the other two are not:
Either $x < 0$ and $x < \frac{1}{2}$ and $x < -1$, so $x < -1$
Or $x < 0$ and $x > \frac{1}{2}$ and $x > -1$ (impossible)
Or $x > 0$ and $x < \frac{1}{2}$ and $x > -1$
Or $x > 0$ and $x > \frac{1}{2}$ and $x < -1$ (impossible)
This gives the ranges $x < -1$ and $0 < x < \frac{1}{2}$.

Q8 a) $y = x^2 + \sqrt{x} = x^2 + x^{\frac{1}{2}} \Rightarrow \frac{dy}{dx} = 2x + \frac{1}{2\sqrt{x}}$
$\frac{dy}{dx} > 0$ for all $x > 0$, so the function is increasing for all $x > 0$.

b) $\frac{dy}{dx} = -x^{-2} + 16 = -\frac{1}{x^2} + 16$.
The function is increasing when $\frac{dy}{dx} > 0$
$\Rightarrow -\frac{1}{x^2} + 16 > 0 \Rightarrow x^2 > \frac{1}{16} \Rightarrow x > \frac{1}{4}$ and $x < -\frac{1}{4}$

c) $y = 4x^2 + \frac{1}{x} = 4x^2 + x^{-1} \Rightarrow \frac{dy}{dx} = 8x - \frac{1}{x^2}$
The function is increasing when $\frac{dy}{dx} > 0$
$\Rightarrow 8x - \frac{1}{x^2} > 0 \Rightarrow x^3 > \frac{1}{8} \Rightarrow x > \frac{1}{2}$

d) $y = -\frac{3x^{-2} + 18x}{2} = -\frac{3}{2}x^{-2} - 9x \Rightarrow \frac{dy}{dx} = \frac{3}{x^3} - 9$
The function is increasing when $\frac{dy}{dx} > 0$
$\Rightarrow \frac{3}{x^3} - 9 > 0 \Rightarrow \frac{1}{x^3} > 3$
When multiplying an inequality by a negative number, you need to flip the inequality sign — but here you don't know if x^3 is positive or negative. So consider the two cases:
If $x > 0$, then $x^3 > 0$, so $\frac{1}{x^3} > 3 \Rightarrow \frac{1}{3} > x^3 \Rightarrow x < \frac{1}{\sqrt[3]{3}}$
If $x < 0$, then $x^3 < 0$, so $\frac{1}{x^3} > 3 \Rightarrow \frac{1}{3} < x^3$
— but this is impossible since $x^3 < 0$ in this case.
So the function is only increasing when x is positive, i.e. for x in the interval $0 < x < \frac{1}{\sqrt[3]{3}}$.

Q9 If the function is decreasing, $\frac{dy}{dx} < 0$ for all x.
$\frac{dy}{dx} = -3 - 5ax^4 \Rightarrow -3 - 5ax^4 < 0 \Rightarrow ax^4 > -\frac{3}{5}$.
The right-hand side is negative, so as $x^4 \geq 0$, a must also be positive to make the LHS > RHS for all x. So $a > 0$.

Q10 If the function is increasing, $\frac{dy}{dx}$ will always be greater than 0.
$\frac{dy}{dx} = kx^{k-1} + 1 \Rightarrow kx^{k-1} + 1 > 0$.
When $k = 1$, $x^0 + 1 > 0$ — true for all x
When $k = 2$, $2x^1 + 1 > 0$ — not true for all x
When $k = 3$, $3x^2 + 1 > 0$ — true for all x
When $k = 4$, $4x^3 + 1 > 0$ — not true for all x, etc.
So k must be an odd number greater than zero.

Exercise 10.3.5 — Curve sketching

Q1 **a)** When $x = 0$, $y = 0^3 - 2(0)^2 = 0$, so the curve crosses the axes at (0, 0). When $y = 0$, $x^3 - 2x^2 = 0$
$\Rightarrow x^2(x - 2) = 0 \Rightarrow x = 0$ and $x = 2$.
So the curve also crosses the axes at (2, 0).
You already knew it crossed the x-axis at x = 0, so you can ignore that one.

b) $\frac{dy}{dx} = 3x^2 - 4x$. When $\frac{dy}{dx} = 0$, $3x^2 - 4x = 0$
$\Rightarrow x(3x - 4) = 0 \Rightarrow x = 0$ and $x = \frac{4}{3}$.
When $x = \frac{4}{3}$, $y = (\frac{4}{3})^3 - 2(\frac{4}{3})^2 = -\frac{32}{27}$.
So the coordinates are (0, 0) and $(\frac{4}{3}, -\frac{32}{27})$.

c) $\frac{d^2y}{dx^2} = 6x - 4$.
At $x = 0$, $\frac{d^2y}{dx^2} = -4$, so it's a maximum.
At $x = \frac{4}{3}$, $\frac{d^2y}{dx^2} = 4$, so it's a minimum.

d) A positive cubic goes from bottom left to top right:

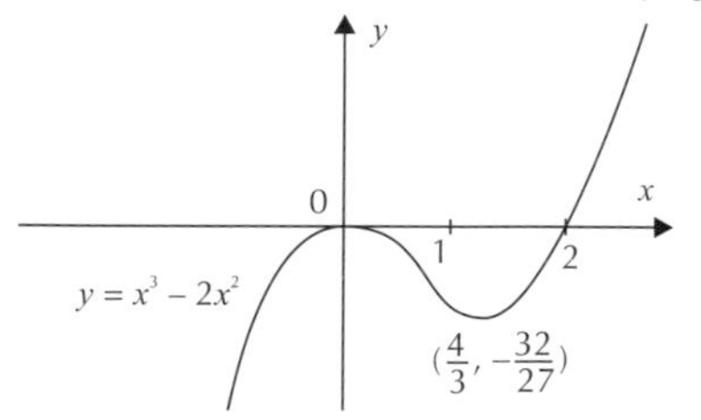

Q2 **a)** $x^3 + x^2 = 0 \Rightarrow x^2(x + 1) = 0 \Rightarrow x = 0$ or $x = -1$.

b) $f'(x) = 3x^2 + 2x$. When $f'(x) = 0$, $3x^2 + 2x = 0$
$\Rightarrow x(3x + 2) = 0 \Rightarrow x = 0$ and $x = -\frac{2}{3}$.
When $x = 0$, $y = 0$. When $x = -\frac{2}{3}$, $y = \frac{4}{27}$,
so the stationary points are at (0, 0) and $\left(-\frac{2}{3}, \frac{4}{27}\right)$.
$f''(x) = 6x + 2$. At (0, 0), $f''(x) = 2$, so it's a minimum.
At $(-\frac{2}{3}, \frac{4}{27})$, $f''(x) = -2$, so it's a maximum.

c)

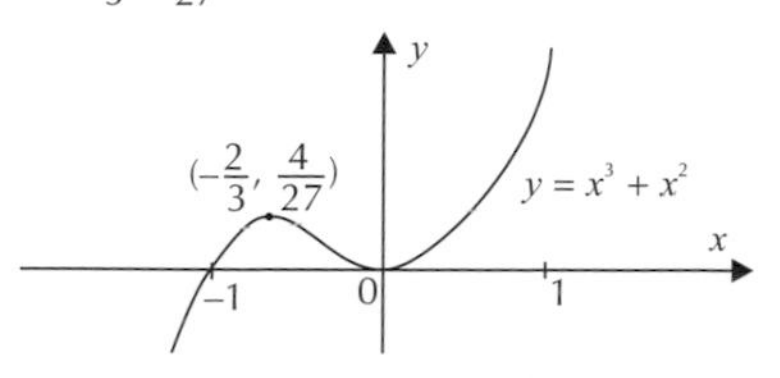

It's a positive cubic, so it goes from bottom left to top right.

Q3 **a)** $f'(x) = 4x^3 - 3x^2$, $f''(x) = 12x^2 - 6x$.

b) f(x) is increasing for $f'(x) > 0 \Rightarrow 4x^3 - 3x^2 > 0$
$\Rightarrow x^2(4x - 3) > 0$, so either:
$x^2 > 0$ and $x > \frac{3}{4}$ ($\Rightarrow x > \frac{3}{4}$), or
$x^2 < 0$ and $x < \frac{3}{4}$ (x^2 can't be less than 0, so this situation is impossible)
So it's increasing when $x > \frac{3}{4}$.
f(x) is decreasing for $f'(x) < 0 \Rightarrow 4x^3 - 3x^2 < 0$
$\Rightarrow x^2(4x - 3) < 0$, so either:
$x^2 < 0$ and $x > \frac{3}{4}$ (x^2 can't be less than 0, so this situation is impossible), or
$x^2 > 0$ and $x < \frac{3}{4}$
So it's decreasing when $x < \frac{3}{4}$, $x \neq 0$.

c) When $x = 0$, $f(x) = 0$. When $f(x) = 0$, $x^4 - x^3 = 0$
$\Rightarrow x^3(x - 1) = 0$, so $x = 0$ or $x = 1$.
So the curve crosses the axes at (0, 0) and (1, 0).
Stationary points occur when $f'(x) = 0$.
$f'(x) = 4x^3 - 3x^2 = 0 \Rightarrow x^2(4x - 3) = 0$, so $x = 0$ or $x = \frac{3}{4}$. When $x = 0$, $y = 0$ and when $x = \frac{3}{4}$, $y = -\frac{27}{256} = -0.11$ (2 d.p.).

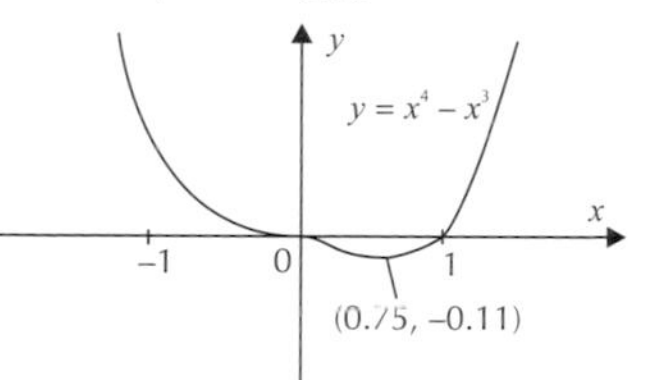

f''(0) = 0 so we cannot say whether the point (0, 0) is a maximum or a minimum — but the function is decreasing for x < 0.75 so (0, 0) must be a point of inflection.

Q4 **a)** When $x = 0$, $y = 0$. When $y = 0$, $3x^3 + 3x^2 = 0$
$\Rightarrow 3x^2(x + 1) = 0$, so $x = 0$ and -1.
When $\frac{dy}{dx} = 0$, $9x^2 + 6x = 0 \Rightarrow 3x(3x + 2) = 0$
so $x = 0$ and $x = -\frac{2}{3}$. When $x = -\frac{2}{3}$, $y = \frac{4}{9}$.
$\frac{d^2y}{dx^2} = 18x + 6$.
When $x = 0$, $\frac{d^2y}{dx^2} = 6$, so it's a minimum.
When $x = -\frac{2}{3}$, $\frac{d^2y}{dx^2} = -6$, so it's a maximum.
It's a positive cubic, so it'll go from bottom left to top right.

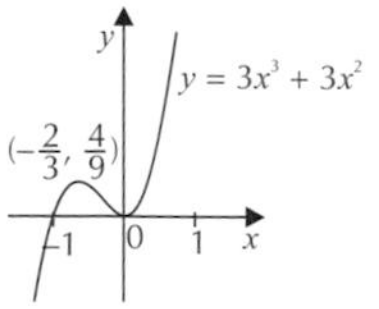

b) When $x = 0$, $y = 0$. When $y = 0$, $-x^3 + 9x = 0$
$\Rightarrow x(9 - x^2) = 0$, so $x = 0$ and $x = \pm 3$.
When $\frac{dy}{dx} = 0$, $-3x^2 + 9 = 0 \Rightarrow x = \pm\sqrt{3}$.
When $x = \sqrt{3}$, $y = 6\sqrt{3}$
and when $x = -\sqrt{3}$, $y = -6\sqrt{3}$, $\frac{d^2y}{dx^2} = -6x$.
When $x = \sqrt{3}$, $\frac{d^2y}{dx^2} = -6\sqrt{3}$, so it's a maximum.
When $x = -\sqrt{3}$, $\frac{d^2y}{dx^2} = 6\sqrt{3}$, so it's a minimum.

It's a negative cubic, so it'll go from top left to bottom right.

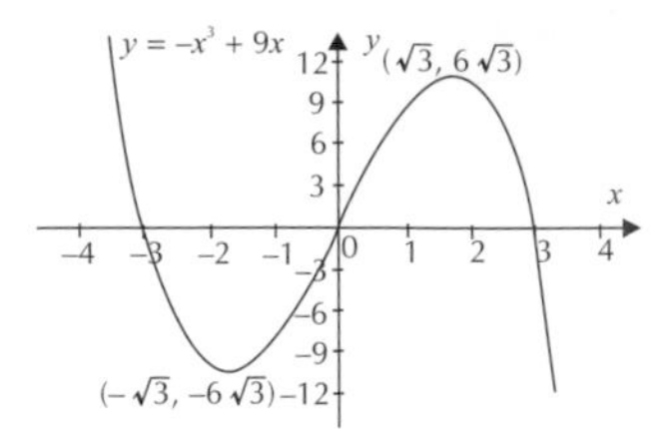

c) When $x = 0$, $y = 0$. When $y = 0$, $x^4 - x^2 = 0$
$\Rightarrow x^2(x^2 - 1) = 0$, so $x = 0$ and ± 1.
When $\frac{dy}{dx} = 0$, $4x^3 - 2x = 0 \Rightarrow x(2x^2 - 1) = 0$.
So $x = 0$ and $\pm\frac{1}{\sqrt{2}}$.
When $x = \frac{1}{\sqrt{2}}$, $y = -\frac{1}{4}$ and
when $x = -\frac{1}{\sqrt{2}}$, $y = -\frac{1}{4}$, $\frac{d^2y}{dx^2} = 12x^2 - 2$.
When $x = 0$, $\frac{d^2y}{dx^2} = -2$, so it's a maximum.
When $x = \frac{1}{\sqrt{2}}$, $\frac{d^2y}{dx^2} = 4$, so it's a minimum.
When $x = -\frac{1}{\sqrt{2}}$, $\frac{d^2y}{dx^2} = 4$, so it's a minimum.

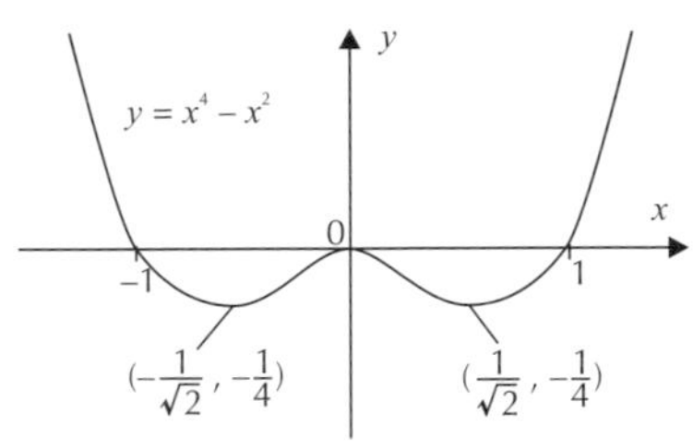

d) When $x = 0$, $y = 0$. When $y = 0$, $x^4 + x^2 = 0$
$\Rightarrow x^2(x^2 + 1) = 0$, so $x = 0$ ($x^2 = -1$ has no solutions).
When $\frac{dy}{dx} = 0$, $4x^3 + 2x = 0$
$\Rightarrow x(2x^2 + 1) = 0$, so $x = 0$.
$\frac{d^2y}{dx^2} = 12x^2 + 2$.
When $x = 0$, $\frac{d^2y}{dx^2} = 2$, so it's a minimum.

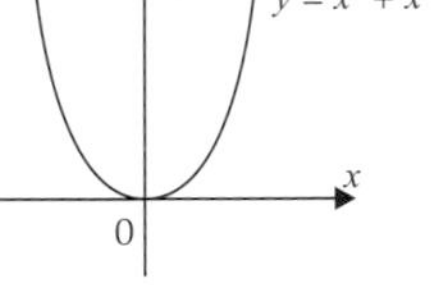

Q5 When $x = 0$, $y = 1$.
When $y = 0$, $(x + 1)(x - 1)^2 = 0$, so $x = 1$ and -1.
When $\frac{dy}{dx} = 0$, $3x^2 - 2x - 1 = 0 \Rightarrow (3x + 1)(x - 1) = 0$
so $x = 1$ and $-\frac{1}{3}$.
When $x = 1$, $y = 0$ and when $x = -\frac{1}{3}$, $y = \frac{32}{27}$.
$\frac{d^2y}{dx^2} = 6x - 2$.
When $x = 1$, $\frac{d^2y}{dx^2} = 4$, so it's a minimum.
When $x = -\frac{1}{3}$, $\frac{d^2y}{dx^2} = -4$, so it's a maximum.

It's a positive cubic, so it'll go from bottom left to top right.

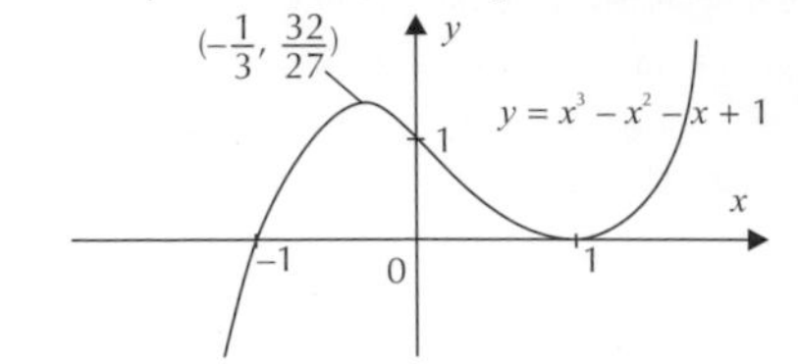

Q6 a) When $x = 2$, $x^3 - 4x = 8 - 8 = 0$.
When $x = -2$, $x^3 - 4x = -8 - (-8) = 0$
When $x = 0$, $x^3 - 4x = 0 - 0 = 0$.

b) $\frac{dy}{dx} = 3x^2 - 4$, $\frac{d^2y}{dx^2} = 6x$.
When $\frac{dy}{dx} = 0$, $3x^2 - 4 = 0 \Rightarrow x = \pm\frac{2\sqrt{3}}{3}$
$= \pm 1.2$ (to 1 d.p.)
When $x = \frac{2\sqrt{3}}{3}$, $y = -3.1$ (1 d.p.)
and when $x = -\frac{2\sqrt{3}}{3}$, $y = 3.1$ (1 d.p.).
So the coordinates of the stationary points to 1 d.p. are (1.2, –3.1) and (–1.2, 3.1).
At $(\frac{2\sqrt{3}}{3}, -3.1)$, $\frac{d^2y}{dx^2} = 6 \times \frac{2\sqrt{3}}{3} = 4\sqrt{3}$,
so it's a minimum.
At $(-\frac{2\sqrt{3}}{3}, 3.1)$, $\frac{d^2y}{dx^2} = 6 \times -\frac{2\sqrt{3}}{3} = -4\sqrt{3}$,
so it's a maximum.

c)

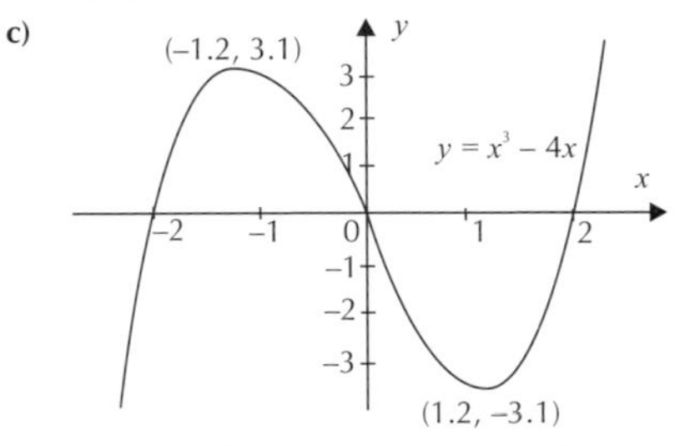

Q7 a) $f'(x) = 1 - \frac{1}{x^2}$. When $f'(x) = 0$, $1 - \frac{1}{x^2} = 0$
$\Rightarrow x^2 = 1 \Rightarrow x = \pm 1$. So the graph of $f(x) = x + \frac{1}{x}$ has stationary points at $x = 1$ and $x = -1$.

b) When $x = 1$, $y = 1 + 1 = 2$, and when $x = -1$, $y = -1 - 1 = -2$. So the coordinates are (1, 2) and (–1, –2).
$f''(x) = \frac{2}{x^3}$. At (1, 2), $f''(x) = 2$, so it's a minimum.
At (–1, –2), $f''(x) = -2$, so it's a maximum.

c) $f(x) = x + \frac{1}{x}$.
As x tends to 0 from below (x is negative), $f(x)$ tends to $-\infty$.
As $x \to 0$ from above (x is positive), $f(x) \to \infty$.

d) As x tends to ∞, $f(x)$ tends to x
i.e. the graph tends towards the line $y = x$.
As x tends to $-\infty$, $f(x)$ tends to x
i.e. the graph tends towards the line $y = x$.

e)

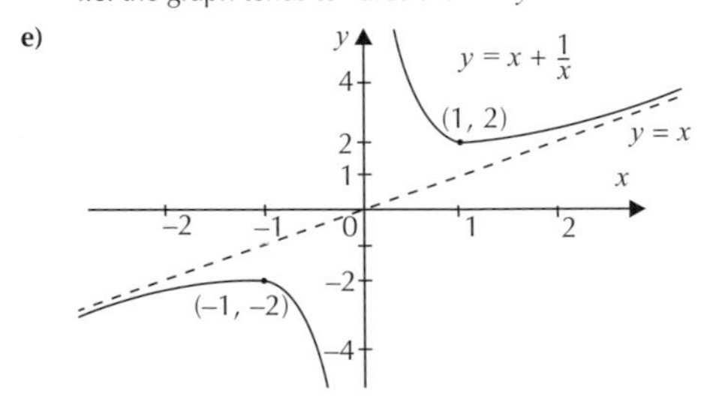

Q8 a) $\frac{dy}{dx} = 4x^3 - \frac{4}{\sqrt{x^3}}$. When $x = 1$, $\frac{dy}{dx} = 4 - 4 = 0$.

b) When $x = 1$, $y = 9$

As $x \to 0$, $y \to \infty$ and as $x \to \infty$, $y \to x^4$.

When $y = 0$, $x^4 + \frac{8}{\sqrt{x}} = 0 \Rightarrow \sqrt{x^9} = -8$. This has no solutions, so the curve doesn't cross the x-axis.

$\frac{d^2y}{dx^2} = 12x^2 + \frac{6}{\sqrt{x^5}}$. When $x = 1$, $\frac{d^2y}{dx^2} = 18$, so it's a minimum.

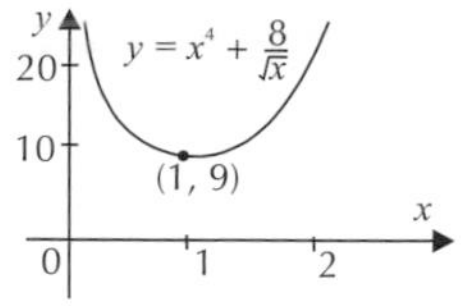

10.4 Real-Life Problems

Exercise 10.4.1 — Speed and acceleration problems

Q1 a) $\frac{dx}{dt} = 6t - 7$

b) (i) $t = 2 \Rightarrow \frac{dx}{dt} = 12 - 7 = 5\ \text{ms}^{-1}$

(ii) $t = 5 \Rightarrow \frac{dx}{dt} = 30 - 7 = 23\ \text{ms}^{-1}$

c) If $\frac{dx}{dt} = 17 \Rightarrow 6t - 7 = 17 \Rightarrow 6t = 24 \Rightarrow t = 4$ s.

d) $\frac{d^2x}{dt^2} = 6\ \text{ms}^{-2}$

Although you were asked for $\frac{d^2x}{dt^2}$ as a function of t, it happens to be constant in this example.

Q2 a) $\frac{dx}{dt} = 6t^2 - 8t$

b) If $\frac{dx}{dt} = 30$ then $6t^2 - 8t = 30 \Rightarrow 6t^2 - 8t - 30 = 0$

$\Rightarrow 3t^2 - 4t - 15 = 0 \Rightarrow (3t + 5)(t - 3) = 0$

$\Rightarrow t = 3$ or $t = -\frac{5}{3}$. But $t > 0$ so $t = 3$.

If $t = 3$, $x = 2t^3 - 4t^2 = 18$.

So $t = 3$ s and $x = 18$ m.

c) $\frac{d^2x}{dt^2} = 12t - 8$.

d) $t = 5 \Rightarrow \frac{d^2x}{dt^2} = 52\ \text{ms}^{-2}$

e) If $\frac{d^2x}{dt^2} = 16 \Rightarrow 12t - 8 = 16 \Rightarrow 12t = 24 \Rightarrow t = 2$.

$\frac{dx}{dt} = 6t^2 - 8t = 8\ \text{ms}^{-1}$.

Q3 a) $\frac{dx}{dt} = \frac{t}{2}\ \text{ms}^{-1}$

b) $\frac{d^2x}{dt^2} = \frac{1}{2}\ \text{ms}^{-2}$

c) $t = 45 \Rightarrow \frac{dx}{dt} = \frac{45}{2} = 22.5\ \text{ms}^{-1}$

d) $\frac{dx}{dt} = 25\ \text{ms}^{-1} \Rightarrow \frac{t}{2} = 25 \Rightarrow t = 50$ seconds

$\Rightarrow x = \frac{50^2}{4} = 625$ m

Exercise 10.4.2 — Length, area and volume problems

Q1 Total length of fence for a rectangular area of length x m and width y m $= 2x + 2y$.

Total area $= 100 = xy \Rightarrow y = \frac{100}{x}$

Sub this into equation for length: $f(x) = 2x + \frac{200}{x}$

You want to minimise the length, so find $f'(x)$:

$f'(x) = 2 - \frac{200}{x^2}$. When $f'(x) = 0$, $2 - \frac{200}{x^2} = 0$

$\Rightarrow x^2 = 100 \Rightarrow x = 10$ (length can't be negative).

Check to see if this gives a minimum value by differentiating again:

$f''(x) = \frac{400}{x^3} \Rightarrow f''(10) = 0.4$, so it's a minimum.

Now find the value of y when $x = 10$:

$100 = xy \Rightarrow y = 10$.

Finally, length $= 2x + 2y = 20 + 20 = 40$ m.

Q2 $\frac{dh}{dt} = 30 - 9.8t$. When $\frac{dh}{dt} = 0$, $30 - 9.8t = 0$

$\Rightarrow t = 3.061... = 3.06$ (3 s.f.).

Check this gives a maximum value:

$\frac{d^2h}{dt^2} = -9.8$, so it's a maximum. So the maximum value of $h = 30(3.061...) - 4.9(3.061...)^2$

$= 45.9$ m (3 s.f.).

Q3 Let the length of fence parallel to the wall be x and the lengths perpendicular to the wall be y.

We want to maximise area, i.e. $A = xy$.

Total length of fence $= 66 = x + 2y \Rightarrow y = \frac{66 - x}{2}$

So we want to maximise $A = x\left(\frac{66 - x}{2}\right) = 33x - \frac{x^2}{2}$.

$\frac{dA}{dx} = 33 - x$, so when $\frac{dA}{dx} = 0$, $x = 33$.

Check that this gives a maximum value by differentiating again:

$\frac{d^2A}{dx^2} = -1$, so it's a maximum.

When $x = 33$, $y = \frac{66 - 33}{2} = 16.5$.

Area $= 33 \times 16.5 = 544.5\ \text{m}^2$.

If you'd labelled the sides the other way round, your working would be different but you'd still get the same answer.

Q4 a) Surface area = area of top and bottom plus area of curved face $= \pi r^2 + \pi r^2 + (2\pi r \times h) = 2\pi r^2 + 2\pi rh$.

To match the question, find an expression for h by thinking about the volume of the tin:

Volume = area of base × height $= \pi r^2 h = 500$

$\Rightarrow h = \frac{500}{\pi r^2}$. So surface area $= 2\pi r^2 + 2\pi r \frac{500}{\pi r^2}$

$= 2\pi r^2 + \frac{1000}{r}$.

b) $\frac{dA}{dr} = 4\pi r - \frac{1000}{r^2}$. When $\frac{dA}{dr} = 0$,

$4\pi r - \frac{1000}{r^2} = 0 \Rightarrow r = \sqrt[3]{\frac{250}{\pi}} = 4.30$ cm (3 s.f.).

Check that this gives a minimum value:

$\frac{d^2A}{dr^2} = 4\pi + \frac{2000}{r^3}$.

When $r = 4.30$, $\frac{d^2A}{dr^2} = 37.7$, so it's a minimum.

You don't really need to work out $\frac{d^2A}{dr^2}$ when r = 4.30 — you can tell it's positive straight away.

c) Surface area $= 2\pi r^2 + \frac{1000}{r} = 349\ \text{cm}^2$ (3 s.f.).

Q5 a) Volume of the box = length × width × height

$= (40 - 2x) \times (40 - 2x) \times x = 4x^3 - 160x^2 + 1600x$

b) $\frac{dV}{dx} = 12x^2 - 320x + 1600$. When $\frac{dV}{dx} = 0$,

$12x^2 - 320x + 1600 = 0 \Rightarrow 3x^2 - 80x + 400 = 0$

$\Rightarrow (3x - 20)(x - 20) = 0 \Rightarrow x = 20$ or $x = \frac{20}{3}$.

Differentiate again to find which of these is a maximum:

$\frac{d^2V}{dx^2} = 24x - 320$. When $x = 20$,

$\frac{d^2V}{dx^2} = 160$, and when $x = \frac{20}{3}$, $\frac{d^2V}{dx^2} = -160$,

so V is a maximum when $x = \frac{20}{3}$.

Note that if x = 20, the volume of the box V = 0, so this can't be the maximum of V — it's always worth checking that answers are sensible in the context of the question.

So the maximum volume is:

$4\left(\frac{20}{3}\right)^3 - 160\left(\frac{20}{3}\right)^2 + 1600\left(\frac{20}{3}\right) = 4740\ \text{cm}^3$ (3 s.f.)

Q6 **a)** The prism is made up of 5 shapes:
2 triangles with base x and height x (area = $\frac{1}{2}x^2$),
2 rectangles with width x and length l (area = xl) and
1 rectangle with width h and length l (area = hl).
So the total surface area is given by:
$A = x^2 + 2xl + hl$. To get rid of the l, find an expression for l by looking at the volume:
$300 = \frac{1}{2}x^2l \Rightarrow l = \frac{600}{x^2}$
To get rid of the h, form an expression for it in terms of x.
It's the hypotenuse of a right-angled triangle, so $h^2 = x^2 + x^2$
$\Rightarrow h = \sqrt{2x^2} = \sqrt{2}\,x$.
Now put these into the original formula for A:
$A = x^2 + 2x(\frac{600}{x^2}) + \sqrt{2}\,x(\frac{600}{x^2}) = x^2 + \frac{600(2+\sqrt{2})}{x}$

b) $\frac{dA}{dx} = 2x - \frac{600(2+\sqrt{2})}{x^2}$. When $\frac{dA}{dx} = 0$,
$2x - \frac{600(2+\sqrt{2})}{x^2} = 0 \Rightarrow x^3 = 300(2 + \sqrt{2})$
$\Rightarrow x = \sqrt[3]{600+300\sqrt{2}}$
Check that this gives a minimum value:
$\frac{d^2A}{dx^2} = 2 + \frac{1200(2+\sqrt{2})}{x^3}$.
When $x = \sqrt[3]{600+300\sqrt{2}}$, $\frac{d^2A}{dx^2} = 6$,
so it's a minimum.
You don't really need to work out $\frac{d^2A}{dx^2}$ when $x = \sqrt[3]{600+300\sqrt{2}}$ — you can see it's positive.

Q7 **a)** There are 6 rectangular sides of the box which make up 3 identical pairs of sides. These sides have areas of xy, $2x^2$ and $2xy$, so the total surface area is:
$A = 2(xy + 2x^2 + 2xy) = 4x^2 + 6xy$
To get rid of y, rearrange the formula for the volume of the box: $V = x \times y \times 2x = 2x^2y$
$\Rightarrow 200 = 2x^2y \Rightarrow y = \frac{100}{x^2}$
so $A = 4x^2 + 6x\left(\frac{100}{x^2}\right) = 4x^2 + \frac{600}{x}$

b) $A = 4x^2 + 600x^{-1} \Rightarrow \frac{dA}{dx} = 8x - 600x^{-2} = 8x - \frac{600}{x^2}$
When $\frac{dA}{dx} = 0$, $8x - \frac{600}{x^2} = 0 \Rightarrow x^3 = 75$
$\Rightarrow x = 4.217...$
$= 4.22$ cm (3 s.f.)

Check that this gives a minimum value:
$\frac{d^2A}{dx^2} = 8 + 1200x^{-3}$.
When $x = 4.217...$, $\frac{d^2A}{dx^2} = 24$, so it's a minimum.

c) $x = 4.21...$ cm $\Rightarrow A = 4(4.217...)^2 + 600(4.217...)^{-1}$
$= 213.4... = 213$ cm^2 (3 s.f.)

Review Exercise — Chapter 10

Q1 **a)** $\frac{dy}{dx} = \lim_{h\to 0}\left[\frac{(x+h)+1-(x+1)}{(x+h)-x}\right]$
$= \lim_{h\to 0}\left[\frac{h}{h}\right] = 1$

b) $\frac{dy}{dx} = \lim_{h\to 0}\left[\frac{4(x+h)^2-4x^2}{(x+h)-x}\right]$
$= \lim_{h\to 0}\left[\frac{4x^2+8xh+4h^2-4x^2}{(x+h)-x}\right]$
$= \lim_{h\to 0}\left[\frac{8xh+4h^2}{h}\right] = \lim_{h\to 0}[8x+4h] = 8x$

c) $\frac{dy}{dx} = \lim_{h\to 0}\left[\frac{3-(x+h)^3-(3-x^3)}{(x+h)-x}\right]$
$= \lim_{h\to 0}\left[\frac{3-x^3-3x^2h-3xh^2-h^3-3+x^3}{(x+h)-x}\right]$
$= \lim_{h\to 0}\left[\frac{-3x^2h-3xh^2-h^3}{h}\right]$
$= \lim_{h\to 0}[-3x^2-3xh-h^2] = -3x^2$

Q2 **a)** $\frac{dy}{dx} = 2x$

b) $y = x^4 + x^{\frac{1}{2}} \Rightarrow \frac{dy}{dx} = 4x^3 + \frac{1}{2}x^{-\frac{1}{2}} = 4x^3 + \frac{1}{2\sqrt{x}}$

c) $y = 7x^{-2} - 3x^{-\frac{1}{2}} + 12x^3$
$\Rightarrow \frac{dy}{dx} = -14x^{-3} + \frac{3}{2}x^{-\frac{3}{2}} + 36x^2 = -\frac{14}{x^3} + \frac{3}{2\sqrt{x^3}} + 36x^2$

Q3 **a)** $\frac{dy}{dx} = 4x$, $x = 2 \Rightarrow \frac{dy}{dx} = 8$

b) $\frac{dy}{dx} = 8x - 1$, $x = 2 \Rightarrow \frac{dy}{dx} = 15$

c) $\frac{dy}{dx} = 3x^2 - 14x$, $x = 2 \Rightarrow \frac{dy}{dx} = -16$

Q4 **a)** Initially $t = 0$, so $v = 3(0)^2 + 4 = 4$ ml.

b) $\frac{dv}{dt} = 6t$, $t = 4 \Rightarrow \frac{dv}{dt} = 24$ ml/s

Q5 $y = x^{\frac{3}{2}} - 3x - 10 \Rightarrow \frac{dy}{dx} = \frac{3x^{\frac{1}{2}}}{2} - 3$
$x = 16 \Rightarrow y = 64 - 48 - 10 = 6$ and $\frac{dy}{dx} = 6 - 3 = 3$
Gradient of the tangent = 3,
so $y = 3x + c \Rightarrow 6 = 3(16) + c \Rightarrow c = -42$
So the equation of the tangent is $y = 3x - 42$,
which can be rearranged to $3x - y - 42 = 0$.
Gradient of the normal = $-\frac{1}{3}$,
so $y = -\frac{1}{3} + c \Rightarrow 6 = -\frac{1}{3}(16) + c \Rightarrow c = \frac{34}{3}$
So the equation of the normal is $y = -\frac{x}{3} + \frac{34}{3}$,
which can be rearranged to $x + 3y - 34 = 0$.

Q6 On the first curve, when $x = 4$
$\Rightarrow y = \frac{(4)^3}{3} - 2(4)^2 - 4(4) + \frac{86}{3} = \frac{64}{3} - 32 - 16 + \frac{86}{3}$
$= \frac{150}{3} - 48 = 2$
On the second curve, when $x = 4$, $y = \sqrt{4} = \pm 2$.
Both curves go through (4, 2) so they meet at this point.
Differentiating the first curve gives:
$\frac{dy}{dx} = x^2 - 4x - 4$, $x = 4 \Rightarrow \frac{dy}{dx} = 4^2 - 4(4) - 4 = -4$
Differentiating the second curve $y = x^{\frac{1}{2}}$ gives:
$\frac{dy}{dx} = \frac{1}{2}x^{-\frac{1}{2}} = \frac{1}{2\sqrt{x}}$, $x = 4 \Rightarrow \frac{dy}{dx} = \frac{1}{2\sqrt{4}} = \frac{1}{4}$
Multiplying the gradients of the curves at (4, 2) gives $-4 \times \frac{1}{4} = -1$,
so the two curves are perpendicular at this point.

Q7 **a)** $y = x^2 - 6$ and $y = 3 \Rightarrow x^2 - 6 = 3 \Rightarrow x^2 = 9 \Rightarrow x = -3$ or 3
So C and L intersect at $A(3, 3)$ and $B(-3, 3)$.
A and B could be the other way round.

b) $\frac{dy}{dx} = 2x$. At A, when $x = 3$, $\frac{dy}{dx} = 6$,
and at B, when $x = -3$, $\frac{dy}{dx} = -6$.

c) The normal at A has a gradient of $-1 \div 6 = -\frac{1}{6}$
and an equation in the form $y = -\frac{1}{6}x + c$.
At $A(3, 3)$, $3 = -\frac{3}{6} + c \Rightarrow c = 3 + \frac{1}{2} = \frac{7}{2}$
So the equation of the normal is $y = -\frac{1}{6}x + \frac{7}{2}$,
which can be rearranged as $x + 6y - 21 = 0$.
The normal at B has a gradient of $-1 \div -6 = \frac{1}{6}$
and an equation in the form $y = \frac{1}{6}x + c$.
At $B(-3, 3)$, $3 = -\frac{3}{6} + c \Rightarrow c = 3 + \frac{1}{2} = \frac{7}{2}$
So the equation of the normal is $y = \frac{1}{6}x + \frac{7}{2}$,
which can be rearranged as $x - 6y + 21 = 0$.

d) Both lines have the same y-intercept, so will intersect each other at $(0, \frac{7}{2})$.

Q8 a) $y = 1,\ \frac{dy}{dx} = 0$

b) $y = x^3 - 2x^2 + 1,\ \frac{dy}{dx} = 3x^2 - 4x$

$\frac{dy}{dx} = 0 \Rightarrow 3x^2 - 4x = 0 \Rightarrow x(3x - 4) = 0$
$\Rightarrow x = 0 \text{ or } x = \frac{4}{3}$

$x > 0$, so the solution for x must be $\frac{4}{3}$.

$y = x^3 - 2x^2 + 1 = \left(\frac{4}{3}\right)^3 - 2\left(\frac{4}{3}\right)^2 + 1$

$= \frac{64}{27} - \frac{32}{9} + 1 = -\frac{5}{27}$

So the coordinates of the point are $\left(\frac{4}{3}, -\frac{5}{27}\right)$.

c) The tangent at this point has a gradient of 0 and an equation in the form $y = (0)x + c$, so $c = -\frac{5}{27}$ and the equation is $y = -\frac{5}{27}$.

Q9 $y = x^3 + 4x^{-1} + 2x^{\frac{1}{2}} \Rightarrow \frac{dy}{dx} = 3x^2 - 4x^{-2} + x^{-\frac{1}{2}}$

When $x = 1$, $y = (1)^3 + 4(1)^{-1} + 2(1)^{\frac{1}{2}} = 7$,

and $\frac{dy}{dx} = 3(1)^2 - 4(1)^{-2} + (1)^{-\frac{1}{2}} = 0$

The tangent to the curve at $x = 1$, has an equation in the form $y = (0)x + c$, so $c = 7$ and the equation is $y = 7$.

Q10 $y = 1 + x^{\frac{3}{2}} \Rightarrow \frac{dy}{dx} = \frac{3}{2}x^{\frac{1}{2}} = \frac{3\sqrt{x}}{2}$

When $x = 16$, $y = 1 + (16)^{\frac{3}{2}} = 65$, and $\frac{dy}{dx} = \frac{3\sqrt{16}}{2} = 6$

The tangent to the curve at $x = 16$ has an equation in the form $y = 6x + c$. Using the point (16, 65):

$65 = 6(16) + c \Rightarrow c = 65 - 96 = -31$

So the tangent has the equation $y = 6x - 31$, which can be rearranged as $6x - y - 31 = 0$.

The normal has a gradient of $-1 \div 6 = -\frac{1}{6}$ and an equation in the form $y = -\frac{1}{6}x + c$.

Using the point (16, 65): $65 = -\frac{16}{6} + c \Rightarrow c = 65 + \frac{16}{6} = \frac{203}{3}$

So the normal has the equation $y = -\frac{1}{6}x + \frac{203}{3}$,

which can be rearranged as $6y = -x + 406 \Rightarrow x + 6y - 406 = 0$

Q11 a) $f(x) = x^3 - 3x \Rightarrow f'(x) = 3x^2 - 3$

$x = -1 \Rightarrow f'(x) = 3(-1)^2 - 3 = 0$

b) $f''(x) = 6x$

$2f''(x) - 3f'(x) + f(x) = 2(6x) - 3(3x^2 - 3) + (x^3 - 3x)$
$= 12x - 9x^2 + 9 + x^3 - 3x$
$= x^3 - 9x^2 + 9x + 9$
$= x^3 + 9(1 + x - x^2)$

Q12 $f(x) = x^4 \Rightarrow f'(x) = 4x^3 \Rightarrow f''(x) = 12x^2$

$f''(x) + 2f'(x) - 4f(x) = 12x^2 + 2(4x^3) - 4(x^4)$
$= 12x^2 + 8x^3 - 4x^4$
$= -4x^2(x + 1)(x - 3)$

Q13 a) $y = x^3 + 3x^{-1} \Rightarrow \frac{dy}{dx} = 3x^2 - 3x^{-2} = 3x^2 - \frac{3}{x^2}$

When $\frac{dy}{dx} = 0$, $3x^2 - \frac{3}{x^2} = 0 \Rightarrow 3x^4 = 3$
$\Rightarrow x^4 = 1$
$\Rightarrow x = -1 \text{ or } x = 1$

When $x = -1$, $y = -1 - 3 = -4$, and when $x = 1$, $y = 1 + 3 = 4$.

So the stationary points are $(-1, -4)$ and $(1, 4)$.

b) $\frac{d^2y}{dx^2} = 6x + 6x^{-3}$

When $x = -1$, $\frac{d^2y}{dx^2} = -6 - 6 = -12 < 0$, so $(-1, -4)$ is a maximum.

When $x = 1$, $\frac{d^2y}{dx^2} = 6 + 6 = 12 > 0$, so $(1, 4)$ is a minimum.

Q14 $y = 2x^4 - x^2 + 4 \Rightarrow \frac{dy}{dx} = 8x^3 - 2x$

When $\frac{dy}{dx} = 0$, $8x^3 - 2x = 0 \Rightarrow 2x(4x^2 - 1) = 0$
$\Rightarrow 2x(2x - 1)(2x + 1)$
$\Rightarrow x = 0 \text{ and } -\frac{1}{2} \text{ and } \frac{1}{2}$

When $x = 0$, $y = 0 - 0 + 4 = 4$,

when $x = -\frac{1}{2}$, $y = \frac{1}{8} - \frac{1}{4} + 4 = \frac{31}{8}$, and

when $x = \frac{1}{2}$, $y = \frac{1}{8} - \frac{1}{4} + 4 = \frac{31}{8}$.

So the stationary points are $(0, 4)$, $\left(-\frac{1}{2}, \frac{31}{8}\right)$, $\left(\frac{1}{2}, \frac{31}{8}\right)$.

$\frac{d^2y}{dx^2} = 24x^2 - 2$.

At $(0, 4)$, $\frac{d^2y}{dx^2} = -2 < 0$, so the point is a maximum.

At $\left(-\frac{1}{2}, \frac{31}{8}\right)$ and $\left(\frac{1}{2}, \frac{31}{8}\right)$, $\frac{d^2y}{dx^2} = 6 - 2 = 4$,

so the points are minimums.

Q15 a) $y = 6(x^2 - x - 6) = 6x^2 - 6x - 36$

$\frac{dy}{dx} = 12x - 6$. If the function is increasing,

$\frac{dy}{dx} > 0 \Rightarrow 12x - 6 > 0 \Rightarrow 12x > 6 \Rightarrow x > \frac{1}{2}$.

If the function is decreasing, $\frac{dy}{dx} < 0 \Rightarrow x < \frac{1}{2}$.

b) $y = x^{-2} \Rightarrow \frac{dy}{dx} = -2x^{-3} = -\frac{2}{x^3} \Rightarrow x \neq 0$

If $x < 0$, $x^3 < 0$, so $-\frac{2}{x^3} > 0$, and if $x > 0$, $x^3 > 0$ so $-\frac{2}{x^3} < 0$.

x cannot equal 0, so the function is increasing when $x < 0$ and decreasing when $x > 0$.

Q16 $y = 3x^3 - 16x \Rightarrow \frac{dy}{dx} = 9x^2 - 16$

When $\frac{dy}{dx} = 0$, $9x^2 - 16 = 0 \Rightarrow (3x + 4)(3x - 4) = 0$
$\Rightarrow x = -\frac{4}{3} \text{ and } x = \frac{4}{3}$

When $x = -\frac{4}{3}$, $y = -\frac{64}{9} + \frac{64}{3} = \frac{128}{9}$, and

when $x = \frac{4}{3}$, $y = \frac{64}{9} - \frac{64}{3} = -\frac{128}{9}$.

When $y = 0$, $3x^3 - 16x = 0 \Rightarrow x(3x^2 - 16) = 0$
$\Rightarrow x = 0 \text{ and } \pm\frac{4\sqrt{3}}{3}$

So the curve crosses the x-axis at $x = -\frac{4\sqrt{3}}{3}$, $x = 0$ and $x = \frac{4\sqrt{3}}{3}$.

The x^2 coefficient is positive, so sketch a bottom-left to top-right curve passing through all these points.

(−1.3, 14.2)
$y = 3x^3 - 16x$
(1.3, −14.2)

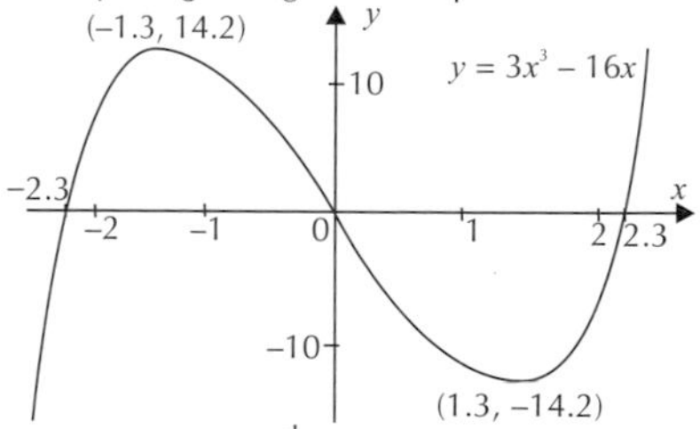

Q17 $y = -3x^3 + 6x^2 \Rightarrow \frac{dy}{dx} = -9x^2 + 12x$

When $\frac{dy}{dx} = 0$, $-9x^2 + 12x = 0 \Rightarrow 3x(-3x + 4) = 0$
$\Rightarrow x = 0 \text{ and } x = \frac{4}{3}$

When $x = 0$, $y = 0$, and when $x = \frac{4}{3}$, $y = -\frac{64}{9} + \frac{32}{3} = \frac{32}{9}$.

When $y = 0$, $-3x^3 + 6x^2 = 0 \Rightarrow -3x^2(x - 2) = 0 \Rightarrow x = 0 \text{ and } 2$

So the curve meets the x-axis at $x = 0$ and $x = 2$.

The x^2 coefficient is negative, so sketch a top-left to bottom-right curve passing through all these points.

$\left(\frac{4}{3}, \frac{32}{9}\right)$
$y = -3x^3 + 6x^2$

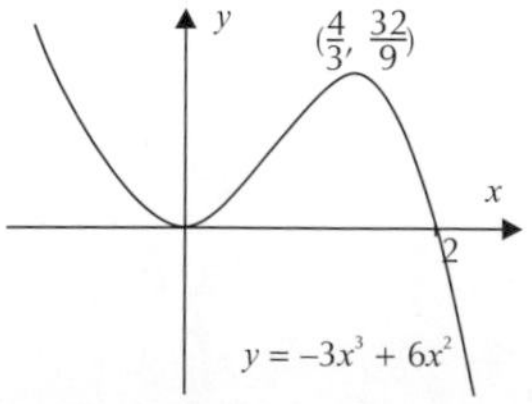

Q18 $xy = 20 \Rightarrow y = \frac{20}{x} \Rightarrow x^2 + y^2 = x^2 + \left(\frac{20}{x}\right)^2 = x^2 + \frac{400}{x^2}$

$f(x) = x^2 + \frac{400}{x^2} = x^2 + 400x^{-2}$,

$f'(x) = 2x - 800x^{-3} = 2x - \frac{800}{x^3}$

If $f'(x) = 0$, $2x - \frac{800}{x^3} = 0 \Rightarrow 2x = \frac{800}{x^3}$

$\Rightarrow 2x^4 = 800$

$\Rightarrow x^4 = 400 \Rightarrow x = \sqrt{20}$

The question says x and y are positive so you can ignore the negative square root of 20.

$f''(x) = 2 + 2400x^{-4}$, $x = \sqrt{20}$

$\Rightarrow f''(x) = 2 + 6 = 8 > 0$, so $x = \sqrt{20}$ is a minimum.

$y = \frac{20}{x} = \frac{20}{\sqrt{20}} = \sqrt{20}$,

so the minimum value of $x^2 + y^2 = 20 + 20 = 40$.

Q19 a) $x = t^3 - 8t \Rightarrow \frac{dx}{dt} = 3t^2 - 8$

b) When $\frac{dx}{dt} = 19$, $3t^2 - 8 = 19 \Rightarrow 3t^2 = 27$

$\Rightarrow t^2 = 9 \Rightarrow t = 3$

When $t = 3$ seconds, $x = 27 - 24 = 3$ metres.

t is a time and cannot be negative, so ignore the negative square root of 9.

c) $\frac{d^2x}{dt^2} = 6t$

d) $t = 2$, $\frac{d^2x}{dt^2} = 12\text{ ms}^{-2}$

e) $\frac{d^2x}{dt^2} = 18$, $6t = 18 \Rightarrow t = 3$

$\Rightarrow \frac{dx}{dt} = 27 - 8 = 19\text{ ms}^{-1}$

Q20 $h = \frac{1}{10}m^2 - \frac{1}{800}m^3 \Rightarrow \frac{dh}{dm} = \frac{1}{5}m - \frac{3}{800}m^2$

When $\frac{dh}{dm} = 0$, $\frac{1}{5}m - \frac{3}{800}m^2 = 0 \Rightarrow \frac{1}{5}m(1 - \frac{3}{160}m)$

$\Rightarrow m = 0$ or $m = \frac{160}{3}$

$\frac{d^2h}{dm^2} = \frac{1}{5} - \frac{3}{400}m$.

When $m = 0$, $\frac{d^2h}{dm^2} = \frac{1}{5} > 0$, so the point is a minimum — there is no fuel when $m = 0$, so this cannot be the mass of fuel needed to reach the maximum height.

When $m = \frac{160}{3}$, $\frac{d^2h}{dm^2} = \frac{1}{5} - \frac{2}{5} = -\frac{1}{5} < 0$,

so the point is a maximum.

So the mass of fuel needed to reach maximum height is $\frac{160}{3}$ g, which is 53.3 g (3 s.f.).

$h = \frac{(53.3...)^2}{10} - \frac{(53.3...)^3}{800} = 94.81... \approx 94.8$ m (3 s.f.)

Exam-Style Questions — Chapter 10

Q1 Using $\frac{dy}{dx} = \lim_{h\to 0}\left[\frac{f(x+h) - f(x)}{(x+h) - x}\right]$:

$$\frac{dy}{dx} = \lim_{h\to 0}\left[\frac{(x+h)^3 + 4(x+h) - (x^3 + 4x)}{(x+h) - x}\right]$$

$$= \lim_{h\to 0}\left[\frac{x^3 + 3x^2h + 3xh^2 + h^3 + 4x + 4h - x^3 - 4x}{h}\right]$$

$$= \lim_{h\to 0}\left[\frac{3x^2h + 3xh^2 + h^3 + 4h}{h}\right]$$

$$= \lim_{h\to 0}[3x^2 + 3xh + h^2 + 4] = 3x^2 + 4$$

[5 marks available — 1 mark for using and substituting into the formula, 2 marks for correctly expanding brackets (otherwise 1 mark for one mistake), 1 mark for simplifying and cancelling h, 1 mark for h tending toward zero to give correct answer]

Q2 $y = \frac{x-1}{x^2} = \frac{x}{x^2} - \frac{1}{x^2} = x^{-1} - x^{-2}$

$\frac{dy}{dx} = -x^{-2} + 2x^{-3} = \frac{-x+2}{x^3}$

$\frac{d^2y}{dx^2} = 2x^{-3} - 6x^{-4} = \frac{2x-6}{x^4}$

Substitute in these expressions:

$\frac{d^2y}{dx^2} + \frac{4}{x}\left(\frac{dy}{dx}\right) + \frac{2}{x^2}y = \frac{2x-6}{x^4} + \frac{4}{x}\left(\frac{-x+2}{x^3}\right) + \frac{2}{x^2}\left(\frac{x-1}{x^2}\right)$

Expand the brackets to get separate terms with a common denominator of x^4:

$\frac{2x-6}{x^4} + \frac{-4x+8}{x^4} + \frac{2x-2}{x^4}$

Simplify the numerator:

$\frac{2x - 6 - 4x + 8 + 2x - 2}{x^4} = \frac{0}{x^4} = 0$

[5 marks available — 1 mark for rewriting y as two separate terms to be differentiated, 1 mark for correctly differentiating to get $\frac{dy}{dx}$, 1 mark for correctly differentiating to get $\frac{d^2y}{dx^2}$, 1 mark for correct substitution into left side of equation, 1 mark for expanding and simplifying to show that it equals zero]

Q3 a) Using Pythagoras' theorem:

$r^2 + (5h)^2 = 15^2 \Rightarrow r^2 + 25h^2 = 225$

$\Rightarrow r^2 = 225 - 25h^2$

$\Rightarrow r = \sqrt{225 - 25h^2}$

[2 marks available — 1 mark for use of Pythagoras' theorem and correct substitution, 1 mark for correct r in terms of h]

b) $V = \pi r^2h = 10\pi h(225 - 25h^2) = 2250\pi h - 250\pi h^3$

[2 marks available — 1 mark for substituting part a) into the volume of a cylinder formula, 1 mark for correct formula for V in terms of h]

c) $\frac{dV}{dh} = 2250\pi - 750\pi h^2$

When $\frac{dV}{dh} = 0$, $2250\pi - 750\pi h^2 = 0$

$\Rightarrow 750\pi(3 - h^2) = 0 \Rightarrow h^2 = 3 \Rightarrow h = \pm\sqrt{3}$

h is a length, so you can ignore the negative solution.

$\frac{d^2V}{dh^2} = -1500\pi h$.

At $h = \sqrt{3}$, $\frac{d^2V}{dh^2} = -8162 < 0$, so it's a maximum.

When $h = \sqrt{3}$, $V = 2250\pi\sqrt{3} - 250\pi(\sqrt{3})^3$

$= 8162.09...\text{ cm}^3$

$= 8162\text{ cm}^3$ (to nearest cm^3)

[5 marks available — 1 mark for correct $\frac{dV}{dh}$, 1 mark for making $\frac{dV}{dh} = 0$, 1 mark for correctly solving for h, 1 mark for substituting h value into part b), 1 mark for correct answer rounded to the nearest cm^3]

Q4 a) $y = x^2(x-3)^2 = x^2(x^2 - 6x + 9) = x^4 - 6x^3 + 9x^2$

$\frac{dy}{dx} = 4x^3 - 18x^2 + 18x$

When $\frac{dy}{dx} = 0$, $4x^3 - 18x^2 + 18x = 0$

$\Rightarrow 2x(2x^2 - 9x + 9) = 0 \Rightarrow 2x(2x - 3)(x - 3) = 0$

$\Rightarrow x = 0$ and $x = \frac{3}{2}$ and $x = 3$

When $x = 0$, $y = 0$. When $x = \frac{3}{2}$, $y = \frac{81}{16}$, and when $x = 3$, $y = 0$.

The coordinates of the stationary points are $(0, 0)$, $\left(\frac{3}{2}, \frac{81}{16}\right)$ (allow (1.5, 5.1)) and $(3, 0)$.

[4 marks available — 1 mark for correct $\frac{dy}{dx}$, 1 mark for setting $\frac{dy}{dx}$ equal to zero, 1 mark for one correct pair of coordinates, 1 mark for the other two correct pairs of coordinates]

b) $\frac{d^2y}{dx^2} = 12x^2 - 36x + 18$.

When $x = 0$, $\frac{d^2y}{dx^2} = 0 - 0 + 18 = 18 > 0$,

so the point (0, 0) is a minimum.

When $x = \frac{3}{2}$, $\frac{d^2y}{dx^2} = 27 - 54 + 18 = -9 < 0$,

so the point $\left(\frac{3}{2}, \frac{81}{16}\right)$ is a maximum.

When $x = 3$, $\frac{d^2y}{dx^2} = 108 - 108 + 18 = 18 > 0$,

so the point (3, 0) is a minimum.

[3 marks available — 1 mark for $\frac{d^2y}{dx^2}$, 1 mark for correctly substituting one x-value from part a) to determine the nature of one point, 1 mark for correctly substituting the other two x-values to determine the nature of the other two points]

c)

y

$\left(\frac{3}{2}, \frac{81}{16}\right)$ $y = x^2(x-3)^2$

x

(0, 0) (3, 0)

[3 marks available — 1 mark for correct shape, 1 mark for one stationary point correct and labelled, 1 mark for the other two stationary points correct and labelled]

Q5 $f(x) = x^4 - 14x^2 + 24x \Rightarrow f'(x) = 4x^3 - 28x + 24$

When $f'(x) = 0$, $4x^3 - 28x + 24 = 0 \Rightarrow x^3 - 7x + 6 = 0$

The coefficients add up to $1 + (-7) + 6 = 0$, so $(x - 1)$ must be a factor. Use algebra or division to find the quadratic factor:

$(x-1)(x^2 + x - 6) \Rightarrow (x-1)(x+3)(x-2) = 0$

$\Rightarrow x = 1$ or $x = -3$ or $x = 2$

If the function is increasing $f'(x) > 0$

$\Rightarrow 4x^3 - 28x + 24 > 0 \Rightarrow (x-1)(x+3)(x-2) > 0$

For this to be true, there are 4 possibilities — all brackets are more than zero, or one is more than zero and two are less than zero.

Either $x > 1$ and $x > -3$ and $x > 2$, so $x > 2$

Or $x > 1$ and $x < -3$ and $x < 2$ (impossible)

Or $x < 1$ and $x > -3$ and $x < 2$

Or $x < 1$ and $x < -3$ and $x > 2$ (impossible)

So the function is increasing when $-3 < x < 1$ and $x > 2$, and is decreasing when $x < -3$ and $1 < x < 2$.

[7 marks available — 1 mark for finding f′(x), 1 mark for f(x) = 0, f(x) > 0 or f(x) < 0, 1 mark for at least one factor and hence root, 1 mark for other factors and hence roots, 1 mark for a correct method to find increasing and decreasing ranges (e.g. solving inequalities, determining nature of stationary points, sketching curve), 1 mark for partially correct ranges, 1 mark for fully correct ranges]

Q6 a) $y = x^2 + 4x - 2 \Rightarrow \frac{dy}{dx} = 2x + 4$

When $x = 2$, $y = 10$ and $\frac{dy}{dx} = 8$.

The tangent has a gradient of 8 and an equation of the form $y = 8x + c$. Using the point (2, 10):

$10 = 16 + c \Rightarrow c = -6$

So the tangent has the equation $y = 8x - 6$, which can be rearranged as $8x - y - 6 = 0$.

[3 marks available — 1 mark for finding $\frac{dy}{dx}$, 1 mark for substituting x = 2 and finding gradient of 8, 1 mark for correct equation of tangent in any appropriate form]

b) When $x = -1$, $y = -5$ and $\frac{dy}{dx} = 2$.

So the normal has a gradient of $-1 \div 2 = -\frac{1}{2}$ and an equation of the form $y = -\frac{1}{2}x + c$.

Using the point (–1, –5):

$-5 = \frac{1}{2} + c \Rightarrow c = -\frac{11}{2}$

So the normal has the equation $y = -\frac{1}{2}x - \frac{11}{2}$, which can be rearranged as $2y = -x - 11$

$\Rightarrow x + 2y + 11 = 0$

[3 marks available — 1 mark for substituting x = –1 and finding gradient of 2, 1 mark for correct gradient of normal, 1 mark for correct equation of normal in any appropriate form]

c)

①	$8x - y = 6$
②	$x + 2y = -11$
① × 2	$16x - 2y = 12$ ③
② + ③	$17x = 1 \Rightarrow x = \frac{1}{17}$
$x = \frac{1}{17}$ in ①	$\frac{8}{17} - y = 6 \Rightarrow y = \frac{8}{17} - 6 = -\frac{94}{17}$

So the tangent at $x = 2$ and normal at $x = -1$ intersect at point $V\left(\frac{8}{17}, -\frac{94}{17}\right)$.

[2 marks available — 1 mark for attempt to solve using simultaneous equations to find V, 1 mark for correct coordinates for V]

Q7 $f(x) = 4\sqrt{x} + \frac{a}{x^2} - 6x + b = 4x^{\frac{1}{2}} + ax^{-2} - 6x + b$

$f'(x) = 2x^{-\frac{1}{2}} - 2ax^{-3} - 6$

$f'(1) = -64 \Rightarrow 2 - 2a - 6 = -64 \Rightarrow 2a = 60 \Rightarrow a = 30$

$f(4) = -\frac{97}{8} \Rightarrow 8 + \frac{a}{16} - 24 + b = -\frac{97}{8}$

$\Rightarrow \frac{a}{16} + b = \frac{31}{8} \Rightarrow a + 16b = 62$

$a = 30$, so $30 + 16b = 62 \Rightarrow b = 2$

[4 marks available — 1 mark for correct f′(x), 1 mark for substituting x = 1 into f′(x) to get a = 30, 1 mark for substituting x = 4 into f(x) to get an equation in a and b, 1 mark for substituting a = 30 into the equation to get b = 2]

Q8 $s = 4 + 12t^2 - t^3 \Rightarrow \frac{ds}{dt} = 24t - 3t^2 \Rightarrow \frac{d^2s}{dt^2} = 24 - 6t$

At $t = 4.5$ seconds, $\frac{d^2s}{dt^2} = 24 - 27 = -3$ ms^{-2}

[4 marks available — 1 mark for finding $\frac{ds}{dt}$, 1 mark for finding $\frac{d^2s}{dt^2}$, 1 mark for substituting t = 4.5 into $\frac{d^2s}{dt^2}$, 1 mark for correct answer]

Chapter 11: Integration

11.1 Indefinite Integration

Exercise 11.1.1 — Integrating x^n

Q1 a) $y = \int \frac{dy}{dx}\,dx = \int x^7\,dx = \frac{x^8}{8} + C$

b) $y = \int \frac{dy}{dx}\,dx = \int 2x^3\,dx = 2\int x^3\,dx$

$= 2\left(\frac{x^4}{4}\right) + C = \frac{x^4}{2} + C$

c) $y = \int \frac{dy}{dx}\,dx = \int 8x\,dx = 8\int x\,dx$

$= 8\left(\frac{x^2}{2}\right) + C = 4x^2 + C$

d) $y = \int \frac{dy}{dx}\,dx = \int -5x^4\,dx = -5\int x^4\,dx$

$= -5\left(\frac{x^5}{5}\right) + C = -x^5 + C$

e) $y = \int \frac{dy}{dx}\,dx = \int 200x^{99}\,dx = 200\left(\frac{x^{100}}{100}\right) + C = 2x^{100} + C$

f) $y = \int \frac{dy}{dx}\,dx = \int x^{-3}\,dx = \frac{x^{-2}}{-2} + C = -\frac{1}{2x^2} + C$

g) $y = \int \frac{dy}{dx}\,dx = \int 4x^{-4}\,dx = 4\int x^{-4}\,dx$
$= 4\left(\frac{x^{-3}}{-3}\right) + C = \frac{4x^{-3}}{-3} + C = -\frac{4}{3x^3} + C$

h) $y = \int \frac{dy}{dx}\,dx = \int -6x^{-5}\,dx = -6\int x^{-5}\,dx$
$= -6\left(\frac{x^{-4}}{-4}\right) + C = \frac{3x^{-4}}{2} + C = \frac{3}{2x^4} + C$

i) $y = \int \frac{dy}{dx}\,dx = \int -12\,dx = \frac{-12x}{1} + C = -12x + C$

j) $y = \int \frac{dy}{dx}\,dx = \int x^{\frac{1}{2}}\,dx = \frac{x^{\frac{3}{2}}}{\left(\frac{3}{2}\right)} + C = \frac{2x^{\frac{3}{2}}}{3} + C$

Don't forget that dividing by a fraction is the same as multiplying by the flipped fraction.

k) $y = \int \frac{dy}{dx}\,dx = \int x^{\frac{1}{3}}\,dx = \frac{x^{\frac{4}{3}}}{\left(\frac{4}{3}\right)} + C = \frac{3x^{\frac{4}{3}}}{4} + C$

l) $y = \int \frac{dy}{dx}\,dx = \int 12x^{\frac{3}{2}}\,dx = \frac{12x^{\frac{5}{2}}}{\left(\frac{5}{2}\right)} + C = \frac{24x^{\frac{5}{2}}}{5} + C$

Q2 a) $f(x) = \int f'(x)\,dx = \int x^5\,dx = \frac{x^6}{6} + C$

b) $f(x) = \int f'(x)\,dx = \int -6x\,dx = -6\int x\,dx$
$= (-6)\left(\frac{x^2}{2}\right) + C = -3x^2 + C$

c) $f(x) = \int f'(x)\,dx = \int \frac{4}{3}\,dx = \left(\frac{4}{3}\right)\left(\frac{x^1}{1}\right) + C = \frac{4x}{3} + C$

d) $f(x) = \int f'(x)\,dx = \int 2x^{-2}\,dx = 2\int x^{-2}\,dx$
$= 2\left(\frac{x^{-1}}{-1}\right) + C = -2x^{-1} + C$

e) $f(x) = \int f'(x)\,dx = \int -9x^{-19}\,dx = -9\int x^{-19}\,dx$
$= (-9)\left(\frac{x^{-18}}{-18}\right) + C = \frac{x^{-18}}{2} + C$

f) $f(x) = \int f'(x)\,dx = \int \frac{x}{5}\,dx = \frac{1}{5}\int x\,dx$
$= \left(\frac{1}{5}\right)\left(\frac{x^2}{2}\right) + C = \frac{x^2}{10} + C$

g) $f(x) = \int f'(x)\,dx = \int -x^{-6}\,dx = -\int x^{-6}\,dx$
$= -\left(\frac{x^{-5}}{-5}\right) + C = \frac{x^{-5}}{5} + C$

h) $f(x) = \int f'(x)\,dx = \int \frac{5}{\sqrt{x}}\,dx = 5\int x^{-\frac{1}{2}}\,dx$
$= 5\left(\frac{x^{\frac{1}{2}}}{\left(\frac{1}{2}\right)}\right) + C = 10\sqrt{x} + C$

i) $f(x) = \int f'(x)\,dx = \int x^{-\frac{4}{3}}\,dx = \frac{x^{-\frac{1}{3}}}{\left(-\frac{1}{3}\right)} + C$
$= -3x^{-\frac{1}{3}} + C = -\frac{3}{\sqrt[3]{x}} + C$

j) $f(x) = \int f'(x)\,dx = \int 30x^{-0.8}\,dx = 30\int x^{-0.8}\,dx$
$= 30\left(\frac{x^{0.2}}{(0.2)}\right) + C = 150x^{0.2} + C$

k) $f(x) = \int f'(x)\,dx = \int -\frac{2}{3}x^{\frac{7}{3}}\,dx = -\frac{2}{3}\int x^{\frac{7}{3}}\,dx$
$= -\frac{2}{3}\left(\frac{x^{\frac{10}{3}}}{\left(\frac{10}{3}\right)}\right) + C = -\frac{1}{5}x^{\frac{10}{3}} + C$

l) $f(x) = \int f'(x)\,dx = \int -\frac{3}{4}x^{-\frac{5}{2}}\,dx = -\frac{3}{4}\int x^{-\frac{5}{2}}\,dx$
$= -\frac{3}{4}\left(\frac{x^{-\frac{3}{2}}}{\left(-\frac{3}{2}\right)}\right) + C = \frac{1}{2}x^{-\frac{3}{2}} + C$

Q3 a) $\int x^{\frac{2}{3}}\,dx = \frac{x^{\frac{5}{3}}}{\left(\frac{5}{3}\right)} + C = \frac{3x^{\frac{5}{3}}}{5} + C$

b) $\int 7x^{\frac{4}{3}}\,dx = 7\int x^{\frac{4}{3}}\,dx = 7\left(\frac{x^{\frac{7}{3}}}{\left(\frac{7}{3}\right)}\right) + C = 3x^{\frac{7}{3}} + C$

c) $\int x^{-\frac{1}{2}}\,dx = \frac{x^{\frac{1}{2}}}{\left(\frac{1}{2}\right)} + C = 2x^{\frac{1}{2}} + C$

d) $\int 2x^{-\frac{1}{3}}\,dx = 2\int x^{-\frac{1}{3}}\,dx = 2\frac{x^{\frac{2}{3}}}{\left(\frac{2}{3}\right)} + C = 3x^{\frac{2}{3}} + C$

e) $\int 14x^{0.4}\,dx = 14\int x^{0.4}\,dx = 14\left(\frac{x^{1.4}}{1.4}\right) + C = 10x^{1.4} + C$

f) $\int -1.2x^{-0.6}\,dx = -1.2\int x^{-0.6}\,dx = -1.2\left(\frac{x^{0.4}}{0.4}\right) + C = -3x^{0.4} + C$

g) $\int -2x^{-\frac{5}{4}}\,dx = -2\int x^{-\frac{5}{4}}\,dx = -2\frac{x^{-\frac{1}{4}}}{\left(-\frac{1}{4}\right)} + C = 8x^{-\frac{1}{4}} + C$

h) $\int -\frac{3}{2}x^{-\frac{1}{2}}\,dx = -\frac{3}{2}\int x^{-\frac{1}{2}}\,dx = -\frac{3}{2}\left(\frac{x^{\frac{1}{2}}}{\left(\frac{1}{2}\right)}\right) + C = -3x^{\frac{1}{2}} + C$

i) $\int -\frac{4}{3}x^{-\frac{4}{3}}\,dx = -\frac{4}{3}\int x^{-\frac{4}{3}}\,dx = -\frac{4}{3}\left(\frac{x^{-\frac{1}{3}}}{\left(-\frac{1}{3}\right)}\right) + C = 4x^{-\frac{1}{3}} + C$

j) $\int \frac{1}{2}x^{\frac{5}{2}}\,dx = \frac{1}{2}\int x^{\frac{5}{2}}\,dx = \frac{1}{2}\left(\frac{x^{\frac{7}{2}}}{\left(\frac{7}{2}\right)}\right) + C = \frac{1}{7}x^{\frac{7}{2}} + C$

k) $\int \frac{2}{3}x^{\frac{7}{5}}\,dx = \frac{2}{3}\int x^{\frac{7}{5}}\,dx = \frac{2}{3}\left(\frac{x^{\frac{12}{5}}}{\left(\frac{12}{5}\right)}\right) + C = \frac{5}{18}x^{\frac{12}{5}} + C$

l) $\int 50.5x^{0.01}\,dx = 50.5\int x^{0.01}\,dx = 50.5\left(\frac{x^{1.01}}{(1.01)}\right) + C$
$= 50x^{1.01} + C$

Q4 $\int 4ax^{a-1}\,dx = 4a\int x^{a-1}\,dx = 4a\left(\frac{x^a}{a}\right) + C = 4x^a + C$
$\int (a+1)x^a\,dx = (a+1)\int x^a\,dx = (a+1)\left(\frac{x^{a+1}}{a+1}\right) + C = x^{a+1} + C$
When $\int 4ax^{a-1}\,dx = \int (a+1)x^a\,dx$ and $C = 0$ for both,
$4x^a = x^{a+1} \Rightarrow 4 = \frac{x^{a+1}}{x^a} \Rightarrow x = 4$

Exercise 11.1.2 — Integrating functions

Q1 a) $f(x) = \int f'(x)\,dx = \int (5x + 3x^{-4})\,dx$
$= 5\int x\,dx + 3\int x^{-4}\,dx$
$= 5\left(\frac{x^2}{2}\right) + 3\left(\frac{x^{-3}}{-3}\right) + C = \frac{5x^2}{2} - x^{-3} + C$

b) $f(x) = \int f'(x)\,dx = \int 4x(x^2 - 1)\,dx$
$= \int (4x^3 - 4x)\,dx = 4\int x^3\,dx - 4\int x\,dx$
$= 4\left(\frac{x^4}{4}\right) - 4\left(\frac{x^2}{2}\right) + C = x^4 - 2x^2 + C$

c) $f(x) = \int f'(x)\,dx = \int (x-3)^2\,dx$
$= \int (x^2 - 6x + 9)\,dx$
$= \int x^2\,dx - 6\int x\,dx + 9\int 1\,dx$
$= \frac{x^3}{3} - 6\left(\frac{x^2}{2}\right) + 9\left(\frac{x^1}{1}\right) + C = \frac{x^3}{3} - 3x^2 + 9x + C$

d) $f(x) = \int f'(x)\,dx = \int x\left(6x + \frac{4}{x^4}\right)dx$
$= \int \left(6x^2 + \frac{4}{x^3}\right)dx = \int (6x^2 + 4x^{-3})\,dx$
$= 6\int x^2\,dx + 4\int x^{-3}\,dx$
$= 6\left(\frac{x^3}{3}\right) + 4\left(\frac{x^{-2}}{-2}\right) + C$
$= 2x^3 - 2x^{-2} + C = 2x^3 - \frac{2}{x^2} + C$

e) $f(x) = \int f'(x)\,dx = \int \left(x + \frac{2}{x}\right)^2 dx$
$= \int \left(x^2 + 4 + \frac{4}{x^2}\right)dx = \int (x^2 + 4 + 4x^{-2})\,dx$
$= \int x^2\,dx + 4\int 1\,dx + 4\int x^{-2}\,dx$
$= \frac{x^3}{3} + 4\left(\frac{x^1}{1}\right) + 4\left(\frac{x^{-1}}{-1}\right) + C = \frac{x^3}{3} + 4x - \frac{4}{x} + C$

f) $f(x) = \int f'(x)\,dx = \int x\left(3x^{\frac{1}{2}} - \frac{2}{x^{\frac{4}{3}}}\right)dx$

$= \int \left(3x^{\frac{3}{2}} - \frac{2}{x^{\frac{1}{3}}}\right)dx = \int (3x^{\frac{3}{2}} - 2x^{-\frac{1}{3}})\,dx$

$= 3\int x^{\frac{3}{2}}\,dx - 2\int x^{-\frac{1}{3}}\,dx$

$= 3\left(\frac{x^{\frac{5}{2}}}{\left(\frac{5}{2}\right)}\right) - 2\frac{x^{\frac{2}{3}}}{\left(\frac{2}{3}\right)} + C = \frac{6}{5}x^{\frac{5}{2}} - 3x^{\frac{2}{3}} + C$

g) $f(x) = \int f'(x)\,dx = \int \left(6\sqrt{x} - \frac{1}{x^2}\right)dx$

$= 6\int x^{\frac{1}{2}}\,dx - \int x^{-2}\,dx$

$= 6\left(\frac{x^{\frac{3}{2}}}{\left(\frac{3}{2}\right)}\right) - \frac{x^{-1}}{-1} + C$

$= \frac{12}{3}x^{\frac{3}{2}} + \frac{1}{x} + C = 4x^{\frac{3}{2}} + \frac{1}{x} + C$

$= 4x\sqrt{x} + \frac{1}{x} + C$

h) $f(x) = \int f'(x)\,dx = \int \left(\frac{2}{\sqrt{x}} - 7x^2\sqrt{x}\right)dx$

$= \int (2x^{-\frac{1}{2}} - 7x^2x^{\frac{1}{2}})\,dx = \int (2x^{-\frac{1}{2}} - 7x^{\frac{5}{2}})\,dx$

$= 2\int x^{-\frac{1}{2}}\,dx - 7\int x^{\frac{5}{2}}\,dx$

$= 2\left(\frac{x^{\frac{1}{2}}}{\left(\frac{1}{2}\right)}\right) - 7\frac{x^{\frac{7}{2}}}{\left(\frac{7}{2}\right)} + C$

$= 4x^{\frac{1}{2}} - 2x^{\frac{7}{2}} + C = 4\sqrt{x} - 2(\sqrt{x})^7 + C$

i) $f(x) = \int f'(x)\,dx = \int \left(5(\sqrt{x})^3 - \frac{3x}{\sqrt{x}}\right)dx$

$= \int (5(x^{\frac{1}{2}})^3 - 3xx^{-\frac{1}{2}})\,dx$

$= 5\int x^{\frac{3}{2}}\,dx - 3\int x^{\frac{1}{2}}\,dx$

$= 5\left(\frac{x^{\frac{5}{2}}}{\left(\frac{5}{2}\right)}\right) - 3\left(\frac{x^{\frac{3}{2}}}{\left(\frac{3}{2}\right)}\right) + C$

$= 2x^{\frac{5}{2}} - 2x^{\frac{3}{2}} + C = 2(\sqrt{x})^5 - 2(\sqrt{x})^3 + C$

j) $f(x) = \int f'(x)\,dx = \int (\sqrt{x}(1-x))\,dx$

$= \int (x^{\frac{1}{2}} - xx^{\frac{1}{2}})\,dx = \int x^{\frac{1}{2}}\,dx - \int x^{\frac{3}{2}}\,dx$

$= \frac{x^{\frac{3}{2}}}{\left(\frac{3}{2}\right)} - \frac{x^{\frac{5}{2}}}{\left(\frac{5}{2}\right)} + C$

$= \frac{2x^{\frac{3}{2}}}{3} - \frac{2x^{\frac{5}{2}}}{5} + C = \frac{2\sqrt{x^3}}{3} - \frac{2\sqrt{x^5}}{5} + C$

k) $f(x) = \int f'(x)\,dx = \int \left(3\sqrt[6]{x} - \frac{\sqrt{x}}{\sqrt[3]{x}}\right)dx$

$= \int (3x^{\frac{1}{6}} - x^{\frac{1}{2}}x^{-\frac{1}{3}})\,dx = \int (3x^{\frac{1}{6}} - x^{\frac{1}{6}})\,dx$

$= \int (2x^{\frac{1}{6}})\,dx = 2\int x^{\frac{1}{6}}\,dx$

$= 2\left(\frac{x^{\frac{7}{6}}}{\left(\frac{7}{6}\right)}\right) + C = \frac{12x^{\frac{7}{6}}}{7} + C$

l) $f(x) = \int f'(x)\,dx = \int \left(\frac{x^3 - 2x^2}{\sqrt{x}}\right)dx$

$= \int (x^{\frac{5}{2}} - 2x^{\frac{3}{2}})\,dx = \int x^{\frac{5}{2}}\,dx - 2\int x^{\frac{3}{2}}\,dx$

$= \frac{x^{\frac{7}{2}}}{\left(\frac{7}{2}\right)} - \frac{2x^{\frac{5}{2}}}{\left(\frac{5}{2}\right)} + C = \frac{2x^{\frac{7}{2}}}{7} - \frac{4x^{\frac{5}{2}}}{5} + C$

Q2 a) $\int \left(\frac{5}{7}x^4 + \frac{2}{3}x + \frac{1}{4}\right)dx$

$= \frac{5}{7}\int x^4\,dx + \frac{2}{3}\int x^1\,dx + \frac{1}{4}\int x^0\,dx$

$= \frac{5}{7}\left(\frac{x^5}{5}\right) + \frac{2}{3}\left(\frac{x^2}{2}\right) + \frac{1}{4}\left(\frac{x^1}{1}\right) + C$

$= \frac{x^5}{7} + \frac{x^2}{3} + \frac{x}{4} + C$

b) $\int (x^{-\frac{1}{2}} + x^{\frac{1}{2}})\,dx = \int x^{-\frac{1}{2}}\,dx + \int x^{\frac{1}{2}}\,dx$

$= \frac{x^{\frac{1}{2}}}{\left(\frac{1}{2}\right)} + \frac{x^{\frac{3}{2}}}{\left(\frac{3}{2}\right)} + C = 2\sqrt{x} + \frac{2\sqrt{x^3}}{3} + C$

c) $\int (3x^{-2} + 3x^{-\frac{1}{3}})\,dx = 3\int x^{-2}\,dx + 3\int x^{-\frac{1}{3}}\,dx$

$= 3\left(\frac{x^{-1}}{(-1)}\right) + 3\left(\frac{x^{\frac{2}{3}}}{\frac{2}{3}}\right) + C = -\frac{3}{x} + \frac{9\sqrt[3]{x^2}}{2} + C$

d) $\int (x^{\frac{5}{4}} + x^{\frac{4}{5}})\,dx = \int x^{\frac{5}{4}}\,dx + \int x^{\frac{4}{5}}\,dx$

$= \left(\frac{x^{\frac{9}{4}}}{\frac{9}{4}}\right) + \left(\frac{x^{\frac{9}{5}}}{\frac{9}{5}}\right) + C = \frac{4x^{\frac{9}{4}}}{9} + \frac{5x^{\frac{9}{5}}}{9} + C$

e) $\int \left(\frac{1}{2}x^{-\frac{1}{2}} + 3x^{-\frac{2}{3}} + 4x^{-\frac{3}{4}}\right)dx$

$= \frac{1}{2}\int x^{-\frac{1}{2}}\,dx + 3\int x^{-\frac{2}{3}}\,dx + 4\int x^{-\frac{3}{4}}\,dx$

$= \frac{1}{2}\left(\frac{x^{\frac{1}{2}}}{\frac{1}{2}}\right) + 3\left(\frac{x^{\frac{1}{3}}}{\frac{1}{3}}\right) + 4\left(\frac{x^{\frac{1}{4}}}{\frac{1}{4}}\right) + C$

$= x^{\frac{1}{2}} + 9x^{\frac{1}{3}} + 16x^{\frac{1}{4}} + C = \sqrt{x} + 9\sqrt[3]{x} + 16\sqrt[4]{x} + C$

f) $\int \left(-\frac{4}{x^3} + \frac{1}{\sqrt{x^3}}\right)dx = -4\int x^{-3}\,dx + \int x^{-\frac{3}{2}}\,dx$

$= -4\left(\frac{x^{-2}}{(-2)}\right) + \left(\frac{x^{-\frac{1}{2}}}{-\frac{1}{2}}\right) + C = 2x^{-2} - 2x^{-\frac{1}{2}} + C$

$= \frac{2}{x^2} - \frac{2}{\sqrt{x}} + C$

Q3 a) $\int (0.55x^{0.1} - 3x^{-1.5}x)\,dx = \int (0.55x^{0.1} - 3x^{-0.5})\,dx$

$= 0.55\int x^{0.1}\,dx - 3\int x^{-0.5}\,dx$

$= 0.55\left(\frac{x^{1.1}}{1.1}\right) - 3\left(\frac{x^{0.5}}{0.5}\right) + C$

$= 0.5x^{1.1} - 6x^{0.5} + C$

b) $\int \left(8x^3 - \frac{2}{\sqrt{x}} + \frac{5}{x^2}\right)dx = \int (8x^3 - 2x^{-\frac{1}{2}} + 5x^{-2})\,dx$

$= 8\int x^3\,dx - 2\int x^{-\frac{1}{2}}\,dx + 5\int x^{-2}\,dx$

$= 8\left(\frac{x^4}{4}\right) - 2\left(\frac{x^{\frac{1}{2}}}{\left(\frac{1}{2}\right)}\right) + 5\left(\frac{x^{-1}}{-1}\right) + C$

$= 2x^4 - 4x^{\frac{1}{2}} - 5x^{-1} + C = 2x^4 - 4\sqrt{x} - \frac{5}{x} + C$

c) $\int \left((\sqrt{x})^5 + \frac{1}{2\sqrt{x}}\right)dx = \int \left((x^{\frac{1}{2}})^5 + \frac{1}{2}x^{-\frac{1}{2}}\right)dx$

$= \int x^{\frac{5}{2}}\,dx + \frac{1}{2}\int x^{-\frac{1}{2}}\,dx$

$= \left(\frac{x^{\frac{7}{2}}}{\left(\frac{7}{2}\right)}\right) + \frac{1}{2}\left(\frac{x^{\frac{1}{2}}}{\left(\frac{1}{2}\right)}\right) + C$

$= \frac{2x^{\frac{7}{2}}}{7} + x^{\frac{1}{2}} + C = \frac{2}{7}(\sqrt{x})^7 + \sqrt{x} + C$

d) $\int \left(\sqrt{x}\left(7x^2 - 1 - \frac{2}{x}\right)\right)dx$

$= \int (x^{\frac{1}{2}}(7x^2 - 1 - 2x^{-1}))\,dx$

$= \int (7x^{\frac{5}{2}} - x^{\frac{1}{2}} - 2x^{-\frac{1}{2}})\,dx$

$= 7\int x^{\frac{5}{2}}\,dx - \int x^{\frac{1}{2}}\,dx - 2\int x^{-\frac{1}{2}}\,dx$

$= 7\left(\frac{x^{\frac{7}{2}}}{\left(\frac{7}{2}\right)}\right) - \left(\frac{x^{\frac{3}{2}}}{\left(\frac{3}{2}\right)}\right) - 2\left(\frac{x^{\frac{1}{2}}}{\left(\frac{1}{2}\right)}\right) + C$

$= 2x^{\frac{7}{2}} - \frac{2}{3}x^{\frac{3}{2}} - 4x^{\frac{1}{2}} + C$

$= 2(\sqrt{x})^7 - \frac{2}{3}(\sqrt{x})^3 - 4\sqrt{x} + C$

e) $\int (3x - 5\sqrt{x})^2\,dx = \int (9x^2 - 30x\sqrt{x} + 25x)\,dx$

$= \int (9x^2 - 30x^{\frac{3}{2}} + 25x)\,dx$

$= 9\int x^2\,dx - 30\int x^{\frac{3}{2}}\,dx + 25\int x\,dx$

$= 9\left(\frac{x^3}{3}\right) - 30\left(\frac{2}{5}x^{\frac{5}{2}}\right) + 25\left(\frac{1}{2}x^2\right) + C$

$= 3x^3 - 12(\sqrt{x})^5 + \frac{25}{2}x^2 + C$

f) $\int \left(\frac{2x^3 - \sqrt{x}}{x}\right) dx = \int \left(\frac{2x^3}{x} - \frac{\sqrt{x}}{x}\right) dx$
$= \int (2x^2 - x^{-\frac{1}{2}})\, dx$
$= 2\int x^2\, dx - \int x^{-\frac{1}{2}}\, dx$
$= 2\left(\frac{x^3}{3}\right) - \left(\frac{x^{\frac{1}{2}}}{\left(\frac{1}{2}\right)}\right) + C = \frac{2}{3}x^3 - 2\sqrt{x} + C$

g) $\int \left(\frac{(5x-3)^2}{\sqrt{x}}\right) dx = \int \left(\frac{(25x^2 - 30x + 9)}{\sqrt{x}}\right) dx$
$= \int \left(\frac{25x^2}{\sqrt{x}} - \frac{30x}{\sqrt{x}} + \frac{9}{\sqrt{x}}\right) dx$
$= \int (25x^{\frac{3}{2}} - 30x^{\frac{1}{2}} + 9x^{-\frac{1}{2}})\, dx$
$= 25\int x^{\frac{3}{2}}\, dx - 30\int x^{\frac{1}{2}}\, dx + 9\int x^{-\frac{1}{2}}\, dx$
$= 25\left(\frac{x^{\frac{5}{2}}}{\left(\frac{5}{2}\right)}\right) - 30\left(\frac{x^{\frac{3}{2}}}{\left(\frac{3}{2}\right)}\right) + 9\left(\frac{x^{\frac{1}{2}}}{\left(\frac{1}{2}\right)}\right) + C$
$= 10x^{\frac{5}{2}} - 20x^{\frac{3}{2}} + 18x^{\frac{1}{2}} + C$
$= 10(\sqrt{x})^5 - 20(\sqrt{x})^3 + 18\sqrt{x} + C$

h) $\int (x^{\frac{1}{2}} + 1)(x^{-\frac{1}{2}} - 3)\, dx = \int (1 - 3x^{\frac{1}{2}} + x^{-\frac{1}{2}} - 3)\, dx$
$= \int (x^{-\frac{1}{2}} - 3x^{\frac{1}{2}} - 2)\, dx$
$= \int x^{-\frac{1}{2}}\, dx - 3\int x^{\frac{1}{2}}\, dx - 2\int 1\, dx$
$= \left(\frac{x^{\frac{1}{2}}}{\left(\frac{1}{2}\right)}\right) - 3\left(\frac{x^{\frac{3}{2}}}{\left(\frac{3}{2}\right)}\right) - 2\left(\frac{x^1}{1}\right) + C$
$= 2x^{\frac{1}{2}} - 2x^{\frac{3}{2}} - 2x + C$

i) $\int (x(2x + \sqrt{x})^2)\, dx = \int (x(4x^2 + 4x^{\frac{3}{2}} + x))\, dx$
$= \int (4x^3 + 4x^{\frac{5}{2}} + x^2)\, dx$
$= 4\int x^3\, dx + 4\int x^{\frac{5}{2}}\, dx + \int x^2\, dx$
$= 4\left(\frac{x^4}{4}\right) + 4\left(\frac{x^{\frac{7}{2}}}{\left(\frac{7}{2}\right)}\right) + \frac{x^3}{3} + C$
$= x^4 + \frac{8x^{\frac{7}{2}}}{7} + \frac{x^3}{3} + C$

j) $\int (\sqrt{x} + 2\sqrt[3]{x})^2\, dx = \int (x^{\frac{1}{2}} + 2x^{\frac{1}{3}})^2\, dx$
$= \int (x + 4x^{\frac{5}{6}} + 4x^{\frac{2}{3}})\, dx$
$= \int x\, dx + 4\int x^{\frac{5}{6}}\, dx + 4\int x^{\frac{2}{3}}\, dx$
$= \frac{x^2}{2} + 4\left(\frac{x^{\frac{11}{6}}}{\left(\frac{11}{6}\right)}\right) + 4\left(\frac{x^{\frac{5}{3}}}{\left(\frac{5}{3}\right)}\right) + C$
$= \frac{x^2}{2} + \frac{24x^{\frac{11}{6}}}{11} + \frac{12x^{\frac{5}{3}}}{5} + C$

Q4 $y = \int \frac{dy}{dx}\, dx = \int \left(1.5x^2 - \frac{4}{x^3}\right) dx$
$= \int (1.5x^2 - 4x^{-3})\, dx = 1.5\int x^2\, dx - 4\int x^{-3}\, dx$
$= 1.5\left(\frac{x^3}{3}\right) - 4\left(\frac{x^{-2}}{-2}\right) + C = \frac{x^3}{2} + \frac{2}{x^2} + C$

Q5 $f(x) = \int f'(x)\, dx = \int \left(\frac{4}{3(x^{\frac{1}{3}})^4} + 5x^{\frac{3}{2}}\right) dx$
$= \int \left(\frac{4}{3x^{\frac{4}{3}}} + 5x^{\frac{3}{2}}\right) dx = \int \left(\frac{4}{3}x^{-\frac{4}{3}} + 5x^{\frac{3}{2}}\right) dx$
$= \frac{4}{3}\int x^{-\frac{4}{3}}\, dx + 5\int x^{\frac{3}{2}}\, dx = \frac{4}{3}\left(\frac{x^{-\frac{1}{3}}}{\left(-\frac{1}{3}\right)}\right) + 5\frac{x^{\frac{5}{2}}}{\left(\frac{5}{2}\right)} + C$
$= -4x^{-\frac{1}{3}} + 2x^{\frac{5}{2}} + C \left(= -\frac{4}{\sqrt[3]{x}} + 2(\sqrt{x})^5 + C\right)$

Q6 a) $\int \left(4x^2 + \frac{3}{\sqrt{x}} - 2\right) dx$
$= \int (4x^2 + 3x^{-\frac{1}{2}} - 2)\, dx$
$= 4\int x^2\, dx + 3\int x^{-\frac{1}{2}}\, dx - 2\int x^0\, dx$
$= 4\left(\frac{x^3}{3}\right) + 3\left(\frac{x^{\frac{1}{2}}}{\frac{1}{2}}\right) - 2\left(\frac{x^1}{1}\right) + C$
$= \frac{4x^3}{3} + 6x^{\frac{1}{2}} - 2x + C = \frac{4x^3}{3} + 6\sqrt{x} - 2x + C$

b) $\int (3\sqrt{x} + 3)^2\, dx$
$= \int (9x + 18\sqrt{x} + 9)\, dx$
$= 9\int x\, dx + 18\int x^{\frac{1}{2}}\, dx + 9\int x^0\, dx$
$= 9\left(\frac{x^2}{2}\right) + 18\left(\frac{x^{\frac{3}{2}}}{\frac{3}{2}}\right) + 9\left(\frac{x^1}{1}\right) + C$
$= \frac{9x^2}{2} + 12x^{\frac{3}{2}} + 9x + C = \frac{9x^2}{2} + 12\sqrt{x^3} + 9x + C$

c) $\int \left(\frac{(\sqrt{x} + 3)(\sqrt{x} - 1)}{\sqrt{x}}\right) dx = \int \left(\frac{x + 2\sqrt{x} - 3}{\sqrt{x}}\right) dx$
$= \int \left(\frac{x}{\sqrt{x}} + \frac{2\sqrt{x}}{\sqrt{x}} - \frac{3}{\sqrt{x}}\right) dx$
$= \int (x^{\frac{1}{2}} + 2 - 3x^{-\frac{1}{2}})\, dx$
$= \int x^{\frac{1}{2}}\, dx + 2\int 1\, dx - 3\int x^{-\frac{1}{2}}\, dx$
$= \left(\frac{x^{\frac{3}{2}}}{\left(\frac{3}{2}\right)}\right) + 2\left(\frac{x^1}{1}\right) - 3\left(\frac{x^{\frac{1}{2}}}{\left(\frac{1}{2}\right)}\right) + C$
$= \frac{2}{3}x^{\frac{3}{2}} + 2x - 6x^{\frac{1}{2}} + C$
$= \frac{2}{3}(\sqrt{x})^3 + 2x - 6\sqrt{x} + C$

d) $\int \left(\sqrt{x}\left(\sqrt{x} - \frac{1}{\sqrt{x}}\right)^2\right) dx = \int \left(\sqrt{x}\left(x - 2 + \frac{1}{x}\right)\right) dx$
$= \int \left(x\sqrt{x} - 2\sqrt{x} + \frac{\sqrt{x}}{x}\right) dx$
$= \int (x^{\frac{3}{2}} - 2x^{\frac{1}{2}} + x^{-\frac{1}{2}})\, dx$
$= \int x^{\frac{3}{2}}\, dx - 2\int x^{\frac{1}{2}}\, dx + \int x^{-\frac{1}{2}}\, dx$
$= \left(\frac{x^{\frac{5}{2}}}{\left(\frac{5}{2}\right)}\right) - 2\left(\frac{x^{\frac{3}{2}}}{\left(\frac{3}{2}\right)}\right) + \left(\frac{x^{\frac{1}{2}}}{\left(\frac{1}{2}\right)}\right) + C$
$= \frac{2}{5}x^{\frac{5}{2}} - \frac{4}{3}x^{\frac{3}{2}} + 2x^{\frac{1}{2}} + C$
$= \frac{2}{5}(\sqrt{x})^5 - \frac{4}{3}(\sqrt{x})^3 + 2\sqrt{x} + C$

e) $\int \left(\frac{(\sqrt{x^3} + 6)(\sqrt{x^3} - 6)}{\sqrt{x^3}}\right) dx = \int \left(\frac{x^3 - 36}{x^{\frac{3}{2}}}\right) dx$
$= \int (x^{\frac{3}{2}} - 36x^{-\frac{3}{2}})\, dx$
$= \int x^{\frac{3}{2}}\, dx - 36\int x^{-\frac{3}{2}}\, dx$
$= \left(\frac{x^{\frac{5}{2}}}{\frac{5}{2}}\right) - 36\left(\frac{x^{-\frac{1}{2}}}{\left(-\frac{1}{2}\right)}\right) + C$
$= \frac{2x^{\frac{5}{2}}}{5} + 72x^{-\frac{1}{2}} + C$
$= \frac{2\sqrt{x^5}}{5} + \frac{72}{\sqrt{x}} + C$

f) $\int \left(\frac{3x - \sqrt{x} - 2}{\sqrt{x} - 1}\right) dx = \int \left(\frac{(3\sqrt{x} + 2)(\sqrt{x} - 1)}{\sqrt{x} - 1}\right) dx$
$= \int (3x^{\frac{1}{2}} + 2)\, dx$
$= 3\int x^{\frac{1}{2}}\, dx + 2\int x^0\, dx$
$= 3\left(\frac{x^{\frac{3}{2}}}{\frac{3}{2}}\right) + 2\left(\frac{x^1}{1}\right) + C$
$= 2x^{\frac{3}{2}} + 2x + C$
$= 2\sqrt{x^3} + 2x + C$

Q7 $\int \left(\frac{2n^2x^2 + 3x^3}{nx}\right) dx = \int \left(\frac{2n^2x^2}{nx} + \frac{3x^3}{nx}\right) dx$
$= \int \left(2nx + \frac{3x^2}{n}\right) dx$
$= 2n\int x\, dx + \frac{3}{n}\int x^2\, dx$
$= 2n\left(\frac{x^2}{2}\right) + \frac{3}{n}\left(\frac{x^3}{3}\right) + C$
$= nx^2 + \frac{x^3}{n} + C$

Exercise 11.1.3 — Integrating to find equations of curves

Q1 a) $f(x) = \int f'(x)\,dx = \int 4x^3\,dx = 4\int x^3\,dx$

$= 4\left(\frac{x^4}{4}\right) + C = x^4 + C$

At the point (0, 5), $x = 0$ and $f(x) = y = 5$, so $5 = 0^4 + C$. So $C = 5$ and $f(x) = x^4 + 5$.

b) $f(x) = \int f'(x)\,dx = \int (3x^2 - 4x + 3)\,dx$

$= 3\int x^2\,dx - 4\int x\,dx + 3\int 1\,dx$

$= 3\left(\frac{x^3}{3}\right) - 4\left(\frac{x^2}{2}\right) + 3\left(\frac{x^1}{1}\right) + C = x^3 - 2x^2 + 3x + C$

At the point (1, –3) $x = 1$ and $f(x) = y = -3$, so $-3 = 1^3 - 2(1^2) + 3(1) + C = 2 + C$. So $C = -5$ and $f(x) = x^3 - 2x^2 + 3x - 5$.

c) $f(x) = \int f'(x)\,dx = \int 6x(x + 2)\,dx$

$= \int (6x^2 + 12x)\,dx = 6\int x^2\,dx + 12\int x\,dx$

$= 6\left(\frac{x^3}{3}\right) + 12\left(\frac{x^2}{2}\right) + C = 2x^3 + 6x^2 + C$

At the point (–1, 1) $x = -1$ and $f(x) = y = 1$, so $1 = 2(-1)^3 + 6(-1)^2 + C = 4 + C$. So $C = -3$ and $f(x) = 2x^3 + 6x^2 - 3$.

d) $f(x) = \int f'(x)\,dx = \int \left(\frac{5}{x^2} + 2x\right)dx$

$= \int (5x^{-2} + 2x)\,dx$

$= 5\int x^{-2}\,dx + 2\int x\,dx$

$= 5\left(\frac{x^{-1}}{-1}\right) + 2\left(\frac{x^2}{2}\right) + C = -\frac{5}{x} + x^2 + C$

At the point (5, 4) $x = 5$ and $f(x) = y = 4$, so $4 = -\frac{5}{5} + 5^2 + C = 24 + C$. So $C = -20$ and $f(x) = -\frac{5}{x} + x^2 - 20$.

e) $f(x) = \int f'(x)\,dx = \int 3x^2(x - 4)\,dx$

$= \int (3x^3 - 12x^2)\,dx$

$= 3\int x^3\,dx - 12\int x^2\,dx$

$= 3\left(\frac{x^4}{4}\right) - 12\left(\frac{x^3}{3}\right) + C = \frac{3}{4}x^4 - 4x^3 + C$

At the point (2, –10) $x = 2$ and $f(x) = y = -10$, so $-10 = \frac{3}{4}(2^4) - 4(2^3) + C = -20 + C$. So $C = 10$ and $f(x) = \frac{3}{4}x^4 - 4x^3 + 10$.

f) $f(x) = \int f'(x)\,dx = \int (3x + 1)(x - 1)\,dx$

$= \int (3x^2 - 2x - 1)\,dx$

$= 3\int x^2\,dx - 2\int x\,dx - \int 1\,dx$

$= 3\left(\frac{x^3}{3}\right) - 2\left(\frac{x^2}{2}\right) - \left(\frac{x^1}{1}\right) + C = x^3 - x^2 - x + C$

At the point (3, –3) $x = 3$ and $f(x) = y = -3$, so $-3 = 3^3 - 3^2 - 3 + C = 15 + C$. So $C = -18$ and $f(x) = x^3 - x^2 - x - 18$.

g) $f(x) = \int f'(x)\,dx = \int x\left(x + \frac{3}{x^3}\right)dx$

$= \int \left(x^2 + \frac{3}{x^2}\right)dx = \int x^2\,dx + 3\int x^{-2}\,dx$

$= \frac{x^3}{3} + 3\left(\frac{x^{-1}}{-1}\right) + C = \frac{x^3}{3} - \frac{3}{x} + C$

At the point (–3, 5) $x = -3$ and $f(x) = y = 5$, so $5 = \frac{(-3)^3}{3} - \frac{3}{-3} + C = -8 + C$. So $C = 13$ and $f(x) = \frac{x^3}{3} - \frac{3}{x} + 13$.

h) $f(x) = \int f'(x)\,dx = \int \frac{9x^3 + 2x^{-2}}{x}\,dx$

$= \int \left(\frac{9x^3}{x} + \frac{2x^{-2}}{x}\right)dx = \int (9x^2 + 2x^{-3})\,dx$

$= 9\int x^2\,dx + 2\int x^{-3}\,dx$

$= 9\left(\frac{x^3}{3}\right) + 2\left(\frac{x^{-2}}{-2}\right) + C = 3x^3 - \frac{1}{x^2} + C$

At the point (–1, 2) $x = -1$ and $f(x) = y = 2$, so $2 = 3(-1)^3 - \frac{1}{(-1)^2} + C = -4 + C$. So $C = 6$ and $f(x) = 3x^3 - \frac{1}{x^2} + 6$.

i) $f(x) = \int f'(x)\,dx = \int \sqrt{x}(3 + x)\,dx$

$= \int (3\sqrt{x} + x\sqrt{x})\,dx = \int (3x^{\frac{1}{2}} + x^{\frac{3}{2}})\,dx$

$= 3\int x^{\frac{1}{2}}\,dx + \int x^{\frac{3}{2}}\,dx$

$= 3\left(\frac{x^{\frac{3}{2}}}{\frac{3}{2}}\right) + \frac{x^{\frac{5}{2}}}{\frac{5}{2}} + C = 2x^{\frac{3}{2}} + \frac{2x^{\frac{5}{2}}}{5} + C$

At the point (1, 4), $x = 1$ and $f(x) = y = 4$, so $4 = 2(1)^{\frac{3}{2}} + \frac{2(1)^{\frac{5}{2}}}{5} + C = \frac{12}{5} + C$. So $C = \frac{8}{5}$ and $f(x) = 2x^{\frac{3}{2}} + \frac{2x^{\frac{5}{2}}}{5} + \frac{8}{5}$

j) $f(x) = \int f'(x)\,dx = \int \frac{2 - 3\sqrt[6]{x}}{\sqrt{x}}\,dx$

$= \int \left(\frac{2}{\sqrt{x}} - \frac{3x^{\frac{1}{6}}}{\sqrt{x}}\right)dx = \int (2x^{-\frac{1}{2}} - 3x^{-\frac{1}{3}})\,dx$

$= 2\int x^{-\frac{1}{2}}\,dx - 3\int x^{-\frac{1}{3}}\,dx$

$= 2\left(\frac{x^{\frac{1}{2}}}{\frac{1}{2}}\right) - 3\left(\frac{x^{\frac{2}{3}}}{\frac{2}{3}}\right) + C = 4x^{\frac{1}{2}} - \frac{9x^{\frac{2}{3}}}{2} + C$

At the point (64, –33), $x = 64$ and $f(x) = y = -33$, so $-33 = 4(64)^{\frac{1}{2}} - \frac{9(64)^{\frac{2}{3}}}{2} + C = -40 + C$. So $C = 7$ and $f(x) = 4x^{\frac{1}{2}} - \frac{9x^{\frac{2}{3}}}{2} + 7$

Q2 a) $y = f(x) = \int f'(x)\,dx = \int (6x^2 + 6x - 5)\,dx$

$= 6\int x^2\,dx + 6\int x\,dx - 5\int 1\,dx$

$= 6\left(\frac{x^3}{3}\right) + 6\left(\frac{x^2}{2}\right) - 5\left(\frac{x^1}{1}\right) + C$

$= 2x^3 + 3x^2 - 5x + C$

At the origin, $x = 0$ and $y = 0$, so $0 = 2(0)^3 + 3(0)^2 - 5(0) + C = 0 + C$. So $C = 0$ and $y = 2x^3 + 3x^2 - 5x$.

b) $y = 2x^3 + 3x^2 - 5x = x(2x^2 + 3x - 5) = x(2x + 5)(x - 1)$

So when $y = 0$, $x = 0$ or $x = -\frac{5}{2}$ or $x = 1$.

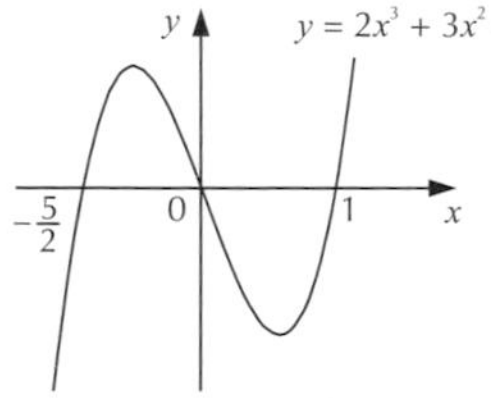

Q3 $y = f(x) = \int f'(x)\,dx = \int \left(\frac{3}{\sqrt{x}} + 2x\right)dx$

$= \int (3x^{-\frac{1}{2}} + 2x)\,dx = 3\int x^{-\frac{1}{2}}\,dx + 2\int x\,dx$

$= 3\left(\frac{x^{\frac{1}{2}}}{\left(\frac{1}{2}\right)}\right) + 2\left(\frac{x^2}{2}\right) + C = 6x^{\frac{1}{2}} + x^2 + C = 6\sqrt{x} + x^2 + C$

At the point (4, 9) $x = 4$ and $y = 9$, so $9 = 6\sqrt{4} + 4^2 + C = 28 + C$. So $C = -19$ and $y = 6\sqrt{x} + x^2 - 19$.

Q4 $y = \int \frac{dy}{dx}\,dx = \int \left(3\sqrt{x} + \frac{1}{x^2}\right) dx = \int (3x^{\frac{1}{2}} + x^{-2})\,dx$

$= 3\int x^{\frac{1}{2}}\,dx + \int x^{-2}\,dx = 3\left(\frac{x^{\frac{3}{2}}}{\left(\frac{3}{2}\right)}\right) + \left(\frac{x^{-1}}{-1}\right) + C$

$= 2x^{\frac{3}{2}} - \frac{1}{x} + C = 2(\sqrt{x})^3 - \frac{1}{x} + C$

At the point (1, 7), $x = 1$ and $y = 7$, so
$7 = 2((\sqrt{1})^3) - \frac{1}{1} + C = 1 + C$.
So C = 6 and $y = 2(\sqrt{x})^3 - \frac{1}{x} + 6$.

Q5 $y = \int \frac{dy}{dt}\,dt = \int (\sqrt{t} - 3)^2\,dt = \int (t - 6\sqrt{t} + 9)\,dt$

$= \int t\,dt - 6\int t^{\frac{1}{2}}\,dt + 9\int 1\,dt$

$= \frac{t^2}{2} - 6\left(\frac{t^{\frac{3}{2}}}{\left(\frac{3}{2}\right)}\right) + 9\left(\frac{t^1}{1}\right) + C$

$= \frac{t^2}{2} - 4t^{\frac{3}{2}} + 9t + C = \frac{t^2}{2} - 4(\sqrt{t})^3 + 9t + C$

When $t = 4$, $y = 9$ so $9 = \frac{4^2}{2} - 4(\sqrt{4})^3 + 9(4) + C = 12 + C$.
So C = –3 and $y = \frac{t^2}{2} - 4(\sqrt{t})^3 + 9t - 3$.

Q6 $f(x) = \int f'(x)\,dx = \int (\sqrt{x}(5x - 1))\,dx$

$= \int (5x\sqrt{x} - \sqrt{x})\,dx = \int (5x^{\frac{3}{2}} - x^{\frac{1}{2}})\,dx$

$= 5\int x^{\frac{3}{2}}\,dx - \int x^{\frac{1}{2}}\,dx = 5\left(\frac{x^{\frac{5}{2}}}{\left(\frac{5}{2}\right)}\right) - \left(\frac{x^{\frac{3}{2}}}{\left(\frac{3}{2}\right)}\right) + C$

$= 2x^{\frac{5}{2}} - \frac{2}{3}x^{\frac{3}{2}} + C = 2(\sqrt{x})^5 - \frac{2}{3}(\sqrt{x})^3 + C$

When $x = 1$, $f(x) = y = \frac{1}{3}$ so $\frac{1}{3} = 2(\sqrt{1})^5 - \frac{2}{3}(\sqrt{1})^3 + C = \frac{4}{3} + C$.
So C = –1 and $f(x) = 2(\sqrt{x})^5 - \frac{2}{3}(\sqrt{x})^3 - 1$

Q7 $y = f(x) = \int f'(x)\,dx = \int \left(x^2 + \frac{2}{x^{\frac{3}{2}}}\right) dx$

$= \int (x^2 + 2x^{-\frac{3}{2}})\,dx = \int x^2\,dx + 2\int x^{-\frac{3}{2}}\,dx$

$= \frac{x^3}{3} + 2\left(\frac{x^{-\frac{1}{2}}}{\left(-\frac{1}{2}\right)}\right) + C = \frac{x^3}{3} - \frac{4}{\sqrt{x}} + C$

When $x = 1$, $y = -\frac{5}{3}$ so $-\frac{5}{3} = \frac{1^3}{3} - \frac{4}{\sqrt{1}} + C = -\frac{11}{3} + C$
So C = 2 and $y = \frac{x^3}{3} - \frac{4}{\sqrt{x}} + 2$

Q8 $y = \int \frac{dy}{dx}\,dx = \int \left(\frac{x - 6}{x^3} + 2\right) dx$

$= \int \left(\frac{x}{x^3} - \frac{6}{x^3} + 2\right) dx = \int (x^{-2} - 6x^{-3} + 2)\,dx$

$= \int x^{-2}\,dx - 6\int x^{-3}\,dx + 2\int 1\,dx$

$= \left(\frac{x^{-1}}{-1}\right) - 6\left(\frac{x^{-2}}{-2}\right) + 2\left(\frac{x^1}{1}\right) + C = -\frac{1}{x} + \frac{3}{x^2} + 2x + C$

When $x = 3$, $y = -1$ so $-1 = -\frac{1}{3} + \frac{3}{3^2} + 2(3) + C = 6 + C$.
So C = –7 and $y = -\frac{1}{x} + \frac{3}{x^2} + 2x - 7$.

Q9 **a)** $\frac{dy}{dx} = \frac{(x + 2)(x - 2)}{\sqrt{x}} = \frac{x^2 - 4}{\sqrt{x}} = \frac{x^2}{x^{\frac{1}{2}}} - \frac{4}{x^{\frac{1}{2}}} = x^{\frac{3}{2}} - 4x^{-\frac{1}{2}}$

b) $y = \int \frac{dy}{dx}\,dx = \int (x^{\frac{3}{2}} - 4x^{-\frac{1}{2}})\,dx$

$= \int x^{\frac{3}{2}}\,dx - 4\int x^{-\frac{1}{2}}\,dx = \left(\frac{x^{\frac{5}{2}}}{\frac{5}{2}}\right) - 4\left(\frac{x^{\frac{1}{2}}}{\frac{1}{2}}\right) + C$

$= \frac{2x^{\frac{5}{2}}}{5} - 8x^{\frac{1}{2}} + C = \frac{2\sqrt{x^5}}{5} - 8\sqrt{x} + C$

At point $\left(1, \frac{7}{5}\right)$, $\frac{7}{5} = \frac{2\sqrt{(1)^5}}{5} - 8\sqrt{1} + C = -\frac{38}{5} + C$

So C = 9 and $y = \frac{2\sqrt{x^5}}{5} - 8\sqrt{x} + 9$

Q10 $v = \int \frac{dv}{dt}\,dt = \int (6t - t^2)\,dt = 6\int t\,dt - \int t^2\,dt$

$= 6\left(\frac{t^2}{2}\right) - \frac{t^3}{3} + C = 3t^2 - \frac{t^3}{3} + C$

When $v = 0$, $t = 10$, so $0 = 3(10)^2 - \frac{10^3}{3} + C = -\frac{100}{3} + C$,

$\Rightarrow C = \frac{100}{3} \Rightarrow v = 3t^2 - \frac{t^3}{3} + \frac{100}{3}$

So when $t = 0$, $v = \frac{100}{3} = 33.3$ ms^{-1} (3 s.f.).

Q11 When $x = 2$, $f'(x) = 18$, so $3(2)^2 + 2k - k = 18$
$\Rightarrow 12 + k = 18 \Rightarrow k = 6$, so $f'(x) = 3x^2 + 6x - 6$.

$f(x) = \int f'(x)\,dx = \int (3x^2 + 6x - 6)\,dx$

$= 3\int x^2\,dx + 6\int x\,dx - 6\int 1\,dx$

$= 3\left(\frac{x^3}{3}\right) + 6\left(\frac{x^2}{2}\right) - 6x + C$

$= x^3 + 3x^2 - 6x + C$

When $x = 2$, $y = 7$, so $7 = 2^3 + 3(2)^2 - 6(2) + C$
$\Rightarrow 7 = 8 + C \Rightarrow C = -1$
So $f(x) = x^3 + 3x^2 - 6x - 1$.

11.2 Definite Integration

Exercise 11.2.1 — Evaluating definite integrals

Q1 **a)** $\int_{-2}^{0} (4x^3 + 2x)\,dx = [x^4 + x^2]_{-2}^{0}$

$= (0^4 + 0^2) - ((-2)^4 + (-2)^2)$

$= -(16 + 4) = -20$

b) $\int_{-2}^{5} (x^3 + x)\,dx = \left[\frac{x^4}{4} + \frac{x^2}{2}\right]_{-2}^{5}$

$= \left(\frac{5^4}{4} + \frac{5^2}{2}\right) - \left(\frac{(-2)^4}{4} + \frac{(-2)^2}{2}\right)$

$= \frac{625}{4} + \frac{25}{2} - \frac{16}{4} - \frac{4}{2} = \frac{651}{4}$

c) $\int_{-5}^{-2} (x + 1)^2\,dx = \int_{-5}^{-2} (x^2 + 2x + 1)\,dx$

$= \left[\frac{x^3}{3} + x^2 + x\right]_{-5}^{-2}$

$= \left(\frac{(-2)^3}{3} + (-2)^2 + (-2)\right) - \left(\frac{(-5)^3}{3} + (-5)^2 + (-5)\right)$

$= \left(\frac{-8}{3} + 4 - 2\right) - \left(\frac{-125}{3} + 25 - 5\right) = 21$

d) $\int_{3}^{4} (6x^{-4} + x^{-2})\,dx = \left[\frac{6x^{-3}}{-3} + \frac{x^{-1}}{-1}\right]_{3}^{4}$

$= \left[-\frac{2}{x^3} - \frac{1}{x}\right]_{3}^{4} = \left(-\frac{2}{4^3} - \frac{1}{4}\right) - \left(-\frac{2}{3^3} - \frac{1}{3}\right)$

$= -\frac{2}{64} - \frac{1}{4} + \frac{2}{27} + \frac{1}{3} = \frac{109}{864}$

e) $\int_{1}^{2} \left(x^2 + \frac{1}{x^2}\right) dx = \int_{1}^{2} (x^2 + x^{-2})\,dx = \left[\frac{x^3}{3} + \frac{x^{-1}}{-1}\right]_{1}^{2}$

$= \left[\frac{x^3}{3} - \frac{1}{x}\right]_{1}^{2} = \left(\frac{2^3}{3} - \frac{1}{2}\right) - \left(\frac{1^3}{3} - \frac{1}{1}\right)$

$= \frac{8}{3} - \frac{1}{2} - \frac{1}{3} + 1 = \frac{17}{6}$

f) $\int_{1}^{4} (3x^{-4} + \sqrt{x})\,dx = \int_{1}^{4} (3x^{-4} + x^{\frac{1}{2}})\,dx$

$= \left[\frac{3x^{-3}}{-3} + \frac{x^{\frac{3}{2}}}{\left(\frac{3}{2}\right)}\right]_{1}^{4} = \left[-\frac{1}{x^3} + \frac{2}{3}(\sqrt{x})^3\right]_{1}^{4}$

$= \left(-\frac{1}{4^3} + \frac{2}{3}(\sqrt{4})^3\right) - \left(-\frac{1}{1^3} + \frac{2}{3}(\sqrt{1})^3\right)$

$= \left(-\frac{1}{64} + \frac{2}{3} \times 2^3\right) - \left(-1 + \frac{2}{3}\right) = \frac{1085}{192}$

g) $\int_0^1 (2x+3)(x+2)\,dx = \int_0^1 (2x^2+7x+6)\,dx$

$= \left[\frac{2x^3}{3} + \frac{7x^2}{2} + 6x\right]_0^1$

$= \left(\frac{2\times1^3}{3} + \frac{7\times1^2}{2} + (6\times1)\right) - \left(\frac{2\times0^3}{3} + \frac{7\times0^2}{2} + (6\times0)\right)$

$= \left(\frac{2}{3} + \frac{7}{2} + 6\right) - 0 = \frac{61}{6}$

h) $\int_1^4 \frac{x^2+2}{\sqrt{x}}\,dx = \int_1^4 (x^{\frac{3}{2}} + 2x^{-\frac{1}{2}})\,dx$

$= \left[\frac{x^{\frac{5}{2}}}{\left(\frac{5}{2}\right)} + 2\frac{x^{\frac{1}{2}}}{\left(\frac{1}{2}\right)}\right]_1^4 = \left[\frac{2}{5}(\sqrt{x})^5 + 4\sqrt{x}\right]_1^4$

$= \left(\frac{2}{5}(\sqrt{4})^5 + 4\sqrt{4}\right) - \left(\frac{2}{5}(\sqrt{1})^5 + 4\sqrt{1}\right)$

$= \left(\frac{2}{5}\times2^5 + 8\right) - \left(\frac{2}{5} + 4\right)$

$= \frac{64}{5} + 8 - \frac{2}{5} - 4 = \frac{82}{5}$

i) $\int_4^9 \left(\frac{1}{x} + \sqrt{x}\right)^2 dx = \int_4^9 \left(\frac{1}{x^2} + 2\frac{\sqrt{x}}{x} + x\right) dx$

$= \int_4^9 (x^{-2} + 2x^{-\frac{1}{2}} + x)\,dx$

$= \left[\frac{x^{-1}}{-1} + \frac{2x^{\frac{1}{2}}}{\left(\frac{1}{2}\right)} + \frac{x^2}{2}\right]_4^9$

$= \left[-\frac{1}{x} + 4\sqrt{x} + \frac{x^2}{2}\right]_4^9$

$= \left(-\frac{1}{9} + 4\sqrt{9} + \frac{9^2}{2}\right) - \left(-\frac{1}{4} + 4\sqrt{4} + \frac{4^2}{2}\right)$

$= -\frac{1}{9} + 12 + \frac{81}{2} + \frac{1}{4} - 8 - 8 = \frac{1319}{36}$

j) $\int_4^{16} \left(\frac{\sqrt{x}-1}{x^2}\right) dx = \int_4^{16} (x^{-\frac{3}{2}} - x^{-2})\,dx$

$= \left[\frac{x^{-\frac{1}{2}}}{\left(-\frac{1}{2}\right)} - \frac{x^{-1}}{(-1)}\right]_4^{16} = [-2x^{-\frac{1}{2}} + x^{-1}]_4^{16} = \left[-\frac{2}{\sqrt{x}} + \frac{1}{x}\right]_4^{16}$

$= \left(-\frac{2}{\sqrt{16}} + \frac{1}{16}\right) - \left(-\frac{2}{\sqrt{4}} + \frac{1}{4}\right)$

$= -\frac{1}{2} + \frac{1}{16} + 1 - \frac{1}{4} = \frac{5}{16}$

k) $\int_1^9 x^{\frac{1}{2}}(5x - x^{-\frac{1}{2}})\,dx = \int_1^9 (5x^{\frac{3}{2}} - 1)\,dx$

$= \left[\frac{5x^{\frac{5}{2}}}{\frac{5}{2}} - \frac{x^1}{1}\right]_1^9 = [2x^{\frac{5}{2}} - x]_1^9$

$= (2(9)^{\frac{5}{2}} - 9) - (2(1)^{\frac{5}{2}} - 1)$

$= 486 - 9 - 2 + 1 = 476$

l) $\int_1^4 (1+x^2)(1+\sqrt{x})\,dx = \int_1^4 (1 + x^{\frac{5}{2}} + x^2 + x^{\frac{1}{2}})\,dx$

$= \left[\frac{x^1}{1} + \frac{x^{\frac{7}{2}}}{\frac{7}{2}} + \frac{x^3}{3} + \frac{x^{\frac{3}{2}}}{\frac{3}{2}}\right]_1^4 = \left[x + \frac{2x^{\frac{7}{2}}}{7} + \frac{x^3}{3} + \frac{2x^{\frac{3}{2}}}{3}\right]_1^4$

$= \left(4 + \frac{2(4)^{\frac{7}{2}}}{7} + \frac{4^3}{3} + \frac{2(4)^{\frac{3}{2}}}{3}\right)$

$- \left(1 + \frac{2(1)^{\frac{7}{2}}}{7} + \frac{1^3}{3} + \frac{2(1)^{\frac{3}{2}}}{3}\right)$

$= 4 + \frac{256}{7} + \frac{64}{3} + \frac{16}{3} - 1 - \frac{2}{7} - \frac{1}{3} - \frac{2}{3} = \frac{1364}{21}$

Q2 $\int_{-1}^3 (4x - 5x^3 + 7)\,dx = \left[4\left(\frac{x^2}{2}\right) - 5\left(\frac{x^4}{4}\right) + 7\left(\frac{x^1}{1}\right)\right]_{-1}^3$

$= \left[2x^2 - \frac{5x^4}{4} + 7x\right]_{-1}^3 = \left(18 - \frac{405}{4} + 21\right) - \left(2 - \frac{5}{4} - 7\right)$

$= -56$

Q3 $\int_0^1 (3 - 4\sqrt{x} + \frac{1}{2}x^2)\,dx = \left[3\left(\frac{x^1}{1}\right) - 4\left(\frac{x^{\frac{3}{2}}}{\frac{3}{2}}\right) + \frac{1}{2}\left(\frac{x^3}{3}\right)\right]_0^1$

$= \left[3x - \frac{8x^{\frac{3}{2}}}{3} + \frac{x^3}{6}\right]_0^1 = \left(3 - \frac{8}{3} + \frac{1}{6}\right) - (0 - 0 + 0) = \frac{1}{2}$

Q4 a) $\int_0^a x^3\,dx = \left[\frac{x^4}{4}\right]_0^a = \left(\frac{a^4}{4}\right) - \left(\frac{0^4}{4}\right) = \frac{a^4}{4}$

So $\frac{a^4}{4} = 64 \Rightarrow a^4 = 64 \times 4 = 256 \Rightarrow a = 4$

a can't be –4 since the question tells you that a > 0.

b) $\int_0^a 2x^4\,dx = \left[\frac{2x^5}{5}\right]_0^a = \left(\frac{2a^5}{5}\right) - \left(\frac{2(0)^5}{5}\right) = \frac{2a^5}{5}$

So $\frac{2a^5}{5} = 4a^4 \Rightarrow a^5 = 10a^4 \Rightarrow a = 10$

Q5 a) The area is all above the x-axis so just integrate:

$\int_0^4 (x + \sqrt{x})\,dx = \int_0^4 (x + x^{\frac{1}{2}})\,dx$

$= \left[\frac{x^2}{2} + \frac{x^{\frac{3}{2}}}{\left(\frac{3}{2}\right)}\right]_0^4 = \left[\frac{1}{2}x^2 + \frac{2}{3}(\sqrt{x})^3\right]_0^4$

$= \left(\frac{1}{2}(4)^2 + \frac{2}{3}(\sqrt{4})^3\right) - \left(\frac{1}{2}(0)^2 + \frac{2}{3}(\sqrt{0})^3\right)$

$= \left(\frac{16}{2} + \frac{2}{3}\times2^3\right) - 0 = \frac{40}{3}$

b) The limits aren't shown on the graph, but they are just the roots of the equation $0 = 4 - x^2$.

Set $y = 0$: $4 - x^2 = 0 \Rightarrow x^2 = 4 \Rightarrow x = 2$ or -2.

So the limits of integration are –2 and 2:

$\int_{-2}^2 (4 - x^2)\,dx = \left[4x - \frac{x^3}{3}\right]_{-2}^2$

$= \left((4\times2) - \frac{2^3}{3}\right) - \left((4\times(-2)) - \frac{(-2)^3}{3}\right)$

$= \left(8 - \frac{8}{3}\right) - \left(-8 - \frac{-8}{3}\right)$

$= 8 - \frac{8}{3} + 8 - \frac{8}{3} = \frac{32}{3}$

c) This area lies above and below the x-axis so you'll have to integrate the bits above and below the axis separately.

First you need to find the points where the curve crosses the axis: $y = x(x-1)(x-3)$ is already factorised, so it's easy.

If $x(x-1)(x-3) = 0$ then either $x = 0$, $x = 1$ or $x = 3$. So these are the three points where the curve crosses the axis.

The area above the x-axis is between 0 and 1 so integrate:

$\int_0^1 x(x-1)(x-3)\,dx = \int_0^1 (x^3 - 4x^2 + 3x)\,dx$

$= \left[\frac{x^4}{4} - \frac{4x^3}{3} + \frac{3x^2}{2}\right]_0^1$

$= \left(\frac{1^4}{4} - \frac{4\times1^3}{3} + \frac{3\times1^2}{2}\right) - \left(\frac{0^4}{4} - \frac{4\times0^3}{3} + \frac{3\times0^2}{2}\right)$

$= \frac{1}{4} - \frac{4}{3} + \frac{3}{2} - 0 = \frac{5}{12}$

The area below the x-axis is between 1 and 3, so integrate:

$\int_1^3 x(x-1)(x-3)\,dx = \int_1^3 (x^3 - 4x^2 + 3x)\,dx$

$= \left[\frac{x^4}{4} - \frac{4x^3}{3} + \frac{3x^2}{2}\right]_1^3$

$= \left(\frac{3^4}{4} - \frac{4\times3^3}{3} + \frac{3\times3^2}{2}\right) - \left(\frac{1}{4} - \frac{4}{3} + \frac{3}{2}\right)$

$= \left(\frac{81}{4} - \frac{108}{3} + \frac{27}{2}\right) - \frac{5}{12} = -\frac{8}{3}$

Areas cannot be negative so the area of the bit below the x-axis is $\frac{8}{3}$. So the total area is $\frac{5}{12} + \frac{8}{3} = \frac{37}{12}$.

d) This area lies above and below the x-axis so you'll have to integrate the bits above and below the axis separately.
$y = x^2 + x - 6 = (x + 3)(x - 2)$, so the curve crosses the x-axis at $x = -3$ and $x = 2$. The area below the x-axis is between $x = -3$ and $x = 2$, so integrate:

$$\int_{-3}^{2} (x^2 + x - 6)\,dx = \left[\frac{x^3}{3} + \frac{x^2}{2} - 6x\right]_{-3}^{2}$$

$$= \left(\frac{2^3}{3} + \frac{2^2}{2} - 6 \times 2\right) - \left(\frac{(-3)^3}{3} + \frac{(-3)^2}{2} - 6 \times (-3)\right)$$

$$= \left(\frac{8}{3} + 2 - 12\right) - \left(-9 + \frac{9}{2} + 18\right) = -\frac{125}{6}$$

Areas cannot be negative so the area below the x-axis is $\frac{125}{6}$.

The area above the x-axis is between $x = 2$ and $x = 3$, so integrate:

$$\int_{2}^{3} (x^2 + x - 6)\,dx = \left[\frac{x^3}{3} + \frac{x^2}{2} - 6x\right]_{2}^{3}$$

$$= \left(\frac{3^3}{3} + \frac{3^2}{2} - 6 \times 3\right) - \left(\frac{2^3}{3} + \frac{2^2}{2} - 6 \times 2\right)$$

$$= \left(9 + \frac{9}{2} - 18\right) - \left(\frac{8}{3} + 2 - 12\right) = \frac{17}{6}$$

So the total area is $\frac{125}{6} + \frac{17}{6} = \frac{71}{3}$.

Q6 The graph of $y = (x - 1)(3x + 9)$ crosses the x-axis at $x = 1$ and $x = -3$.

a)

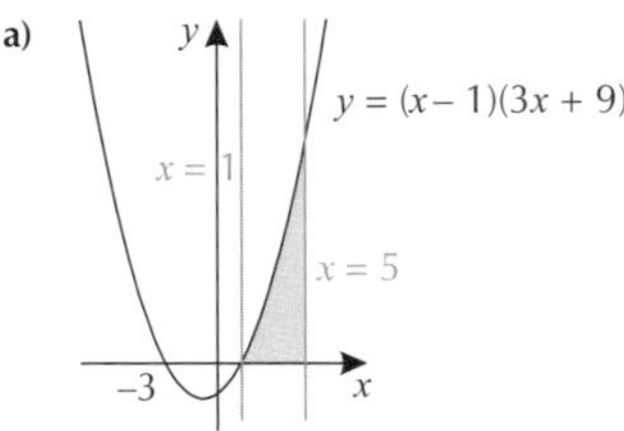

Work out the area above the x-axis between $x = 1$ and $x = 5$:

$$\int_{1}^{5} (x - 1)(3x + 9)\,dx = \int_{1}^{5} (3x^2 + 6x - 9)\,dx$$

$$= [x^3 + 3x^2 - 9x]_1^5$$

$= (5^3 + 3(5)^2 - 9(5)) - (1^3 + 3(1)^2 - 9(1)) = 155 + 5 = 160$

b)

Use the integration from part a) to work out the area above the x-axis between $x = -3$ and $x = -4$:

$$\int_{-4}^{-3} (x - 1)(3x + 9)\,dx = [x^3 + 3x^2 - 9x]_{-4}^{-3}$$

$= ((-3)^3 + 3(-3)^2 - 9(-3)) - ((-4)^3 + 3(-4)^2 - 9(-4))$
$= 27 - 20 = 7$

c)

Between $x = -2$ and $x = 2$, the graph crosses the x-axis at $x = 1$ which means the area lies both above and below the x-axis.

Work out the area between $x = -2$ and $x = 1$:

$$\int_{-2}^{1} (x - 1)(3x + 9)\,dx = [x^3 + 3x^2 - 9x]_{-2}^{1}$$

$= (1^3 + 3(1)^2 - 9(1)) - ((-2)^3 + 3(-2)^2 - 9(-2)) = -5 - 22 = -27$
So the area below the x-axis is 27.

Now work out the area between $x = 1$ and $x = 2$:

$$\int_{1}^{2} (x - 1)(3x + 9)\,dx = [x^3 + 3x^2 - 9x]_1^2$$

$= (2^3 + 3(2)^2 - 9(2)) - (1^3 + 3(1)^2 - 9(1)) = 2 - (-5) = 7$
So the area above the x-axis is 7.
Therefore the total area is $27 + 7 = 34$.

d)

Between $x = -6$ and $x = 0$, the graph crosses the x-axis at $x = -3$ which means the area lies both above and below the x-axis.

Work out the area between $x = -6$ and $x = -3$:

$$\int_{-6}^{-3} (x - 1)(3x + 9)\,dx = [x^3 + 3x^2 - 9x]_{-6}^{-3}$$

$= ((-3)^3 + 3(-3)^2 - 9(-3)) - ((-6)^3 + 3(-6)^2 - 9(-6))$
$= 27 - (-54) = 81$
So the area above the x-axis is 81.

Now work out the area between $x = -3$ and $x = 0$:

$$\int_{-3}^{0} (x - 1)(3x + 9)\,dx = [x^3 + 3x^2 - 9x]_{-3}^{0}$$

$= ((0)^3 + 3(0)^2 - 9(0)) - ((-3)^3 + 3(-3)^2 - 9(-3)) = 0 - 27 = -27$
Area cannot be negative, so the area below the x-axis is 27.
So the total area is $81 + 27 = 108$.

Q7 $y = \frac{20}{x^5}$ is positive between $x = 1$ and $x = 2$, so:

$$\int_{1}^{2} \frac{20}{x^5}\,dx = \int_{1}^{2} 20x^{-5}\,dx = \left[\frac{20x^{-4}}{-4}\right]_1^2 = \left[-\frac{5}{x^4}\right]_1^2$$

$$= \left(-\frac{5}{2^4}\right) - \left(-\frac{5}{1^4}\right) = -\frac{5}{16} + 5 = \frac{75}{16}$$

So the area is $\frac{75}{16}$.

Q8 In order to find this area, you need to split it into two sections and find the area of each section separately:

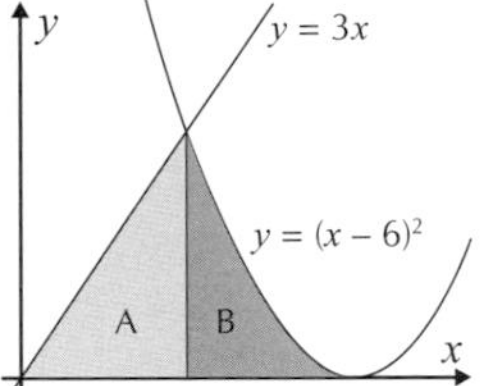

First, find the places where the graphs meet the x-axis.
$y = 3x$ crosses the x-axis at $x = 0$
$y = (x - 6)^2$ touches the x-axis at $x = 6$
The point of intersection is where the two lines meet, i.e. where $3x = (x - 6)^2$. So solve this to find x:
$3x = (x - 6)^2 = x^2 - 12x + 36$
$x^2 - 15x + 36 = 0$
$(x - 12)(x - 3) = 0$
So $x = 3$ or $x = 12$, and since you're looking for the intersection between $x = 0$ and $x = 6$, you want the solution $x = 3$.
Now the graph looks like this:

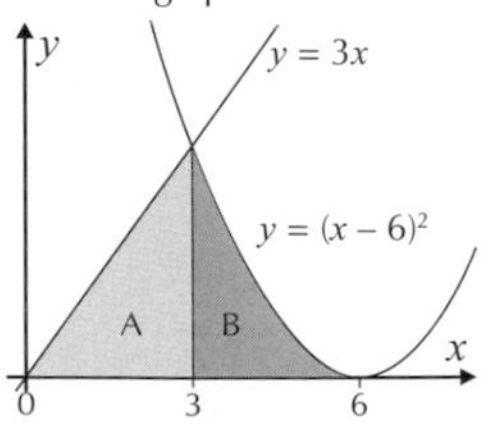

Now find the areas of the two sections.
Area A is a triangle with base length 3.
For the height, substitute $x = 3$ into the equation of the line. $y = 3 \times 3 = 9$
So the area of the triangle is $\frac{1}{2} \times 3 \times 9 = \frac{27}{2}$

Area B $= \int_3^6 (x-6)^2\, dx = \int_3^6 (x^2 - 12x + 36)\, dx$
$= \left[\frac{1}{3}x^3 - 6x^2 + 36x\right]_3^6$
$= (72 - 216 + 216) - (9 - 54 + 108) = 72 - 63 = 9$

So the total area is $\frac{27}{2} + 9 = \frac{45}{2}$.

Q9 $y = x^3 + 4x^2 + 3x = x(x^2 + 4x + 3) = x(x + 3)(x + 1)$

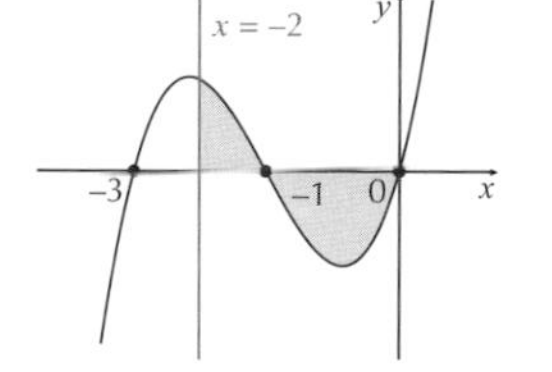

The graph of $y = x^3 + 4x^2 + 3x$ crosses the x-axis at $x = -3$, $x = -1$ and $x = 0$, so between $x = -2$ and $x = 0$ (the y-axis) the graph crosses the x-axis at $x = -1$ which means the area lies above and below the x-axis.
Work out the area between $x = -2$ and $x = -1$:
$\int_{-2}^{-1} (x^3 + 4x^2 + 3x)\, dx = \left[\frac{x^4}{4} + \frac{4x^3}{3} + \frac{3x^2}{2}\right]_{-2}^{-1}$
$= \left(\frac{(-1)^4}{4} + \frac{4(-1)^3}{3} + \frac{3(-1)^2}{2}\right) - \left(\frac{(-2)^4}{4} + \frac{4(-2)^3}{3} + \frac{3(-2)^2}{2}\right)$
$= \frac{5}{12} - \left(-\frac{2}{3}\right) = \frac{13}{12}$

So the area above the x-axis is $\frac{13}{12}$.
Work out the area between $x = -1$ and $x = 0$:
$\int_{-1}^{0} (x^3 + 4x^2 + 3x)\, dx = \left[\frac{x^4}{4} + \frac{4x^3}{3} + \frac{3x^2}{2}\right]_{-1}^{0}$
$= \left(\frac{(0)^4}{4} + \frac{4(0)^3}{3} + \frac{3(0)^2}{2}\right) - \left(\frac{(-1)^4}{4} + \frac{4(-1)^3}{3} + \frac{3(-1)^2}{2}\right)$
$= 0 - \frac{5}{12} = -\frac{5}{12}$

So the area below the x-axis is $\frac{5}{12}$.
Therefore the total area is $\frac{13}{12} + \frac{5}{12} = \frac{18}{12} = \frac{3}{2}$.

Q10 a) First, evaluate the integral, treating A as a constant.
$\int_2^3 (1 - 2Ax)\, dx = [x - Ax^2]_2^3 = (3 - 9A) - (2 - 4A) = 1 - 5A$
From the question, you know that this is equal to $6A^2$, so set up and solve the quadratic.
$1 - 5A = 6A^2 \Rightarrow 6A^2 + 5A - 1 = 0$
$(6A - 1)(A + 1) = 0$, so $A = \frac{1}{6}$ or $A = -1$

b) Again, integrate the function with constant A:
$\int_{-2}^{2} \left(\frac{21}{8}x^2 + \frac{A}{x^2}\right) dx = \int_{-2}^{2} \left(\frac{21}{8}x^2 + Ax^{-2}\right) dx$
$= \left[\frac{21}{8} \times \frac{1}{3}x^3 - Ax^{-1}\right]_{-2}^{2}$
$= \left[\frac{7}{8}x^3 - \frac{A}{x}\right]_{-2}^{2}$
$= \left(\frac{7}{8}(8) - \frac{A}{2}\right) - \left(\frac{7}{8}(-8) - \frac{A}{-2}\right)$
$= \left(7 - \frac{A}{2}\right) - \left((-7) + \frac{A}{2}\right) = 14 - A$
Set this equal to $3A^2$ from the question to form a quadratic in A.
$14 - A = 3A^2 \Rightarrow 3A^2 + A - 14 = 0$
$(3A + 7)(A - 2) = 0$, so $A = -\frac{7}{3}$ or $A = 2$

c) Evaluate the integral, treating A as a constant.
$\int_4^5 (2A^2 - 6x^2)\, dx = [2A^2x - 2x^3]_4^5 = (10A^2 - 250) - (8A^2 - 128)$
$= 2A^2 - 122$
From the question, you know that this is equal to $120A$, so set up and solve the quadratic.
$2A^2 - 122 = 120A \Rightarrow 2A^2 - 120A - 122 = 0$
$\Rightarrow 2(A + 1)(A - 61) = 0$
So $A = -1$ or $A = 61$.

d) Evaluate the integral, treating A as a constant.
$\int_1^4 (A - 2\sqrt{x})^2\, dx = \int_1^4 (A^2 - 4Ax^{\frac{1}{2}} + 4x)\, dx$
$= \left[A^2x - \frac{8Ax^{\frac{3}{2}}}{3} + 2x^2\right]_1^4$
$= (4A^2 - \frac{64A}{3} + 32) - (A^2 - \frac{8A}{3} + 2) = 3A^2 - \frac{56A}{3} + 30$
From the question, you know that this is equal to $10 - A^2$, so set up and solve the quadratic.
$3A^2 - \frac{56A}{3} + 30 = 10 - A^2$
$4A^2 - \frac{56A}{3} + 20 = 0 \Rightarrow 12A^2 - 56A + 60 = 0$
$\Rightarrow 3A^2 - 14A + 15 = 0 \Rightarrow (3A - 5)(A - 3) = 0$,
so $A = \frac{5}{3}$ or $A = 3$

Q11 a) $\int_1^3 (10t - t^2)\, dt = \left[5t^2 - \frac{1}{3}t^3\right]_1^3$
$= (45 - 9) - \left(5 - \frac{1}{3}\right) = \frac{94}{3}$ m (or $31\frac{1}{3}$ m)

b) First work out at what times the object's velocity is 0:
$10t - t^2 = 0 \Rightarrow t(10 - t) = 0$
So $t = 0$ or $t = 10$
This means the object starts at time $t = 0$ and comes to rest at $t = 10$. To find the total distance travelled, integrate between these limits:
$\int_0^{10} (10t - t^2)\, dt = \left[5t^2 - \frac{1}{3}t^3\right]_0^{10}$
$= \left(500 - \frac{1000}{3}\right) - 0$
$= \frac{500}{3}$ m (or $166\frac{2}{3}$ m)

Review Exercise — Chapter 11

Q1 a) $\int (x^{-\frac{1}{2}} + 4 - 5x^3)\, dx$
$= \int x^{-\frac{1}{2}}\, dx + 4\int x^0\, dx - 5\int x^3\, dx$
$= \frac{x^{\frac{1}{2}}}{\left(\frac{1}{2}\right)} + 4\left(\frac{x^1}{1}\right) - 5\left(\frac{x^4}{4}\right) + C$
$= 2\sqrt{x} + 4x - \frac{5x^4}{4} + C$

b) $\int (2x + 3x^{-2})\, dx = 2\int x\, dx + 3\int x^{-2}\, dx$
$= 2\left(\frac{x^2}{2}\right) + 3\left(\frac{x^{-1}}{-1}\right) + C = x^2 - \frac{3}{x} + C$

c) $\int \left(6x^2 - \frac{1}{3}x^{-\frac{1}{2}}\right) dx = 6\int x^2\, dx - \frac{1}{3}\int x^{-\frac{1}{2}}\, dx$
$= 6\left(\frac{x^3}{3}\right) - \frac{1}{3}\left(\frac{x^{\frac{1}{2}}}{\frac{1}{2}}\right) + C = 2x^3 - \frac{2\sqrt{x}}{3} + C$

Q2 $y = \int \left(3x^2 - \frac{7}{\sqrt{x}}\right) dx = 3\int x^2\, dx - 7\int x^{-\frac{1}{2}}\, dx$
$= 3\left(\frac{x^3}{3}\right) - 7\left(\frac{x^{\frac{1}{2}}}{\frac{1}{2}}\right) + C = x^3 - 14x^{\frac{1}{2}} + C = x^3 - 14\sqrt{x} + C$
At (1, 0), $x = 1$ and $y = 0 \Rightarrow 0 = (1)^3 - 14\sqrt{1} + C \Rightarrow C = 13$
So the equation of the curve is $y = x^3 - 14\sqrt{x} + 13$.

Q3 a) $y = \int 2(3x - 6.5)\,dx = \int (6x - 13)\,dx$
$= 6\int x\,dx - 13\int x^0\,dx$
$= 6\left(\frac{x^2}{2}\right) - 13\left(\frac{x^1}{1}\right) + C = 3x^2 - 13x + C$

At $(1, 2)$, $x = 1$ and $y = 2 \Rightarrow 2 = 3(1)^2 - 13(1) + C$
$\Rightarrow C = 12$

So the equation of the curve is $y = 3x^2 - 13x + 12$.

b) $y = 3x^2 - 13x + 12 = (3x - 4)(x - 3)$
When $y = 0$, $x = \frac{4}{3}$ or $x = 3$.
When $x = 0$, $y = 12$.

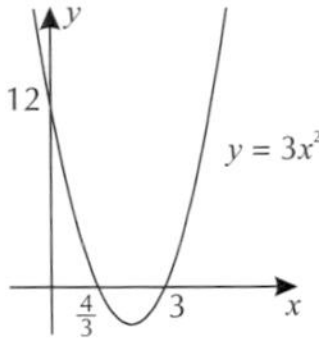

Q4 $y = \int\left(6x^2 - 12 - \frac{8}{x^2}\right)dx = 6\int x^2\,dx - 12\int x^0\,dx - 8\int x^{-2}\,dx$
$= 6\left(\frac{x^3}{3}\right) - 12\left(\frac{x^1}{1}\right) - 8\left(\frac{x^{-1}}{-1}\right) + C = 2x^3 - 12x + \frac{8}{x} + C$
At $(-2, 5)$, $x = -2$ and $y = 5$, so
$5 = 2(-2)^3 - 12(-2) + \frac{8}{(-2)} + C = 4 + C \Rightarrow C = 1$
So the curve has the equation $y = 2x^3 - 12x + \frac{8}{x} + 1$.

Q5 $f'(x) = \frac{5x^2 + 1}{x^{\frac{1}{2}}} - 10 = \frac{5x^2}{x^{\frac{1}{2}}} + \frac{1}{x^{\frac{1}{2}}} - 10 = 5x^{\frac{3}{2}} + x^{-\frac{1}{2}} - 10$

$f(x) = \int(5x^{\frac{3}{2}} + x^{-\frac{1}{2}} - 10)\,dx = 5\left(\frac{x^{\frac{5}{2}}}{\frac{5}{2}}\right) + \left(\frac{x^{\frac{1}{2}}}{\frac{1}{2}}\right) - 10\left(\frac{x^1}{1}\right) + C$
$= 2x^{\frac{5}{2}} + 2x^{\frac{1}{2}} - 10x + C$
$= 2\sqrt{x^5} + 2\sqrt{x} - 10x + C$
At $(1, -9)$, $x = 1$ and $y = -9$,
so $-9 = 2\sqrt{1^5} + 2\sqrt{1} - 10(1) + C = -6 + C \Rightarrow C = -3$
So the curve has the equation $y = 2\sqrt{x^5} + 2\sqrt{x} - 10x - 3$.

Q6 a) $\int_0^1 (4x^3 + 3x^2 + 2x + 1)\,dx = [x^4 + x^3 + x^2 + x]_0^1$
$= (1^4 + 1^3 + 1^2 + 1) - (0^4 + 0^3 + 0^2 + 0) = 4$

b) $\int_1^6 \frac{3}{x^2}\,dx = \int_1^6 3x^{-2}\,dx = \left[3\left(\frac{x^{-1}}{-1}\right)\right]_1^6 = \left[-\frac{3}{x}\right]_1^6$
$= \left(-\frac{3}{6}\right) - \left(-\frac{3}{1}\right) = -\frac{1}{2} - (-3) = \frac{5}{2}$

c) $\int_1^2 \left(\frac{8}{x^5} + \frac{3}{\sqrt{x}}\right)dx = \int_1^2 (8x^{-5} + 3x^{-\frac{1}{2}})\,dx$
$= \left[8\left(\frac{x^{-4}}{-4}\right) + 3\left(\frac{x^{\frac{1}{2}}}{\frac{1}{2}}\right)\right]_1^2 = [-2x^{-4} + 6x^{\frac{1}{2}}]_1^2$
$= \left[-\frac{2}{x^4} + 6\sqrt{x}\right]_1^2 = \left(-\frac{2}{2^4} + 6\sqrt{2}\right) - \left(-\frac{2}{1^4} + 6\sqrt{1}\right)$
$= -\frac{33}{8} + 6\sqrt{2}$

d) $\int_4^9 \sqrt{x}(1 - 2\sqrt{x})\,dx = \int_4^9 (x^{\frac{1}{2}} - 2x)\,dx$
$= \left[\left(\frac{x^{\frac{3}{2}}}{\frac{3}{2}}\right) - 2\left(\frac{x^2}{2}\right)\right]_4^9 = \left[\frac{2x^{\frac{3}{2}}}{3} - x^2\right]_4^9$
$= \left(\frac{2(9)^{\frac{3}{2}}}{3} - 9^2\right) - \left(\frac{2(4)^{\frac{3}{2}}}{3} - 4^2\right)$
$= 18 - 81 - \frac{16}{3} + 16 = -\frac{157}{3}$

e) $\int_1^4 \frac{3\sqrt{x} + 2x^4}{x^2}\,dx = \int_1^4 \left(\frac{3\sqrt{x}}{x^2} + \frac{2x^4}{x^2}\right)dx$
$= \int_1^4 3x^{-\frac{3}{2}} + 2x^2\,dx = \left[3\left(\frac{x^{-\frac{1}{2}}}{-\frac{1}{2}}\right) + 2\left(\frac{x^3}{3}\right)\right]_1^4$
$= \left[-6x^{-\frac{1}{2}} + \frac{2x^3}{3}\right]_1^4 = \left[-\frac{6}{\sqrt{x}} + \frac{2x^3}{3}\right]_1^4$
$= \left(-\frac{6}{\sqrt{4}} + \frac{2(4)^3}{3}\right) - \left(-\frac{6}{\sqrt{1}} + \frac{2(1)^3}{3}\right)$
$= -3 + \frac{128}{3} + 6 - \frac{2}{3} = 45$

f) $\int_{-1}^{0.5} \frac{2x^5 + 4x^4}{5x^3}\,dx = \int_{-1}^{0.5} \frac{2x^2}{5} + \frac{4x}{5}\,dx$
$= \left[\frac{2}{5}\left(\frac{x^3}{3}\right) + \frac{4}{5}\left(\frac{x^2}{2}\right)\right]_{-1}^{0.5} = \left[\frac{2x^3}{15} + \frac{2x^2}{5}\right]_{-1}^{0.5}$
$= \left(\frac{2(0.5)^3}{15} + \frac{2(0.5)^2}{5}\right) - \left(\frac{2(-1)^3}{15} + \frac{2(-1)^2}{5}\right)$
$= \frac{1}{60} + \frac{1}{10} + \frac{2}{15} - \frac{2}{5} = -\frac{3}{20}$

Q7 a) $\int_{-3}^3 (9 - x^2)\,dx = \left[9\left(\frac{x^1}{1}\right) - \frac{x^3}{3}\right]_{-3}^3$
$= \left[9x - \frac{x^3}{3}\right]_{-3}^3 = \left(9(3) - \frac{3^3}{3}\right) - \left(9(-3) - \frac{(-3)^3}{3}\right)$
$= 27 - 9 + 27 - 9 = 36$

b)

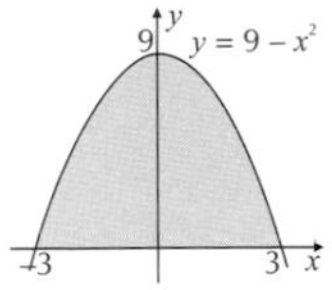

Q8 a) $A = \int_1^8 x^{-\frac{1}{3}}\,dx = \left[\frac{3x^{\frac{2}{3}}}{2}\right]_1^8 = \frac{3(8)^{\frac{2}{3}}}{2} - \frac{3(1)^{\frac{2}{3}}}{2} = 6 - \frac{3}{2} = \frac{9}{2}$

b) Factorise the equation to find where the graph crosses the x-axis:
$y = x^2 + 8x + 7 = (x + 7)(x + 1)$
$\Rightarrow$ where $y = 0$, $x = -7$ or $x = -1$
A is split into an area below the x-axis between $x = -7$ and $x = -1$ and an area above the x-axis between $x = -1$ and $x = 0$, so integrate separately for each area:
$\int_{-7}^{-1} (x^2 + 8x + 7)\,dx = \left[\frac{x^3}{3} + 8\left(\frac{x^2}{2}\right) + 7\left(\frac{x^1}{1}\right)\right]_{-7}^{-1}$
$= \left[\frac{x^3}{3} + 4x^2 + 7x\right]_{-7}^{-1}$
$= \left(\frac{(-1)^3}{3} + 4(-1)^2 + 7(-1)\right) - \left(\frac{(-7)^3}{3} + 4(-7)^2 + 7(-7)\right)$
$= -\frac{1}{3} + 4 - 7 + \frac{343}{3} - 196 + 49 = -36$
Area cannot be negative, so the area below the x-axis is 36.
$\int_{-1}^0 (x^2 + 8x + 7)\,dx = \left[\frac{x^3}{3} + 4x^2 + 7x\right]_{-1}^0$
$= \left(\frac{(0)^3}{3} + 4(0)^2 + 7(0)\right) - \left(\frac{(-1)^3}{3} + 4(-1)^2 + 7(-1)\right)$
$= 0 + \frac{1}{3} - 4 + 7 = \frac{10}{3}$
So the total area of A is $36 + \frac{10}{3} = \frac{118}{3}$

Q9 a) You want to find the area between the curve, the x-axis and the lines $x = 0$ and $x = 2$.
$A = \int_0^2 (x^3 - 5x^2 + 6x)\,dx = \left[\frac{x^4}{4} - 5\left(\frac{x^3}{3}\right) + 6\left(\frac{x^2}{2}\right)\right]_0^2$
$= \left[\frac{x^4}{4} - \frac{5x^3}{3} + 3x^2\right]_0^2$
$= \left(\frac{2^4}{4} - \frac{5(2)^3}{3} + 3(2)^2\right) - \left(\frac{(0)^4}{4} - \frac{5(0)^3}{3} + 3(0)^2\right)$
$= 4 - \frac{40}{3} + 12 - 0 = \frac{8}{3}$

b) $A = \int_1^4 2\sqrt{x}\,dx = \int_1^4 2x^{\frac{1}{2}}\,dx$
$= \left[2\left(\frac{x^{\frac{3}{2}}}{\frac{3}{2}}\right)\right]_1^4 = \left[\frac{4x^{\frac{3}{2}}}{3}\right]_1^4$
$= \left(\frac{4(4)^{\frac{3}{2}}}{3}\right) - \left(\frac{4(1)^{\frac{3}{2}}}{3}\right) = \frac{32}{3} - \frac{4}{3} = \frac{28}{3}$

c) Integrate $y = 2x^2$ between $x = 0$ and $x = 2$:

$$\int_0^2 2x^2\,dx = \left[2\left(\frac{x^3}{3}\right)\right]_0^2 = \left[\frac{2x^3}{3}\right]_0^2$$

$$= \left(\frac{2(2)^3}{3}\right) - \left(\frac{2(0)^3}{3}\right) = \frac{16}{3}$$

Integrate $y = 12 - 2x$ between $x = 2$ and $x = 6$:

$$\int_2^6 (12 - 2x)\,dx = \left[12\left(\frac{x^1}{1}\right) - 2\left(\frac{x^2}{2}\right)\right]_2^6$$

$$= [12x - x^2]_2^6 = (12(6) - 6^2) - (12(2) - 2^2)$$

$$= 72 - 36 - 24 + 4 = 16$$

So the total area of A is $\frac{16}{3} + 16 = \frac{64}{3}$.

You could have used the formula for the area of a triangle to work out the second area, taking 8 as the height and 6 – 2 = 4 as the base length.

d) Integrate $y = x + 3$ between $x = 1$ and $x = 4$:

$$\int_1^4 (x + 3)\,dx = \left[\frac{x^2}{2} + 3\left(\frac{x^1}{1}\right)\right]_1^4$$

$$= \left[\frac{x^2}{2} + 3x\right]_1^4 = \left(\frac{4^2}{2} + 3(4)\right) - \left(\frac{1^2}{2} + 3(1)\right)$$

$$= 8 + 12 - \frac{1}{2} - 3 = \frac{33}{2}$$

You could have found this area using the trapezium formula.

Integrate $y = x^2 - 4x + 7$ between $x = 1$ and $x = 4$:

$$\int_1^4 (x^2 - 4x + 7)\,dx = \left[\frac{x^3}{3} - 4\left(\frac{x^2}{2}\right) + 7\left(\frac{x^1}{1}\right)\right]_1^4$$

$$= \left[\frac{x^3}{3} - 2x^2 + 7x\right]_1^4$$

$$= \left(\frac{4^3}{3} - 2(4)^2 + 7(4)\right) - \left(\frac{1^3}{3} - 2(1)^2 + 7(1)\right)$$

$$= \frac{64}{3} - 32 + 28 - \frac{1}{3} + 2 - 7 = 12$$

Subtract the area below $y = x^2 - 4x + 7$ between $x = 1$ and $x = 4$ from the area below $y = x + 3$ between $x = 1$ and $x = 4$:

$$A = \frac{33}{2} - 12 = \frac{9}{2}$$

Q10 a) Factorise the equation to find where the graph crosses the x-axis:

$y = -x^3 + 7x^2 - 10x = -x(x - 2)(x - 5)$

$\Rightarrow$ where $y = 0$, $x = 0$ or $x = 2$ or $x = 5$

The shaded area is split into an area below the x-axis between $x = 0$ and $x = 2$ and an area above the x-axis between $x = 2$ and $x = 5$, so integrate separately for each area:

$$\int_0^2 (-x^3 + 7x^2 - 10x)\,dx = \left[-\left(\frac{x^4}{4}\right) + 7\left(\frac{x^3}{3}\right) - 10\left(\frac{x^2}{2}\right)\right]_0^2$$

$$= \left[-\frac{x^4}{4} + \frac{7x^3}{3} - 5x^2\right]_0^2$$

$$= \left(-\frac{2^4}{4} + \frac{7(2)^3}{3} - 5(2)^2\right) - \left(-\frac{(0)^4}{4} + \frac{7(0)^3}{3} - 5(0)^2\right)$$

$$= -4 + \frac{56}{3} - 20 - 0 = -\frac{16}{3}$$

Area cannot be negative, so the area below the x-axis is $\frac{16}{3}$.

$$\int_2^5 (-x^3 + 7x^2 - 10x)\,dx = \left[-\frac{x^4}{4} + \frac{7x^3}{3} - 5x^2\right]_2^5$$

$$= \left(-\frac{5^4}{4} + \frac{7(5)^3}{3} - 5(5)^2\right) - \left(-\frac{2^4}{4} + \frac{7(2)^3}{3} - 5(2)^2\right)$$

$$= -\frac{625}{4} + \frac{875}{3} - 125 + 4 - \frac{56}{3} + 20 = \frac{63}{4}$$

So the total area shaded is $\frac{16}{3} + \frac{63}{4} = \frac{253}{12}$

b) The shaded area is split into an area below the x-axis between $x = -1$ and $x = 0$, above the x-axis between $x = 0$ and $x = 2$ and below the x-axis between $x = 2$ and $x = 3$. Integrate for these three areas separately:

$$\int_{-1}^0 (x^4 - 3x^3 - 4x^2 + 12x)\,dx$$

$$= \left[\left(\frac{x^5}{5}\right) - 3\left(\frac{x^4}{4}\right) - 4\left(\frac{x^3}{3}\right) + 12\left(\frac{x^2}{2}\right)\right]_{-1}^0$$

$$= \left[\frac{x^5}{5} - \frac{3x^4}{4} - \frac{4x^3}{3} + 6x^2\right]_{-1}^0$$

$$= \left(\frac{(0)^5}{5} - \frac{3(0)^4}{4} - \frac{4(0)^3}{3} + 6(0)^2\right)$$

$$- \left(\frac{(-1)^5}{5} - \frac{3(-1)^4}{4} - \frac{4(-1)^3}{3} + 6(-1)^2\right)$$

$$= (0 - 0 - 0 + 0) - \left(-\frac{1}{5} - \frac{3}{4} + \frac{4}{3} + 6\right) = -\frac{383}{60}$$

Area cannot be negative, so the area below the x-axis between $x = -1$ and $x = 0$ is $\frac{383}{60}$.

$$\int_0^2 (x^4 - 3x^3 - 4x^2 + 12x)\,dx = \left[\frac{x^5}{5} - \frac{3x^4}{4} - \frac{4x^3}{3} + 6x^2\right]_0^2$$

$$= \left(\frac{2^5}{5} - \frac{3(2)^4}{4} - \frac{4(2)^3}{3} + 6(2)^2\right)$$

$$- \left(\frac{(0)^5}{5} - \frac{3(0)^4}{4} - \frac{4(0)^3}{3} + 6(0)^2\right)$$

$$= \left(\frac{32}{5} - 12 - \frac{32}{3} + 24\right) - (0 - 0 - 0 + 0) = \frac{116}{15}$$

$$\int_2^3 (x^4 - 3x^3 - 4x^2 + 12x)\,dx = \left[\frac{x^5}{5} - \frac{3x^4}{4} - \frac{4x^3}{3} + 6x^2\right]_2^3$$

$$= \left(\frac{3^5}{5} - \frac{3(3)^4}{4} - \frac{4(3)^3}{3} + 6(3)^2\right)$$

$$- \left(\frac{2^5}{5} - \frac{3(2)^4}{4} - \frac{4(2)^3}{3} + 6(2)^2\right)$$

$$= \left(\frac{243}{5} - \frac{243}{4} - 36 + 54\right) - \left(\frac{32}{5} - 12 - \frac{32}{3} + 24\right)$$

$$= \frac{117}{20} - \frac{116}{15} = -\frac{113}{60}$$

Area cannot be negative, so the area below the x-axis between $x = 2$ and $x = 3$ is $\frac{113}{60}$.

So the total area shaded is $\frac{383}{60} + \frac{116}{15} + \frac{113}{60} = 16$

Q11 Factorise $y = 2x^3 - 15x^2 + 36x - 28$ using the Factor Theorem:

If $x = 2$, $y = 2(2)^3 - 15(2)^2 + 36(2) - 28$
$= 16 - 60 + 72 - 28 = 0$
$\Rightarrow$ $(x - 2)$ is a factor

Using e.g. algebraic division:

$y = 2x^3 - 15x^2 + 36x - 28 = (x - 2)(2x^2 - 11x + 14)$
$= (x - 2)^2(2x - 7)$

The double root is at $x = 2$, so $a = 2$.

Integrate the function between $x = 0$ and $x = 2$:

$$\int_0^2 (2x^3 - 15x^2 + 36x - 28)\,dx$$

$$= \left[2\left(\frac{x^4}{4}\right) - 15\left(\frac{x^3}{3}\right) + 36\left(\frac{x^2}{2}\right) - 28\left(\frac{x^1}{1}\right)\right]_0^2$$

$$= \left[\frac{x^4}{2} - 5x^3 + 18x^2 - 28x\right]_0^2$$

$$= \left(\frac{2^4}{2} - 5(2)^3 + 18(2)^2 - 28(2)\right)$$

$$- \left(\frac{(0)^4}{2} - 5(0)^3 + 18(0)^2 - 28(0)\right)$$

$$= 8 - 40 + 72 - 56 - 0 = -16$$

Area cannot be negative so the area is 16.

Q12 a) Integrate $y = 9 - \frac{1}{3}\sqrt{x^3}$ between $x = 0$ and $x = 9$:

$$\int_0^9 \left(9 - \frac{1}{3}x^{\frac{3}{2}}\right)dx$$

$$= \left[9\left(\frac{x^1}{1}\right) - \frac{1}{3}\left(\frac{x^{\frac{5}{2}}}{\frac{5}{2}}\right)\right]_0^9 = \left[9x - \frac{2}{15}x^{\frac{5}{2}}\right]_0^9$$

$$= \left(9(9) - \frac{2}{15}(9)^{\frac{5}{2}}\right) - \left(9(0) - \frac{2}{15}(0)^{\frac{5}{2}}\right)$$

$$= 81 - \frac{162}{5} - 0 = \frac{243}{5} = 48.6 \text{ m}^2$$

So the plan meets the requirement of at least 40 m^2 floor space.

b) Integrate $y = 2 - \frac{1}{2}x^2$ between $x = 0$ and $x = 2$:

$$\int_0^2 \left(2 - \frac{1}{2}x^2\right)dx$$

$$= \left[2\left(\frac{x^1}{1}\right) - \frac{1}{2}\left(\frac{x^3}{3}\right)\right]_0^2 = \left[2x - \frac{1}{6}x^3\right]_0^2$$

$$= \left(2(2) - \frac{1}{6}(2)^3\right) - \left(2(0) - \frac{1}{6}(0)^3\right)$$

$$= 4 - \frac{4}{3} - 0 = \frac{8}{3} \text{ m}^2$$

So there is $48.6 \text{ m}^2 - \frac{8}{3} \text{ m}^2 = 45.9 \text{ m}^2$ (3 s.f.) of floor space left.

Exam-Style Questions — Chapter 11

Q1 $\frac{d^2y}{dx^2} = 12x$, $\frac{dy}{dx} = \int 12x\,dx = 12\int x\,dx$

$= 12\left(\frac{x^2}{2}\right) + C_1 = 6x^2 + C_1$

When $x = -2$, $\frac{dy}{dx} = -8$,

so $6(-2)^2 + C_1 = -8 \Rightarrow C_1 = -32 \Rightarrow \frac{dy}{dx} = 6x^2 - 32$

So $y = \int (6x^2 - 32)\,dx = 6\int x^2\,dx - 32\int x^0\,dx$

$= 6\left(\frac{x^3}{3}\right) - 32\left(\frac{x^1}{1}\right) + C_2$

$= 2x^3 - 32x + C_2$

When $x = -2$ and $y = -13$, $-13 = 2(-2)^3 - 32(-2) + C_2$

$\Rightarrow -13 = -16 + 64 + C_2 \Rightarrow C_2 = -61$

So the equation of the curve $y = f(x)$ is $y = 2x^3 - 32x - 61$.

[7 marks available — 1 mark for correctly integrating $\frac{d^2y}{dx^2}$, 1 mark for setting $\frac{dy}{dx}$ equal to -8 and attempting to substitute in $x = -2$, 1 mark for correct C_1 value, 1 mark for equation for $\frac{dy}{dx}$ using C_1 or calculated value as coefficient of x, 1 mark for substituting $x = -2$, $y = -13$ and $C_1 = -32$ into equation for y, 1 mark for correct C_2 value, 1 mark for writing equation for y using values found]

Q2 $\int \left(\frac{5}{3}x^{\frac{2}{3}} - \frac{4}{\sqrt{x}} + (2x)^3 - \frac{3}{2x^2}\right)dx$

$= \int \left(\frac{5}{3}x^{\frac{2}{3}} - 4x^{-\frac{1}{2}} + 8x^3 - \frac{3}{2}x^{-2}\right)dx$

$= \frac{5}{3}\int x^{\frac{2}{3}}\,dx - 4\int x^{-\frac{1}{2}}\,dx + 8\int x^3\,dx - \frac{3}{2}\int x^{-2}\,dx$

$= \frac{5}{3}\left(\frac{x^{\frac{5}{3}}}{\frac{5}{3}}\right) - 4\left(\frac{x^{\frac{1}{2}}}{\frac{1}{2}}\right) + 8\left(\frac{x^4}{4}\right) - \frac{3}{2}\left(\frac{x^{-1}}{-1}\right) + C$

$= x^{\frac{5}{3}} - 8x^{\frac{1}{2}} + 2x^4 + \frac{3}{2}x^{-1} + C = \sqrt[3]{x^5} - 8\sqrt{x} + 2x^4 + \frac{3}{2x} + C$

[5 marks available — 1 mark for writing the original function as powers of x, 1 mark for attempting to raise powers of each term by one and dividing by result, 1 mark for two correct terms, 1 mark for other two correct terms, 1 mark for C with at least two terms correct]

Q3 a)

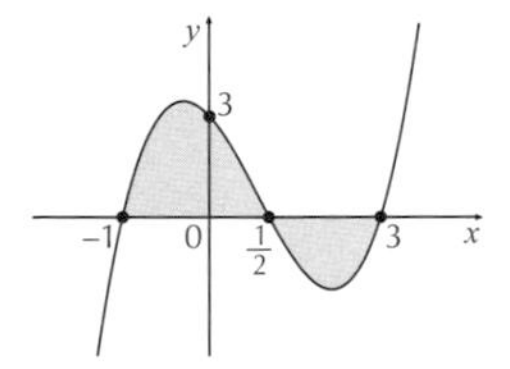

[3 marks available — 1 mark for correct curve shape, 1 mark for correct y-intercept, 1 mark for correct x-intercepts]

b) Between $x = -1$ and $x = 3$, the graph crosses the axes at $x = \frac{1}{2}$ which means the area lies both above and below the x-axis (as shown on the diagram for part a)).

$y = (x + 1)(2x - 1)(x - 3) = (x + 1)(2x^2 - 7x + 3)$

$= 2x^3 - 5x^2 - 4x + 3$

So integrate $y = 2x^3 - 5x^2 - 4x + 3$ between $x = -1$ and $x = \frac{1}{2}$ to find the area above the x-axis:

$\int_{-1}^{\frac{1}{2}} (2x^3 - 5x^2 - 4x + 3)\,dx$

$= \left[2\left(\frac{x^4}{4}\right) - 5\left(\frac{x^3}{3}\right) - 4\left(\frac{x^2}{2}\right) + 3\left(\frac{x^1}{1}\right)\right]_{-1}^{\frac{1}{2}}$

$= \left[\frac{1}{2}x^4 - \frac{5}{3}x^3 - 2x^2 + 3x\right]_{-1}^{\frac{1}{2}}$

$= \left(\frac{1}{2}\left(\frac{1}{2}\right)^4 - \frac{5}{3}\left(\frac{1}{2}\right)^3 - 2\left(\frac{1}{2}\right)^2 + 3\left(\frac{1}{2}\right)\right)$

$- \left(\frac{1}{2}(-1)^4 - \frac{5}{3}(-1)^3 - 2(-1)^2 + 3(-1)\right)$

$= \left(\frac{1}{32} - \frac{5}{24} - \frac{1}{2} + \frac{3}{2}\right) - \left(\frac{1}{2} + \frac{5}{3} - 2 - 3\right)$

$= \frac{79}{96} - \left(-\frac{17}{6}\right) = \frac{117}{32}$

Integrate $y = 2x^3 - 5x^2 - 4x + 3$ between $x = \frac{1}{2}$ and $x = 3$ to find the area below the x-axis:

$\int_{\frac{1}{2}}^{3} (2x^3 - 5x^2 - 4x + 3)\,dx = \left[\frac{1}{2}x^4 - \frac{5}{3}x^3 - 2x^2 + 3x\right]_{\frac{1}{2}}^{3}$

$= \left(\frac{1}{2}(3)^4 - \frac{5}{3}(3)^3 - 2(3)^2 + 3(3)\right)$

$- \left(\frac{1}{2}\left(\frac{1}{2}\right)^4 - \frac{5}{3}\left(\frac{1}{2}\right)^3 - 2\left(\frac{1}{2}\right)^2 + 3\left(\frac{1}{2}\right)\right)$

$= \left(\frac{81}{2} - 45 - 18 + 9\right) - \left(\frac{1}{32} - \frac{5}{24} - \frac{1}{2} + \frac{3}{2}\right)$

$= -\frac{27}{2} - \frac{79}{96} = -\frac{1375}{96}$

Area cannot be negative so the area below the x-axis is $\frac{1375}{96}$.

So the total area enclosed is $\frac{117}{32} + \frac{1375}{96} = \frac{863}{48}$.

[8 marks available — 1 mark for writing integral with limits of $x = -1$ and $x = \frac{1}{2}$, 1 mark for writing integral with limits of $x = \frac{1}{2}$ and $x = 3$, 1 mark for correctly integrating y, 1 mark for attempting to substitute the limits into integrated function, 1 mark for correct area above x-axis, 1 mark for correct area below x-axis, 1 mark for finding negative area below x-axis and omitting negative sign, 1 mark for correct total area]

Q4 $I = \int_a^6 (4x - 5)\,dx = \left[4\left(\frac{x^2}{2}\right) - 5\left(\frac{x^1}{1}\right)\right]_a^6$

$= [2x^2 - 5x]_a^6 = (2(6)^2 - 5(6)) - (2a^2 - 5a)$

$= 42 - 2a^2 + 5a$

If $I = 24$, $42 - 2a^2 + 5a = 24 \Rightarrow 2a^2 - 5a - 18 = 0$

$\Rightarrow (2a - 9)(a + 2) = 0$

$\Rightarrow a = \frac{9}{2}$ or $a = -2$

[6 marks available — 1 mark for correct integration, 1 mark for correctly substituting in limits, 1 mark for making equation in terms of a equal to 24, 1 mark for simplifying quadratic function and making equal to zero, 1 mark for factorising quadratic, 1 mark for correct a-values]

Q5 a) Use the Factor Theorem to find a linear factor of $y = x^3 - 2x^2 - 9x + 18$:

When $x = 2$, $y = (2)^3 - 2(2)^2 - 9(2) + 18 = 0$, so $(x - 2)$ is a factor.

So $y = (x - 2)(x^2 + nx - 9) \Rightarrow -2x^2 = nx^2 - 2x^2$

$\Rightarrow n = 0 \Rightarrow y = (x - 2)(x^2 - 9)$

$= (x - 2)(x - 3)(x + 3)$

So the curve crosses the x-axis where $x = 2$, and $x = 3$ and $x = -3$.

[3 marks available — 1 mark for one correct linear factor, 1 mark for two other correct linear factors, 1 mark for all correct x-values]

b) Split the area into two sections and find the areas separately.

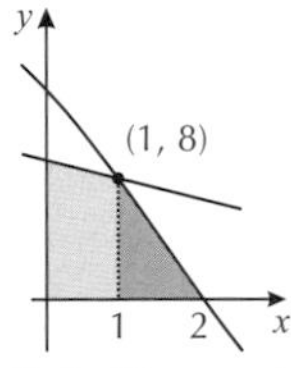

To find the area to the left, integrate $y = 12 - 4x$ between $x = 0$ and $x = 1$.

$\int_0^1 (12 - 4x)\,dx = \left[12\left(\frac{x^1}{1}\right) - 4\left(\frac{x^2}{2}\right)\right]_0^1 = [12x - 2x^2]_0^1$

$= (12(1) - 2(1)^2) - (12(0) - 2(0)^2) = 10$

To find the area to the right, integrate $y = x^3 - 2x^2 - 9x + 18$ between $x = 1$ and $x = 2$.

$$\int_1^2 (x^3 - 2x^2 - 9x + 18)\,dx$$
$$= \left[\left(\frac{x^4}{4}\right) - 2\left(\frac{x^3}{3}\right) - 9\left(\frac{x^2}{2}\right) + 18\left(\frac{x^1}{1}\right)\right]_1^2$$
$$= \left[\frac{x^4}{4} - \frac{2x^3}{3} - \frac{9x^2}{2} + 18x\right]_1^2$$
$$= \left(\frac{(2)^4}{4} - \frac{2(2)^3}{3} - \frac{9(2)^2}{2} + 18(2)\right) - \left(\frac{(1)^4}{4} - \frac{2(1)^3}{3} - \frac{9(1)^2}{2} + 18(1)\right)$$
$$= \frac{50}{3} - \frac{157}{12} = \frac{43}{12}$$

So the total area is $10 + \frac{43}{12} = \frac{163}{12}$.

[7 marks available — 1 mark for correctly splitting the area in two and attempting to integrate the two areas separately, 1 mark for each correctly integrated function, 1 mark each for correctly substituting the limits into the integrated functions, 1 mark for at least one correct positive area, 1 mark for correct total area]

Chapter 12: Vectors

12.1 Vectors

Exercise 12.1.1 — Introducing vectors

Q1 **a)** vector **b)** scalar **c)** vector

Q2 **a)** **b)** **c)** **d)**

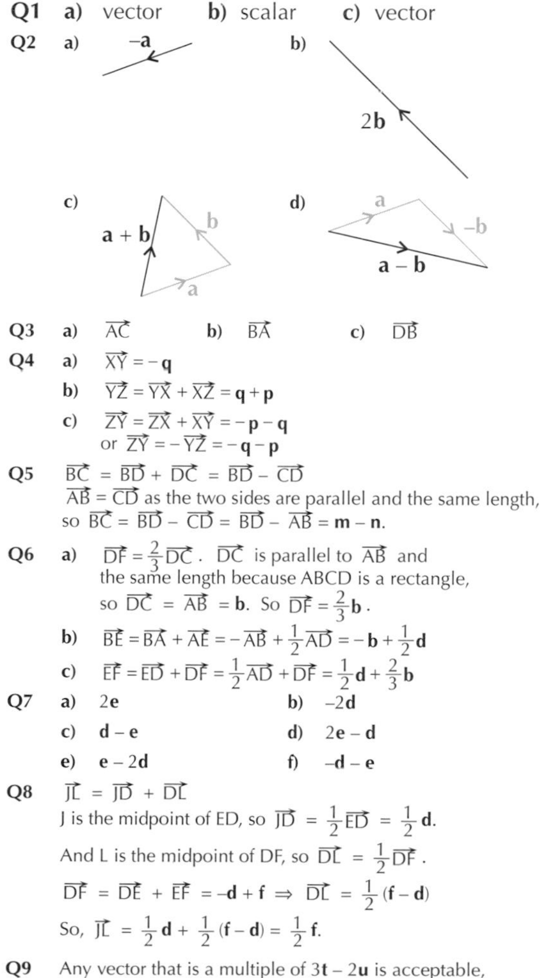

Q3 **a)** $\overrightarrow{AC}$ **b)** $\overrightarrow{BA}$ **c)** $\overrightarrow{DB}$

Q4 **a)** $\overrightarrow{XY} = -\mathbf{q}$

b) $\overrightarrow{YZ} = \overrightarrow{YX} + \overrightarrow{XZ} = \mathbf{q} + \mathbf{p}$

c) $\overrightarrow{ZY} = \overrightarrow{ZX} + \overrightarrow{XY} = -\mathbf{p} - \mathbf{q}$
or $\overrightarrow{ZY} = -\overrightarrow{YZ} = -\mathbf{q} - \mathbf{p}$

Q5 $\overrightarrow{BC} = \overrightarrow{BD} + \overrightarrow{DC} = \overrightarrow{BD} - \overrightarrow{CD}$
$\overrightarrow{AB} = \overrightarrow{CD}$ as the two sides are parallel and the same length, so $\overrightarrow{BC} = \overrightarrow{BD} - \overrightarrow{CD} = \overrightarrow{BD} - \overrightarrow{AB} = \mathbf{m} - \mathbf{n}$.

Q6 **a)** $\overrightarrow{DF} = \frac{2}{3}\overrightarrow{DC}$. $\overrightarrow{DC}$ is parallel to $\overrightarrow{AB}$ and the same length because ABCD is a rectangle, so $\overrightarrow{DC} = \overrightarrow{AB} = \mathbf{b}$. So $\overrightarrow{DF} = \frac{2}{3}\mathbf{b}$.

b) $\overrightarrow{BE} = \overrightarrow{BA} + \overrightarrow{AE} = -\overrightarrow{AB} + \frac{1}{2}\overrightarrow{AD} = -\mathbf{b} + \frac{1}{2}\mathbf{d}$

c) $\overrightarrow{EF} = \overrightarrow{ED} + \overrightarrow{DF} = \frac{1}{2}\overrightarrow{AD} + \overrightarrow{DF} = \frac{1}{2}\mathbf{d} + \frac{2}{3}\mathbf{b}$

Q7 **a)** $2\mathbf{e}$ **b)** $-2\mathbf{d}$
c) $\mathbf{d} - \mathbf{e}$ **d)** $2\mathbf{e} - \mathbf{d}$
e) $\mathbf{e} - 2\mathbf{d}$ **f)** $-\mathbf{d} - \mathbf{e}$

Q8 $\overrightarrow{JL} = \overrightarrow{JD} + \overrightarrow{DL}$
J is the midpoint of ED, so $\overrightarrow{JD} = \frac{1}{2}\overrightarrow{ED} = \frac{1}{2}\mathbf{d}$.
And L is the midpoint of DF, so $\overrightarrow{DL} = \frac{1}{2}\overrightarrow{DF}$.
$\overrightarrow{DF} = \overrightarrow{DE} + \overrightarrow{EF} = -\mathbf{d} + \mathbf{f} \Rightarrow \overrightarrow{DL} = \frac{1}{2}(\mathbf{f} - \mathbf{d})$
So, $\overrightarrow{JL} = \frac{1}{2}\mathbf{d} + \frac{1}{2}(\mathbf{f} - \mathbf{d}) = \frac{1}{2}\mathbf{f}$.

Q9 Any vector that is a multiple of $3\mathbf{t} - 2\mathbf{u}$ is acceptable, e.g. $6\mathbf{t} - 4\mathbf{u}$, $6\mathbf{u} - 9\mathbf{t}$, $\frac{3}{2}\mathbf{t} - \mathbf{u}$, etc.

Q10 $\overrightarrow{YZ} = -\mathbf{a} + \mathbf{b}$, so $\overrightarrow{YP} = \frac{1}{2}\overrightarrow{YZ} = \frac{1}{2}(-\mathbf{a} + \mathbf{b})$.
This means, $\overrightarrow{PQ} = \overrightarrow{YP} + \overrightarrow{YQ} = -\frac{1}{2}(-\mathbf{a} + \mathbf{b}) + -\frac{1}{2}(-\mathbf{a}) = -\frac{1}{2}\mathbf{b}$.
$\overrightarrow{PQ}$ is a scalar multiple of $\overrightarrow{XZ}$, so they're parallel.

Q11 $\overrightarrow{US} = \overrightarrow{UT} + \overrightarrow{TS} = -\overrightarrow{TU} - \overrightarrow{ST} = -(\mathbf{v} - \mathbf{w}) - (2\mathbf{v} - \mathbf{w}) = -3\mathbf{v}$.
This is a scalar multiple of $\mathbf{v}$, so is parallel to $\mathbf{v}$.

Q12 $\overrightarrow{YZ} = \overrightarrow{YX} + \overrightarrow{XZ} = -\overrightarrow{XY} + \overrightarrow{XZ}$
$= -(3\mathbf{a} - 4\mathbf{b} + 2\mathbf{c}) + \mathbf{a} - 2\mathbf{b} - 2\mathbf{c}$
$= -2\mathbf{a} + 2\mathbf{b} - 4\mathbf{c} = 2(\mathbf{a} - \mathbf{b} + 2\mathbf{c})$
This is a scalar multiple of $\mathbf{a} - \mathbf{b} + 2\mathbf{c}$, so they're parallel.

Q13 $\overrightarrow{YZ} = \overrightarrow{AZ} - \overrightarrow{AY} = \mathbf{u} - (-\mathbf{t}) = \mathbf{t} + \mathbf{u} = \overrightarrow{XY}$.
These are the same vector, so X, Y and Z are collinear.

Q14 $4\mathbf{b} + 8\mathbf{a} = 4(2\mathbf{a} + \mathbf{b}) = -4(-\mathbf{b} - 2\mathbf{a})$,
so $4\mathbf{b} + 8\mathbf{a}$, $2\mathbf{a} + \mathbf{b}$ and $-\mathbf{b} - 2\mathbf{a}$ are parallel.

$2\mathbf{p} + \mathbf{q} = 2\left(\frac{1}{2}\mathbf{q} + \mathbf{p}\right)$, so $2\mathbf{p} + \mathbf{q}$ and $\frac{1}{2}\mathbf{q} + \mathbf{p}$ are parallel.
$10\mathbf{a} - 5\mathbf{b} = 5(2\mathbf{a} - \mathbf{b})$, so $10\mathbf{a} - 5\mathbf{b}$ and $2\mathbf{a} - \mathbf{b}$ are parallel.

Q15 $\overrightarrow{AB} = \overrightarrow{OB} - \overrightarrow{OA} = \mathbf{b} - \mathbf{a}$
$\overrightarrow{BC} = \overrightarrow{OC} - \overrightarrow{OB} = (5\mathbf{a} - 4\mathbf{b}) - \mathbf{b} = 5(\mathbf{a} - \mathbf{b})$
So $\overrightarrow{BC} = -5\overrightarrow{AB}$, so A, B & C lie on the same straight line — i.e. they are collinear.

Q16 $\overrightarrow{QR} = 2\mathbf{m} + 3\mathbf{n}$ and $\overrightarrow{RS} = \mathbf{m} + \frac{3}{2}\mathbf{n} = \frac{1}{2}\overrightarrow{QR}$.
This shows they're scalar multiples, so they are parallel. Therefore, Q, R and S are collinear.

Q17 $\overrightarrow{OA} = 4\mathbf{a} - 2\mathbf{b}$
$\overrightarrow{AC} = \overrightarrow{AB} + \overrightarrow{BD} + \overrightarrow{DC} = 2\mathbf{b} + 4\mathbf{a} - \mathbf{b} - \frac{5}{2}\mathbf{b} - \mathbf{a}$
$= 3\mathbf{a} - \frac{3}{2}\mathbf{b} = \frac{3}{4}\overrightarrow{OA}$
This shows that $\overrightarrow{OA}$ and $\overrightarrow{AC}$ are scalar multiples of one another, so they're parallel. Therefore, O, A and C are collinear and OAC is a straight line.

Q18 $\overrightarrow{BC} = \overrightarrow{AC} - \overrightarrow{AB} = \mathbf{p} - (\mathbf{q} - \frac{1}{2}\mathbf{p}) = \frac{3}{2}\mathbf{p} - \mathbf{q}$
$\overrightarrow{CD} = \overrightarrow{AD} - \overrightarrow{AC} = (-5\mathbf{p} + 4\mathbf{q}) - \mathbf{p} = -6\mathbf{p} + 4\mathbf{q}$
$-6\mathbf{p} + 4\mathbf{q} = -4(\frac{3}{2}\mathbf{p} - \mathbf{q}) = -4\overrightarrow{BC}$
$\overrightarrow{BC}$ and $\overrightarrow{CD}$ are scalar multiples of one another, so they're parallel. They also meet at point C, so B, C and D are collinear.

Q19 **a)** $\overrightarrow{PS} = \overrightarrow{PQ} + \overrightarrow{QR} + \overrightarrow{RS}$
$= (-\mathbf{a}) + (-\frac{1}{2}\mathbf{a} + \mathbf{b}) + (\frac{5}{2}\mathbf{a} - 3\mathbf{b}) = \mathbf{a} - 2\mathbf{b}$

b) $\mathbf{a} - 2\mathbf{b} = -2(-\frac{1}{2}\mathbf{a} + \mathbf{b}) = -2\overrightarrow{QR}$, so $\overrightarrow{PS}$ and $\overrightarrow{QR}$ are parallel. Since $\overrightarrow{PQ}$ and $\overrightarrow{RS}$ are not scalar multiples of each other, they are not parallel, so PQRS must be a trapezium.

Q20 X is the midpoint of AC, so $\overrightarrow{AX} = \overrightarrow{XC} = \mathbf{a}$.
So, $\overrightarrow{DX} = \mathbf{a} - \mathbf{b} + \mathbf{a} = 2\mathbf{a} - \mathbf{b}$ and $\overrightarrow{XB} = -\mathbf{a} + 3\mathbf{a} - 2\mathbf{b} = 2\mathbf{a} - 2\mathbf{b}$.
This shows $\overrightarrow{DX}$ and $\overrightarrow{XB}$ are not scalar multiples of one another, so are not parallel. Therefore, D, X and B are not collinear and DXB is not a straight line.

Q21 **a)** $\overrightarrow{CD} = \overrightarrow{CB} + \overrightarrow{BA} + \overrightarrow{AD} = \overrightarrow{AD} - \overrightarrow{BC} - \overrightarrow{AB} = \frac{3}{2}\mathbf{a} - \mathbf{a} - (\mathbf{b} - \mathbf{a})$
$= \frac{3}{2}\mathbf{a} - \mathbf{b}$

$\overrightarrow{BE} = \overrightarrow{BA} + \overrightarrow{AE} = \overrightarrow{AE} - \overrightarrow{AB} = (2\mathbf{a} - \mathbf{b}) - (\mathbf{b} - \mathbf{a}) = 3\mathbf{a} - 2\mathbf{b}$
$\overrightarrow{BE} = 2\overrightarrow{CD}$, so they are parallel.

b) $\overrightarrow{AO} = \overrightarrow{AB} + \overrightarrow{BO} = \overrightarrow{AB} + \frac{1}{3}\overrightarrow{BE} = (\mathbf{b} - \mathbf{a}) + \frac{1}{3}(3\mathbf{a} - 2\mathbf{b}) = \frac{1}{3}\mathbf{b}$
$\overrightarrow{OC} = \overrightarrow{OB} + \overrightarrow{BC} = \overrightarrow{BC} - \frac{1}{3}\overrightarrow{BE} = \mathbf{a} - \frac{1}{3}(3\mathbf{a} - 2\mathbf{b}) = \frac{2}{3}\mathbf{b}$
$\overrightarrow{AO}$ and $\overrightarrow{OC}$ are scalar multiples of one another, so they are parallel and therefore A, O and C are collinear — they lie on the same straight line.

Exercise 12.1.2 — Position vectors

Q1 a) $\mathbf{a}+\mathbf{b}+\mathbf{c} = \begin{pmatrix}-1\\-2\end{pmatrix}+\begin{pmatrix}3\\-2\end{pmatrix}+\begin{pmatrix}4\\3\end{pmatrix}=\begin{pmatrix}6\\-1\end{pmatrix}$

b) $\mathbf{c}-2\mathbf{b} = \begin{pmatrix}4\\3\end{pmatrix}-2\begin{pmatrix}3\\-2\end{pmatrix}=\begin{pmatrix}-2\\7\end{pmatrix}$

c) $3\mathbf{a}-\mathbf{b}+2\mathbf{c} = 3\begin{pmatrix}-1\\-2\end{pmatrix}-\begin{pmatrix}3\\-2\end{pmatrix}+2\begin{pmatrix}4\\3\end{pmatrix}=\begin{pmatrix}2\\2\end{pmatrix}$

d) $5\mathbf{a}-5\mathbf{c} = 5\begin{pmatrix}-1\\-2\end{pmatrix}-5\begin{pmatrix}4\\3\end{pmatrix}=\begin{pmatrix}-25\\-25\end{pmatrix}$

Q2 $\mathbf{p}+3\mathbf{q}+\mathbf{r} = \begin{pmatrix}-1\\-2\end{pmatrix}+3\begin{pmatrix}3\\-2\end{pmatrix}+\begin{pmatrix}-4\\5\end{pmatrix}=\begin{pmatrix}4\\-3\end{pmatrix} = 4\mathbf{i}-3\mathbf{j}$

$12\mathbf{i}-9\mathbf{j} = 3(4\mathbf{i}-3\mathbf{j})$ — this is a scalar multiple of $\mathbf{p}+3\mathbf{q}+\mathbf{r}$, so they are parallel.

Q3 Position vector for Jack's house: $\begin{pmatrix}2\\3\end{pmatrix}$

Position vector for Jack's school: $\begin{pmatrix}4\\-5\end{pmatrix}$

Q4 a) $2\mathbf{i}-4\mathbf{j}$ b) $-\mathbf{i}-2\mathbf{j}$
c) $5\mathbf{i}+7\mathbf{j}$ d) $3\mathbf{i}-7\mathbf{j}$

Q5 a) C (–1, 2), D (4, –3)

b) $\overrightarrow{CD}=\overrightarrow{OD}-\overrightarrow{OC}=(4\mathbf{i}-3\mathbf{j})-(-\mathbf{i}+2\mathbf{j})=5\mathbf{i}-5\mathbf{j}$
$\overrightarrow{DC}=-\overrightarrow{CD}=-(5\mathbf{i}-5\mathbf{j})=-5\mathbf{i}+5\mathbf{j}$

Q6 $\overrightarrow{OA}=\begin{pmatrix}2\\4\end{pmatrix}$, $\overrightarrow{OB}=\begin{pmatrix}0\\1\end{pmatrix}$, $\overrightarrow{OC}=\begin{pmatrix}-1\\3\end{pmatrix}$

You could use unit form instead of column vectors to answer this question if you prefer.

$\overrightarrow{AB}=\overrightarrow{OB}-\overrightarrow{OA}=\begin{pmatrix}0\\1\end{pmatrix}-\begin{pmatrix}2\\4\end{pmatrix}=\begin{pmatrix}-2\\-3\end{pmatrix}$

$\overrightarrow{BC}=\overrightarrow{OC}-\overrightarrow{OB}=\begin{pmatrix}-1\\3\end{pmatrix}-\begin{pmatrix}0\\1\end{pmatrix}=\begin{pmatrix}-1\\2\end{pmatrix}$

$\overrightarrow{CA}=\overrightarrow{OA}-\overrightarrow{OC}=\begin{pmatrix}2\\4\end{pmatrix}-\begin{pmatrix}-1\\3\end{pmatrix}=\begin{pmatrix}3\\1\end{pmatrix}$

Q7 Diagonal $\overrightarrow{AC} = \begin{pmatrix}3\\-1\end{pmatrix}-\begin{pmatrix}-4\\6\end{pmatrix}=\begin{pmatrix}7\\-7\end{pmatrix} = 7\mathbf{i}-7\mathbf{j}$

i.e. the **i**- and **j**- components are equal length.
So the sides of the square ABCD have length 7.

E.g. A: $-4\mathbf{i}+6\mathbf{j}$ B

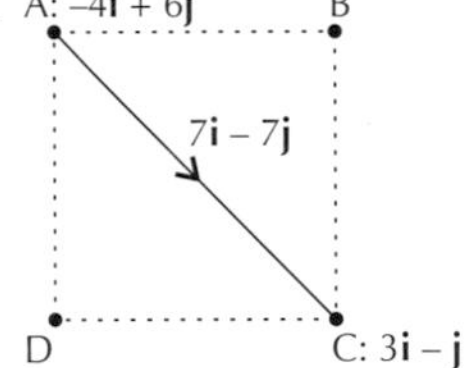

B has coordinates (3, 6) or position vector $3\mathbf{i}+6\mathbf{j}$ and D has coordinates (–4, –1) or position vector $-4\mathbf{i}-\mathbf{j}$.
You might have B and D the other way around, depending on how you labelled the points.

Q8 $\overrightarrow{PM}=\overrightarrow{MQ}$ because M is the midpoint of PQ. This is because the lines are the same length and point in the same direction.

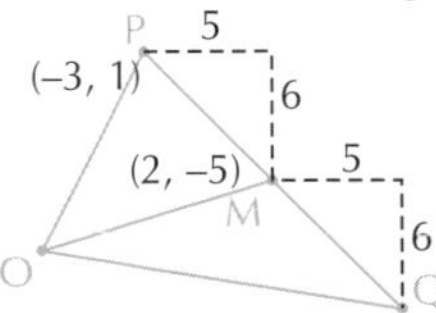

$\overrightarrow{MQ}=\overrightarrow{PM}=\overrightarrow{OM}-\overrightarrow{OP}=2\mathbf{i}-5\mathbf{j}-(-3\mathbf{i}+\mathbf{j})=5\mathbf{i}-6\mathbf{j}$
So $\overrightarrow{OQ}=\overrightarrow{OM}+\overrightarrow{MQ}=2\mathbf{i}-5\mathbf{j}+(5\mathbf{i}-6\mathbf{j})=7\mathbf{i}-11\mathbf{j}$

Q9 a) $\overrightarrow{OP}=\overrightarrow{OA}+\overrightarrow{AP}=\overrightarrow{OA}+\frac{2}{3}\overrightarrow{AB}=\begin{pmatrix}-2\\4\end{pmatrix}+\frac{2}{3}\begin{pmatrix}4-(-2)\\-5-4\end{pmatrix}$

$=\begin{pmatrix}-2\\4\end{pmatrix}+\frac{2}{3}\begin{pmatrix}6\\-9\end{pmatrix}=\begin{pmatrix}-2\\4\end{pmatrix}+\begin{pmatrix}4\\-6\end{pmatrix}=\begin{pmatrix}2\\-2\end{pmatrix}$

b) $\overrightarrow{PB}=\overrightarrow{OB}-\overrightarrow{OP}=\begin{pmatrix}4\\-5\end{pmatrix}-\begin{pmatrix}2\\-2\end{pmatrix}=\begin{pmatrix}2\\-3\end{pmatrix}$

Q10 a) $\overrightarrow{OD}=\begin{pmatrix}-7\\-2\end{pmatrix}$, $\overrightarrow{OE}=\begin{pmatrix}-3\\-1\end{pmatrix}$, $\overrightarrow{OF}=\begin{pmatrix}-1\\5\end{pmatrix}$, $\overrightarrow{OG}=\begin{pmatrix}-3\\10\end{pmatrix}$

$\overrightarrow{DE}=\overrightarrow{OE}-\overrightarrow{OD}=\begin{pmatrix}-3\\-1\end{pmatrix}-\begin{pmatrix}-7\\-2\end{pmatrix}=\begin{pmatrix}4\\1\end{pmatrix}$

$\overrightarrow{EF}=\overrightarrow{OF}-\overrightarrow{OE}=\begin{pmatrix}-1\\5\end{pmatrix}-\begin{pmatrix}-3\\-1\end{pmatrix}=\begin{pmatrix}2\\6\end{pmatrix}$

$\overrightarrow{FG}=\overrightarrow{OG}-\overrightarrow{OF}=\begin{pmatrix}-3\\10\end{pmatrix}-\begin{pmatrix}-1\\5\end{pmatrix}=\begin{pmatrix}-2\\5\end{pmatrix}$

$\overrightarrow{GD}=\overrightarrow{OD}-\overrightarrow{OG}=\begin{pmatrix}-7\\-2\end{pmatrix}-\begin{pmatrix}-3\\10\end{pmatrix}=\begin{pmatrix}-4\\-12\end{pmatrix}$

b) The vacuum cleaner could have travelled along vector $\overrightarrow{DE}+\overrightarrow{EF}=\overrightarrow{DF}=\begin{pmatrix}4\\1\end{pmatrix}+\begin{pmatrix}2\\6\end{pmatrix}=\begin{pmatrix}6\\7\end{pmatrix}$

Q11 E.g. $\overrightarrow{PR}=\overrightarrow{OR}-\overrightarrow{OP}$, so $\overrightarrow{OR}=\overrightarrow{PR}+\overrightarrow{OP}=\begin{pmatrix}4\\7\end{pmatrix}+\begin{pmatrix}3\\0\end{pmatrix}=\begin{pmatrix}7\\7\end{pmatrix}$

So the coordinates of R are (7, 7).

$\overrightarrow{QR}=\overrightarrow{OR}-\overrightarrow{OQ}=\begin{pmatrix}7\\7\end{pmatrix}-\begin{pmatrix}3\\3\end{pmatrix}=\begin{pmatrix}4\\4\end{pmatrix}$

Q12 $\overrightarrow{AC}=\overrightarrow{OC}-\overrightarrow{OA}=(4\mathbf{i}-2\mathbf{j})-(\mathbf{i}-\mathbf{j})=3\mathbf{i}-\mathbf{j}$
$\overrightarrow{BD}=\overrightarrow{OD}-\overrightarrow{OB}=(3\mathbf{i}+\mathbf{j})-(-4\mathbf{j})=3\mathbf{i}+5\mathbf{j}$

12.2 Calculating with Vectors

Exercise 12.2.1 — Calculating with vectors

Q1 a) (i) $\sqrt{6^2+8^2}=\sqrt{36+64}=\sqrt{100}=10$
(ii) $\theta=\tan^{-1}\frac{8}{6}=53.13°$
Both components are positive, so direction = 53.13°

b) (i) $\sqrt{12^2+(-5)^2}=13$
(ii) $\theta=\tan^{-1}\frac{-5}{12}=-22.62°$
The horizontal component is positive and the vertical component is negative, so direction = 360° – 22.62° = 337.38°

c) (i) $\sqrt{2^2+4^2}=\sqrt{20}=2\sqrt{5}$
(ii) $\theta=\tan^{-1}\frac{4}{2}=63.43°$
Both components are positive, so direction = 63.43°

d) (i) $\sqrt{(-3)^2+(-1)^2}=\sqrt{10}$
(ii) $\theta=\tan^{-1}\frac{-1}{-3}=18.43°$
Both components are negative, so direction = 180° + 18.43° = 198.43°

e) (i) $\sqrt{(24)^2+(-7)^2}=25$
(ii) $\theta=\tan^{-1}\frac{-7}{24}=-16.26°$
The horizontal component is positive and the vertical component is negative, so direction = 360° – 16.26° = 343.74°

f) (i) $\sqrt{(-\sqrt{13})^2+6^2}=\sqrt{13+36}=\sqrt{49}=7$
(ii) $\theta=\tan^{-1}\frac{6}{-\sqrt{13}}=59.00°$
The horizontal component is negative and the vertical component is positive, so direction = 180° – 59.00° = 121.00°

g) (i) $\sqrt{3^2+(\sqrt{7})^2}=4$
(ii) $\theta=\tan^{-1}\frac{\sqrt{7}}{3}=41.41°$
Both components are positive so direction = 41.41°

h) (i) $\sqrt{0^2+(-7)^2}=7$
(ii) The horizontal component is 0 and the vertical component is negative, so direction = 270.00°

Q2 $|\overrightarrow{OS}|=\sqrt{10^2+5^2}=\sqrt{100+25}=\sqrt{125}=5\sqrt{5}$

Q3 **a)** (0, 1) has the position vector $\mathbf{j}$, and (2, 2) has the position vector $2\mathbf{i} + 2\mathbf{j}$. Find the vector to get from one point to the other: $(2 - 0)\mathbf{i} + (2 - 1)\mathbf{j} = 2\mathbf{i} + \mathbf{j}$.
The distance between the points is the magnitude of the resultant vector: $\sqrt{2^2 + 1^2} = \sqrt{5}$.

b) (–3, 2) has position vector $-3\mathbf{i} + 2\mathbf{j}$, and (4, 3) has position vector $4\mathbf{i} + 3\mathbf{j}$. Find the vector to get from one point to the other: $(4 - (-3))\mathbf{i} + (3 - 2)\mathbf{j} = 7\mathbf{i} + \mathbf{j}$.
The distance between the points is the magnitude of the resultant vector: $\sqrt{7^2 + 1^2} = \sqrt{50} = 5\sqrt{2}$

c) (–1, –1) has the position vector $-\mathbf{i} - \mathbf{j}$, and (0, 4) has the position vector $4\mathbf{j}$. Find the vector to get from one point to the other: $(0 - (-1))\mathbf{i} + (4 - (-1))\mathbf{j} = \mathbf{i} + 5\mathbf{j}$.
The distance between the points is the magnitude of the resultant vector: $\sqrt{1^2 + 5^2} = \sqrt{26}$

Q4 **a)** $\mathbf{a} + \mathbf{b} = (2\mathbf{i} + \mathbf{j}) + (2\mathbf{i} - 4\mathbf{j}) = 4\mathbf{i} - 3\mathbf{j}$
The magnitude of the resultant is $\sqrt{4^2 + (-3)^2} = 5$

b) $\mathbf{u} + \mathbf{v} = 4\mathbf{i} - 4\mathbf{j}$, $|4\mathbf{i} - 4\mathbf{j}| = \sqrt{4^2 + (-4)^2} = \sqrt{32} = 4\sqrt{2}$

c) $\mathbf{f} + \mathbf{g} = \begin{pmatrix} 24 \\ -10 \end{pmatrix}$, $\left|\begin{pmatrix} 24 \\ -10 \end{pmatrix}\right| = \sqrt{24^2 + (-10)^2} = 26$

d) $\mathbf{d} + \mathbf{e} = \begin{pmatrix} 3 \\ -6 \end{pmatrix}$, $\left|\begin{pmatrix} 3 \\ -6 \end{pmatrix}\right| = \sqrt{3^2 + (-6)^2} = \sqrt{45} = 3\sqrt{5}$

e) $\mathbf{s} + \mathbf{t} = -5\mathbf{j}$, $|-5\mathbf{j}| = \sqrt{(-5)^2} = 5$

f) $\mathbf{w} + \mathbf{x} = \begin{pmatrix} -9 \\ 0 \end{pmatrix}$, $\left|\begin{pmatrix} -9 \\ 0 \end{pmatrix}\right| = \sqrt{(-9)^2} = 9$

Q5 **a)** $\dfrac{3\mathbf{i}}{|3\mathbf{i}|} = \dfrac{3\mathbf{i}}{\sqrt{3^2}} = \mathbf{i}$

b) $\begin{pmatrix} 1 \\ 2 \end{pmatrix} \times \dfrac{1}{\sqrt{1^2 + 2^2}} = \dfrac{1}{\sqrt{5}}\begin{pmatrix} 1 \\ 2 \end{pmatrix}$

c) $5\mathbf{i} - 4\mathbf{j} \times \dfrac{1}{\sqrt{5^2 + (-4)^2}} = \dfrac{5}{\sqrt{41}}\mathbf{i} - \dfrac{4}{\sqrt{41}}\mathbf{j}$

d) $\begin{pmatrix} -2 \\ -6 \end{pmatrix} \times \dfrac{1}{\sqrt{(-2)^2 + (-6)^2}} = \begin{pmatrix} -2 \\ -6 \end{pmatrix} \times \dfrac{1}{2\sqrt{10}} = \dfrac{1}{\sqrt{10}}\begin{pmatrix} -1 \\ -3 \end{pmatrix}$

Q6 $\overrightarrow{AC} = \overrightarrow{AB} + \overrightarrow{BC} = 3\mathbf{i} - 2\mathbf{j} + \mathbf{i} + 5\mathbf{j} = 4\mathbf{i} + 3\mathbf{j}$
$|\overrightarrow{AC}| = \sqrt{4^2 + 3^2} = 5$
So unit vector $= \frac{4}{5}\mathbf{i} + \frac{3}{5}\mathbf{j}$

Q7 $\overrightarrow{BA} = (2\mathbf{i} - \mathbf{j}) - (7\mathbf{i} - 13\mathbf{j}) = -5\mathbf{i} + 12\mathbf{j}$,
$|\overrightarrow{BA}| = \sqrt{(-5)^2 + 12^2} = \sqrt{169} = 13$
So unit vector $= -\frac{5}{13}\mathbf{i} + \frac{12}{13}\mathbf{j}$

Q8 **a)** $\mathbf{a} = \sqrt{2}\cos 45\,\mathbf{i} + \sqrt{2}\sin 45\,\mathbf{j}$
$= \left(\sqrt{2} \times \frac{1}{\sqrt{2}}\right)\mathbf{i} + \left(\sqrt{2} \times \frac{1}{\sqrt{2}}\right)\mathbf{j} = \mathbf{i} + \mathbf{j}$

b) $\mathbf{b} = \sqrt{7}\cos 60\,\mathbf{i} + \sqrt{7}\sin 60\,\mathbf{j}$
$= \sqrt{7} \times \frac{1}{2}\,\mathbf{i} + \left(\sqrt{7} \times \frac{\sqrt{3}}{2}\right)\mathbf{j} = \frac{\sqrt{7}}{2}\mathbf{i} + \frac{\sqrt{21}}{2}\mathbf{j}$

c) $3\cos 33°\,\mathbf{i} + 3\sin 33°\,\mathbf{j}$

d) $5\cos 76°\,\mathbf{i} + 5\sin 76°\,\mathbf{j}$
c) and d) don't give exact values when evaluated, so leave them in terms of cos and sin.

Q9 $|\mathbf{d}| = \sqrt{8^2 + (-6)^2} = \sqrt{100} = 10$
The magnitude of $\mathbf{c}$ is seven times the magnitude of $\mathbf{d}$, so $\mathbf{c} = 7\mathbf{d} = 7(8\mathbf{i} - 6\mathbf{j}) = 56\mathbf{i} - 42\mathbf{j}$

Q10 $\mathbf{v}$ makes an angle $\tan^{-1}\left(\frac{4}{a}\right) = 51°$ below the x-axis,
so $\left(\frac{4}{a}\right) = \tan(51°) = 1.234... \Rightarrow a = \frac{4}{1.234...} = 3.239...$
So $|\mathbf{v}| = \sqrt{3.239...^2 + 4^2} = 5.147... = 5.15$ (2 d.p.)

Q11 $|\overrightarrow{AB}|$ and $|\overrightarrow{BC}|$ form two sides of a triangle.

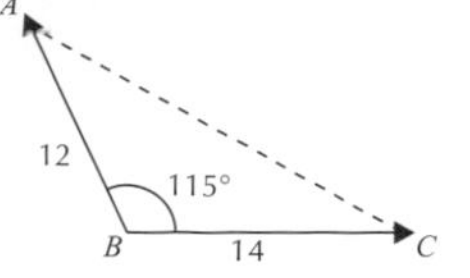

Use the cosine rule to find the remaining side:
$|\overrightarrow{CA}|^2 = 12^2 + 14^2 - (2 \times 12 \times 14 \times \cos 115°)$
$|\overrightarrow{CA}| = \sqrt{481.99...} = 21.954... = 21.95$ (2 d.p.)

Q12 $|\mathbf{a}| = \sqrt{3^2 + 3^2} = \sqrt{18}$, $|\mathbf{b}| = \sqrt{(-2)^2 + 5^2} = \sqrt{29}$
To be able to use the cosine rule to find θ, you also need to know the magnitude of the resultant vector
$\mathbf{a} - \mathbf{b} = \begin{pmatrix} 5 \\ -2 \end{pmatrix}$. $|\mathbf{a} - \mathbf{b}| = \sqrt{5^2 + (-2)^2} = \sqrt{29}$

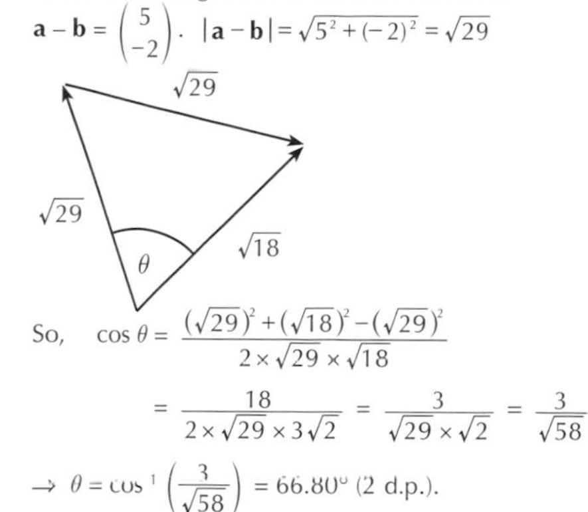

So, $\cos\theta = \dfrac{(\sqrt{29})^2 + (\sqrt{18})^2 - (\sqrt{29})^2}{2 \times \sqrt{29} \times \sqrt{18}}$
$= \dfrac{18}{2 \times \sqrt{29} \times 3\sqrt{2}} = \dfrac{3}{\sqrt{29} \times \sqrt{2}} = \dfrac{3}{\sqrt{58}}$

$\rightarrow \theta = \cos^{-1}\left(\frac{3}{\sqrt{58}}\right) = 66.80°$ (2 d.p.).

Q13 $\overrightarrow{WY}$ is horizontal, so you can make a right-angled triangle:

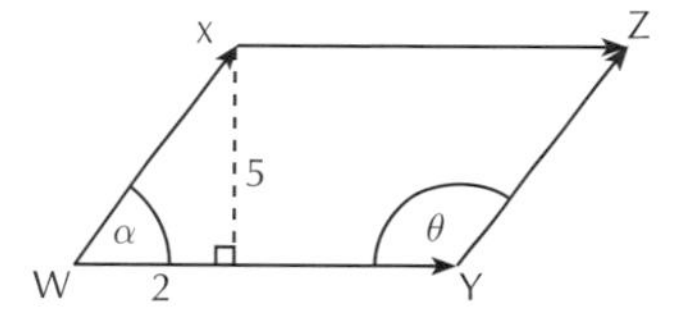

So, find the angle α between $\overrightarrow{WX}$ and $\overrightarrow{WY}$,
$\tan\alpha = \frac{5}{2} \Rightarrow \alpha = \tan^{-1}\frac{5}{2}$
$\overrightarrow{WX}$ and $\overrightarrow{YZ}$ are parallel, so $\theta = 180° - \alpha$
$= 180° - \tan^{-1}\frac{5}{2}$
$= 111.8°$ (1 d.p.)

You could also find θ by using the cosine rule with the lengths of $\overrightarrow{WY}$, $\overrightarrow{YZ}$ and $\overrightarrow{WZ}$ — you should get the same answer.

12.3 Modelling with Vectors

Exercise 12.3.1 — Modelling with vectors

Q1 **a)** $|\mathbf{a}| = \sqrt{1^2 + 2^2} = \sqrt{5}$ ms^{-2}

b) $|\mathbf{b}| = \sqrt{(-1)^2 + (-1)^2} = \sqrt{2}$ ms^{-2}

c) $|\mathbf{c}| = \sqrt{3^2 + 2^2} = \sqrt{13}$ ms^{-2}

d) $|\mathbf{d}| = \sqrt{(-2)^2 + 3^2} = \sqrt{13}$ ms^{-2}

Q2 From the diagram you can see that
$\overrightarrow{AB} = \begin{pmatrix} 10 \\ 0 \end{pmatrix}$, $\overrightarrow{BC} = \begin{pmatrix} -3 \\ 6 \end{pmatrix}$, $\overrightarrow{CD} = \begin{pmatrix} -7 \\ 2 \end{pmatrix}$.
Now, calling the start point of the path S and the end point T:
$\mathbf{p} = \overrightarrow{ST} = \overrightarrow{SB} + \overrightarrow{BC} + \overrightarrow{CT} = \frac{3}{10}\overrightarrow{AB} + \overrightarrow{BC} + \frac{1}{2}\overrightarrow{CD}$
$= \frac{3}{10}\begin{pmatrix} 10 \\ 0 \end{pmatrix} + \begin{pmatrix} -3 \\ 6 \end{pmatrix} + \frac{1}{2}\begin{pmatrix} -7 \\ 2 \end{pmatrix} = \begin{pmatrix} 3 \\ 0 \end{pmatrix} + \begin{pmatrix} -3 \\ 6 \end{pmatrix} + \begin{pmatrix} -3.5 \\ 1 \end{pmatrix}$
$= \begin{pmatrix} -3.5 \\ 7 \end{pmatrix}$, so $x = -3.5$ and $y = 7$.

Q3 $|\mathbf{v}_1| = \sqrt{3^2 + (-2)^2} = \sqrt{13}$, $|\mathbf{v}_2| = \sqrt{1^2 + 3^2} = \sqrt{10}$.
To be able to use the cosine rule, you also need to know the length of the resultant of these two vectors,
$\mathbf{v}_1 + \mathbf{v}_2 = \begin{pmatrix} 4 \\ 1 \end{pmatrix}$, so $|\mathbf{v}_1 + \mathbf{v}_2| = \sqrt{4^2 + 1^2} = \sqrt{17}$

$\sqrt{17}$
$\mathbf{v}_2$
$\sqrt{13}$
$\sqrt{10}$
θ
$\mathbf{v}_1$

Using the cosine rule:
$$\cos\theta = \frac{(\sqrt{13})^2 + (\sqrt{10})^2 - (\sqrt{17})^2}{2 \times \sqrt{13} \times \sqrt{10}} = \frac{13 + 10 - 17}{2\sqrt{130}} = \frac{3}{\sqrt{130}}$$
$$\Rightarrow \theta = \cos^{-1}\left(\frac{3}{\sqrt{130}}\right) = 74.74° \text{ (2 d.p.)}$$

Q4 **a)** When $t = 2$, $\mathbf{v} = 2 \times 2(2\mathbf{i} - \mathbf{j}) = 8\mathbf{i} - 4\mathbf{j}$ ms^{-1}

b) When $t = 3$, $\mathbf{v} = 12\mathbf{i} - 6\mathbf{j}$ ms^{-1}.
Speed = $|\mathbf{v}|$ at $t = 3$: $\sqrt{12^2 + (-6)^2} = 6\sqrt{5}$ ms^{-1}.

c) $\theta = \tan^{-1}\left(\frac{1}{2}\right) = 26.56...°$ below the positive horizontal.
So the direction of $\mathbf{v}$ (anticlockwise from the positive horizontal) is $360° - 26.56...° = 333.43...° = 333.4°$ (1 d.p.)

Q5 **a)** When $t = 1.5$, $\mathbf{s} = 2 \times 1.5^2(-\mathbf{i} + 7\mathbf{j})$ m $= -\frac{9}{2}\mathbf{i} + \frac{63}{2}\mathbf{j}$ m

b) $\theta = \tan^{-1}\left(\frac{7}{1}\right) = 81.86...°$ above the negative horizontal.
So the direction of $\mathbf{s}$ (anticlockwise from the positive horizontal) is $180° - 81.86...° = 98.13... = 98.1°$ (3 s.f.).

c) When $t = 2.5$, $\mathbf{s} = 2 \times 2.5^2(-\mathbf{i} + 7\mathbf{j})$ m $= -12.5\mathbf{i} + 87.5\mathbf{j}$ m
You need the vertical component of the overall displacement (i.e. the $\mathbf{j}$ component), so the firework is 87.5 m from the ground.

Q6

30 ms^{-2}
y
35°
x

Use trigonometry to find x and y:
$\cos 35° = \frac{x}{30}$ and $\sin 35° = \frac{y}{30}$
$\Rightarrow x = 30\cos 35° = 24.574... = 24.57$ (2 d.p.)
$y = 30\sin 35° = 17.207... = 17.21$ (2 d.p.)
The x-component is negative, so $\mathbf{a} = -24.57\mathbf{i} + 17.21\mathbf{j}$ (2 d.p.)

Q7 $\overrightarrow{EF}$ is parallel to $\overrightarrow{GH}$, as $-2(-\mathbf{i} - \frac{3}{2}\mathbf{j}) = 2\mathbf{i} + 3\mathbf{j}$.
$\overrightarrow{HE}$ has two negative components, while $\overrightarrow{FG}$ has one positive and one negative, so they cannot be parallel.
So this quadrilateral has one and only one pair of parallel sides. This means it must be a trapezium.

Q8 **a)**

600
θ
75

The two vectors form a right-angled triangle.
$\tan\theta = \frac{600}{75} = 8 \Rightarrow \theta = \tan^{-1} 8 = 82.87...°$. So the aircraft's actual bearing is $90° - 82.87...° = 007°$

b) Speed is the magnitude of the resultant vector. Using Pythagoras, speed $= \sqrt{600^2 + 75^2} = \sqrt{365\,625}$
$= 604.67$ km/hr (2 d.p.)

Q9 Drawing the vectors end to end:

f
210 N
80°
100°
250 N

The triangle isn't a right-angled triangle, so you need to use the cosine rule:
$|\mathbf{f}|^2 = 210^2 + 250^2 - (2 \times 210 \times 250 \times \cos 80°)$
$|\mathbf{f}| = \sqrt{88366.94...} = 297.27$ N (2 d.p.)

Q10 **a)**

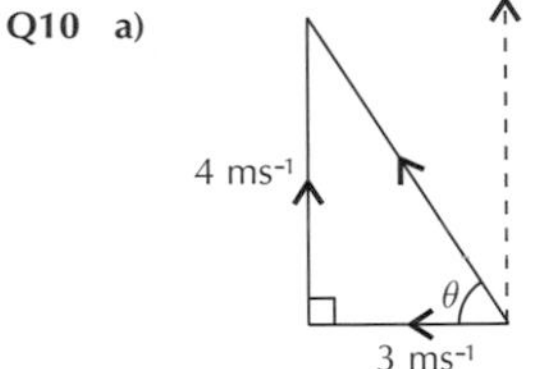

The two vectors form a right-angled triangle.
$\tan\theta = \frac{4}{3} \Rightarrow \theta = \tan^{-1}\frac{4}{3} = 53.13...°$.
Bearings are measured clockwise from North, so Ray travels on the bearing $270° + 53.13...° = 323°$ (3 s.f.)

b) Speed is the magnitude of the resultant vector, so Ray's speed is $\sqrt{3^2 + 4^2} = 5$ ms^{-1}.

Q11 $\overrightarrow{PW} = \overrightarrow{PQ} + \overrightarrow{QW}$. Because W divides QR in the ratio $a:b$, we get that: $\overrightarrow{QW} = \frac{a}{a+b}\overrightarrow{QR}$
$\overrightarrow{QR} = \overrightarrow{QP} + \overrightarrow{PR} = (-\mathbf{s}) + \mathbf{t} = \mathbf{t} - \mathbf{s}$, $\Rightarrow \overrightarrow{QW} = \frac{a}{a+b}(\mathbf{t} - \mathbf{s})$
We also know that $\overrightarrow{PQ} = \mathbf{s}$, so if we substitute these back into the formula for $\overrightarrow{PW}$ we get:
$\overrightarrow{PW} = \mathbf{s} + \frac{a}{a+b}(\mathbf{t} - \mathbf{s}) = \left(1 - \frac{a}{a+b}\right)\mathbf{s} + \frac{a}{a+b}\mathbf{t}$.
The question tells us that $\overrightarrow{PW} = \frac{5}{9}\mathbf{s} + \frac{4}{9}\mathbf{t}$
$\Rightarrow \frac{5}{9}\mathbf{s} + \frac{4}{9}\mathbf{t} = \left(1 - \frac{a}{a+b}\right)\mathbf{s} + \frac{a}{a+b}\mathbf{t}$.
By equating the coefficients of $\mathbf{t}$ this gives you
$\frac{4}{9} = \frac{a}{a+b} \Rightarrow a = 4$ and $b = 5$
Equating the coefficients just means setting the numbers in front of the same variable equal to one another.

Q12 $\overrightarrow{PR} = \overrightarrow{OR} - \overrightarrow{OP} = (6\mathbf{i} + 6\mathbf{j}) - (2\mathbf{i} + 3\mathbf{j}) = 4\mathbf{i} + 3\mathbf{j}$
$|\overrightarrow{PR}| = \sqrt{4^2 + 3^2} = 5$
$\overrightarrow{QS} = \overrightarrow{OS} - \overrightarrow{OQ} = (\mathbf{i} + 5\mathbf{j}) - (7\mathbf{i} + 4\mathbf{j}) = -6\mathbf{i} + \mathbf{j}$
$|\overrightarrow{QS}| = \sqrt{(-6)^2 + 1^2} = \sqrt{37}$

Review Exercise — Chapter 12

Q1

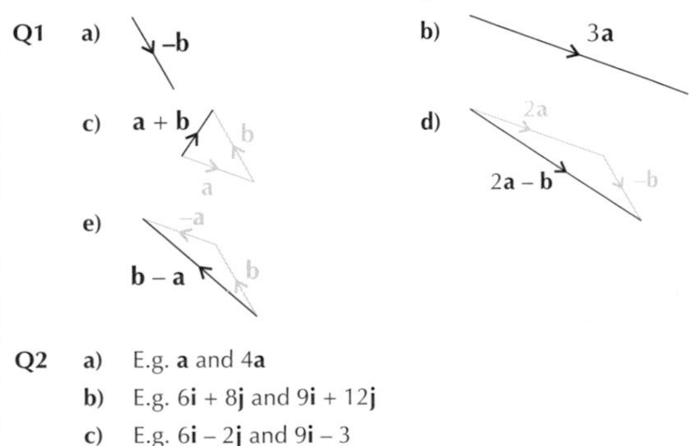

Q2 **a)** E.g. $\mathbf{a}$ and $4\mathbf{a}$

b) E.g. $6\mathbf{i} + 8\mathbf{j}$ and $9\mathbf{i} + 12\mathbf{j}$

c) E.g. $6\mathbf{i} - 2\mathbf{j}$ and $9\mathbf{i} - 3$

d) E.g. $\begin{pmatrix} 6 \\ 10 \end{pmatrix}$ and $\begin{pmatrix} 9 \\ 15 \end{pmatrix}$

Q3 a) $\overrightarrow{YZ} = 2(\mathbf{a} - \mathbf{b}) - 2\mathbf{b} = 2\mathbf{a} - 4\mathbf{b} = -2(2\mathbf{b} - \mathbf{a}) = -2\,\overrightarrow{XY}$.

This shows they're scalar multiples, so $\overrightarrow{YZ}$ is parallel to $\overrightarrow{XY}$. Therefore X, Y and Z are collinear.

b) $\overrightarrow{AB} = \mathbf{q} - 3\mathbf{p}$,
$\overrightarrow{BC} = 4\mathbf{q} - 9\mathbf{p} - \mathbf{q} = 3\mathbf{q} - 9\mathbf{p} = 3(\mathbf{q} - 3\mathbf{p}) = 3\,\overrightarrow{AB}$.

This shows they're scalar multiples, so $\overrightarrow{AB}$ is parallel to $\overrightarrow{BC}$. Therefore A, B and C are collinear.

Q4 a) $\overrightarrow{AB} = -\overrightarrow{OA} + \overrightarrow{OB} = \mathbf{b} - \mathbf{a}$

b) $\overrightarrow{BA} = -\overrightarrow{OB} + \overrightarrow{OA} = \mathbf{a} - \mathbf{b}$

c) $\overrightarrow{CB} = -\overrightarrow{OC} + \overrightarrow{OB} = \mathbf{b} - \mathbf{c}$

d) $\overrightarrow{AC} = -\overrightarrow{OA} + \overrightarrow{OC} = \mathbf{c} - \mathbf{a}$

Q5 $\overrightarrow{XO} = -6\mathbf{i} + \mathbf{j} = \begin{pmatrix}-6\\1\end{pmatrix}$ and $\overrightarrow{YO} = 4\mathbf{i} - 7\mathbf{j} = \begin{pmatrix}4\\-7\end{pmatrix}$

Q6 a) $2\mathbf{d} - \mathbf{f} = 2\begin{pmatrix}3\\2\end{pmatrix} - \begin{pmatrix}3\\-1\end{pmatrix} = \begin{pmatrix}3\\5\end{pmatrix}$

b) $\mathbf{d} - \mathbf{e} - \mathbf{f} = \begin{pmatrix}3\\2\end{pmatrix} - \begin{pmatrix}-1\\-2\end{pmatrix} - \begin{pmatrix}3\\-1\end{pmatrix} = \begin{pmatrix}1\\5\end{pmatrix}$

c) $\mathbf{f} - 3\mathbf{d} + \mathbf{e} = \begin{pmatrix}3\\-1\end{pmatrix} - 3\begin{pmatrix}3\\2\end{pmatrix} + \begin{pmatrix}-1\\-2\end{pmatrix} = \begin{pmatrix}-7\\-9\end{pmatrix}$

d) $-2\mathbf{e} + \mathbf{d} - \mathbf{f} = -2\begin{pmatrix}-1\\-2\end{pmatrix} + \begin{pmatrix}3\\2\end{pmatrix} - \begin{pmatrix}3\\-1\end{pmatrix} = \begin{pmatrix}2\\7\end{pmatrix}$

Q7 $\overrightarrow{RS} = \begin{pmatrix}-5\\-7\end{pmatrix} - \begin{pmatrix}3\\-1\end{pmatrix} = \begin{pmatrix}-8\\-6\end{pmatrix}$

So $|\overrightarrow{RS}| = \sqrt{(-8)^2 + (-6)^2} = \sqrt{100} = 10$

Q8 $\mathbf{q}$ is parallel to $\mathbf{p}$, so $\mathbf{q} = n(5\mathbf{i} - 12\mathbf{j}) = 5n\mathbf{i} - 12n\mathbf{j}$ for some scalar n.

$|\mathbf{q}| = 65$, so $65 = \sqrt{(5n)^2 + (-12n)^2}$
$= \sqrt{25n^2 + 144n^2} = \sqrt{169n^2} = 13n$

$13n = 65 \Rightarrow n = 5$, so $\mathbf{q} = 5(5\mathbf{i} - 12\mathbf{j}) = 25\mathbf{i} - 60\mathbf{j}$.

Q9 $\overrightarrow{AB} = \begin{pmatrix}x\\1\end{pmatrix} - \begin{pmatrix}2\\-3\end{pmatrix} = \begin{pmatrix}x-2\\4\end{pmatrix}$

$|\overrightarrow{AB}| = \sqrt{(x-2)^2 + 16} = 5 \Rightarrow (x-2)^2 + 16 = 25$
$\Rightarrow (x-2)^2 = 9$
$\Rightarrow (x-2) = \pm 3$
$\Rightarrow x = 5$ or $x = -1$

The direction of $\overrightarrow{OB}$ is less than 90°, so both the **i** and **j** component of B must be positive. This means you can disregard the negative solution — so $x = 5$.

Q10

$|\mathbf{v}| = \sqrt{4^2 + (-6)^2} = \sqrt{52}$
$|\mathbf{u}| = \sqrt{(-1)^2 + (-3)^2} = \sqrt{10}$
$|\mathbf{v} - \mathbf{u}| = |5\mathbf{i} - 3\mathbf{j}| = \sqrt{5^2 + (-3)^2} = \sqrt{34}$

Using the cosine rule: $\cos\theta = \dfrac{\sqrt{52}^2 + \sqrt{10}^2 - \sqrt{34}^2}{2 \times \sqrt{52} \times \sqrt{10}}$

$\Rightarrow \theta = \cos^{-1}\left(\dfrac{\sqrt{52}^2 + \sqrt{10}^2 - \sqrt{34}^2}{2 \times \sqrt{52} \times \sqrt{10}}\right) = 52.1250...° = 52.13°$ (2 d.p)

Q11 The speed of the current and the resultant speed are perpendicular, so use Pythagoras' theorem to find the required speed for the swimmer:

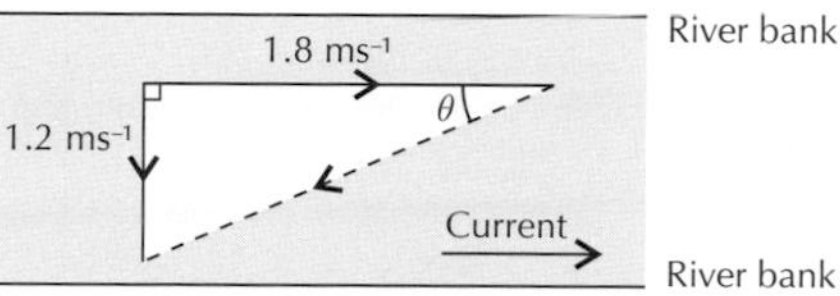

Speed $= \sqrt{1.8^2 + 1.2^2} = 2.163... = 2.16$ ms^{-1} (3 s.f.)
Then also using the right-angled triangle, he should swim at an upstream angle θ to the river bank, where
$\theta = \tan^{-1}\left(\frac{1.2}{1.8}\right) = 33.69... = 33.7°$ (3 s.f.).

Q12 Bearings are measured from North, so the angle the velocity makes with **i** is 171° – 90° = 81°.

The **i** component of velocity is:
$16\cos(81°) = 2.502... = 2.50$ (3 s.f.).
The **j** component of velocity is:
$16\sin(81°) = 15.803... = 15.8$ (3 s.f.).
From the diagram, the **j** component is negative, so the girl's velocity is $2.50\mathbf{i} - 15.8\mathbf{j}$ km/h.

Q13 $\overrightarrow{PM} = \frac{3}{4}\overrightarrow{US} + \frac{1}{2}\overrightarrow{ST}$

$\overrightarrow{US} = \overrightarrow{UT} - \overrightarrow{ST} = (3\mathbf{b} - \mathbf{a}) - 4\mathbf{a} = 3\mathbf{b} - 5\mathbf{a}$
So $\overrightarrow{PM} = \frac{3}{4}(3\mathbf{b} - 5\mathbf{a}) + \frac{1}{2}(4\mathbf{a})$
$= \frac{9}{4}\mathbf{b} - \frac{15}{4}\mathbf{a} + 2\mathbf{a} = \frac{9}{4}\mathbf{b} - \frac{7}{4}\mathbf{a}$

Exam-Style Questions — Chapter 12

Q1 a) $-5\mathbf{a} + 2\mathbf{b} = -5(-3\mathbf{i} + 4\mathbf{j}) + 2(\mathbf{i} + 2\mathbf{j}) = 15\mathbf{i} - 20\mathbf{j} + 2\mathbf{i} + 4\mathbf{j}$
$= 17\mathbf{i} - 16\mathbf{j}$

[2 marks available — 1 mark for correct substitution, 1 mark for correct answer]

b) Magnitude $= |\mathbf{a}| = \sqrt{(-3)^2 + 4^2} = 5$

Angle above the negative horizontal $\alpha = \tan^{-1}\left(\frac{4}{3}\right) =$ 53.13...°

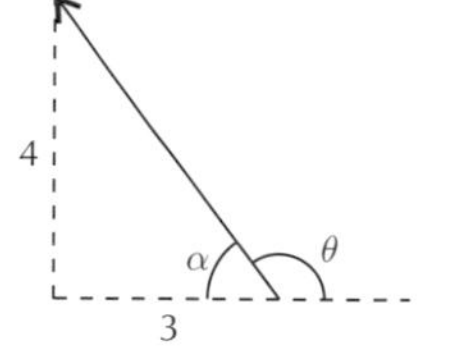

So the direction $\theta = 180 - \alpha = 126.9°$ (1 d.p.)

[4 marks available — 1 mark for using Pythagoras to find magnitude, 1 mark for correct magnitude, 1 mark for finding the angle α, 1 mark for the correct direction of the vector]

Q2 **c** is parallel to **d**, so $\mathbf{c} = n(6\mathbf{i} - 9\mathbf{j}) = 6n\mathbf{i} - 9n\mathbf{j}$ for some scalar n.
$|\mathbf{c}| = \sqrt{13}$, so $\sqrt{13} = \sqrt{(6n)^2 + (-9n)^2}$
$= \sqrt{36n^2 + 81n^2} = \sqrt{117n^2} = 3n\sqrt{13}$

$3n\sqrt{13} = \sqrt{13} \Rightarrow n = \frac{1}{3}$, so $\mathbf{c} = \frac{1}{3}(6\mathbf{i} - 9\mathbf{j}) = 2\mathbf{i} - 3\mathbf{j}$.

*[4 marks available — 1 mark for writing **c** as a scalar multiple of **d**, 1 mark for finding an expression for the length of **c**, 1 mark for finding the scalar multiple, 1 mark for the correct vector]*

Q3 a) $\overrightarrow{MN} = \overrightarrow{ON} - \overrightarrow{OM} = (3\mathbf{i} + 3\mathbf{j}) - (-7\mathbf{i} + 6\mathbf{j}) = 10\mathbf{i} - 3\mathbf{j}$

[2 marks available — 1 mark for subtracting the position vectors, 1 mark for correct answer]

b) $|\overrightarrow{MN}| = \sqrt{10^2 + (-3)^2} = \sqrt{109} = 10.4$ (3 s.f.)

[2 marks available — 1 mark for correctly using Pythagoras, 1 mark for the correct answer]

Q4 Draw the vectors end to end:
Then the angle opposite R is
$180° - 80° - 65° = 35°$.

R
10 N
11 N
35°
P
65°
80°

Using the cosine rule:
$R^2 = 10^2 + 11^2 - 2 \times 10 \times 11 \times \cos 35°$
$R^2 = 40.786... \Rightarrow R = 6.386... = 6.39$ N (3 s.f.)

[3 marks available — 1 mark for finding the angle 35° between the vectors, 1 mark for attempting to use the cosine rule, 1 mark for the correct answer]

Q5 **a)** $\overrightarrow{PR} = \begin{pmatrix}1\\5\end{pmatrix} - \begin{pmatrix}-4\\-5\end{pmatrix} = \begin{pmatrix}5\\10\end{pmatrix}$ so $|\overrightarrow{PR}| = \sqrt{5^2 + 10^2} = \sqrt{125} = 5\sqrt{5}$

$\overrightarrow{QS} = \begin{pmatrix}8\\3\end{pmatrix} - \begin{pmatrix}-5\\1\end{pmatrix} = \begin{pmatrix}13\\2\end{pmatrix}$ so $|\overrightarrow{QS}| = \sqrt{13^2 + 2^2} = \sqrt{173}$

[4 marks available — 1 mark for finding the vectors $\overrightarrow{PR}$ and $\overrightarrow{QS}$, 1 mark for attempting to use Pythagoras to find the length, 1 mark for both lengths correct]

b) $\overrightarrow{QR} = \begin{pmatrix}1\\5\end{pmatrix} - \begin{pmatrix}-5\\1\end{pmatrix} = \begin{pmatrix}6\\4\end{pmatrix}$

$\overrightarrow{PS} = \begin{pmatrix}8\\3\end{pmatrix} - \begin{pmatrix}-4\\-5\end{pmatrix} = \begin{pmatrix}12\\8\end{pmatrix}$

$\overrightarrow{PS} = 2\overrightarrow{QR}$, i.e. they are scalar multiples of each other, so the vectors are parallel.

$|\overrightarrow{PS}| = 2|\overrightarrow{QR}|$, i.e. $PQ = 2QR$, so the parallel sides are not equal in length. Therefore $PQRS$ must be a trapezium.

[4 marks available — 1 mark for finding $\overrightarrow{PS}$ and $\overrightarrow{QR}$, 1 mark for showing they are parallel, 1 mark for stating $PQ = 2QR$, 1 mark for using this to show that $PQRS$ is a trapezium]

You could also find the vectors $\overrightarrow{PQ}$ and $\overrightarrow{RS}$ and show they are not parallel.

Practice Paper

Q1 $3\mathbf{p} - 2\mathbf{q} = 3\begin{pmatrix}-3\\2\end{pmatrix} - 2\begin{pmatrix}6\\10\end{pmatrix} = \begin{pmatrix}-9\\6\end{pmatrix} - \begin{pmatrix}12\\20\end{pmatrix} = \begin{pmatrix}-9-12\\6-20\end{pmatrix}$

$= \begin{pmatrix}-21\\-14\end{pmatrix} = -7\begin{pmatrix}3\\2\end{pmatrix} = -7\mathbf{r}$

Since $3\mathbf{p} - 2\mathbf{q}$ is a scalar multiple of $\mathbf{r}$, $3\mathbf{p} - 2\mathbf{q}$ is parallel to $\mathbf{r}$.
[2 marks available — 1 mark for expressing $3\mathbf{p} - 2\mathbf{q}$ as a single vector, 1 mark for showing $3\mathbf{p} - 2\mathbf{q} = -7\mathbf{r}$]

Q2 **a)** Any non-zero negative value of x proves the statement is incorrect, e.g.:
When $x = -4$, $x^2 = 16$ and $\sqrt{x^2} = \sqrt{16} = 4 \neq x$.
[1 mark for correct counter-example with demonstration that $\sqrt{x^2} \neq x$ for that value]

b) Let $n = 2k$ represent a positive even number.
Then $2n^2 + 2n + 6 = 2(2k)^2 + 2(2k) + 6 = 8k^2 + 4k + 6$
$= 4(2k^2 + k + 1) + 2$
So $2n^2 + 2n + 6$ is not exactly divisible by 4 for any positive number, as dividing by 4 leaves a remainder of 2.
[3 marks available — 1 mark for using $n = 2k$ or equivalent, 1 mark for writing part of expression as multiple of 4, 1 mark for correct interpretation]

Q3 $f'(x) = \lim_{h \to 0}\left[\frac{f(x+h) - f(x)}{(x+h) - x}\right]$

$= \lim_{h \to 0}\left[\frac{4(x+h)^2 - 4(x+h) + 3 - (4x^2 - 4x + 3)}{(x+h) - x}\right]$

$= \lim_{h \to 0}\left[\frac{4x^2 + 8xh + 4h^2 - 4x - 4h + 3 - 4x^2 + 4x - 3}{(x+h) - x}\right]$

$= \lim_{h \to 0}\left[\frac{8xh + 4h^2 - 4h}{h}\right] = \lim_{h \to 0}[8x + 4h - 4] = 8x - 4$

[4 marks available — 1 mark for correct substitution into expression for f′(x), 1 mark for correct expansion of brackets, 1 mark for correct simplification of expression, 1 mark for correct answer]

Q4 $2\log_5(2x - 1) = 1 + \log_5(3 - x)$
$\Rightarrow \log_5(2x - 1)^2 = \log_5 5 + \log_5(3 - x)$
$\Rightarrow \log_5(2x - 1)^2 = \log_5(5(3 - x)) \Rightarrow (2x - 1)^2 = 5(3 - x)$
$\Rightarrow 4x^2 - 4x + 1 = 15 - 5x \Rightarrow 4x^2 + x - 14 = 0$
$\Rightarrow (4x - 7)(x + 2) = 0 \Rightarrow x = \frac{7}{4}$ or $x = -2$
But $x = -2$ is not a valid solution since $2x - 1 = -5 < 0$ and $\log_5 n$ isn't defined for $n \leq 0$, so $\frac{7}{4}$ is the only solution.
[4 marks available — 1 mark for suitable manipulation of logs, 1 mark for correct elimination of logs, 1 mark for both correct solutions resulting from quadratic, 1 mark for identifying correct valid solution]

Q5 **a)** Using the Factor Theorem, if $(x - 3)$ and $(x + 2)$ are factors of $p(x)$, then $p(3) = 0$ and $p(-2) = 0$.
So $p(3) = 2(3)^3 + a(3)^2 + b(3) + 18 = 0$
$\Rightarrow 54 + 9a + 3b + 18 = 0$
$\Rightarrow 9a + 3b = -72 \Rightarrow 3a + b = -24$
And $p(-2) = 2(-2)^3 + a(-2)^2 + b(-2) + 18 = 0$
$\Rightarrow -16 + 4a - 2b + 18 = 0 \Rightarrow 4a - 2b = -2 \Rightarrow 2a - b = -1$
Solve these equations simultaneously:

(1) $3a + b = -24$
(2) $2a - b = -1$
(1) + (2) $5a = -25 \Rightarrow a = -5$

$a = -5$ in (1) $3(-5) + b = -24 \Rightarrow b = -9$
[3 marks available — 1 mark for use of p(3) and p(−2), 1 mark for correct equations derived from p(3) and p(−2), 1 mark for correct values of both a and b]

b) Using part a) substitute a and b, then factorise:
$p(x) = 2x^3 - 5x^2 - 9x + 18 = (x - 3)(x + 2)(cx - d)$
By comparing coefficients:
$2x^3 = x \times x \times cx \Rightarrow c = 2$
$18 = -3 \times 2 \times -d \Rightarrow d = 3$
So $p(x) = (x - 3)(x + 2)(2x - 3)$
[1 mark for correct factorisation]

Q6 **a)** $\int \frac{9x^2 - 3x^3}{\sqrt{x^3}}\,dx = \int\left(\frac{9x^2}{x^{\frac{3}{2}}} - \frac{3x^3}{x^{\frac{3}{2}}}\right)dx$

$= 9\int \frac{x^2}{x^{\frac{3}{2}}}\,dx - 3\int \frac{x^3}{x^{\frac{3}{2}}}\,dx$

$= 9\int x^{\frac{1}{2}}\,dx - 3\int x^{\frac{3}{2}}\,dx$

$= 9\left(\frac{x^{\frac{3}{2}}}{\frac{3}{2}}\right) - 3\left(\frac{x^{\frac{5}{2}}}{\frac{5}{2}}\right) + C$

$= 6x^{\frac{3}{2}} - \frac{6x^{\frac{5}{2}}}{5} + C$

[3 marks available — 1 mark for writing each term in function as a power of x, 1 mark for correct powers in integral, 1 mark for fully correct expression]

b) Area $= \int_0^2 \frac{9x^2 - 3x^3}{\sqrt{x^3}}\,dx = \left[6x^{\frac{3}{2}} - \frac{6}{5}x^{\frac{5}{2}}\right]_0^2$

$= \left(6(2)^{\frac{3}{2}} - \frac{6}{5}(2)^{\frac{5}{2}}\right) - \left(6(0)^{\frac{3}{2}} - \frac{6}{5}(0)^{\frac{5}{2}}\right)$

$= 6(2\sqrt{2}) - \frac{6}{5}(4\sqrt{2}) - 0 = \frac{36}{5}\sqrt{2}$

[3 marks available — 1 mark for correct limits on definite integral, 1 mark for attempting to evaluate at limits of integration, 1 mark for answer in correct form (accept $\frac{36\sqrt{2}}{5}$)]

Q7 a) $\overrightarrow{AB} = \begin{pmatrix}3\\7\end{pmatrix} - \begin{pmatrix}1\\1\end{pmatrix} = \begin{pmatrix}2\\6\end{pmatrix}$

$|\overrightarrow{AB}| = \sqrt{2^2 + 6^2} = \sqrt{40} = 2\sqrt{10}$

[3 marks available — 1 mark for correct $\overrightarrow{AB}$, 1 mark for correct magnitude in any form, 1 mark for fully simplified magnitude]

b) Let the angle at vertex A be θ:

Then using the cosine rule:

$\cos\theta = \frac{(2\sqrt{10})^2 + (2\sqrt{3})^2 - 4^2}{2 \times 2\sqrt{10} \times 2\sqrt{3}}$

$\Rightarrow \theta = \cos^{-1}\left(\frac{(2\sqrt{10})^2 + (2\sqrt{3})^2 - 4^2}{2 \times 2\sqrt{10} \times 2\sqrt{3}}\right) = 34.75...$

$= 34.8°$ (1 d.p.)

[2 marks available — 1 mark for attempting to use cosine rule, 1 mark for correct answer to one decimal place]

c) Area $= \frac{1}{2} \times 2\sqrt{10} \times 2\sqrt{3} \times \sin 34.75...°$

$= 6.244... = 6.2$ (1 d.p.)

[2 marks available — 1 mark for using formula for area of triangle, 1 mark for correct answer to one decimal place]

Q8 a) $f(x) = 2 - 6x + 8x^2 - 2x^3$

$f'(x) = -6 + 8(2x) - 2(3x^2) = -6 + 16x - 6x^2$

When $x = 1$, $f'(x) = -6 + 16(1) - 6(1)^2 = 4$

The gradient of the curve at P is 4, so the tangent has a gradient of 4 and an equation in the form $y = 4x + c$.

At P, $x = 1$ and $y = 2$, so $2 = 4(1) + c \Rightarrow c = -2$

So the equation of the line is $y = 4x - 2$.

[4 marks available — 1 mark for one correct term of f′(x), 1 mark for other correct terms of f′(x), 1 mark for correct gradient of f(x) at x = 1, 1 mark for correct equation of line]

b) At Q, $2 - 6x + 8x^2 - 2x^3 = 4x - 2$

$\Rightarrow 0 = 2x^3 - 8x^2 + 10x - 4 \Rightarrow 0 = x^3 - 4x^2 + 5x - 2 = g(x)$

Since the coefficients of $g(x)$ add up to 0, $(x - 1)$ is a factor of $g(x)$. Divide $g(x)$ by $(x - 1)$ to factorise the function:

$$\begin{array}{r} x^2 - 3x + 2 \\ x - 1 \overline{\smash{)}\, x^3 - 4x^2 + 5x - 2} \\ -\ (x^3 - x^2) \\ \hline -3x^2 + 5x \\ -\ (-3x^2 + 3x) \\ \hline 2x - 2 \\ -\ (2x - 2) \\ \hline 0 \end{array}$$

So $g(x) = (x - 1)(x^2 - 3x + 2) = (x - 1)(x - 1)(x - 2)$

This means l intersects the curve again at $x = 2$.

When $x = 2$, $y = 4(2) - 2 = 6$, so Q has the coordinates $(2, 6)$.

[3 marks available — 1 mark for setting equations to each other and rearranging to equal zero, 1 mark for correct factorisation of g(x), 1 mark for correct coordinates of Q]

Q9 a) $(2x + 3) = \binom{5}{0}(2x)^5(3)^0 + \binom{5}{1}(2x)^4(3)^1 + \binom{5}{2}(2x)^3(3)^2 + \binom{5}{3}(2x)^2(3)^3 + \binom{5}{4}(2x)^1(3)^4 + \binom{5}{5}(2x)^0(3)^5$

$= (1 \times 32x^5 \times 1) + (5 \times 16x^4 \times 3) + (10 \times 8x^3 \times 9) + (10 \times 4x^2 \times 27) + (5 \times 2x \times 81) + (1 \times 1 \times 243)$

$= 32x^5 + 240x^4 + 720x^3 + 1080x^2 + 810x + 243$

[4 marks available — 1 mark for substituting into the binomial formula correctly, 1 mark for one correct simplified term, 1 mark for a further three correct simplified terms, 1 mark for a fully correct expansion]

b) $(2x - 3)^5 = 32x^5 - 240x^4 + 720x^3 - 1080x^2 + 810x - 243$

[1 mark for correct answer]

c) $(2x + 3)^5 - (2x - 3)^5 = 32x^5 + 240x^4 + 720x^3 + 1080x^2 + 810x + 243 - (32x^5 - 240x^4 + 720x^3 - 1080x^2 + 810x - 243)$

$= 480x^4 + 2160x^2 + 486$

So $480x^4 + 2160x^2 + 486 = 475x^4 + 2138x^2 + 501$

$\Rightarrow 5x^4 + 22x^2 - 15 = 0$

Let $u = x^2 \Rightarrow 5u^2 + 22u - 15 = 0 \Rightarrow (5u - 3)(u + 5) = 0$

$\Rightarrow u = \frac{3}{5}$ or $u = -5$

$u = x^2 > 0$, so $\frac{3}{5}$ is the only solution

$\Rightarrow x = \sqrt{\frac{3}{5}} = \frac{\sqrt{15}}{5}$

[3 marks available — 1 mark for correct substitution of binomial expansions and rearranging to get an equation equal to zero, 1 mark for correct values of x^2, 1 mark for correct x-value]

Q10 a) $f(x) = \frac{2x - 4}{4 - x} = \frac{2x - 8 + 4}{4 - x} = \frac{2(x - 4) + 4}{4 - x}$

$= \frac{2(x - 4)}{4 - x} + \frac{4}{4 - x} = -2 + \frac{4}{4 - x}$

[2 marks available — 1 mark for each of a and b]

You could also use algebraic division to find the answer.

b) When $x = 0$, $y = \frac{2(0) - 4}{4 - 0} = -1$,

so the curve intersects the y-axis at $(0, -1)$.

When $y = 0$, $0 = \frac{2x - 4}{4 - x} \Rightarrow 0 = -2 + \frac{4}{4 - x}$

$\Rightarrow 2 = \frac{4}{4 - x}$

$\Rightarrow 2(4 - x) = 4$

$\Rightarrow 4 - x = 2 \Rightarrow x = 2$

So the curve intersects the x-axis at $(2, 0)$.

$y = \frac{1}{x}$ has asymptotes at $x = 0$ and $y = 0$.

If $g(x) = \frac{1}{x}$, then:

$\frac{1}{4 - x} = g(4 - x)$, i.e. $g(x)$ has been translated 4 units left and reflected in the y-axis, so the asymptotes have been translated 4 units left and reflected — the horizontal asymptote isn't changed by the translation, and neither asymptote is changed by the reflection, so the asymptotes are $x = 4$ and $y = 0$.

$\frac{4}{4 - x} = 4g(4 - x)$, i.e. $g(4 - x)$ has been stretched vertically by a scale factor of 4, so the asymptotes aren't changed — they are still $x = 4$ and $y = 0$.

$-2 + \frac{4}{4-x} = 4g(4-x) - 2$, i.e. $4g(4-x)$ has been translated down 2 units, so the asymptotes are translated down 2 units.
The vertical asymptote doesn't change, so the asymptotes are
$x = 4$ and $y = -2$.

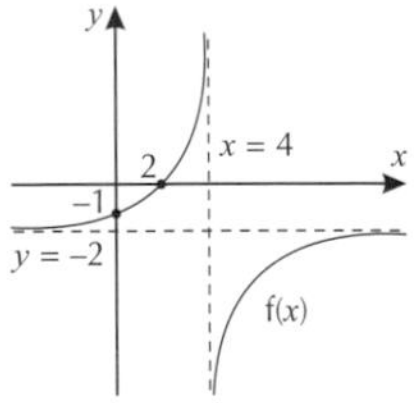

[3 marks available — 1 mark for correct curve shape, 1 mark for both correct axis intercepts labelled, 1 mark for both correct asymptotes labelled]

Q11 a) $q(0) = a + b(0) = 90 \Rightarrow a = 90$
$q(7) = a + b(7) = 34 \Rightarrow 90 + 7b = 34$
$\Rightarrow 7b = -56 \Rightarrow b = -8$
[2 marks available — 1 mark for each correct value]

b) Pressure in tank $P \geq$ pressure in tank Q
$\Rightarrow p(t) \geq q(t) \Rightarrow 72 + 10t - 2t^2 \geq 90 - 8t$
$\Rightarrow 0 \geq 2t^2 - 18t + 18 \Rightarrow 0 \geq t^2 - 9t + 9$
Use the quadratic formula to find t where the above quadratic is equal to 0:

$$t = \frac{-(-9) \pm \sqrt{(-9)^2 - 4 \times 1 \times 9}}{2 \times 1} = \frac{9 \pm 3\sqrt{5}}{2}$$

$t^2 - 9t + 9 < 0$ between these two points, so $\frac{9-3\sqrt{5}}{2} \leq t \leq \frac{9+3\sqrt{5}}{2}$. So the time the pressure in $P \geq$ pressure in Q is:
$\frac{9+3\sqrt{5}}{2} - \frac{9-3\sqrt{5}}{2} = 6.708...$ minutes
$= 6$ minutes 42 seconds
[4 marks available — 1 mark for recognising $p(t) \geq q(t)$, 1 mark for rearranging to get correct quadratic inequality with 0 on one side, 1 mark for correctly solving quadratic inequality, 1 mark for time period given to nearest second]

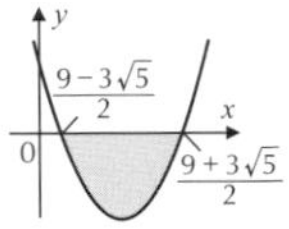

c) E.g. The models predict that the pressure in both tank P and Q will eventually become negative, which is impossible.
[1 mark for any correct reason]

Q12 a) $\frac{\sin^4 x - \cos^4 x}{\cos^2 x} + \tan^2 x = \frac{(\sin^2 x + \cos^2 x)(\sin^2 x - \cos^2 x)}{\cos^2 x} + \tan^2 x$
$= \frac{\sin^2 x - \cos^2 x}{\cos^2 x} + \tan^2 x = \frac{\sin^2 x}{\cos^2 x} - \frac{\cos^2 x}{\cos^2 x} + \tan^2 x$
$= \tan^2 x - 1 + \tan^2 x = 2\tan^2 x - 1$
[3 marks available — 1 mark for use of trig identity $\sin^2 x + \cos^2 x = 1$, 1 mark for use of trig identity $\tan x = \frac{\sin x}{\cos x}$, 1 mark for correct algebraic manipulation]

You may have used different steps to prove the identity, but as long as they're correct you'll get the marks.

b) $\frac{\sin^4 x - \cos^4 x}{\cos^2 x} + \tan^2 x = 3\tan x - 1$
$\Rightarrow 2\tan^2 x - 1 = 3\tan x - 1 \Rightarrow 2\tan^2 x - 3\tan x = 0$
$\Rightarrow \tan x(2\tan x - 3) = 0 \Rightarrow \tan x = 0$ or $\tan x = \frac{3}{2}$
$\tan x = 0 \Rightarrow x = -180°, 0°, 180°$
$\tan x = \frac{3}{2} \Rightarrow x = -123.690...°, 56.309...°$
So the solutions to the equation are
$x = -180°, -123.7°, 0°, 56.3°, 180°$.
[4 marks available — 1 mark for substituting result from part a) and rearranging equation equal to zero, 1 mark for solving quadratic to give correct values of $\tan x$, 1 mark for using $\tan^{-1}$ to get correct values of x, 1 mark for all answers in given domain]

Q13 a) $M = \left(\frac{0+(-2)}{2}, \frac{0+4}{2}\right) = (-1, 2)$
Gradient of $l_1 = \frac{4-2}{4-(-1)} = \frac{2}{5}$
So l_1 has an equation in the form $y = \frac{2}{5}x + c$.
At M, $x = -1$, $y = 2$, so $2 = \frac{2}{5}(-1) + c \Rightarrow c = \frac{12}{5}$
So the equation of l_1 is $y = \frac{2}{5}x + \frac{12}{5}$.
l_2 is perpendicular to l_1, so has a gradient of $-1 \div \frac{2}{5} = -\frac{5}{2}$.
So l_2 has an equation in the form $y = -\frac{5}{2}x + d$.
At D, $x = 6$, $y = 0$, so $0 = -\frac{5}{2}(6) + d \Rightarrow d = 15$
so the equation of l_2 is $y = -\frac{5}{2}x + 15$.
At T, $y = \frac{2}{5}x + \frac{12}{5}$ and $y = -\frac{5}{2}x + 15$,
so $\frac{2}{5}x + \frac{12}{5} = -\frac{5}{2}x + 15 \Rightarrow \frac{29}{10}x = \frac{63}{5} \Rightarrow x = \frac{126}{29}$
When $x = \frac{126}{29}$, $y = \frac{2}{5}\left(\frac{126}{29}\right) + \frac{12}{5} = \frac{120}{29}$
So the coordinates of T are $\left(\frac{126}{29}, \frac{120}{29}\right)$.
[6 marks available — 1 mark for correct coordinates of M, 1 mark for correctly calculating gradient of l_1, 1 mark for correct equation of l_1, 1 mark for use of negative reciprocal to find gradient of l_2, 1 mark for setting equations equal to each other to find coordinates of T, 1 mark for coordinates of T]

b) Length of $AD = 6$
Length of $AT = \sqrt{\left(\frac{126}{29} - 0\right)^2 + \left(\frac{120}{29} - 0\right)^2} = 6$
Length of $DT = \sqrt{\left(\frac{126}{29} - 6\right)^2 + \left(\frac{120}{29} - 0\right)^2} = 4.46$ (3 s.f.)
AD and AT have equal lengths, and DT has a different length, so ADT is isosceles.
[2 marks available — 1 mark for finding AT and DT, 1 mark for correct justification that ADT is isosceles]

Q14 a) Volume of cylinder $= \pi r^2 h \Rightarrow \pi r^2 h = 128\pi \Rightarrow h = \frac{128}{r^2}$
Surface area of cylinder $(S) = 2\pi r^2 + 2\pi rh$
$= 2\pi r^2 + 2\pi r\left(\frac{128}{r^2}\right)$
$= 2\pi r^2 + \frac{256\pi}{r}$ cm²
[2 marks available — 1 mark for rearranging formula for volume to find h, 1 mark for correct surface area expression]

b) $S = 2\pi r^2 + \frac{256\pi}{r} = 2\pi r^2 + 256\pi r^{-1}$
$\Rightarrow \frac{dS}{dr} = 2(2\pi r) + (-1)(256\pi r^{-2}) = 4\pi r - 256\pi r^{-2}$
$\Rightarrow \frac{d^2S}{dr^2} = 4\pi - (-2)(256\pi r^{-3}) = 4\pi + 512\pi r^{-3}$
When $\frac{dS}{dr} = 0$, $4\pi r - 256\pi r^{-2} = 0$
$\Rightarrow 4\pi r = 256\pi r^{-2} \Rightarrow r^3 = 64 \Rightarrow r = 4$
When $r = 4$, $\frac{d^2S}{dr^2} = 4\pi + 512\pi(4)^{-3} = 12\pi > 0$,
so $r = 4$ is minimum point.
When $r = 4$, $S = 2\pi(4)^2 + \frac{256\pi}{4} = 32\pi + 64\pi = 96\pi$ cm²
[5 marks available — 1 mark for correct expression for $\frac{dS}{dr}$, 1 mark for correct expression for $\frac{d^2S}{dr^2}$, 1 mark for correct solution for r using $\frac{dS}{dr} = 0$, 1 mark for using $\frac{d^2S}{dr^2}$ to determine r = 4 is a minimum, 1 mark for correct exact value of S with correct units]

Q15 a) $\log_{10}P = \log_{10}a + t\log_{10}b$,
which has the form of a straight line.
Gradient $= \frac{5.012 - 4.762}{20} = 0.0125$
At $t = 0$, $y = 4.762$, so the line has the equation
$\log_{10}P = 0.0125t + 4.762$
[3 marks available — 1 mark for forming linear equation for $\log_{10}P$ in terms of t, 1 mark for correct gradient, 1 mark for correct equation]

b) $\log_{10}a = 4.762 \Rightarrow a = 10^{4.762} = 57\ 809.60...$

$\log_{10}b = 0.0125 \Rightarrow b = 10^{0.0125} = 1.02920...$

In 2007, $t = 12$, so $P = 10^{4.762}(10^{0.0125})^{12} = 81\ 658.23...$
$= 81\ 700$ (to nearest hundred)

[3 marks available — 1 mark for correct value of a, 1 mark for correct value of b, 1 mark for correct predicted value to nearest hundred]

c) From part a) you know that $\log_{10}P = 0.0125t + 4.762$, so:

$\log_{10}100\ 000 = 0.0125t + 4.762$

$\Rightarrow 5 = 0.0125t + 4.762 \Rightarrow 0.0125t = 0.238 \Rightarrow t = 19.04$

$t = 19.04$ corresponds to the year 2014,
so the population first reaches 100 000 in 2014

[2 marks available — 1 mark for substituting into equation from part a) and rearranging to get t, 1 mark for correct year]

Q16 a) E.g. A and B lie on the vertical line $x = 10$, and B and C lie on the horizontal line $y = 10$, so the sides AB and BC form a right angle and ABC is a right-angled triangle.

You could also use Pythagoras' Theorem to show it's a right-angled triangle.

[1 mark for correct explanation]

b) The hypotenuse of the triangle AC must be the diameter of the circle D_1 since the angle in a semicircle is a right-angle.

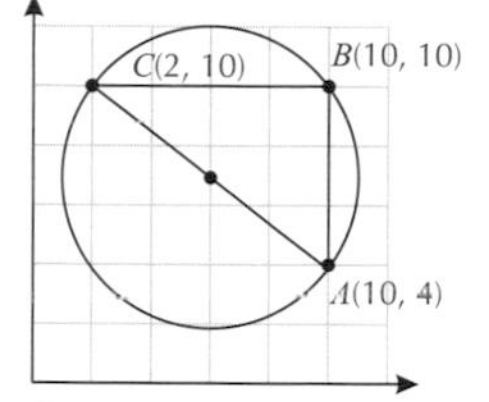

The midpoint of AC is the centre of D_1.

Midpoint of $AC = \left(\frac{10+2}{2}, \frac{4+10}{2}\right) = (6, 7)$

Length of $AC = \sqrt{(10-2)^2 + (4-10)^2} = 10$

Radius of $D_1 = \frac{1}{2} \times AC = 5$

So D_1 has equation $(x - 6)^2 + (y - 7)^2 = 25$.

[3 marks available — 1 mark for correct centre, 1 mark for correct radius, 1 mark for correct circle equation]

c) When $x = 9$ and $y = 11$,

$(x - 6)^2 + (y - 12)^2 = (9 - 6)^2 + (11 - 12)^2$
$= 3^2 + (-1)^2 = 9 + 1 = 10$

$(x - 6)^2 + (y - 7)^2 = (9 - 6)^2 + (11 - 7)^2$
$= 3^2 + 4^2 = 9 + 16 = 25$

So M lies on the circumference of D_1 and D_2 and is a point of intersection of the circles. The line $x = 6$ lies across the centres of both circles, so is a line of symmetry.

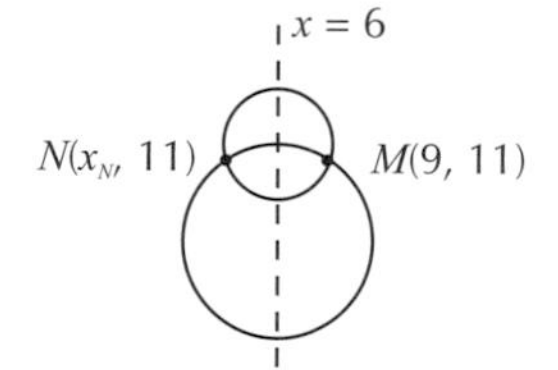

M is 3 right from $x = 6$, so N is 3 left from $x = 6$ and has coordinates (3,11).

[2 marks available — 1 mark for showing M is point of intersection of circles with correct explanation, 1 mark for correct coordinates of N]

d) Work out the gradient of the radius from centre of $D_2(6, 12)$ to $M(9,11)$:

Gradient of radius $(D_2M) = \frac{12-11}{6-9} = -\frac{1}{3}$

So the tangent to D_2 at M has the gradient $-1 \div -\frac{1}{3} = 3$ and an equation in the form $y = 3x + c$.

At M, $x = 9$ and $y = 11$, so $11 = 3(9) + c \Rightarrow c = -16$

So the tangent to D_2 at M has the equation $y = 3x - 16$.

By symmetry, you know the tangent to D_2 at N has the gradient -3 and an equation in the form $y = -3x + c$.

At N, $x = 3$ and $y = 11$, so $11 = -3(3) + c \Rightarrow c = 20$

So the tangent to D_2 at N has the equation $y = -3x + 20$.

The tangents intersect where $3x - 16 = -3x + 20$
$\Rightarrow 6x = 36 \Rightarrow x = 6$ and $y = 3(6) - 16 = 2$

When $x = 6$, $y = 2$, $(x - 6)^2 + (y - 7)^2 = (6 - 6)^2 + (2 - 7)^2 = 25$

So the tangent lines to D_2 intersect on the circumference of D_1.

You could also have started by working out the gradient of the radius from the centre of D_2 to N.

[6 marks available — 1 mark for correct gradient of radius, 1 mark for use of negative reciprocal to find gradient of one tangent to D_2 from gradient radius, 1 mark each for correct equation of tangents to D_2, 1 mark for correct intersection point of tangents, 1 mark for clearly showing that intersection point lies on circumference of D_1]

Glossary

A

Acceleration
The rate of change of an object's **velocity** with respect to time.

Algebraic division
Dividing one **algebraic expression** by another.

Algebraic expression
An **expression** which contains **constants** and / or **variables**.

Assumption
A simplification of a real-life situation used in a **model**.

Asymptote
A line that a curve gets infinitely closer to, but never touches.

B

Bearing
A direction, given as an angle measured clockwise from north.

Binomial
A **polynomial** with only two terms e.g. $a + bx$.

Binomial coefficient
The **coefficients** of the terms of a **binomial expansion**.
For the binomial expansion $(1 + x)^r$, the coefficient of x^r is: $\frac{n!}{r!(n-r)!}$

Binomial expansion
The result of expanding a **binomial** raised to a **power** — e.g. $(a + bx)^n$.

Binomial formula
A **formula** that describes the general terms of a **binomial expansion**.

Chord
A line joining two points that lie on the circumference of a circle.

Coefficient
The **constant** multiplying the **variable**(s) in an algebraic **term** e.g. 4 in the term $4x^2y$.

Collinear points
Three or more points are collinear if they all lie on the same straight line.

Common denominator
A denominator (i.e. bottom of a fraction) that is shared by all fractions in an **expression**.

Common factor
A **factor** that is shared by all the **terms** in an **expression**.

Completing the square
Rewriting a **quadratic** function as: $p(x + q)^2 + r$. Useful for solving **equations** or sketching curves.

Component
The effect of a **vector** in a given direction.

Constant
A fixed numerical value in an **expression**.

Constant of integration
A **constant** term coming from an indefinite **integration** representing any number.

Cosine rule
A rule for finding the missing sides or angles in a triangle when you know all of the sides, or two sides and the angle between them.

Cubic equation
An **equation** that can be written $ax^3 + bx^2 + cx + d = 0$ (where $a \neq 0$).

Decreasing function
A **function** for which the **gradient** is always less than zero.

Definite integral
An **integral** that is evaluated over an interval given by two **limits**, representing the area under the curve between those limits.

Derivative
The result you get when you **differentiate** something.

Differentiation
A method of finding the rate of change of a **function** with respect to a **variable**.
$\frac{dy}{dx}$ is 'derivative of y with respect to x'.

Direct proportion
A relationship between two **variables** where multiplying one of them by any **constant** has the same effect on the other one.

Discriminant
The discriminant of a **quadratic** function $ax^2 + bx + c$ is the value of $b^2 - 4ac$.

Displacement
A **vector** measurement of an object's distance from a particular point.

Disproof by counter-example
Finding one example of where a statement doesn't hold, hence showing that it is false.

Divisor
The number or **expression** you're dividing by in a division.

Scalar
A quantity that has a **magnitude** but not a direction.

Second order derivative
The result of **differentiating** a **function** twice — it tells you the rate of change of the **gradient** of a function.
$\frac{d^2y}{dx^2}$ means 'second order derivative of y with respect to x'.

Set
A collection of objects or numbers (called elements).

Simultaneous equations
A set of **equations** containing two or more unknown quantities, often x and y, for which the same set of values satisfy each equation.

Sine rule
A rule for finding missing sides or angles in a triangle. It can be used if you know any two angles and a side, and in some cases, if you know two sides and an angle that isn't between them.

Solution
The value or values (usually of a **variable**) that satisfy a problem, e.g. an **equation** or **inequality**.

Speed
The **magnitude** of an object's **velocity**.

Stationary point
A point on a curve where the **gradient** is zero.

Substitution
Method for solving **simultaneous equations**, where you replace each occurrence of one unknown with an **expression** in terms of the other unknown.

Surd
A number that can only be expressed precisely by using a square root sign.

T

Tangent
A straight line that just touches a curve at a point. Its **gradient** is the same as the curve's gradient at that point.

Term
A collection of numbers, **variables** and brackets all multiplied or divided together.

Turning point
A **stationary point** that is a (local) **maximum** or **minimum** point of a curve.

Unit vector
A **vector** of **magnitude** one unit.

Variable
A letter in an **expression** representing an unknown which, unlike a **constant**, can take on different values.

Vector
A quantity that has both a **magnitude** and a direction.

Velocity
The rate of change of an object's **displacement** with respect to time.

Vertex
Turning point of a graph — the **maximum** or **minimum** point for a **quadratic graph**.

Index

Formula Sheet

These are the formulas you'll be given in the exam, but make sure you know exactly **when you need them** and **how to use them**.

Binomial Series

$$(a+b)^n = a^n + \binom{n}{1}a^{n-1}b + \binom{n}{2}a^{n-2}b^2 + \ldots + \binom{n}{r}a^{n-r}b^r + \ldots + b^n \qquad (n \in \mathbb{N})$$

$$\text{where } \binom{n}{r} = {}^nC_r = \frac{n!}{r!(n-r)!}$$

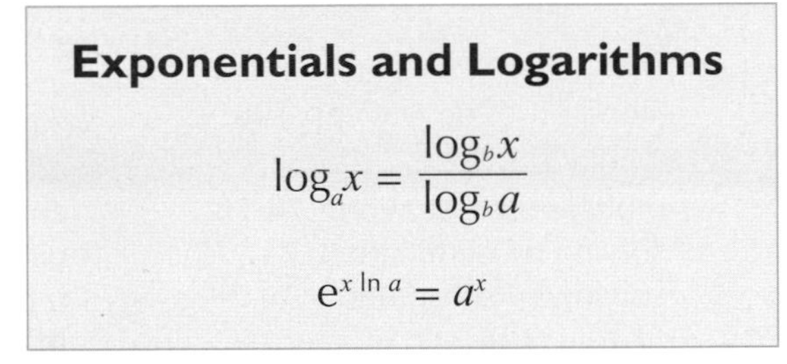

Exponentials and Logarithms

$$\log_a x = \frac{\log_b x}{\log_b a}$$

$$e^{x \ln a} = a^x$$

Mensuration

Surface area of sphere = $4\pi r^2$

Area of curved surface of cone = $\pi r \times$ slant height

Differentiation from First Principles

$$f'(x) = \lim_{h \to 0} \frac{f(x+h) - f(x)}{h}$$